World Sustainability Series

Series Editor

Walter Leal Filho, Research and Transfer Centre "Sustainable Development and Climate Change Management", Hamburg University of Applied Sciences, Hamburg, Germany

Due to its scope and nature, sustainable development is a matter which is very interdisciplinary, and draws from knowledge and inputs from the social sciences and environmental sciences on the one hand, but also from physical sciences and arts on the other. As such, there is a perceived need to foster integrative approaches, whereby the combination of inputs from various fields may contribute to a better understanding of what sustainability is, and means to people. But despite the need for and the relevance of integrative approaches towards sustainable development, there is a paucity of literature which address matters related to sustainability in an integrated way.

Notes on the quality assurance and peer review of this publication.

Prior to publication, the works published in this book are initially assessed and reviewed by an in-house editor. If suitable for publication, manuscripts are sent for further review, which includes a combined effort by the editorial board and appointed subject experts, who provide independent peer-review. The feedback obtained in this way was communicated to authors, and with manuscripts checked upon return before finally accepted. The peer-reviewed nature of the books in the "World Sustainability Series" means that contributions to them have, over many years, been officially accepted for tenure and promotion purposes.

Walter Leal Filho ·
Jaluza Maria Lima Silva Borsatto ·
Carla Bonato Marcolin · Amanda Lange Salvia ·
João Eustachio · Ayyoob Sharifi
Editors

Sustainability in the Curriculum and Operations of Universities

Springer

Editors
Walter Leal Filho
Hamburg University of Applied Sciences
Hamburg, Germany

Carla Bonato Marcolin
School of Management and Business
Federal University of Uberlândia
(FAGEN-UFU)
Uberlândia, Brazil

João Eustachio
Sustainable Development and Climate
Change Management Research
and Transfer Centre (FTZ-NK)
Hamburg University of Applied Sciences
Hamburg, Germany

Jaluza Maria Lima Silva Borsatto
School of Management and Business
Federal University of Uberlândia
(FAGEN-UFU)
Uberlândia, Brazil

Amanda Lange Salvia
Sustainable Development and Climate
Change Management Research
and Transfer Centre (FTZ-NK)
Hamburg University of Applied Sciences
Hamburg, Germany

Ayyoob Sharifi
IDEC Institute
Hiroshima University
Hiroshima, Japan

ISSN 2199-7373 ISSN 2199-7381 (electronic)
World Sustainability Series
ISBN 978-3-032-12677-1 ISBN 978-3-032-12678-8 (eBook)
https://doi.org/10.1007/978-3-032-12678-8

This Springer imprint is published by the registered company Springer Nature Switzerland AG
The registered company address is: Gewerbestrasse 11, 6330 Cham, Switzerland

Preface

The need to tackle global sustainability challenges is more urgent than ever. Climate change, resource depletion, and social inequities require significant action from all parts of society. Higher education institutions have an essential role in this shift. Universities not only create and share knowledge but also serve as active examples of sustainable practices. They have a dual responsibility: to incorporate sustainability into their teaching, which helps shape a new generation of environmentally and socially aware leaders, and to implement sustainable practices on their campuses. This way, they can model broader societal change.

This book, *Sustainability in the Curriculum and Operations of Universities*, arises from the understanding that while many institutions endorse sustainability in theory, the practical application often varies and is frequently isolated. There is a strong need to connect academic discussions on sustainability with real actions within university systems. By gathering insights from scholars, practitioners, and policymakers, this volume aims to explore how sustainability can be effectively integrated into both the educational purpose and daily operations of universities.

The book is divided into two distinct but related parts, reflecting the two key areas of sustainability in higher education. The first part highlights sustainability education, focusing on innovative teaching methods, curriculum design, and interdisciplinary partnerships that prepare students with the knowledge, skills, and values necessary to confront complex sustainability issues. These chapters include case studies, theoretical frameworks, and practical strategies for incorporating sustainability across various subjects, ensuring that graduates are ready to make meaningful contributions to a sustainable future.

The second part shifts attention to campus initiatives, presenting real examples of how universities are decreasing their environmental impact, promoting social equity, and encouraging sustainable governance. From improving energy efficiency and reducing waste to green building practices and community involvement, these chapters demonstrate the significant impact that institutional dedication to sustainability can have. They also address the challenges in implementation, providing insights into governance structures, collaboration with stakeholders, and measuring outcomes.

Together, these parts emphasise the interconnected relationship between sustainability in education and operations. A university that teaches sustainability but does not practice it risks losing credibility. Likewise, one that undertakes green initiatives without integrating sustainability into its curriculum misses the chance to foster deeper cultural change. This book is aimed at educators, administrators, students, and policymakers who are committed to promoting sustainability in higher education. It serves as both a call to action and a practical resource, offering motivation and research-based strategies for making universities true leaders in the global sustainability movement.

We thank all authors for sharing their experiences and knowledge and for promoting their work in this book which, thanks to their efforts, is a truly interdisciplinary publication.

We hope this book will encourage collaboration within and across institutions and accelerate the transition towards more sustainable universities and, by extension, a more sustainable world.

Hamburg, Germany — Walter Leal Filho
Uberlândia, Brazil — Jaluza Maria Lima Silva Borsatto
Uberlândia, Brazil — Carla Bonato Marcolin
Hamburg, Germany — Amanda Lange Salvia
Hamburg, Germany — João Eustachio
Hiroshima, Japan — Ayyoob Sharifi

Contents

Sustainability and Curricular Issues

Innovative Outreach Teaching Models: Sustainability and Community Development Through the 'Integrated Design Process' at the University of Brasília

Paula Lelis Rabelo Albala and Caio Frederico e Silva

Abstract Integrating teaching and academic outreach projects is a fundamental and innovative strategy to continuous pedagogical development, benefiting academia and society. In this context, the article aims to examine the didactic experience of a university course, and the process adopted to integrate teaching, academic outreach, and sustainability. This integration was guided by active learning methodologies and the application of the Integrated Design Process (IDP) framework. As a case study, the article presents the experience of the course *Integrated Environmental Project* (PAI), offered at the Faculty of Architecture and Urbanism of the University of Brasília. The course engaged students in responding to a real-world challenge: the development of a Memory and Cultural Center in Chapada dos Veadeiros, located in Alto Paraíso, in the state of Goiás, Brazil. The pedagogical approach incorporated the IDP concept to implement active teaching methodologies. The method of this integration was structured in three interconnected stages. The first stage focused on implementing the IDP, promoting a collaborative and interdisciplinary approach. This practice has involved collective knowledge construction through lectures, guest talks, training sessions, field visits, and discussion circles, expanding students' design repertoire. The second stage consisted of aligning the course content to the university outreach project, which aimed to develop the design project. Finally, following the IDP approach, the third stage adopted a gamification technique, developing and evaluating the students' projects in an architectural competition format, with design loops to promote feedback. These methods collectively fostered a deeper understanding of sustainability principles and their application in architecture and at the urban scale, promoting interdisciplinarity and critical thinking. The course had significant and positive impacts: students developed enhanced problem-solving skills and a greater

P. L. R. Albala (✉)
Faculty of Architecture, Urbanism and Design, University of São Paulo, São Paulo, SP, Brazil
e-mail: paula.rabelo@unb.br

P. L. R. Albala · C. F. Silva
Faculty of Architecture and Urbanism, University of Brasília, Brasília, DF, Brazil
e-mail: caio.silva@unb.br

W. Leal Filho et al. (eds.), *Sustainability in the Curriculum and Operations of Universities*, World Sustainability Series,
https://doi.org/10.1007/978-3-032-12678-8_1

appreciation for sustainable architectural practices. Beyond the educational scope, the project positively influenced the local community by developing nine project proposals, thereby strengthening the community's identity and heritage. The experience demonstrates that integrating teaching, outreach projects, and sustainability can yield meaningful educational outcomes and community benefits. Looking ahead, this model may serve as a valuable framework for other academic fields and initiatives, advancing in sustainability and multidisciplinarity, and enriching the university's contribution to societal development.

Keywords Sustainability · Integrated design process · Community outreach · Active learning · Architectural education

1 Teaching, Sustainability, and Outreach: An Integrated Approach

Universities, as institutions responsible for educating both professionals and citizens with critical thinking skills, must keep pace with the rapid social, technological, and environmental transformations occurring today. In this context, universities should not merely reproduce existing knowledge but rather challenge prevailing structures, stimulate new debates, and develop innovative solutions to societal problems (Santos 2005; Abu-Khalil 2024).

In the contemporary scenario, particular attention must be given to ongoing climate change and the urgent need for resilience and adaptation in urban environments (IPCC 2018). UNESCO (2011) emphasizes that university education must prepare students to act as agents of change on a planet facing critical ecological limits.

Accordingly, incorporating sustainability as a curricular principle within universities ceases to be optional and emerges as an imperative necessity. This inclusion not only enriches students' educational experience but also reinforces the university's role as a catalyst for social transformation.

The effective integration of sustainability into university curricula is further enhanced when teaching is combined with outreach activities. Such integration bridges the gap between theory and practice, yielding significant benefits for both the academic community and society at large (Abu-Khalil 2024; Caputo et al. 2021).

University outreach facilitates direct engagement with diverse social realities and fosters the development of essential skills such as empathy and social responsibility, thereby preparing students to act with greater ethical awareness and conscientiousness (Freire 1996; Caputo et al. 2021). Education, therefore, must transcend the confines of formal classroom settings, underscoring the transformative potential of university outreach. It is not simply a matter of transmitting content but of fostering a dialogical relationship among educators, students, and society (Freire 1996; Caputo et al. 2021).

This integrative approach, which seeks to educate rather than merely inform, enables students to apply theoretical knowledge in practical contexts, thereby strengthening both their professional and civic formation. Furthermore, it potentiates reflective learning, which occurs when students confront real and complex challenges—a process considered essential for training competent professionals (Schön 1983; Larivee 2008).

In addition, such engagement positively impacts the communities involved, as universities contribute directly to addressing local issues. This co-construction of knowledge enhances inclusion and relevance through knowledge transfer, individual empowerment, increased civic participation, and the strengthening of local social networks (Santos 2005; Abu-Khalil 2024).

Within the field of Architecture and Urbanism education, the synergy between teaching, sustainability, and university outreach assumes a particularly central and vital role, given that architects and urban planners work directly with cities and their constituent elements. These practices enable students to engage with real community challenges—including socio-environmental issues, housing, and urban mobility—thus facilitating learning through field-based application of knowledge to actual demands. Moreover, such experiences reinforce the urban resilience approach, preparing communities to confront and adapt to climate change.

Consequently, the integration of teaching, outreach, and sustainability contributes both to the formation of professionals equipped with critical thinking skills capable of addressing contemporary challenges and to the development of communities better prepared for future uncertainties.

In contemporary education, fostering student involvement in projects connected to pressing social and environmental priorities—such as energy efficiency (Lange Salvia et al. 2020), decarbonization (Ouria and Aníbal 2021), and nature-based solutions (Kabisch et al. 2016)—has become increasingly important.

In the light of the above, this article presents the experience of a university course, and the process adopted to integrate teaching, academic outreach, and sustainability. This integration was guided by active learning methodologies and framed by the Integrated Design Process (IDP). The case study focuses on the course Integrated Environmental Project (PAI), offered at the Faculty of Architecture and Urbanism at the University of Brasília in 2023. The course incorporated an outreach initiative within the curriculum, specifically the development of a Memory and Cultural Center in the Chapada dos Veadeiros region, Goiás State, Brazil.

Therefore, this work aims to discuss the didactic experience of the course, as well as the process undertaken to integrate teaching, outreach, and sustainability through active learning methodologies and the implementation of the IDP framework.

1.1 *Integrated Design Process (IDP) as an Active Learning Technique*

Contemporary education shifts toward pedagogies that are collaborative and socially engaged, based on real-world complexity. In the architecture and urbanism field, this alignment is particularly significant, given the professional's responsibility to address. Within this context, the IDP emerges not only as a professional design methodology but also as a powerful framework for active learning.

1.2 *Integrated Design Process (IDP): A Brief Conceptualization*

The increased performance requirements of a project building cannot be addressed without the efforts of an interdisciplinary design team, from the earliest stages of project development. Traditional design processes, often fragmented and sequential, have failed to address sustainability goals. In contrast, IDP emerges as a collaborative and multidisciplinary approach to conduct the project design process. It involves all project participants from the preliminary stages to optimize performance and sustainability outcomes. In the context of a real building, IDP aims to involve stakeholders, clients, and technical partners from the beginning stages to address energy efficiency, environmental sustainability, or any other relevant criteria defined during the design and construction phases. This approach contrasts with traditional methods, where different professionals work in isolation (Athienitis and O'Brien 2015).

Considering the above, IDP has become a central methodology in sustainable architecture, not only for its capacity to optimize energy and environmental performance but also for its capacity to facilitate dialogue across disciplines and sectors. Its structure emphasizes early-stage collaboration, shared decision-making, and performance evaluation throughout the building lifecycle. Recent studies affirm that the IDP promotes improved building outcomes and strengthens professional collaboration. For example, Wang et al. (2024) show that applying IDP in university courses fosters students' understanding of integrated, sustainable development through collaborative project-based learning. It also highlights the pedagogical potential of IDP as a means of promoting teamwork, reflection, and environmental awareness in architectural education.

It is also important to mention that the Integrated Design Process is not only concerned about the development of a real architectural project. The process also emphasizes the need for educational and institutional support to foster a mindset that values collaboration and sustainability in architectural practice.

By adopting the IDP in an educational context, the course has the potential to enhance students' technical and critical thinking skills, and reinforce the importance of integrated, participatory, and sustainable design in contemporary education.

1.3 Integrated Design Process—Active Learning Technique

Integrating teaching, sustainability, and outreach is significantly enhanced through the adoption of active learning methodologies, which redefine the teacher-student relationship. Active learning methodologies are grounded in the premise of students' active participation in the knowledge-creation and learning process, shifting away from the traditional teacher-centered approach. In other words, the student becomes the protagonist of the process rather than a passive recipient of information. Such methodologies foster autonomy, problem-solving, and collaboration, ensuring that the student takes ownership of their educational development (Guarizzo et al. 2024). Furthermore, research indicates that this approach leads to more effective learning by increasing student engagement (Diesel et al. 2017). It also enhances motivation and retention (Freeman et al. 2014).

Among the most common strategies are problem-based learning (PBL), project-based learning (PjBL), the flipped classroom, and gamification, all of which stimulate critical thinking and self-directed learning (Dresch et al. 2015; Guarizzo et al. 2024). Problem-based learning (PBL), for instance, is centered on solving real-world challenges, requiring students to develop applicable solutions. In project-based learning (PjBL), students engage in extended, real-world projects that demand in-depth research (Simamora 2024). The flipped classroom model reverses conventional instruction: students review content independently while class time is devoted to interactive activities and discussions. Gamification incorporates game-like elements, such as points, leaderboards, and challenges, to enhance motivation. Additionally, these methodologies often promote cross-disciplinary integration, fostering both interdisciplinary and transdisciplinary approaches.

Most importantly, the transition from traditional to active methodologies reflects a broader educational paradigm shift. Students' active participation leads to deeper, more contextualized learning, better preparing them for real-world challenges (Morán 2015; Guarizzo et al. 2024). When connected to pressing global issues such as sustainability, climate change, and urban resilience, the impact of these methods becomes even more pronounced.

When applied as an active learning methodology, the IDP fosters key strategic elements that support student engagement and holistic understanding:

(i) Collaboration and communication: It establishes the need for early and continuous communication among all participants;
(ii) Optimization of building performance: By integrating various disciplines, the design process becomes more holistic, improving its performance in terms of sustainable features, environmental impact, among other;
(iii) Sustainability and Innovation: It enables the incorporation of innovative technologies and sustainable strategies; and
(iv) Performance metrics: The process suggests evaluating building performance throughout its lifecycle, ensuring long-term sustainability goals.

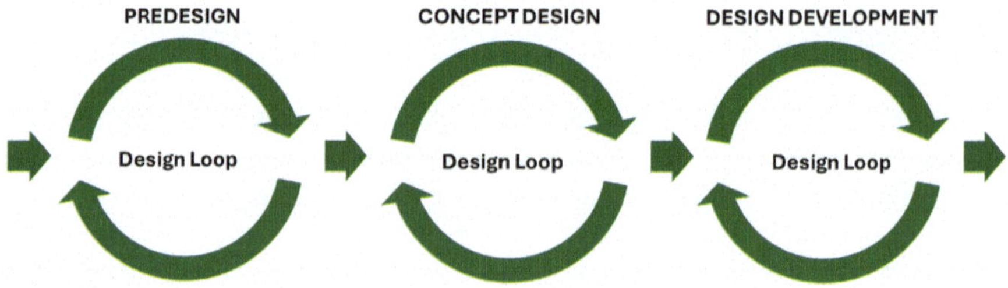

Fig. 1 Design loops—integrated design process. *Source* Authors' elaboration, adapted from International Energy Agency—IEA (2003)

Another key feature of this methodology is its non-linear nature: it is an iterative process that fosters continuous feedback cycles among all project participants. Figure 1 illustrates this process, through interconnected design loops. Each loop signifies a cycle of feedback and refinement involving multiple stakeholders, emphasizing the non-linear and collaborative character of the process. Unlike traditional design models that follow a linear path from concept to execution, the IDP encourages revisiting earlier stages as added information, constraints, or insights emerge. The outermost loop encompasses the broad stages of project development—predesign, concept design, design development—while the inner loops represent constant re-evaluation of performance criteria, stakeholder inputs, and integration of technical disciplines. This iterative dynamic ensures that decisions are continuously tested and optimized, leading to improved outcomes in terms of energy efficiency, environmental performance, and user satisfaction.

By enabling early collaboration and performance feedback, this looped model enhances the design's adaptability to sustainability goals. The figure, adapted from the International Energy Agency (IEA 2003), underscores how IDP aligns design decision-making with evolving project requirements and shared learning across disciplines, making it especially suitable for both professional and educational contexts.

The advantages include more precise timing, ensuring that specialized knowledge and expertise are applied at the appropriate stage of the process. Furthermore, considering a wide range of professional approaches enhances the ability to address complexity effectively. Iterations facilitate the exploration of fundamental design alternatives, allowing for the assessment and optimization of their potential performance. Lastly, the client benefits from having multiple options, avoiding pressure to adopt predefined solutions (International Energy Agency 2003).

In an educational context, applying the IDP helps students grasp the importance of interdisciplinary collaboration and systems thinking. Specifically, teaching IDP in architecture schools enhances the understanding of sustainable and efficient construction processes while strengthening communication and teamwork skills.

Furthermore, by adopting IDP as the foundation of an active learning approach, the course has the potential to encourage students to embrace a holistic design perspective, emphasizing interdisciplinary collaboration as a core value in their work.

2 Methods

This study adopts a qualitative, descriptive, and exploratory approach, aligned with the research objective of analyzing how active learning methodologies—particularly the Integrated Design Process (IDP)—can be implemented within architectural education to integrate teaching, outreach, and sustainability. It presents a didactic case study, based on the PAI (Integrated Environmental Project) course offered in 2023 at a Brazilian public university.

The aim is to detail and critically reflect upon the pedagogical methods employed, enabling replication in similar academic contexts. The research methodology was structured in three stages:

1. Implementation of the IDP concept in a pedagogical context

The IDP was applied as an active learning technique, emphasizing collaborative, interdisciplinary, and iterative design processes. This involved activities such as lectures, guest expert talks, training workshops, site visits, and discussion circles. These tools aim at constructing collective knowledge and enrich students' architectural design repertoires.

2. Integration with a real-world outreach demand

The PAI course was directly aligned with an ongoing university outreach initiative focused on developing a Memory and Cultural Center in the Chapada dos Veadeiros region. This ensured that the design tasks addressed concrete community needs, promoting a socially engaged learning experience.

3. Gamified project development and evaluation

Students worked in teams to develop architectural proposals in the format of a design competition, structured around design loops that allowed for continuous feedback and improvement. This gamification element fostered motivation and creativity in project development.

The results were analyzed based on student feedback, the quality of the design outcomes, and the level of engagement demonstrated throughout the semester.

The following diagrams synthesize all the pedagogical processes (Fig. 2). This format has provided students with an engaging and motivating environment to apply their acquired knowledge while developing technical skills and critical thinking.

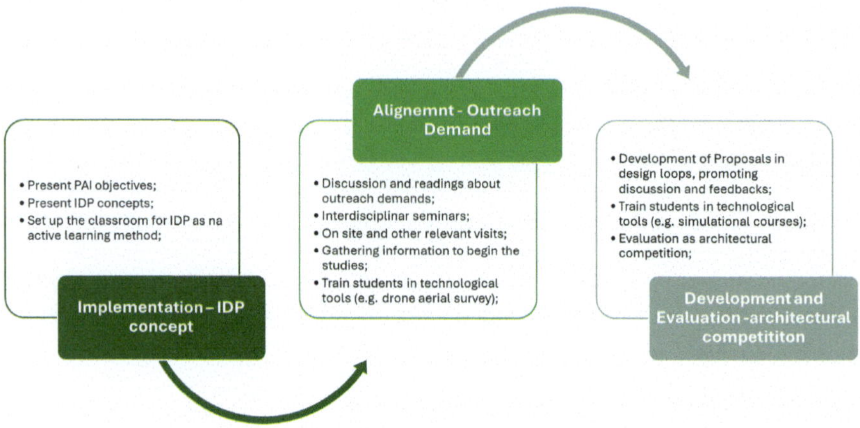

Fig. 2 Pedagogical approach fluxogram used at PAI 2023 course. *Source* Authors' elaboration (2025)

3 Results and Discussion

3.1 Implementing IDP in the Course

As previously mentioned, IDP is a collaborative approach that involves all project agents, working together to share information and make joint decisions. This process enabled the early identification of issues and the development of innovative solutions that meet the needs of all partners.

In this educational setting, applying IDP has enabled students to understand the importance of interdisciplinary collaboration and systemic thinking. During the course, this concept was applied to innovate the teacher-student relationship within the classroom.

Within the concept of the integrated design process, the course was structured into three steps (Fig. 3). The first phase consisted of introductory lectures that outlined the course objectives and introduced the principles of the IDP. This initial stage was essential to establish well-defined objectives, aligned with the team's capacities and objectives. Students were informed early on that they would assume an active role throughout the semester. The second step emphasized a facilitated learning environment, where the presence of a facilitator helped establish a collaborative tone and ensured clear communication among participants. Finally, in the third step, a participatory process was initiated, in which both students and teachers suggested readings and discussions to support project development. In this stage, teachers acted solely as mediators, focusing on the student protagonism in both the learning and development processes. As a result, students understood from the outset that they would play an active role throughout the semester.

Fig. 3 IDP implementation diagram presented in class. *Source* Author's elaboration (2025)

3.2 Alignment Between the Course and the Outreach Project Demands

The demand for a Memory and Cultural Center originated from students engaged with the Network of Outreach Poles (REPE) at the University of Brasília (UnB). REPE is a strategic initiative under the Dean's Office of Outreach (DEX/UnB), aimed at fostering stronger ties between the university and society through integrated outreach, teaching, and research activities. Structured around regional hubs, REPE serves as a platform for dialogue and collaborative knowledge production, connecting local community needs with academic endeavors. The network consists of five outreach poles, one of which is the UnB Chapada dos Veadeiros Pole, encompassing municipalities in the state of Goiás, including Alto Paraíso de Goiás—the project's designated location (UnB 2023).

As previously mentioned, the PAI course aims to integrate sustainability and environmental comfort issues into projects from a practical perspective. Aligning it with the outreach project, which sought to develop a space for memory and culture in Chapada dos Veadeiros, entailed challenges and a revision of pedagogical practices. Another relevant aspect is that, as part of a REPE hub, the outreach project was actively involved in various initiatives focused on the Cerrado biome.

The project's objective was to develop a space for educational practices about the Cerrado Biome, aiming to be a rural ecovillage and to promote public education in the region. This center would also serve as a meeting point for local tourist guides in Chapada dos Veadeiros. Also, it should be conceived as an example of sustainable architecture, employing vernacular and bioclimatic techniques and functioning as a model of resilience to climate change.

Within the IDP (Integrated Design Process) context, the course was structured in various stages. Aiming at interdisciplinary knowledge, lectures, guided readings, discussion circles, and courses were organized to enhance students' training and critical thinking, sensitizing them to issues inherent to the study object. For example, the course hosted discussions with other actors involved in the outreach network,

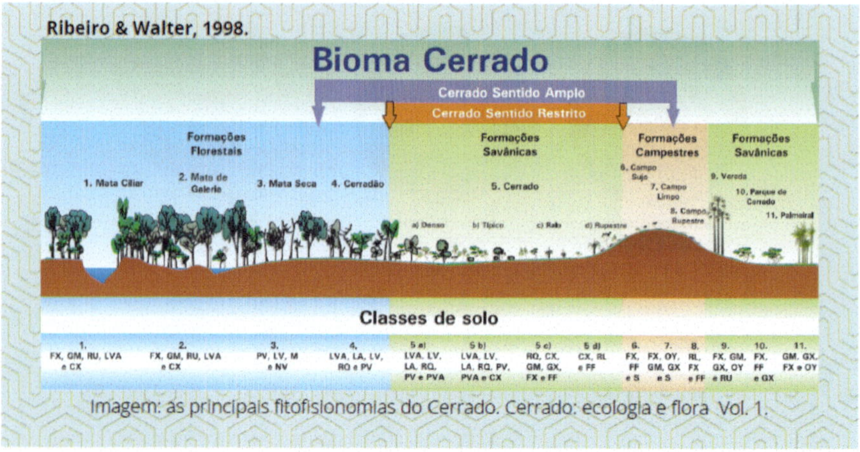

Fig. 4 A slide from the lecture on the phytophysionomies of the Cerrado biome, where knowledge from other fields of study was integrated into the course. *Source* Authors' archive (2023)

such as the lecture on "Cerrado Phytophysionomies," given by agronomy students. This lecture was initiated by the students, who actively participated in this collective knowledge-building process (Fig. 4).

Also, the course adapted the subject of the project proposal, to the actual site of the outreach project. It also conducted field recognition visits (Fig. 5) and organized the schedule based on demands brought forth by outreach participants.

Another interesting aspect was the aerial survey conducted by drone (Fig. 6) as part of the course, which provided technical support to the outreach project, as it lacked an updated cartographic base for the intervention site. In this process, the students worked with new technological tools, also learning how to generate the 3D digital terrain from point cloud data.

The course schedule was also more flexible compared to that of a traditional course, as activities that arose either from student demand or from outreach needs were often incorporated. An example of this were the workshops, which were reorganized or suggested based on visits by outreach members to the community. Finally, the course also encouraged participation in extracurricular events, such as lectures on professional practice, and organized visits to museum spaces in Brasília, such as the Art Museum of Brasília (MAB). During the visit, students had the opportunity to explore the museum's collection, architectural design, and facilities. They engaged in discussions with museum staff and management on programmatic aspects related to exhibition and collection organization, as well as maintenance challenges and the specific requirements of a museum based on concrete experience (Fig. 7). This activity played a crucial role in shaping their design repertoire, which would later be applied to their projects.

Additionally, the study of national and international design references, analyzed and debated in a seminar with all students, along with discussions with architecture

Fig. 5 On-site visits with students and professors. *Source* Authors' archive (2023)

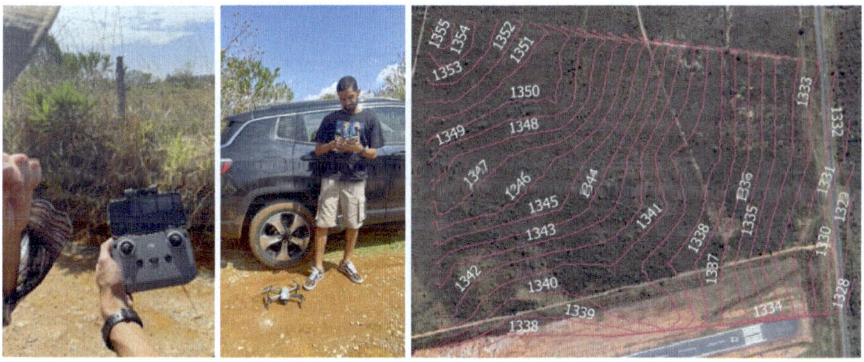

Fig. 6 Drone survey to generate a 3D model of the terrain, showcasing the integration of technological tools in the course. *Source* Authors' archive (2023)

firms, further enriched this process. The course proposed a debate circle with young architects from São Paulo, Hiperstudio, winners of various competitions. This session aimed to foster exchanges between students and professionals regarding the universe

Fig. 7 Visit to explore the collection, technical facilities, and installations of the Brasília Art Museum (MAB). *Source* Authors' archive (2023)

of competitions, bringing design and representation references while discussing challenges. Still regarding the teaching approach, students were also invited to participate in workshops organized by professors, which covered computational simulation techniques for environmental project analysis (e.g., Envi-met and Climate Studio).

Subsequently, they were guided to develop proposals for the center in groups. The work was organized into two stages: an initial submission of a preliminary study and, at the end, the delivery of the consolidated project. The format of presentation boards and the presentation method were also the subject of collective discussion with students.

3.3 Development and Evaluation—Architectural Competition as a Playful Learning Tool

The project development and evaluation, from the IDP perspective, was structured as an architectural competition, where students applied their learning in groups, proposing designs for the Memory and Cultural Center. As stated, the proposals were required to incorporate sustainability, innovation, and context-appropriate design techniques.

Simulating a design competition in an architecture course offers several advantages that enrich students' learning experience. First, this gamification methodology provides a playful and motivating environment where healthy competition stimulates creativity and innovation. In this context, students were encouraged to solve design problems, consider budget and space constraints, and present innovative and viable solutions. This not only improved their technical skills but also developed essential competencies such as teamwork, communication, critical thinking, and decision-making.

During the design process, intermediate deliverables were organized through seminars, where students presented their work and received feedback. The professors also provided a shared document containing observations on the projects, ensuring that everyone had access to all evaluations. This process is inherent to the design loops of the IDP, which establishes cycles of debates among all those involved. The loops also served to incorporate demands arising from the local Cerrado community, which emerged after site visits and further development of outreach work. For the final evaluation, in addition to professors acting as "jurors," architect Nonato Veloso, a renowned architect from Brasília and winner of several architectural competitions, was invited to join the jury (Fig. 8). The presence of a renowned architect as a guest on evaluation provided students with the opportunity to receive valuable feedback from an experienced professional. It also has encouraged them to achieve excellence in their projects.

On the final presentation day, an exhibition of the projects was held at the university, disseminating the content produced in the course to other undergraduate students. After the internal evaluation, all students received feedback on their submitted projects, along with comments from the presentation day. Following the

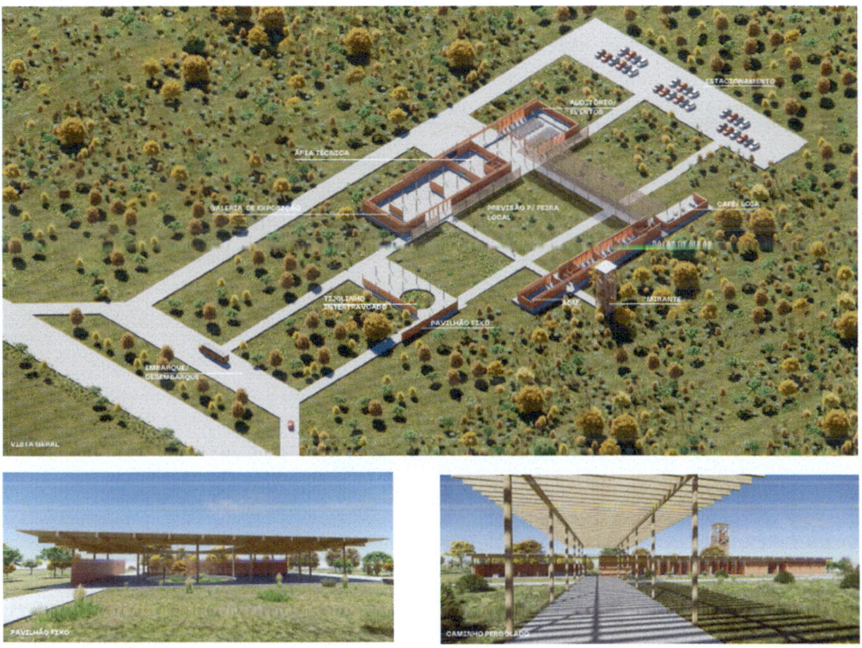

Fig. 8 Examples of a project delivered at the end of the course—perspectives of the first-place winners (students Ferro and Costa). *Source* Authors' archive (2023)

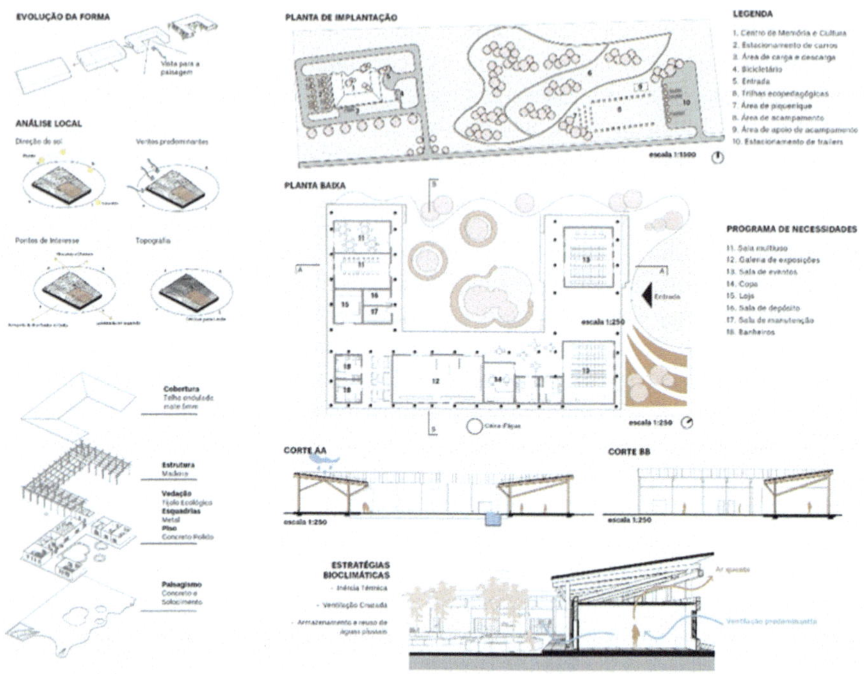

Fig. 9 Example of a project delivered at the end of the course—plans, sections, and project process of the second-place winners (students Macedo and Albuquerque). *Source* Authors' archive (2023)

"judgment" of the proposals, the professors shared the highest-rated proposals with students, providing technical justifications for their selection (Figs. 8, 9 and 10).

This competition structure enabled a practical and theoretical understanding of the IDP, as the process also familiarized students with submission, presentation, and defense procedures while receiving constructive feedback from the jury. This cycle of continuous evaluation and revision is fundamental for professional development, allowing future architects to have a better understanding of the expectations and challenges they will face.

3.4 Learning from the Process and Future Applications

In summary, the implementation of the Integrated Design Process (IDP) in the PAI course brought numerous advantages and benefits. Structured according to Lang's (1974) four design phases—synthesis, prediction, evaluation, and decision-making—and integrated with IDP principles, the course enabled a systemic and interdisciplinary approach from the preliminary stages.

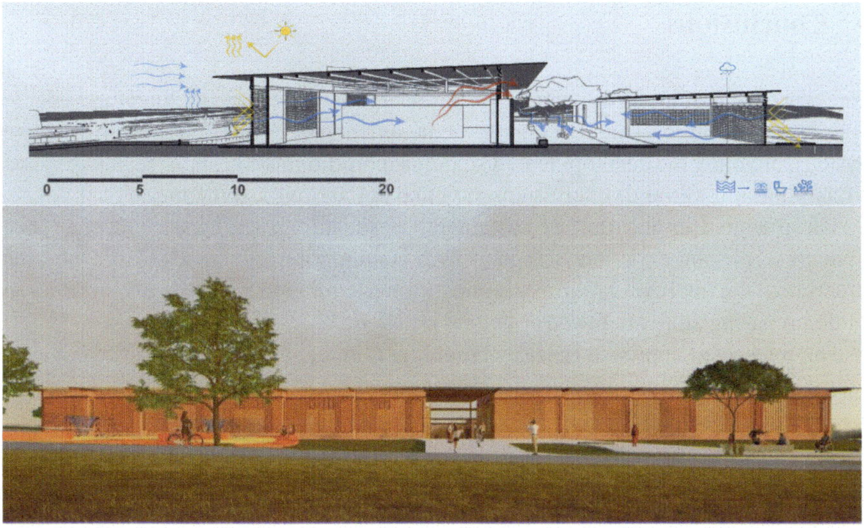

Fig. 10 Example of a project delivered at the end of the course—section and perspective of the third-place winners (students Victoria and Nogueira). *Source* Authors' archive (2023)

Students actively participated in all design phases, from site recognition and to technological explorations such as drone mapping and environmental simulations. IDP promoted a collaborative approach from the start of the project, involving students, professors, outreach participants, and guests in a shared process of decision-making. feedback loops and collaborative discussions supported continuous learning and refinement.

In the educational environment, this methodology allowed students to understand the importance of interdisciplinary collaboration and systemic thinking. Working in teams and sharing responsibilities helped develop communication, teamwork, and critical thinking skills. The active learning approach, where students had a voice, was positively received, and student engagement remained high throughout the semester. In the final discussion, many referred to the course as "the best" they had taken, precisely because it broke away from traditional teaching models.

These qualities were notably reflected in the project's outcomes. High-quality proposals were developed for the Memory and Cultural Center, combining technical depth on sustainability with a critical reflection on the concepts of memory and culture in the context of Chapada dos Veadeiros.

Finally, the project positively influenced the local community by delivering nine proposals that helped strengthen local identity and heritage. This experience offers a replicable model for future editions of the PAI course and other curricular innovations seeking to integrate teaching, outreach, and sustainable design in a coherent and transformative way.

4 Conclusions

The experience of the PAI course was unique and innovative in architectural education. The integration of the IDP into the pedagogical process proved to be an effective strategy, enabling the development of students' communication skills, teamwork, and critical thinking through interdisciplinary and collaborative dynamics.

The practical application of sustainability and urban resilience concepts gave students a relevant experience aligned with contemporary demands. Moreover, the structure of the final evaluation as a competition encouraged creativity and innovation while preparing students for professional practice.

The integrated approach between teaching, outreach, and sustainability not only contributed significantly to students' training but also promoted meaningful engagement with the local community. The use of active learning methodologies, including the flipped classroom, demonstrated positive results, notably in student motivation and participation.

As a future perspective, it is suggested that this methodology be replicated in other contexts and design courses to further enrich the institution's teaching and learning processes.

Acknowledgements We would like to thank the students on the course, who engaged in the process and allowed the methodology to be carried out. We also thank the REPE Outreach Program and the Dean's Office of Extension of the University of Brasília for scholarships. We also acknowledge the support of FAPESP through grant number 24/23425-1.

References

Abo-Khalil, A. G. (2024). Integrating sustainability into higher education: Challenges and opportunities for universities worldwide. *Heliyon, 10*(9), e29946.

Athienitis, A., & O'Brien, W. (2015). *Modeling, design, and optimization of net-zero energy buildings*. Hoboken: John Wiley & Sons.

Caputo, F., Ligorio, L., & Pizzi, S. (2021). The Contribution of Higher Education Institutions to the SDGs – An Evaluation of Sustainability Reporting Practices. *Administrative Sciences, 11*(3), 97.

Diesel, A., Baldez, A. L. S., & Martins, S. N. (2017). Os princípios das metodologias ativas de ensino: uma abordagem teórica [The principles of active learning methodologies: a theoretical approach]. *THEMA, 14*(1), 268-288.

Dresch, A., Lacerda, D. P., & Antunes Junior, J. A. V. (2015). *Design science research: A method for science and technology advancement*. Cham: Springer.

Freire, P. (1996). *Pedagogia da autonomia: Saberes necessários à prática educativa* [Pedagogy of autonomy: Knowledge necessary for educational practice]. São Paulo: Paz e Terra.

Freeman, S., Eddy, S. L., McDonough, M., Smith, M. K., Okoroafor, N., Jordt, H., & Wenderoth, M. P. (2014). Active learning increases student performance in science, engineering, and mathematics. *Proceedings of the National Academy of Sciences, 111*(23), 8410–8415.

Guarizzo, A. B., Silva, A. M. R. da, Araújo, A. F. de, Santos, G. C. dos, Botelho, S. de O., Pereira, S. M. da S., Pereira, T. do R. dos S., & Woodcock, Z. S. P. (2024). Metodologias de aprendizagem

ativa: uma mudança de paradigma no ensino [Active learning methodologies: a paradigm shift in teaching]. *Caderno Pedagógico, 21*(5), e4186.
Intergovernmental Panel on Climate Change (IPCC). (2018). *Global warming of 1.5°C: An IPCC special report on the impacts of global warming of 1.5°C above pre-industrial levels and related global greenhouse gas emission pathways, in the context of strengthening the global response to the threat of climate change, sustainable development, and efforts to eradicate poverty.* Cambridge University Press.
International Energy Agency (IEA). (2003). *Integrated design process guideline.* Berlin: IEA.
Kabisch, N., Frantzeskaki, N., Pauleit, S., Naumann, S., Davis, M., Artmann, M., Haase, D., Knapp, S., Korn, H., Stadler, J., Zaunberger, K., & Bonn, A. (2016). Nature-based solutions to climate change mitigation and adaptation in urban areas: Perspectives on indicators, knowledge gaps, barriers, and opportunities for action. *Ecology and Society, 21*(2), 39.
Lang, J. (1974). *Creating architectural theory: The role of the behavioral sciences in environmental design.* New York: Van Nostrand Reinhold.
Lange Salvia, A., Leal Filho, W., Brandli, L. L., Griebeler, J. S., & Vasconcelos, C. R. (2020). Energy sustainability in teaching and outreach initiatives and the contribution to the 2030 Agenda. *International Journal of Sustainability in Higher Education, 21*(7), 1607–1624.
Larrivee, B. (2008). Development of a tool to assess teachers' level of reflective practice. Reflective Practice, 9(3), 341–360.
Morán, J. (2015). Mudando a educação com metodologias ativas [Changing education with active methodologies]. In C. A. de Souza & O. E. T. Morales (Orgs.), *Convergências midiáticas, educaçao e cidadania: Aproximações jovens* [Media convergences, education and citizenship: Youth approaches]. Ponta Grossa: UEPG.
Ouria, M., & de Almeida, A. T. (2021). How to decarbonize developing cities by 2050: A case from Tabriz-Iran. *Renewable Energy, 178*, 620–638. https://doi.org/10.1016/j.renene.2021.06.065
Santos, B. de S. (2005). A universidade no século XXI: Para uma reforma democrática e emancipatória da universidade [The university in the 21st century: For a democratic and emancipatory university reform]. São Paulo: Cortez.
Schön, D. A. (1983). *The reflective practitioner: How professionals think in action.* New York: Basic Books.
Simamora, A. M. (2024). A decade of Science, Technology, Engineering, and Mathematics (STEM) Project-Based Learning (PjBL): A systematic literature review. Journal of Computers for Science and Mathematics Learning, 1(1), 1–15.
UNESCO (Organização das Nações Unidas para a Educação, a Ciência e a Cultura). (2011). Os sete saberes necessários à educação do futuro (Edgar Morin) [The seven knowledge necessary for education of the future (Edgar Morin)]. Brasília: UNESCO.
Universidade de Brasília (UnB). (2023). *REPE 2023 - Catálogo de Programas e Projetos da Rede de Polos de Extensão da UnB* [REPE 2023 – Catalog of Programs and Projects of the Outreach Network Poles at UnB]. Brasília: UnB.
Wang, X., Short, K. M., & Fiori, C. M. (2024). Board 69: Learning Sustainable Development Through Integrative Design Process (a Case Study). In ASEE Annual Conference & Exposition (Portland, OR). American Society for Engineering Education. https://doi.org/10.18260/1-2--48366

Paula Lelis Rabelo Albala Assistant Professor of the Faculty of Architecture and Urbanism at the University of Brasilia (FAU-UnB). Postdoctoral researcher at Sciences Center for Development—CCD USP Cidades, University of São Paulo (2025). She holds a PhD in Architecture and Urbanism from University of Sao Paulo—FAUUSP (2022), a Master's degree in Architecture, Energy, and Environment from Polytechnical University of Catalonia—UPC (2010), and a Bachelor's degree from FAUUSP (2007). Postdoctoral fellow at FAU-UnB (2024). Researcher at LaSUS and Lapis (FAU-UnB), focusing on sustainable architecture and urbanism, environmental comfort, active mobility, accessibility, nature-based solutions, and social inclusion.

Caio Frederico e Silva Associate Professor and Dean of the Faculty of Architecture and Urbanism at the University of Brasília (FAU-UnB). He is a CNPq Productivity Fellow and conducted postdoctoral research at Harvard Graduate School of Design (2020) and NOVA University Lisbon (2016). He holds a PhD (2013) and a Master's degree (2009) in Architecture and Urbanism from FAU-UnB. Advisor in graduate programs and coordinator of Reabilita Postgraduate Certificate Course. He leads research at LaSUS and SICAC (FAU-UnB), developing projects on sustainable architecture, energy-efficient public buildings, and integrated outreach methodologies.

Universities and Global Transformative Processes to Tackle Climate Change. An Overview of Latin America

Marcelo Knobel, Neri de Barros Almeida, and Mariana Ceci de França e Silva

Abstract Addressing the escalating climate crisis requires transformative societal action. Higher education institutions, as centers of knowledge and vital contributors to society, play an essential role in this effort. This study examines the sustainability initiatives of 24 leading higher education institutions across 11 Latin American countries. It specifically looks at how these institutions publicly showcase their climate change responses on their websites, focusing on policies, research, education, and community engagement. The findings highlight the need for more ambitious and systemic strategies, improved cooperation and communication, and a greater focus on education and community outreach to create meaningful societal change in response to climate challenges.

Keywords Climate change · Universities · Latin America · Transformative processes

1 University Engagement: A Crucial Issue

The concept of sustainable development gained global recognition in 1987 when the United Nations General Assembly adopted the *Our Common Future* report. It was established as a collective commitment to meet the needs of the present without compromising the ability of future generations to meet their own needs (Brundtland et al. 1987). Over the past 38 years, however, concerns regarding the capacity of nations to adhere to this principle have intensified significantly.

M. Knobel (✉) · N. de Barros Almeida · M. C. de França e Silva
University of Campinas, Campinas, Brazil
e-mail: knobel@unicamp.br

N. de Barros Almeida
e-mail: neri@unicamp.br

M. C. de França e Silva
e-mail: m240605@g.unicamp.br

W. Leal Filho et al. (eds.), *Sustainability in the Curriculum and Operations of Universities*, World Sustainability Series,
https://doi.org/10.1007/978-3-032-12678-8_2

Since 2015, with the establishing of the 17 Sustainable Development Goals (SDGs) under the United Nations 2030 Agenda, sustainability has gained unprecedented prominence, becoming part of mainstream discourse and inspiring initiatives and actions across various sectors of society. However, an apparent contradiction remains between the goal to promote prosperity while protecting the planet and the ongoing decline of the conditions needed to achieve this aim.

The Potsdam Institute for Climate Impact Research (PIK) monitors planetary boundaries, which clearly indicates this situation. In October, the Institute observed that in 2024, seven out of nine planetary boundaries had been exceeded: land use change, climate change, biodiversity loss, nitrogen and phosphorus cycles, freshwater use, chemical pollution, and ocean acidification (Ripple et al. 2024). Only the levels of aerosols in the atmosphere and the size of the ozone layer remained within safe limits.

The data also indicates that climate change is directly or indirectly responsible for the adverse outcomes. Indeed, according to the *Copernicus Climate Change Service* (C3S), 2024 was the "warmest year on record globally, and the first year that the average global temperature exceeded 1.5 °C above its pre-industrial level" (Copernicus Climate Change Service 2024). The new planetary reality requires us to consider sustainability in the context of climate change, meaning recognizing the problem as a priority for caring for the planet, which involves speeding up the energy transition and protecting and restoring ecosystems, especially forests and water resources. The necessity of preventing the imminent failure to keep global temperatures below 1.5 °C compared to pre-industrial levels, as outlined in the Paris Agreement,[1] makes it essential to have diverse institutional spaces for information, knowledge and action. These spaces should be organized democratically, connected to citizenship and the common good, and capable of supporting processes of cultural change.

The literature identifies at least four key stakeholders who should recognize climate change as an urgent issue: politicians and policymakers, who control resources and have the authority to regulate green transitions; markets, which dominate the energy sector and play a role in investing in and adopting green solutions; citizens, who represent a powerful political force; and Higher Education Institutions (HEIs), which possess the capacity to generate new knowledge and drive innovation (Kekäle and Pinheiro 2025).

Generally speaking, universities have extensive reach, diversity, and influence both locally and globally. This influence is shown through their ability to produce scientific knowledge on climate change, develop effective technologies and solutions to minimize its impacts, and shape society through the professionals they educate

[1] "Article 2—(1) This Agreement, by strengthening the implementation of the Convention, including its objective, aims to strengthen the global response to the threat of climate change, in the context of sustainable development and poverty eradication efforts, including:

(a) Keep the increase in the global average temperature well below 2 °C above pre-industrial levels, and endeavour to limit this temperature increase to 1.5 °C above pre-industrial levels, recognising that this would significantly reduce the risks and impacts of climate change".

https://www.planalto.gov.br/ccivil_03/_Ato2015-2018/2017/Decreto/D9073.htm.

and the broad international networks they establish. Additionally, universities help promote the behavioral changes needed for society to adopt adaptive practices to face the challenges of the climate crisis. In this context, a key question arises: Are universities addressing the sustainability challenge related to climate change?

2 A Perspective on the Issue in Latin America

To answer this question, between August and November 2024, we investigated 24 higher education institutions—23 universities and one polytechnic—from 11 Latin American countries. The institutions were selected from those top-ranked by the SCIMago Institutions Rankings (SIR). The SIR ranks academic and research institutions using a composite indicator that combines three dimensions: research performance, innovation outputs, and societal impact, with the latter assessed through web visibility. The ranking is based on bibliometric data from over 5000 of the world's most productive research institutions (López-Illescas et al. 2011).

Despite knowing the various criticisms of university rankings, that range from the lack of transparency in the data collection method to their bias towards research and STEM subjects (UNU-IIGH 2023), we chose to use the SIR as a starting point for selecting institutions because, starting in 2024, it included the United Nations Sustainable Development Goals as part of its evaluation criteria. Specifically, the SDGs now make up 5% of the total assessment, reflecting the number of documents related to them produced by each institution (SCIMago n.d.). This is particularly relevant given that the literature highlights the impossibility of separating the objective of measuring quality from that of promoting it. In this context, the criteria and values assessed by rankings not only serve as measures of quality but also actively promote and reinforce those same values (Kulczycki 2023).

We analyzed data from four institutions in Brazil, three in Colombia, two each in Argentina, Chile, Uruguay, Bolivia, Peru, Ecuador, Venezuela, and Mexico, and one institution in Paraguay. The complete list of institutions is shown in Table 1.

To identify initiatives aimed at transforming the institutional climate, the research focused on two areas: (1) central management initiatives related to themes like sustainability, climate change, or environmental change, which, despite their differences, serve as strong indicators of the institutions' engagement with policies in the field of "sustainability" and (2) the mention of these themes in the institutions' strategic plans. To pinpoint efforts to change the institutional climate, we evaluated the scope of the initiatives by considering their potential impact on the institution. This involved assessing the effects on one or more of the institutions' core activities (teaching, research, extension) and support activities (management).

A total of 51 initiatives were identified and sorted into the following categories:

- **Educational**: Includes courses and graduate programs focused on climate change.
- **Teaching**: Considers initiatives aimed at integrating the climate change theme into the overall curriculum of courses and teacher training.

Table 1 Institutions analyzed by country

Country	Institution
Brazil	Universidade de São Paulo
	Universidade Estadual de Campinas
	Universidade Federal do Rio de Janeiro
	Universidade Federal de Minas Gerais
Colombia	Universidad Nacional de Colombia
	Universidad de Antioquia
	Universidad de los Andes
Argentina	Universidad de Buenos Aires
	Universidad Austral
Chile	Pontificia Universidad Católica de Chile
	Universidad de Chile
Uruguay	Universidad Católica del Uruguay
	Universidad ORT Uruguay
Bolivia	Universidad Mayor de San Andrés
	Universidad Mayor de San Simón
Peru	Pontificia Universidad Católica del Perú
	Universidad Peruana Cayetano Heredia
Ecuador	Escuela Superior Politécnica del Litoral
	Universidad de las Américas
Venezuela	Universidad Central de Venezuela
	Universidad de los Andes Venezuela
Mexico	Universidad Nacional Autónoma de México
	Instituto Tecnológico y de Estudios Superiores de Monterrey
Paraguay	Universidad Nacional de Asunción

Source Author's own

- **Students**: Accounts for students' initiatives and committees supported and recognized by the institution.
- **Institutional**: Considers the inclusion in strategic plans and creation of specific management boards or committees.
- **Research**: Includes centers that are officially recognized and highlighted by the university. When identifying research initiatives, terms like "climate change," "environmental change," or "sustainability" in their titles were considered.

Our methodological approach involved a systematic review of publicly available information on each institution's official website, focusing specifically on the homepage and sections dedicated to research, teaching, and strategic planning. Additionally, we included information that, while not directly visible on the homepage, was accessible within a maximum of two clicks from the website's main navigation menus.

The decision to focus on the content displayed on the institution's homepage was based on the understanding that effective communication is a key instrument in managing climate change risks, as it promotes public engagement and raises awareness (Badullovich et al. 2025). Additionally, the literature highlights that communication is often overlooked within higher education institutions, despite its strategic importance in addressing major challenges like the rapid spread of anti-scientific misinformation and the underestimation of education's societal role, issues that threaten the legitimacy of universities themselves. Some authors argue that communication is so vital to the modern mission of higher education that it should be considered the "fourth mission" of the university—alongside teaching, research, and extension—as a means to connect with society, foster public engagement, and actively participate in the public sphere (Knobel and Reisberg 2022). In this context, the website, particularly the homepage, serves as the institution's "virtual face," reflecting its values and goals (Meyer 2008).

We emphasize that the data collected represent only part of the actions taken by institutions related to "sustainability." Local initiatives developed by teaching and research units would definitely make each institution's results more solid. However, we chose to focus solely on initiatives connected to central management, identified through the main menu of their website's research section, from the homepage, or through items that are easily accessible and identifiable by the public.

In view of the categories listed above, we found the following prevalence among the 51 initiatives listed in the 24 institutions:

- **Institutional**: 28 initiatives or 54.9% of the total;
- **Research**: 13 initiatives or 25.49% of the total;
- **Educational**: 6 initiatives or 11.76% of the total;
- **Teaching**: 3 initiatives or 5.88% of the total;
- **Students**: 1 initiative or 1.9% of the total.

It is also worth noting that 34 initiatives were not listed on the website's homepage. Instead, they were identified through the institution's institutional plan, strategic plan, or the 'teaching,' "research," and "extension" tabs. Only 17 initiatives appeared directly on the website's homepage, one of which had an active affiliate link. Analyzing this data requires careful consideration and the incorporation of new information. We might ask ourselves whether this indicates that publicizing this information is not a priority for the institutions' policies. Nonetheless, this institutional data could be more prominently featured to enhance the universities' emphasis on the issue of climate change.

All the institutions analyzed have initiatives in the field of "sustainability" regarding climate change; however, in most cases, their scope does not suggest the existence of projects aimed at transforming the institutional climate by focusing on specific segments. Although these are important—as they can demonstrate effective use of concrete opportunities offered to the community, such as waste collection and the partnership with a cooperative of urban waste pickers carried out by the University of Buenos Aires (UBA Verde)—they overlook other opportunities of significant impact that could be equally leveraged, particularly in the areas of professional and citizen training.

Out of the 24 institutions studied, 15 (62.5%) have sustainability policies overseen by dedicated departments or committees. However, in most cases, these groups mainly focus on campus infrastructure, dealing with issues like energy, waste, and water management, and sometimes on reducing greenhouse gas (GHG) emissions. The scope and effect of these efforts differ widely among institutions, depending on their priorities and how much they impact various aspects of university life. Usually, these policies are developed within the frameworks of "environmental management" and "sustainable campus" approaches.

Only two institutions explicitly expand these efforts to include formal education as part of their sustainability strategies: the Federal University of Minas Gerais, which sets sustainability targets for teaching activities, and the Escuela Superior Politécnica del Litoral in Ecuador, which integrates operational and academic sectors. Even among initiatives limited to environmental management, there are notable differences in scope. For example, the Universidad Peruana Cayetano Heredia stands out by combining waste management and GHG reduction with afforestation programs and the creation of an environmental volunteer organization. The remaining 13 institutional initiatives are embedded in strategic plans and involve varying levels of ambition—ranging from afforestation and recycling to benchmarking and partnerships with universities and government agencies. Notable examples include developing a sustainable innovation district at the State University of Campinas and creating a sustainability and climate change agency at the Universidad de Chile.

Following institutional activities, "Research" is the second most common category, with 13 initiatives, including 12 research centers and one scientific journal. As expected, this category highlights one of the most important aspects of the institutional profile and covers a wide range of topics, such as the Amazon, tropical agriculture, GHG emissions, climate change, biodiversity, marine and atmospheric studies, sustainable development, and energy. Notably, together, research centers and institutional initiatives account for about 80% of all actions identified in the areas of sustainability and responses to environmental and climate change. This focus highlights a clear imbalance compared to another equally vital mission of these institutions: teaching.

The "Educational" category, which includes courses and graduate programs—primarily focused on professional training—is smaller in number but notable for its diverse offerings. Examples include a complementary transdisciplinary certification in sustainability at the Universidad de Chile and multidisciplinary training in responsible business management at the Universidad Católica del Uruguay. Additionally, four other programs were identified in this category, covering areas such as sustainable engineering, GHG emissions management, sustainable architecture, and broader sustainability themes.

The "Teaching" category is the second least represented in terms of the number of initiatives. In fact, only one institution offers an introductory course in a virtual format aimed at academics and teachers, providing concepts and tools to address sustainable development and the climate crisis in their educational practices (Universidad de Chile). Additionally, two institutions have adopted the proposal to include the subject in their curriculum, focusing on a minimum level of training for all students (Universidad Nacional Autónoma de México and Instituto Tecnológico y de Estudios Superiores de Monterrey). This may partly explain the low number of initiatives in the "student" category, where we have only seen one institutional action—the Ecology and Environment Secretariat—organized by the Universidad de Chile Student Federation, which has a presence on the central management website.

The data show that institutional commitment mainly focuses on two areas: (1) updating campus infrastructure to lessen environmental impact and (2) supporting research centers and groups. In contrast, only five institutions had initiatives related to education and training, which appeared in two different forms: (1) voluntary participation of the academic community in basic training offered by central management, and (2) the provision of formal and informal education to help shape the institutional climate.

In the first case, examples include training young leaders (Pontificia Universidad Católica de Chile and Universidad de Chile), offering training courses for students and faculty (Universidad de Chile), and developing educational content and sharing information with the wider community (Pontificia Universidad Católica del Perú). In the second case, a more comprehensive and integrated approach combining operational and educational initiatives was observed, with notable examples being the Escuela Superior Politécnica del Ecuador, the Universidad Nacional Autónoma de México (UNAM), and the Instituto Tecnológico y de Estudios Superiores de Monterrey. These latter institutions are especially distinguished for being the only ones to formally promote broad-based sustainability training for all students within their strategic plans. Specifically, UNAM plans to update its curriculum to ensure that sustainable development is integrated at all educational levels, with particular emphasis on undergraduate programs.

In this scenario, some institutions stand out for having more comprehensive policies and enjoying priority status. They are:

- ***Universidad de Chile***: Of the five data categories used in this study (institutional, educational, teaching, student, and research), the institution met all of them, and these are highlighted on the homepage of the central management website.
- ***Universidad Nacional Autónoma de México***: Combines two major interdisciplinary and intersectoral initiatives in its institutional, research, and teaching areas, emphasizing climate change in the university's strategic plan and recognizing the importance of transforming curricula to include discussions on the issue. It has established a social commitment to sustainability in its Strategic Plan 2023–2027, which links sustainability to five projects: updating study plans and programs, including at least one item on sustainable development at all educational levels; fostering a cross-cutting debate on sustainable development across all degree programs; strengthening the UNAM Environmental Badge as a tool for assessing the environmental performance of university buildings; incorporating the 2030

Agenda into UNAM; and launching a mobility program for bicycles, "*pumabuses y automóviles.*"

- ***Escuela Superior Politécnica del Litoral***: Its Sustainability Programme, established in 2018, seeks to develop, implement, and coordinate initiatives related to sustainability. It includes an academic division (supporting critical thinking development among teachers and students on environmental issues and encouraging research), an operational division (focused on making the campus more environmentally efficient), and a community division (promoting campaigns for students, teachers, and staff to apply sustainability in daily life).
- ***Instituto Tecnológico y de Estudios Superiores de Monterrey***: Has developed the "Blue Route" plan to address climate change, focusing on five strategic areas: culture, mitigation and adaptation, research, education, and networking. These efforts involve over 50 departments and 70 concurrent projects. The plan also aims to integrate climate change and SDG 13 (Climate Action) into all professional curricula across the 14 training units, ensuring that every student in professional programs encounters the topic at least once during their studies. Additionally, sustainable development is incorporated into curriculum design, and at least 15 credits are dedicated to it. Sustainable development should also be embedded into the design of transversal competencies within teaching programs. A key highlight is that the goals are clearly outlined and supported by a predefined evaluation model reviewed annually to ensure the objectives are met by 2025. The *site's homepage* is also emphasized through news updates or direct access to the plan itself.

Based on these results, institutions can be categorized as more or less responsive to the imperatives of climate change using the following criteria: (1) those with a policy that has clear objectives, targets, and a body responsible for promoting, monitoring, and evaluating them; (2) the extent of this policy's impact on different segments of the university community and its surroundings (students, teachers, staff, users, external community); and (3) its influence on the core activities (teaching, research, extension, and management). From this perspective, most institutions respond slowly and sectorally, failing to leverage much of their potential to contribute meaningfully to the transformation needed by climate change.

3 A Crucial Intergenerational Space

Given this situation, it is important to consider another aspect of the challenge: What are the current social conditions for implementing large-scale institutional sustainability projects within universities? The study's findings reveal that although most higher education institutions focus on sustainability efforts in infrastructure and research, initiatives related to education and student engagement remain limited.

This gap is especially significant when considering the attitudes of the main group within these institutions: young adults. The "National Survey on Climate

Change," conducted in Brazil by the non-governmental organization Reconectta and the Schools for Climate movement and published in April 2025, highlights this issue. The survey gathered responses from 6586 people across all regions of Brazil, divided into three age groups. It found that while children aged 7–14 are optimistic and eager to act, and adults over 30 are still willing to take action despite low optimism, the 15–29 age group—which includes university students—shows both low optimism and little motivation to participate in climate efforts (Grandisoli et al. 2025). This paradox highlights the need for universities to go beyond merely implementing structural and research-based sustainability initiatives and to actively develop engagement strategies that inspire and mobilize students—the future educators, professionals, and citizens responsible for advancing these efforts. As spaces dedicated to knowledge creation, sharing, and social change, universities are uniquely positioned to shape social attitudes, foster resilience, and prepare young leaders to face the climate crisis.

To genuinely fulfill this role and connect infrastructure-focused actions with broader educational and social engagement, institutions must move beyond isolated efforts and adopt integrated, ambitious sustainability models that span multiple aspects of university life. This transformation requires a comprehensive approach that includes:

- Management: the sustainable management of energy, water, waste, soil, food, vegetation, and transportation; the setting of clear sustainability goals within ambitious deadlines overseen by a dedicated management team; and the development of assessment systems to evaluate the effectiveness and long-term impact of these initiatives.
- Extension: ongoing community engagement, development of long-term collaborative projects, and active promotion of volunteer involvement in sustainability management.
- Research: enhanced support for interdisciplinary research centers that unite diverse fields to tackle complex environmental challenges through innovative, solution-focused methods.
- Education: investment in teacher and staff training for both formal and informal learning settings, development of technical leadership, creation of new courses and disciplines, and systematic review and update of existing curricula to incorporate sustainability and climate change as central themes.

By progressing in these interconnected areas, universities can not only reinforce their leadership in sustainability but also contribute more effectively to societal change in response to the climate emergency.

To implement this vision, the institutional data collected in this study provides a valuable resource for improving sustainability policies, setting clearer and more ambitious goals, and strengthening accountability systems. Additionally, creating regular channels for information exchange among institutions can accelerate progress and promote greater alignment in tracking and assessing sustainability outcomes. In this context, university rankings can serve as powerful drivers, encouraging institutions to adopt more comprehensive, transparent, and impactful methods for climate action—thereby increasing their societal relevance and their role in addressing one of the most urgent challenges of our time.

4 Conclusion

Based on the data presented, several key conclusions emerge. First, institutional responsibilities are not understood consistently among the analyzed institutions. Second, most of these institutions lack a sense of urgency, and third, a major curriculum review is not imminent for most of them. The variety of initiatives and their partial implementation also led us to conclude that sharing experiences between institutions could accelerate the institutional response to sustainability challenges in the context of climate change.

The variety of initiatives and their partial or fragmented implementation indicate that sharing experiences and best practices among institutions could significantly help accelerate responses to the sustainability challenges posed by climate change. Improved collaboration could help close existing gaps and promote more consistent and effective strategies.

Additionally, the analysis reveals critical issues that remain largely absent from current institutional policies. These include:

1. The organization of actions to support the mental health of university communities, especially in response to the increasing issue of climate anxiety;
2. The recognition of universities as key spaces for fostering intergenerational dialogue and cooperation, which are important both for advancing climate action and for mitigating its psychological impacts; and
3. Enhancing communication strategies and creating ongoing outreach efforts that can foster public trust and widespread social support for institutional environmental agendas.

These findings emphasize what the literature describes as a core weakness in universities' responses to the climate crisis: the difficulty of articulating their contributions in a systematic, coordinated, and interdisciplinary manner. Current organizational structures, curricula, and research programs often do not reflect either the urgency or the cross-cutting nature of the climate emergency. This shortcoming is also evident in the low visibility of initiatives that engage key stakeholders, such as policymakers and businesses—actors vital for scaling up effective climate action.

Universities, with their scientific expertise and societal influence, are uniquely positioned to promote dialogue among academia, government, industry, and civil society. Therefore, publicizing institutional efforts and clearly communicating the central role of universities in addressing the climate crisis is necessary not only to increase societal engagement but also to reinforce the importance of higher education institutions as active players in confronting the most urgent global challenges.

It should be noted that the research conducted offers some indications that require further investigation for confirmation. First, it would be important to conduct direct inquiries with the analyzed institutions to verify the actions actually taken and their performance compared to the initial plan and schedule. A particularly interesting aspect to explore is the verification of extension actions and whether their low profile in the survey is related to the still emerging focus universities have on the general education of their students. This could be assessed by checking whether institutions with plans that involve the transversal education of all their students show more significant extension activities in this area, whether due to institutional policies or the efforts of their members.

Finally, a more detailed analysis of key terms such as "sustainability," "climate change," and "environmental change" could help us better understand progress toward sustainability policies from a climate change perspective. Currently, among the initiatives analyzed, 36 mention one or more of the SDGs, and climate change is referenced in 11 initiatives, primarily in research activities (5 direct mentions). Only the Instituto Tecnológico y de Estudios Superiores de Monterrey (Mexico) includes climate change as a core element of its strategic plan.

This exploratory research aimed to gather straightforward and accessible information on institutional sustainability activities. While acknowledging the existence of dedicated research groups, content development initiatives, and informal community efforts within these institutions, our focus was on publicly visible institutional strategies. We believe this research sheds light on key aspects of the institutional vision, which can help improve sustainability policies and highlight initiatives led by specific academic sectors. Indeed, our findings indicate that most institutions currently provide limited visibility to sustainability efforts and their measurable outcomes. Indirectly, the data analysis suggests that external pressures from governments, agencies, and university rankings—despite their inherent limitations and ongoing critiques regarding their methodologies and impact on institutional priorities—could speed up the transformation of the organizational climate within teaching and research institutions. Moreover, it emphasizes the vital role of actively involving the student community in creating and sharing knowledge essential for building resilience to environmental changes. Encouraging student participation in decision-making and backing their initiatives can act as a strong driver for real progress.

References

Badullovich N, Kotcher J, Myers TA, et al. Understanding six "key truths" about climate change predicts policy support, discussion, and political advocacy. Climatic Change. 2025;178(89). https://doi.org/10.1007/s10584-025-03934-3

Brasil. Decreto nº 9.073, de 5 de junho de 2017 [Decree no. 9.073, of 5 June 2017]. Presidência da República. Available from: https://www.planalto.gov.br/ccivil_03/_Ato2015-2018/2017/Decreto/D9073.htm

Brundtland GH, Khalid M, Agnelli S, Al-Athel SA, Chidzero B, Fadika LM, Hauff V, Lang I, Ma S, Botero MM, Singh N, Nogueira Neto P. Our common future; by World Commission on Environment and Development. 1987

Copernicus Climate Change Service. Global Climate Highlights 2024. 10 Jan 2025. Available from: https://climate.copernicus.eu/global-climate-highlights-2024

Camacho E. La nueva red universitaria que busca soluciones científicas ante la crisis climática [The new university network seeking scientific solutions to the climate crisis]. TecScience. 2025 May 26. Available from: https://tecscience.tec.mx/es/clima-y-sostenibilidad/red-mexicana-de-universidades-accion-climatica/

Copernicus Climate Change Service. 2024: First year to exceed 1.5°C above pre-industrial level. 2024. Available from: https://climate.copernicus.eu/copernicus-2024-first-year-exceed-15degc-above-pre-industrial-level

Dibbern TA, Cristofoletti EC, Serafim MP, Alves DS. SDGs and Latin American university: Impact of scientific knowledge production in policy documents. In: Leal Filho W, et al., editors. SDGs in the Americas and Caribbean region: Implementing the UN Sustainable Development Goals - Regional perspectives. Springer; 2022. p. 1–29. https://doi.org/10.1007/978-3-030-91188-1_37-1

Escuela Superior Politécnica del Litoral. Programa de Sostenibilidad [Sustainability Programme]. Available from: https://sostenibilidad.espol.edu.ec/

Grandisoli E, Ribeiro L, Giglioti D, Fanti A, Marchini S. Pesquisa nacional sobre mudanças climáticas: percepções das crianças e das juventudes [National survey on climate change: perceptions of children and youth]. São Paulo: Movimento Escolas pelo Clima e Reconectta; 2025

Instituto Tecnológico y de Estudios Superiores de Monterrey. Ruta Azul - Reporte de Sostenibilidad [Ruta Azul - Sustainability Report]. Available from: https://tec.mx/sites/default/files/repositorio/sentido-humano/sostenibilidad/ruta-azul/reportes/ReporteRutaAzul2023-2024Ejecutivo.pdf

Knobel M, Reisberg L. Effective communication: the 4th mission of universities—a 21st century challenge [Internet]. Research & Occasional Paper Series: CSHE.6.2022. Berkeley: Center for Studies in Higher Education; 2022 [cited 2025 Jul 9]. Available from: https://cshe.berkeley.edu/publications/rops

Kekäle J, Pinheiro R. Climate change, higher education and the wicked problem of time. In: Kekäle J, Pinheiro R, editors. Higher education policy for tackling climate change. Palgrave Macmillan; 2025. https://doi.org/10.1007/978-3-031-78479-8_1

Kulczycki E. The Evaluation Game: How Publication Metrics Shape Scholarly Communication. Cambridge: Cambridge University Press; 2023.

Leal Filho W, Aina YA, Dinis MAP, Purcell W, Nagy GJ. Climate change: Why higher education matters? Sci Total Environ. 2023;892:164819. https://doi.org/10.1016/j.scitotenv.2023.164819

Leal Filho W, Portela de Vasconcelos CR, editors. Pursuing sustainable development across disciplines. In: Handbook of best practices in sustainable development at university level. Springer; 2022. p. 495–504.

López-Illescas C, de Moya-Anegón F, Moed HF. A ranking of universities should account for differences in their disciplinary specialization. Scientometrics. 2011;88(2):563-74. https://doi.org/10.1007/s11192-011-0398-6

McCowan T. Universities and climate action. UCL Press; 2025. Available from: https://discovery.ucl.ac.uk/id/eprint/10207460/

McCowan T, Leal Filho W, Brandli L, editors. Universities facing climate change and sustainability. Hamburg; 2021.

Meyer KA. The "virtual face" of institutions: what do home pages reveal about higher education? Innov High Educ. 2008;33(3):141–57. https://doi.org/10.1007/s10755-008-9071-2

Pontificia Universidad Católica de Chile. Estrategia de Acción Climática - Carbono Neutral 2038 [Climate Action Strategy - Carbon Neutral to 2038]. Available from: https://sustentable.uc.cl/wp-content/uploads/2023/09/Estrategia-de-Accion-Climatica-UC-2023-VF.pdf

Pontificia Universidad Católica de Chile. Proyecto HEADCET - Alianzas para el Desarrollo Local Sostenible [HEADCET Project - Higher Education Partnerships for Sustainable Local Development]. Available from: https://www.headcet.eu/project/

Pontificia Universidad Católica del Perú. Campus Sostenible [Sustainable Campus]. Available from: https://dars.pucp.edu.pe/que-hacemos/desarrollo-organizacional/medio-ambiente/
Pontificia Universidad Católica del Perú. Clima de Cambios [Climate of Change]. Available from: https://www.pucp.edu.pe/climadecambios/
Pontificia Universidad Católica del Perú. Kawsaypacha: Sociedad y Medio Ambiente [Kawsaypacha: Society and Environment]. Available from: https://revistas.pucp.edu.pe/index.php/Kawsaypacha
Pontificia Universidad Católica del Perú. Red Peruana de Ciclo de Vida y Ecología Industrial [Peruvian Life Cycle Assessment and Industrial Ecology Network]. Available from: https://red.pucp.edu.pe/ciclodevida/en/sobre-la-red-ciclo-de-vida/presentacion/
Ripple WJ, Wolf C, Gregg JW, Rockström J, Mann ME, Oreskes N, Lenton TM, Rahmstorf S, Newsome TM, Xu C, Svenning JC, Pereira CC, Law BE, Crowther TW. The 2024 state of the climate report: Perilous times on planet Earth. BioScience. 2024;biae087. https://doi.org/10.1093/biosci/biae087
SCImago. SJR — SCImago Journal & Country Rank [Internet].. Available from: http://www.scimagojr.com
Silva RLF, Grandisoli E, Jacobi PR. The role of environmental education in tackling the climate emergency and biodiversity decline: Articulating Brazilian experiences. Int J Environ Resil Sci. 2024;6(2):1-25.
TSO Uruguay. Diploma de Especialización en Arquitectura Sostenible [Diploma of Specialisation in Sustainable Architecture]. Available from: https://fa.ort.edu.uy/diploma-de-especializacion-en-arquitectura-sostenible
TSO Uruguay. ORT Recicla [ORT Recycles]. Available from: https://www.ort.edu.uy/recicla
United Nations University International Institute for Global Health. Statement on global university rankings [Internet]. Kuala Lumpur: UNU-IIGH; 2023 [cited 2025 Jul 9]. Available from: https://doi.org/10.37941/PB/2023/2
Universidad Austral. Comité de Sostenibilidad [Sustainability Committee]. Available from: https://www.austral.edu.ar/sostenibilidad/
Universidad Austral. Diplomatura en Gestión de Reducción de Emisiones [Diploma in Emissions Reduction Management]. Available from: https://www.austral.edu.ar/ingenieria/ingenieria-posgrados/energia/diplomatura-en-gestion-de-reduccion-de-emisiones/
Universidad de Buenos Aires. Centro GEO de Desarrollo Sostenible [GEO Sustainable Development Centre]. Available from: https://centrogeo.economicas.uba.ar/
Universidad de Buenos Aires. Centro de Investigación del Mar y la Atmósfera [Sea and Atmosphere Research Centre]. Available from: https://www.cima.fcen.uba.ar/CambioClimatico.php
Universidad de Buenos Aires. UBA Verde [Green UBA]. Available from: https://www.uba.ar/ubasociedad
Universidade de São Paulo. Centro de Estudos da Amazônia Sustentável [Centre for Sustainable Amazon Studies]. Available from: https://ceas.usp.br/
Universidade de São Paulo. Centro de Estudos de Agricultura Tropical Sustentável (STAC) [Centre for the Study of Sustainable Tropical Agriculture]. Available from: https://www.stacusp.com.br/
Universidade de São Paulo. Centro de Pesquisa de Carbono em Agricultura Tropical (CCARBON) [Centre for Carbon Research in Tropical Agriculture]. Available from: https://ccarbon.usp.br/
Universidade de São Paulo. Centro de Pesquisa em Inovação em Gases de Efeito Estufa (RCGI) [Research Centre for Greenhouse Gas Innovation]. Available from: https://sites.usp.br/rcgi/br/home/
Universidade de São Paulo. Superintendência de Gestão Ambiental [Superintendence of Environmental Management]. Available from: https://sga.usp.br/
Universidade Estadual de Campinas. Centro de Pesquisas em Genômica Aplicada às Mudanças Climáticas [Centre for Research in Genomics Applied to Climate Change]. Available from: https://www.gccrc.unicamp.br/
Universidade Estadual de Campinas. Coordenadoria de Sustentabilidade (CSUS) [Sustainability Division Coordinator]. Available from: https://www.depi.unicamp.br/csus/

Universidade Estadual de Campinas. Consórcio Universitário Hemisférico [Hemispheric University Consortium]. Available from: https://thehuc.org/initiatives/
Universidade Estadual de Campinas. HUB Internacional para o Desenvolvimento Sustentável [International HUB for Sustainable Development]. Available from: https://hids.org.br/
Universidade Estadual de Campinas. Relatório de Sustentabilidade [Sustainability Report]. Available from: https://drive.google.com/file/d/1v3zCVV-RTKIsSLBavzQd3GILyjZoHp89/view
Universidade Federal de Minas Gerais. UFMG Sustentável [Sustainable UFMG]. Available from: https://www.ufmg.br/sustentabilidade/
Universidade Federal do Rio de Janeiro. Instituto de Biodiversidade e Sustentabilidade [Institute of Biodiversity and Sustainability]. Available from: https://nupem.ufrj.br/
Universidade Federal do Rio de Janeiro. Plano de Logística Sustentável [Sustainable Logistics Plan]. Available from: https://ufrj.br/acesso-a-informacao/acoes-e-programas/plano-de-logistica-sustentavel/
Universidad de Chile. Acuerdo de Producción Limpia [Clean Production Agreement]. Available from: https://uchile.cl/sustentabilidad/iniciativas-destacadas/apl-i
Universidad de Chile. Certificación Complementaria en Interdisciplina y Transdisciplina para la Sustentabilidad [Complementary Certificate in Interdiscipline and Transdiscipline for Sustainability]. Available from: https://cits.uchile.cl/certificacion-complementaria-cits
Universidad de Chile. Curso 'Educar para la Sustentabilidad' [Educating for Sustainability course]. Available from: https://uchile.cl/sustentabilidad/iniciativas-destacadas/curso-educar-para-la-sustentabilidad
Universidad de Chile. Política de Sustentabilidad [Sustainability Policy]. Available from: https://uchile.cl/sustentabilidad/comite-por-la-sustentabilidad/objetivos
Universidad de Chile. Secretaría de Ecología y Medio Ambiente FECH [Ecology and Environment Secretariat FECH]. Available from: https://www.instagram.com/secma.fech/
Universidad de Antioquia. UdeA Sostenible [Sustainable UdeA]. Available from: https://encurtador.com.br/iYVA0
Universidad de los Andes. Campus Sostenible [Sustainable Campus]. Available from: https://sostenibilidad.uniandes.edu.co/
Universidad de los Andes. Maestría en Sostenibilidad [Master's in Sustainability]. Available from: https://cider.uniandes.edu.co/es/programas/maestria/estudios-sobre-sustentabilidad
Universidad Nacional de Colombia. Campus Sostenible [Sustainable Campus]. Available from: https://minas.medellin.unal.edu.co/campus-sostenible/lineas-de-accion-campus-sostenible/
Universidad de Las Américas. Acciones Ambientales en Campus [Environmental actions on campus]. Available from: https://www.udla.edu.ec/ambiental
Universidad Católica del Uruguay. Curso de Desarrollo Sostenible - Escuela de Negocios UCU [Sustainable Development Course - UCU Business School]. Available from: https://ucubs.edu.uy/programas-focalizados/programa-de-desarrollo-sustentable
Universidad Católica del Uruguay. Observatorio de Energía y Desarrollo Sostenible [Observatory on Energy and Sustainable Development]. Available from: https://www.ucu.edu.uy/categoria/Observatorio-de-Energia-y-Desarrollo-Sus-401
Universidad Central de Venezuela. Plan Estratégico de Gestión Ambiental [Strategic Environmental Management Plan]. Available from: http://www.ucv.ve/sobre-la-ucv/resena-organizacional/plan-estrategico-ucv-en-construcao/lineas-estrategicas/gestion-ambiental.html
Universidad Nacional Autónoma de México. Compromiso social con la sostenibilidad - Plan Estratégico 2023–2027 [Social commitment to sustainability - Strategic Plan 2023–2027]. Available from: https://www.rector.unam.mx/docs/PDI-2023-2027.pdf
Universidad Nacional Autónoma de México. Programa de Investigación en Cambio Climático (PINCC) [Climate Change Research Programme]. Available from: https://www.pincc.unam.mx/
Universidad Nacional de Asunción. Plan de forestación y reciclaje [Afforestation and recycling plan]. Available from: https://www.una.py/wp-content/uploads/2021/11/Experiencias-UNA-con-RS.pdf

Universidad Peruana Cayetano Heredia. Centro de Investigación Ambiental [Centre for Environmental Research]. Available from: https://investigacion.cayetano.edu.pe/catalogo/csa/
Universidad Peruana Cayetano Heredia. Política ambiental institucional [Institutional environmental policy]. Available from: https://responsabilidadsocial.cayetano.edu.pe/desarrollo-sostenible/
World Meteorological Organisation. State of the global climate 2024. Available from: https://wmo.int/publication-series/state-of-global-climate-2024

Marcelo Knobel (International civil servant currently serving as UNESCO The World Academy of Sciences. The ideas and opinions expressed herein are those of the author(s) and do not necessarily reflect the views of UNESCO.) Executive Director of The World Academy of Sciences for the Advancement of Science in Developing Countries (TWAS/UNESCO). He served as Dean of Undergraduate Studies from 2009 to 2013 and was Rector of University of Campinas (Unicamp) from 2017 to 2021. He is also a professor in the Condensed Matter Physics Department at the Gleb Wataghin Institute of Physics (IFGW/Unicamp). Dr. Knobel holds both a Ph.D. and a bachelor's degree in Physics from Unicamp.

Neri de Barros Almeida Professor of History at the Institute of Philosophy and Human Sciences at the University of Campinas (Unicamp), where she oversees the Center for Studies on Ancient Diversity and was the Executive Director of Human Rights (2019–2021). Dr. Barros Almeida holds a Ph.D. in Social History and a bachelor's degree in History from the University of São Paulo (USP).

Mariana Ceci de França e Silva Master's student in Scientific and Cultural Communication at the Laboratory for Advanced Studies in Journalism at the University of Campinas (Labjor/Unicamp). Holds a specialization in Scientific and Technological Information in Health from the Oswaldo Cruz Institute (Fiocruz) and a bachelor's degree in Social Communication from the Federal University of Rio Grande do Norte (UFRN).

Reporting Progress to Sustainability: Contributions of an Interdisciplinary Graduate Program to Agribusiness

Fábio Mosso Moreira, Eduardo Guilherme Satolo, Wagner Luiz Lourenzani, and Ana Elisa Bressan Smith Lourenzani

Abstract This study analyzes the integration of sustainability-oriented education into both the curriculum and institutional practices of the Graduate Program in Agribusiness and Development (PGAD) at the Faculty of Science and Engineering, in São Paulo State University (UNESP), Tupã Campus. The research began with a comprehensive literature review, which provided a theoretical foundation for understanding the essential contributor behind the incorporation of sustainability into the curricula and operations of higher education institutions. Subsequently, a participatory methodology was employed, involving PGAD faculty members as primary sources of information through interviews and questionnaires. In addition, data were collected from strategic institutional documents, such as theses and dissertations, to evaluate the extent and depth of PGAD's engagement with sustainability and the United Nations Sustainable Development Goals (SDGs). The findings were structured around the three fundamental pillars of higher education: Teaching, Research, and Extension. In the Teaching dimension, the program presents a curricular structure that encourages interdisciplinary and critical reflection on the challenges of agribusiness. In the field of Research, the diversity of topics addressed confirms PGAD's relevance in generating knowledge applied to sustainability. Lastly, Extension initiatives demonstrate that the program is commitment to social engagement and the dissemination of sustainable practices within the community.

Keywords Sustainability · Education · Agribusiness · Graduate program · PGAD

F. M. Moreira (✉) · E. G. Satolo · W. L. Lourenzani · A. E. B. S. Lourenzani
Graduate Program in Agribusiness and Development, School of Sciences and Engineering, São Paulo State University, Tupã, São Paulo, Brazil
e-mail: fabio.moreira@unesp.br

E. G. Satolo
e-mail: eduardo.satolo@unesp.br

W. L. Lourenzani
e-mail: w.lourenzani@unesp.br

A. E. B. S. Lourenzani
e-mail: ana.lourenzani@unesp.br

W. Leal Filho et al. (eds.), *Sustainability in the Curriculum and Operations of Universities*, World Sustainability Series,
https://doi.org/10.1007/978-3-032-12678-8_3

1 Introduction

The twenty-first century presents numerous challenges for contemporary society, including population growth, increased consumption of natural resources, and the intensification of poverty (FAO 2020). Within this context, agribusiness emerges as a crucial economic sector with the potential to these challenges, playing a strategic role in ensuring food security while simultaneously preserving environmental integrity.

Agribusiness is a complex system that encompasses the totality of operations related to the supply of agricultural inputs, production on farming units, storage, processing, and distribution of agricultural products and the products and their derivatives (Batalha and Silva 2021). In response to rising global food demand, this sector holds the potential not only to guarantee food availability for future generations but also to drive economic development through job creation, foster social inclusion, and improve living conditions in rural areas.

Brazil is a leading player in the global agribusiness arena, distinguished by its strong performance in the production and exportation of commodities such as soybeans, sugar, and beef, as well as in ensuring domestic food supply through family farming (MAPA 2024). For decades, Brazilian agriculture has been recognized for its productivity and quality. Nevertheless, the country still faces significant challenges to sustainable development, including ongoing deforestation and forest fires driven by agricultural expansion into frontier areas; inefficient use of natural resources; excessive pesticide application that threatens worker health and contaminates water sources; and precarious living conditions in many rural regions (Batalha and Silva 2021).

Sustainability is a critical topic in contemporary discourse, acquiring particular relevance in agribusiness due to the sector's direct dependence on the availability of natural resources. Agricultural production significantly impacts the environment through land use, water consumption, greenhouse gas emissions, and the application of chemical inputs, making the adoption of sustainable practices imperative (Burnquist et al. 2020). In this context, one of the most effective mechanisms for disseminating sustainability principles across agro-industrial chains is through education and training.

Education for sustainable development encompasses a broad array of dimensions, including sustainability, resilience, cognitive and experiential learning, as well as collaborative and transformative learning processes (Didham et al. 2013). Since the introduction of Agenda 21 and the United Nations Decade of Education for Sustainable Development, education has assumed a central role in knowledge generation, capacity building, and the facilitation of knowledge transfer (Jickling and Wals 2012; Lozano et al. 2013). More recently, the 2030 Agenda and the establishment of the Sustainable Development Goals (SDGs) have provided a comprehensive framework for international efforts to build sustainable societies through education (UN 2021).

Education holds substantial potential to shape values, attitudes, and behaviors of individuals and organizations regarding sustainable development. As such, it serves as a fundamental promoter in the advancement of the SDGs (Hallinger and Chatpinyakoop 2019; Jickling and Wals 2012; UNESCO 2014). According to Leal Filho et al. (2025), universities are key institutions currently leading efforts in environmental sustainability through their policies, research, and community engagement, while also serving as reference in professional education and training. These institutions, however, However, these institutions are increasingly subject to regulatory pressures to demonstrate their commitment to sustainability and often encounter challenges such as the lack of robust models for formulating and implementing sustainable actions.

In this context, the research problem explores how the principles of sustainability and the SGDs are being interpreted and implemented within the context of a graduate program specializing in agribusiness. Accordingly, this study aims to analyze the integration of sustainability-oriented education into the curriculum and operations of the Graduate Program in Agribusiness and Development (PGAD, as its acronym in Portuguese) at the Faculty of Science and Engineering (FCE/UNESP), Brazil.

2 Sustainability in Agribusiness: Education at the Forefront

Incorporating sustainability principles into the curriculum and operations of Higher Education Institutions (HEIs) requires promoting interdisciplinary learning, building partnerships with other institutions and organizations, and equipping both faculty and students to act as agents of change (Leal Filho et al. 2018). Thus, it is essential to establish a clear and coherent theoretical framework that defines the key concepts involved and to explain their interrelationships.

The first key concept that underpins this investigation is the definition of sustainable development. It refers to a process of social and economic progress based on poverty eradication, reducing inequality, promoting social inclusion, protecting the environment, and creating effective economic strategies for both present and future generations (Sinakou et al. 2019). Sustainable development is a concept surrounded by multiple discourses and definitions. Since its introduction in the 1980s, it has sparked considerable academic debate, giving rise to different—and sometimes opposing—schools of thought (Ruggerio 2021).

Sustainable development is often mistakenly used as a synonym for sustainability. According to Feil and Schreiber (2017), sustainability represents a goal defined through scientific criteria that help measure and monitor the outcomes of strategies linked to sustainable development. Thus, sustainability relies on guiding principles to ensure human well-being, such as precaution, responsibility, equity, social and environmental justice, participation, and integration.

Vadakkepatt et al. (2021) describe sustainability as a combination of ideas, attitudes, intentions, and behaviors that involve the strategic management of economic, environmental, and social resources for the benefit of current and future generations.

Fig. 1 The 17 SDGs of the 2030 agenda. *Source* UN (2021)

Bravo et al. (2021) add two equally important dimensions, health and culture, arguing that sustainable models applied to food production directly affect public health and must include cultural elements specific to certain groups, such as indigenous peoples, *quilombolas*, and other traditional communities.

To guide the planning of actions aligned with its sustainability principles, the United Nations proposed the 2030 Agenda (UN 2021), a document that outlines a framework with 17 SDGs to guide international efforts between 2016 and 2030 toward advancing sustainable development (see Fig. 1).

According to Abo-Khalil (2024), integrating the SDGs into the planning and design of curriculum and institutional operations in HEIs offers a range of benefits. These include fostering critical thinking and achieving tangible outcomes such as academic publications; enhancing students' competencies through active learning and engagement; and strengthening planning capacities by developing collaboration skills and encouraging reflective, high-quality communication. Student motivation is also positively impacted as they gain problem-solving abilities and greater environmental awareness. Teamwork improves through exposure to collaborative practices and intercultural experiences, providing broader perspectives and a deeper understanding of global issues. Additionally, this approach allows students to acquire local knowledge and practical skills, sharpening their analytical capabilities and sustainability literacy.

In this way, integrating sustainability into curriculum and operations not only enriches students' education but also prepares them to face real-world challenges, contributing to a more sustainable future. However, HEIs around the world are currently at different stages in adopting sustainability education. While some have quickly implemented sustainable practices, many have only recently begun to integrate such principles (Abo-Khalil 2024).

Ruiz-Mallén and Heras (2020) conducted a study involving 167 universities across five continents to examine the extent to which the SDGs have been incorporated into sustainability education. The study found that lectures remain the most commonly used method to introduce these themes. Although, other approaches (e.g. equipping educators to teach and inspire students about sustainable practices or support for interdisciplinary research addressing complex sustainability challenges) could have a more direct societal impact but they appear to be less frequently adopted.

Bradley (2019) explored the challenges and barriers faced in integrating sustainable development into undergraduate economics curriculum. The author identified a range of obstacles, including institutional barriers (e.g. resistance to change, lack of institutional support, limited funding), academic constraints (e.g. scarcity of specific materials, rigid course structures, overemphasis on neoclassical economics), pedagogical issues (e.g. insufficient faculty training, difficulty in connecting theory to practice, student resistance), and cultural/structural limitations (e.g. influence of external policies and regulations, labor market pressures, lack of successful case studies). To overcome these challenges, Bradley (2019) suggests strategies such as developing targeted teaching materials, offering faculty-training workshops, and promoting interdisciplinary approaches in economics education.

Abo-Khalil (2024) further emphasizes the importance of interdisciplinary approaches and active faculty engagement in advancing sustainability education. The author argues that interdisciplinary pedagogical methods are essential for effectively embedding sustainability into teaching and learning processes. In response, many universities are beginning to overcome traditional academic silos by creating interdisciplinary research centers and programs that encourage collaboration and innovation across different fields of knowledge.

Abo-Khalil (2024) also presents a set of strategic recommendations to enhance the effectiveness and reach of interdisciplinary sustainability education, underscoring the essential role of HEIs in preparing future generations to address the complex challenges of sustainable development:

- Interdisciplinary approaches: Promote collaboration across academic disciplines to foster a holistic understanding of sustainability challenges;
- Active faculty involvement: Encourage proactive participation by faculty in the design and implementation of sustainable curriculum, ensuring they are equipped and motivated to integrate sustainability into their teaching;
- Partnerships with industry and the community: Establish collaborations with industry sectors and local communities to align educational programs with real-world sustainability needs and facilitate the practical application of acquired knowledge;
- Adequate infrastructure and resources: Invest in facilities and tools that support sustainable practices, such as specialized laboratories and green technologies, to create a learning and research environment conducive to sustainability;

- Global initiatives and rankings: Participate in international sustainability initiatives and use sustainability rankings as tools to assess and guide institutional practices, promoting continuous improvement and global recognition.

Having outlined the possibilities for integrating sustainability principles into HEIs curriculum and operations, the following section presents an overview of how such strategies are being implemented by PGAD, an interdisciplinary graduate program focused on the agribusiness sector.

3 Methodology

To achieve the study's objectives, the methodological approach began with a literature review to conceptualize 'sustainability', 'sustainable development', and the 'SDGs', and to establish linkages between these concepts and the domains of teaching and education. This review provided a foundational understanding of the contributors behind the integration of sustainability into higher education institutions' curriculum and operations.

The guiding methodological framework of this research was participatory methodology (Bergold and Thomas 2012), involving PGAD faculty members as primary sources of information. The data collection occurred in 2024 and was based on strategic documents to assess the extent and depth of PGAD's integration with sustainability and the SDGs.

The results were organized in accordance with the three core pillars of higher education—'Teaching', 'Research', and 'Extension'. To examine the incorporation of sustainability principles in the Teaching pillar, the PGAD course curriculum was analyzed. Data collection included an unstructured interview conducted with the program coordinator and vice-coordinator, which provided insights into the design of course modules and the interdisciplinary strategies adopted.

In addressing the Research pillar, questionnaires were distributed to all faculty members affiliated with PGAD. These questionnaires gathered information on faculty-led research projects, including the primary research focus and the types of impacts considered (e.g., environmental, social, cultural, economic). Additionally, research-related data were obtained from Theses and Dissertations defended within PGAD between 2021 and 2024 (76 documents). Each of these final academic works indicated its contribution to one of the 17 SDGs. This allowed for an evaluation of the sustainability concerns present in both faculty members and student research endeavors.

Finally, the Extension pillar was analyzed through the identification of outreach projects coordinated by PGAD faculty that demonstrated direct actions and impacts related to sustainability. A questionnaire was distributed to faculty project coordinators to collect descriptive data and document effective initiatives undertaken between 2021 and 2024.

Regarding data analysis, research-related data were processed to generate network maps highlighting key elements from faculty research projects and student theses/dissertations (i.e., research objects, impacts, SDGs), and their interconnections according to the program's research lines. Gephi® v.0.10 software was employed to develop these visual representations. The results pertaining to Teaching and Extension were presented descriptively, based on interviews with program coordinators and faculty members, emphasizing the importance of integrating sustainability principles into the curriculum and operations of graduate programs that address the multifaceted challenges of the agribusiness sector.

The following section of this paper offers a critical reflection on sustainability concepts and their implications for agribusiness, while also establishing the role of education as a central actor in the formation and professional training of agents operating within this strategic economic sector.

4 Results

The analysis carried out a variety of actions aligned with the commitment to integrating sustainability into higher education, with a focus on the strategies adopted by PGAD to contribute to the achievement of the SDGs. The results are organized around the academic pillars: **Teaching** (knowledge transmission in the classroom), **Research** (generation of scientific and technological knowledge), and **Extension** (university-community engagement), which are considered essential both for academic training and for social development.

4.1 Teaching

To examine how sustainability was incorporated into the Teaching pillar within PGAD's activities, this study investigated the teaching methodologies, course content, and pedagogical practices used to promote both professional and civic development among students. This information was gathered through consultation with the Program's coordinators and highlights the role of interdisciplinary as a facilitator in integrating sustainability into the curriculum and operations.

Regarding PGAD's curricular structure, the core courses in the Master's program are organized into modules designed to foster dialogue across different fields of knowledge and promote interdisciplinary practice.

Module I includes two compulsory courses offered during the first semester: "Systemic Approach to Agribusiness" and "Construction of Interdisciplinary Knowledge." These courses are designed in an integrative manner and aim to provide theoretical, conceptual, and methodological foundations in agribusiness and development, along with methodological support for interdisciplinary research.

As part of the interdisciplinary learning experience, students in these courses collectively explore a research problem related to agribusiness and development, guided by the course instructors and with contributions from other PGAD faculty. The broad, interdisciplinary research themes addressed in recent years include:

- 2021—Socioeconomic, technical, and environmental concepts in the agribusiness debate
- 2022—The Sustainable Development Goals (SDGs) from the perspective of agribusiness
- 2023—The SDGs from the perspective of agribusiness: SDG 2—Zero Hunger and Sustainable Agriculture
- 2024—The SDGs from the perspective of agribusiness: SDG 12—Responsible Consumption and Production

At the end of the semester, a workshop is held in which students present their group research projects as part of the evaluation process. The instructors serve as a review panel, engaging with the presenters alongside other students. This process, centered on a complex and interdisciplinary research object, promotes the co-construction of knowledge through autonomy, dialogue, and inquiry. Often, these outcomes are further developed into conference papers, book chapters, and peer-reviewed journal articles.

Module II includes three additional compulsory courses offered across the first and second semesters (Module I is a prerequisite): "Research Methodology I," "Research Methodology II," and "Interdisciplinary Seminars." Continuing the interdisciplinary approach, these courses involve active participation and interaction between students and faculty, supporting the development and refinement of dissertation research methods and techniques, as well as providing opportunities to present, defend, and discuss research projects.

These courses are taught in an integrated manner, and by the end of the module, Master's students present their research projects in the "Interdisciplinary Seminars I" course. Their advisory committee, fellow students, and the course instructors provides feedback aimed to refining the projects and aligning them with the program's objectives, evaluate these presentations. This stage can be considered a form of pre-qualification, allowing for knowledge exchange, methodological discussion, and interdisciplinary contributions to improve the proposed research. It also ensures that all members of the program, including students and faculty, are informed about ongoing research topics.

Module III, which has no prerequisites, includes elective courses that may be taken in any semester. The student's supervisor and advisory committee recommend internal or external electives based on their relevance to the student's research development. Doctoral students are encouraged to follow the same modular structure as the Master's program. If a doctoral student has not completed a Master's program in an area classified as interdisciplinary by CAPES, it is recommended that they enroll in "Systemic Approach to Agribusiness" and "Construction of Interdisciplinary Knowledge" to develop competencies aligned with PGAD's graduate profile.

The same recommendation applies to students from academic backgrounds unrelated to agribusiness.

After completing Modules I, II, and/or III, the doctoral student fulfills the course credit requirements and may proceed to the qualification exam, which is expected to take place between the 18th and 30th months. Because of the Program's commitment to promoting interdisciplinary in both instruction and advising, PGAD has seen a growing number of scientific articles published in reputable journals, as well as increased graduate employability in academia and job market. Furthermore, the research generated often leads to practical managerial, technological, social, and environmental solutions applicable to diverse areas within agribusiness.

4.2 Research

According to Abo-Khalil (2024), there is a general consensus that sustainable development should be central to HEIs curriculum. However, the author emphasizes that sustainability must also be supported by broader research initiatives. Strengthening research contributes not only to scientific knowledge production but is also key to advancing sustainability and social well-being. It fosters a dynamic academic culture and provides evidence to inform practice and policy.

At PGAD, research is structured around two main lines: 'Competitiveness of Agro-Industrial Systems (L1)', and Development and Environment (L2). Both lines address the complexities of agribusiness and development through an interdisciplinary lens that integrates various theoretical and methodological frameworks.

To explore how sustainability is reflected in PGAD's research activities, a questionnaire was distributed to all permanent faculty members. The survey collected information on the main research objects of their integrative projects and the dimensions of impact associated with each one.

Figure 2 presents the main research themes categorized according to the research line to which the projects belong. It shows a predominance of research projects developed by PGAD faculty whose study objects that are directly related to the theme of sustainability.

In Research Line 1 (L1), key concepts include "short supply chains", "climate change", "animal welfare", "animal health", and "energy". The research topics within this line converge on investigations that address ongoing challenges in agricultural production, particularly those intrinsically linked to the environment, such as energy resources, input supply, and responsible management practices.

In Research Line 2 (L2) the identified concepts include "environmental education", "Amazon region", "agro-industrial effluents", "waste treatment", "discarded materials", "wastewater", "By-products", "pesticide degradation", "water pollution", and "traditional communities". The study objects of the projects in this line focus on the environmental effects of agricultural production, including waste management and natural resource recovery.

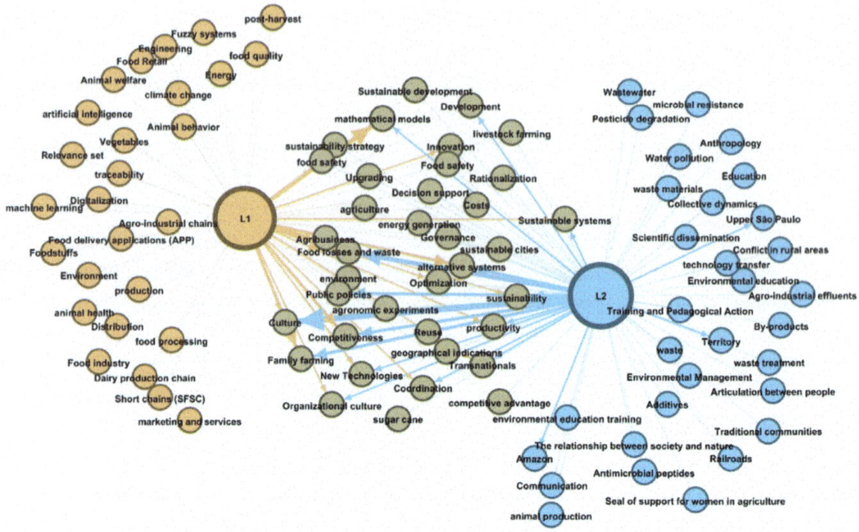

Fig. 2 Research objects of the PGAD faculty member's integrative projects. *Source* Authors

Some research objects appear in projects that span both of PGAD's research lines. These topics reflect a broader, more general perspective on sustainability, and include concepts such as "sustainability", "sustainable systems", "energy generation", "sustainable development", "sustainable cities", "environment", "agroecology", and "food security". This overlap highlights that both research lines host projects engaging with themes strongly tied to sustainable development in the agribusiness context.

In addition to identifying the main objects of study in their PGAD research projects, faculty members were also asked to indicate the potential impacts of their research on society. These impacts were categorized as environmental, social, economic, cultural, technological, scientific, educational, political, health-related, or animal welfare. Figure 3 presents a network diagram showing the distribution of project impacts from PGAD's research lines. The intensity of the connections in the network reflects the number of projects associated with each specific type of impact.

Based on Fig. 3, "environmental", "social", and "economic" impacts emerge as the most prominent within the context of research projects across both of PGAD's research lines. However, in Research Line 1 (L1), there is a stronger emphasis on "economic" impact and "animal welfare", the latter not being associated with any projects in Research Line 2 (L2). Conversely, in L2, there is a slight predominance of "political" impact among the projects.

In addition to analyzing, the main research projects conducted by PGAD's permanent faculty, this study also examined the relationship between SDGs and the research outcomes generated by graduate students during the 2021–2024 period. Each thesis or dissertation submitted during this quadrennial identifies at least one SDG as a guiding

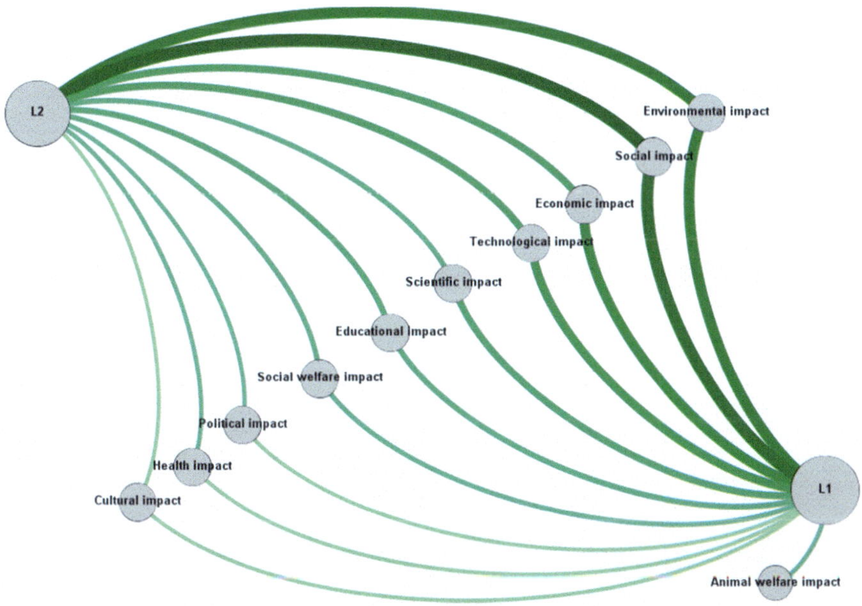

Fig. 3 The research impact of the PGAD faculty member's integrative projects. *Source* Authors

reference for its results, with the option to include secondary and even tertiary SDG connections.

Figure 4 presents a network diagram highlighting the links between the SDGs referenced in the analyzed documents and the intensity of those relationships. Each node in the diagram represents an SDG, with larger nodes indicating higher frequency of mention in theses and dissertations. The links between nodes represent associations among SDGs based on secondary mentions. The thicker the link, the stronger or more frequent the connection between the SDGs (Fig. 4).

Figure 4 reveals that SDG 2—Zero Hunger and Sustainable Agriculture, SDG 4—Quality Education, and SDG 12—Responsible Consumption and Production are the most frequently referenced goals and show strong interconnections both with each other and with other SDGs. These three goals are closely linked and mutually reinforcing, allowing them to span both of the Program's research lines.

Given that, PGAD's research and extension projects are integrative and interdisciplinary, bringing together member from different universities and academic backgrounds, certain themes stand out. For instance, SDG 2 shows strong links to SDGs 1, 8, and 9; SDG 12 is significantly associated with SDGs 4, 8, and 9; while other goals exhibit more uniform distribution, with SDGs 7, 13, and 14 being the least interconnected with other elements.

The analysis also indicates that SDGs 12, 2, 8, 3, 4, 7, and 9 are referenced in nearly 80% of the Master's theses and doctoral dissertations produced in the Program. This thematic focus is subject to change over time as new challenges and

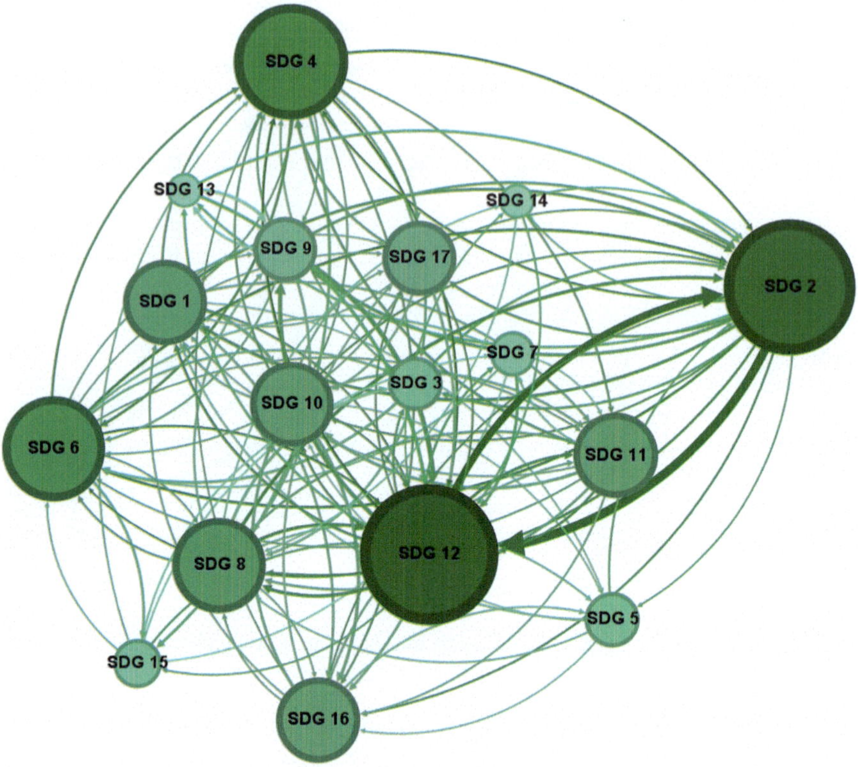

Fig. 4 The relationship between SDGs and the thesis and dissertation of the PGAD. *Source* Authors

opportunities emerge in the agribusiness sector, reflecting shifts in student profiles, evolving societal demands, and updates to PGAD's strategic planning.

As an interdisciplinary program, the two research lines are in continuous dialogue, with some faculty members actively contributing to both areas. One clear indication of this integration is that all 17 SDGs are addressed to some extent across the final academic works defended in the Program. As a result, the SDGs serve as key guiding elements, shaping the Program's research, teaching, and outreach activities, and fostering actions that contribute to sustainable development across environmental, economic, and social dimensions. These findings underscore the interdisciplinary nature of PGAD's activities and their alignment with the SDGs, the Program's mission, and the intended graduate profile.

4.3 *Extension*

In addition to Teaching and Research, Extension forms a crucial foundation for fulfilling the social mission of higher education institutions. In this regard, the study sought to identify university outreach projects coordinated by PGAD's faculty members that are directly engaged in practical actions promoting sustainability.

In December of 2024, PGAD hosted five integrative extension projects involving 14 faculty members (which represents 70% of the total members). These projects are responsible for a significant number of extension activities that generate direct societal impact. Table 1 highlights three out of the five extension projects, specifically those that have direct implications for sustainability-related actions.

According to Abo-Khalil (2024), the integration of sustainability into higher education curriculum and operations is becoming increasingly important on a global level and should be incorporated into all courses, not only those focused on the environment or sustainability. In this context, the results presented demonstrate how actions can be adopted within Agribusiness based on parameters obtained through the analysis of activities related to Teaching, Research, and Extension within the scope of PGAD, a nationally recognized program for training professionals to work in the sector.

5 Conclusions

This study analyzed the integration of sustainable development principles into the PGAD, highlighting the role of interdisciplinary in academic training and in the development of research aligned with the SDGs. The results show that the graduate program has played a fundamental role in preparing professionals capable of addressing the complex challenges of the agribusiness sector, promoting innovative and sustainable solutions.

Among the main conclusions, there is a strong interaction between the research developed and SDGs 2 (Zero Hunger and Sustainable Agriculture) and 12 (Responsible Consumption and Production), reflecting the program's commitment to addressing critical global sustainability issues. The impacts observed cover social, environmental, economic, and technological dimensions, evidencing the contribution of research to the improvement of value chains and regional development.

The analysis of the three academic pillars 'Teaching', 'Research', and 'Extension' revealed a consistent alignment with the guidelines of education for sustainable development:

- In the Teaching dimension, the program presents a curricular structure that encourages interdisciplinary and critical reflection on the challenges of agribusiness.
- In the field of Research, the diversity of topics addressed confirms PGAD's relevance in generating knowledge applied to sustainability.

Table 1 Integrative extension projects and it sustainable actions

Integrative extension projects	Description	Sustainable actions
Empower and enhance: a project for women in agribusiness	The project aims to promote the empowerment of women in agribusiness through the organization and participation in meetings and events. The team believes that female empowerment in modern agriculture is not just a fact—it is a revolution. Diverse environments that foster professional development perform better	• Support in the creation of the 'Tribe Agro Association' and the 'Support for Women in Agribusiness' seal • Active participation in the organization of the Farm of the Future: The Farm of the Future was created with the goal of showcasing to students and the broader community the innovations, technologies, and opportunities offered by the rural sector—including biotechnology, agricultural technology, and environmental sustainability. The event takes place alongside the EXAPIT Fair in the municipality of Tupã, São Paulo

(continued)

Table 1 (continued)

Integrative extension projects	Description	Sustainable actions
REAP Green Room	The REAP Green Room is a space dedicated to the development of educational activities focused on socio-environmental and cultural topics. It aims to contribute to and encourage critical discussion, reflection, and the strengthening of group identities, ultimately fostering the formation of more critical, informed, and engaged citizens. The REAP Green Room is a space dedicated to the development of educational activities focused on socio-environmental and cultural topics. It aims to contribute to and encourage critical discussion, reflection, and the strengthening of group identities, ultimately fostering the formation of more critical, informed, and engaged citizens	• Use of social media (blog, YouTube, Instagram, and Facebook) to share knowledge focused on environmental education; monthly live sessions on Instagram supported by the Research Group on Environmental Management and Education (PGEA), covering various topics to promote scientific outreach to the community • Organization of local and regional environmental events, with live streaming of events and debates on REAP's YouTube channel • Promotion of activities to connect primary school communities with UNESP, such as the environmental circuit for children from the municipal school system • Offering of courses and lectures, for example, on building composting systems and creating biofertilizers for the community • Active participation in the organization of the Farm of the Future: The Farm of the Future was created with the goal of displaying to students and the broader community the innovations, technologies, and opportunities offered by the rural sector—including biotechnology, agricultural technology, and environmental sustainability. The event takes place alongside the EXAPIT Fair in the municipality of Tupã, São Paulo

(continued)

Table 1 (continued)

Integrative extension projects	Description	Sustainable actions
Kamby Extension Program—"Keeping an Eye on Milk"	Since 2015, the Kamby Extension Program has aimed to enhance the external community's understanding of the dairy production chain, with the goal of improving both the quantity and quality of milk production in the Alta Paulista region. This is achieved through the dissemination of scientific and humanistic knowledge and the exchange of experiences among university students, primary and secondary school students, teachers, technicians, dairy farmers, and the public. Through its knowledge-sharing initiatives, the program contributes to the United Nations SDGs, specifically: Zero Hunger and Sustainable Agriculture (SDG 2), Responsible Consumption and Production (SDG 12), and Decent Work and Economic Growth (SDG 8)	• Establishment of partnerships with the municipal governments of Tupã and Lucélia, as well as with the Rural Association of Small and Medium-Sized Producers of Lucélia-SP • Collaborative work with rural producers in Tupã on a herd genetic improvement program. The results gained regional recognition, including an official commendation from the Municipality of Tupã-SP • Annual promotion of the Dairy Cattle Best Practices Meeting, featuring 13 lectures and 689 participants, including undergraduate and graduate students, rural producers, extension agents, rural entrepreneurs, and industry professionals • Organization of the "Milk Morning" event, focused on animal nutrition, pasture management, and reproduction—with the participation of 23 dairy farmers and 16 graduate students • Active participation in the organization of the Farm of the Future: The Farm of the Future was created to display innovations, technologies, and opportunities offered by the rural sector—including biotechnology, agricultural technology, and environmental sustainability—to students and the public. The event is held alongside the EXAPIT Fair in the municipality of Tupã, São Paulo

Source Authors

- Lastly, Extension initiatives demonstrate the program's commitment to social engagement and the dissemination of sustainable practices within the community.

Given this scenario, it is essential for graduate programs in fields related to agribusiness to incorporate sustainability into their curriculum and operations, training professionals capable of balancing productivity, competitiveness, and socio-environmental responsibility.

By integrating sustainability principles into academic training, these programs not only contribute to advancing knowledge and innovation in the sector but also prepare specialists to face the global challenges of sustainable development in agribusiness. The experience of PGAD may serve as a model for other institutions, contributing to the formulation of strategic guidelines that enhance the social and economic impact of higher education in Brazil.

Finally, the inclusion of only one graduate program in this study may be regarded as a limitation, as it confines the analysis to a single institutional context. Nevertheless, this delimitation was necessary to enable a more in-depth investigation and a comprehensive understanding of the various dimensions of teaching. Consequently, the study established analytical parameters that may serve as a foundation for future research aimed at comparing the contexts of other graduate programs focused on agribusiness-related issues.

Thus, the need for public policies that encourage interdisciplinary and the integration of sustainability in higher education is reinforced, consolidating universities as key players in building a more balanced and resilient future.

References

Abo-Khalil, AG (2024) Integrating sustainability into higher education challenges and opportunities for universities worldwide. Heliyon 10(9):e29946. https://pubmed.ncbi.nlm.nih.gov/38707336/

Batalha MO, Silva AL (2021) Gestão e economia dos sistemas agroindustriais: definições, correntes metodológicas e métodos de análise [Management and economics of agribusiness systems: definitions, methodological currents, and methods of analysis]. In: Batalha, MO (Ed.) Gestão Agroindustrial. Atlas, pp. 1–48.

Bergold J, Thomas S (2012) Participatory research methods: A methodological approach in motion. Historical Social Research 37(4): 191–222. https://www.jstor.org/stable/41756482

Bradley P (2019) Integrating sustainable development into economics curriculum: a case study analysis and sector wide survey of barriers. J Clean Prod 209:333–352. https://doi.org/10.1016/j.jclepro.2018.10.184

Bravo PS, Chambers E, Artiaga LN, Sendra E, Barrachina, A (2021) Consumer understanding of sustainability concept in agricultural products. Food Quality and Preference 28(4):e104136. https://doi.org/10.1016/j.foodqual.2020.104136

Burnquist H, Martins M, Campoli J, Costa C (2020). Covid-19 e agroalimentos: recalibrando expectativas [Covid-19 and agrifood: recalibrating expectations]. Revista de Política Agrícola 29(2):88–101. https://seer.sede.embrapa.br/index.php/RPA/article/view/1582.

Didham R, Ofei-Manu P, Atiti A, Yasuda S, Tabucanon M, Vanhala K, Ogihara A (2013) Approaches to Educational Assessment & Application to Education for Sustainable Development. In: Institute for Global Environmental Strategies (Ed.) Monitoring & evaluation of education for sustainable development: a framework of the main factors and important leverage points in the implementation of education for sustainable development in the Asia-Pacific region. Institute for Global Environmental Strategies, pp. 27–82. www.jstor.org/stable/resrep00886.8.

FAO - Food and Agriculture Organization of United Nations (2020) The state of food security and nutrition in the world. Transforming food systems for affordable healthy diets. http://www.fao.org/3/ca9692en/ca9692en.pdf

Feil AA, Schreiber D (2017) Sustentabilidade e desenvolvimento sustentável: desvendando as sobreposições e alcances de seus significados [Sustainability and sustainable development: unraveling the overlaps and scope of their meanings]. Cad EBAPEBR 15(3):667–81. https://doi.org/10.1590/1679-395157473

Hallinger P, Chatpinyakoop C (2019) A bibliometric review of research on higher education for sustainable development 1998–2018. Sustainability 11(8): 2401 – 2421. https://doi.org/10.3390/su11082401

Jickling B, Wals AE (2012) Debating education for sustainable development 20 years after Rio: A conversation between Bob Jickling and Arjen Wals. Journal of Education for Sustainable Development 6(1):49–57. https://doi.org/10.1177/097340821100600111

Leal Filho W, Pallant E, Enete A, Richter B, Brandli LL (2018) Planning and implementing sustainability in higher education institutions: an overview of the difficulties and potentials. International Journal of Sustainable Development and World Ecology 25(8):713–721. https://e-space.mmu.ac.uk/621884/1/Planning.SD.HE.Evaluation.Challenges.IJSDWE..pdf

Leal Filho W, Lopes G, Trevisan L, Bella, RL, Rampasso I, Brunozi Júnior AC, Bichueti R, Quelhas O, Anholon R, Galleli B (2025). Green metrics: how are Brazilian universities doing? Discover Sustainability 6(90) 1-15. https://doi.org/10.1007/s43621-025-00864-z

Lozano R, Lozano FJ, Mulder K, Huisingh D, Waas T (2013) Advancing Higher Education for Sustainable Development: international insights and critical reflections. Journal of Cleaner Production 48(6):3–9. hdoi.org/https://doi.org/10.1016/j.jclepro.2013.03.034.

MAPA - Ministério da Agricultura e Pecuária (2024). Exportações do agronegócio fecham 2023 com US$ 166,55 bilhões em vendas [Agribusiness exports close 2023 with US$ 166.55 billion in sales]. https://www.gov.br/agricultura/pt-br/assuntos/noticias/exportacoes-do-agronegocio-fecham-2023-com-us-166-55-bilhoes-em-vendas?utm_source=chatgpt.com. Acesso: 24 fev. 2025.

Ruggerio CA (2021) Sustainability and sustainable development: A review of principles and definitions. Science of The Total Environment 786:1-11. https://doi.org/10.1016/j.scitotenv.2021.147481

Ruiz-Mallén I, Heras M (2020). What sustainability? Higher education institutions' pathways to reach the agenda 2030 goals. Sustainability 12(4):1290. https://www.mdpi.com/2071-1050/12/4/1290

Sinakou E, Pauw, JB, Van Petegem P (2019). Exploring the concept of sustainable development within education for sustainable development: implications for ESD research and practice. Environment, Development and Sustainability: A Multidisciplinary Approach to the Theory and Practice of Sustainable Development 21(1): 1-10. https://doi.org/10.1007/s10668-017-0032-8

UN - United Nations (2021) Transforming our world: the 2030 Agenda for Sustainable Development. https://sdgs.un.org/2030agenda

UNESCO - United Nations Educational, Scientific and Cultural Organization (2014) Shaping the future we want: UN decade of education for sustainable development. https://sustainabledevelopment.un.org/content/documents/1682Shaping%20the%20future%20we%20want.pdf (accessed 13 Aug 2021).

Vadakkepatt GG, Winterich KP, Mittal V, Zinn W, Beitelspacher L, Aloysius L, Ginger J, Reilman J (2021) Sustainable Retailing. Journal of Retailing 97(1):62-80. https://doi.org/10.1016/j.jretai.2020.10.008.

Fábio Mosso Moreira Bachelor's degree in Business Administration from São Paulo State University (UNESP/Tupã). Master and Ph.D. in Information Science from São Paulo State University (UNESP/Marília). Postdoctoral research in Agribusiness and Development at São Paulo State University (UNESP/Tupã). Works in research in the area of public policies, data access, small farming and Information and Communication Technology. Tupã, São Paulo, Brazil. Telephone + 55 (14) 3404–4200.

Eduardo Guilherme Satolo Bachelor's degree, Master, and Ph.D in Production Engineering from the Methodist University of Piracicaba (UNIMEP). Postgraduate in Management and Production of Alcohol and Sugar (MBA) from IAT/Fatep. Postdoctoral research in Production Engineering and Computer Systems (MESC-ICT) from the Universidade Federal Fluminense (UFF). Associate Professor at the Faculty of Science and Engineering (UNESP/Tupã). Permanent Professor of the Graduate Program in Agribusiness and Development (UNESP/Tupã). Leader of the research group Group for Multicriteria Analysis in Agribusiness (GAMA). Works on research in the area of production management and supply chain management, using the theory of lean production systems with agribusiness as his object of study. Tupã, São Paulo, Brazil. Telephone + 55 (14) 3404–4249.

Wagner Luiz Lourenzani Bachelor's degree in Food Engineering from the Federal University of Viçosa (UFV), Master's degree in Food Science and Technology from UFV, and Ph.D. in Production Engineering from the Federal University of São Carlos (UFSCar). Completed a Postdoctoral Fellowship at Kansas State University (K-State). Currently serves as an Associate Professor at the Faculty of Science and Engineering, São Paulo State University (UNESP). Permanent faculty member of the Graduate Program in Agribusiness and Development. Research areas include agribusiness, agroindustrial systems, and regional development. Tupã, São Paulo, Brazil. Telephone + 55 (14) 3404–4249.

Ana Elisa Bressan Smith Lourenzani Bachelor's degree in Agronomy from the Federal University of Viçosa (UFV), and a Master's and Ph.D. in Production Engineering from the Federal University of São Carlos (UFSCar). She was a visiting scholar at Kansas State University. Currently, she is an Associate Professor at the Faculty of Science and Engineering, São Paulo State University (UNESP), and a permanent faculty member of the Graduate Program in Agribusiness and Development at UNESP/Tupã. She leads the research group on Family Farming and Sustainability (GEAFS/UNESP). Her main research areas include Agribusiness Systems Management, family farming, market access, institutional procurement, certification, and geographical indications. Tupã, São Paulo, Brazil. Telephone + 55 (14) 3404–4249.

Integrating SDGs and LCSA: Enhancing Sustainability in a University Research Laboratory

João Victor Encide Salla, Abner Fernandes Souza da Silva, Jayson Wilson Barretti, Virgínia Aparecida Silva Moris, Juliana Veiga Mendes, and Diogo Aparecido Lopes Silva

Abstract Universities exert a considerable impact on the environment, economy, and society due to their complex infrastructures and multifaceted activities. This study applies the Life Cycle Sustainability Assessment (LCSA) methodology within the Sustainable Production Laboratory at the Federal University of São Carlos—Sorocaba campus, São Paulo, Brazil, to evaluate and enhance the sustainability performance of the laboratory. The LCSA framework integrates Environmental Life Cycle Assessment (LCA), Social Life Cycle Assessment (S-LCA), and Life Cycle Costing (LCC) to identify critical sustainability hotspots. The findings reveal that transportation constitutes the primary contributor to both environmental and social impacts, whereas expenditures related to human resources represent the predominant component of economic costs. To broaden the analytical scope, the study correlates LCSA outcomes with the Sustainable Development Goals (SDGs). The results underscore the necessity of implementing targeted strategies such as optimizing transportation systems, improving energy efficiency, and advancing resource management practices to mitigate adverse impacts. This research highlights the applicability and relevance of LCSA in higher education institutions, offering valuable insights for embedding sustainability principles into university operations. Furthermore, it contributes to the broader discourse on aligning institutional practices with global sustainable development agendas.

Keywords Environmental impacts · Social impacts · Economic impacts · Research laboratory

---

J. V. E. Salla (✉) · A. F. S. da Silva · J. W. Barretti · V. A. S. Moris · J. V. Mendes · D. A. L. Silva
Research Group On Sustainability Engineering (EngS Group), Department of Production Engineering, School of Management and Technology, Federal University of São Carlos, Sorocaba, SP, Brazil
e-mail: vesjoao@estudante.ufscar.br

W. Leal Filho et al. (eds.), *Sustainability in the Curriculum and Operations of Universities*, World Sustainability Series,
https://doi.org/10.1007/978-3-032-12678-8_4

1 Introduction

Universities play a fundamental role in society, grounded in the pillars of education and research. However, due to the complexity of their infrastructure and the diversity of activities carried out, these institutions generate significant environmental impacts, including greenhouse gas emissions (Robinson et al. 2018). Furthermore, their operation demands substantial human effort and financial resources, underscoring the importance of addressing the three dimensions of sustainability: environmental, economic, and social (Zhang et al. 2024).

In response to global challenges such as environmental degradation and social inequality, the United Nations (UN) launched the 2030 Agenda in 2015, which sets forth 17 Sustainable Development Goals (SDGs) and 169 associated targets to be achieved by 2030 (UNEP 2015). As signatories to this agenda, countries and their respective institutions, including higher education institutions (HEIs), are expected to align their strategies and operations with these global objectives. To monitor and assess such alignment, it is essential to adopt robust quantitative tools capable of evaluating environmental, economic, and social impacts (Genta et al. 2022).

One well-established and comprehensive methodological framework for this purpose is the Life Cycle Sustainability Assessment (LCSA), which integrates three complementary approaches: Environmental Life Cycle Assessment (LCA), Social Life Cycle Assessment (S-LCA), and Life Cycle Costing (LCC) (Wafa et al. 2022). The LCA methodology enables the quantification of potential environmental impacts of products and services throughout their life cycle, from raw material extraction to final disposal (De Oliveira et al. 2021). According to ISO 14040 and ISO 14044 (2006a, b), LCA encompasses four distinct phases: goal and scope definition, life cycle inventory, impact assessment, and result interpretation. S-LCA assesses social impacts across all stages of the life cycle, considering various stakeholders, including workers, local communities, and society as a whole (UNEP 2020). LCC, in turn, provides an economic perspective by analyzing the costs incurred throughout the life cycle of a product or service (Huang et al. 2018).

Integrating LCSA results with the SDG framework enables HEIs to develop informed strategies and make decisions that enhance the sustainability of their processes. A methodological approach for this integration was proposed by Weidema et al. (2020), enabling a systematic assessment of the contributions of production systems to achieving specific SDG targets.

Several studies have explored the application of LCA in the context of university operations. Clabeaux et al. (2020) identified electricity generation, transportation, and heating as the primary sources of environmental impact at a university campus in Clemson, United States. In the Brazilian context, Barros et al. (2020) proposed sustainability enhancement strategies for a university campus, such as replacing disposable cups with reusable alternatives distributed to students across its 13 campuses. Beyond environmental impacts, Erauskin-Tolosa et al. (2021) analyzed the social footprint of an HEI in Spain, concluding that labor-related and indirect transportation activities were the main contributors to social impacts. Similarly,

Giusti et al. (2023) demonstrated that social impacts in universities are significantly associated with food consumption and mobility, with particular emphasis on risks linked to energy and material supply chains.

The relationship between LCSA and the SDGs has also been the subject of study. Giusti et al. (2024) examined this nexus within a university setting, comparing three scenarios—remote, hybrid, and in-person instruction. Their findings identified transportation and energy consumption as the principal environmental hotspots across scenarios, with the hybrid model presenting the most sustainable alternative. In another context, Ugaia et al. (2020) conducted an LCA of the Life Cycle Sustainability Assessment Center (Gyro) in Paraná, Brazil, identifying construction, equipment uses, and food as major contributors to environmental impacts, while personnel costs dominated the economic analysis.

In light of this context, the present study applies the LCSA framework to a Sustainable Production Laboratory located at a university in Sorocaba city, São Paulo, Brazil. The objective is to identify key environmental, economic, and social hotspots. Subsequently, the LCSA results are integrated with the SDG framework to evaluate the laboratory's alignment with global sustainability goals and to identify opportunities for enhancing its performance and advancing the institution's broader sustainability agenda.

2 Literature Review

Numerous authors from diverse perspectives have explored the theme of sustainability within universities. In recent years, Life Cycle Assessment (LCA)—and, at the organizational level, Organizational LCA (O-LCA)—has become a crucial tool for quantifying environmental impacts, informing decision-making, and demonstrating institutional alignment with the United Nations Sustainable Development Goals (SDGs) (Filho et al. 2024).

Operational LCA studies conducted across entire university campuses reveal recurring patterns. At Kharazmi University, Iran, fossil-based electricity and natural gas heating accounted for more than 80% of daily environmental damages in the categories of climate change, human health, and resource depletion (Asadollahfardi et al. 2024). Kocaeli University obtained a similar result at its Umuttepe campus in Turkey, where O-LCA, employing the EF 3.0 method, showed that grid electricity and natural gas account for approximately 80% of the annual environmental footprint. A scenario involving 24% photovoltaic electricity reduced all impact categories, except for mineral use, and outperformed the baseline in 60% of Monte Carlo iterations (Çetin Doğruparmak et al. 2024). These studies reinforce that low-carbon energy interventions are the primary environmental hotspots in universities, which is especially relevant for SDGs 7 and 13.

By incorporating economic and social dimensions, the Life Cycle Sustainability Assessment (LCSA) expands the traditional environmental scope of LCA. At the Sümer campus of Abdullah Gül University, the integration of E-LCA, Life Cycle

Costing (LCC), and Social LCA (S-LCA) revealed an average emission of 2.92 t CO_2 eq per capita, with annual energy costs comprising 86% of total expenditures and indicating fragile social performance with the local community (Gulcimen et al. 2023). Along the same lines, a study at the University of the Basque Country combined O-LCA and S-LCA for 46,813 users and found that transport was responsible for approximately 60% of environmental impacts. At the same time, global supply chain footprints related to child labor and illiteracy have emerged as significant social issues, most of which are outside the Basque region (Bueno et al. 2021). These findings underscore the need for integrated strategies that transcend emissions inventories and encompass governance, cost, and social justice—core pillars of SDGs 8, 12, and 16.

Carbon-focused analyses further support this diagnosis. At Clemson University, USA, the campus GHG inventory totaled 95 kt CO_2 eq in 2014 (4.4 t CO_2 eq per student), with electricity (41%), automotive transport (18%), and steam (16%) as the major contributors (Clabeaux et al. 2020). Moreover, comparisons between face-to-face and remote classes during the COVID-19 pandemic indicated that online teaching reduced 10 out of 13 impact categories and could potentially cut up to 57% of the total environmental burden in an ideal hybrid model (Silva et al. 2021), highlighting the relevance of behavioral and mobility-related measures for SDGs 11 and 13.

From a building perspective, Xue et al. (2019) proposed an integrated LCA–LCC model for university buildings in China, monetizing impacts through willingness-to-pay and assigning 73% of the weight to environmental performance versus 27% to costs. Their application to a teaching building revealed that the operational phase had the most significant impact on both environmental impacts—particularly GWP and fossil ADP—and costs, suggesting that policies focused on electrical efficiency yield the best socio-environmental returns (Xue et al. 2019).

In summary, the recent evidence demonstrates that (i) energy consumption (electricity and heating) is the principal environmental hotspot in university campuses; (ii) renewable energy scenarios and demand management measures offer high environmental and economic returns; (iii) the inclusion of cost and social aspects via LCSA enhances the capacity to prioritize actions; and (iv) education based on LCA links academic competencies to sustainable practice. Nonetheless, a gap remains in the integrated assessment—environmental, economic, and social—of university production laboratories, which present unique consumption and emission profiles. The present study addresses this gap by proposing an LCSA applied to a Sustainable Production Laboratory, aligning operational metrics with SDG targets and the best practices identified in the literature.

3 Methodology

The environmental assessment adhered to the methodological guidelines established by ISO 14040 and ISO 14044 (2006a, b), which structure the Life Cycle Assessment (LCA) into four main phases: (i) Goal and Scope definition, (ii) Life Cycle Inventory (LCI), (iii) Life Cycle Impact Assessment (LCIA), and (iv) Interpretation. In the first phase, the functional unit was defined as one day of laboratory operation, ensuring a standardized and representative basis for quantifying and comparing average daily impacts. We selected this functional unit to enable extrapolation of the findings and facilitate comparisons with similar studies in the literature.

The analyzed system comprises an anteroom (used for meetings, measuring 4.5 m × 3.2 m) and a main room (4.5 m × 6.8 m), containing eight individual workstations and ten seats at shared tables, as depicted in Fig. 1. The system boundaries were delineated from cradle to grave, encompassing the full life cycle of inputs required for laboratory operations—from the production of energy, food, furniture, and electronic equipment to their use and eventual disposal. As such, we included the environmental impacts associated with the manufacturing, long-term use, and end-of-life treatment of essential goods and services.

In the second phase, the Life Cycle Inventory (LCI) was developed based on primary data collected through direct monitoring of the laboratory's activities. This LCI included information on electricity consumption, use of office and laboratory supplies, waste generation and disposal practices, maintenance activities, and the expected lifespan of furniture and equipment. Researchers sourced secondary data from databases such as Ecoinvent and relevant scientific literature for processes where primary data were unavailable. Table 1 presents the main input and output flows.

Fig. 1 Sustainable Production Laboratory overview at UFSCar—Sorocaba campus

Table 1 Main input/output inventory for the Sustainable Production Laboratory operation at UFSCar Sorocaba campus

Stage/item		Amount (items)	useful life	Consumption details (per day)	Amount per F.U. (per week)
Electricity	Computer—Core i3	4	5 years	280 Wh per day	0.0021 items
	Monitor	4	5 years	2.52 Wh per day	0.0019 items
	Air conditioner 36,000 BTU	1	10/15 years	22,320 Wh per day	0.0001 items
	Air conditioner 12,000 BTU	1	10/15 years	4912 Wh per day	0.0001 items
	Electric coffee maker	1	5 years	91.7 Wh per day	0.0005 items
	Electric kettle	1	5 years	87.5 Wh per day	0.0005 items
	LED tube light	10	1000 h	1.728 Wh per day	0.0006 items
	Landline phone	1	5 years	72 Wh per day	0.0005 items
	Printer	1	5 years	0.15 Wh per day	0.0005 items
	Laptops	4	5 years	0.0019 Wh per day	0.0019 items
Furniture and equipment	Individual L-shaped desk	4	10 years	–	0.0010 items
	Round meeting table	1	10 years	–	0.0002 items
	Individual rectangular desk	6	10 years	–	0.0016 items
	Large rectangular table	3	10 years	–	0.0008 items
	Swivel office chair	12	10 years	–	0.0032 items
	Fixed ISO chair	4	10 years	–	0.0010 items
	Closed cabinet with two doors	4	10 years	–	0.0010 items
	Open shelf cabinet	1	10 years	–	0.0002 items
Consumables	Coffee	–	–	37.5 g	37.5 g
	Water	–	–	5100.75 ml	5100.75 ml
	Tea	–	–	3.5 ml	3.5 ml
	Coffee grounds	–	–	37.5 g	37.5 g
	Effluent	–	–	5100.75 ml	5100.75 ml

(continued)

Table 1 (continued)

Stage/item		Amount (items)	useful life	Consumption details (per day)	Amount per F.U. (per week)
	Tea grounds	–	–	3.5 ml	3.5 ml
Transp.	Transportation (3.5 students)—distance	–	–	Car (2.5 students): 42.52 km Bus (1 student): 25.812 km	341.68 km (diesel consumption: 11 L and gasoline consumption: 5 L)

To estimate consumption, an average of 3.5 students per day was considered over a five-day academic week. The assessment of energy consumption included not only the direct use of electrical equipment but also the allocation from the production impact (as indicated in the penultimate column of the table). For example, experts apportioned the environmental impact of a laptop to one week of use based on its expected lifespan.

For durable goods, such as furniture and electronic devices, we considered only the production impact, factoring in their useful life. We estimated consumables based on weekly observations of usage patterns. Finally, transportation impacts were calculated based on the daily commute of 3.5 students to the laboratory, taking into account their modes of transportation (e.g., bus or car).

The third phase consisted of the Life Cycle Impact Assessment (LCIA), which utilized the Impact World+ Footprint method. This method was selected due to its comprehensive and spatially differentiated characterization factors, enabling greater accuracy in regionalized analyses, including within the Brazilian context (Impact World+ 2025). The method considers five impact categories: (i) climate change, (ii) ecosystem quality, (iii) depletion of non-renewable energy resources, (iv) human health, and (v) water use. Researchers assessed climate change based on greenhouse gas emissions from energy consumption, as well as the indirect impacts of manufacturing and waste disposal. They attributed environmental degradation resulting from material extraction and processing to a decline in ecosystem quality. The non-renewable energy resources category accounted for the dependence on fossil fuels and minerals for equipment manufacturing and operation. Human health impacts were based on air pollutant emissions, while water use encompassed both direct consumption in the laboratory and indirect water demands throughout the production chain of the materials.

In the fourth phase, Interpretation, we analyzed the results to identify the most critical environmental hotspots—the life cycle stages with the most significant adverse impacts. This analysis supported the proposal of mitigation strategies, including the substitution of materials with lower environmental impacts, the optimization of energy use, and the enhancement of waste management practices.

Subsequently, a social sustainability assessment was conducted, following the same structural phases and utilizing the same functional unit and system boundaries as the environmental assessment. The Social Organizational Context Analysis (SOCA) method was employed to evaluate the social implications of the laboratory's operations across different stakeholder groups. According to the UNEP and Life Cycle Initiative framework (Traverso et al. 2021), researchers considered six stakeholder categories: children, consumers, the local community, society at large, value chain actors, and workers. The framework examined social impacts through indicators related to working conditions, community engagement, and societal influence.

Finally, for the economic dimension, the Life Cycle Costing (LCC) methodology was applied to quantify the total operational costs of the laboratory and to identify the primary cost drivers. Researchers collected data on infrastructure needs (G1 2023), procurement and maintenance of equipment and furniture, material and food consumption, transportation, research scholarships, and electricity usage. The analysis enabled a comparative evaluation of expenditure categories and supported the development of strategies to optimize resource allocation.

Weidema et al. (2020) integrated LCSA findings with the Sustainable Development Goals (SDGs) using their proposed methodological framework. This approach allowed us to assess how the laboratory's impacts align with specific SDG targets. We used a benchmarking reference value for each indicator (details in Table 6) and applied a standardized scale to classify the results.

+ 2: Significantly lower impact (environmental, social, or economic) than the reference value (difference > 10% or two performance levels above);
+ 1: Considerably lower impact than the reference value (difference between 5 and 10% or one level above);
0: No significant difference in impact compared to the reference value;
− 1: Considerably higher impact than the reference value (difference between 5 and 10% or one level below);
− 2: Significantly higher impact than the reference value (difference > 10% or two levels below).

This classification enabled the identification of critical areas requiring attention to reduce environmental, social, and economic impacts throughout the laboratory's life cycle. The adopted methodology enabled an integrated and detailed analysis of the laboratory's sustainability, providing support for improvements aligned with sustainability principles.

4 Results and Discussions

4.1 Environmental LCA

The study evaluated the results from an environmental standpoint according to the impact categories defined in the ImpactWorld+ method (ImpactWorld+ 2025), using the established functional unit of one day of laboratory operation (see Table 2 and Fig. 2).

The results underscore the significant contribution of transportation within the analyzed system, as it represents the most substantial impact in four out of the five assessed categories. Electricity consumption then drives the primary production and use of furniture and equipment.

These findings are consistent with those reported by Ugaya et al. (2020) in their Life Cycle Sustainability Assessment (LCSA) for the Gyro laboratory in Brazil, where they similarly identified transportation and electricity usage as the most critical impact sources.

In the climate change category, transportation emerged as the primary contributor, mainly due to greenhouse gas (GHG) emissions associated with ethanol-powered vehicles and diesel-fueled buses used by students commuting to the laboratory. Similarly, electricity consumption was the most significant contributor in the water

Table 2 Results for the environmental LCA

Life cycle stages	Impact categories (impact unit)				
	Climate change (kg de CO_2 eq.)	Ecosystem quality (PDF m^2 y)	Energy resources: non-renewable (MJ deprived)	Human health (DALY)	Water use (m^3 world eq)
Transportation	22.91	2.62	310.99	1.27×10^{-05}	2.05
Electricity	2.15	0.33	33.26	3.80×10^{-06}	8.19
Furniture and equipment	2.12	0.57	30.08	2.68×10^{-06}	0.86
Building Infrastructure	0.96	0.27	10.39	1.38×10^{-06}	0.24
Consumables	0.52	0.29	2.99	3.35×10^{-07}	2.95
Human resources (HR)	0.28	0.05	2.99	2.68×10^{-07}	0.08
Total (per day)	28.94	4.14	390.7	2.12×10^{-05}	14.38
Total (per week)	144.70	20.68	1953.51	1.06×10^{-4}	71.88

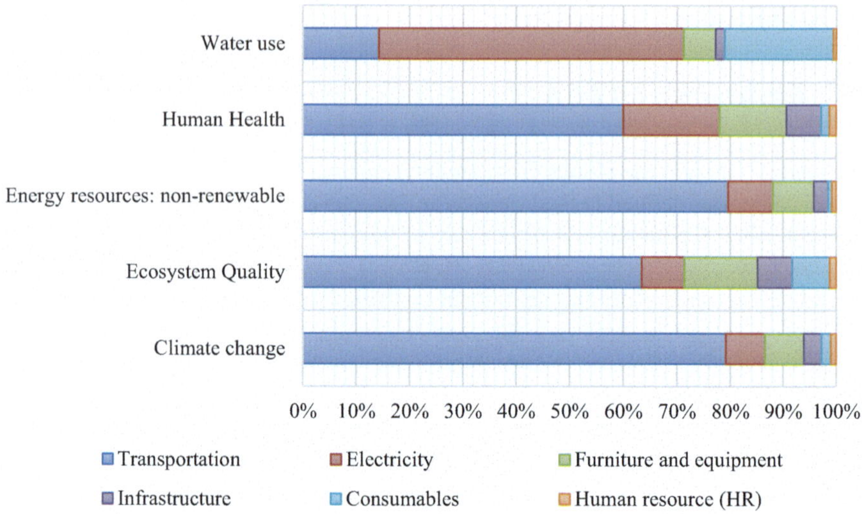

Fig. 2 Results for the environmental LCA contribution analysis of the evaluated lab

use category, owing to Brazil's predominantly hydroelectric energy matrix, which, although renewable, exerts considerable pressure on water resources.

Production and use of furniture and equipment, along with building infrastructure, were notable contributors to the ecosystem quality and non-renewable energy categories. Extracting raw materials like minerals and fossil fuels for manufacturing laptops, desktop computers, and air conditioning units, using timber for desks and chairs, and the impacts of building construction attributed to this.

Conversely, material consumption and human resource activities showed the lowest absolute impacts. This result occurred because the study excluded student food consumption, considering only water, tea, and coffee intake. Furthermore, the modelled activities by students resembled administrative computer-based tasks, which are relatively low-impact in nature.

4.2 Economic LCA

The economic assessment revealed that the operational cost of the Sustainability Engineering Laboratory for one working day with 3.5 students (comprising 2.625 master's and 0.875 doctoral students) was approximately R$ 622 (≈ 115.69 dollars (USD) – exchange rate on 01/22/2026). As illustrated in Table 3 and Fig. 3, the majority of this cost (91.56%) is attributable to human resources, primarily consisting of graduate student scholarships.

These findings corroborate those of Ugaya et al. (2020), who observed that human resource expenses dominate laboratory budgets due to the specialized

Table 3 Results for the economic LCA

Life cycle stages	Life cycle cost
Transportation	R$ 28.53
Electricity	R$ 17.60
Furniture and equipment	R$ 2.60
Infrastructure	R$ 0.79
Consumables	R$ 3.00
Human resources (HR)	R$ 569.43
Total (per day)	R$ 621.95
Total (per week)	R$ 3109.75

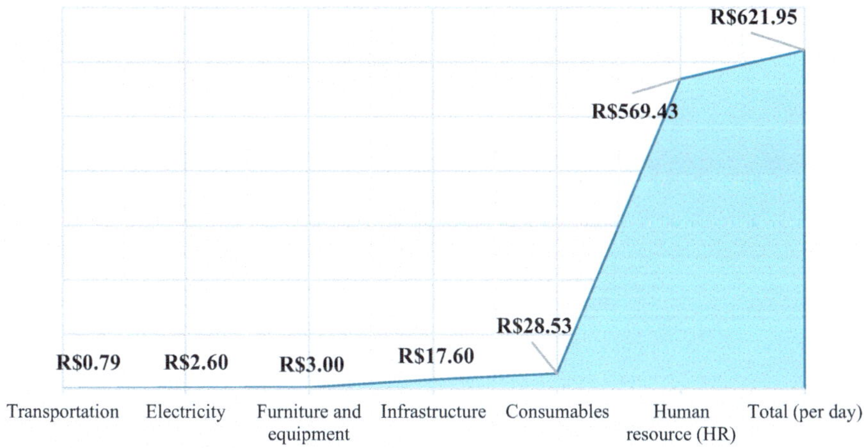

Fig. 3 Results for the economic LCA contribution analysis of the evaluated lab

nature of academic work, which requires adequately compensated labor. Secondary contributors to total costs were transportation and electricity.

Transportation costs encompass fuel purchases (such as ethanol) and bus fare expenditures by students. Vehicle depreciation was not included in the analysis, as its relevance to laboratory activities was deemed negligible and excluded via the cutoff criterion.

Electricity costs were also significant, mainly due to the high energy consumption of two air conditioning units. Infrastructure-related expenses represented the lowest absolute costs, a result explained by the long lifespan assumed for buildings (50 years), the annual depreciation rate of 4% (MM Contabilidade 2025), and the local average construction cost per square meter in Sorocaba (R$ 8046.00) (G1 2023).

In absolute terms, the weekly life cycle cost per researcher was approximately R$ 888. This result aligns with Ugaya et al. (2020), who estimated a weekly cost of R$7000 for nine researchers at the Gyro laboratory, equating to approximately R$777 per researcher per week.

4.3 Social LCA

The study utilized the 55 impact subcategories from the SOCA 3.7.1 database to conduct the social LCA. To enhance interpretability, results were normalized to identify the main contributors at each life cycle stage (Fig. 4).

The analysis of the Fig. 3 indicates that transportation is the dominant source of social impacts, leading across all subcategories. Its contributions range from 36% in "Embodied Agricultural Area Footprints" to 88% in "Net Migration". These outcomes are primarily due to the production phase of vehicles, which is closely linked to the mining and petroleum sectors—highlighting transportation as a critical factor in social sustainability.

Fig. 4 Results for the social LCA contribution analysis of the evaluated lab

Furniture, equipment, and infrastructure ranked as the second and third most significant contributors, respectively, followed by energy use, administrative and human resource activities, and consumables. However, it is essential to acknowledge that the SOCA database primarily relies on global-scale data (e.g., RoW: rest-of-world; GLO: global), which may not accurately reflect the Brazilian context.

Ugaya et al. (2020), using the PSILCA database, observed that a substantial portion of social risks at the Gyro lab originated from its human resources. This difference stems from the PSILCA model's cost-based structure, which inherently translates financial data into social risk estimations.

Additionally, several SOCA subcategories exhibit environmental bias in their analyses, such as biomass consumption; certified environmental management system; DALYs (Disability-adjusted life years) due to indoor and outdoor air and water pollution; embodied (agricultural area footprint, biodiversity footprints, forest area footprints, and water footprints); fossil fuel consumption; GHG footprints; industrial water depletion; minerals consumption and pollution. Therefore, these categories were excluded from the LCSA integration stage with the SDGs, along with the categories of youth illiteracy, female and youth illiteracy, male; illiteracy, female and illiteracy, male; and child labor, male and child labor, female, as they have corresponding total categories (Youth illiteracy, total; Illiteracy, total; and Child Labour, total), resulting in 37 subcategories for final evaluation.

These were subsequently grouped into six stakeholder categories following the UNEP framework by Traverso et al. (2021): children, consumers, local community, society, value chain actors, and workers (see Table 4).

Table 5 presents the normalized data, expressed as a percentage of 100%, within each of the six categories, highlighting the contributions of each life cycle stage in the categories defined by Traverso et al. (2021).

Consumers' social impacts included 37% from transportation, 34% from furniture and equipment, and 17% from infrastructure, resulting from the laboratory's construction. Electricity consumption generated 8% of the impacts, with human resources (2%) and water, tea, and coffee consumption (1%) accounting for the remaining 4%. Overall, the results indicate that transportation remains the primary contributor, accounting for more than 60% of the impacts related to children, local community, society, value chain actors, and workers.

Table 4 Correlation of categories and subcategories SOCA 3.7.1

.Categories (Traverso et al. 2021)	Subcategories (Traverso et al. 2021)	Subcategories (SOCA 3.7.1)	Categories (Traverso et al. 2021)	Subcategories (Traverso et al. 2021)	Subcategories (SOCA 3.7.1)
Children	Education provided in the local community	Expenditures on education	Value chain actors	Promoting social responsibility	Promoting social responsibilty
		Illiteracy, total	Value chain actors	Respect of intellectual property rights	Violations of employment laws and regulations
		Youth illiteracy, total	Worker	Child labor	Child Labour, total
Consumer	Health and safety	Safety measures	Worker	Employment relationship	Trade unionism
Local community	Delocalization and migration	Trafficking in persons	Worker	Equal opportunities/ discrimination	Gender wage gap
		International migrant workers			Men in the sectoral labour force
		Migration flows			Women in the sectoral labour force
		Net migration	Worker	Fair salary	Fair salary
		International migrant stock	Worker	Forced labor	Frequency of forced labour
Local community	Safe and healthy living conditions	Drinking water coverage			Goods produced by forced labour
		Health expenditure	Worker	Freedom of association and collective bargaining	Association and bargaining rights
		Life expectancy at birth	Worker	Health and safety	Fatal accidents
		Sanitation coverage			Non-fatal accidents
Local community	Local employment	Unemployment			Workers affected by natural disasters
Local community	Respect of indigenous rights	Indigenous rights	Worker	Social benefits/ social security	Social security expenditures

(continued)

Table 4 (continued)

.Categories (Traverso et al. 2021)	Subcategories (Traverso et al. 2021)	Subcategories (SOCA 3.7.1)	Categories (Traverso et al. 2021)	Subcategories (Traverso et al. 2021)	Subcategories (SOCA 3.7.1)
Society	Contribution to economic develop	Contribution of the sector to economic development	Worker	Working hours	Weekly hours of work per employee
Society	Corruption	Active involvement of enterprises in corruption	Society	Poverty alleviation	Value added (total)
		Anti-competitive behaviour or violation of anti-trust and monopoly legislation	Society	Prevention and mitigation of armed conflicts	Risk of conflicts
		Public sector corruption			

Table 5 Normalized results for the social stakeholder categories

Categories (Traverso et al. 2021)	Transportation (%)	Furniture and equipment (%)	Infrastructure (%)	Electricity (%)	HR (%)	Consumables (%)
Children	64	21	6	5	3	2
Consumer	37	34	17	8	2	1
Local Community	67	16	9	4	3	1
Society	62	17	9	6	3	2
Value chain actors	68	14	9	5	3	1
Worker	62	18	10	5	3	2

4.4 Linking LCA and SDGs

To evaluate the system's contribution to the SDGs, LCSA results were benchmarked using a five-level scoring scale proposed by Weidema et al. (2020), ranging from − 2 (significantly worse) to + 2 (significantly better). A total of 12 categories—five environmental, six social, and one economic—were mapped to relevant Sustainable Development Goals (SDGs) as presented in Table 6.

Table 6 Linking SGDs with LCSA categories

Categories	Score					Sustainable Development Goals
	-2	-1	0	1	2	
Climate Change						2 Zero Hunger; 9 Industry, Innovation and Infrastructure; 13 Climate Action; 15 Life on Land
ecosystem Quality						2 Zero Hunger; 3 Good Health and Well-Being; 6 Clean Water and Sanitation; 9 Industry, Innovation and Infrastructure; 12 Responsible Consumption and Production; 13 Climate Action; 14 Life Below Water; 15 Life on Land
energy resources: non-renewable						12 Responsible Consumption and Production
Human Health						3 Good Health and Well-Being; 11 Sustainable Cities and Communities; 12 Responsible Consumption and Production
Water use						2 Zero Hunger; 3 Good Health and Well-Being; 6 Clean Water and Sanitation; 12 Responsible Consumption and Production; 15 Life on Land
Children						1 No Poverty; 3 Good Health and Well-Being; 4 Quality Education; 8 Decent Work and Economic Growth; 10 Reduced Inequalities

(continued)

Table 6 (continued)

Workers	1 No poverty; 2 Zero hunger; 3 Good health and well-being; 4 Quality education; 5 Gender equality; 6 Clean water and sanitation; 8 Decent work and economic growth; 9 Industry, innovation and infrastructure; 10 Reduced inequalities; 11 Sustainable cities and communities; 16 Peace, justice and strong institutions
Local community	3 Good health and well-being; 6 Clean water and sanitation; 8 Decent work and economic growth; 9 Industry, innovation and infrastructure; 10 Reduced inequalities; 11 Sustainable cities and communities; 12 Responsible consumption and production
Society	1 No poverty; 2 Zero hunger; 3 Good health and well-being; 4 Quality education; 8 Decent work and economic growth; 9 Industry, innovation and infrastructure; 1 No poverty; 11 Sustainable cities and communities; 12 Responsible consumption and production; 16 Peace, justice and strong institutions; 17 Partnerships for the goals
Consumer	2 Zero hunger; 3 Good health and well-being; 12 Responsible consumption and production
Value chain actors	9 Industry, innovation and infrastructure; 12 Responsible consumption and production
Laboratory cost	8 Decent work and economic growth; 9 Industry, innovation and infrastructure; 1 No poverty; 12 Responsible consumption and production; 17 Partnerships for the goals

Giusti et al. (2024) obtained benchmark values for environmental indicators related to climate change, non-renewable energy resources, and water use from their study, which assessed 2-h classes with 60 students at the UFSCar Sorocaba campus.

Extrapolated to a scenario involving 3.5 students over an 8-h day for one week, the present study reported impacts that were 353%, 1607%, and 565% higher, respectively—resulting in a score of − 2 in all three categories. Since we found no comparative data for the ecosystem quality and human health categories, we assigned a neutral score of 0.

For social categories, no equivalent benchmarks were available; however, a qualitative assessment suggested a positive social impact of the laboratory's research activities. Contributions include advances in pedagogical practices, special education, ethnic-racial relations, affectivity in teaching, teacher identity, public education policy, and historical and educational analysis—particularly within the Brazilian context (Wood Jr et al. 2016; Borges 2019). The study team, therefore, rated these impacts as + 2.

Details for Benchmarking

Environmental categories: (adapted from Giusti et al. 2024)

Social Categories: (adapted from Traverso et al. 2021; Borges 2019 and Wood Jr et al. 2016)

Economic categories: (adapted from Parvathy et al. 2025)

5 Conclusions

In terms of environmental impacts, the findings indicated that transportation was the most significant contributor, followed by electricity consumption. Strategies such as carpooling among research group members can mitigate these effects by reducing individual trips and distributing associated impacts across multiple users. To minimize environmental burdens associated with electricity consumption, we can adopt energy-efficient equipment and implement basic operational practices, such as turning off devices and lights when not in use and maintaining moderate indoor temperature settings.

From an economic perspective, human resource expenditures—specifically the stipends provided to master's and doctoral students attending the laboratory under study—were identified as the primary cost drivers. Transportation-related expenses, notably fuel consumption and electricity usage, also represented relevant cost categories. In this context, promoting carpooling practices and enhancing energy efficiency would not only reduce environmental impacts but also contribute to cost optimization.

Regarding social impacts, transportation again emerged as the predominant factor, mainly due to its dependency on the mining and petroleum sectors involved in vehicle production. Other relevant stages included the use of furniture, equipment, and building infrastructure, followed by energy consumption, administrative activities, and the use of consumables. A limitation of this assessment stems from the use of globalized data from the Ecoinvent and the SOCA databases, which may not fully capture the specific socio-environmental context of the Brazilian laboratory under study.

The correlation between the LCSA results and the Sustainable Development Goals (SDGs) was evident throughout the analysis. Each evaluated category was mapped to its corresponding SDGs, highlighting the relevance of LCSA as a tool for sustainability assessment and for aligning institutional practices with international sustainability frameworks. Notably, our study covered 16 out of the 17 SDGs in at least one impact category, except for SDG 7—"Affordable and Clean Energy," for which we found no direct evidence.

The integration of sustainability into higher education curricula demands methodological approaches that allow students to apply theoretical knowledge in real-world contexts, thereby enhancing both critical thinking and practical skills. In this respect, the application of the LCSA methodology to assess the sustainability of the Sustainable Production Laboratory demonstrates its potential as a replicable, quantitative, and pedagogically valuable tool in sustainability education. By analyzing their institutional infrastructure, students can develop key competencies without relying on external resources, thereby fostering an accessible, applied, and holistic educational model that aligns with the principles of academic sustainability.

It is important to note that the benchmarks used to connect LCA with the SDGs are fundamental assumptions of this study. Consequently, proposing new benchmarks would alter the score obtained. Among the limitations, the lack of a social benchmark for this work stands out, given the specificity of the research.

For future research, we recommend expanding the underlying database to incorporate additional research laboratories, thereby enhancing the robustness and generalizability of the findings. Furthermore, conducting sensitivity and uncertainty analyses is also encouraged to improve data reliability. Another avenue for advancement involves the regionalization of assessments, enabling the inclusion of local specificities—particularly relevant to social and environmental analyses LCSA, which in this study were based on globally averaged datasets from Ecoinvent and SOCA. The standardization of methodological procedures, alongside the systematic documentation of all analytical steps, is essential to ensure the reproducibility of results and to facilitate the continuity of research activities by future members of the Sustainability Engineering Laboratory.

Acknowledgements We would like to thank the São Paulo Research Foundation (FAPESP), process no. 2022/15134-1; the National Council for Scientific and Technological Development (CNPq, grant number 303343/2022-2); and the Coordination for the Improvement of Higher Education Personnel—Brazil (CAPES)—Funding Code 001.

References

Asadollahfardi G, Alipour M, Panahandeh A, Karimi Ardestani MH (2024) Environmental Life Cycle Assessment of University Campus in Operation Phase, a Case Study of Kharazmi University in Iran. Process Integr Optim Sustain 8:1035–1049. https://doi.org/10.1007/s41660-024-00411-9

Barros, M.V., Puglieri, F.N., Tesser, D.P., Kuczynski, O. and Piekarski, C.M. (2020), "Sustainability at a Brazilian university: developing environmentally sustainable practices and a LCA case study", International Journal of Sustainability in Higher Education, Vol. 21 No. 5, pp. 841-859, https://doi.org/10.1108/IJSHE-10-2019-0309.

Borges, D. (2019, July 31). O impacto social da pesquisa científica em educação [The Social Impact of Scientific Research in Education]. https://comunica.ufu.br/noticias/2019/07/o-impacto-social-da-pesquisa-cientifica-em-educacao

Bueno G, De Blas M, Pérez-Iribarren E, Zuazo I, Torre-Pascual E, Erauskin A, Etxano I, Tamayo U, García M, Akizu-Gardoki O, León I, Marieta C, Zulueta G, Barrio I (2021) The environmental and social footprint of the university of the Basque Country UPV/EHU. Journal of Cleaner Production 315:128019. https://doi.org/10.1016/j.jclepro.2021.128019

Clabeaux, R., Carbajales-Dale, M., Ladner, D. and Walker, T. (2020), "Assessing the carbon footprint of a university campus using a life cycle assessment approach", Journal of Cleaner Production, Vol. 273, p. 122600, https://doi.org/10.1016/j.jclepro.2020.122600.

Çetin Doğruparmak Ş, Çankaya S, Kibar ME, Pekey B (2025) Assessing the environmental performance of a university campus using a life cycle assessment approach. Environ Dev Sustain 27:645–664. https://doi.org/10.1007/s10668-024-05730-y

De Oliveira, J.A., Silva, D.A.L., Puglieri, F.N. and Saavedra, Y.M.B. (2021), "Introduction to life cycle engineering and management (LCEM)", in de Oliveira, J.A., Lopes Silva, D.A., Puglieri, F.N., Saavedra, Y.M.B., (Eds) Life Cycle Engineering and Management of Products, Springer, Cham, https://doi.org/10.1007/978-3-030-78044-9_1.

Erauskin-Tolosa, A., Bueno, G., Etxano, I., Tamayo, U., García, M., de Blas, M., Perez-Iribarren, E., Zuazo, I., Torre-Pascual, E. and Akizu-Gardoki, O. (2021), "Social organisational LCA for the academic activity of the university of the Basque Country UPV/EHU", The International Journal of Life Cycle Assessment, Vol. 26 No. 8, pp. 1648-1669, https://doi.org/10.1007/s11367-021-01940-y.

Filho, W.L., Trevisan, L.V., Dinis, M.A.P. *et al.* Fostering students' participation in the implementation of the sustainable development goals at higher education institutions. *Discov Sustain* **5**, 22 (2024). https://doi.org/10.1007/s43621-024-00204-7

G1. (2023, September 19). Pesquisa aponta que Sorocaba teve aumento de mais de 10% no preço médio do metro quadrado [Research shows Sorocaba saw over a 10% increase in the average price per square meter]. https://g1.globo.com/sp/sao-jose-do-rio-preto-aracatuba/mercado-imobiliario-do-interior/noticia/2023/09/19/pesquisa-aponta-que-sorocaba-teve-aumento-de-mais-de-10percent-no-preco-medio-do-metro-quadrado.ghtml

Genta, C., Favaro, S., Sonetti, G., & Lombardi, P. L. (2022). Quantitative assessment of environmental impacts at the urban scale: The ecological footprint of a university campus. Environmental Development and Sustainability, 24(12), 2501-2522. https://doi.org/10.1007/s10668-021-01686-5

Giusti, G., Farrapo, A.C., Jr, Sigahi, T.F.A.C., Saltorato, P., Vieira, J.G.V. and Silva, D.A.L. (2023), "A multicriteria approach for consumption footprint in higher education institutions before and during the COVID-19 pandemic", International Journal of LCA, Vol. 28, pp. 1735-1755.

Giusti, G., Farrapo Junior, A. C., Silva, D. V., & Silva, D. A. L. (2024). "Linking the UN sustainable development goals to life cycle impact applied to a university campus", International Journal of Sustainability in Higher Education, https://doi.org/10.1108/IJSHE-08-2023-0389.

Gulcimen S, Qadri S, Donmez RO, Uzal N (2023) A holistic sustainability assessment of a university campus using life cycle approach. Int J Environ Sci Technol 20:3309–3322. https://doi.org/10.1007/s13762-022-04214-8

Huang, L., Liu, Y., Krigsvoll, G., & Johansen, F. (2018). "Life cycle assessment and life cycle cost of university dormitories in the southeast China: Case study of the university town of Fuzhou", Journal of Cleaner Production, Vol. 173, pp. 151–159, https://doi.org/10.1016/j.jclepro.2017.06.021.

Impact World+. (2025). IMPACT World+. Retrieved March 20, 2025, from https://www.impactworldplus.org/?utm_source

ISO 14040 (2006a), Environmental Management - LCA - Principles and Framework.

ISO 14044 (2006b), Environmental Management – LCA – Requirements and Guidelines.

MM Contabilidade. (2025). Taxas de depreciação. https://www.mmcontabilidade.com.br/flash/taxasdepreciacao.htm?utm_source=chatgpt.com

Parvathy, S. U., Kolil, V. K., Raman, R., Vinuesa, R., & Achuthan, K. (2025). Integrating sustainable development goals into life cycle thinking: a multidimensional approach for advancing sustainability. Environment, Development and Sustainability. https://doi.org/10.1007/s10668-024-05810-z

Robinson, O. J., Tewkesbury, A., Kemp, S., & Williams, I. D. (2018). "Towards a universal carbon footprint standard: A case study of carbon management at universities", Journal of Cleaner Production, Vol. 172, pp. 4435–4455, https://doi.org/10.1016/j.jclepro.2017.02.147.

Silva DAL, Giusti G, Rampasso IS, Junior ACF, Marins MAS, Anholon R (2021) The environmental impacts of face-to-face and remote university classes during the COVID-19 pandemic. Sustainable Production and Consumption 27:1975–1988. https://doi.org/10.1016/j.spc.2021.05.002

Traverso, M., Luthin, A., & Roche, L. (2021). Methodological Sheets for Subcategories in Social life cycle assessment (S-LCA) 2021. https://www.researchgate.net/publication/356810860

Ugaya, C. M. L., de Loreto, A. C., Sturm, G., Savioli, J. P. P. D., Crippa, J., Marin, L. E., Juchen, R. T., & de Araújo, J. B. (2020). Faço o que eu digo: ACV ambiental, social e econômica do Gyro [Walk the Talk: Environmental, Social, and Economic LCA of Gyro]. Anais de conferência.

United Nations (2015), Transforming Our World: The 2030 Agenda for Sustainable Development, UN Publishing, New York, NY.

UNEP. (2020). Guidelines for Social Life Cycle Assessment of Products and Organizations 2020. Benoît Norris, C., Traverso, M., Neugebauer, S., Ekener, E., Schaubroeck, T., Russo Garrido, S., Berger, M., Valdivia, S., Lehmann, A., Finkbeiner, M., & Arcese, G. (Eds.). United Nations Environment Programme (UNEP).

Wafa, W., Sharaai, A. H., Matthew, N. K., Ho, S. A., & Akhundzada, N. A. (2022). Organizational Life Cycle Sustainability Assessment (OLCSA) for a Higher Education Institution as an Organization: A Systematic Review and Bibliometric Analysis. Sustainability, 14(5), 2616.

Weidema, B., Goedkoop, M., Meijer, E., & Harmens, R. (2020). LCA-based assessment of the Sustainable Development Goals: Development update and preliminary findings of the Project "Linking the UN Sustainable Development Goals to life cycle impact pathway frameworks". PRé Sustainability & 2.-0 LCA consultants.

Wood Jr, T., Costa, C. C. M., Lima, G. de M. R., & Guimarães, R. C. (2016). Impacto Social: Estudo sobre Programas Brasileiros Selecionados de Pós-graduação em Administração de Empresas [Social Impact: A Study on Selected Brazilian MBA Programs]. Revista de Administração Contemporânea, 20(1), 21–40. https://doi.org/10.1590/1982-7849rac2016184?

Xue Z, Liu H, Zhang Q, Wang J, Fan J, Zhou X (2019) The Impact Assessment of Campus Buildings Based on a Life Cycle Assessment–Life Cycle Cost Integrated Model. Sustainability 12:294. https://doi.org/10.3390/su12010294

Zhang, G., Chen, P., & Xu, S. (2024). "Development and validation of a college organisational sustainability scale", International Journal of Sustainability in Higher Education, https://doi.org/10.1108/IJSHE-10-2023-0493.

João Victor Encide Salla He's currently a Ph.D. candidate in Production Engineering at UFSCar, deepening his studies in sustainable Production, product life cycle assessment, and artificial intelligence. Since April 2021, he's been a researcher at UFSCar's Sustainability Engineering Research Group (EngS). João also has professional experience in consulting projects, including an internship in the Research and Development sector at EMBRAPA (Brazilian Agricultural Research Corporation) Environment; leading LCA projects; and contributing to the UN Environment's GO4SDGs platform.

Abner Fernandes Souza da Silva He is a Production Engineer graduated from the Federal University of São Carlos (UFSCar), Sorocaba campus, Brazil. He is currently a Ph.D. candidate in Production Engineering at UFSCar, focusing his research on Life Cycle Assessment, Circular Economy, and Waste Management. Since 2023, he has been a member of UFSCar's Sustainability Engineering Research Group (EngS). Abner also has academic and professional experience in Discrete-Event Simulation, Optimization, and Applied Mathematics.

Jayson Wilson Barretti holds a master's degree in production engineering from the Federal University of São Carlos, Department of Industrial Engineering. He holds a degree in Industrial Engineering from Centro Universitário FEI and an MBA in Business Management from USP/ESALQ. He is a member of the Intersectoral Committee on Artificial Intelligence of the Assemblée Nationale du Québec. He is a Lean Manufacturing Specialist with Black Belt certifications in Lean Six Sigma. His current research interests are Artificial Intelligence, Industry 4.0, models and solution methods based on information technology and applied to the integration of strategic and tactical decisions in the management of sustainable operations. He currently works at the Ministry of the Government of Quebec, Canada.

Virgínia Aparecida Silva Moris Associate professor at the Federal University of São Carlos in the Department of Production Engineering. She is a member of CNPq's Research Group on Sustainability Engineering (Engs Group—https://www.grupoengs.com.br/). She has experience in the areas of: Industrial Process Monitoring, Industrial Waste Treatment, Cleaner Production, Life Cycle Assessment, Circular Economy, Reverse Logistics, Sustainability Indicators, Emerging Technologies.

Juliana Veiga Mendes Associate Professor in Operations Management at the Federal University of São Carlos (UFSCar) -Sorocaba campus, Brazil. She is part of the Center for Management Sciences and Technologies, Department of Industrial Engineering. She is a member of Supply Chain 4.0 and a member of CNPq's Research Group on Sustainability Engineering. She heads the Studies Group on Information Technology applied to reverse logistics. Her research focuses on Operations Management with an interface in Sustainability. Key areas of study include: Information technologies in reverse logistics systems, Industry 4.0, Sustainability Indicators, Production planning, Lean manufacturing, and ERP systems. Currently, her research interests revolve around emergent technologies and practices for sustainable operations in reverse logistics and production systems.

Diogo Aparecido Lopes Silva Professor and researcher at Federal University of São Carlos (UFSCar) - Sorocaba campus, in Brazil. He performs research in Life Cycle Engineering and Management of products (Life Cycle Assessment—LCA, Sustainable Manufacturing, Circular Economy, Absolute Sustainability and Ecodesign) and Lean & Green Production (Cleaner Production, Sustainable VSM, Sustainable Production Planning and Control, and Ecoefficiency). He is head of the CNPq's Research Group on Sustainability Engineering (Engs Group), and is editor of the textbook "Life Cycle Engineering and Management of products: theory and practice" published by Springer Nature.

IFSP Sustentare: Promoting Urban Agriculture and Sustainable Gastronomy for Food Education and Conscious Citizenship

Simone Mendes Delphino, Fabiana Souza Ferreira, Rodrigo De Benedictis Delphino, Marcelo Bernardino Araújo, and Luciano Delmondes De Alencar

Abstract This chapter explores some of the means which may be deployed to pursue and implement sustainable development across disciplines. The increasing need to address urban challenges—such as food waste, food insecurity, and environmental degradation—has prompted the investigation of sustainable practices, with urban agriculture and conscious gastronomy emerging as viable strategies. This study focuses on the Sustentare Project developed by the Federal Institute of São Paulo (IFSP), which integrates education, technology, and community engagement to promote the circular economy, waste management, and food security, in alignment with selected United Nations Sustainable Development Goals (SDGs). Adopting an interdisciplinary approach, the research combined literature review, data analysis, and case studies to evaluate initiatives including urban gardens, composting systems, and the application of IoT sensors. Additionally, food reuse workshops were conducted with IFSP students and members of the surrounding community. The findings indicate a significant reduction in organic waste disposal, increased awareness regarding sustainable food practices, and improved cultivation efficiency through the use of technology. Nonetheless, challenges related to energy consumption and financial

S. M. Delphino · F. S. Ferreira · L. D. De Alencar
Federal Institute of São Paulo, São Paulo, Brazil
e-mail: simone.delphino@ifsp.edu.br

F. S. Ferreira
e-mail: ferreira.souza@ifsp.edu.br

L. D. De Alencar
e-mail: luciano.delmondes@ifsp.edu.br

R. D. B. Delphino (✉)
University of São Paulo, São Paulo, Brazil
e-mail: rodrigo.delphino@ifsp.edu.br; rodrigodelphino@usp.br

M. B. Araújo
Fundação Instituto Tecnológico de Osasco, Osasco, Brazil
e-mail: marcelo.bernardino@fito.br

W. Leal Filho et al. (eds.), *Sustainability in the Curriculum and Operations of Universities*, World Sustainability Series,
https://doi.org/10.1007/978-3-032-12678-8_5

sustainability persist. The project also contributed to strengthening social bonds and fostering income generation through entrepreneurship, illustrating that the integration of research, community outreach, and innovation can play a transformative role in urban food systems. However, subjective impacts—such as behavioral change—require more refined methodologies for future measurement. In conclusion, these practices are deemed essential for building resilient cities. The findings advocate for their incorporation into public policies and the expansion of participatory models to ensure continuity and scalability. Sustentare serves as a compelling example of how educational institutions can function as agents of change, bridging theory and practice in pursuit of a more equitable and sustainable future.

Keywords Urban agriculture · Food waste reduction · Sustainable development · Environmental education · Interdisciplinary practices

1 Introduction

Urban agriculture has increasingly been recognized as a key component in advancing urban sustainability, both socially and environmentally. It contributes to the promotion of recycling practices, resource conservation, and the development of environmental awareness—elements that are critical to addressing the complex challenges of the twenty-first century (Deelstra and Girardet 2000). Within this context, periods of economic crisis often serve as catalysts for the emergence of collaborative and sustainable solutions (Rinaldi et al. 2020), underscoring the importance of integrated approaches that combine innovation, education, and socio-environmental responsibility.

The Federal Institute of Education, Science and Technology of São Paulo (IFSP), through the Sustentare Project—an institutional initiative that encompasses both outreach activities and academic research—exemplifies a systemic response aligned with the Sustainable Development Goals (SDGs) outlined in the United Nations 2030 Agenda. The project develops and implements actions centered on the circular economy, waste management, and sustainable gastronomy. These include the selective collection and reuse of materials such as plastic caps, blister packs, textiles, and organic residues, while also promoting engagement from both the academic community and surrounding population in responsible consumption and waste disposal practices (Projeto Sustentare—IFSP).

As part of its integrated approach, the Sustentare Project (2024) adopts the concept of sustainable gastronomy, which encompasses not only responsible consumption practices but also the strategic planning of urban gardens as a foundation for local food production. This framework connects environmental education, circular economy, and food sovereignty, promoting low-impact diets and resilient food systems in urban settings. As an outgrowth of this work, one of the project's researchers was invited to coordinate the Quintais IFSP Project, a state-level initiative carried out in partnership with the Ministry of Agrarian Development (MDA). The Quintais Project focuses on the design and implementation of productive backyards and urban gardens in vulnerable territories, including favelas and agrarian reform settlements across São Paulo State, thus scaling up the methodologies and principles developed under Sustentare to broader socio-environmental contexts.

The IFSP São Paulo campus, which receives over 10,000 people daily, faces considerable logistical and environmental challenges due to the high volume of waste produced—particularly in the student restaurant and cafeteria areas—further emphasizing the urgency of sustainable interventions (ABRELPE 2020). Beyond waste collection efforts, Sustentare promotes interdisciplinary integration by supporting student-led research on the application of digital and technological innovations in agriculture. Advanced technologies such as IoT (Internet of Things) sensors and autonomous systems have been shown to enhance agricultural productivity and sustainability, particularly in urban contexts where resource constraints are pronounced (Elbeheiry and Balog 2023).

Within this framework, the project incorporates practices such as composting, the development of urban gardens, and the transformation of agricultural by-products into valuable resources, thereby operationalizing the principles of the circular economy to reduce waste and foster more resilient food systems (Aziz et al. 2023). These practices are not only implemented on campus, but also serve as educational tools to raise awareness about the ecological and economic significance of waste recovery, encouraging the community to adopt more sustainable behaviors in line with circular economy models.

Additionally, the valorization of local agricultural products and their integration into sustainable gastronomic practices is a strategic component in the design of resilient urban food systems. Figures 1, 2, 3 and 4 illustrate practical applications of the Sustentare Project in fostering circular economy principles and environmentally responsible behavior, including the separation and disposal of plastic waste (Fig. 1), the collection of used cooking oil (Fig. 2), and the organization of a food reuse fair (Fig. 3). The project also includes a series of online workshops on food reuse (Fig. 4), which serve as tools for environmental education and civic engagement. These virtual activities promote sustainable gastronomy while expanding access to knowledge on food recovery techniques, especially in urban households and school environments.

In addition, valuing local agricultural resources and integrating them into gastronomic practices can strengthen sustainable urban food systems and foster conscious citizenship, especially when addressing challenges such as seasonality and sustainability in the supply chain (Kuang and Bhat 2017). The Sustentare Project reflects this comprehensive approach by combining applied research, extension outreach, and social entrepreneurship, thereby empowering communities and generating measurable environmental and socioeconomic benefits.

This article aims to investigate and critically assess urban agriculture and sustainable gastronomy. It examines their role as educational tools and catalysts for the development of environmentally conscious citizenship. It examines their social, economic, and environmental impacts within the urban context, with a specific focus on the Sustentare Project as a case study. By integrating theory and practice, the study

Fig. 1 Collectors of waste (Sustentare Project)

Fig. 2 The collection of used cooking oil (Sustentare Project)

seeks to contribute to the advancement of more inclusive, resilient, and intelligent urban environments, aligned with the core principles of global sustainability.

Finally, this investigation also addresses how excessive waste generation and inadequate disposal practices undermine biodiversity in urban and peri-urban areas. These actions contribute to habitat loss, soil and water pollution, and the disruption of ecological functions. Therefore, raising awareness of the ecological impacts of organic and food waste is essential for promoting responsible consumption and strengthening urban ecosystems. In this regard, practices such as composting, food

Fig. 3 Food reuse fair (Sustentare Project)

Fig. 4 Online workshops on food reuse (Sustentare Project)

recovery, and integrated waste management—central to Sustentare's methodology—represent critical strategies for mitigating biodiversity loss and fostering sustainable urban living.

2 Sustainable Gastronomy Across Urban Agriculture

Urban agriculture and sustainable gastronomy have emerged as fundamental pillars for building more resilient and socially just cities, integrating both traditional practices and technological innovations. Sgroi (2023) emphasizes that historic markets—such as the Ballarò Market in Palermo—are not only centers of commerce but also serve as spaces for social engagement, the preservation of culinary traditions, and the promotion of local products. These markets act as forms of resistance to global mass standardization, valuing food diversity and the relationship between producers and consumers. The study highlights that consumers frequent these spaces not only for product quality and affordability but also for the cultural and social experience, aligning with the principles of the civil economy, which prioritizes human relationships and sustainability over short-term profit (Sgroi 2023).

In line with this perspective, the Sustentare Project promotes awareness initiatives that connect food systems—from production, transportation, and storage to consumption—with broader issues such as climate change, hunger, and food waste. Food inflation has become a global concern in the post-pandemic context and amid increasingly evident signs of climate instability.

In this scenario, educational institutions play a transformative role by collaborating with local stakeholders to address complex challenges, such as those posed by sustainable development. This highlights the importance of multisectoral participation in fostering food literacy and environmentally conscious citizenship (Rinaldi et al. 2020). Developed by the Federal Institute of Education, Science and Technology of São Paulo (IFSP), the Sustentare Project exemplifies this model by integrating academic research, outreach, and community actions aimed at waste management, circular economy practices, and food security.

Shifting consumption habits begins with food, which can serve as a powerful driver for sustainable development by integrating social, economic, and environmental dimensions (Rinaldi et al. 2020). This approach reinforces the idea that sustainable gastronomy not only preserves cultural identity but also raises awareness about local food systems and sustainability. The integration of cultural globalization into urban food tourism, for instance, can foster sustainability by balancing the tension between global homogenization and local diversity, ultimately reinforcing cultural identity and generating positive socio-economic impacts (Stalmirska and Ali 2023).

Research such as that by Csurgó et al. (2016) shows that urban agriculture initiatives and alternative food networks—like AMAP (Associations for the Maintenance of Peasant Agriculture) in Paris—not only support sustainability but also promote social cohesion and food education. These models underscore the significance of participatory governance and the valorization of local crops in the pursuit of sustainable urban development. At IFSP, practices including urban gardens, composting systems, and food reuse workshops reflect this vision by fostering circular economy principles and minimizing waste (ABRELPE 2020).

Technological innovation also plays a critical role in this transformation. Automated irrigation systems and data analytics platforms that monitor crop needs can contribute directly to sustainable gastronomy by enabling the cultivation of fresh, local food while reducing the carbon footprint associated with transportation. However, challenges persist, including energy consumption in sensor networks and the demand for robust communication infrastructure. Solutions such as solar-powered systems and low-energy algorithms have been proposed to mitigate these challenges (Elbeheiry and Balog 2023). These considerations are crucial for evaluating the technical and financial feasibility of urban agriculture in resource-constrained environments, such as university campuses.

Deelstra and Girardet (2000) argue that urban agriculture plays a significant role in improving resource management, including nutrient recycling, waste reduction, and soil conservation. These practices contribute to lowering ecological footprints and promoting more responsible food systems. The Sustentare Project incorporates these principles by conducting research on composting, fabric reuse in construction, and partnerships with recycling cooperatives. This reinforces IFSP's commitment to sustainability, citizen education, and local economic development, while also seeking to address hunger. Hunger continues to hinder national development and imposes severe social consequences. Kimani-Murage et al. (2014, p. 1099) state that food insecurity is a defining feature of many crises, with declining access to food leading to increased morbidity and mortality. They note that "the nature, extent, and duration of coping strategies employed by the urban poor in the face of food crises and consequent food insecurity will determine the magnitude and severity of their suffering."

On the other hand, Thomas Malthus argued that hunger acts as a natural check on population growth, suggesting that overpopulation would result in scarcity and increased pressure on food systems (Ziegler 2012). However, this perspective does not fully reflect today's reality. The current food supply crisis is not solely the result of population growth, but rather of unsustainable environmental practices, particularly food waste. In 2019, food waste accounted for approximately 40% of all food produced, and by 2022, this figure had exceeded 1 billion tons globally—equivalent to 132 kg per capita, or nearly 25% of total food production (Food Waste Index Report 2024).

The environmental impact of food waste is immense. It includes wasted irrigation water, soil degradation, agrochemical use, and greenhouse gas emissions from decomposition. In Brazil, the situation is further exacerbated by the fact that organic waste is often sent to landfills without treatment, intensifying the environmental burden.

Figure 5 illustrates the lines of action of the Sustentare Project, which strives, through actions in partnership with the community, as well as with applied research, to contribute in some way to the reduction of environmental and social impacts, while strengthening the tripod of sustainability (environmental, social, and economic).

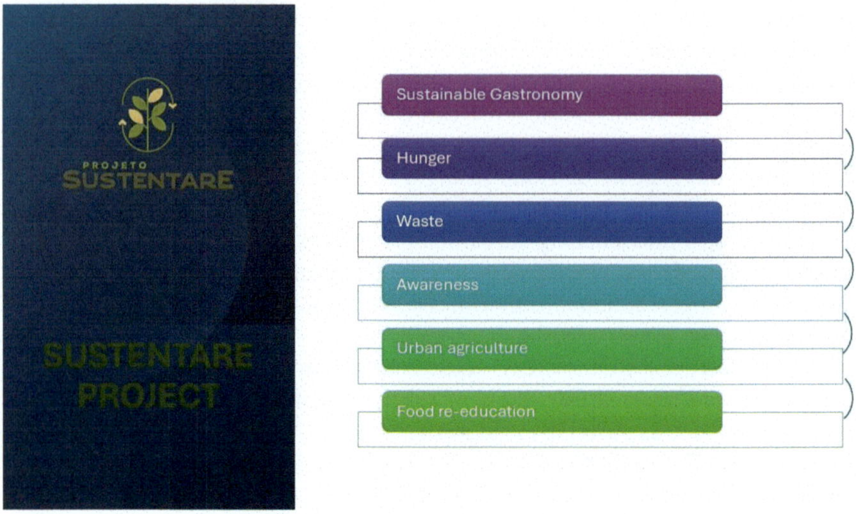

Fig. 5 Sustentare Project research and outreach activities

3 Future Trends in Sustainability

Extension actions in higher education institutions play a role in facilitating educational processes and social transformation, especially in the context of promoting food education and conscious citizenship (Rinaldi et al. 2020). Within the scope of the Sustentare Project, this function is evidenced through initiatives that integrate sustainable urban gardens, composting and sustainable gastronomy, aligning with the Sustainable Development Goals of the UN 2030 Agenda. The multidimensional approach to sustainability, which encompasses social, economic, and ecological aspects (Csurgó et al. 2016), is essential to discussing how public policies and citizen participation can strengthen sustainable urban food systems.

The appreciation of traditional foods and the holding of gastronomic events, such as the wine festivals in Etyek, Hungary (Csurgó et al. 2016), demonstrate how cultural practices can strengthen links between rural and urban communities. In the context of the IFSP, sustainable gastronomy not only promotes food education, but also rescues and values local practices, contributing to the formation of a collective awareness of responsible consumption. Figure 6 illustrates one of our sustainable gastronomy workshops, demonstrating the ease of preparing novel dishes using commonly discarded vegetable parts.

By reusing parts of food that are typically discarded, we can reduce the volume of organic waste sent to landfills and promote healthier eating habits without increasing costs. The sustainability of projects after funding ends is a critical challenge (European Commission 2006), and urban agriculture and sustainable gastronomy initiatives are not exempt. To ensure continuity, it is essential to develop

Fig. 6 Workshop of sustainable gastronomy (Sustentare Project)

management models that foster active community participation, empower individuals, and integrate these practices into institutional policies (Rinaldi et al. 2020). The Sustentare Project addresses this challenge through community workshops and income generation activities based on social entrepreneurship.

The adoption of smart agriculture technologies, such as sensors and automated irrigation systems, can serve as educational tools—demonstrating the importance of sustainable food production and encouraging citizen engagement in urban gardens (Elbeheiry and Balog 2023). Carneiro and Carvalho (2013) highlight the need for Brazil to revisit its approach to technological innovation, emphasizing that such knowledge could be introduced as early as elementary school to develop human capital and contribute to national development. Similarly, Carvalho et al. (2020) argue that technological innovation is a key enabler in addressing urbanization challenges, improving urban management, and enhancing quality of life—thereby supporting the creation of smart, inclusive, and sustainable cities.

In addition, effective supply chain management of local agricultural products is crucial to overcome challenges such as seasonality and material shortages while generating employment and income in both urban and rural areas (Kuang and Bhat 2017).

The use of agricultural by-products—such as pumpkin seeds, peels, and leaves—in the food industry (Aziz et al. 2023) reinforces the importance of minimizing food waste, a core tenet of urban agriculture and sustainability. In addition to transforming ingredients into value-added products, the Sustentare Project works directly with home cooks and food handlers to develop new menus using unconventional food parts. Examples include banana peel steaks (a vegan alternative), watermelon rind stroganoff, and stalk-based pies. These recipes are nutritious, easy to prepare with ingredients already available at home, and reduce waste disposal.

This circular approach to food emphasizes responsible consumption, sustainable production, and new ways of using food, thereby reinforcing sustainability practices and valuing natural capital. Costanza et al. (1997) underscores the importance of ecological systems for life support on Earth, reminding us that human well-being is directly linked to how we relate to and impact the environment.

The strategic use of social media can further raise environmental awareness. Although this article focuses on healthy eating, sustainability, and alternative agricultural practices, it also aligns with the broader goals of food education and the promotion of conscious citizenship (Unay-Gailhard et al. 2025). Similarly, urban agriculture practices have the potential to transform citizens' perceptions of the impact of their food choices, promoting greater environmental engagement (Deelstra and Girardet 2000).

Finally, valuing local food practices not only educates the academic community but also enriches the experience of tourists, deepening their understanding of regional culture and traditions. These efforts encourage more conscious and sustainable choices (Kuang and Bhat 2017). When integrated into projects such as Sustentare, they illustrate how universities can act as transformative agents, aligning teaching, research, and extension to promote a more sustainable and inclusive future.

4 Conclusions

Urban agriculture, as implemented through the IFSP Sustentare Project, has demonstrated substantial potential for generating employment and supporting livelihoods across diverse urban settings. Moreover, the integration of local food practices with sustainable gastronomy initiatives illustrates the feasibility of reconciling regional culinary traditions with global sustainability imperatives, thereby fostering culturally diverse and nutritionally adequate diets. Despite the observable benefits—such as improved dietary habits and enhanced environmental awareness among participants—many of these outcomes remain subjective and difficult to quantify. Accordingly, future research should adopt longitudinal and mixed-method approaches to assess the extent to which educational programs based on urban agriculture and sustainable gastronomy influence behavioral change, particularly across different demographic cohorts.

Farley (2010) reinforces that the main drivers of environmental degradation are of economic origin and that the global economy depends on the extraction of raw materials, and this has led to a deterioration of ecosystems, mainly by the generation of waste that is discarded in life support systems, such as aquifers. The resulting accumulation of waste in vital ecological infrastructures poses a direct threat to biodiversity and human health. In this regard, excessive food and organic waste contribute significantly to the degradation of urban ecosystems, reinforcing the need for integrated waste management practices within sustainability frameworks.

This context underscores the strategic importance of initiatives such as Sustentare, which align urban agriculture with broader environmental objectives, including biodiversity conservation, climate resilience, and circular economy principles, in accordance with the United Nations 2030 Agenda. By incorporating actions such as composting, IoT-enabled efficiency in food production, and community-based educational interventions, the project exemplifies how systemic approaches to sustainability can be operationalized in urban environments.

According to Daly and Farley (2005) most universities educate students within the confines of traditional disciplines, often ignoring real problems. The Sustentare Project, by combining education, research and extension, reinforces the potential of educational institutions, in extension actions, as agents of social change, demonstrating that the integration between theory and practice is essential to promote a fairer, healthier and more environmentally responsible future and engage society in the protection of natural capital—a resource essential for biodiversity conservation and the continuity of life on Earth.

References

ABRELPE. (2020). *Overview of solid waste in Brazil, 2020*. Brazilian Association of Public Cleaning and Waste Companies. https://www.abrema.org.br/panorama/

Aziz, M. A., Sarfraz, S., Ashraf, I., & Iqbal, M. (2023). Valorization of pumpkin by-products for nutritional and functional food applications. *Journal of Food Science and Technology, 60*(2), 412–425. https://doi.org/10.1007/s13197-022-05534-9

Carneiro, A., & Carvalho, L. (2013). Determinants of innovation in Brazilian industries: Empirical evidence for the food, automotive, and electronics sectors. In *Annals of the XVI Symposium on Production Management, Logistics and International Operations*. São Paulo, Brazil.

Carvalho, S. M. S., Martin, A. R., Carneiro, A. G., Santos, E. R., & Barbosa, A. P. T. (2020). Smart cities: Evaluation of the characteristics of the innovation ecosystems of two Brazilian smart cities. *Cadernos de Prospecção, 13*(3), 693–706. https://doi.org/10.9771/cp.v13i3.32928

Costanza, R., d'Arge, R., de Groot, R., Farber, S., Grasso, M., Hannon, B., Limburg, K., Naeem, S., O'Neill, R. V., Paruelo, J., Raskin, R. G., Sutton, P., & van den Belt, M. (1997). The value of the world's ecosystem services and natural capital. *Nature, 387*, 253–260. https://doi.org/10.1038/387253a0

Csurgó, B., Kovách, I., & Mathieu, C. (2016). Exploring innovation and sustainability in the metropolitan rural areas of Budapest and Paris. *Metropolitan Ruralities, 23*, 275–301. https://doi.org/10.1108/S1057-192220160000023012

Daly, H., Farley, J. (2005). Ecological Economics: Principles And Applications. Ecological Economics. https://doi.org/10.1016/J.ECOLECON.2005.08.005

Deelstra, T., & Girardet, H. (2000). Urban agriculture and sustainable cities. In N. Bakker, M. Dubbeling, S. Gündel, U. Sabel-Koshella, & H. de Zeeuw (Eds.), *Growing cities, growing food: Urban agriculture on the policy agenda* (pp. 43–66). Zentralstelle für Ernährung und Landwirtschaft.

Elbeheiry, M., & Balog, R. S. (2023). Technologies driving the shift to smart farming: A review. *IEEE Sensors Journal, 23*(3), 1752–1789. https://doi.org/10.1109/JSEN.2022.3225183

European Commission. (2006). *Sustainability of EU-funded projects: Guidelines for long-term impact*. Publications Office of the European Union.

Farley, J. (2010). Conservation through the economics lens. *Environmental Management, 45*(1), 26–38. https://doi.org/10.1007/s00267-008-9232-1

Kimani-Murage, E. W., Schofield, L., Wekesah, F., Mohamed, S., Mberu, B., Ettarh, R., Egondi, T., Kyobutungi, C., & Ezeh, A. (2014). Vulnerability to food insecurity in urban slums: Experiences from Nairobi, Kenya. *Journal of Urban Health, 91*(6), 1094–1107. https://doi.org/10.1007/s11524-014-9894-3

Kuang, L., & Bhat, G. A. (2017). Local food supply chains: Socioeconomic impacts and resilience strategies. *Journal of Rural Studies, 52*, 123–135. https://doi.org/10.1016/j.jrurstud.2017.04.005
Rinaldi, C., Cavicchi, A., & Spigarolo, R. (2020). University contributions to co-creating sustainable tourism destinations. *Journal of Sustainable Tourism, 28*(1), 1–20. https://doi.org/10.1080/09669582.2020.1797056
Sgroi, F. (2023). Sustainability and culinary traditions? Understand the role of historical markets in the development of agrifood and local gastronomy from the perspective of behavioral economics. *International Journal of Gastronomy and Food Science, 34*, 100809. https://doi.org/10.1016/j.ijgfs.2023.100809
Stalmirska, A., & Ali, F. (2023). Glocal gastronomy: Balancing standardization and cultural identity in urban tourism. *Tourism Management, 94*, 104630. https://doi.org/10.1016/j.tourman.2022.104630
Sustentare Project – IFSP. (2024). *Annual activity report 2023*. Federal Institute of São Paulo.
Unay-Gailhard, I., Simões, F., & Lança, M. (2025). Social media as a tool for food citizenship education. *Journal of Agricultural Education and Extension, 31*(1), 1–18. https://doi.org/10.1080/1389224X.2024.2292106
United Nations Environment Programme. (2024). *Food waste index report 2024*. Nairobi, Kenya.
Ziegler, J. (2012). *Massive destruction: Geopolitics of hunger*.

Simone Mendes Delphino holds a degree in Administration from Centro Universitário SantAnna (2002) and a specialization in Public Management from Universidade Católica Dom Bosco (UCDB). She is an administrator and currently serves as a "pregoeira" (auctioneer/bidding agent) for the Federal Institute of São Paulo. She coordinates the Sustentare Project: Intelligent, Inclusive, and Sustainable Communities. With experience in Administration, her focus is on Government Policy and Planning, particularly in the areas of sustainability, waste management, textile reuse and transformation, climate change, and ESG.

Fabiana Souza Ferreira is a Ph.D. in Geography from the Federal University of Pernambuco (UFPE). She works with geological risk mapping, territorial justice, and climate emergency communication. She also coordinates projects that integrate research, art, and outreach in peripheral areas and community libraries.

Rodrigo de Benedictis Delphino is a Ph.D. Candidate in Sustainability at the University of São Paulo, Master's in Strategic Planning in Hospitality from Anhembi Morumbi University, Postgraduate Specialization (Lato Sensu) in Tourism and Hospitality Planning from Universidade Paulista, and a Bachelor's degree in Business Administration and Hospitality from Centro Universitário SantAnna. Currently, he serves as a Director of Cooperation and Sustainability at the Pro-Rectory of Extension and Culture, and as a Professor in the Technology in Tourism Management program at the Federal Institute of Education, Science and Technology of São Paulo—IFSP—Campus São Paulo. He hold the position of General Coordinator of the IFSP/MDA Productive Backyards Project and are the leader of the Sustentare Project—IFSP/CNPQ research group. His areas of expertise include: Sustainability, Sustainable Gastronomy, Urban and Peri-urban Agriculture, Climate Change, and ESG.

Marcelo Bernardino Araújo Ph.D. in Production Engineering, Master's in Accounting and Actuarial Sciences, Specialist in Controllership and Finance, Bachelor's in Accounting, and a Technical degree in Accounting. He is a university professor in undergraduate and postgraduate courses across in-person, blended, and distance learning (EaD) modalities, and also serve as a content developer for EaD and blended learning disciplines for the Fundação Instituto Tecnológico de Osasco.

Luciano Delmondes de Alencar Ph.D. in Genetics and Molecular Biology (UNICAMP), with degrees in Agronomy (IAC) and Agroecology (UFSCar), researches the genetics of natural populations to inform conservation policies. He is a public servant at IFSP, where he integrates molecular biology and agroecology to promote biodiversity and sustainable agriculture. His work spans research, extension, academic management, and public policy grounded in social justice and food sovereignty.

Teaching Sustainability in Postgraduate Studies in Accounting: An Analysis of Subjects and Syllabuses

Viviana dos Reis Nunes and Edvalda Araújo Leal

Abstract Sustainability has been the focus of study in various spheres of the educational process, making it essential to promote a social, environmental and economic awareness that seeks to balance human needs with the limits of the planet. Education plays a fundamental role in academic and professional training. The aim of this study is therefore to identify the subjects and content related to sustainability offered in postgraduate courses in Accounting Sciences in Brazil. The research is descriptive and documental, analyzing course files and institutional documents available on open data platforms and on the websites of stricto sensu postgraduate accounting programs. The results show that the issue of sustainability is addressed in 41 of the 53 courses investigated, either through specific subjects or topics included in subjects already offered. However, it was noted that most of the subjects on sustainability are optional. Based on this result, it is suggested that academic managers review the curricula to include compulsory subjects on sustainability. As the job market increasingly demands competencies related to sustainability, Accountancy training must evolve to meet these demands, both for academic work and in other areas.

Keywords Graduate studies in accounting · Sustainable development · Education · Sustainability

1 Introduction

Reflecting on environmental complexity opens the door to a deeper understanding of the dynamics surrounding the interaction between human beings and the natural environment. In this sense, stimulating opportunities arise for the emergence of new social actors engaged in the responsible appropriation of nature and the promotion

V. R. Nunes (✉) · E. A. Leal
Federal University of Uberlândia, Uberlândia, Brazil
e-mail: vivianareisnunes@gmail.com

E. A. Leal
e-mail: edvalda@ufu.br

W. Leal Filho et al. (eds.), *Sustainability in the Curriculum and Operations of Universities*, World Sustainability Series,
https://doi.org/10.1007/978-3-032-12678-8_6

of sustainability (Jacobi 2003). Sustainability, according to Gadotti (2008, p. 75), is "the dynamic balance with the other and with the environment, it is harmony between the different".

Associated with sustainability, it is also important to address the issue of sustainable development, which is also one of the mandatory topics at global meetings (Guimarães et al. 2023). Sustainable development is not just an environmental issue, but a holistic approach that seeks to create a balance between human needs and the limits of the planet. This implies a profound change in the way we live, produce and consume (Jacobi 2005).

Associated with sustainability, it is also important to address the issue of sustainable development, which is also one of the mandatory topics at global meetings (Guimarães et al. 2023). Sustainable development is not just an environmental issue, but a holistic approach that seeks to create a balance between human needs and the limits of the planet. This implies a profound change in the way we live, produce and consume (Jacobi 2005 sustainable throughout the world, considering not only environmental issues, but also social and economic aspects.

2030 Agenda has 17 Sustainable Development Goals (SDGs) that cover various areas of global concern. These goals are designed to be achieved by the year 2030 and aim to end poverty, protect the planet and ensure that all people have peace and prosperity. Among the goals, SDG 4 focuses on ensuring that all people, regardless of their ethnic, economic, social or disability background, have access to quality education and lifelong learning opportunities (United Nations Brazil 2015).

Jacobi (2003) points out that an educational process that is articulated and committed to sustainability is crucial to promoting a deeper and more lasting environmental awareness. The author considers that the educator plays a fundamental role in environmental education and in promoting awareness of environmental issues. The teacher, in this context, acts as a facilitator who helps students understand the importance of the natural environment and develop a deeper connection with nature (Jacobi 2003).

Integrating sustainability into the higher education sector can be a challenge, but it is actually an extremely important and necessary initiative (Hazelton and Haigh 2014). Thus, the role assigned to Higher Education Institutions (HEIs), which is that of transforming agents of society, is recognized for its ability to positively influence the social, economic and cultural development of a nation. It should be noted that HEIs have an important role to play in promoting the SDGs established by the UN, as pointed out by Marques et al. (2020).

The study by Nascimento et al. (2022) provides evidence that scholars in the field of organizational studies and accounting can play a significant role in promoting sustainability in the organizational environment in order to contribute to achieving the goals of the 2030 Agenda. Guimarães et al. (2023) point out that the relationship between accounting and companies is relevant to sustainability, since the accounting sciences provide the information needed for decisions, and companies are included in the economic pillar of sustainability.

Gomes et al. (2012) report on the teaching model with an educational approach that divides topics into isolated disciplines, such as environmental financial accounting and environmental management accounting, and that this approach can lead to a lack of integration and holistic understanding of socio-environmental issues. Thus, there is a lack of preparation of accounting professionals to meet the market demand for skills related to sustainability (Gomes et al. 2012).

At master's level, Serafini's (2022) research reinforces the scarcity of studies on sustainability in the area of accounting, especially those dealing with the SDGs, with the author having identified 06 theses and dissertations on the SDGs in accounting from 2015 to 2021. It is therefore important to encourage research that covers this topic. According to Botomé and Kubo (2002), master's and doctoral programs play a fundamental role in the advancement of knowledge, the training of researchers and the practical application of knowledge for the benefit of society in general.

Paisey and Paisey's (2004) point of view suggests that while micro-teaching is instructive and valuable, one should not neglect concerns related to the bigger picture of accounting education, including the definition of its objectives and the construction of comprehensive curricula. Addressing these broader issues can be crucial to the evolution and adaptation of accounting education to the changing demands of society.

Given the above, this study poses the following research question: How do the syllabuses and bibliographies of the subjects offered in postgraduate courses in Accounting present the theme of Sustainability? The aim of this article is to investigate the subjects and content offered in postgraduate courses in Accounting, in relation to the theme of Sustainability.

The justification for this research is the relevance of the subject and the role of postgraduate courses in accounting education. It was identified that few researches were carried out involving the curricular structure of the postgraduate course in Accounting Sciences (Hazelton and Haigh 2014; Paisey and Paisey 2004). Thus, it is relevant to research the theme of sustainability in Postgraduate Programs in Accounting Sciences (PPGCC), especially with the awareness that one of the objectives of Postgraduate Studies is the training of teachers (Nganga et al. 2016).

Furthermore, the Coordination for the Improvement of Higher Education Personnel—CAPES, a Foundation of the Ministry of Education (MEC) whose mission is to expand and consolidate *stricto sensu* postgraduate studies in Brazil, proposed for the four-year period from 2021 to 2024 a restructuring of the evaluation system, which should analyze the quality of postgraduate programs in terms of promotion and assessment of impacts, "their cultural, social and economic dimensions, focusing on strategic, regional, national and international challenges" and "the paths that consolidate sustainability" (CAPES, 2019, pp. 9–10). Now, on May 20, 2025, CAPES Board of Directors approved the VII National Postgraduate Plan (PNPG) with guidelines until 2029. The preliminary version of this document (CAPES, 2023) includes an axis of internationalization of Brazilian postgraduate studies oriented towards achieving the Sustainable Development Goals (SDGs).

2 Literature Review

Education for Sustainability

According to Freire (1996, p. 51), "education is a way of intervention in the world". Tacca and Rey (2008) point out that learning plays a crucial role in our lives, both on a personal and social level, and learning goes beyond mere survival. For the authors, education is a process that allows us to explore our potential, enrich our emotional and intellectual lives and contribute to building a more vibrant and informed society.

Pazos et al. (2022) add that when discussing human formation, it is crucial to consider the central role of the social and historical context, as well as to recognize that education plays a vital role in preparing individuals to become active and engaged members of their communities and society as a whole. Through learning and education, researchers and practitioners are looking for more innovative approaches to find ways to promote sustainability (Aboytes and Barth 2020).

Involving the UN Member States, various institutions and civil society gathered in Rio de Janeiro, Brazil, in June 2012, at a conference to identify the future we want, called Rio+20, and developed the concept of Sustainable Development. Among the objectives, number 230 addresses the need to improve education systems by developing sustainability curricula and thus prepare people for the pursuit of sustainable development (United Nations 2012, p. 60). Education plays a key role in creating a more sustainable world and engaging people in sustainability-related issues, as outlined in the Sustainable Development Goals (SDGs) established by the United Nations. Education for sustainable development aims to equip people with the knowledge, skills, values and attitudes needed to promote sustainable practices and decisions in their daily lives and in their communities (United Nations Educational, Scientific and Cultural Organization [UNESCO], 2018).

The SDGs give reason for transformation, Baden and Higgs (2015) mention that this challenges the idea that profit maximization is the ultimate business solution and sets new benchmarks for management education, with greater emphasis on social responsibility and sustainability. For Thomassen and Jorgensen (2021) the SDGs meet the demands of learning through discrepancies and multiple competing agendas, and seek innovative solutions in areas that are constantly evolving.

With regard to integrating the SDGs into academic research, Serafini et al. (2023) identified the perception of students and teachers at the Federal University of Rio Grande do Norte (UFRN) about the SDGs. The authors portray a scenario that indicates the need to train both teachers and students on the subject in order to raise awareness and encourage knowledge and implementation of the SDGs. They also recommend that the UFRN should encourage the creation of a program that addresses the SDGs not only by in terms of management, as a mission and development plan, but also so that the students' research reflects this pattern of sustainable development.

Setzer (2010) points out that acquiring knowledge is essential if a person is to develop both professionally and socially. In this way, the incorporation of the Sustainable Development Goals (SDGs) into master's degree courses is a very relevant approach and in line with the current demands of society and the market, with observance of the "inclusion of a specific topic on the SDGs in some disciplines or the development in a transversal way in different disciplines of the course" (Zamora-Polo and Sánchez-Martín 2019, p. 7).

It is necessary to highlight the contributions of Botomé and Kubo (2002), who point out that the demand for highly qualified professionals in specific areas of knowledge is a crucial factor for the relevance and validity of a postgraduate program. These programs not only train individuals to work in academic institutions, but also to make a significant contribution to various sectors of society.

Mochizuki and Fadeeva (2010) recognize the need for professionals, including teachers, who can deal effectively with immediate sustainability challenges or who can actually facilitate their students' participation in processes of positive social change, in the case of teachers. Thus, stricto sensu programs play a fundamental role in training teachers capable of acting as multipliers of principles and values linked to sustainability (Lima and Amancio-Vieira 2017).

According to Fariñas (2006), the development of postgraduate curricula requires quality criteria for constant improvement, and the skills of students are crucial, as they must be in line with the demands of the socio-cultural progress of the time in which they live (present and future). In this sense, Malheiros et al. (2013) identified relevant elements to make postgraduate programs more relevant to sustainability, such as: (i) ecosystem vision; (ii) interdisciplinary perspective; active involvement of different social actors; (iii) cultural diversity; (iv) new learning methodologies, among others.

Studies by Dillard and Vinnari (2017) show that academics have a responsibility to exercise their role consciously, critically and as advisors to society. In fulfilling this public interest responsibility as accounting academics, they are especially concerned with how accounting, accountants and accountability regimes facilitate more democratic institutions and processes that contribute to improving economic, social and environmental justice. Sustainability accounting must not only assess the environmental impacts of human activities, but also understand how these activities affect people and communities (Larrinaga 2023). In this way, the union between integrated thinking and action is fundamental to implementing sustainable economic activities in a way that is successful for all the communities involved (Tettamanzi et al. 2022).

As a force for transformative education, Sheehan et al. (2022) recommend that accounting academics discuss with students that companies have a personality of their own, which implies recognizing that they have a responsibility towards multiple stakeholders, not only to generate immediate profits, but also to ensure their long-term sustainability. In this sense, the authors indicate that directors have a duty to consider not only the interests of shareholders, but also the impact of the company's decisions on the community, the environment and employees.

For Gray and Collison (2002) it is necessary to understand and address the profound challenges presented by sustainability in accounting, which requires a significant change not only in teaching and research, but also in the very definition and way the profession is perceived and taught. For the authors, a continuous commitment to updating and evolving educational curricula is needed, as well as the need for a cultural change that values, in addition to financial efficiency, responsibility and sustainability from a long-term perspective.

3 Methodology

The research seeks to identify the disciplines and content offered in postgraduate courses in Accounting that are closely linked to the theme of Sustainability. The research had a qualitative approach, whose main focus is the creation of precise representations of social phenomena and which identifies, in this precision, its main added value, as stated by Cardano (2017). The procedure was documentary research. Analysis from a qualitative perspective is a method that uses specific techniques to apprehend and understand different types of documents and uses a careful process of selecting, collecting, analyzing and interpreting data, according to Lima et al. (2021).

Data was collected from the course files, based on institutional documents found on open data platforms and HEI websites. For the study, all the HEIs that offer *stricto sensu* postgraduate courses in Accounting Sciences in Brazil were selected, according to the spreadsheet of results from CAPES' 2021 Quadrennial Evaluation. The 53 postgraduate courses in Accounting Sciences—PPGCC existing in 2023 were considered, of which 29 were academic master's courses (ME), 7 were professional master's courses (MP), 15 were academic doctorates (DO) and 2 were professional doctorates (DP), evaluated by CAPES (Coordination for the Improvement of Higher Education Personnel 2022).

Data was collected and analyzed from July and September 2023. The following data was analyzed: the number of subjects offered related to sustainability and the content covered in the respective subjects. Based on these analyses, the aim was to identify the integration of sustainability into academic and professional training in the subjects offered in the Postgraduate Program in Accounting Sciences. Content analysis was used to interpret the data, and the categories for analysis were as follows: Lines of research in sustainability; disciplines on sustainability and the workload; content of the syllabuses on sustainability; and main authors and bibliographical references covering sustainability.

4 Presentation of Results

4.1 Lines of Research and Disciplines in Sustainability

Initially, a search was carried out to identify *stricto sensu* postgraduate programs in Accounting Sciences in Brazil, with lines of research that contain the word "sustainability and/or sustainable" in their name. Table 1 shows the presented programs.

Table 1 Postgraduate programs with a research line in sustainability

IES	State	Program	Line of research
UFC	CE	MP	Strategy and sustainability
UFC	CE	ME/DO	Organizations, strategy and sustainability
UFSM	RS	ME	Governance and organizational sustainability

According to the information presented, it can be seen that only three Postgraduate Accounting programs in Brazil have lines of research specifically related to sustainability, indicating the beginning of the concentration of research involving the subject and showing the integration of sustainable knowledge into accounting, but this data is only available in the states of Ceará and Rio Grande do Sul.

Table 2 lists the existing courses in the Master's and Doctorate programs in Science Accounting, both academic and professional, that include in their name (denomination) the word "sustainability" or "sustainable".

It can be seen that the number of courses on sustainability is greater than the number of lines of research and they are offered in a greater number of Brazilian states. Three programs are in Santa Catarina, two programs are concentrated in Ceará and São Paulo and one program is in other Brazilian states, namely: DF, ES, MS, PB, PE, PR, RJ and RS. The total workload varies among institutions, with the highest recorded at University of São Paulo, Ribeirão Preto campus (USP/RP), at 120 h.

About the Federal University of Ceará, it was found that, in the professional master's degree, the subject "Innovation and Sustainability Management" and, at the Federal University of Santa Maria, the subjects "Accounting and Sustainable Innovation, Accounting and Organizational Sustainability, Research Seminars in Governance and Organizational Sustainability" are electives. At all the other institutions, the subjects are optional. Therefore, access to Governance and Organizational Sustainability in postgraduate courses is subject to student interest.

4.2 *Bibliographies and Content on Sustainability*

When analyzing the bibliographies of the courses shown in Table 2, following the same search format, which was based on the nomenclature "sustainability and/or sustainable", it was possible to identify that, of the 18 courses on offer, 4 of them do not publish bibliographical references. In general, the references were quite diverse, with 104 different references identified. Figure 1 shows the most cited authors in the bibliographies, considering the first author.

The analysis shows that the leading authors in the subjects' syllabuses account for approximately 19% of the total of 104 references as first author. Among the names in Fig. 1, three authors have published 03 articles/books each: Schaltegger, Searcy and Bebbington. The other five authors published 02 articles/books each: Bonacchi, Da Veiga, Gray, Kemp, Robertson, totaling 12 scientific articles and seven books.

Table 2 Postgraduate courses named with sustainable and/or sustainability

IES	State	Program	Discipline	WL
UFC	CE	MP	Innovation management and sustainability	48
UFC	CE	ME/DO	Innovation and sustainability	64
			Sustainable development, economics and evaluation	64
UnB	DF	ME/DO	Accounting and sustainability	*
UFES	ES	ME/DO	Ethics, corporate social responsibility and sustainability	60
UFMS	MS	ME	Sustainability in organizations	60
UFPB-JP	PB	ME/DO	Accounting and sustainable development	60
UFRPE	PE	ME	Environmental performance and sustainability	60
UFPR	PR	ME/DO	Accounting and sustainability	45
UFSM	RS	ME	Accounting and sustainable innovation	45
			Accounting and organizational sustainability	45
			Research seminars on governance and organizational sustainability	45
FURB	SC	ME/DO	Sustainability and corporate governance	45
UFSC	SC	ME/DO	Accounting and sustainability	60
UNOCHAPECÓ	SC	ME	Environmental accounting and sustainability	45
USP/RP	SP	ME/DO	Sustainability accounting	120
UPM	SP	MP/DP	Sustainable development and corporate sustainability	48
UERJ	RJ	MP	Special topics in controllership and public management II—public administration and sustainability	30

*The teaching plan does not include the workload and mentions that the course offers 04 credits

The author José Eli da Veiga is a senior professor at USP's Institute for Advanced Studies and has been cited for his books: 'The Global Disgovernance of Sustainability' (Veiga 2013) and 'Sustainable Development: The Challenge of the twenty-first Century' (Veiga 2005). The other most cited authors are members of international institutions located in the United Kingdom, the United States, the Netherlands, Germany and Canada.

With regard to the 81% of bibliographies cited only once, it is worth highlighting the references in the syllabus of the Sustainability Accounting course at USP/RP, which has the highest workload, 120 h. The course is offered entirely in the non-face-to-face system, which is justified by the importance of promoting interdisciplinary communication within and outside USP. The compulsory bibliography includes only two books: (i) Matias Laine, Helen Tregidga and Jeffrey Unerman (Laine et al. 2021)

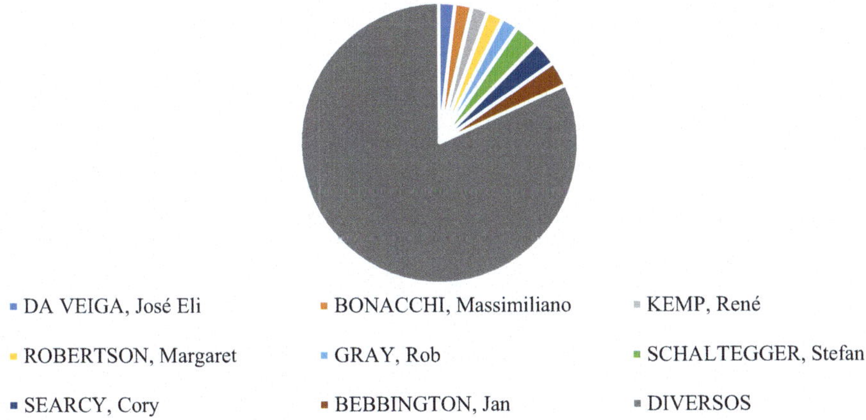

Fig. 1 Authors and citations in sustainability courses

Sustainability Accounting and Accountability. New York: Routledge, 3ed. (ii) Adams (2022).

Research Handbooks in Accounting series. The complementary reading concerns a single publication selected in the disciplines with the expressions sustainable and/or sustainability in the year 2023, which was by the authors Nicolas Garcia-Torea, Carlos Larrinaga and Mercedes Luque-Vílchez with the title Bridging the Understanding of Sustainability Accounting and Organizational Change (Garcia-Torea et al. 2023). The aim of the work was to review the Accounting associated with sustainability reports and sustainable organizational change. Next, the sheets for all disciplines offered in postgraduate programs. By checking the syllabuses and bibliographies that include the words "sustainability and sustainable", the information shown in Figs. 2 and 3 was obtained.

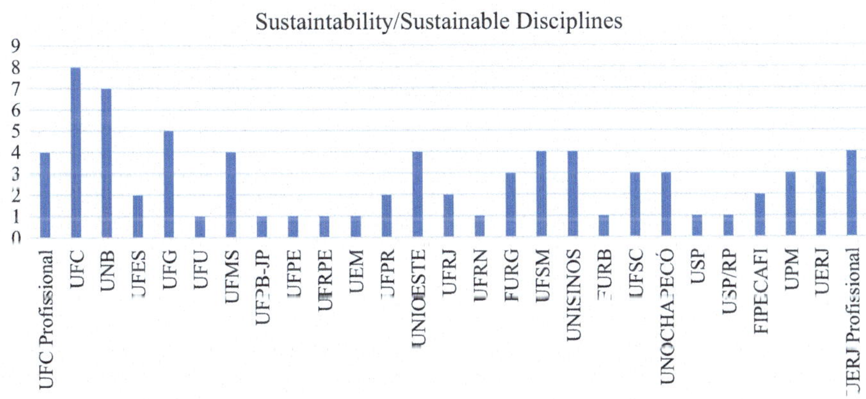

Fig. 2 Contents and bibliographies on sustainability

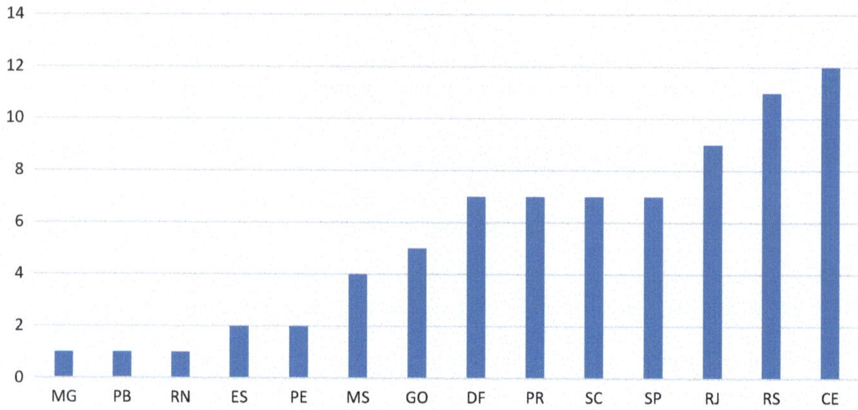

Fig. 3 Subjects with a sustainability theme

Figure 2 shows all the postgraduate programs that address sustainability in their syllabuses and bibliographies. It can be seen that Federal University of Ceará (UFC) has 8 subjects that deal with sustainability and the University of Brasília (UNB) has 7 subjects. In short, these two postgraduate programs have the largest number of courses in which the terms sustainability or sustainable appear.

At UFC, among the courses mentioned above, only the Multivariate Data Analysis course is compulsory and includes sustainability in its bibliographical references, the title of which is "Perception of sustainability attributes in a supermarket retailer" (Cunha et al. 2011).

UnB also has a compulsory subject called 'Teaching Methodology'. The article is "Peru, mountains and los niños: Dialogic action, Accounting and sustainable transformation" (Contrafatto et al. 2015). There are another 04 subjects, but it is not stated whether they are compulsory or optional.

The fact there is a specific line of research on "sustainability" has partly contributed to fostering interest in the subject in the courses offered in the programs. It was found the Postgraduate Accounting programs at the Universities of Ceará (MP) and Santa Maria had a total of 04 specific subjects in each program.

Education for Sustainable Development involves system change and implies respect for life, prevention for the planet and care for all of humanity, as Gadotti (2008) asserts. Tregidga et al. (2018) emphasize the importance of plurality and academic involvement in the sustainable development discourse, especially in relation to the dominant perspectives in the business context.

Figure 3 illustrates the subjects that cover sustainability by federative unit.

When analyzing the contributions by state, CE and RS have 12 and 11 courses respectively. It can be seen that the lines of research in sustainability in these states have contributed to discussions on sustainability, as can be seen in the respective curriculum sheets.

In Ceará, the contribution of UFC's professional master's degree is seen in subjects such as: Organizational Theory, Socio-environmental Management and Strategic People Management. The academic postgraduate course collaborates with the following subjects: Organizational Network Analysis, Multivariate Data Analysis, Socio-environmental Accounting, Corporate Finance, Environmental Management and Corporate Social Responsibility. Thus, there is a wide coverage of the subject in the course.

The institutions in Rio Grande do Sul that make up the sample are: (i) UFSM, which is represented by the subject Environmental Accounting and Social Reports; (ii) FURG, with the subjects Management Control, Information Systems in Organizations, as well as Accounting and Governance in the Public Sector; (iii) UNISINOS, which presents: Special Topics in Controllership and Finance I—Emerging Themes in Governance—ESG—Environmental, Social and Corporate Governance; Special Topics in Controllership and Finance II—Advanced Cost Studies; Special Topics in Controllership and Finance I—Governance in Cooperatives; and Accounting Theory.

The analysis shows that subjects and/or content on the subject of sustainability are present in certain postgraduate programs, especially those that use a specific line of research. It is noteworthy that the other programs did not show any content and/or bibliographical references that could instigate discussions on the subject of sustainability in the training of postgraduate students.

Thus, the development and continuous improvement of methods and tools to evaluate transformative learning in sustainability becomes relevant, which is fundamental for understanding the impact of educational and awareness-raising initiatives and for promoting effective changes towards a more sustainable future (Aboytes and Barth 2020).

5 Final Considerations

The purpose of the research was to identify the subjects and content offered in postgraduate courses in Accounting that make reference to the theme of Sustainability. The role attributed to Higher Education Institutions in the training of future professionals, a responsibility and social development, is relevant, especially when it comes to post-graduation.

The need for training in sustainability is evidenced by the responsibility that 2030 Agenda expresses to humanity, especially SDG 4, which emphasizes education should promote learning opportunities. Based on the analyses carried out on the subject files of the Postgraduate Accounting courses, it was found that 41 of the 53 programs include sustainability, which represents a percentage of 77%, either by offering subjects, syllabuses and/or bibliographical references. When we look at the specific courses offered with the sustainability nomenclature, the number is reduced to 24 and they are not in the compulsory format.

It also emerged from the analysis that the courses dealing specifically with sustainability are optional, with an average workload of 40–60 h. It was also found that the predominant references are scientific articles, with an emphasis on international authors.

This study contributes to highlighting the subjects that deal with issues related to sustainability and that are offered in postgraduate courses in Accounting. The results could encourage postgraduate programs that do not offer specific subjects and/or content in the course syllabus to review and implement such discussions, which are extremely relevant to the academic and professional training of the individual. Thus, teaching Accounting students about sustainable development is fundamental to ensuring the future well-being of the planet (Sheehan et al. 2022).

For future research, we suggest deepening the analysis by conducting interviews with postgraduate coordinators and postgraduates in order to understand the relevance of teaching about sustainability in the academic environment.

References

Aboytes, J. G. R., & Barth, M. (2020). Transformative learning in the field of sustainability: A systematic literature review (1999–2019). *International Journal of Sustainability in Higher Education, 21*(5), 993–1013. https://doi.org/10.1108/IJSHE-05-2019-0168

Adams, C. A. (Ed.). (2022). *Handbook of Accounting and Sustainability*. Edward Elgar Publishing. https://doi.org/10.4337/9781800373518

Baden, D., & Higgs, M. (2015). Challenging the perceived wisdom of management theories and practice. *Academy of Management Learning & Education, 14*(4), 539–555. https://doi.org/10.5465/amle.2014.0170

Botomé, S. P., & Kubo, O. M. (2002). Responsabilidade social dos programas de pós-graduação e formação de novos cientistas e professores de nível superior. *Interação em Psicologia, 6*(1), 81–110. https://doi.org/10.5380/psi.v6i1.3196

Cardano, M. (2017). *Manual de pesquisa qualitativa: A contribuição da teoria da argumentação* (E. R. Conill Trad.). Vozes.

Contrafatto, M., Thomson, I., & Monk, E. A. (2015). Peru, mountains and los niños: Dialogic action, accounting and sustainable transformation. *Critical Perspectives on Accounting, 33*, 117–136. https://doi.org/10.1016/j.cpa.2015.04.009

Coordenação de Aperfeiçoamento de Pessoal de Nível Superior. (2019). Documento de área – Área 27: Administração pública e de empresas. *ciências contábeis e turismo*. https://www.gov.br/capes/pt-br/centrais-deconteudo/adm-pdf

Coordenação de Aperfeiçoamento de Pessoal de Nível Superior. (2022). *Áreas de Avaliação*. https://www.gov.br/capes/pt-br/centrais-de-conteudo/adm-pdf

Cunha, C. F. D., Spers, E. E., & Zylbersztajn, D. (2011). Percepção sobre atributos de sustentabilidade em um varejo supermercadista. Revista de Administração de Empresas, 51, 542–552. https://doi.org/10.1590/S0034-75902011000600004

Dillard, J., & Vinnari, E. (2017). A case study of critique: Critical perspectives on critical accounting. *Critical Perspectives on Accounting, 43*, 88–109. https://doi.org/10.1016/j.cpa.2016.09.004

Fariñas, G. (2006). Desafios do currículo na educação de pós-graduação e o desenvolvimento do pensamento complexo. *Revista E-curriculum, 1*(2), 1–17.

Freire, P. (1996). *Pedagogia da autonomia: Saberes necessários à prática educativa*. Paz e Terra.

Gadotti, M. (2008). Educar para a sustentabilidade. *Inclusão social, 3*(1), 75–78.

Garcia-Torea, N., Larrinaga, C., & Luque-Vílchez, M. (2023). Bridging the understanding of sustainability accounting and organizational change. *Organization & Environment, 36*(1), 17–38. https://doi.org/10.1177/10860266221083339

Gomes, S. M. S., Sampaio, M. S., Azevedo, T. C., & Slomski, V. G. (2012). Proposta para o ensino da controladoria ambiental nos cursos de graduação de Ciências Contábeis nas IES's brasileiras. *Revista de Gestão Social e Ambiental, 6*(1), 177–189. https://repositorio.ufba.br/handle/ri/27922

Gray, R., & Collison, D. (2002). Can't see the wood for the trees, can't see the trees for the numbers? Accounting education, sustainability and the public interest. *Critical Perspectives on Accounting, 13*(5–6), 797–836. https://doi.org/10.1006/cpac.2002.0554

Guimarães, A. A. B., Damasceno, A. F. B.. Pereira, S. S., Martins, V. F., & Barbosa, J. E. C. (2023). Valores de sustentabilidade na visão dos acadêmicos do curso de Ciências Contábeis numa instituição de ensino em Minas Gerais. *Revista Científica E-Locução, 23*, 51–76. https://doi.org/10.57209/e-locucao.v1i23.525

Hazelton, J., & Haigh, M. (2014). Incorporating sustainability into accounting curricula: Lessons learnt from an action research study. In A. Sangster, & R. M. S. Wilson (Eds.), *Liberalising the Accounting Curriculum in University Education* (pp. 126–145). Routledge.

Jacobi, P. R. (2005). Educação ambiental: O desafio da construção de um pensamento crítico, complexo e reflexivo. *Educação e pesquisa, 31*, 233–250. https://doi.org/10.1590/S1517-97022005000200007

Jacobi, P. (2003). Educação ambiental, cidadania e sustentabilidade. *Cadernos de Pesquisa, 118*, 189–205. https://doi.org/10.1590/S0100-15742003000100008

Laine, M., Tregidga, H., & Unerman, J. (2021). *Sustainability accounting and accountability.* Routledge.

Larrinaga, C. (2023). Contabilidade sustentável: As abordagens ESG não são suficientes. *Revista Contabilidade & Finanças, 34*, 1–4. https://doi.org/10.1590/1808-057x20239042.en

Lima, C. E., & Amancio-Vieira, S. F. (2017). A institucionalização da temática da sustentabilidade no programa de pós-graduação em administração da Universidade Estadual de Londrina. *Revista Pensamento Contemporâneo em Administração, 11*(5), 20–36.

Lima, E. B., Jr., Oliveira, G. S., Santos, A. C. O., & Schnekenberg, G. F. (2021). Análise documental como percurso metodológico na pesquisa qualitativa. *Cadernos da FUCAMP, 20*(44), 1–16.

Malheiros, T. F., Pérez, M. A., Sampaio, C. A. C., & Zuñiga, C. H. (2013). Os desafios do tema sustentabilidade no ensino da pós-graduação. *Revista Brasileira de Pós-Graduação, 10*(21), 537–552. https://doi.org/10.21713/2358-2332.2013.v10.597

Marques, J. F. S., Santos, Â. V., & Aragão, J. M. C. (2020). Planejamento e sustentabilidade em Instituições de Ensino Superior à luz dos objetivos do Desenvolvimento Sustentável. *REUNIR Revista de Administração Contabilidade e Sustentabilidade, 10*(1), 14–29.

Mochizuki, Y., & Fadeeva, Z. (2010). Competences for sustainable development and sustainability: Significance and challenges for ESD. *Int. J. Sustain. High. Educ., 11*, 391–403. https://doi.org/10.1108/14676371011077603

Nações Unidas Brasil (2015). *Agenda 2030 para o Desenvolvimento Sustentável.* https://brasil.un.org/pt-br/91863-agenda-2030-para-o-desenvolvimento-sustentável

Nascimento, Í. C. S., Oliveira, M. C., & Rodrigues, M. S., Jr. (2022). Diversidade nas organizações: Contribuições para o alcance dos objetivos de desenvolvimento sustentável da Agenda 2030. *Revista de Gestão e Secretariado, 13*(3), 1033–1058. https://doi.org/10.7769/gesec.v13i3.1390

Nganga, C. S. N., Botinha, R. A., Miranda, G. J., & Leal, E. A. (2016). Mestres e doutores em contabilidade no Brasil: Uma análise dos componentes pedagógicos de sua formação inicial. *REICE: Revista Iberoamericana sobre Calidad, Eficacia y Cambio en Educación, 14*(1), 83–99. https://doi.org/10.15366/reice2016.14.1.005

Organização das Nações Unidas para a Educação, a Ciência e a Cultura. (2018). *Educação para os Objetivos de Desenvolvimento Sustentável: Objetivos de aprendizagem.* https://unesdoc.unesco.org/ark:/48223/pf0000252197

Paisey, C., & Paisey, N. J. (2004). An analysis of accounting education research in accounting education: An international journal–1992–2001. *Accounting Education, 13*(1), 69–99. https://doi.org/10.1080/0963928042000207396

Pazos, V. C., Santos, C. H. M., Andrade, L. M. S., & de Moraes, R. A. (2022). Educação para a sustentabilidade no ensino remoto emergencial na graduação e pós-graduação: Uma visão do processo de formação continuada. *International Journal of Development Research, 12*(4), 55444–55449. https://doi.org/10.37118/ijdr.24390.04.2022

Serafini, P. (2022). Os objetivos de desenvolvimento sustentável nas instituições de ensino superior: Iniciativas e barreiras no processo de implementação. [Dissertação de Mestrado]. Universidade Federal do Rio Grande do Norte.

Serafini, P. G., Gomes, S.M.S, Moura, J.M., Nobre, A.C.S, Rezende, F.D.R. (2023). Sustainable development goals (SDGS) in the University context: The perception of UFRN students and professors. *Anais da oitava Conferência Sul-amaericana de Contabilidade Ambiental, Governança, Accountability e Sustentabilidade* (pp.107–123). Belo Horizonte, MG, Brasil. Disponível em: www.csca.net.br

Setzer, V. W. (2010). *Consequências do materialismo.* São Paulo: USP.

Sheehan, N. T., Fox, K. A., Klassen, M., & Vaidyanathan, G. (2022). Threshold concepts and ESG performance: Teaching accounting students reconceptualized fundamentals to drive future ESG advocacy. *Accounting Education, 33*(1), 1–25. https://doi.org/10.1080/09639284.2022.2122727

Tacca, M. C. V. R., & Rey, F. L. G. (2008). Produção de sentido subjetivo: As singularidades dos alunos no processo de aprender. *Psicologia: ciência e profissão, 28*, 138–161. https://doi.org/10.1590/S1414-98932008000100011

Tettamanzi, P., Venturini, G., & Murgolo, M. (2022). Sustainability and financial accounting: A critical review on the ESG dynamics. *Environmental Science and Pollution Research, 29*(11), 16758–16761. https://doi.org/10.1007/s11356-022-18596-2

Thomassen, A. O., & Jørgensen, K. M. (2021). John Dewey and continuing management education: Problem-based learning for organizational sustainability. *Journal of workplace learning, 33*(3), 229–242. https://doi.org/10.1108/JWL-05-2020-0080

Tregidga, H., Milne, M. J., & Kearins, K. (2018). Ramping up resistance: Corporate sustainable development and academic research. *Business & Society, 57*(2), 292–334. https://doi.org/10.1177/0007650315611459

United Nations. (2012). *The future we want: Outcome document of the United Nations Conference on Sustainable Development.* https://sustainabledevelopment.un.org/content/documents/733FutureWeWant.pdf

Veiga, J. E. (2013). *A desgovernança mundial da sustentabilidade.* Editora 34.

Veiga, J. E. (2005). *Desenvolvimento sustentável o desafio do século XXI.* Garamond.

Zamora-Polo, F., & Sánchez-Martín, J. (2019). Teaching for a better world. Sustainability and sustainable development goals in the construction of a change-maker university. *Sustainability, 11*(15), 1–15. https://doi.org/10.3390/su11154224

Viviana dos Reis Nunes works as an accounting technician at the Instituto Federal do Triângulo Mineiro. Master in Accounting Sciences from the Federal University of Uberlândia (UFU). She is a special student of the doctorate at the Faculty of Administration of UFU.

Edvalda Araújo Leal is a professor at the Federal University of Uberlândia, teaching in both undergraduate and graduate programs in Accounting Sciences. She holds a Ph.D. in Administration from Fundação Getulio Vargas (FGV) and is the Editor-in-Chief of the journal Advances in Scientific and Applied Accounting (ASAA). She has served as the coordinator of the Graduate Program in Accounting Sciences (PPGCC/UFU) and as a tutor for the Tutorial Education Program—PET Accounting. Her research focuses on Accounting Education and Managerial Accounting, with nationally recognized academic output. She is also a CNPq Research Productivity Fellow.

Green Innovation and Artificial Intelligence: A Systematic Literature Review

Gabriel Augusto Sampaio Soares, Helena Farias Ribeiro, João Victor Gomes Ribeiro, Carla Bonato Marcolin, and Jaluza Maria Lima Silva Borsatto

Abstract The search for technological solutions that can mitigate environmental impacts has become a global priority, driven by climate change challenges and the need for sustainable practices across economic sectors. The integration of artificial intelligence (AI) with green innovation represents a field of study that combines technological advancement with environmental sustainability, where AI serves as a catalyst for developing new technologies and processes that enhance energy efficiency, optimize resource utilization, and promote sustainable management systems. Universities play a fundamental role in this integration through sustainable education initiatives, serving as catalysts for developing AI-driven sustainability solutions while addressing challenges related to ethical considerations, technological barriers, and implementation complexities. The objective of this review is to map the relationship between green innovation and artificial intelligence (AI), analyzing the relevant scientific production in specialized journals. The adopted methodology followed a structured approach in stages. Initially, a search was conducted in the Web of Science and Scopus databases using a specific query that combined terms such as "Green Innovation", "Artificial Intelligence", "Machine Learning", among others. The search was restricted to articles published from 2015 onwards, focusing on journals in the administration field. A total of 8065 articles were collected, which were analyzed through natural language processing (NLP) techniques, including n-grams and topic modeling, using Latent Dirichlet Allocation (LDA), to group the articles into different thematic categories. As a result, the articles address central themes related to sustainability, innovation, and digital transformation, with emphasis on areas such as supply chain management, sustainable development, and circular economy. The most cited article explores the potential of blockchain technology to improve sustainability

This work was supported by the National Council for Scientific and Technological Development (CNPq) of Brazil.

G. A. S. Soares · H. F. Ribeiro · J. V. G. Ribeiro · C. B. Marcolin (✉) · J. M. L. S. Borsatto
Faculty of Management and Business (FAGEN), Federal University of Uberlândia (UFU), Uberlândia, Minas Gerais, Brazil
e-mail: carla@ufu.br

W. Leal Filho et al. (eds.), *Sustainability in the Curriculum and Operations of Universities*, World Sustainability Series,
https://doi.org/10.1007/978-3-032-12678-8_7

in global supply chain management, addressing transparency and traceability challenges. The implementation of Sustainable Development Goals (SDGs) was also a highlighted theme, with proposals for transformations encompassing education, health, energy, sustainable cities, and the digital revolution. The structuring of topics through LDA modeling revealed five main thematic areas: (1) research in sustainability and emerging technologies; (2) technical approaches to problem solving; (3) how green innovation and environmental performance are shaping companies; (4) the role of companies in the sustainable development process; and, finally, (5) concern about the impact of climate change. These results show the growing importance of integration between AI and green innovation to face global environmental challenges, evidenced both in article citations and in the structure of identified topics.

Keywords Green innovation · Artificial intelligence · Sustainability · Sustainable education · Machine learning · Digital transformation · Environmental management

1 Introduction

The search for technological solutions that can mitigate environmental impacts has become a global priority, especially given the challenges posed by climate change and the need to adopt more sustainable practices in various economic sectors (Racz 2019). In this context, green innovation, defined as the development and implementation of new technologies, processes, and practices that promote environmental sustainability, has gained prominence on the agenda of governments, companies, and research institutions (Sharma 2017; Schiederig et al. 2012).

Simultaneously, the advancement of artificial intelligence (AI) has transformed how data is analyzed and decisions are made in different areas, offering new possibilities to address environmental issues more efficiently and innovatively. The integration of AI in green innovation promotes sustainability and requires a collaborative approach between different sectors to maximize benefits and mitigate associated risks (El Koshiry et al. 2025; Hojageldiyev 2019).

This integration presents a promising field for exploring solutions that combine technology and sustainability. AI can be used to optimize production processes, reduce waste, improve energy efficiency, monitor environmental impact, and develop more sustainable management systems (Willenbacher et al. 2018). Furthermore, the application of techniques such as machine learning and deep learning allows for the analysis of large volumes of environmental data, identifying patterns and trends that assist in strategic decision-making aimed at environmental preservation (Hsieh 2009; Galaz et al. 2021).

1.1 *AI as a Catalyst for Green Innovation Across Sectors*

Recent studies have demonstrated the role of AI as a catalyst for green innovation across various economic sectors. In the manufacturing sector, AI technologies have been applied to optimize production processes, reduce waste, and promote sustainable supply chains (Li et al. 2024; Liu and Zhou 2024). Machine learning algorithms and predictive analytics are being used to predict equipment failures, reducing downtime and energy consumption while facilitating the development of more efficient production methods (Chun 2024; Wang et al. 2024).

The energy sector has emerged as a relevant area for AI-driven green innovation. AI applications are enhancing the efficiency and reliability of renewable energy systems, with algorithms being used to predict energy demand and optimize production from renewable sources such as wind and solar power (Tao 2024; Chen et al. 2024). These technologies enable better integration of renewable energy sources into existing grids, reducing dependence on fossil fuels and contributing to greenhouse gas emission reductions (Yang et al. 2024).

In the transportation sector, AI is driving green innovation by improving traffic flow efficiency, promoting electric vehicle adoption, and enabling the development of autonomous systems that can lead to more efficient use of transportation resources (Biggi et al. 2025; Chen et al. 2024). These applications demonstrate how AI can address complex environmental challenges while creating opportunities for sustainable development.

The impact of AI on green innovation is mediated through several mechanisms that enhance organizations' technological capabilities. AI facilitates knowledge sharing and collaboration between different departments and organizations, leading to the creation of new green technologies (Du et al. 2024; Cheng et al. 2024). This is particularly evident in small and medium enterprises (SMEs), where AI can bridge the gap between technological advancement and environmental sustainability.

Furthermore, AI enables the optimization of industrial processes to reduce environmental impact, including supply chain optimization that reduces carbon footprints associated with transportation and logistics (Wang et al. 2024; Fu and Rasiah 2024). The technology also plays a role in promoting circular economy models, where resources are reused and recycled, reducing waste and pollution (Biggi et al. 2025; Lin et al. 2024).

Recent research has highlighted AI's potential to address environmental challenges by providing innovative solutions to complex problems. AI systems are being used to monitor and mitigate climate change effects by analyzing large datasets related to environmental conditions (Biggi et al. 2025; Lin et al. 2024). Additionally, AI can enhance the effectiveness of environmental policies by providing policymakers with data-driven insights for more informed decision-making, particularly in the context of achieving sustainable development goals (Nahar 2024).

Despite the potential, several challenges need to be addressed to maximize AI's impact on green innovation. Data privacy and security concerns arise from the collection and analysis of large datasets required for AI applications (Cheng et al. 2024; Zhang 2024). Ethical considerations related to AI deployment in environmental monitoring and its impact on local communities and ecosystems must also be carefully addressed (Nahar 2024). Additionally, the evolving regulatory frameworks governing AI use in green innovation requires continuous attention to ensure effective and responsible implementation (Lin et al. 2024; Yang et al. 2024).

Future research directions include the integration of AI with other technologies such as blockchain and Internet of Things (IoT) to enhance green innovation impact, the development of AI-driven green innovation models for organizational adoption and increasing public awareness and engagement in AI-driven sustainability initiatives (Fu and Rasiah 2024; Tian 2023; Vezzani 2023).

1.2 AI, Green Innovation, and Sustainable Education

Universities play a fundamental role in advancing the integration of AI and green innovation through sustainable education initiatives. Higher education institutions serve as catalysts for developing new technologies and processes that enhance energy efficiency, optimize resource utilization, and promote sustainable management systems. However, this integration presents both opportunities and challenges that require careful consideration (Khan et al. 2025; Al-Omari et al. 2025).

The implementation of AI in sustainable education faces several practical and theoretical challenges. Ethical considerations are central to this integration, including data privacy and security concerns, algorithmic bias and fairness issues, and digital divide considerations that may exacerbate existing inequalities (Khan et al. 2025; Adel et al. 2025; Ofem and Chukwujama 2024). Additionally, technological barriers such as infrastructure constraints, lack of expertise and training, and resource allocation challenges can impede the effective integration of AI in sustainability initiatives (Al-Zahrani 2024; Filho et al. 2024; Juanda et al. 2024).

Universities must address practical implementation challenges including financial constraints, resistance to change from stakeholders, implementation complexity, and policy and regulatory obstacles (Al-Zahrani 2024; Filho et al. 2024; Adel et al. 2025). Theoretical challenges encompass the need for interdisciplinary collaboration, balancing innovation with responsibility, and developing robust methodologies to measure the impact and effectiveness of AI-driven sustainability solutions (Gupta and Nyamapfene 2025; Chemlal and Azzouazi 2024; Tripathi et al. 2024).

The role of universities extends beyond implementation to include fostering collaboration between computer scientists, sustainability experts, and other stakeholders to maximize the impact of AI-driven initiatives (Zönnchen et al. 2024; Saxena et al. 2024). This requires the development of ethical frameworks and governance structures that ensure transparency, fairness, and accountability in AI deployment while promoting sustainable practices across academic and administrative operations.

Given this scenario, this systematic review aims to map the relationship between green innovation and artificial intelligence, analyzing the relevant scientific production in specialized journals in the administration field. It seeks to identify the main thematic approaches and research trends, as well as to evidence the potential of AI in promoting sustainable practices in organizations. The analysis covers articles published from 2015, considering the increased interest in emerging technologies and sustainable practices in recent years, and uses natural language processing (NLP) techniques to categorize studies into relevant thematic areas.

Through this analysis, it is expected to contribute to the understanding of synergies between green innovation and artificial intelligence, evidencing how these two areas can complement each other to face global environmental challenges. The systematic review also aims to identify gaps in literature and propose future research directions, highlighting the role of emerging technologies in building a more sustainable future.

2 Methodology

2.1 Research Design

This systematic review followed a structured process to ensure comprehensiveness, relevance, and reproducibility of the study. The methodology was designed to map the relationship between green innovation and artificial intelligence through bibliometric analysis of scientific publications. The review protocol was developed following established guidelines for systematic reviews and bibliometric studies.

2.2 Data Sources and Search Strategy

The review was conducted in two major scientific databases: Web of Science (WoS) and Scopus, selected for their broad coverage of scientific journals and provision of updated and reliable information about academic publications. The search was performed in January 2025, covering publications from 2015 to 2024.

A comprehensive search query was developed to identify articles exploring the intersection between green innovation and artificial intelligence. The search string combined keywords using boolean operators to capture variations of core concepts: ALL = (green innovation OR environmental innovation OR sustainable innovation OR eco-innovation OR inovação verde OR inovação ambiental OR inovação sustentável OR ecoinovação) AND ALL = (artificial intelligence OR machine intelligence OR smart support OR smart virtual reality OR chat bot* OR large language model OR llm OR chatgpt OR machine learning OR intelligent agent* OR neural network* OR natural language processing OR ensemble algorithm* OR regression algorithm* OR decision tree* OR digital transformation OR inteligência artificial OR inteligência de máquina OR suporte inteligente OR realidade virtual inteligente OR aprendizado de máquina OR agente inteligente* OR rede neural* OR processamento de linguagem natural OR algoritmo de conjunto* OR algoritmo de regressão* OR árvore de decisão* OR processamento de linguagem natural OR transformação digital).

The search string incorporated both English and Portuguese terms to ensure comprehensive coverage, considering the international scope of the research field and the potential inclusion of studies from Portuguese speaking countries (Table 1).

To maintain thematic alignment with Business, Management, and Organizational Studies, the search was refined by including only relevant subject areas:

Table 1 Inclusion and exclusion criteria

Inclusion criteria	Exclusion criteria
Articles published from 2015 onwards	Non-article publications (e.g., conference papers, book chapters, reviews, editorials)
Studies focusing on the intersection of green innovation and artificial intelligence	Articles not aligned with the research scope
Articles published in journals classified within relevant subject areas	Duplicate publications
Peer-reviewed articles only	Articles published before 2015

Source Authors' own work

Web of Science Categories: Engineering, Environmental Sciences Ecology, Science Technology Other Topics, Computer Science, Business Economics, Remote Sensing, Agriculture, Telecommunications, Imaging Science Photographic Technology, Public Environmental Occupational Health, Automation Control Systems, Transportation, Operations Research Management Science, Mathematics, Public Administration, Social Sciences Other Topics, Food Science Technology, Forestry, Education Educational Research, Urban Studies, Robotics, Mathematical Computational Biology, Optics, Information Science Library Science, Life Sciences Biomedicine Other Topics, Psychology, Health Care Sciences Services, Development Studies, Government Law, Social Issues, Architecture, Arts Humanities Other Topics, Cardiovascular System Cardiology, Communication, Mining Mineral Processing, Physiology, Behavioral Sciences, Sociology, History Philosophy Of Science, Mathematical Methods In Social Sciences, International Relations, Linguistics, Audiology Speech Language Pathology, Area Studies, Art, Anthropology, Philosophy, Demography, Sport Sciences, History, Music, Religion, Substance Abuse, Criminology Penology, Cultural Studies, Emergency Medicine, Social Work, Transplantation.

Scopus Subject Areas: Environmental Science, Business Management and Accounting, Social Sciences, Engineering, Computer Science, Energy, Economics Econometrics and Finance, Decision Sciences, Mathematics, Multidisciplinary, Arts and Humanities.

2.3 Study Selection Process

In accordance with the PRISMA guidelines, the study selection process was conducted. Following the initial search, the query identified twenty-eight thousand nine hundred eighty-eight documents across the Web of Science and Scopus databases. Subsequently, nine hundred twenty-two duplicate records were detected and eliminated through a combination of automated deduplication tools and manual verification, resulting in twenty-eight thousand sixty-six unique records for further screening. A document type filter was then applied, leading to the exclusion of six thousand four hundred eighty-three records classified as non-articles, such as conference papers, book chapters, reviews, and editorials; this yielded twenty-one thousand five hundred eighty-three articles.

To ensure source relevance, a journal quality filter based on the Area 27 classification (Public and Business Administration, Accounting Sciences, and Tourism) established by the Coordination for the Improvement of Higher Education Personnel (CAPES)—a Brazilian government agency responsible for evaluating graduate programs and academic journals—was implemented. This filtering step resulted in eight thousand sixty-five articles being retained for the final analysis.

2.4 *Data Analysis*

The analysis began with a descriptive overview of selected articles, focusing on key bibliometric indicators including number of publications per year; most frequent journals; most influential authors; most cited articles, among others. We conducted textual frequency analysis based on bigrams and trigrams extracted from article titles and abstracts. These n-grams represent sequences of consecutive words, allowing identification of frequent word combinations and recurring thematic expressions related to green innovation and artificial intelligence.

Natural Language Processing (NLP) techniques were applied to uncover deeper semantic patterns. Specifically, Topic Modeling using the Latent Dirichlet Allocation (LDA) algorithm was employed to group articles into coherent thematic clusters. This approach enabled identification of dominant research topics, emerging trends, and recurring approaches across the selected literature (Lee et al. 2021; Shin et al. 2015).

To ensure reproducibility, all search queries, inclusion/exclusion criteria, and filtering procedures were documented. The search strategy was designed to be transparent and replicable, with clear documentation of database selection, search terms, and filtering criteria. Data extraction and analysis procedures were standardized to minimize bias and ensure consistency across the research process.

3 Results and Discussion

The results obtained through bibliometric analysis corroborate the empirical evidence about the relationship between artificial intelligence and green innovation presented in literature. The exponential growth of publications, which went from 87 articles in 2015 to 2784 in 2024, reflects the growing recognition of AI as a catalyst for the development of green technologies, as observed by Biggi et al. (2025) when demonstrating that AI acts directly in increasing the efficiency of green technologies, particularly in the transport and energy sectors (Fig. 1).

The analysis of the most frequent bigrams, which identified terms such as "supply chain", "digital transformation", "sustainable development" and "carbon emission", aligns with the indirect effects of AI on green innovation described by Yang et al. (2023), who highlight how industrial modernization and human capital enhancement constitute pathways through which AI stimulates green innovation. This thematic convergence suggests that academic research has focused precisely on the areas where AI impacts are most evident (Fig. 2).

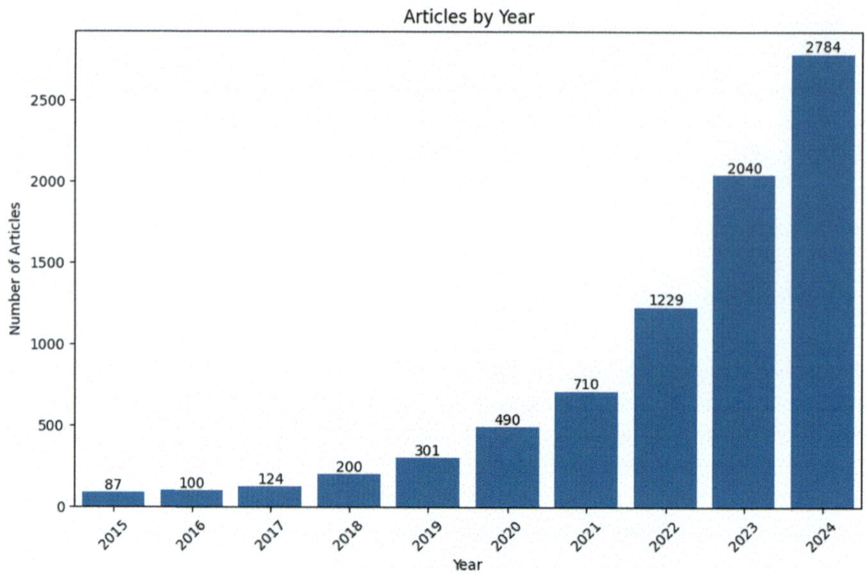

Fig. 1 Articles by year

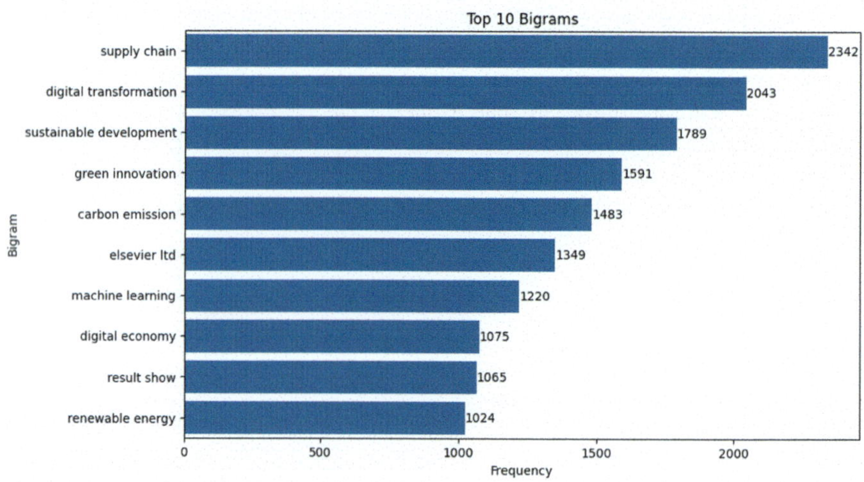

Fig. 2 Most frequent bigrams

The prominence of the most cited articles for themes related to blockchain in sustainable supply chain management (Saberi et al. 2019) and SDGs resonates with the observations of Du et al. (2024) about the importance of integrating AI in green innovation to achieve SDGs, especially those related to industry, innovation, and infrastructure.

Topic modeling revealed five thematic areas that reflect both direct and indirect effects of AI on green innovation.

The first thematic (Fig. 3) area encompasses sustainability and emerging technologies, evidencing the interconnection between technological innovation and sustainability in industrial development. This theme addresses how emerging technologies can contribute to more sustainable practices, emphasizing the importance of aligning innovation and environmental responsibility. The analysis reveals concrete examples of how emerging technologies address environmental challenges.

Dinh et al. (2018) developed CO_2 electroreduction systems that convert atmospheric pollutants into valuable chemicals, achieving 70% faradaic efficiency in ethylene production. This research illustrates how advances in electrochemistry can transform environmental waste into economic resources. Saberi et al. (2019) examine how blockchain technology can revolutionize sustainable supply chain management through enhanced transparency, traceability, and security.

The research addresses governmental, community, and consumer pressures for sustainability goals, demonstrating how emerging digital technologies can facilitate environmental compliance. Paustian et al. (2016) highlight the potential of soils to mitigate greenhouse gas emissions, emphasizing how developments in information technology can expand the inclusion of soil-based practices in climate policies. This thematic area corroborates the findings of Biggi et al. (2025) about the originality and generality of AI-driven innovations.

The second thematic area (Fig. 4) refers to methodologies and technical models used to address sustainability challenges. It includes approaches such as networks, systems, and data analysis, suggesting technical solutions to solve environmental problems and improve the efficiency of organizational processes. The methodological approaches identified are exemplified by various studies that employ technical models to solve sustainability problems. Luthra et al. (2017) propose an integrated framework using Analytical Hierarchy Process (AHP) and VIKOR for sustainable supplier evaluation, identifying 22 selection criteria across three dimensions: economic, environmental, and social.

This methodological approach demonstrates how multi-criteria decision-making techniques can be applied to sustainability challenges. Govindan et al. (2015) analyze multi-criteria decision-making approaches for green supplier evaluation and selection, finding that applied techniques are mainly based on single fuzzy models. Di Vaio et al. (2020) conduct bibliometric analysis of 73 publications on artificial intelligence and sustainable business models, employing quantitative methods to map the research field. These studies illustrate how rigorous methodologies can be applied to understand and optimize sustainability practices.

The third thematic area (Fig. 5) highlights the relationship between green innovation and environmental performance, showing how companies are being shaped by digital technologies and sustainable practices. The studies emphasize the impact of digital transformation on improving corporate performance and adopting more responsible practices. Digital transformation and its impact on corporate performance are evidenced in several analyzed studies. Saberi et al. (2019) identify blockchain as a disruptive technology in early stages of evolution, categorizing adoption barriers into four types: inter-organizational, intra-organizational, technical, and external. The research demonstrates how blockchain implementation can improve organizational capabilities in transparency and security, responding to sustainability pressures from multiple stakeholders.

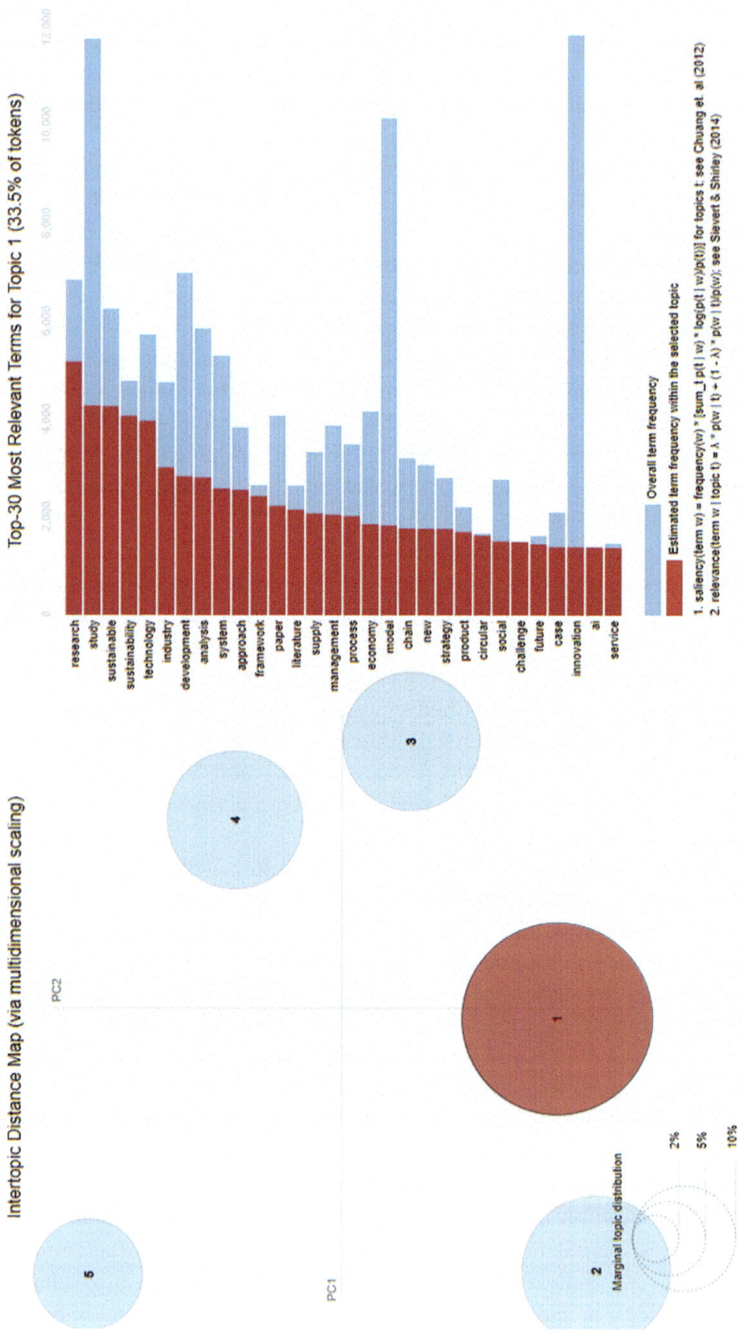

Intertopic Distance Map (via multidimensional scaling)
PC2
PC1
1
2
3
4
5
Marginal topic distribution
2%
5%
10%
Top-30 Most Relevant Terms for Topic 1 (33.5% of tokens)
0
2,000
4,000
6,000
8,000
10,000
12,000
research
study
sustainable
sustainability
technology
industry
development
analysis
system
approach
framework
paper
literature
supply
management
process
economy
model
chain
new
strategy
product
circular
social
challenge
future
case
innovation
ai
service
Overall term frequency
Estimated term frequency within the selected topic
1. saliency(term w) = frequency(w) * [sum_t p(t | w) * log(p(t | w)/p(t))] for topics t; see Chuang et. al (2012)
2. relevance(term w | topic t) = λ * p(w | t) + (1 - λ) * p(w | t)/p(w); see Sievert & Shirley (2014)

Fig. 3 Topic 1

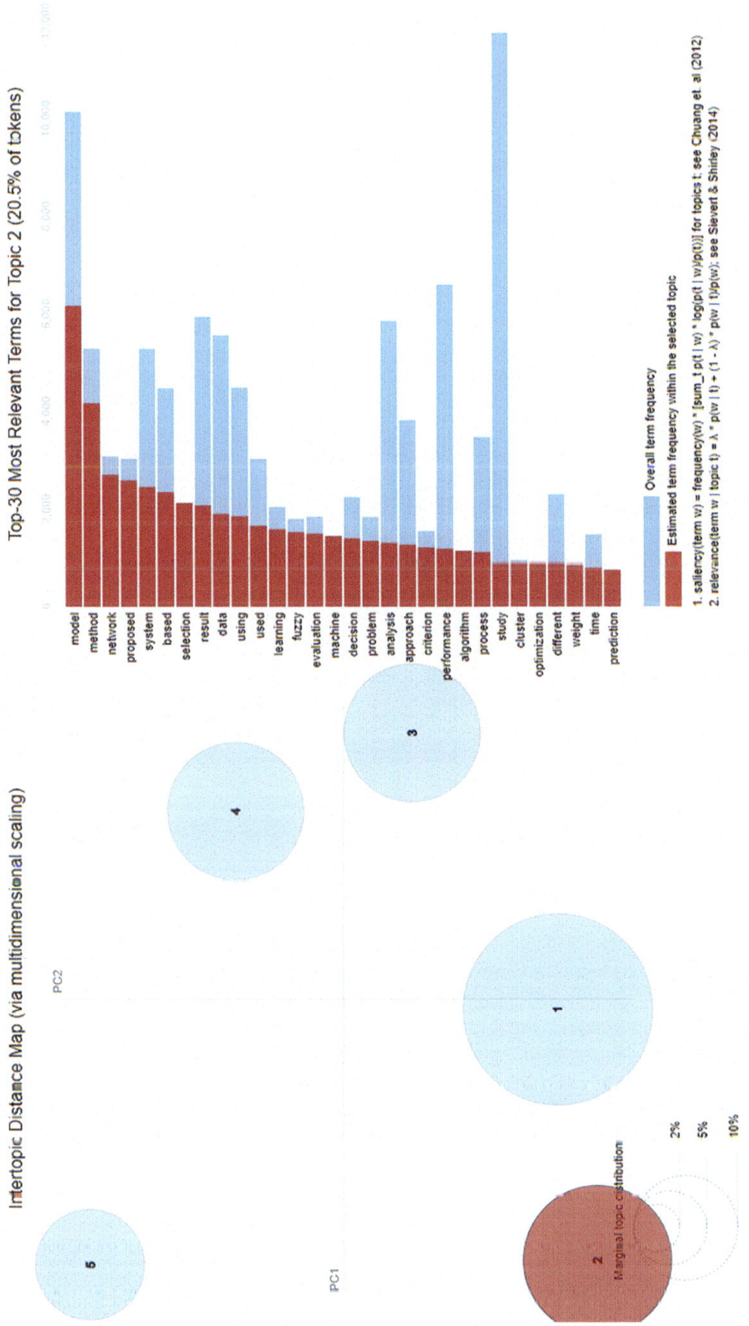
Intertopic Distance Map (via multidimensional scaling)
PC2
PC1
1
2
3
4
5
Marginal topic distribution
2%
5%
10%
Top-30 Most Relevant Terms for Topic 2 (20.5% of tokens)
0
2,000
4,000
6,000
8,000
10,000
12,000
model
method
network
proposed
system
based
selection
result
data
using
used
learning
fuzzy
evaluation
machine
decision
problem
analysis
approach
criterion
performance
algorithm
process
study
cluster
optimization
different
weight
time
prediction
Overall term frequency
Estimated term frequency within the selected topic
1. saliency(term w) = frequency(w) * [sum_t p(t | w) * log(p(t | w)/p(t))] for topics t; see Chuang et. al (2012)
2. relevance(term w | topic t) = λ * p(w | t) + (1 - λ) * p(w | t)/p(w); see Sievert & Shirley (2014)

Fig. 4 Topic 2

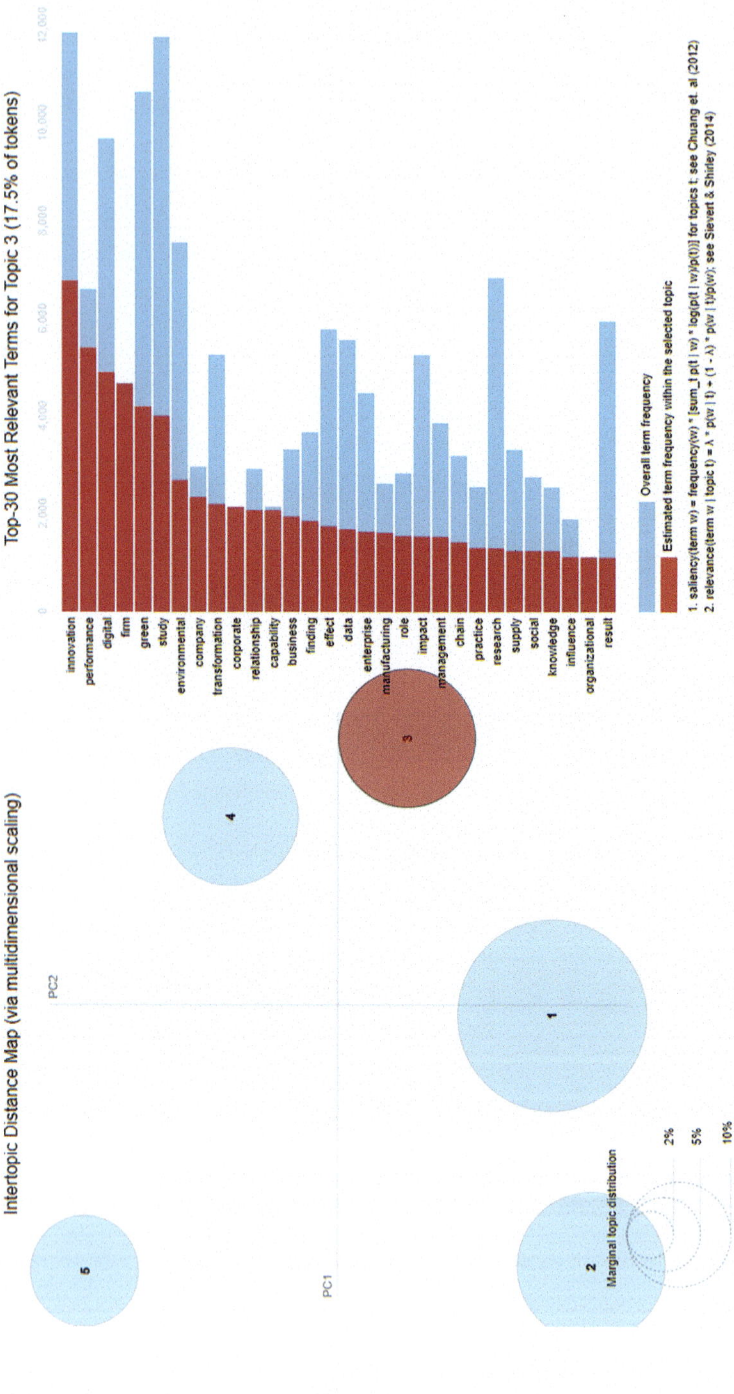
Intertopic Distance Map (via multidimensional scaling)
PC2
PC1
1
2
3
4
5
Marginal topic distribution
2%
5%
10%
Top-30 Most Relevant Terms for Topic 3 (17.5% of tokens)
0
2,000
4,000
6,000
8,000
10,000
12,000
innovation
performance
digital
firm
green
study
environmental
company
transformation
corporate
relationship
capability
business
finding
effect
data
enterprise
manufacturing
role
impact
management
chain
practice
research
supply
social
knowledge
influence
organizational
result
Overall term frequency
Estimated term frequency within the selected topic
1. saliency(term w) = frequency(w) * [sum_t p(t | w) * log(p(t | w)/p(t))] for topics t; see Chuang et. al (2012)
2. relevance(term w | topic t) = λ * p(w | t) + (1 - λ) * p(w | t)/p(w); see Sievert & Shirley (2014)

Fig. 5 Topic 3

Di Vaio et al. (2020) investigate how artificial intelligence can influence production and consumption patterns to achieve sustainable resource management according to the Sustainable Development Goals. The study highlights the role of Knowledge Management Systems in cultural change toward AI dissemination for sustainable business models. Luthra et al. (2017) demonstrate how integrated evaluation frameworks can help managers distinguish important supplier selection criteria and evaluate more efficient suppliers for supply chain sustainability. This theme aligns with the results of Zhu and Zhu (2024), who demonstrated how AI applications promote green innovation at the corporate level, especially in companies with high administrative shareholding indices.

The fourth thematic area (Fig. 6) emphasizes the role of companies in the sustainable development process, with focus on regions like China. The articles analyze how green innovation and digital transformation are driving sustainable economic growth, highlighting policies and practices adopted by different countries. This topic is broadly represented by articles that connect green innovation with sustainable development objectives. Sachs et al. (2019) introduce six SDG transformations as modular building blocks to achieve the goals, specifically including the "digital revolution for sustainable development" as one of the central transformations. This research establishes direct connections between technological advances and global sustainability goals. D'Amato et al. (2017) comparatively analyze the concepts of Circular Economy, Green Economy, and Bioeconomy, finding that Green Economy acts as an "umbrella" concept, including elements of the other two concepts.

The study reveals that Circular Economy and Bioeconomy are resource-focused, while Green Economy recognizes the fundamental role of all ecological processes. Dinh et al. (2018) contribute to this topic by demonstrating how CO_2 conversion to ethylene addresses multiple sustainability challenges simultaneously: reducing atmospheric CO_2 concentrations and producing valuable chemical raw materials. The economic and regional variations identified in the literature find correspondence in this thematic area, which is consistent with the observations of Hussain et al. (2024) about the association between AI adoption and increased green innovation in China, particularly among private companies.

The fifth thematic area (Fig. 7) addresses issues related to sustainable resource use, energy consumption, and concerns about climate change. The studies gathered in this theme indicate a strong concern about the impacts of climate change and the need to develop innovative solutions to mitigate environmental risks. Climate change concerns and environmental risks are directly addressed by several studies. Paustian et al. (2016) examine the potential of soils to mitigate greenhouse gas emissions, identifying gaps in data and understanding about soil-based mitigation practices. The research suggests ways to close these gaps through new research, technology, and collaboration, highlighting developments in information technology as facilitators. Dinh et al. (2018) directly address climate challenges through the development of CO_2 utilization technologies, providing a pathway to reduce atmospheric CO_2 concentrations while producing valuable chemicals. The demonstrated operational stability (150 h of continuous operation) suggests that such technologies can be implemented at industrial scale to address climate change challenges.

Sachs et al. (2019) include "energy decarbonization and sustainable industry" as one of the six necessary transformations, directly connecting energy and climate

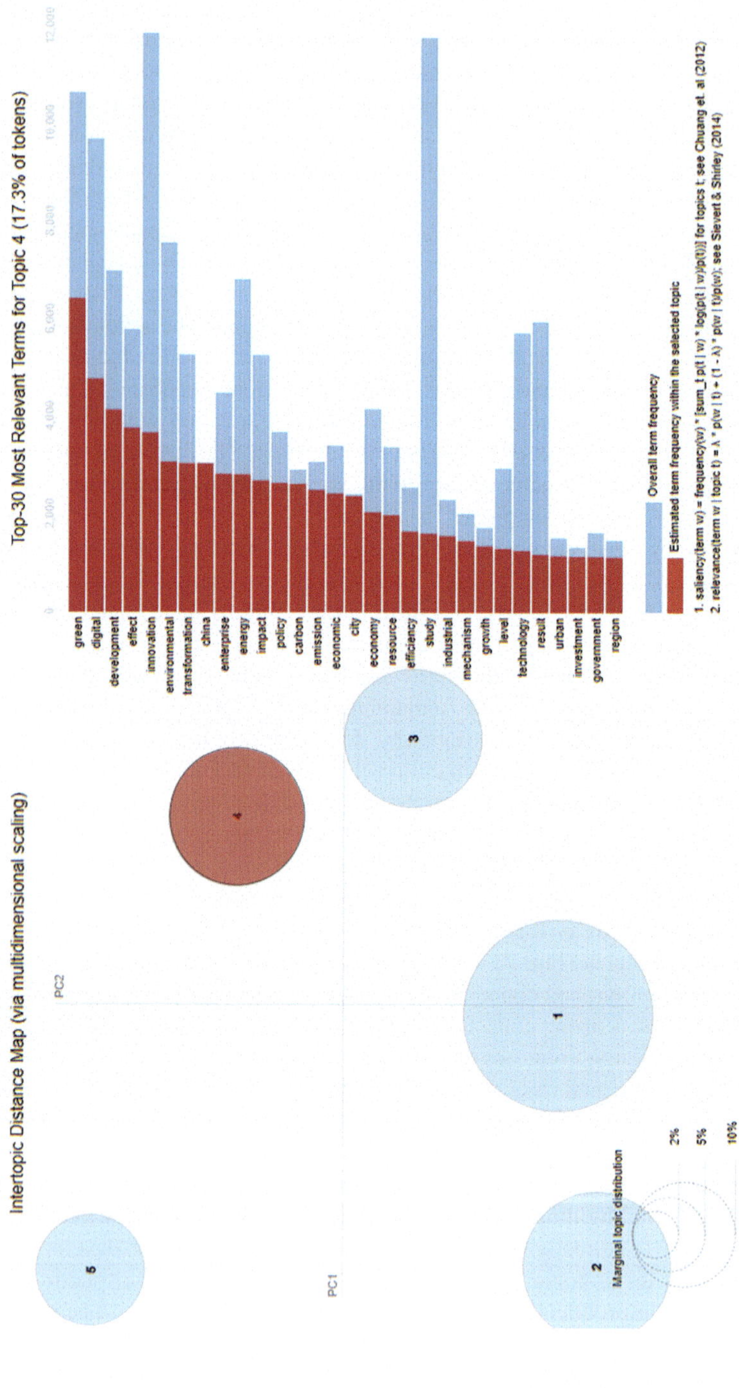
Intertopic Distance Map (via multidimensional scaling)
PC2
PC1
Marginal topic distribution
2%
5%
10%
Top-30 Most Relevant Terms for Topic 4 (17.3% of tokens)
green
digital
development
effect
innovation
environmental
transformation
china
enterprise
energy
impact
policy
carbon
emission
economic
city
economy
resource
efficiency
study
industrial
mechanism
growth
level
technology
result
urban
investment
government
region
Overall term frequency
Estimated term frequency within the selected topic
1. saliency(term w) = frequency(w) * [sum_t p(t | w) * log(p(t | w)/p(t))] for topics t; see Chuang et. al (2012)
2. relevance(term w | topic t) = λ * p(w | t) + (1 - λ) * p(w | t)/p(w); see Sievert & Shirley (2014)

Fig. 6 Topic 4

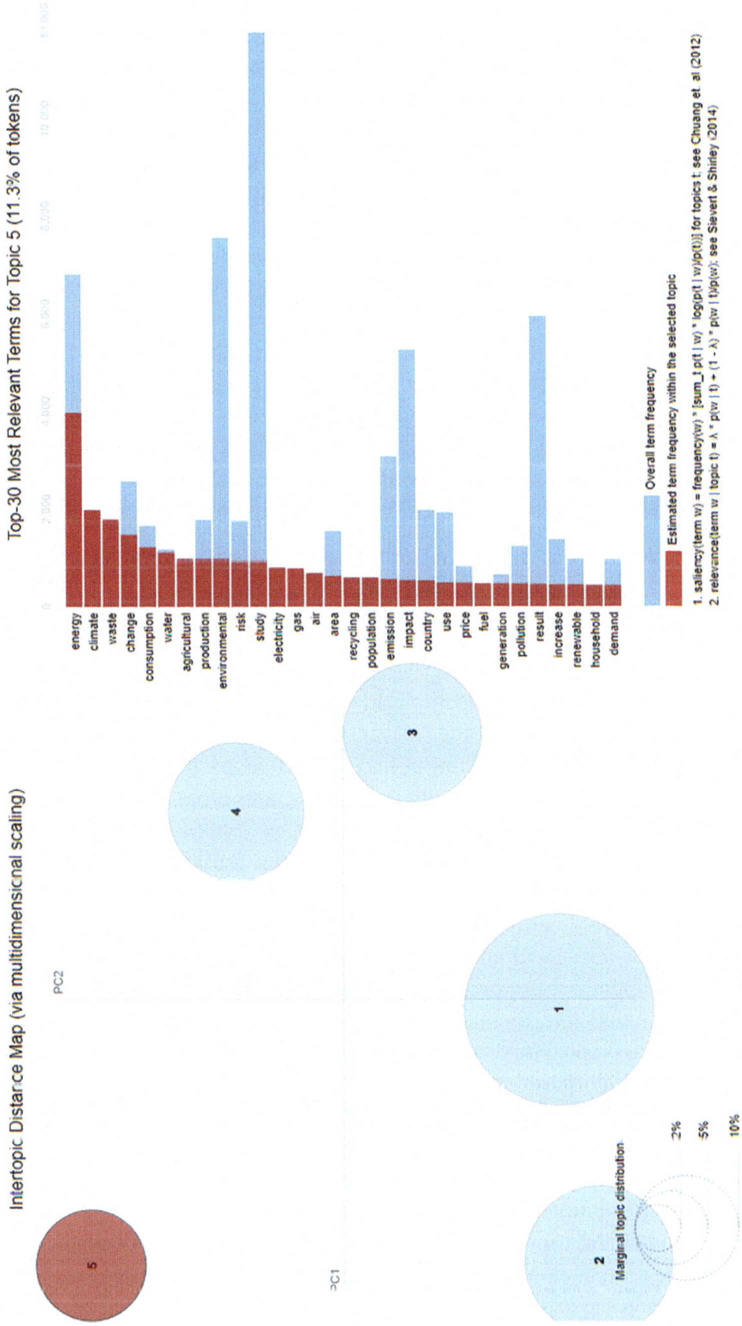
Intertopic Distance Map (via multidimensional scaling)
PC2
PC1
Marginal topic distribution
2%
5%
10%
Top-30 Most Relevant Terms for Topic 5 (11.3% of tokens)
energy
climate
waste
change
consumption
water
agricultural
production
environmental
risk
study
electricity
gas
air
area
recycling
population
emission
impact
country
use
price
fuel
generation
pollution
result
increase
renewable
household
demand
Overall term frequency
Estimated term frequency within the selected topic

Fig. 7 Topic 5

action with sustainable development. This thematic area reflects the central concerns that motivate the integration between AI and green innovation, corroborating the perspective of Yang et al. (2023) about the differences in AI impact between developed and developing countries, where advanced technological capabilities result in more pronounced relationships between these variables.

The challenges identified in the literature, particularly the incompatibility between the technological and market potential of green and AI technologies due to uncertainties in risk-return profiles (Biggi et al. 2025), may partially explain the thematic concentration observed in the bibliometric analysis. The predominance of studies in areas such as methodologies and technical models suggests that the academic community is seeking solutions to overcome these technological and economic barriers.

The convergence between the patterns identified in the bibliometric analysis and the empirical evidence from the literature indicates that the research field is evolving coherently, with focus on areas where AI demonstrates greater potential to drive green innovation. However, the need for strategic management and policy interventions to fully leverage AI capabilities, as highlighted by various authors, remains as a gap that requires continued attention from the academic community and policymakers.

4 Conclusion

The objective of this review was to map the relationship between green innovation and artificial intelligence (AI), analyzing the relevant scientific production in specialized journals. This objective was achieved through a comprehensive bibliometric analysis that examined publication trends, thematic structures, and research patterns in the field from 2015 to 2024.

The results demonstrate that the relationship between artificial intelligence and green innovation has evolved into a consolidated research area, as evidenced by the exponential growth in publications from 87 articles in 2015 to 2784 in 2024. This growth pattern reflects the academic community's recognition of AI as a catalyst for sustainable technological development and environmental problem-solving.

The analysis revealed five thematic areas that encompass the current state of research: sustainability and emerging technologies, methodologies and technical models, innovation and corporate performance through digital transformation, green innovation and sustainable development, and energy, climate change and environmental risks. These areas demonstrate that AI applications in green innovation extend beyond direct technological interventions to include methodological frameworks, organizational transformations, and policy implications.

The examination of the most cited articles and frequent bigrams confirmed that research has concentrated on areas where AI demonstrates measurable impact on sustainability outcomes, particularly in supply chain management, carbon emission reduction, and sustainable development goal achievement. The prominence of blockchain technology, multi-criteria decision-making approaches, and electrochemical innovations in CO_2 utilization illustrates the diversity of technological solutions being explored.

The bibliometric findings align with empirical evidence from the literature regarding both direct and indirect effects of AI on green innovation. The thematic concentration observed in methodologies and technical models suggests that researchers are addressing practical barriers to implementation, while the focus on corporate performance and digital transformation indicates growing attention to organizational adoption mechanisms.

The study identified that research in this field has evolved coherently, with academic focus aligning with areas where AI demonstrates clear potential for driving green innovation. However, gaps remain in strategic management approaches and policy frameworks needed to fully leverage AI capabilities for sustainability outcomes.

The trends identified in scientific production suggest continued growth in this research area, with implications for public policy development, business strategy formulation, and technological innovation pathways. The mapping provided by this review offers a foundation for future research directions and practical applications in the integration of artificial intelligence and green innovation.

References

Adel, M. A. A., Abouelnour, M. M., Alhourani, M. I., & Awad, A. (2025). Towards intelligent universities enhanced with artificial intelligence (AI). *Journal of Infrastructure, Policy and Development*. https://doi.org/10.24294/jipd10412

Al-Omari, O., Alyousef, A., Fati, S. M., Shannaq, F., & Omari, A. (2025). Governance and ethical frameworks for AI integration in higher education: Enhancing personalized learning and legal compliance. *Journal of Ecohumanism*. https://doi.org/10.62754/joe.v3i8.5781

Al-Zahrani, A. M. (2024). Harnessing AI for sustainable university practices: Toward a greener campus. *International Journal of Sustainability in Higher Education*. https://doi.org/10.1108/ijshe-08-2024-0566

Biggi, G., Iori, M., Mazzei, J., & Mina, A. (2025). Green intelligence: The AI content of green technologies. *Eurasian Business Review*. https://doi.org/10.1007/s40821-024-00288-1

Chemlal, Y., & Azzouazi, M. (2024). A methodology for evaluating and reporting the integration of artificial intelligence for sustainability in higher education: New insights and opportunities. In *Research and Innovation Forum 2023* (pp. 69-82). Springer. https://doi.org/10.1007/978-3-031-44721-1_7

Chen, M., Wang, S., & Wang, X. (2024). How does artificial intelligence impact green development? Evidence from China. *Sustainability, 16*(3), 1260. https://doi.org/10.3390/su16031260

Cheng, J., Xu, N., Khan, N. U., & Singh, H. (2024). The impacts of artificial intelligence literacy, green absorptive capacity, and green information system on green innovation. *Corporate Social Responsibility and Environmental Management*. https://doi.org/10.1002/csr.3017

Chun, Y. (2024). The nexus of artificial intelligence and green innovation: A cross-density analysis at the country level. *Journal of the Knowledge Economy*. https://doi.org/10.1007/s13132-024-02076-8

D'Amato, D., Droste, N., Allen, B., Kettunen, M., Lähtinen, K., Korhonen, J., Leskinen, P., Matthies, B. D. & Toppinen, A. (2017). Green, circular, bio economy: A comparative analysis of sustainability avenues. Journal of Cleaner Production, 168, 716-734. https://doi.org/10.1016/j.jclepro.2017.09.053

Di Vaio, A., Palladino, R., Hassan, R. & Escobar, O. (2020). Artificial intelligence and business models in the sustainable development goals perspective: A systematic literature review. Journal of Business Research, 121, 283–314. https://doi.org/10.1016/j.jbusres.2020.08.019

Dinh, C. T., Burdyny, T., Kibria, M. G., Seifitokaldani, A., Gabardo, C. M., de Arquer, F. P. G., Kiani, A., Edwards, J.P., De Luna, P., Bushuyev, O. S., Zou, CQ., Quintero-Bermudez, R., Pang, Y. J., Sinton, D. & Sargent, E. H. (2018). CO2 electroreduction to ethylene via hydroxide-mediated copper catalysis at an abrupt interface. Science, 360, 783–787. https://doi.org/10.1126/science.aas9100

Du, J., Cai, H., & Jin, X. (2024). Exploring the association between artificial intelligence management and green innovation: Expanding the research field for sustainable outcomes. *Sustainability, 16*(21), 9315. https://doi.org/10.3390/su16219315

El Koshiry, A. M., Eliwa, E. H., Abdel Mohsen, N. A., & Khalil, S. S. (2025). The impact of AI-based cloud network management on Microsoft Azure in promoting green technology awareness. *Sustainability, 17*(3), 1065.

Filho, W. L., Ribeiro, P. C. C., Mazutti, J., Salvia, A. L., Marcolin, C. B., Borsatto, J. M. L. S., Sharifi, A., Sierra, J. D., Luetz, J. M., Pretorius, R., & Trevisan, L. V. (2024). Using artificial intelligence to implement the UN sustainable development goals at higher education institutions. *Environment, Development and Sustainability*. https://doi.org/10.1080/13504509.2024.2327584

Fu, H., & Rasiah, R. (2024). Fostering inclusive green growth in Chinese cities: Investigating the role of artificial intelligence. *Sustainability, 16*(22), 9809. https://doi.org/10.3390/su16229809

Galaz, V., Centeno, M. A., Callahan, P. W., Causevic, A., Patterson, T. S., Baum, S. D., Farber, D., Fischer, J., Garcia, D., McPhearson, T., Jiménez, D., King, B., & Levy, K. (2021). Artificial intelligence, systemic risks, and sustainability. *Technology in Society, 67*, 101741. https://doi.org/10.1016/j.techsoc.2021.101741

Govindan, K., Rajendran, S., Sarkis, J. & Murugesan, P. (2015). Multi criteria decision making approaches for green supplier evaluation and selection: A literature review. Journal of Cleaner Production, 98, 66–83. https://doi.org/10.1016/j.jclepro.2013.06.046

Gupta, V., & Nyamapfene, A. (2025). Generative AI in universities: Practices at UCL and other institutions, and the path forward. *Internet Reference Services Quarterly*. https://doi.org/10.1080/10875301.2025.2453461

Hojageldiyev, D. (2019, October 21). *Artificial intelligence opportunities for environmental protection*. Society of Petroleum Engineers. https://doi.org/10.2118/198616-MS

Hsieh, W. W. (2009). *Machine learning methods in the environmental sciences: Neural networks and kernels*. Cambridge University Press.

Hussain, M., Yang, S., Maqsood, U. S., & Zahid, R. M. A. (2024). Tapping into the green potential: The power of artificial intelligence adoption in corporate green innovation drive. *Business Strategy and the Environment*. https://doi.org/10.1002/bse.3710

Juanda, J., Riansyah, R. J., Arsadi, A., & Bethany, L. (2024). Toward entrepreneurial campus sustainability: Integrating artificial intelligence for resource allocation in business management. *Aptisi Transactions on Technopreneurship, 6*(3). https://doi.org/10.34306/att.v6i3.410

Khan, S., Mazhar, T., Shahzad, T., Khan, M. A., Rehman, A. U., Saeed, M. M., & Hamam, H. (2025). Harnessing AI for sustainable higher education: Ethical considerations, operational efficiency, and future directions. *Discover Sustainability*. https://doi.org/10.1007/s43621-025-00809-6

Lee, J. H., Wood, J., & Kim, J. (2021). Tracing the trends in sustainability and social media research using topic modeling. *Sustainability, 13*(3), 1269. https://doi.org/10.3390/su13031269

Li, D., Xiao, J., & Yang, F. (2024). Artificial intelligence and enterprise green innovation: Intrinsic mechanisms and heterogeneous effects. *Sustainability, 16*(21), 9246. https://doi.org/10.3390/su16219246

Lin, X., Hu, X., Qu, Y., & Wei, Z. (2024). The role of institutional innovation and artificial intelligence in corporate eco-innovation: A case study of high-tech enterprises in Liaoning province, China. *Journal of Asian Business Strategy, 14*(2), 1–12. https://doi.org/10.55493/5006.v14i2.5188

Liu, T., & Zhou, B. (2024). The impact of artificial intelligence on the green and low-carbon transformation of Chinese enterprises. *Managerial and Decision Economics*. https://doi.org/10.1002/mde.4164

Luthra, S., Govindan, K., Kannan, D., Mangla, S. K. & Garg, C. P. (2017). An integrated framework for sustainable supplier selection and evaluation in supply chains. Journal of Cleaner Production, 140(3), 1686–1698. https://doi.org/10.1016/j.jclepro.2016.09.078

Nahar, S. (2024). Modeling the effects of artificial intelligence (AI)-based innovation on sustainable development goals (SDGs): Applying a system dynamics perspective in a cross-country setting. *Technological Forecasting and Social Change, 200*, 123203. https://doi.org/10.1016/j.techfore.2023.123203

Ofem, U. J., & Chukwujama, G. (2024). Sustainable artificial intelligence-driven classroom assessment in higher institutions: Lessons from Estonia, China, the USA, and Australia for Nigeria. *European Journal of Interactive Multimedia and Education*. https://doi.org/10.30935/ejimed/15265

Paustian, K., Lehmann, J., Ogle, S., Reay, D., Robertson, GP. & Smith, P. (2016). Climate-smart soils. Nature, 532, 49–57. https://doi.org/10.1038/nature17174

Racz, L. (2019). Mitigation in the industrial sector, CO_2 trade. In W. Leal Filho (Ed.), *Encyclopedia of sustainability in higher education* (pp. 1–8). Springer. https://doi.org/10.1007/978-3-030-11352-0_22

Saberi, S., Kouhizadeh, M., Sarkis, J., & Shen, L. (2019). Blockchain technology and its relationships to sustainable supply chain management. *International Journal of Production Research, 57*(7), 2117–2135. https://doi.org/10.1080/00207543.2018.1533261

Sachs, J. D., Schmidt-Traub, G., Mazzucato, M., Messner, D., Nakicenovic, N. & Rockström, J. (2019). Six Transformations to achieve the Sustainable Development Goals. Nature Sustainability, 2, 805–814. https://doi.org/10.1038/s41893-019-0352-9

Saxena, A., Pancholi, G., Yadav, B. P., & Saini, R. P. (2024). Exploring the role of artificial intelligence in advancing higher education for sustainable development. In *Handbook of research on innovative approaches to information technology in library and information science* (pp. 184–201). IGI Global. https://doi.org/10.4018/979-8-3693-2440-0.ch010

Schiederig, T., Tietze, F., & Herstatt, C. (2012). Green innovation in technology and innovation management – an exploratory literature review. *R&D Management, 42*(2), 180–192. https://doi.org/10.1111/j.1467-9310.2011.00672.x

Sharma, N. (2017). Innovation in green practices: A tool for environment sustainability and competitive advantage. *SSRN Electronic Journal*. https://doi.org/10.2139/ssrn.2861530

Shin, K., Choi, H., & Lee, H.-C. (2015). Topic model analysis of research trend on renewable energy. *Journal of the Korea Academia-Industrial Cooperation Society, 16*(9), 6411–6417. https://doi.org/10.5762/KAIS.2015.16.9.6411

Tao, M. (2024). Digital brains, green gains: Artificial intelligence's path to sustainable transformation. *Journal of Environmental Management, 362*, 122679. https://doi.org/10.1016/j.jenvman.2024.122679

Tian, C. (2023). Research on impact of artificial intelligence orientation on enterprise green innovation. *International Journal of Business and Economics Research, 12*(1), 1–7. https://doi.org/10.11648/j.ijber.20231201.16

Tripathi, S. K., Bachmann, N., Brunner, M., Rizk, Z., & Jodlbauer, H. (2024). Assessing the current landscape of AI and sustainability literature: Identifying key trends, addressing gaps and challenges. *Journal of Big Data, 11*, 65. https://doi.org/10.1186/s40537-024-00912-x

Vezzani, A. (2023). Digital technologies and eco-innovation: Evidence of the twin transition from Italian firms. *Industry and Innovation*. https://doi.org/10.1080/13662716.2023.2213179

Wang, J., Wang, X., Sun, F., & Li, X. (2024). The functional mechanisms through which artificial intelligence influences the innovation of green processes of enterprises. *Systems, 12*(9), 378. https://doi.org/10.3390/systems12090378

Willenbacher, M., Kunisch, C., & Wohlgemuth, V. (2018). Application of methods of artificial intelligence for sustainable production of manufacturing companies. In *Advances and new trends in environmental and energy informatics* (pp. 313–328). Springer. https://doi.org/10.1007/978-3-319-65687-8_20

Yang, Q., Sun, T., & Li, R. (2023). Does artificial intelligence promote green innovation? An assessment based on direct, indirect, spillover, and heterogeneity effects. *Energy & Environment*. Advance online publication. https://doi.org/10.1177/0958305X231220520

Yang, S., Wang, J., Dong, K., Dong, X., Wang, K.-J., & Fu, X. (2024). Is artificial intelligence technology innovation a recipe for low-carbon energy transition? A global perspective. *Energy, 300*, 131539. https://doi.org/10.1016/j.energy.2024.131539

Zhang, D. (2024). The pathway to curb greenwashing in sustainable growth: The role of artificial intelligence. *Energy Economics, 134*, 107562. https://doi.org/10.1016/j.eneco.2024.107562
Zhu, Y., & Zhu, G. (2024). Research on the influence mechanism of artificial intelligence application on corporate green innovation. *South Asian Research Journal of Engineering and Technology, 6*(6), 77–84. https://doi.org/10.36346/sarjet.2024.v06i06.003
Zönnchen, B., Böhm, C., & Socher, G. (2024). Bridging disciplines in higher education: The convergence of AI and sustainability. In *Proceedings of the 8th International Conference on Higher Education Advances (HEAd'24)*. https://doi.org/10.4995/head24.2024.17278

Gabriel Augusto Sampaio Soares holds a Bachelor's degree in Information Management from the Federal University of Uberlândia (UFU), Brazil. He is affiliated with UFU's Faculty of Management and Business (FAGEN), where his academic focus spans information systems, data analytics, and business intelligence. His research interests include the application of information management strategies to enhance organizational decision-making. As an emerging scholar, he actively contributes to academic projects exploring the intersection of technology and business operations. His work supports evidence-based approaches in information governance and digital transformation.

Helena Farias Ribeiro is a doctoral candidate in Administration at the Federal University of Uberlândia (UFU), funded by a CNPq scholarship and affiliated with the Faculty of Management and Business (FAGEN). Her doctoral research focuses on applied artificial intelligence (AI) and sustainability, specifically investigating the alignment of academic research with the Sustainable Development Goals (SDGs) of the 2030 Agenda. She holds a Master's degree in Administration (UFU), where her research emphasized organizational management and regional dynamics, exploring AI's role in institutional transformation and global sustainability targets.

João Victor Gomes Ribeiro is an undergraduate student in Information Management at the Federal University of Uberlândia (UFU), affiliated with the Faculty of Management and Business (FAGEN). His academic pursuits center on information systems, data science, and business process optimization. He participates in research initiatives exploring digital governance and technology-driven business models, gaining hands-on experience in academic collaborations. His emerging interests include the role of big data in organizational efficiency and innovation. As a junior scholar, he contributes to projects aimed at enhancing information management practices in diverse sectors.

Carla Bonato Marcolin is a Professor at UFU's Faculty of Management and Business (FAGEN). She holds a Ph.D. in Administration (Operational Research) and a Master's in Administration (IT Management) from the Federal University of Rio Grande do Sul (UFRGS), complemented by a specialization in Finance. She completed postdoctoral research at the Getúlio Vargas Foundation (FGV-SP). Her expertise spans operational research, information systems, financial technology, and IT governance. She publishes extensively on quantitative decision-making models and digital innovation. With industry experience and academic leadership, she drives projects integrating technology, finance, and strategic management.

Jaluza Maria Lima Silva Borsatto is a Professor of Finance at UFU's Faculty of Management and Business (FAGEN). She holds a Ph.D. in Administration from the University of São Paulo (FEA-RP/USP), a Master's in Financial Management from UFU, and a specialization in Business Management from FGV-RJ. Her research focuses on corporate finance, financial risk, and control systems, with publications on fiscal policy and organizational economics. She combines academic rigor with practical insights, mentoring students and leading projects that address financial sustainability in emerging markets. Her work informs both scholarly discourse and industry practices.

Are We Training Nutritionists for Food Sustainability? Analysis of the Educational Planning of Federal Universities

Lidiane Lieseski, Gabriele Gomes Da Silva, and Rozane Marcia Triches

Abstract Climate crises and resource scarcity reveal planetary boundaries, requiring a multidimensional approach to Sustainable Development. After all, Food highlights the crucial role of the university and the training of nutritionists in this scenario. The objective of this study was to investigate the integration of food and sustainability training within the Nutrition Bachelor's degrees offered by Brazilian federal universities. A total of forty-two Educational Planning documents (PPCs, as per the Portuguese abbreviation) were systematically analyzed and summarized through the application of both keyword identification and qualitative as well as quantitative analytical methods. The findings revealed that only one-third of the Nutrition programs at Federal Higher Education Institutions (IFES, as per the Portuguese abbreviation) incorporated keywords associated with sustainability in their PPCs. Furthermore, while subjects pertaining to sustainability were present in 61.9% of the curricula, only half of these featured the term explicitly in their titles. In terms of guiding principles, 64.3% of the PPCs included one of the examined terms. However, there was a notable scarcity of information regarding faculty profiles within these documents. These results suggest a marginalization of the sustainability theme, albeit indicative of an increasing concern reflected in more recent PPCs. Consequently, there is an imperative to reevaluate the epistemological foundations of Nutrition Science and to refine the PPCs to ensure comprehensive inclusion of sustainability across all facets of the program.

Keywords Sustainable Development · Food System · Nutrition · Guidelines

L. Lieseski (✉) · G. G. Da Silva · R. M. Triches
Federal University of Fronteira Sul, Chapecó, Brazil
e-mail: lidianelieseski@gmail.com

W. Leal Filho et al. (eds.), *Sustainability in the Curriculum and Operations of Universities*, World Sustainability Series,
https://doi.org/10.1007/978-3-032-12678-8_8

1 Introduction

The increasing discussion surrounding the effects of the food system on both human and environmental health has gained significant attention within global agendas that intersect food systems with sustainability (Jacob and Araújo 2020). Moreover, the challenge of aligning population nutrition with sustainability has emerged as a fundamental concern in food policy for the twenty-first century. This raises a pertinent question: how can we feed large populations in a manner that is equitable and healthy, while also ensuring the preservation of a functional ecosystem for future generations? (Lang and Barling 2013; Marchioni et al. 2021).

The 2030 Agenda adopted by the United Nations General Assembly is a comprehensive framework focusing on human health and environmental sustainability. It encompasses 17 ambitious and interconnected goals outlined in its Sustainable Development Goals (SDGs), which address the main challenges prevalent in Brazil and globally. These goals include the elimination of poverty, the protection of the planet, the promotion of prosperity, and eradicating hunger and malnutrition. The Agenda aspires to establish a more equitable and sustainable future, highlighting the critical interrelations between global food security, food systems, and overall well-being (UN 2015).

The interplay between the SDGs and the food system is significant, particularly given that many environmental issues originate from the prevailing food system characterized by monoculture practices of land misuse, the application of pesticides, and intensive animal husbandry. These practices have led not only to increased greenhouse gas emissions but also adversely impacted food quality, contributing to a decline in the consumption of diverse and fresh foods and an increase in the intake of ultra-processed items that are high in sodium, fats, and sugars. Consequently, public health has suffered, as evidenced by a rise in obesity and chronic noncommunicable diseases, alongside a notable prevalence of malnutrition and hunger (FAO 2018; Sonnino 2019; Machado et al. 2021). The concurrent challenges posed by malnutrition, obesity, and climate change constitute a phenomenon known as the global syndemic, which has both direct and indirect implications for the health of populations (Swinburn et al. 2019).

In this context, a debate was initiated during the early 1980s and 1990s, led by Gussow and Clancy (1986), who introduced the concept of sustainable regional diets in relation to sustainable agriculture. From the 2000s onwards, this dialogue has escalated, due to increased food security issues and environmental concerns, which have prompted an integrated perspective that connects food production with consumption practices (Triches 2021). In 2010, the Food and Agriculture Organization (FAO) articulated the definition of "Sustainable Diets" as those that exhibit a low environmental impact, support food and nutritional security, and promote a healthy life for current and future populations, while safeguarding biodiversity and ecosystems. Furthermore, sustainable diets must be culturally acceptable, economically equitable, accessible, nutritionally adequate, safe, and healthy, while simultaneously optimizing both natural and human resources (FAO 2010; Freitas and Garcia 2018).

Sustainable diets are designed to enhance population health while ensuring the sustainability of food production, processing, transportation, storage, sales, and consumption. The core principles of these diets are established to minimize the use of natural resources and reduce environmental footprints. This is achieved through the increased consumption of plant-based foods, moderation in the intake of animal-based proteins, encouragement of efficient use of food, and minimization of waste. Additionally, sustainable diets advocate for the optimal use of available resources and endorse agricultural practices that conserve biodiversity, protect soil quality, and promote responsible water usage (Willet et al. 2019).

In contrast, Triches (2022) posits that a universally sustainable diet does not exist; instead, there are various eating patterns that exhibit varying degrees of sustainability. Furthermore, the notion of healthy eating does not inherently equate to sustainability. In this context, the author asserts that it cannot be presumed a nutritionally adequate diet will inherently result in lower greenhouse gas emissions, reduced land utilization, or diminished water and energy consumption. The sustainability of a diet is contingent upon the specific food selections made. Certain foods may exhibit greater sustainability than others, influenced by a range of factors including the nature of the supply chain, methods of production, processing, transportation, preparation, consumption, and disposal.

Given this scenario, the involvement of academia and the field of nutrition science in addressing the challenges related to food and nutrition have become pertinent. Nutritionists will become agents of the transformation of food systems, ensuring they are developed with a focus on health and sustainability (Fanzo et al. 2015; Naves and Recine 2014). However, studies show that the performance of nutritionists is limited in terms of sustainability (Bianchini et al. 2020; Edemundo et al. 2022). Therefore, it is essential to prioritize the training of professionals equipped with the necessary knowledge, skills, and attitudes to engage with policies, conduct research, and provide services aimed to promote food sustainability (Fanzo et al. 2015).

A limited number of studies have addressed the training of professionals in Brazil concerning sustainability (Jacob and Araújo 2020; Triches and Brito 2021), revealing significant gaps in this area. Teixeira (2021) aimed to assess the integration of sustainability and environmental education within the Educational Planning documents (PPCs, as per the Portuguese abbreviation) of Nutrition Bachelor's degrees in the Northeast region. In a related inquiry, Jeronimo (2015) examined the incorporation of this theme within the Nutrition Bachelor's degrees that participated in the National Student Performance Examination (ENADE, as per the Portuguese abbreviation) in 2010. Both authors concluded that the emphasis on sustainability in Nutrition education is minimal and lacks a transversal approach, being evident only in select curricular matrices observed.

As of now, there exists no requirement for the integration of sustainability topics within Nutrition Bachelor's degrees, as the prevailing National Curriculum Guidelines (Diretrizes Curriculares Nacionais—DCNs, as per the Portuguese abbreviation), outlined in the CNE/CES Resolution No. 5/2001, do not explicitly mandate the inclusion of sustainability in the educational curricula for training programs.

The recently approved National Curriculum Guidelines of 2024 by the Higher Education Chamber of the National Education Council has defined a significant shift in vocational training by prioritizing sustainability. In light of this regulatory change, it is imperative to evaluate the extent to which sustainability is incorporated within the curricula of Nutrition Bachelor's programs. This examination is crucial for determining the alignment of higher education institutions with the new guidelines and for facilitating the necessary revisions of curricula to more effectively address the contemporary social and environmental challenges.

The purpose of this study is to conduct a comprehensive analysis of the Nutrition programs at federal universities in Brazil, particularly focusing on their integration of food and sustainability training and the preparation of professionals for sustainable development. Specifically, the study seeks to ascertain the most recent year in which the curricula were revised, to identify both the compulsory and elective components of the curriculum and their associated syllabuses related to these topics. Furthermore, it aims to understand how these programs cultivate the competencies in graduates by examining the descriptions provided in the graduate profile. Additionally, the research will investigate the presence of references to sustainability within the epistemological and methodological frameworks of the programs and will assess whether faculty training encompasses relevant data on this subject.

2 Methodology

This research adopts both qualitative and quantitative methodologies, focusing specifically on Brazil's federal higher education institutions (IFES, as per the Portuguese abbreviation) that are duly registered and operational with the Ministry of Education, and which offer undergraduate degrees in Nutrition. The study primarily utilizes secondary data, with the Educational Planning documents (PPCs) of the degrees serving as the foundational material for the investigation. In this regard, the PPCs were identified and accessed through the official websites of the respective degrees. In instances where the documents were not available online, direct communication was established with the relevant degree coordinators or institutions via e-mail.

Brazil is home to 69 federal higher education institutions, with 42 of these offering programs in nutrition (ANDIFES 2024). These institutions are situated throughout the country's five geographical regions: there are five in the Midwest, eleven in the Northeast, four in the North, fourteen in the Southeast, and eight in the South, as illustrated in Table 1. Additionally, detailed information regarding each university, including the degrees' title, the respective city or state, the region, and the year of the most recent update can be found in the PPCs on the initial pages.

In order to collect data, keywords were systematically searched within the PPC, resulting in a tool that aided in the aggregation and synthesis of information related to the topic under study. Subsequent analysis indicated the necessity to categorize the

Table 1 Distribution of pedagogical projects for the nutrition programme at federal higher education institutions in Brazil by region, city and year. 2024

Region of Brazil	City/State	Universities	PPC year
South	Chapecó/SC	Universidade Federal da Fronteira Sul (UFFS)	2023
	Porto Alegre/RS	Universidade Federal do Rio Grande do Sul (UFRGS)	2018
	Florianópolis/SC	Universidade Federal de Santa Catarina (UFSC)	2018
	Curitiba/PR	Universidade Federal do Paraná (UFPR)	2017
	Santa Maria/RS	Universidade Federal de Santa Maria (UFSM)	2023
	Pelotas/RS	Universidade Federal de Pelotas (UFPel)	2015
	Itaqui/RS	Universidade Federal do Pampa (UNIPAMPA)	2023
	Porto Alegre/RS	Universidade Federal de Ciências da Saúde de Porto Alegre (UFCSPA)	2015
Midwest	Brasília/DF	Universidade de Brasília (UNB)	2022
	Goiânia/GO	Universidade Federal de Goiás (UFG)	2013
	Cuiabá/MT	Universidade Federal de Mato Grosso (UFMT)	2010
	Campo Grande/MS	Universidade Federal de Mato Grosso do Sul (UFMS)	2021
	Dourados/MS	Universidade Federal da Grande Dourados (UFGD)	2023
Southeast	Belo Horizonte/MG	Universidade Federal de Minas Gerais (UFMG)	2019
	Macaé/RJ	Universidade Federal do Rio de Janeiro (UFRJ)	2019
	Viçosa/MG	Universidade Federal de Viçosa (UFV)	2023
	Santos/SP	Universidade Federal de São Paulo (UNIFESP)	2023
	Niterói/RJ	Universidade Federal Fluminense (UFF)	2018
	Rio de Janeiro/RJ	Universidade Federal do Estado do Rio de Janeiro (UNIRIO)	2009
	Ouro Preto/MG	Universidade Federal de Ouro Preto (UFOP)	2018
	Governador Valadares/MG	Universidade Federal de Juiz de Fora (UFJF)	2015
	Vitória/ES	Universidade Federal do Espírito Santo (UFES)	2013
	Uberaba/MG	Universidade Federal do Triângulo Mineiro (UFTM)	2023
	Lavras/MG	Universidade Federal de Lavras (UFLA)	2021
	Uberlândia/MG	Universidade Federal de Uberlândia (UFU)	2022
	Alfenas/MG	Universidade Federal de Alfenas (UNIFAL)	2023
	Diamantina/MG	Universidade Federal dos Vales do Jequitinhonha e Mucuri (UFVJM)	2007
North	Belém/PA	Universidade federal do Pará (UFPA)	2010
	Coari/AM	Universidade Federal do Amazonas (UFAM)	2018
	Rio Branco/AC	Universidade Federal do Acre (UFAC)	2015

(continued)

Table 1 (continued)

Region of Brazil	City/State	Universities	PPC year
	Palmas/TO	Universidade Federal do Tocantins (UFT)	2019
North east	Vitória de Santo Antão/PE	Universidade Federal de Pernambuco (UFPE)	2012
	Vitória da Conquista/BA	Universidade Federal da Bahia (UFBA)	2018
	Santa Cruz/RN	Universidade Federal do Rio Grande do Norte (UFRN)	2020
	Maceió/AL	Universidade Federal de Alagoas (UFAL)	2010
	Lagarto/SE	Universidade Federal de Sergipe (UFS)	2019
	João Pessoa/PB	Universidade Federal da Paraíba (UFPB)	2006
	Teresina/PI	Universidade Federal do Piauí (UFPI)	2011
	Cuité/PB	Universidade Federal de Campina Grande (UFCG)	2010
	Cruz das Almas/BA	Universidade Federal do Recôncavo da Bahia (UFRB)	2021
	Barreiras/BA	Universidade Federal do Oeste da Bahia (UFOB)	2014
	São Luiz/MA	Universidade Federal do Maranhão (UFMA)	2007

Source Authors (2024)

findings into four principal themes: curriculum components, faculty profile, graduate profile/skills/competencies, and primary approaches/guiding principles. It is worth mentioning that the process of gathering information for the section on "primary approaches/guiding principles" revealed significant variation among the PPCs, attributable to differing nomenclature employed across the programs.

For the investigation, specific criteria were established to streamline data collection across various topics. In terms of curriculum components, the following keywords were targeted: "sustain" ("sustent" in Portuguese); "envir" ("ambi" in Portuguese); food system; food security; food sovereignty; food insecurity; and development (the term "development" had to be contextually relevant to be considered). Other than that, a distinction was made between compulsory and elective subjects. Concerning faculty profiles, graduate profiles/skills/competencies, and core guiding principles, an initial assessment was made to determine their inclusion in the PPC, their qualifications and their relevance to the concept of "sustainable development" using the same keywords as mentioned above.

Searches were conducted across the four specified sections of the document, namely curriculum components, faculty profiles, graduate profiles/skills/competencies, and main approaches/guiding principles. Utilizing the "Search" function in PDF readers, each section was examined using relevant keywords. Subsequently, a comprehensive search was performed across the entire document to ascertain the presence of any significant information that might reside outside the designated sections of the PPCs. This task was executed by a master's student in nutrition and an undergraduate student in Nutrition, both of whom received appropriate training for this analytical process. The 42 PPCs were allocated evenly between the two researchers for examination.

The qualitative data was examined through documentary analysis, aimed at systematically organizing and interpreting the information in alignment with the established objectives. This analytical approach is particularly beneficial, given the richness of information contained within the documents, as well as its cost-effectiveness and the absence of necessity for direct interaction with the subjects (Gil 2008).

3 Results

42 PPCs were analysed, reaching all (100%) federal universities offering the Nutrition Bachelor's degree. Of the total represented, 39 PPCs (92%) were found on the institutions' websites and only three (8%) required contacting the coordinator by e-mail as they were not available on the website at the time of the search. Figure 1 shows the distribution of these documents according to the year in which they were formulated or last revised.

Looking at the year of the PPCs and their relationship to sustainable development, the oldest ones are represented in between the years 2006 and 2010, and only three (37.50) of the total of eight PPCs mention the term sustainability in one of the topics. In the period between 2011 and 2015, it is mentioned in seven (77.78%) of the nine documents. In the years 2016–2020, 10 (83.33%) of 12 projects mentioned the term "sustainability" at some point, and in the most recent projects, which were created in the years 2021–2025, the term is mentioned in 12 (92.31%) of the 13 PPCs in the entire document.

Of the 42 universities offering the Nutrition programme, 26 of them (61.9%) offer at least one subject that mentions sustainable development either in its name or in

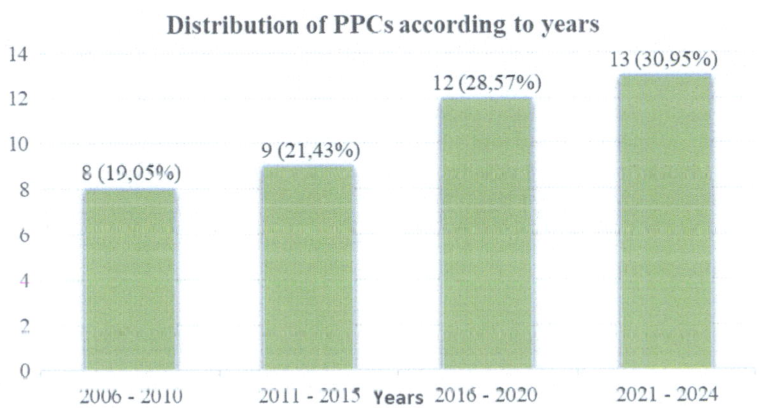

Fig. 1 Distribution of PPCs by year of formulation or last review. 2024. *Source* Authors (2024)

its curriculum using keywords. In total, there are 53 components, of which 41 are compulsory and 12 are elective.

The university with the most subjects (compulsory and elective) dealing with sustainability is the UFMS (Table 2) in the Midwest. Seven specific subjects were found to have the term sustainability/sustainable in their name, four of which were compulsory subjects. UFFS also stands out with six subjects, four of which are compulsory.

These disciplines can be found in the following IFES: UFFS, UFSM, UNB, UFMS, UFMG, UFV, UFTM, UFU, UNIFAL, UFRN, UFS and UFRB.

In the subjects where the curriculum was presented, it was possible to observe how the topic was treated in its context, especially in those subjects where it was treated secondarily, i.e. sustainability was not presented as the main topic of the component.

In terms of graduate profile, all 42 IFES list this topic in the plan, however, only 13 (30.95%) mention the concept of sustainability as something to be considered in the training of nutritionists, as shown in Table 3. The other 29 (69.05%) do not mention it.

This analysis found that 27 PPCs (64.29%) mention terms related to sustainable development in the Guiding Principles and 15 (35.71%) do not include the topic or do not mention sustainability as a priority. Of the 27 PPCs in which they present the topic, 20 (74.07%) actually mention the words "sustainable development" or "sustainability"; and seven (25.93%) show that they indirectly address the topic by including topics focusing on the environment and economy.

In addition, it can be highlighted that among the 27 PPCs that use sustainability-oriented terms in their guiding principles, two IFES stand out: (1) UNIPAMPA, which states that: "the course aims to contribute to the Sustainable Development of the region through… curricular components and projects…" and (2) UFMS, which states that "training must be linked… to sustainable food production, with a reduced waste of raw materials, water and energy, as well as a correct destination of solid and liquid waste".

Regarding the teacher's profile, 37 (88.09%) did not attend the PPC. Of the five PPCs that presented, only one of them did not have a professor with training in postgraduate programs (PPG) dealing with the topic of sustainable or related development. However, there may be teachers with training in this subject in other PPGs, although this information was not verified in other sources. As for the other four (9.52%), the following stand out as the most adherent formations: UFFS with a professor trained in Rural development; UFAC with a PhD in Sanitation, environment and water resources; UFRB with a PhD in Development and environment; and UFOB with professors trained in Health, environment and society, such as in Food and nutrition security.

Table 2 Compulsory and elective subjects with a focus on sustainability/sustainable development at state higher education institutions in the field of nutrition in Brazil. 2024

Region of Brazil	Universities	Compulsory/Elective subjects
South	Universidade Federal Da Fronteira Sul (UFFS)—2023	Environment, economy and society (compulsory) Food and nutrition security (compulsory) Management of food and nutrition I (compulsory) Management of food and nutrition II (compulsory) Sustainable development and its relationship to food and nutrition (elective) Sustainability in the production of community catering (elective)
	Universidade Federal Do Rio Grande Do Sul (UFRGS)—2018	Sustainability in meal production (compulsory)
	Universidade Federal De Santa Maria (UFSM)—2023	Sustainability in meal production (compulsory)
	Universidade Federal De Pelotas (UFPel)—2015	Rural extension (elective)
	Universidade Federal Do Pampa (UNIPAMPA)—2023	Food sociology (compulsory) Food quality control (compulsory) Health and environment (elective)
	Universidade Federal de Santa Catarina (UFSC)—2018	Food system (compulsory) Ethics and bioethics (compulsory)
Midwest	Universidade De Brasília (UNB)—2022	Meal production management II—practice (compulsory) Introduction to sustainable development (elective)
	Universidade Federal De Mato Grosso (UFMT)—2010	Health and environment (compulsory)
	Universidade Federal De Mato Grosso Do Sul (UFMS)—2021	Community catering III (compulsory) Food and nutrition education (compulsory) Compulsory practical training in public health nutrition (compulsory) Nutrition and dietetics III (compulsory) Society, environment and sustainability (elective) Dietetic technique II (compulsory) New technologies in food preservation (elective)

(continued)

Table 2 (continued)

Region of Brazil	Universities	Compulsory/Elective subjects
	Universidade Federal Da Grande Dourados (UFGD)—2023	Community catering I (compulsory) Phytotherapy (compulsory) Dietetic technique I (compulsory)
Southeast	Universidade Federal De Minas Gerais (UFMG)—2019	Ecological principles of sustainable development (elective)
	Universidade Federal Do Rio De Janeiro (UFRJ)—2019	Community catering II (compulsory)
	Universidade Federal De Viçosa (UFV)—2023	Sustainable engineering (elective)
	Universidade Federal De São Paulo (UNIFESP)—2023	Nutrition and food preparation II (compulsory) Food systems II (compulsory)
	Universidade Federal Fluminense (UFF)—2018	Nutrition and environment (elective) Food management for the community I (compulsory) Integrated practice in early childhood education I (compulsory) Integrated practice in food marketing I (compulsory)
	Universidade Federal De Ouro Preto (UFOP)—2018	Planning in food service (compulsory)
	Universidade Federal Do Espírito Santo (UFES)—2013	Planning and management in food II (compulsory)
	Universidade Federal Do Triângulo Mineiro (UFTM)—2023	Community catering I (compulsory) Nutrition and sustainability (compulsory)
	Universidade Federal De Uberlândia (UFU)—2022	Curricular extension activities: nutrition and sustainability (compulsory)
	Universidade Federal De Alfenas (UNIFAL) —2023	Menu planning (compulsory) Food and nutrition management I (compulsory) Legislation and food quality (compulsory) Food technology (compulsory) Food and sustainability (compulsory)
North	Universidade Federal Do Acre (UFAC)—2015	Environment, health and nutrition (compulsory)
North east	Universidade Federal Do Rio Grande Do Norte (UFRN)—2020	Sustainable food systems for food and nutrition security (elective)

(continued)

Table 2 (continued)

Region of Brazil	Universities	Compulsory/Elective subjects
	Universidade Federal De Sergipe (UFS)—2019	Food and nutrition unit management (compulsory) Organic food and sustainability (elective)
	Universidade Federal Do Piauí (UFPI)—2011	Experimental food study (compulsory)
	Universidade Federal Do Recôncavo Da Bahia (UFRB)—2021	University, society and environment (compulsory)
	Universidade Federal Do Oeste Da Bahia (UFOB)—2014	Nutrition in public health I: environment, health and community (compulsory)

The presented table of contents does not take into account distinctive terms such as "food security", "food insecurity" and "food sovereignty" that was found in subjects curriculum
Source Authors (2024)

4 Discussion

It was found that over the last few decades, the percentage of PPCs that take the sustainability approach into account, at least to some extent, has increased slightly. It was also found that the majority of PPCs updated or formulated in recent years (12 out of 13) mention sustainable development/sustainability, which differs from the older ones.

This statement is in line with the historical moment in which the concept of sustainable development emerged as a relatively recent global concern and gained prominence in the last decades of the 20th century (Triches 2021). Prior to this, the focus was on economic and industrial growth without considering the long-term environmental and social impacts. However, sustainable development has evolved from a peripheral issue to a global priority, driven by international agreements such as the United Nations Conference (Rio +20) in 2012 becoming deeply rooted in global policy agendas such as the United Nations' SDGs (Martinelli and Cavalli 2019).

With the growing concern, studies on sustainability have also increased in recent years. When Triches (2021) addresses the topic of sustainable nutrition, she notes that the topic has gained prominence since 2010 and especially in recent years, in line with the advance of these concerns in science and society.

In terms of guiding principles, most PPCs were found to show some interest in sustainability as these discussions were included as principles throughout the degree. However, there is no mention of the term in 35.71% of PPCs. Furthermore, in the curricula where it is mentioned, the topic is only marginally addressed, confirming the idea that biomedical fields still predominate and that dialog with social fields is only rudimentary. As early as 1992, Mennell, Murcott and Oterloo pointed out that nutrition science has become specialized and strives to create uniform nutritional

Table 3 Federal higher education institutions in Brazil that emphasize sustainability/sustainable development in the profile of nutrition graduates. 2024

Region of Brazil	IFES	Graduate profile
South	Universidade Federal da Fronteira Sul (UFFS)	In this way, the programme aims to contribute to the graduate becoming an active subject in the promotion of sustainable and supportive economic alternatives, especially for micro and small enterprises in rural areas, the economic base of the southern border Mesoregion
	Universidade Federal do Rio Grande do Sul (UFRGS)	The promotion of healthy and sustainable eating habits with the aim of promoting and maintaining health and preventing disease
	Universidade Federal de Santa Catarina (UFSC)	Working interdisciplinary in a committed, ethical and sustainable way
	Universidade Federal de Ciências da Saúde de Porto Alegre (UFCSPA)	You'll become aware of the importance of continuing education and your commitment to people and social and environmental sustainability
Midwest	Universidade Federal de Goiás (UFG)	You will be entrepreneurs aware of your role in transforming the world and committed to environmental sustainability
	Universidade Federal de Mato Grosso (UFMT)	To be able to act as an agent of transformation, in a committed, ethical and sustainable way focusing on nutrition
Southeast	Universidade Federal de Viçosa (UFV)	To act respecting nature and the search for ecological balance and sustainable development
	Universidade Federal de São Paulo (UNIFESP)	To recognize adequate and healthy food as a human right, based on sustainable and nutrition-sensitive food systems The promotion of sustainable food systems To help promote a sustainable environment, conserve natural resources and manage waste production in order to improve people's quality of life and health
	Universidade Federal Fluminense (UFF)	To act to ensure environmental and social sustainability
	Universidade Federal de Uberlândia (UFU)	To plan menus with the concept of sustainability in mind

(continued)

Table 3 (continued)

Region of Brazil	IFES	Graduate profile
	Universidade Federal de Alfenas (UNIFAL)	To promote health and healthy eating practices based on scientific, ethical and environmentally sustainable knowledge
North	Universidade Federal do Amazonas (UFAM)	To use the principles of rationality, economic efficiency and sustainability
North east	Universidade Federal do Rio Grande do Norte (UFRN)	To commit to global sustainability and environmental protection by supporting and promoting local and sustainable production

Source Authors (2024)

standards for health based on biology, microbiology, biochemistry and physiology, which is referred to as biological reductionism.

This type of diet, based on what Lang, Barling and Caraher (2009) call Life Sciences Nutrition, attaches little importance to issues related to the impact on the environment and the use of natural resources. In a Brazilian discussion, Prado et al. (2011) noted that from the beginning, Nutrition Bachelor's degrees have focused on the individual, emphasizing biomedical aspects and downplaying social aspects. In addition, the social aspects have improved, e.g. through discussions such as Food and Nutrition Security, but as far as the ecological approach is concerned, it is still in its infancy.

In line with this, Teixeira (2021) analyzed the PPCs of universities in the North-east and found that few of them mentioned sustainability in their curricula, which shows the deficit in terms of environmental education in the Nutrition curricula, as well as the imminent need to reformulate the PPCs to have a transversal and transdisciplinary approach to the subject. According to the still valid 2001 National Curriculum Guidelines (DCNs) for Nutrition Bachelor's degrees, there is no obligation to consider sustainability in program and the terms 'sustainable/sustainability' are not even mentioned. Furthermore, the guidelines do not include environmental aspects, but only emphasize the social and biological binomial (Brasil 2001).

According to Soares and Aguiar (2010), the DCNs need to clarify health promotion, which should be seen in a broader framework and linked to sustainable development, taking into account social and economic inequalities. Although there are no direct references to the teaching of sustainability in the 2001 DCNs, the National Environmental Education Policy (PNEA, as per the Portuguese abbreviation), established by Law No. 9,795/1999, stipulates that the subject should be addressed transversally and as a whole and not only in specific, fragmented subjects. Articles 1 and 2 state that environmental education is an essential and permanent part of national education and must be articulated at all levels and in all modalities of the educational process. It also emphasizes that environmental education means an attitude aimed at preserving the environment and striving for its sustainability (Brasil 1999).

Of the 25 PPCs that include sustainability in their curriculum (59.5%), it was found that the topic is rarely addressed across the board, but mostly within subjects, particularly in the fields of Social nutrition and Community catering (Table 2). Some universities have included specific subjects on sustainability in their curricula, with a distinction between elective and compulsory subjects (Table 2). According to the study by Jeronimo (2015), although most courses are elective, universities are making efforts to include the topic in their curricula. For Jeronimo (2015) and Preuss (2009), however, the topic of sustainability permeates all areas of nutrition and is not as fragmented as in the PPCs.

On the other hand, in a 2009 reading of Nutrition Science, authors Lang, Barling and Caraher proposed three strands/approaches to nutrition: Life science nutrition, which is more related to the natural sciences, or what might be called a clinical nutrition-related approach; Social nutrition, which is more related to the social sciences; and Ecological nutrition, which considers environmental foundations and impacts. The latter is concerned with simple and sustainable diets, conscious consumption, respect for natural limits and sustainable production (Lang et al. 2009). It is important to consider these authors' epistemological discussion of nutrition science when they note that this third approach is still far from being emphasized in Nutrition Bachelor's degrees in Brazil.

Another point analyzed was the degrees' profile. The 2001 DCNs state that the profile of Nutrition Bachelor's degrees should aim to ensure food safety, promote health and be based on ethical principles. Although professional nutritionists and sustainability are not mentioned in the guidelines, some universities have highlighted a profile focused on this topic in their PPCs, as shown in Table 3. The documents highlight profiles that focus on sustainable food practices, social and environmental sustainability, sustainable food systems and the relationship with rural areas and conservation. The PPCs of these universities encourage graduates to work with practices that take sustainability into account, respect the environment and seek alternatives for healthier and more sustainable food.

Currently on the agenda are the new DCNs provided for in Resolution No. 704 of October 20, 2022 of the National Health Council, which represent an important step for the professional training of nutritionists (Brasil 2022). One of the principles for the generalist training of nutritionists, mentioned in the third article of the new proposal for guidelines, refers to the action of professionals who respect the "human right to adequate and healthy food and healthy and sustainable food systems" (Brasil 2022, 7). For nutritionists to develop these skills, they must also understand that food is only adequate if it is part of a sustainable food system and work towards sustainable development that promotes health in all aspects of nutrition (Brasil 2022).

In Baungaard et al's (2023) study of nutrition students in the UK, 76% of students surveyed reported that sustainability content was briefly introduced, covered in a module or throughout the degree, and 24% of students could not recall having received any teaching on the subject. The majority of students surveyed reported low to medium confidence (between three and seven out of ten) in advising future clients on sustainable eating.

In the study conducted by Lopes and Maynard (2020) at the University Center of Brasília (UniCEUB, as per the Portuguese abbreviation), students in the Nutrition program showed that although the topic of sustainability is well known, it is still little covered in the degrees and that there is a demand for more lectures and conferences that encourage students not only to think about sustainability, but also to integrate sustainable practices into their daily lives.

According to a study by Triches and Brito (2021) in Brazil with nutritionists in schools, 87.5% of professionals stated that they did not understand sustainable diets. They also reported that they had not heard the topic of sustainability in any curriculum during their academic training.

Finally, in order to train a professional capable of working, considering issues related to sustainability, there must be a better training in these topics. In this work, the low presence of teachers with training in sustainability areas was observed. In this way, and considering that teaching that encompasses these concerns requires discussions and criticism, it may be harmed (Borgonove 2022). There must be training aimed at teachers (Borges et al. 2020), covering all areas of the profession (Clinical, Public Food, Public Health). Therefore, this affects not only undergraduate courses, but also postgraduate courses (PPGs). PPGs with an emphasis on Nutrition Science also need to direct their research to raise more professors that undertake Masters and Ph.D. theses on the amount of topics related to Sustainability. It is also necessary to highlight that few documents actually presented the teachers' training, so there is a possible observation bias, as there may be teachers trained based on sustainability who are not mentioned in the PPCs.

5 Conclusions

The examination of the Educational Planning documents (PPCs) of federal universities in Brazil that offer the Nutrition Bachelor's degree indicates a gradual integration of sustainability themes into academic curricula, particularly evident in the most recent documentation. Nonetheless, the data also highlight persistent and notable shortcomings; the incorporation of sustainability remains largely sporadic, predominantly found in select disciplines that are often elective and exhibiting limited influence on the overall profile of graduates and faculty development.

The findings of this study suggest that, despite the increasing prominence of the sustainable development paradigm in recent decades, its incorporation into the curricula of Nutrition remains in a state of gradual consolidation. The research reveals that a majority of educational programs do not embrace a transversal and critical approach to sustainability, resulting in an educational model that predominantly emphasizes the biomedical dimensions of the profession, at the expense of a broader ecological and social perspective.

The implications of these findings hold significant relevance for the revision of the National Curriculum Guidelines, as well as for the reformulation of the curricular matrices of various academic programs. The education of nutritionists who are dedicated to sustainable food practices and adept at addressing contemporary socio-environmental challenges necessitates not only the incorporation of specific content but also an epistemological reformation of the subjects in Nutrition, as previously advocated by both national and international scholars.

Among the limitations of this study, it is important to emphasize that the analysis is confined to PPCs and relies solely on publicly available data. This approach does not allow for an in-depth exploration of the curriculum subjects or the practical performance of educators. Furthermore, the absence of an investigation into student perceptions represents a missed opportunity to enhance the comprehension of the impact that these professional development initiatives may have.

Future perspectives encompass the expansion of research to include additional higher education institutions, a more profound qualitative analysis of curricular content, and the ongoing assessment of the new Diretrizes Curriculares Nacionais (DCNs) established by Resolution No. 704/2022. These guidelines, which explicitly incorporate a commitment to sustainable food systems and the human right to adequate food, present a significant opportunity for the transformation of nutrition education within the country. This transformation is essential for aligning educational practices with the pressing demands of the global environmental and social agenda.

References

ANDIFES (Associação Nacional dos Dirigentes das Instituições Federais de Ensino Superior) [National Association of Presidents of Federal Higher Education Institutions] (2024). https://www.andifes.org.br/

Baungaard C, Lane KE, Richardson L (2023) Understanding nutrition students' knowledge, perceived barriers and their views on the future role of nutritionists regarding sustainable diets. Nutr Bull 48(4): 572–86. https://doi.org/10.1111/nbu.12649.

Bianchini VU, Martinelli SS, Soares P, Fabri RK & Cavalli SB (2020) Criteria adopted for school menu planning within the framework of the Brazilian School Feeding Program. Revista de Nutrição, 33. https://doi.org/10.1590/1678-9865202033e190197

Borges CLP, Silva LC, Carniatto I (2020) Ambientalização curricular no ensino superior: uma revisão integrativa da literatura [Curriculum greening in higher education: na integrative review of the literature]. RSD 9(11): 1–23. https://rsdjournal.org/index.php/rsd/article/view/9734

Borgonove CM (2022) A Educação Ambiental no ensino superior: análise dos processos de ambientalização curricular e sua contribuição para o pensamento crítico [An Analysis of Curricular Environmentalization Processes and Their Contribution to Critical Thinking in Higher Education Environmental Education]. Doctoral dissertation, Universidade Federal de São Carlos.

Brasil (1999). Lei nº 9.795, de 27 de abril de 1999. Presidência da República [Law No. 9.795, of April 27th, 1999. Republic Presidency]. https://www.planalto.gov.br/ccivil_03/leis/l9795.htm.

Brasil (2001). Resolução CNE/CES nº 5, de 7 de novembro de 2001 [CNE/CES Resolution no 5, of November 7th, 2001]. http://portal.mec.gov.br/cne/arquivos/pdf/CES05.pdf https://www.gov.br/conselho-nacional-de-saude/pt-br/acesso-a-informacao/legislacao/resolucoes/2022/resolucao-no-704.pdf.

Brasil (2022). Resolução CNE/CES nº 704, de 20 de outubro de 2022 [CNE/CES Resolution No. 704, of October 20th, 2022]. http://portal.mec.gov.br/cne/arquivos/pdf/CES05.pdf

Edemundo AS, Pinto E, Baltazar AL, Fialho S, Lima J (2022) Sustentabilidade alimentar em alimentação coletiva e restauração: percepção dos nutricionistas [Sustainability in food service: dietitians' perceptions]. Acta portuguesa de nutrição 31:44–49.

Fanzo JC, Graziose MM, Kraemer K, Gillespie S, Johnston JL, de Pee S, et al (2015) Educating and training a workforce for nutrition in a post-2015 world. Adv Nutr 6(6): 639–47. https://doi.org/10.3945/an.115.010041

FAO (2010) Report of the international symposium on Biodiversity and Sustainable Diets. Rome, FAO.

FAO (2018) The State of Food Security and Nutrition in the World 2018. Building climate resilience for food security and nutrition. Rome, FAO.

Freitas JF, Garcia JC (2018) Evolução conceitual do princípio da sustentabilidade [Conceptual evolution of the principle of sustainability]. Dir Sem Front 2(4): 13–26. https://e-revista.unioeste.br/index.php/direitosemfronteiras/article/view/20229.

Gil AC (2008) Como elaborar projetos de pesquisa [How to develop research projects]. 4. ed. Atlas.

Gussow JD, Clancy KL (1986) Dietary guidelines for sustainability. J Nutr Educ 18(1): 1–5.

Jacob MCM, Araújo FR (2020) Desenvolvimento de competências para Nutrição no contexto de Sistemas Alimentares Sustentáveis [Competency building in Nutrition in the context of Sustainable Food Systems]. Ciência & Saúde Coletiva 25(11): 4369–78. https://doi.org/10.1590/1413-812320202511.31652018

Jeronimo AC (2015) O Ensino da Sustentabilidade na formação do nutricionista [The Instruction of Sustainability within the Professional Education of Nutritionists]. Undergraduate thesis, Universidade Federal do Rio Grande do Sul.

Lang T, Barling D, Caraher (2009) Food Policy: Integrating health, environment and society. Oxford University Press.

Lang T, Barling D (2013) Nutrition and sustainability: an emerging food policy discourse. Proc Nutr Soc 72(1): 1–12. https://doi.org/10.1017/S002966511200290X

Lopes BNF, Maynard DC (2020) Sustentabilidade: percepção e hábitos sustentáveis de estudantes de nutrição [Sustainability: Insights and Eco-conscious Practices Among Nutrition Students]. Undergraduate thesis, Centro Universitário De Brasília.

Machado AD, Marchioni DML, Carvalho AM (2021) A insustentabilidade dos sistemas alimentares atuais deve ser integrada no entendimento da COVID-19 como uma sindemia [The unsustainability of current food systems must be integrated into the understanding of COVID-19 as a syndemic]. Cadernos de Saúde Pública 37(12): e00253221. https://doi.org/10.1590/0102-311x00253221

Marchioni DM, Carvalho AM, Villar BS (2021) Dietas sustentáveis e sistemas alimentares: novos desafios da nutrição em saúde pública [Sustainable Diets and Food Systems: Emerging Challenges in Public Health Nutrition]. Revista USP (128): 61–76. https://doi.org/10.11606/issn.2316-9036.i128p61-76

Martinelli SS, Cavalli SB (2019). Alimentação saudável e sustentável: uma revisão narrativa sobre desafios e perspectivas [Healthy and sustainable diet: a narrative review of the challenges and perspectives]. Ciência & Saúde Coletiva 24(11): 4251–62. https://doi.org/10.1590/1413-81232018241 1.30572017

Naves CC, Recine E (2014) A atuação profissional do nutricionista no contexto da sustentabilidade [The Role of Nutritionists in Promoting Sustainable Practices in Professional Performance]. Demetra 9(1): 121–36. https://doi.org/10.12957/demetra.2014.6246

Prado SD, Bosi MLM, Carvalho MCVS de, Gugelmin SÂ, Mattos RA de, Camargo Junior KR, et al (2011) Alimentação e nutrição como campo científico autônomo no Brasil: conceitos, domínios e projetos políticos [Food and nutrition as scientific field in brazil: Concepts, domains and political projects]. Revista de Nutrição 24(6): 927–38. https://doi.org/10.1590/s1415-52732011000600013

Preuss K (2009) Integrando Nutrição e desenvolvimento sustentável: atribuições e ações do nutricionista [Integrating Nutrition and Sustainable Development: attributions and actions of the nutritionist]. Nutrição em Pauta.

Soares NT, Aguiar AC (2010) Diretrizes curriculares nacionais para os cursos de nutrição: avanços, lacunas, ambiguidades e perspectivas [Curricular guidelines for nutrition courses: advances, ambiguities, omissions and perspectives]. Rev Nutr 23(5): 895–905. https://doi.org/10.1590/s1415-52732010000500019

Sonnino R (2019) Translating sustainable diets into practice: the potential of public food procurement. Redes 24(1): 14–29. https://doi.org/10.17058/redes.v24i1.13036

Swinburn BA, Kraak VI, Allender S, Atkins VJ, Baker PI, Bogard JR, Brinsden H, Calvillo A, De Schutter O, Devarajan R., Ezzati M, Friel S, Goenka S, Hammond RA, Hastings G, Hawkes C, Herrero M, Hovmand PS, Howden M, ... Dietz WH (2019) The global syndemic of obesity, undernutrition, and climate change: The lancet commission report. Lancet, 393(10173), 791–846. https://doi.org/10.1016/S0140-6736(18)32822-8

Teixeira GA (2021) Sustentabilidade e educação ambiental nos cursos de nutrição do nordeste brasileiro: análise dos Projetos Pedagógicos de Curso [Sustainability and Environmental Education in Nutrition Degrees in Northeastern Brazil: An Analysis of Course Pedagogical Projects]. Undergraduate thesis, Universidade Federal do Rio Grande do Norte.

Triches RM (2022) Alimentação e sustentabilidade: o que são dietas sustentáveis? [Food and sustainability: what are sustainable diets?] In: PARRADO-BARBOSA, A; RUIZ, E. N; TRICHES, R. M. Sustentabilidade, circuitos curtos de abastecimento e compras públicas de alimentos [Sustainability, short supply circuits and public food purchases]. 1. ed. Chapecó: UFFS, 2022.

Triches RM, De Brito IC (2021) Conhecimento e atuação de nutricionistas da alimentação escolar sobre dietas sustentáveis [Assessment of School Nutritionists' Knowledge and Performance Regarding Sustainable Diets]. Demetra 16: e60571.

Triches RM (2021) Sustainable diets: definition, state of the art and perspectives for a new research agenda in Brazil. Ciência & Saúde Coletiva 26(5): 1833–46. https://doi.org/10.1590/1413-81232021265.09742019

UN (2015) Transforming our world: the 2030 Agenda for Sustainable Development. https://sdgs.un.org/2030agenda.

Willett W, Rockström J, Loken B, Springmann M, Lang T, Vermeulen S, Garnett T, Tilman D, DeClerck F, Wood A, Jonell M, Clark M, Gordon LJ, Fanzo J, Hawkes C, Zurayk R, Rivera JA, De Vries W, Sibanda LM ... Murray CJL (2019) Food in the Anthropocene: the EAT-Lancet Commission on healthy diets from sustainable food systems. Lancet 393(10170): 447–92. https://doi.org/10.1016/S0140-6736(18)31788-4

Lidiane Lieseski Graduated in Nutrition from Regional University of the Midwest of Brazil (Universidade Estadual do Centro-Oeste—UNICENTRO, as per the Portuguese abbreviation), Master in Agroecology and Sustainable rural development from Federal University of Fronteira Sul (Universidade Federal da Fronteira Sul—UFFS, as per the Portuguese abbreviation) and, currently, nutritionist of school feeding in the City Hall of Mangueirinha-Paraná, Brazil.

Gabriele Gomes Da Silva Nutrition student at the Federal University of Fronteira Sul (UFFS).

Rozane Marcia Triches Graduated in Nutrition from the Regional University of the Northwest of the State of Rio Grande do Sul, Master in Epidemiology and PhD in Rural development, both from the Federal University of Rio Grande do Sul (UFRGS, as per the Portuguese abbreviation). Currently a professor of the Undergraduate Course in Nutrition and the Stricto Sensu Postgraduate Course in Agroecology and sustainable rural development at the Federal University of Fronteira Sul.

Incorporating Sustainable Development Goals into Accounting Subjects: Reflections and Strategies

Bruno Barbosa de Souza and Deivson Vinicius Barroso

Abstract From the perspective that Accounting goes beyond technical procedures and is understood as a social and moral practice, it becomes evident that accounting education should contribute to the promotion of sustainable development. Therefore, it is essential to consider how socio-environmental practices can be incorporated into accounting curricula and, more specifically, how the Sustainable Development Goals (SDGs) can be integrated into accounting education. This concern leads to the primary objective of this study: to propose strategies for integrating the SDGs into introductory, intermediate, and advanced accounting courses in undergraduate Accounting programs. To achieve this, the study analyzed the objectives and syllabi of these three courses based on the curriculum framework proposed by the Brazilian Accounting Foundation (FBC) and associated their content with the SDGs through a coding process. As a result, we found that it is possible to integrate the SDGs into all three analyzed courses, as SDGs 5, 8, 9, 12, 16, and 17 were identified as applicable within their syllabi. The theoretical contribution of this research lies in expanding the understanding of the relationship between the SDGs and accounting education, highlighting how course content can effectively integrate sustainability. From a practical standpoint, the study provides guidelines for instructors on incorporating the SDGs into accounting education, facilitating the adoption of pedagogical practices aligned with this objective. Socially, the research reinforces the importance of educating individuals and accounting professionals committed to sustainability, thereby contributing to a more inclusive and socially responsible accounting profession.

Keywords Accounting education · Curriculum · Sustainability · Sustainable development goals · SDGs

---

B. B. de Souza (✉) · D. V. Barroso
Accounting Sciences, UFU, Uberlândia, Brazil
e-mail: bruno06bs@gmail.com

D. V. Barroso
e-mail: deivson.vinicius07@gmail.com

W. Leal Filho et al. (eds.), *Sustainability in the Curriculum and Operations of Universities*, World Sustainability Series,
https://doi.org/10.1007/978-3-032-12678-8_9

1 Introduction

Historically, the development of accounting as a field of knowledge, technique, and practice has been directed toward meeting the demands of shareholders and investors, often at the expense of social and environmental concerns (Brown and Dillard 2019). This perception of accounting as a purely technical discipline, detached from ethical and political dimensions, has long constrained the field, leading academics and professionals to overlook issues such as corporate responsibility, ecological sustainability, and social justice (Power 1992).

However, in recent decades, academic discussions on the need to transform accounting have intensified, highlighting its ethical responsibility and role in advancing sustainability (Boyce 2004; Carnegie et al. 2021, 2022; Chabrak and Craig 2013). Within this context, and in alignment with the United Nations' global agenda, scholars advocate for integrating the Sustainable Development Goals (SDGs) into higher education, highlighting their importance in fostering values, knowledge, and skills aimed at sustainability (Dean et al. 2018; Falivena and Gulluscio 2023; Powell and McGuigan 2024).

Training accountants with a strong understanding of sustainability has been recognized as a crucial factor in driving sustainable practices within organizations (Botes et al. 2014; Lee et al. 2017). Moreover, universities play a fundamental role in preparing future professionals to make socially and ethically responsible decisions by incorporating principles of corporate social responsibility and sustainability into business practices (Alonso-Almeida et al. 2015; Godemann et al. 2014).

Given this context, there is growing concern about integrating sustainability-related topics into business and accounting curricula, either through dedicated courses or a cross-disciplinary approach (Christensen et al. 2007; Wright and Bennett 2011). Studies show that universities that adopt sustainability initiatives encourage students to engage actively with these topics and apply them in their professional careers (Cooray et al. 2022; Fanea-Ivanovici and Baber 2022).

Despite recent advancements, accounting education has not yet significantly incorporated the SDGs. Although the topic has gained more attention, as evidenced by recent academic discussions (Abhayawansa et al. 2021; Sobkowiak et al. 2020), a substantial gap remains. There is a lack of structured debate among educators on pedagogical practices and effective ways to integrate the SDGs into accounting curricula. Addressing this gap is essential to training accountants who are not only technically proficient but also agents of social transformation committed to the public interest (International Federation of Accountants 2024; Sytnik et al. 2021).

Against this backdrop, the present study aims to propose strategies for integrating the SDGs into introductory, intermediate, and advanced accounting courses in undergraduate Accounting programs. This initiative seeks to institutionalize sustainability education within accounting curricula and provide practical guidelines for instructors of these courses. Strengthening the integration of sustainability in accounting education is fundamental to preparing citizens and professionals who are willing and capable of aligning accounting practices with the SDGs (Ratnaningsih et al. 2024; Dagiliute et al. 2018).

2 Theoretical Framework

2.1 Sustainable Development Goals (SDGs)

The Sustainable Development Goals (SDGs) represent a universal framework proposed by the United Nations (UN) in 2015, comprising 17 goals aimed at addressing critical global challenges such as poverty, inequality, climate change, and peace (Roma 2019). These goals are interconnected, emphasizing the need for integrated approaches to achieving sustainable development (Roma 2019). The SDGs were unanimously adopted by 193 UN member states as part of the 2030 Agenda for Sustainable Development (Bak 2024).

These goals not only focus on environmental sustainability but also promote social equity and economic growth (Ferreira et al. 2023). Figure 1 presents the 17 goals to be achieved.

The United Nations' 2024 Sustainable Development Goals Report highlights that progress on the SDGs has encountered significant challenges in recent years, with many indicators exhibiting stagnation or decline (United Nations 2024). The report highlights that despite reaffirmed commitments, multiple areas of progress have been disrupted or reversed. The lingering effects of the COVID-19 pandemic, coupled with

Fig. 1 The 17 sustainable development goals. *Source* United Nations (2025)

conflicts, climate shocks, and economic instability, have exacerbated pre-existing inequalities.

Among the consequences of this scenario, extreme poverty has increased, affecting an additional 23 million people, while food insecurity impacted over 100 million individuals between 2019 and 2022. Moreover, education, a fundamental pillar of sustainable development (Mohanty et al. 2024), is under threat, with declining math and reading skills among students in various countries, jeopardizing essential competencies for the future (United Nations 2024).

Bak (2024) explains that progress in one SDG often influences others, a dynamic clearly reflected in the 2024 UN Report findings. Among the established SDG targets, only 17% show sufficient progress to be achieved by 2030, while 48% exhibit moderate or severe deviations from the desired trajectory (United Nations 2024). Of these, 18% are stagnating, and 17% have regressed to levels lower than those of 2015 (United Nations 2024).

This diagnosis reinforces the urgent need to intensify efforts to bring the SDGs back on track and prevent the failure of the global agenda. Consequently, initiatives from various actors—such as governments, third-sector organizations, private companies, and society at large (Roma 2019)—are required. Furthermore, strengthening academic curricula with sustainability-driven practices becomes even more crucial.

2.2 *Curriculum and SDGs*

A curriculum is understood as a sequence of established and necessary content for learning (Santos and Coimbra 2018), defined by the educational institution. It explicitly outlines the aspects that shape the educational experience a particular course aims to provide (Perim et al. 2020).

The way curriculum planning is structured in each institution can sometimes distinguish the programs offered by different universities (Santos and Coimbra 2018). In this context, Carneiro et al. (2017), with the support of the Brazilian Accounting Foundation (FBC) and several Brazilian higher education institutions in the field, proposed a curriculum framework for accounting programs. The aim of this proposal is not to impose a fully standardized curriculum but rather to contribute to improving curriculum structures while respecting the specific characteristics of each institution, ensuring broad applicability (Carneiro et al. 2017).

The discussion on the curriculum also has a regulatory aspect. The Ministry of Education (2024) updated the National Curriculum Guidelines for the Undergraduate Accounting Program, establishing principles and criteria to be observed by educational institutions in the organization, planning, development, and evaluation of undergraduate programs in the field, including curriculum structure.

The curriculum reflects the institution's philosophy (Santos and Coimbra 2018). According to Silva (2021), learning outcomes extend beyond what is formally stated in course plans and proposals, encompassing attitudes, values, rituals, practices, rules, procedures, and ways of organizing space and time. These aspects are transmitted subliminally through social interactions and daily routines within the educational environment. Thus, the incorporation of the SDGs into curricula can occur through subject content as well as through institutional practices that promote sustainability awareness among students and faculty. Studies have increasingly examined these dynamics.

Analyzing how a public university in Sri Lanka integrated the SDGs into its Accounting program, Cooray et al. (2022) identified sustainability-related concepts embedded in the curriculum, covering topics such as management accounting, financial reporting, corporate governance, and ethics. The authors highlighted various institutional initiatives contributing to multiple SDGs.

Similarly, Ferreira et al. (2023) investigated the incorporation of sustainability into Accounting programs based on the adaptation of Brazilian Indicators for the SDGs. Their research analyzed the curricula of public higher education institutions (HEIs) in the state of Goiás, revealing that sustainability, when considering the adaptation of national indicators to the SDGs, is not included in the mandatory courses of the analyzed programs. Instead, its inclusion is limited to elective courses. These findings highlight the gap in accounting education regarding the systematic integration of sustainability, reinforcing the need for initiatives that promote a more structured and permanent approach to sustainability within curricula.

The study by Twyford et al. (2024) demonstrated how four accounting professors from different countries incorporated SDGs through specific pedagogical strategies within their courses. In the context of SDG 4 (Quality Education), one of the approaches involved creating a more inclusive classroom environment, ensuring that students from diverse cultural and socioeconomic backgrounds could actively participate in discussions and academic activities. To achieve this, one professor encouraged students to contribute in their native languages, fostering greater inclusion and engagement. This initiative also contributed to SDG 10 (Reduced Inequalities) by helping to overcome linguistic and cultural barriers that could hinder learning for certain groups.

In the same autoethnographic study, another strategy related to SDG 16 (Peace, Justice, and Strong Institutions) involved the introduction of case studies that explored accounting's role in transparency, governance, and social justice. Through these cases, students were encouraged to reflect on the ethical and moral implications of accounting practices. Another example from taxation education emphasized the role of taxes in financing public policies and reducing inequalities, reinforcing the importance of accounting in promoting collective well-being (Twyford et al. 2024).

It is observed that previous studies have made efforts to address the topic of the accounting curriculum in alignment with the SDGs. However, an opportunity for further investigation arises in the effective association of the SDGs with course subjects, as well as in the systematization of strategies that can be effectively employed to incorporate the SDGs into students' education.

3 Methodology

This study adopts a qualitative approach within the interpretivist epistemological paradigm and is conducted through documentary research. The interpretivist stance of this study aligns with Chua's (1986) perspective, which argues that this research paradigm seeks to understand social reality as a phenomenon constructed by individuals, emphasizing the meaning and context of the practices analyzed. In this sense, the study aims to interpret and construct meaning regarding how accounting curriculum content can be connected to the Sustainable Development Goals (SDGs).

The research focuses on three professional training courses outlined in the Curriculum Proposal of the Brazilian Accounting Foundation (FBC). This proposal establishes guidelines for the training of accounting professionals in Brazil and aims to guide higher education institutions (HEIs) in structuring undergraduate programs in the field, ensuring that the content covered aligns with market demands, professional regulations, and industry trends (Carneiro et al. 2017).

The FBC proposal divides curricular components into two groups: basic training, which includes broader knowledge areas such as administration, law, ethics, economics, psychology, and mathematics, and professional training, which refers to courses specific to the accounting field (Carneiro et al. 2017). Due to the need to delimit the scope of the investigation, as well as the propositional nature of the study, we chose to analyze introductory accounting, intermediate accounting, and advanced accounting, as they represent the starting point for students' professional training in the field.

To conduct the data analysis, the study initially collected the following information regarding the selected courses: course name, objectives, and syllabus. Subsequently, a coding process was performed to associate the SDGs with the course objectives and syllabi. This step was first conducted by one of the researchers and then validated by another, ensuring that discrepancies and disagreements were resolved collaboratively. Finally, to discuss the results, strategies for teaching and learning course content were proposed based on the established associations with the SDGs.

4 Results and Discussions

In presenting and discussing the results of this research, it is essential to frame the integration of the SDGs into Accounting education within a broader conceptual perspective. Carnegie et al. (2021, p. 69) define accounting as "a technical, social, and moral practice concerned with the sustainable utilization of resources and proper accountability to stakeholders to enable the flourishing of organizations, people, and nature". This perspective underscores the necessity of embedding sustainability within accounting curricula, aligning educational frameworks with the discipline's broader societal and ethical responsibilities.

From this perspective, accounting goes beyond a set of techniques, concepts, and procedures. It is not limited to a purely technical or instrumental practice, as it plays a fundamental role in structuring organizational actions and human activities. As a social practice, accounting guides behaviors, influences organizational culture, and establishes parameters for decision-making. Furthermore, it is a moral practice, whose actions and omissions impact individuals and institutions in both the present and the future, contributing to the construction of ethical order within organizations and society (Carnegie et al. 2021).

Building on this definition, the integration of the SDGs into accounting education should not occur in isolation, but rather as a foundational principle embedded across all subjects within the program. According to Carnegie (2022), each accounting topic, beyond its technical nature, influences organizational dynamics, business culture, and decisions that affect both individuals and society. Thus, the professor, in different subjects, has the possibility of adopting a pedagogical approach that enables reflection on the social and ethical impacts of accounting practices, and incorporating one or more SDGs into the covered content, promoting accounting education aligned with contemporary challenges and the commitment to sustainable development. It is from this perspective that strategies and proposals for the incorporation of SDGs into teaching practice are presented below.

The authors of this study start from the principle that the professor, when planning and organizing their lesson, already contributes significantly to the achievement of SDG 4 (Quality Education). After all, when lesson structuring, content selection, methodologies, and pedagogical approaches are innovative and inclusive, accessible, equitable, and quality education is promoted. This planning, as explained by Procópio and Cortez (2023), makes the teaching process more flexible and adaptable to students' needs. In this sense, the professor must consider the diversity of students, taking into account different social, cultural, and economic contexts, as well as the inclusion of people with disabilities and historically underrepresented groups.

By preparing their subject with intentionality, the professor stimulates critical thinking, interdisciplinarity, and students' ethical commitment, aiming for accounting education to be a tool for social and professional transformation.

The discussion begins with the introductory accounting course, as shown in Table 1.

In the Introductory Accounting course, the authors of this study highlight the integration of four key SDGs. As shown in Table 1, for Introductory Accounting, opportunities to address SDGs 5, 8, 12, and 16 are observed.

SDG 8 (Decent Work and Economic Growth) and SDG 5 (Gender Equality) can be explored when presenting the scope of the accounting profession, emphasizing the growing participation of women in the field and the importance of inclusion and wage equity. At this early stage of training for future accountants, it is essential to highlight that, although women represent a significant portion of accounting professionals, they still face challenges in reaching leadership and decision-making positions in the market (Santos et al. 2021; Silva et al. 2023). Classroom discussions can address female representation in the profession, data analysis on strategic positions held by women, and debates on corporate policies that promote equity

Table 1 Introductory accounting

Objective: To provide introductory accounting knowledge, with an emphasis on the preparation, interpretation, and analysis of financial reports, assisting managers in the company's decision-making process		
Syllabus: Scope of the accounting professional. Basic accounting concepts. Study of equity and basic conceptual framework. Study of equity variations. Chart of accounts. Basic bookkeeping procedures. Commercial transactions. Financial transactions. Transactions with fixed and intangible assets. Financial statements		
SDG	**Target**	
5	**5.5**	Ensure women's full and effective participation and equal opportunities for leadership at all levels of decision-making in political, economic and public life
8	**8.5**	By 2030, achieve full and productive employment and decent work for all women and men, including for young people and persons with disabilities, and equal pay for work of equal value
12	**12.2**	By 2030, achieve the sustainable management and efficient use of natural resources
16	**16.6**	Develop effective, accountable and transparent institutions at all levels

Source Research Data

and inclusion. Additionally, case studies of minority groups who have excelled in accounting can serve as inspiration for students, demonstrating the importance of diversity in strengthening the profession.

In Introductory Accounting, SDG 12 (Responsible Consumption and Production) can also be integrated when introducing and defining the accounts that compose equity, including the concept of environmental assets and liabilities. Furthermore, when discussing inventory and fixed assets (and variations in these accounts), emphasis can be placed on business practices that minimize waste and promote the circular economy. There are studies that highlight circular economy practices as a topic to be addressed by accounting (Rizos et al. 2019; Wang et al. 2014). Through this approach, accounting is directly linked to sustainable management and the efficient use of natural resources, which also relates to SDG 14 (Life Below Water) and SDG 15 (Life on Land).

Additionally, in Introductory Accounting, SDG 16 (Peace, Justice, and Strong Institutions) can be incorporated by discussing the importance of sustainability reports within financial statements and accounting standards, aligning with the findings of Cooray et al. (2022), who also identified the relevance of this SDG in accounting education, among others. In this context, students should understand the role of accounting in ensuring transparency and accountability for external stakeholders, such as investors, regulatory bodies, and society at large.

Beyond the SDGs mentioned above, the newly introduced SDG 18 (Ethnic-Racial Equality) was created in Brazil and presented in November 2024 (Ministry of Racial Equality 2024). Although it has not yet been incorporated into corporate sustainability reports at the time of this research, it can already be integrated into accounting education. This inclusion fosters awareness and recognition of racial and ethnic diversity within the accounting profession.

On this topic, a study by Alston et al. (2023) investigated the representation of underrepresented minorities (such as Black and Latino individuals) in Introductory Accounting textbooks, seeking to understand how these materials influence inclusion and retention in the accounting profession. The study found that the lack of representation in textbooks can negatively impact the interest of students from these minority groups. Alston et al. (2023) highlight that, despite initiatives over the past decades to increase minority participation in accounting, progress has been limited. Therefore, addressing this gap should also be considered a responsibility for accounting educators (Table 2).

Cooray et al. (2022), in their study, advocate for the gradual introduction of sustainability concepts into the accounting program curriculum, which constitutes an educational intervention that can directly contribute to achieving SDG 4 (Quality Education), SDG 8 (Decent Work and Economic Growth), SDG 9 (Industry, Innovation and Infrastructure), SDG 12 (Responsible Consumption and Production), SDG 13 (Climate Action), and SDG 16 (Peace, Justice, and Strong Institutions). In this context, incorporating SDG-related aspects into Introductory Accounting (Table 1) establishes a foundation that can be further explored in Intermediate Accounting, allowing for a more in-depth examination of the topic.

Initially, it is important to highlight that the syllabus for this course, proposed by the FBC (Carneiro et al. 2017), includes the study of the concept and structure of all financial statements. However, companies are increasingly pressured to disclose their social and environmental impacts, leading to the development of new accounting

Table 2 Intermediate accounting

Objective: To understand the content, prepare, and interpret financial statements, as well as the main explanatory notes that complement these statements, considering the applicable regulations and the needs of users		
Syllabus: Accounting process. Balance sheet. Income statement and comprehensive income statement. Statement of retained earnings. statement of changes in equity. Cash flow statement. Value added statement. Explanatory notes		
SDG	**Target**	
9	**9.4**	By 2030, upgrade infrastructure and retrofit industries to make them sustainable, with increased resource-use efficiency and greater adoption of clean and environmentally sound technologies and industrial processes, with all countries taking action in accordance with their respective capabilities
12	**12.6**	Encourage companies, especially large and transnational companies, to adopt sustainable practices and to integrate sustainability information into their reporting cycle
17	**17.19**	By 2030, build on existing initiatives to develop measurements of progress on sustainable development that complement gross domestic product, and support statistical capacity-building in developing countries

Source Research Data

structures that support sustainable practices (Chugumbaev 2024). Thus, instructors can address sustainability reports and integrated reporting alongside financial statements, as these initiatives focus on social and environmental responsibility. Unlike financial statements, which primarily present financial data, sustainability reports combine both financial and non-financial information.

Therefore, just as the structure and examples of financial statements are discussed, the same approach can be applied to sustainability reports and integrated reports. This directly links the course content to SDG 12 (Responsible Consumption and Production), which, in its target 12.6, requires the integration of sustainability information into corporate reporting cycles. The growing adoption of sustainability reporting within financial statements highlights the need for accountants trained to integrate socio-environmental information into traditional financial reporting (Fota et al. 2024; Institute of Singapore Chartered Accountants et al. 2022). Consequently, Intermediate Accounting can help prepare students to meet global sustainability requirements.

When teaching sustainability report preparation, instructors should emphasize key qualitative characteristics, including accuracy, clarity, conciseness, and reliability. This ensures that the reported information reaches not only investors and shareholders but also the broader public, including civil society, communities directly affected by the company's activities, NGOs, and public authorities.

Another topic covered in Intermediate Accounting is the financial analysis of companies, which, among other possibilities, allows for assessing their capacity to invest in innovation and sustainability. Al-Muhanadi et al. (2020) mention that the accounting profession can contribute to reducing carbon emission levels by participating in corporate decision-making processes and providing high-quality information that impacts both society and the environment. In line with this, Chugumbaev (2024) states that accounting methods should be developed to structure, summarize, and analyze information regarding initiatives, processes, and resource utilization to enhance corporate sustainability, taking into account the social and environmental impact of business decisions. We reckon that these topics should be incorporated into classroom discussions through case studies and/or seminars, for example.

Additionally, the role of accountants in managing information and indicators that assist companies in decision-making, including sustainable practices, is reinforced. The Cash Flow Statement (CFS) can serve as a tool to assess the feasibility of investments by analyzing available cash flow. Through the CFS, managers can determine whether there is sufficient cash available to modernize production processes, invest in more efficient equipment, and reduce the environmental impact of industrial operations. These examples can be incorporated into teaching this financial statement. In this way, accounting can facilitate the measurement of innovation and infrastructure development, which are essential for progress toward SDG 9 (Industry, Innovation and Infrastructure) (Al-Muhanadi et al. 2020).

Beyond these connections, the course can also be linked to SDG 17 (Partnerships for the Goals), which, under target 17.19, emphasizes the development of alternative measures to complement GDP in assessing sustainable development progress. The

Value Added Statement (VAS) provides insights into a company's economic contribution to society, detailing the generation and distribution of wealth within an organization. This report presents information on salaries, taxes, and community investments, serving as a metric for calculating a company's contribution to GDP. Makhura (2024) highlights that SDG 17 aims to foster partnerships among multiple stakeholders, and the VAS allows investors, governments, and other interested parties to evaluate how organizations contribute to sustainable economic development, adding value to these agents. This is one of the topics covered in Intermediate Accounting.

Table 3 presents the content and associations related to the Advanced Accounting course.

The syllabus for Advanced Accounting under analysis covers topics such as mergers, spin-offs, and acquisitions of organizations, business combinations, provisions, accounting estimates, exchange rate changes, and financial statement consolidation (Carneiro et al. 2017). Regarding SDG 12 (Responsible Consumption and Production), the course can delve into more advanced aspects of sustainability reporting, particularly in relation to estimates and provisions associated with environmental liabilities and socio-environmental impacts. Proper measurement and disclosure of these elements in financial statements ensure greater transparency for stakeholders (Tegofack and Kamgang 2021; Xu et al. 2024).

It is also possible to associate Advanced Accounting with SDG 8 (Decent Work and Economic Growth), especially when addressing topics such as financial statement consolidation and business combinations. The focus of the course, as structured, is to understand and record the accounting impacts of these operations. However, these

Table 3 Advanced accounting

Objective: To understand advanced accounting topics, using conceptual and regulatory issues currently in force in Brazil as a reference, in order to critically position oneself and produce information that supports the decision-making process of organizations		
Syllabus: Capital structure of companies. Corporate restructuring. Equity investments. Business combinations. Consolidation of financial statements. Equity compensation. Various transactions		
SDG	**Target**	
8	8.10	Strengthen the capacity of domestic financial institutions to encourage and expand access to banking, insurance and financial services for all
9	9.3	Increase the access of small-scale industrial and other enterprises, in particular in developing countries, to financial services, including affordable credit, and their integration into value chains and markets
12	12.6	Encourage companies, especially large and transnational companies, to adopt sustainable practices and to integrate sustainability information into their reporting cycle
17	17.1	Strengthen domestic resource mobilization, including through international support to developing countries, to improve domestic capacity for tax and other revenue collection

Source Research Data

transactions contribute to business expansion and strengthening, increasing productive capacity and market competitiveness, including the role of financial institutions in facilitating mergers and acquisitions, which is also related to target 9.3 of SDG 9 (Industry, Innovation, and Infrastructure). This target aims to ensure companies have access to financial services and affordable credit. Additionally, when discussing capital structure, it is relevant to reflect on public policies and best practices related to raising third-party capital, such as bank loans and financing, which stimulate business growth. In this context, accounting plays a key role in strategic decision-making, balancing debt and sustainable investments, thereby ensuring responsible economic development.

Finally, the subject can be linked to SDG 17 (Partnerships for the Goals), as the internationalization of companies is affected by exchange rate fluctuations. The adoption of international accounting standards facilitates the harmonization of financial statements across different countries, making global transactions smoother and attracting foreign investment (Mensah 2020). Exchange rate volatility can significantly impact the financial results of multinational companies. In this regard, target 17.1 suggests that companies seek international support to enhance national financial capacity, which directly affects exchange rate variations.

Figure 2 presents a summary of the results.

Following the analysis of these subjects, it is understood that there are strategies and methods to incorporate the SDGs into accounting courses in a specific and effective manner, aligning them with the syllabi and program content of curricular components. This perspective reflects an approach to be adopted by accounting educators, recognizing the field as one that can and should contribute to the global sustainable development agenda. In this regard, the effort of this research was to

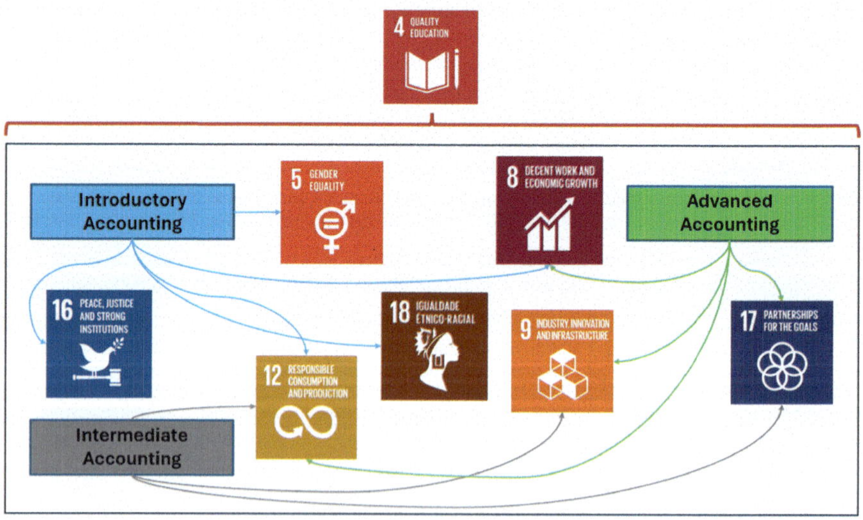

Fig. 2 Summary of results. *Source* Prepared by the Authors

provide strategies and possibilities to make the integration of SDGs increasingly common in the education of accounting students.

5 Conclusions

This study proposed strategies for integrating the SDGs into the three examined subjects, ensuring their alignment with the existing syllabus. Thus, the study contributes theoretically, practically, and socially to the field of investigation. Theoretically, it expands the understanding of the relationship between the SDGs and accounting education, highlighting how the curricular subjects and their syllabi can incorporate SDGs more directly. From a practical perspective, the study suggests that educators can utilize these proposed strategies to guide their teaching, making it easier to integrate SDG-related topics into lessons while maintaining a connection with disciplinary content. Socially, the study reinforces the importance of training accounting professionals who are more engaged with sustainability, contributing to a more responsible practice aligned with the challenges of achieving the UN's goals.

However, it is important to acknowledge the limitations of this study. First, the methodological choice to focus on three disciplines within the FBC curriculum proposal represents a deliberate scope. Beyond this selection, future studies could explore basic training courses or other professional training subjects, considering different curricula and institutional contexts. Another limitation arises from the fact that different higher education institutions face unique challenges in implementing these strategies due to the specific characteristics of each course and university. Nonetheless, this research serves as a starting point for such implementation. Future research is encouraged to test these strategies in classroom settings and assess their actual effectiveness. Finally, considering the recent implementation of new curricular guidelines for accounting programs by the Ministry of Education (2024), future studies could analyze how SDGs align with the competencies and skills outlined in these new directives.

Above all, this study represents an invitation—a call to reflection for accounting educators who believe in a teaching practice committed to sustainability, incorporating critical reflections on the socio-environmental impacts of accounting and encouraging practices that go beyond economic reductionism. It is an invitation to celebrate diversity, to foster an inclusive educational environment open to dialogue between different sociocultural perspectives. It is an invitation to create learning spaces that promote critical thinking and active participation, forming individuals and professionals who can contribute to a more pluralistic, sustainable, and socially just accounting practice.

References

Abhayawansa, S., Adams, C. A., & Neesham, C. (2021). Accountability and governance in pursuit of Sustainable Development Goals: Conceptualising how governments create value. *Accounting, Auditing and Accountability Journal*, 34(4), 923–945. https://doi.org/10.1108/AAAJ-07-2020-4667

Al-Muhanadi, M., Al-Fadhel, H., & Al-Jalahma, A. (2020). How can the Accounting Profession Contribute to the Reduction of CO2 Emissions in the GCC Region. *2020 Second International Sustainability and Resilience Conference: Technology and Innovation in Building Designs.* https://doi.org/10.1109/IEEECONF51154.2020.9319946

Alonso-Almeida, M. M., Navarrete, F. C. F. de, & Rodríguez-Pomeda, J. (2015). Corporate social responsibility perception in business students as future managers: A multifactorialanalysis. *Business Ethics: A European Review*, 24(1), 1–17. https://doi.org/10.1111/beer.12060

Alston, C., Flint, L., & Manley, S. (2023). A content analysis of the portrayal of under-represented minorities in introductory accounting textbooks. *Issues in Accounting Education* (in press). https://ssrn.com/abstract=4665069

Bak, I. (2024). Sustainable Development Goals (SDGs): Concept and measurement (pp. 15–41). In: Ziolo, M. (Ed.). *The Role of Financial Markets in Achieving the Sustainable Development Goals*. Edward Elgar Publishing. https://doi.org/10.4337/9781035323265.00009

Botes, V., Low, M., & Chapman, J. (2014). Is accounting education sufficiently sustainable? *Sustainability Accounting, Management and Policy Journal*, 5(1), 95–124. https://doi.org/10.1108/SAMPJ-11-2012-0041

Boyce, G. (2004). Critical accounting education: Teaching and learning outside the circle. *Critical Perspectives on Accounting*, 15(4/5), 565–586.https://doi.org/10.1016/S1045-2354(03)00047-9

Brown, J., & Dillard, J. (2019). Accounting education, democracy and sustainability: Taking divergent perspectives seriously. *International Journal of Pluralism and Economics Education*, 10(1), 24–45. https://doi.org/10.1504/IJPEE.2019.098168

Carnegie, G. D. (2022) Accounting 101: Redefining accounting for tomorrow. *Accounting Education*, 31(6), 615–628. https://doi.org/10.1080/09639284.2021.2014915

Carnegie, G., Parker, L., & Tsahuridu, E. (2021). It's 2020: What is accounting today? *Australian Accounting Review*, 31(1), 65–73. https://doi.org/10.1111/auar.12325

Carnegie, G., Ferri, P., Parker, L. D., Sidaway, S. I., & Tsahuridu, E. E. (2022). Accounting as technical, social and moral practice: The monetary valuation of public cultural, heritage and scientific collections in financial reports. *Australian Accounting Review*, 32(4), 460–472. https://doi.org/10.1111/auar.12371

Carneiro, J. D. (Coord.), Rodrigues, A. T. L., Silva, A. C. R. da, França, J. A. de, Almeida, J. E. F. de, & Morais, M. L. S. de. (2017). *Matriz curricular para cursos de ciências contábeis [Curriculum matrix for accounting science courses]*. Fundação Brasileira de Contabilidade [Brazilian Accounting Foundation].

Chabrak, N., & Craig, R. (2013). Student imaginings, cognitive dissonance and critical thinking. *Critical Perspectives on Accounting*, 24(2), 10–91. https://doi.org/10.1016/j.cpa.2011.07.008

Christensen, L. J., Peirce, E., Hartman, L. P., Hoffman, H. W., & Carrier, J. (2007). Ethics, CSR, and sustainability education in the Financial Times top 50 global business schools: Baseline data and future research directions. *Journal of Business Ethics*, 73(4), 347–368. https://doi.org/10.1007/s10551-006-9211-5

Chua, W. F. (1986). Radical developments in accounting thought. *The Accounting Review*, 61(4), 601–632.

Chugumbaev, R. R. (2024). Conceptual Aspects of Accounting Transformation in the Context of the Imperatives of Sustainable Development. *Journal of Volgograd State University. Economics*, 26(2), 176–186. https://doi.org/10.15688/ek.jvolsu.2024.2.14

Cooray, T., Senaratne, S., & Gunarathne, N. (2022). Engagement with Sustainable Development Goals in accounting education: The case of a public university in Sri Lanka. In: Öztürk, M. (Ed.).

Engagement with Sustainable Development in Higher Education. Sustainable Development Goals Series. Springer.

Dagiliute, R., Liobikien, G., & Minelgait, A. (2018). Sustainability at universities: Students' perceptions from green and non-green universities. *Journal of Cleaner Production*, 181, 473–482. https://doi.org/10.1016/j.jclepro.2018.01.213

Dean, B., Gibbons, B., & Perkiss, S. (2018). An experiential learning activity for integrating the United Nations Sustainable Development Goals into business education. *Social Business*, 8(4), 387–409. https://doi.org/10.1362/204440818X15445231830058

Falivena, C., & Gulluscio, C. (2023). Integrating SDGs in accounting education: Evidence from Italian universities. In: Idowu, S. O., & Zu, L. (Eds). *The Elgar Companion to Corporate Social Responsibility and the Sustainable Development Goals*. Edward Elgar Publishing. https://doi.org/10.4337/9781803927367.00013

Fanea-Ivanovici, M., & Baber, H. (2022). Sustainability at universities as a determinant of entrepreneurship for sustainability. *Sustainability*, 14(1). https://doi.org/10.3390/su14010454

Ferreira, C. D., Silva, E. G. C., & Crispim, G. (2023). Inserção do tema sustentabilidade na grade curricular dos cursos de ciências contábeis a partir dos parâmetros dos ODS [Inclusion of the sustainability theme in the curriculum of accounting courses based on SDG parameters]. *Ambiente & Educação*, 28(2), 1–29. https://doi.org/10.14295/ambeduc.v28i2.15279

Fota, M.-S., Cristea, N.-E., Ureche, A., & Albu, N. (2024). Accountants' competencies for sustainability reporting: An exploratory study. *Journal of Accounting and Management Information Systems*, 23(2), 446–460. https://doi.org/10.24818/jamis.2024.02007

Godemann, J., Haertle, J., Herzig, C., & Moon, J. (2014). United Nations supported principles for responsible management education: Purpose, progress and prospects. *Journal of Cleaner Production*, 62, 16–23. https://doi.org/10.1016/j.jclepro.2013.07.033

International Federation of Accountants (2024). *Strategic plan*. https://ifacweb.blob.core.windows.net/publicfiles/2024-01/IFAC-Strategic-Plan-2024_0.pdf

Institute of Singapore Chartered Accountants, Ernst & Young, Singapore Management University, & Singapore Accountancy Commission. (2022). *Sustainability: Jobs and skills for the accountancy profession*. https://isca.org.sg/docs/default-source/i-p--our-future-together/sustainability-jobs-and-skills-for-the-accountancy-sector.pdf?sfvrsn=f6af2dae_0

Lee, W. E., Birkey, R. N., & Patten, D. M. (2017). Exposing students to environmental sustainability in accounting: An analysis of its impacts in a US setting. *Social and Environmental Accountability Journal*, 37(2), 81–96. https://doi.org/10.1080/0969160X.2016.1270225

Makhura, M. N. (2024). *Sustainable Development Goal 17: Partnerships for the goals*. Edward Elgar Publishing. https://doi.org/10.4337/9781839105449.00043

Mensah, E. (2020). The effect of IFRS adoption on financial reporting quality: Evidence from listed manufacturing firms in Ghana. *Economic Research-Ekonomska Istraživanja, 33*(1), 2890–2905. https://doi.org/10.1080/1331677X.2020.1860109

Ministry of Education. (2024). *Resolução CNE/CES nº 1, de 27 de março de 2024 [CNE/CES Resolution No. 1, of March 27, 2024]*. Institui as Diretrizes Curriculares Nacionais do Curso de Graduação em Ciências Contábeis [Establishes the National Curricular Guidelines for the Undergraduate Course in Accounting Sciences]. http://portal.mec.gov.br/index.php?option=com_docman&view=download&alias=257031 rces001 24&category_slug=marco 2024&Itemid=30192

Ministry of Racial Equality. (2024). *ODS 18—Igualdade Étnico-Racial [SDG 18—Ethnic-Racial Equality]*. https://www.gov.br/igualdaderacial/pt-br/assuntos/ods18

Mohanty, A., Alam, A., & Mohanty, A. K. (2024). Education for Sustainable Development (ESD) and Global Citizenship for India. *Journal of Education for Sustainable Development Studies*, 1(2), 134–159. https://doi.org/10.70232/jesds.v1i2.17

Perim, L. F., Lima, C. A. de, Ventura, J., Scarton, J., Brum, A. N., & Paula, S. F. de. (2020). The hidden curriculum and its relevance in vocational education. *Research, Society and Development*, 9(2), e101922050. https://doi.org/10.33448/rsd-v9i2.2050

Power M. (1992). After calculation? Reflections on critique of economic reason by André Gorz. *Accounting, Organizations and Society*, 17(5), 477–499. https://doi.org/10.1016/0361-3682(92)90043-R

Powell, L., & McGuigan, N. (2024). Looking within: Cultivating compassion for shaping sustainable mindsets in accounting education. *Meditari Accountancy Research*, 32(5), 1787–1808. https://doi.org/10.1108/MEDAR-07-2023-2082

Procópio, R. B., & Cortez, G. T. (2023). O importante papel do plano de aula na prática pedagógica [The important role of the lesson plan in pedagogical practice]. *Instrumento: Revista de Estudos e Pesquisas em Educação, 25*(3), 620–632.https://doi.org/10.34019/1984-5499.2023.v25.44937

Ratnaningsih, S. D. A., Ghozali, I., & Harto, P. (2024). Antecedents of students' intention to be sustainable accountants: Evidence from Indonesia. *Arab Gulf Journal of Scientific Research*, 42(4), 1950–1968. https://doi.org/10.1108/AGJSR-08-2023-0366.

Rizos, V., Elkerbout, M., & Egenhofer, C. (2019). C*ircular economy for climate neutrality: Setting the priorities for the EU*. CEPS Policy Brief.

Roma, J. C. (2019). Os Objetivos de Desenvolvimento do Milênio e sua Transição para os Objetivos de Desenvolvimento Sustentável [The Millennium Development Goals and their Transition to the Sustainable Development Goals]. *Ciência & Cultura*, 71(1). https://doi.org/10.21800/2317-66602019000100011

Santos, I. O., & Coimbra, C. L. (2018). Currículo. In: Miranda, G. J., Leal, E. A., Casa Nova, S. P. de C. *Revolucionando a docência universitária: Orientações, experiências e teorias para a prática docente em negócios [Revolutionizing university teaching: Guidelines, experiences and theories for teaching practice in business]*. 1. ed. Atlas.

Santos, M. A., Melo, M. C. O. L., & Batinga, G. L. (2021). Representatividade da mulher contadora em escritórios de contabilidade e a desigualdade de gênero na prática contábil: Uma questão ainda em debate? [Representation of female accountants in accounting firms and gender inequality in accounting practice: An issue still under debate?]. *Sociedade, Contabilidade e Gestão, 16*(1), 149–163. https://doi.org/10.21446/scg_ufrj.v0i0.30679

Silva, A. R. P. da, Nonato, K. E. C., Nascimento, I. C. S. do, Miranda, M. C. de, & Costa, W. P. L. B. da. (2023). Mulheres em evidência: desafios e perspectivas da mulher contabilista [Women in the spotlight: challenges and perspectives for female accountants]. *Qualitas Revista Eletrônica, 21*(1), 1–21. https://revista.uepb.edu.br/QUALITAS/article/view/2597

Silva, F. X. (2021). Cultura Escolar e as Especificidades dos Currículos Ocultos [School Culture and the Specificities of Hidden Curricula]. *Revista Tópicos Educacionais*, 27(1), 195–209. https://doi.org/10.51359/2448-0215.2021.249424

Sobkowiak, M., Cuckston, T., & Thomson, I. (2020). Framing sustainable development challenges: accounting for SDG-15 in the UK. *Accounting, Auditing and Accountability Journal*, 33(7), 1671–1703. https://doi.org/10.1108/AAAJ-01-2019-3810

Sytnik, O. E., Kulish, N. V., Tunin, S. A., Frolov, A. V., & Germanova, V. S. (2021). Accounting as a tool for achieving global sustainable development goals. In: Bogoviz, A. V. (Ed.). *The Challenge of Sustainability in Agricultural Systems*. Springer. https://doi.org/10.1007/978-3-030-72110-7_75

Tegofack, B. C., & Kamgang Fodjo, G. J. (2021). Stakeholders' engagement in the sustainability reporting of companies: A study within the context of Cameroon. *Revue du Contrôle, de la Comptabilité et de l'Audit, 5*(3). Retrieved from https://www.revuecca.com/index.php/home/article/view/728

Twyford, E. J., Musundwa, S., Tanima, F. A., & George, S. (2024). Bridging the gap: Sustainable development goals as catalysts for change in accounting education and society. *Meditari Accountancy Research*, 32(5), 1758–1786. https://doi.org/10.1108/MEDAR-02-2024-2375

United Nations. (2024). *The Sustainable Development Goals Report 2024*. United Nations. https://unstats.un.org/sdgs/report/2024/

United Nations. (2025). *Teach SDGs*. https://www.un.org/pt/teach/SDGs

Wang, P., Che, F., Fan, S., & Gu, C. (2014). Ownership governance, institutional pressures and circular economy accounting information disclosure. *Chinese Management Studies, 8*(3), 487–501. https://doi.org/10.1108/CMS-10-2013-0192

Wright, N. S., & Bennett, H. (2011). Business ethics, CSR, sustainability and the MBA. *Journal of Management & Organization*, 17(5), 641–655. https://doi.org/10.5172/jmo.2011.17.5.641

Xu, L., Xie, L., Mei, S., Hao, J., Zhang, Y., & Song, Y. (2024). Corporate sustainability reporting and stakeholders' interests: Evidence from China. *Sustainability, 16*(8), 1–19. https://doi.org/10.3390/su16083443

Bruno Barbosa de Souza is a Ph.D. student in Accounting Sciences at UFU, where he also earned his master's degree in the same field (2023). He holds a postgraduate certificate in Higher Education Teaching with Active Learning Methodologies from Faculdade Descomplica (2022), and a bachelor's degree in Accounting Sciences from UNIFESP (2019). He has teaching experience in vocational and higher education programs in the business field and is currently a substitute professor in the Administration and Accounting Sciences at UFU. He also works independently with critical review, Portuguese language editing, and formatting of academic papers. Passionate about academic research and scientific writing, his current research interests include accounting education (in general), focusing in active learning methodologies, financial reporting, and sustainability-related issues.

Deivson Vinicius Barroso is a Ph.D. student in Accounting Sciences at UFU. Holds a master's degree in Accounting from UFBA and a bachelor's degree in accounting sciences from UFVJM. Also holds a bachelor's degree in theology from Faculdade Unida de Vitória and a postgraduate specialization in Higher Education Teaching from UCAM. Has experience as a Higher Education Professor in managerial-related courses, having taught at Centro Universitário Doctum (2015), IFNMG (2017), UNIFACS (2020), UNESPAR (2022), and UNIPAC (2023). Experienced in distance education as a mediator professor (UFBA), content developer (UFBA, Universidade La Salle), and distance tutor (UFBA). Author of textbooks in the areas of Accounting Theory, Management Control, Public Finance, and Governmental Accounting. Research interests include Accounting Education, Accounting Education for Sustainability, Environmental Accounting, Integrated Reporting, Sustainability, Accounting Theory, and the History of Accounting.

Sustainability Throughout the Curriculum of the Chemistry and Chemical Engineering Courses at the Institute of Chemistry—UNESP

Lorena Oliveira Pires, Kelly Johana Dussan Medina, and Cintia Duarte de Freitas Milagre

Abstract This chapter aims to critically analyze how sustainability is integrated throughout the Chemistry and Chemical Engineering curricula at the Institute of Chemistry, São Paulo State University (UNESP) in Araraquara, Brazil. It aims to identify the main strategies adopted, evaluate their expected impact on student training, and discuss the potential for replicating this curricular model in other higher education institutions. Implemented in 2023, the Chemical Engineering curriculum at UNESP in Araraquara marks a transformative step toward sustainability, positioning the university at the forefront of training future engineers to address global challenges. This restructured pedagogical approach goes beyond traditional coursework, embedding sustainability as a core value in practical activities, outreach projects and an enriched list of subjects. Core subjects such as Environment and Sustainability I, II and III will offer an in-depth exploration of sustainable practices and resource management. However, even before these specialized courses become available in 2026, sustainability principles have already been integrated throughout the curriculum. From the outset, students engage with the United Nation's Sustainable Development Goals (SDGs) and learn about renewable materials, recycling and optimizing production processes. By integrating these themes early on, the course fosters a critical and innovative mindset, enabling future chemical engineers to develop sustainable solutions as they progress through their studies. This proactive approach reflects UNESP's commitment not only to technical excellence, but also to developing professionals capable of driving meaningful change toward a more sustainable world.

L. O. Pires (✉) · K. J. D. Medina
Department of Chemical Engineering, Institute of Chemistry, São Paulo State University (UNESP), Araraquara, Brazil
e-mail: lorena.pires@unesp.br

K. J. D. Medina
e-mail: kelly.medina@unesp.br

C. D. de Freitas Milagre
Department of Biochemistry and Organic Chemistry, Institute of Chemistry, São Paulo State University (UNESP), Araraquara, Brazil
e-mail: cintia.milagre@unesp.br

W. Leal Filho et al. (eds.), *Sustainability in the Curriculum and Operations of Universities*, World Sustainability Series,
https://doi.org/10.1007/978-3-032-12678-8_10

The emphasis on sustainability throughout the curriculum ensures that graduates are equipped not only for success in a rapidly evolving job market, but also for leadership in addressing the environmental and social challenges of the future. The chapter concludes by highlighting the main findings, implications for curriculum development in higher education, and the challenges faced in implementing sustainability as a cross-cutting theme.

Keywords Curriculum innovation · Global challenges · Interdisciplinary learning

1 Introduction

Sustainability is one of the main challenges of the twenty-first century and has become an indispensable topic in the training of professionals in several fields, particularly in Chemical Sciences and Engineering (Allen et al. 2021; Anastas et al. 2021). Global population growth, coupled with an increased demand for natural resources, calls for an urgent reassessment of production and consumption models to minimize environmental impacts and promote sustainable development (United Nations 2015). The main objective of this chapter is to provide a comprehensive overview of the curricular changes, analyze their alignment with global sustainability goals, and discuss their implications for training future professionals.

In this context, universities play a crucial role, as they are responsible for training future professionals who will work in industry, research and policymaking. Integrating sustainability into higher education has been widely discussed as a strategy to align academic training with global needs (Lozano et al. 2013; Wals and Jickling 2002).

The Institute of Chemistry in Araraquara (IQAr) at São Paulo State University (UNESP) has demonstrated its commitment to this agenda by incorporating sustainability principles into the curricula of both the Chemistry and Chemical Engineering undergraduate programs (Milagre et al 2024). The recent curricular restructuring, initiated in 2023, reflects this commitment by promoting the inclusion of content and methodologies that emphasize sustainable practices from the first semester, in both mandatory and elective disciplines.

Thus, this paper aims to present the curricular structure of these courses and highlights how sustainability is integrated into their content, methodologies and practical activities. It also discusses the expected impact of this approach on student training and the potential for replicating this model in other educational institutions.

2 Sustainability in Chemistry and Chemical Engineering Education

The demand for professionals with sustainability-focused training has increased significantly in recent decades, driven by factors such as the depletion of natural resources, climate change and the need to reduce the environmental impacts of industrial activities (Rockström et al. 2009). Sectors such as chemicals, energy, and technology face major challenges in balancing economic growth and environmental responsibility, requiring trained professionals to have not only solid technical knowledge but also a critical understanding of sustainable practices and their long-term impacts.

Companies in various sectors are currently adopting stricter environmental guidelines, driven by regulatory requirements and increasing pressure from markets and consumers. Innovative solutions are being sought to reduce carbon emissions, minimize waste, optimize the use of raw materials, and increase energy efficiency. Chemical engineers and chemists with a background in sustainability are increasingly valued in this scenario (Azapagic et al. 2006).

The United Nations 2030 Agenda and the 17 Sustainable Development Goals (SDGs) provide guidelines that directly impact the field of Chemistry and Chemical Engineering (UNEP 2024). Among the most relevant SDGs for these areas are:

- SDG 7—Affordable and clean energy: Encouraging the development of more efficient, sustainable energy processes, such as biofuels and renewable energy sources.
- SDG 9—Industry, innovation and infrastructure: Fostering innovations in the chemical industry that reduce environmental impacts through cleaner production technologies.
- SDG 12—Responsible consumption and production: Promoting strategies for reducing raw material waste, minimizing industrial waste, and encouraging the reuse and recycling of chemical products.
- SDG 13—Climate action: Emphasizing the adoption of practices that reduce greenhouse gas emissions through cleaner and more efficient production processes (United Nations 2015).

Incorporating these principles into higher education has become essential to prepare future professionals for contemporary challenges. Prestigious universities, such as the Massachusetts Institute of Technology (MIT), Harvard University, the Federal Institute of Technology (ETH Zürich), and the University of Cambridge, have already integrated the SDGs into their curricula, preparing engineers and scientists to address complex environmental problems (Cortese 2003).

In Brazil, some universities are also adapting their curricula to include sustainability-related content, although this process still faces challenges (Silva and Freire 2023). Integration goes beyond adding specific courses it involves a broader methodological shift, including active, problem-based learning (PBL), which allows students to address real-world sustainability problems, thus fostering a proactive approach to environmental solutions (Thomas 2004).

3 Restructuring the Chemistry and Chemical Engineering Undergraduate Courses at UNESP Araraquara

UNESP Araraquara underwent a major curriculum restructuring in 2023, aligning academic training with sustainability and the institutionalization of outreach education. The reform aimed to modernize teaching and embed environmental and social dimensions into academic training, reinforcing UNESP's commitment to the SDGs (United Nations 2015). The initiative seeks to prepare students for a global scenario defined by environmental challenges, such as climate change, resource scarcity, and a push toward cleaner industrial practices.

Figure 1 illustrates the integration of sustainability principles into the curricula of UNESP's Chemical Engineering and Chemistry programs. It highlights how the main curricular components are aligned with institutional goals and expected learning outcomes.

The restructured curricula promote comprehensive education by combining advanced technical knowledge with a critical understanding of sustainability. This

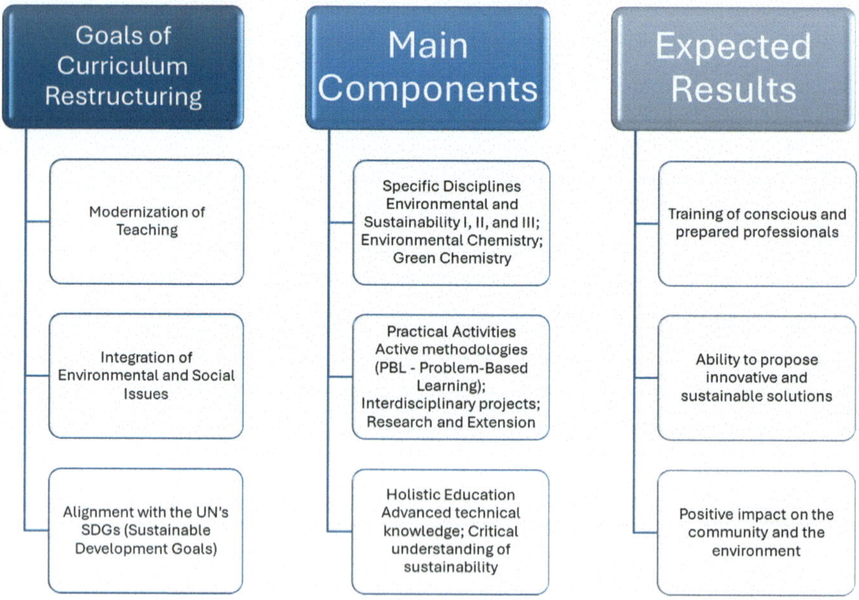

Fig. 1 Integration of sustainability into the chemical engineering and chemistry curricula—UNESP Araraquara. *Source* Author (2025)

includes specific subjects, interdisciplinary projects, and practical activities that immerse students in real-world industry and societal challenges. The curricular inclusion of outreach activities fosters collaborative, dialogical knowledge construction between the university and the wider community, supported by active teaching methodologies. In addition, integrating innovative methodologies, such as project-based learning (PBL), strengthens the connection between theory and practice, encouraging students to develop creative and sustainable solutions (Thomas 2004).

3.1 Integration of Sustainability into the Chemical Engineering Curriculum

The Chemical Engineering program at the Institute of Chemistry, UNESP Araraquara, aims to train highly qualified professionals to work in strategic sectors of the chemical and transformation industries. These sectors include petrochemicals, pharmaceuticals, food, cosmetics, biotechnology, and energy. In addition to providing a solid foundation in the exact sciences and engineering, the program is designed to develop skills that enable future chemical engineers to address contemporary challenges, such as reducing environmental impacts, optimizing industrial processes, and implementing sustainable practices in various production sectors.

In recent years, growing concern about environmental issues and the transition to a more sustainable economy have underscored the need to revise academic curricula to incorporate sustainability more comprehensively. In response, UNESP underwent a major reform of the Chemical Engineering program, leading to a new curriculum being implemented in 2023. This reform includes a more explicit and in-depth focus on environmental issues and sustainable practices, preparing students to meet the demands of a job market that increasingly values social and environmental responsibility.

One of the most significant changes in the new curriculum is the introduction of a three-part course entitled Environment and Sustainability I, II, and III. These courses, scheduled for the final years of the course, are designed to provide students with comprehensive training in fundamental sustainability concepts, environmental regulations, clean technologies, circular economy, and applying sustainable methodologies in the chemical industry. However, due to the gradual implementation of the new curriculum, these courses will be offered for the first time in the second semester of 2026.

Although there is not yet concrete data on the impact of these changes on student training, expectations are positive regarding the development of sustainability-oriented competencies. It is anticipated that, upon graduation, chemical engineers from UNESP Araraquara will be better equipped to work in environmental management, develop more efficient processes, and propose innovative solutions that minimize negative environmental impacts.

In addition to specialized sustainability courses, sustainability principles are also integrated transversally across various subjects throughout the program. From the early semesters, students engage with environmental content, such as in Introduction to Chemical Engineering, where challenges within the profession and its relationship with environmental and social issues are discussed. Analytical Chemistry introduces essential techniques for environmental monitoring and pollution control, while Industrial Processes explores strategies for optimizing chemical processes and reducing waste.

To enhance practical learning, the program promotes active methodologies and hands-on activities that immerse students in real-world industrial challenges. Interdisciplinary projects, case studies, and participation in research and community outreach initiatives are encouraged to help students apply classroom concepts in practical settings. Moreover, students are increasingly encouraged to engage in sustainable innovation efforts, such as developing technologies for waste reuse, improving energy efficiency, and designing new, less polluting and safer materials.

The reform of the Chemical Engineering curriculum at UNESP reflects a continuing commitment to aligning academic training with the demands of the twenty-first century. While implementation is still in progress, the new curriculum is expected to help prepare professionals who are equipped to face global environmental challenges and contribute to a more sustainable and responsible chemical industry.

3.2 Integration of Sustainability in the Chemistry Curriculum

The Chemistry program at the Institute of Chemistry, UNESP Araraquara, offers three distinct degree tracks: the Bachelor's Degree in Chemistry, the Bachelor's Degree in Technological Chemistry and the Bachelor's Degree in Chemistry focused on Teacher Education. Each path is tailored to meet different professional profiles. Regardless of the specific program, one of the foundational pillars is Green Chemistry. It is integrated into both the curriculum and the teaching methodologies adopted.

Green Chemistry is an emerging field that aims to reduce the environmental impact of chemical processes by developing cleaner and more efficient technologies. This approach aligns with the SDGs, particularly SDG 12—Responsible Consumption and Production—and has been increasingly embraced by higher education institutions worldwide. At UNESP, integrating Green Chemistry into the curriculum reflects the institution's commitment to training professionals who can contribute to a circular economy with reduced environmental harm.

Furthermore, the Institute of Chemistry became a signatory of the Green Chemistry Commitment in 2019, a global initiative encouraging higher education institutions to embed sustainability into their curricula through Green Chemistry. This reinforces UNESP's dedication to fostering educational practices grounded in environmental and social responsibility.

Among the changes introduced in the 2023 curriculum reform is the inclusion of subjects such as an Introduction to the Study of Chemistry, Environmental Chemistry, Circular Economy, and Green Chemistry. These courses offer students a critical understanding of the environmental impacts of the chemical industry, along with strategies to mitigate them. The Chemistry program is particularly notable for focusing on teacher education, preparing future educators to disseminate sustainability concepts in secondary schools. Courses focused on environmental education and the didactics of Green Chemistry equip future teachers to act as change agents in their communities, fostering a culture of socio-environmental responsibility from an early stage.

In addition to these specific subjects, the new curriculum for both the Bachelor's Degree in Chemistry and the Bachelor's Degree in Technological Chemistry includes the Integrative Teaching, Research, and Extension Project (PIEPEX). This initiative is designed to develop students' theoretical and practical skills for solving real-world problems, consolidating the foundational knowledge acquired by the third year of study. It also promotes integration with research and community outreach efforts for the benefit of society.

PIEPEX is divided into three modules. Module III includes a semester-long extension project (4 credits) focused on environmental awareness developed for and with the community. It also features an 11-credit, year-long extension project entitled Chemistry for Society, which addresses real issues raised by the community and must align with one or more of the 17 UN SDGs within the field of Chemistry. Due to the phased implementation of the new curriculum, PIEPEX will be offered for the first time in the second semester of 2025.

4 Sustainability Integrated into the Curriculum

4.1 Subjects and Content Related to Sustainability

Sustainability is integrated into the curriculum of the Chemical Engineering and Chemistry courses at UNESP Araraquara through specific subjects and a transversal approach that permeates various areas of knowledge. In the Chemical Engineering course, the subjects Environment and Sustainability I, II and III play a central role, covering topics such as environmental management, pollution control, waste treatment and applying sustainable methodologies in the chemical industry. These subjects are complemented by practical activities that bring students closer to real problems faced by industry, such as waste reuse.

In the Chemistry course, sustainability is addressed even more explicitly through subjects such as Environmental Chemistry, Circular Economy and Green Chemistry. These subjects provide students with a critical understanding of the environmental impacts of chemical processes and encourage adopting innovative practices to minimize these impacts. In addition, the integration of sustainability occurs transversally in subjects such as Introduction to the Study of Chemistry and Industrial Processes, where students learn to apply sustainability principles in different contexts.

4.2 Integration with Research and Outreach

Integration between teaching, research, and outreach is one of the pillars of the approach adopted by IQAr. Extension projects such as "Implementation and Benefits of Composting Processes in Urban and/or Rural Gardens" and "Compost and Cultivate: Environmental Education for Sustainability" allow students to apply classroom knowledge to address real environmental challenges. These initiatives not only help reduce environmental impact but also raise community awareness of sustainable practices.

Figures 2 and 3 illustrate the practical implementation of these extension projects, showing the development of a compost bin in a community garden and environmental education activities conducted with local residents. These images highlight the interaction between students and community members.

Throughout these projects, students had the opportunity to work directly with the community. As shown in Fig. 3, environmental education activities were also carried out in local schools.

These initiatives exemplify how the practical application of sustainability concepts can generate positive impacts on both the community and the environment, reinforcing IQAr's commitment to promoting sustainable practices.

Another notable example is the project entitled "Disposal of Cooking Oil: Diagnosis, Environmental Education, and Recycling," which helps to address the pressing issue of environmental pollution and proposes practical solutions, such as producing

Fig. 2 Chemical Engineering students and community members while implementing the community garden. *Source* Personal Collection (2023)

Fig. 3 Environmental education activity with students and children at a local school. *Source* Personal Collection (2024)

soap from recycled oil. This project demonstrates how circular economy can be applied in a simple and effective way, encouraging the adoption of sustainable practices at a local level.

5 Impact and Future Prospects

The integration of sustainability into the Chemistry and Chemical Engineering curricula at UNESP Araraquara represents a significant advancement in preparing professionals to address contemporary environmental challenges. It is expected that, as new courses and methodologies are implemented, students will develop essential skills to propose innovative and sustainable solutions within their areas of expertise.

Moreover, the model adopted by IQAr can serve as a reference for other higher education institutions aiming to incorporate sustainability into their curricula. Replicating this approach across different academic contexts can contribute to more globally aligned training, supporting the transition to a more sustainable and resilient economy.

To evaluate the impact of the curriculum reform, several indicators will be used, including the level of student involvement in sustainability-related research and outreach projects, graduates' perceptions of the relevance of the content covered, and the performance of trained professionals in the job market. These indicators will help identify strengths and areas for improvement, ensuring that the curriculum continues to evolve in response to the needs of both society and the labor market.

6 Conclusion

The restructuring of the Chemistry and Chemical Engineering programs at the Institute of Chemistry, UNESP Araraquara, is a significant milestone within the broader movement to integrate sustainability into higher education. By embedding concepts such as green chemistry, environmental management, and the circular economy throughout the curricula, the institute demonstrates its commitment to preparing professionals equipped to address global sustainability challenges. This comprehensive reform not only reflects current educational trends but also highlights the expected impact on student training: fostering a new generation of professionals who are environmentally aware, capable of critical thinking, and ready to contribute to more sustainable development at local and global levels.

Although the new curricula are still in the early stages of implementation, expectations are high for the development of professionals who are more environmentally aware and better prepared to address global challenges. UNESP's approach—rooted in active learning methodologies, interdisciplinarity, and community engagement—serves as an inspiring model for other higher education institutions. The implications of this reform extend beyond technical training: they promote interdisciplinary learning, active methodologies, and community engagement, which are essential for developing critical and innovative thinking. However, the process is not without limitations. The full impact of these changes will only be measurable after several years of implementation, and ongoing monitoring will be necessary to identify areas for improvement.

Looking ahead, monitoring the outcomes of this curricular reform will be essential for assessing its impact on student training and the university's broader contribution to a more sustainable world. Sustainability must remain a central pillar of academic education, reflecting the urgent demands of our time and fostering a cultural shift toward more responsible and innovative practices.

Future prospects include continuously evaluating student outcomes, expanding partnerships with local communities, and sharing best practices with other institutions. These efforts will contribute to the ongoing evolution of sustainability education in Brazil and internationally.

Acknowledgements The authors gratefully acknowledge the financial support provided by the National Council for Scientific and Technological Development (CNPq), grant number #316230/2023-5, which enabled the professional English language revision of this chapter.

References

Allen D T, Carrier D J, Chen J, Gathergood N, Garcia J M, Gong J, Han H, Kuok (Mimi) Hii K, Hwang B-J, James A L, Jha M, Licence P, Marr A C, Meier M, Moores A, Nakamura R, Pradeep T, Rossi L, Sels B, Subramaniam B, Tam M K C, Zhuang L, Serrano J F (2021) Expectations

for perspectives. ACS Sustain Chem Eng 9(49):16528–16530 https://doi.org/10.1021/acssuschemeng.1c07865
Anastas P, Nolasco M, Kerton F, Kirchhoff M, Licence P, Pradeep T, Subramaniam B, Moores A (2021) The power of the United Nations Sustainable Development Goals. ACS Sustain Chem Eng 9(24):8015–8017. https://doi.org/10.1021/acssuschemeng.1c03762
Azapagic A, Millington A, Collett A (2006) A methodology for integrating sustainability considerations into process design. Chemical Engineering Research and Design, 84(6A), 439–452.
Cortese A D (2003) The critical role of higher education in creating a sustainable future. Planning for Higher Education 31(3):15–22.
Lozano R, Lukman R, Lozano F J, Huisingh D, Lambrecht W (2013) Declarations for sustainability in higher education: Becoming better leaders, through sustaining leadership. J Clean Prod 48:10–19. https://doi.org/10.1016/j.jclepro.2012.10.030
Milagre C D F, Milagre H M S, Silva D H S, Cavalcanti V H O, Ganeo Neto L (2024) Paths and synergies in accelerating the UN 17 SDGs through the lens of green chemistry: Contributions from a Brazilian university and its Institute of Chemistry. Pure Appl Chem 94(9):1279–1290. https://doi.org/10.1515/pac-2024-0232
Rockström J, Steffen W, Noone K, Persson A, Stuart Chapin III F, Lambin E F, Lenton T M, Scheffer M, Folke C, Schellnhuber H J, Nykvist B, de Wit C A, Hughes T, van der Leeuw S, Rodhe H, Sorlin S, Snyder P K, Costanza R, Svedin U, Falkenmark M, Karlberg L, Corell R W, Fabry V J, Hansen J, Walker B, Liverman D, Richardson K, Crutzen P, Foley J A (2009) A safe operating space for humanity. *Nature* 461(7263):472–475. https://doi.org/10.1038/461472a
Silva N O, Freire F S (2023) Teaching for sustainability in Brazilian higher education from the perspective of the Sustainable Development Goals [Ensino para a sustentabilidade no ensino superior brasileiro sob a perspectiva dos Objetivos de Desenvolvimento Sustentável]. *Rev Bras Educ* 28:e280080. https://doi.org/10.1590/S1413-24782023280080
Thomas I (2004) Sustainability in tertiary curricula: What is stopping it happening? Int J Sustain High 5(1):33–47. https://doi.org/10.1108/14676370410517947
United Nations (2015) Transforming our world: The 2030 agenda for sustainable development. Nova York: United Nations. https://sdgs.un.org/2030agenda
United Nations Environment Program (UNEP) (2024) Green and sustainable chemistry. https://www.unep.org/topics/chemicals-and-pollution-action/circularity-sectors/green-and-sustainable-chemistry
Wals A E J, Jickling B (2002) "Sustainability" in higher education: From doublethink and newspeak to critical thinking and meaningful learning. Int J Sustain High Edu 3(3):221–232. https://doi.org/10.1108/14676370210434673

Lorena Oliveira Pires holds a degree in Chemical Engineering from the Federal University of São Carlos (2003), a master's in Hydraulics and Sanitation from the University of São Paulo (2006), and a PhD in Sciences from the São Carlos School of Engineering, USP (2010). She is currently a professor at the Institute of Chemistry in Araraquara, São Paulo State University (UNESP), where she teaches Unit Operations I, II, and III. She coordinates the undergraduate Chemical Engineering program and is a faculty member in both the Graduate Program in Chemistry and the National Professional Master's in Chemistry (PROFQUI). Her research lies in Environmental Engineering, especially biological wastewater treatment, focusing on surfactants, glycerol, biogas, anaerobic degradation, and bioreactors. She is also actively involved in outreach programs and projects aimed at engaging undergraduate students in sustainability and outreach activities.

Kelly Johana Dussan Medina holds a Ph.D. in Sciences specializing in Applied Microbiology from the School of Engineering of Lorena (USP, 2013), and an M.Sc. and Bachelor's degree in Chemical Engineering from the National University of Colombia. She is an Assistant

Professor at the Department of Chemical Engineering, Institute of Chemistry, UNESP Araraquara. Her academic and scientific career is dedicated to developing innovative and sustainable solutions for the chemical and biotechnological sectors, with experience in lignocellulosic biomass valorization and the production of advanced biofuels and bioproducts. Her research covers yeast and enzyme immobilization, magnetic nanostructures, reaction-extraction processes, and thermal biomass treatments for biochar production. She is also involved in process modeling and simulation and actively participates in outreach projects promoting sustainability and the circular economy.

Cintia Duarte de Freitas Milagre holds a degree in Pharmacy from the Federal University of Minas Gerais (Brazil) and earned both her M.Sc. and Ph.D. in Chemistry from the University of Campinas (Brazil), where she also completed postdoctoral research. She is currently an Associate Professor in the Department of Biochemistry and Organic Chemistry at the Institute of Chemistry, UNESP Araraquara, where she also advises graduate students in the Chemistry program. She co-leads the Milagreen Lab, which conducts research in Organic Chemistry and Biotechnology, strongly focusing on Green Chemistry, Sustainable Chemistry, and Biocatalysis. Her main research interests include enzymatic catalysis for the synthesis of bioactive chiral molecules and the enzymatic degradation of organic compounds. She held the position of vice-coordinator of the Bachelor's Degree in Chemistry, the Bachelor's Degree in Technological Chemistry and the Bachelor's Degree in Chemistry focused on Teacher Education between 2020 and 2022. Dr. Milagre serves as the Green Chemistry Commitment representative at the UNESP Institute of Chemistry and is an ambassador for the Green Chemistry Teaching and Learning Community (GCTLC) by Beyond Benign. She is also an active member of the Brazilian Chemical Society and the American Chemical Society.

A Theoretical Sustainability SWOT Analysis for Aligning the University of São Paulo's Operations and Curriculum with Sustainable Development Goals

Christian Eric Barrantes-Briceño and Marcelo Seido Nagano

Abstract Universities serve as engines of social and cultural development, acting as centers of teaching and learning and influencing regional growth and environmental transformation. They also play a critical role in advancing sustainable development, including the Sustainable Development Goals (SDGs) and broader sustainability agendas. However, university operations, management, and curriculum are often misaligned with sustainability goals established, thus requiring analyses of such a disconnection and exploration of potential solutions. This study identified and categorized internal and external factors that facilitate or hinder the alignment of university operations and curriculum with SDGs, focusing on the University of São Paulo (USP). It is based on a comprehensive literature review that identified sustainability-related factors in the context of university operations and curriculum. The factors were filtered and structured through a Sustainability SWOT analysis (S-SWOT), which integrates sustainability considerations into the traditional SWOT framework. S-SWOT framework proved a valuable tool for identifying key internal and external elements that influence USP's sustainability initiatives and offered a structured approach for evaluations of how the initiatives align with global sustainability goals. The analysis identified 109 factors across the categories of strengths, weaknesses, opportunities, and threats mapped to USP's 11 Thematic Environmental Policies (TEPs). According to the results, whereas some environmental policies are exclusive to operational domains (e.g., water and effluents), others are curriculum-centered (e.g., environmental education). The study also highlighted the way SDGs intersect with TEPs and showed their distribution across university operations, curriculum, or both.

Keywords Sustainability SWOT · Sustainable development goals · University operations · University curriculum · High education institution

C. E. Barrantes-Briceño (✉) · M. S. Nagano
Department of Production Engineering, São Carlos School of Engineering, University of São Paulo, São Carlos, SP, Brazil
e-mail: eric.barrantes@usp.br

W. Leal Filho et al. (eds.), *Sustainability in the Curriculum and Operations of Universities*, World Sustainability Series,
https://doi.org/10.1007/978-3-032-12678-8_11

1 Introduction

According to Rebelatto et al. (2019) and Awuzie et al. (2021), universities are considered living laboratories, for they provide environments where experiments are conducted, and tests are performed on a routine basis. Innovative strategies and solutions grounded in sustainable development (SD) are implemented in those settings towards transforming education, organizational structure, operations, and infrastructure, establishing new standards.

Although Brazilian universities have been moving from vision to execution on the "smart-campus" agenda, the picture is still one of early-stage breakthroughs against a backdrop of latent potential. A national framework study with stakeholders at the Federal University of Campina Grande rated energy and technology as the most critical smart-campus pillars, yet overall performance sits at only 2.6 from 5, evidencing a sizeable implementation gap (Silva-Da-Nóbrega et al. 2022). Flagship institutions have started to close that gap—the State University of Campinas (Unicamp) has blended sub-metering, electric-bus charging, and IoT-driven HVAC optimization since 2018 and, in November 2024, switched on "CampusGrid," a 565 kW-PV + 1 MW/1.27 MWh microgrid that saves approximately R$ 450,000 per year and can island critical buildings for two hours (Hitchens 2024; Neris 2024; da Silva et al. 2018).

At the University of São Paulo, a students' audit mapped twelve smart solutions across energy, water, and waste, illustrating how Brazil's largest campus could scale data-driven efficiency (Sá et al. 2015). Besides those flagships, a multiple-case study in Ceará's Cariri region reported 22% of higher-education institutions already running PV systems that generate up to 115000 kWh/month with an average 4.9-year payback, confirming the financial viability of solar energy on campus, even for medium-sized HEIs (Soares et al. 2023). Together, those cases portray a sector that recognizes energy sustainability as the backbone of smart-campus ambitions and is beginning to scale concrete solutions, yet national surveys remind widespread performance lags behind the pioneers, signaling ample room for policy support, investment, and governance reform.

Budihardjo et al. (2021) claimed several institutions implement rating programs and utilize green marketing tools to measure sustainability achievements and to adhere to Sustainable Development Goal (SDG) targets and indicators. Such instruments evaluate environmental and political aspects, thereby facilitating integration of environmental concerns, management practices, performance metrics, and their broader impacts on campus and society.

Consequently, Strength, Weakness, Opportunity, Threat (SWOT) analysis serves as a valuable instrument for addressing sustainability challenges within higher education institutions (HEIs). Mian et al. (2020) described SWOT as a strategic planning methodology that identifies and assesses an organization's internal strengths and weaknesses, alongside its external opportunities and threats. Its effectiveness is often enhanced by its inclusive approach, typically engaging a wider community to gathering a range of perspectives.

According to Awuzie et al. (2021), the contribution of SWOT analysis to strategic management within the university context has been well established. It supports the development of strategies by systematically assessing an institution's internal strengths and weaknesses alongside external opportunities and threats (Eifert and Julmi 2022). Those insights enable decision-makers to leverage organizational strengths and opportunities while making informed decisions to mitigating potential weaknesses and threats (Mian et al. 2020). However, the limitations inherent to SWOT include lack of statistical rigor and no supply of a mechanism that objectively compares the relative importance of various factors (Wang et al. 2022).

In the context of sustainability, the primary purpose of SWOT analysis is two-fold, namely, to identify opportunities that support sustainable development and to anticipate threats that may hinder progress, thus ultimately contributing to a competitive and responsible strategic positioning (Eifert and Julmi 2022).

Pesonen and Horn (2013) introduced "Sustainability SWOT" (S-SWOT) as an empirical tool to assess how well sustainability is understood in the business sector and its potential to drive meaningful changes in strategic and operational practices. In contrast, Saric et al. (2021) highlighted sustainable SWOT frameworks had already been applied in decision-making contexts where strategic recommendations were critical. Their literature review integrated SWOT analysis with Analytical Hierarchy Process (AHP) and entropy methods.

This study adopted an approach similar to that proposed by Pesonen and Horn (2013), in which they utilize S-SWOT as an analytical tool that builds upon the conventional SWOT model by incorporating elements of Life Cycle Analysis. The integration enables the framework to assess sustainability at the level of individual products or groups of products within an organization.

Certain aspects of the original S-SWOT were adjusted to evaluations of strengths, weaknesses, opportunities, and threats through a sustainability lens that covers environmental, economic, and social dimensions. The adapted framework focuses on how the University of São Paulo's (USP) operations and curriculum align with the Sustainable Development Goals (SDGs).

The theoretical background of this study includes a discussion on universities, their operations, and curriculum, and the way such elements intersect with sustainability. It also provides a more detailed examination of SWOT analysis and its application in the sustainability context followed by a description of the research methodology, presentation and discussion of the findings, and a final section addressing the study's broader implications and key considerations.

2 Theoretical Background

2.1 Universities and Sustainability

The concept of Education for Sustainable Development (ESD) was first introduced in 1992, at the United Nations Conference on Environment and Development (UNCED), commonly referred to as Rio-92. Several initiatives were proposed during the event, including "Educating for a Sustainable Future" (Saric et al. 2021). In 2005, the United Nations Educational, Scientific, and Cultural Organization (UNESCO) launched the Decade of Education for Sustainable Development (DESD), emphasizing equitable access to education and promotion of sustainable lifestyles (Aznar Minguet et al. 2014). In 2015, the SDGs were adopted as the successor framework to DESD, with Target 4.7 focusing on fostering sustainability competencies among students (Saric et al. 2021).

Within universities, ESD—now often referred to as ESDU—aims at building institutional capacity towards addressing global challenges by integrating sustainability into teaching, research, operations, and governance (Hernandez-Diaz et al. 2021). Higher education institutions (HEIs) have been identified as key agents of transformation, tasked with developing future leaders and promoting practices aligned with environmental stewardship (Lozano et al. 2013; Pacheco et al. 2019). According to Leiber (2018), HEIs are responsible for (i) preservation, advancement, and dissemination of freedom and democracy, (ii) generation of knowledge and innovation, and (iii) promotion of sustainable behaviors and environmental conservation.

As claimed by Lozano et al. (2013), institutionalizing sustainable development in universities requires an integrated model that encompasses curriculum design, pedagogy, research, campus operations, assessment, outreach, and partnerships (Lozano 2009; Pacheco et al. 2019; Budihardjo et al. 2021). For instance, the creation of green campuses encourages the adoption of sustainable practices both academically and in everyday university life (Rebelatto et al. 2019). A sustainable HEI minimizes its environmental, economic, and social impacts by aligning its educational, research, and administrative activities with a broader transition towards sustainable societies (Bautista-Puig and Sanz-Casado 2021).

In terms of operations and curriculum, Lozano et al. (2013) identified four foundational pillars, namely, teaching, research, campus operations, and community engagement. Operational efficiency has shown to reduce environmental impacts and generate cost savings (Smith 2011) and the integration of sustainability into academic curriculum is essential for an effective environmental management (Jain and Pant 2010). Leal Filho et al. (2021) categorized "campus operations initiatives" as one of the three primary domains through which Latin American HEIs incorporate sustainability, with curricular innovation being a key element under the broader category of "teaching initiatives."

University Operations

A comprehensive review of the literature identified multiple ways through which universities integrate sustainability, including institutional governance, campus operations, teaching, research, community engagement, and transparency. The integration of sustainability across those domains contributes to an institution being classified as a "sustainable" or "green" university (Fissi et al. 2020).

Such institutions are widely regarded as key drivers of transformation in advancing sustainability and expected to lead efforts towards urgent environmental challenges (Bautista-Puig and Sanz-Casado 2021). Campus operations play a central role in that endeavor, with initiatives focused on water, energy, and waste management, as well as sustainable procurement practices (Leal Filho et al. 2021). Strategic projects include green infrastructure, sustainable mobility, and energy-efficiency measures (Fissi et al. 2020; Hernandez-Diaz et al. 2021).

A variety of tools are available for assessments of sustainability in HEIs. Assessment Instrument for Sustainability in Higher Education (AISHE) 2.0, developed by Roorda et al. (2009), defines operations as the functional dimensions of an institution, encompassing infrastructure, environmental management systems, and human resource policies.

Sustainability Tracking, Assessment and Rating System™ (STARS), developed by the Association for the Advancement of Sustainability in Higher Education (AASHE), enables institutions to evaluate their sustainability performance. Within the Operations category, STARS outlines the following five key impact areas (AASHE 2024):(I)(I)(I)(I)

(I) Buildings and Grounds
(II) Energy and Climate
(III) Food and Dining
(IV) Procurement and Waste
(V) Transportation

Sustainability Assessment Questionnaire (SAQ) for Colleges and Universities, developed by the Association of University Leaders for a Sustainable Future (ULSF), covers practices such as sustainable construction, energy conservation, waste reduction, and green transportation (ULSF 2009).

The aforementioned tools support institutions in measuring sustainability performance and setting strategic goals for continued improvement.

University Curriculum

The primary goal of sustainability education programs is to prepare students to contribute to a sustainable future across industry, business, and society (Franco et al. 2018). Nevertheless, the integration of sustainable development into the university curriculum remains limited, despite increasing global relevance and awareness (Leal Filho et al. 2018). Faculty members play a critical role in the process, for they are responsible for embedding sustainability principles into course content and linking those concepts to global challenges and societal values (Franco et al. 2018).

Despite growing awareness of sustainability, curricular reform has advanced slowly, often due to the lack of explicit frameworks and the limited adoption of transdisciplinary approaches (Franco et al. 2018; Tasdemir and Gazo 2020).

According to Fissi et al. (2020), integration can occur through two main ways, namely, vertically, via dedicated academic programs such as master's degrees in sustainability, and horizontally, by embedding sustainability concepts within existing disciplinary structures. An example is the University of Florence, which offers specialized courses in sustainability across fields including economics and engineering, while also incorporating sustainability themes into traditional course offerings (Fissi et al. 2020).

The implementation of transdisciplinary curriculum has shown to enhance students' understanding and align academic preparation with labor market demands by combining foundational knowledge with innovative, systems-oriented thinking (Tasdemir and Gazo 2020). However, universities must tailor their strategies to local contexts rather than rely on standardized one-size-fits-all models (Leal Filho et al. 2018).

Lozano (2009) proposed evaluation criteria for sustainability-oriented curriculum across four domains, namely, economic (e.g., productivity, consumption), environmental (e.g., biodiversity, energy efficiency), social (e.g., employment, poverty), and transversal themes (e.g., governance, responsibility).

Assessment tools such as AISHE 2.0 recommend introducing sustainability education early in academic programs, either through standalone courses, or by integrating the topic into existing subjects (Roorda et al. 2009). STARS framework (AASHE 2024) emphasizes the importance of academic programs that equip students so that they can face sustainability challenges, whereas SAQ (ULSF 2009) evaluates the extent to which institutions offer courses on themes such as social justice and sustainable production.

2.2 SWOT

SWOT (Strengths, Weaknesses, Opportunities, and Threats) analysis, a widely used strategic management tool, was initially developed at Harvard Business School in the 1950s. Its application to business strategy was further refined and popularized in the 1960s by Albert Humphrey, who conducted research with Fortune 500 companies (Benzaghta et al. 2021). Since then, SWOT has evolved into a flexible framework for scenario analysis across various fields, including education, industry, and agriculture (Ghazinoory et al. 2011; Leiber et al. 2018).

SWOT distinguishes between internal factors—strengths and weaknesses, which are controllable—and external ones—opportunities and threats, which are not controllable (Jain and Pant 2010; Aznar Minguet et al. 2014). Its matrix outlines

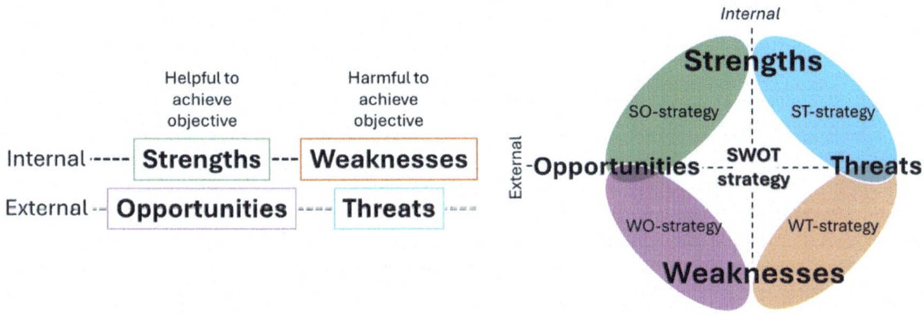

Fig. 1 SWOT analysis matrix and SWOT strategies SO: Leverage strengths to capitalize on opportunities ST: Use strengths to minimize or neutralize threats. WO: Address weaknesses by taking advantage of opportunities. WT: Develop defensive strategies to mitigate weaknesses and threats.

four strategic options, enabling a structured and comprehensive analysis of internal and external environments (Vlados 2019; Benzaghta et al. 2021) (see Fig. 1):

SWOT framework offers several advantages, including simplicity and accessibility (Agyekum 2020). Moreover, it provides a systemic overview of internal and external conditions and enables integration with other strategic planning tools such as PESTEL and AHP (Benzaghta et al. 2021).

However, the tool faces criticisms. Scholars have pointed to its potential superficiality (Pickton and Wright 1998), tendency to generate an excess of non-prioritized data, and limited empirical validation (Leiber et al. 2018). SWOT should ideally be combined with complementary methodologies to enhancing its effectiveness in strategic planning (Eifert and Julmi 2022).

In the context of higher education, SWOT analysis is particularly effective for identifying institutional capacities (strengths), diagnosing implementation barriers (weaknesses), recognizing favorable opportunities, and anticipating environmental or contextual risks (Aznar Minguet et al. 2014; Budihardjo et al. 2021).

3 Methodology

This study aimed at identifying and categorizing internal and external factors that facilitate or hinder the alignment of university operations and curriculum with SDGs, with a specific focus on the USP. It was guided by the conceptual framework displayed in Fig. 2.

3.1 Case Study

A case study methodology that enables an in-depth and contextualized examination of phenomena within their real-world setting was adopted in the study. As claimed

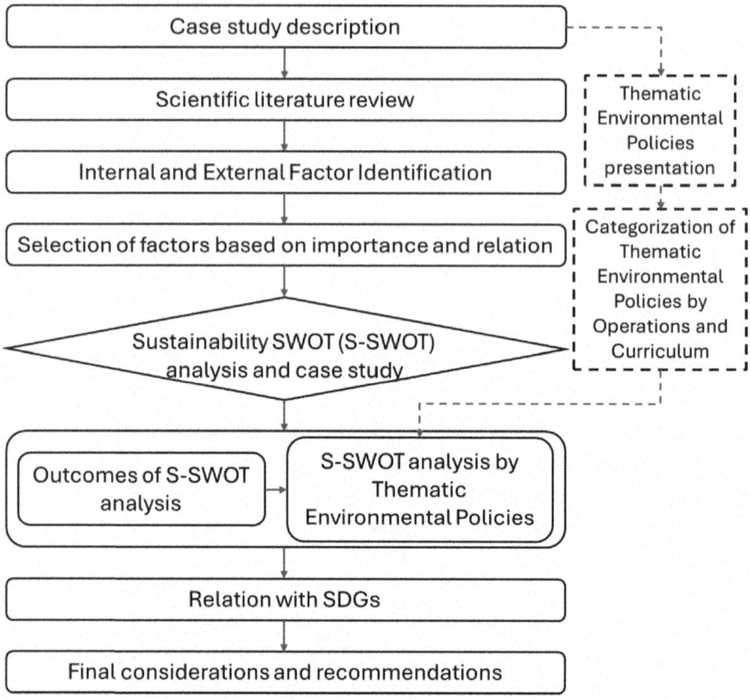

Fig. 2 Conceptual scheme of the methodology

by Yin (2014), such an approach is particularly effective in addressing "how" and "why" research questions and provides rich insights into complex systems.

It also offers notable methodological flexibility, which makes it a valuable approach across disciplines such as social sciences, business, education, and public health (Stake 1995). Its strength lies in the triangulation of diverse data sources, including interviews, observations, and document analysis which, together enable a holistic and nuanced understanding of the research subject while accounting for contextual variables.

Building on that foundation, Flyvbjerg (2006) underscored the case study's unique ability to generate context-sensitive, empirically grounded knowledge. The methodological approach not only supports detailed analyses, but also provides a solid basis for theory development, ensuring academic insights remain rooted in observed reality.

Accordingly, the selection of a case study in this study is justified by the need to investigate, through the lens of S-SWOT, the factors that promote or hinder the alignment of university operations and curriculum with SDGs, with a specific emphasis on USP's institutional characteristics and dynamics. The approach promotes capture of institutional nuances and interrelationships that may be overlooked by broader, more generalized methodologies.

The University of São Paulo is a leading higher education institution in both Latin America and the global academic landscape. It is a tuition-free public university funded by the State of São Paulo and widely recognized for its academic excellence. Established on January 25, 1934, USP is Brazil's largest research university, operating across eight campuses and maintaining facilities in 15 additional municipalities (USP 2023). It comprises 43 teaching and research units and, in 2023, supported a student population of more than 97,000, along with over 12,000 technical and administrative staff and more than 5000 faculty members. The university offers 324 undergraduate programs and 264 graduate ones, accounting for approximately 22% of Brazil's scientific output indexed in Clarivate Analytics. It is ranked among the world's top 10 institutions in terms of research publications, with over 6.3 million citations on Google Scholar and more than 123,000 publications listed in Elsevier's Scopus (USP 2023). Such indicators reinforce its leadership in both regional and international research and education.

In October 2024, USP maintained its position as the highest-ranked university in the QS Latin America and The Caribbean Ranking, published by the British consultancy Quacquarelli Symonds (Jornal da USP 2024). According to THE World Reputation Ranking 2025, USP is the only Latin American institution listed among the 150 universities with highest academic reputation worldwide. The ranking is established by Times Higher Education (THE), a UK-based higher education consultancy (CNN Brasil 2025).

USP has also been recognized as one of the five most sustainable universities globally and holds the top position in Latin America, according to UI GreenMetric World University Ranking 2024 (SGA 2025a).

In 2009, USP established a working group tasked with developing strategies for implementing environmental management, in accordance with Regulation GR-4.448 of September 29, 2009 (USP 2009). As a result, Resolution 6062 of February 27, 2012, created the Superintendence for Environmental Management (SGA, its acronym in Portuguese), with the mission of planning, implementing, maintaining, and promoting environmental sustainability across USP's campuses (SGA 2025b).

In 2014, SGA began developing USP's Environmental Management Plan, which encompasses the following 11 Thematic Environmental Policies (TEPs):

- Water and Effluents
- Green Areas and Ecological Reserves
- Sustainable Buildings
- Environmental Education
- Energy
- Fauna
- Mobility
- Greenhouse Gas and Air Pollutant Emission Reduction
- Solid Waste
- Sustainability in Administration
- Land Use and Occupation

Fig. 3 Representation of USP's thematic environmental policies

According to SGA, "USP's Environmental Policy, published in the Official Gazette on January 12, 2018, by Resolution No. 7465 of January 11, 2018, was created to promote more efficient and integrated environmental management in alignment with the university's institutional principles" (USP 2018; SGA 2025c) (Fig. 3).

3.2 Analytical Frame of USP's Environmental Policies

Since university operations refer to activities related to management and infrastructure (Sect. 2.1.1), the curriculum is associated with teaching and learning processes (Sect. 2.1.2), and the TEPs may be part of one or both domains, this study proposes the following classification (Fig. 4):

Regarding only Operations:

- **Water and Effluents**: Management of water consumption and treatment in campus infrastructure.
- **Sustainable Buildings**: Design, construction, and maintenance of environmentally responsible buildings.
- **Sustainability in Administration**: Implementation of sustainable practices within administrative operations.
- **Land Use and Occupation**: Planning and management of the institution's physical space and land resources.

Regarding only Curriculum:

- **Environmental Education:** Courses, disciplines, programs, and academic initiatives focused on environmental awareness, education, and sustainability-oriented projects.

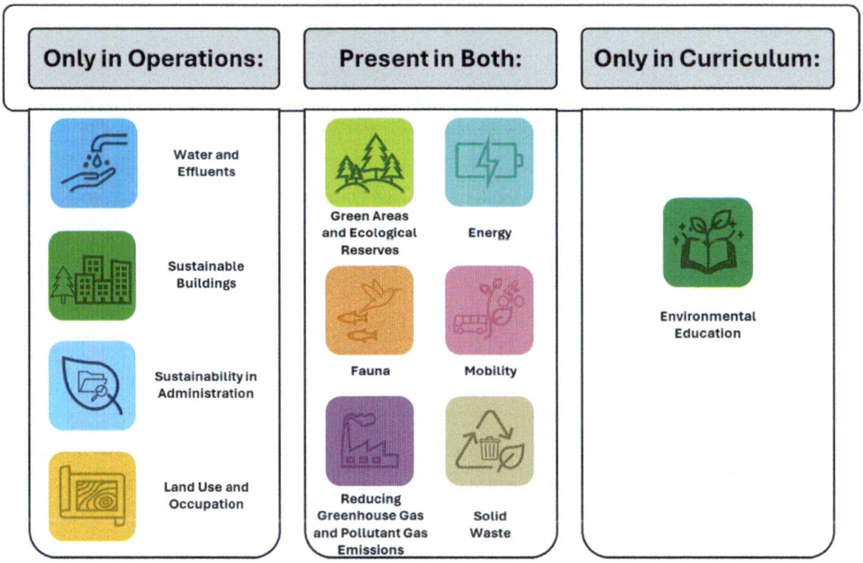

Fig. 4 Classification of thematic environmental policy

Regarding both:

- **Green Areas and Ecological Reserves**: Environmental preservation on campus (operations) and academic research into biodiversity and ecosystem conservation (curriculum).
- **Energy**: Efficient and renewable energy use on campus (operations) combined with teaching and research into sustainable energy and energy efficiency (curriculum).
- **Fauna**: On-campus biodiversity protection (operations) and research into species conservation and wildlife management (curriculum).
- **Mobility**: Development of sustainable transportation policies on campus (operations) and research into urban planning and sustainable mobility systems (curriculum).
- **Reduction of Greenhouse Gas and Pollutant Gas Emissions**: Implementation of environmentally responsible practices on campus (operations) and academic instruction on environmental impact mitigation (curriculum).
- **Solid Waste**: Campus waste management and recycling initiatives (operations), alongside curricular content related to waste reduction, recycling, and circular economy (curriculum).

3.3 Identification of Sustainable SWOT Factors

The development of the S-SWOT framework applied in this study was based on the methodology proposed by Pesonen and Horn (2013), who defined S-SWOT as "an analytical tool that combines two well-known methods, namely, the basic SWOT tool, a strategic business planning process in matrix form, and life cycle assessment, which calculates the environmental impacts of a given product system. It is able to integrate all aspects of sustainability into one assessment matrix efficiently."

However, life cycle assessment was not applied in the present study. Instead, an alternative interpretation of S-SWOT identified in the literature was adopted and selected elements from that approach were incorporated. World Resources Institute (WRI) developed its own version of a sustainability-oriented SWOT, known as sSWOT, designed to encourage actions and foster collaboration around environmental issues that pose real business risks and opportunities. It also aims to engage and motivate stakeholders—particularly those with limited familiarity with environmental or business sustainability concepts (WRI 2012).

The model was later adopted by Livzeniece et al. (2021), who described it as a revised version of the traditional SWOT framework, applied in private-sector contexts such as renewable energy planning. A key distinction between the classic SWOT and sSWOT lies in their respective orientations—the former emphasizes internal strengths and weaknesses, whereas the latter introduces a global sustainability perspective, focusing on long-term environmental and social challenges (Livzeniece et al. 2021).

Accordingly, the S-SWOT developed in this study aims to analyze various internal and external factors through a sustainability lens. It compiles a set of factors derived from existing literature that correspond to each of the 11 TEPs outlined in USP's Environmental Management Plan and then examines their alignment with the 17 SDGs.

Factors were excluded according to the following criteria:

1. Strategic-level corporate factors not related to USP's TEPs;
2. Sustainability-related factors overly specific to individual case studies or highly localized scenarios; and
3. Sustainability-related factors with no linkage to any of the TEPs, such as those limited to cultural dimensions.

A systematic literature review using Web of Science and Scopus databases between September and October 2024 identified internal and external factors. A variety of search strings and their variants including "SWOT," "Sustainability," "Higher Education Institution," and "University" were applied. Figures 5, 6, 7 and 8 illustrate an adapted PRISMA Statement that guided the literature review, with variations in the search strategies adopted.

The initial review used search terms "SWOT" and *Sustainab** to identify papers that explicitly addressed SWOT analysis in conjunction with sustainability-related concepts, including variations such as "sustainable" and "sustainability." The aim

		Web of Science	Scopus
Identification	Strings used:	"SWOT" AND Sustainab*	
	In:	TOPIC	Article title, Abstract, Keywords
	Documents found:	1,714	2,352
Screening	**Refined by**		
	Publication Years:	2020 to 2024	
	Document Type:	Article	
	Languages:	English	
	Optional Keywords:	N/A	Circular Economy; SDGs; SWOT; SWOT Analysis; Strategic Planning; Strength, Weakness, Opportunity, And Threat; Sustainability; Sustainable Development; Sustainable Development Goal; Sustainable Development Goals; TOWS Matrix
	Documents found:	882	655
Eligibility	**Title and Abstract analysis**		
	Eligible documents:	57	
Included	Studies included in qualitative synthesis	57	

Fig. 5 Adapted PRISMA diagram with 'SWOT'—'Sustainab*' as strings

		Web of Science	Scopus	Web of Science	Scopus
Identification	Strings used:	"SWOT" AND High* Education* Institution*		"SWOT" AND Universit*	
	In:	TOPIC	Article title, Abstract, Keywords	TOPIC	Article title, Abstract, Keywords
	Documents found:	103	187	509	749
Screening		**Refined by**		**Refined by**	
	Publication Years:	N/A		2020 to 2024	
	Document Type:	Article		Article	
	Languages:	English or Spanish		English or Spanish or Portuguese	
	Documents found:	97	112	175	219
		Title and Abstract analysis		**Title and Abstract analysis**	
Eligibility	Eligible documents:	27		12	
Included	Studies included in qualitative synthesis	27		12	

Fig. 6 Adapted PRISMA diagram with 'SWOT'—'High* Education* Institution*' and 'SWOT'—'Universit*' as strings

was to locate documents containing those terms in title or abstract. The preliminary search yielded 4066 documents. Subsequent refinements—filtering by publication years (2020–2024), document type (article), language (English), and optional Scopus keywords resulted in 1537 articles. After removal of duplicates across databases and reviewing of titles and abstracts, 57 documents were deemed eligible and fully accessed for inclusion in the qualitative synthesis (Fig. 5).

		Web of Science	Scopus	Web of Science	Scopus
Identification	Strings used:	strength*, weakn*, opportunit*, threat* AND High* Education* Institution*		strength*, weakn*, opportunit*, threat* AND Universit*	
	In:	TOPIC	Article title, Abstract, Keywords	TOPIC	Article title, Abstract, Keywords
	Documents found:	61	109	280	412
Screening		**Refined by**		**Refined by**	
	Publication Years:	N/A		N/A	
	Document Type:	Article		Article	
	Languages:	English or Spanish		English or Spanish or Portuguese	
	Documents found:	46	75	188	292
		Title and Abstract analysis		**Title and Abstract analysis**	
Eligibility	Eligible documents:	8		9	
Included	Studies included in qualitative synthesis	8		9	

Fig. 7 Adapted PRISMA diagram with 'strength*, weakn*, opportunit*, threat*'—'High* Education* Institution*' and 'Universit*' as strings

		Web of Science	Scopus
Identification	Strings used:	"SWOT" AND sustainab* AND Universit*	
	In:	TOPIC	Article title, Abstract, Keywords
	Documents found:	87	115
Screening		**Refined by**	
	Publication Years:	N/A	
	Document Type:	Article	
	Languages:	English, Spanish or Portuguese	
	Documents found:	64	73
		Title and Abstract analysis	
Eligibility	Eligible documents:	5	
Included	Studies included in qualitative synthesis	5	

Fig. 8 Adapted PRISMA diagram with 'SWOT'—'sustainab*'—'Universit*' as strings

In a subsequent review (Fig. 6), two sets of search terms, namely, *(i) "SWOT" and High* Education* Institution**, and *(ii) "SWOT" and Universit** were used for addressing variations in terminology and expanding the pool of studies. The strategy aimed at identifying studies explicitly addressing SWOT in the context of higher education institutions or universities. Terms *High* Education* Institution** and *Universit** were used for capturing variations such as "higher," "educational," "institutions," and "universities."

The first set of search terms returned 290 documents. Given the volume, filtering by publication year was impractical. Instead, refinement based on document type (article) and language (English and Spanish) was conducted for ensuring a broader inclusion. After removal of duplicate records and previously reviewed studies and upon evaluation of titles and abstracts, 27 articles met the inclusion criteria and were added to the qualitative synthesis.

The second set of search terms—*"SWOT" and Universit**—yielded 1258 results; therefore, filters for publication year (2020–2024), document type (article), and language (English, Spanish, and Portuguese) were applied, reducing the set to 394 articles. After elimination of duplicates from both databases and earlier reviews and assessment of titles and abstracts, 12 articles were selected for the synthesis.

In a third review (Fig. 7), two new sets of combined strings based on prior title and abstract analysis were tested. *(i)* The full SWOT terminology, i.e., *strength*, weakn**, *opportunit**, and *threat** was used in the string in combination with *High* Education* Institution** and *(ii)* the same full SWOT terminology—*strength*, weakn**, *opportunit**, and *threat**—and *Universit**. The first searching set returned 170 documents, which was reduced to 121 after application of filters by document type (article) and language (English and Spanish). After the removal of duplicates, articles already screened, and analysis of the remaining ones, 8 articles were identified and included in the qualitative synthesis.

The second variation of the strategy retrieved 692 documents. After refinement by document type (article) and language (English, Spanish, and Portuguese), 480 results remained and after exclusion of previously reviewed or duplicate articles and analysis of titles and abstracts, 9 papers were ultimately included in the synthesis.

Finally, towards broadening the scope and according to the findings of the previous three reviews, an additional search was conducted using terms *"SWOT"*,*sustainab**, and *Universit**. The combination aimed at identifying documents explicitly mentioning "SWOT" in title or abstract and capturing term variations such as "sustainability," "sustainable," "university," and "universities." The search returned 202 articles. After refining by document type (article) and language (English, Spanish, and Portuguese), 137 articles remained and after exclusion of duplicates and articles already reviewed and a detailed evaluation of titles and abstracts, 5 documents met the inclusion criteria and were incorporated into the qualitative synthesis (Fig. 8).

118 articles were included across all literature reviews. They were fully analyzed towards identification of internal and external factors relevant to the construction of S-SWOT. Synthesis, organization, and analysis of the qualitative data were facilitated by MAXQDA Plus 2022 software (version 22.7.0), which supported efficient data exploration, coding, classification, and pattern recognition. The process enabled the development of a structured list of articles and the extraction of relevant internal and external factors.

97 articles were excluded from the 118 ones after analysis for falling under the first exclusion criterion, i.e., they addressed purely strategic corporate factors unrelated to the TEPs. The remaining 21 articles contained relevant information; however, only 17 were ultimately used. The 4 excluded papers fit either the second exclusion category—factors with a sustainability focus that were overly specific to localized case studies—or the third one, involving factors unrelated to the TEPs, such as cultural dimensions.

Table 1 shows the 17 selected articles from which 109 internal and external factors were extracted for the formulation of S-SWOT. The Results and Discussion section provides the factors, classified as strengths, weaknesses, opportunities, or threats, and organized according to the 11 TEPs established by USP.

Table 1 List of authors and studies used for the construction of S-SWOT

Author	Title
Agyekum (2020)	Energy poverty in energy rich Ghana: A SWOT analytical approach for the development of Ghana's renewable energy
Ahonen et al. (2024)	Carbon neutral higher education institutions: a reality check, challenges and solutions
Alsharari (2020)	Internationalization market and higher education field: institutional perspectives
Alves et al. (2024)	Strategic analysis of the forest carbon market in Brazil
Awuzie et al. (2021)	Facilitating successful smart campus transitions: a systems thinking-SWOT analysis approach
Aznar Minguet et al. (2014)	Sustainability in higher education: challenges and opportunities
Budihardjo et al. (2021)	Towards sustainability in higher-education institutions: analysis of contributing factors and appropriate strategies
Camilleri (2019)	Higher education marketing: opportunities and challenges in the digital era
Cheng et al. (2023)	Towards sustainable construction through better construction and demolition waste management practices: a SWOT analysis of Suzhou, China
Díaz et al. (2019)	Strategic actions of environmental education in the community La Majagua for its local development
Fahim et al. (2021)	Sustainable higher education reform quality assessment using SWOT analysis with integration of AHP and entropy models: a case study of Morocco
Gill (2009)	Knowledge management initiatives at a small university
Nobrega et al. (2021)	Logistics 4.0 in Brazil: critical analysis and relationships with SDG 9 targets
Ramdan et al. (2023)	Addressing sustainability through waste management: a perspective from higher education institutions in Southeast Asia
Rebelatto et al. (2019)	Energy efficiency actions at a Brazilian university and their contribution to sustainable development Goal 7
Smith (2011)	Developing a sustainability plan at a large U.S. college of education
Soutullo et al. (2024)	Sustainable transformation in the Latin American and Caribbean districts through the implementation of a qualitative methodology. Challenges and key aspects to be addressed

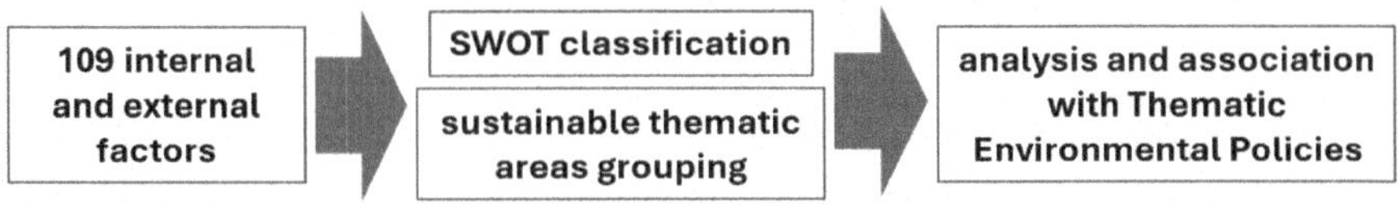

Fig. 9 Diagram for S-SWOT construction

3.4 Coding, Classification, and Mapping Procedures

The development of S-SWOT analysis involved classification of 109 internal and external factors, identified from the studies listed in Table 1. Each text segment referring to a sustainability issue was coded as an internal or external factor and then classified into four distinct categories, namely, (S) strengths, (W) weaknesses, (O) opportunities, and (T) threats. Those factors were systematically organized according to the three pillars of sustainability—economic, environmental, and social. The codes were then cross-referenced with the 11 TEPs and subsequently analyzed. Such a type of analysis allowed multidimensional queries. The findings were mapped to each of USP's TEPs. Figure 9 illustrates the general process.

4 Results and Discussion

4.1 Sustainable SWOT Construction and Analysis

A comprehensive list of S-SWOT factors based on the structure described in Sect. 3.4 was compiled and organized by TEP, as shown in Table 2.

4.2 S-SWOT Results and Sustainable Development Goals

The 17 SDGs were linked to the 11 TEPs through the S-SWOT analysis of each policy. The policies were then evaluated according to type of influence they might exert on the SDGs within the university context. Therefore, each influence was categorized as Strength (S), Weakness (W), Opportunity (O), or Threat (T).

As an example, the university has an opportunity to advance SDG 1 (No Poverty) through the TEP "Environmental Education," represented by O4.6, which refers to the creation of vocational sustainability courses for vulnerable communities. Another opportunity is provided through the "Sustainability in Administration" policy, represented by O10.9 to O10.12, which includes initiatives such as green internships and programs promoting social inclusion. However, the achievement of those goals is challenged by institutional barriers such as bureaucratic inefficiencies (W10.4) and budgetary constraints (T10.15) (see Table 2).

Table 2 S-SWOT factors compiled and organized by thematic environmental policy

1. Water and Effluents	
Strengths: 1.1. Presence of advanced infrastructure 1.2. Technical capacity for water management 1.3. Available technologies for efficient water management	Weaknesses: 1.4. Liquid waste management issues 1.5. Incomplete legal framework for environmental regulation 1.6. Insufficient educational programs on water resources 1.7. Lack of conservation culture for natural resources
Opportunities: 1.8. Optimization of water use through circular economy practices 1.9. Increase in global awareness of climate change 1.10. Development of internal USP policies for water resource management 1.11. Potential development of innovative technologies for wastewater treatment	Threats: 1.12. High costs of development of conservation systems
2. Green Areas and Ecological Reserves	
Strengths: 2.1. Development of sustainable tourism projects 2.2. Preservation of forest biodiversity 2.3. Traditional knowledge of forest management 2.4. High number of open green areas and forest coverage 2.5. Recognition of biodiversity importance	Weaknesses: 2.6. Deforestation and illegal logging 2.7. Introduction of invasive exotic plant species 2.8. Lack of reliable biodiversity data on flora 2.9. Weak integration between environmental stakeholders
Opportunities: 2.10. Implementation of training and capacity-building programs 2.11. Creation of ecological corridors 2.12. Enhancement of flora biodiversity conservation 2.13. Expansion of sustainable tourism	Threats: 2.14. Economic pressures encouraging illegal exploitation 2.15. Reduction of protected areas due to public policies converting them into agricultural or exploitable land
3. Sustainable Buildings	
Strengths: 3.1. Improved environmental performance of buildings 3.2. Use of bioclimatic elements in architectural designs 3.3. Availability of innovative technologies for green buildings	Weaknesses: 3.4. Outdated and inefficient infrastructure 3.5. Low energy efficiency standards in buildings
Opportunities: 3.6. Cost reduction through sustainable practices 3.7. Promotion of sustainable urbanization and energy rehabilitation	Threats: 3.8. High costs of implementation or energy rehabilitation 3.9. Lack of governmental regulations

(continued)

Table 2 (continued)

4. Environmental Education	
Strengths: 4.1. High intellectual capacity and interdisciplinary studies 4.2. Growing awareness of sustainability 4.3. Inclusion of sustainable development within institutional structures	Weaknesses: 4.4. Lack of consistent educational strategies 4.5. Low sustainability culture
Opportunities: 4.6. Development of training and capacity-building programs 4.7. Integration of sustainable curriculum 4.8. Engagement of professionals in educational processes 4.9. Participation of national and international media	Threats: 4.10. Low interest and motivation among students and faculty
5. Energy	
Strengths: 5.1. Internal policies for energy efficiency 5.2. Government support for renewable energy projects 5.3. Diversified energy matrix within USP 5.4. Specialized human resources and technical support	Weaknesses: 5.5. High initial costs for renewable energy projects 5.6. Low development of thermal storage technologies 5.7. Highly bureaucratic licensing processes
Opportunities: 5.8. Reduction in costs of renewable energy technology 5.9. Development of new sources of local renewable energy 5.10. Strengthening of regulatory frameworks within USP for energy efficiency	Threats: 5.11. Climate change 5.12. Unstable national economy
6. Fauna	
Strengths: 6.1. Conservation of endangered species 6.2. Recognition of biodiversity as a natural heritage	Weaknesses: 6.3. Lack of reliable biodiversity data 6.4. Introduction of invasive exotic species
Opportunities: 6.5. Strengthening partnerships for conservation projects	Threats: 6.6. Lack of incentives for conservation projects
7. Mobility	
Strengths: 7.1. Advanced transportation infrastructure 7.2. Geography favorable for multimodal transportation (within USP and the city) 7.3. Available technologies for efficient transportation	Weaknesses: 7.4. Inefficient public transportation fleet

(continued)

Table 2 (continued)

Opportunities: 7.5. Promotion of sustainable public transportation and bike lanes 7.6. Development of sustainable logistics technologies	Threats: 7.7. High costs of sustainable transportation technologies 7.8. Traffic congestion in critical areas (within USP and the city)
8. Reducing Greenhouse Gas and Pollutant Gas Emissions	
Strengths: 8.1. Implementation of public policies 8.2. Commitment to carbon neutrality goals 8.3. Well-structured carbon footprint calculations	Weaknesses: 8.4. Lack of standardization in emissions calculations
Opportunities: 8.6 Technological innovation for emission reduction	Threats: 8.7. Changes in international carbon policies 8.8. Dependence on external funding
9. Solid Waste	
Strengths: 9.1. Established recycling systems 9.2. Well-structured waste collection system with segregated, visible, and accessible bins	Weaknesses: 9.3. Lack of effective waste collection management 9.4. Lack of awareness about 3R (Reduce, Reuse, Recycle) practices
Opportunities: 9.5. Development of recycling markets 9.6. Promotion and implementation of internal waste policies	Threats: 9.7. Lack of support from local companies for segregated waste collection
10. Sustainability in Administration	
Strengths: 10.1. Implemented environmental management system 10.2. Existing Information and Communication Technology infrastructure at the university 10.3. Administrative motivation to advance the environmental agenda	Weaknesses: 10.4. Complex decision-making processes, bureaucratic hurdles, and slow decision-making 10.5. Weak integration of existing electronic platforms 10.6. Lack of knowledge of current environmental legislation 10.7. Low multisectoral integration 10.8. Governance and team management issues
Opportunities: 10.9. Strengthening of public–private partnerships 10.10. Partnerships between universities and the market 10.11. Improvement of environmental governance quality 10.12. Potential expansion of partnership networks and involvement of national and international external stakeholders	Threats: 10.13. Lack of non-governmental support 10.14. Political and economic instability 10.15. Decrease in external subsidies
11. Land Use and Occupation	

(continued)

Table 2 (continued)

Strengths: 11.1. Availability and planning of land use for sustainable projects 11.2. Strategic planning with environmental integration	Weaknesses: 11.3. Lack of integration in territorial planning 11.4. Lack of standardization in sustainable urban projects
Opportunities: 11.5. Improved land management 11.6. Harmonization of urban interventions with the natural environment	Threats: 11.7. Ineffective regulations that do not consider local context 11.8. Pressure to convert forested areas into urban zones

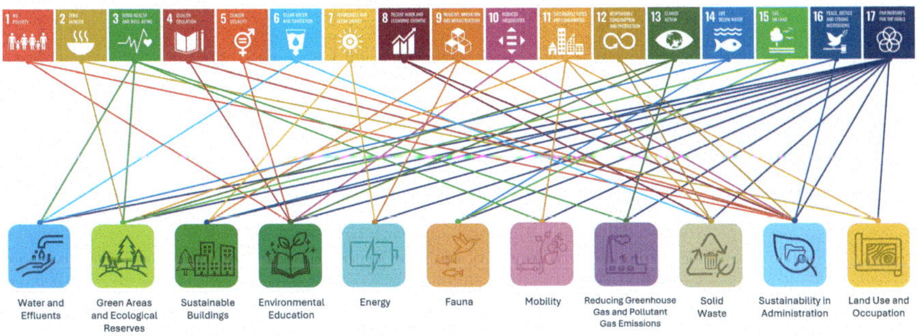

Fig. 10 Relationships between SDGs and TEPs

The analysis evaluated the way each policy's S-SWOT elements influenced the SDGs. Although the study does not aim to provide a detailed account of each influence—whether positive or negative—the groundwork laid here can serve as a basis for future research. Figure 10 shows an overview of the connections between the SDGs and the TEPs, without distinguishing between types of influence.

The following several key insights emerge when those relationships are viewed holistically:

- SDG 17 (Partnerships for the Goals) is connected to all TEPs, reflecting the university's reliance on networks to implement its environmental agenda, which includes collaboration with current students, alumni, faculty, and staff, as well as partnerships with local, regional, and national governments, other universities, and public and private sector organizations.
- "Sustainability in Administration" policy is associated with nine SDGs, the highest number among all policies.
- 12 out of the 17 SDGs are linked to at least two TEPs, indicating a strong degree of interconnection.
- Five of the eleven TEPs are associated with three or more SDGs, highlighting their cross-cutting nature.

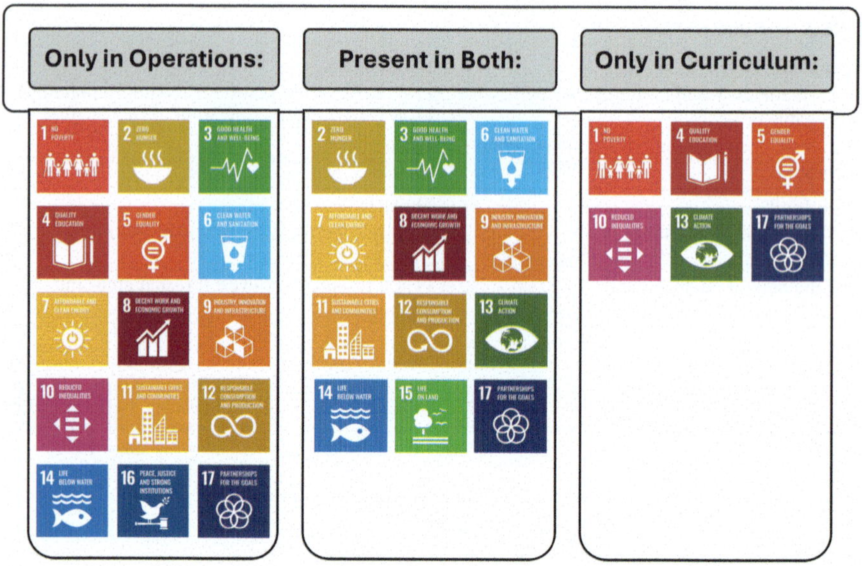

Fig. 11 Classification of the SDGs by operations, curriculum or both

An attempt was made to cross-reference Fig. 4, which categorizes TEPs by university domain (Operations, Curriculum, or Both) with Fig. 10, which maps the SDG relationships. The resulting Fig. 11 illustrates the distribution of SDGs across those three classifications.

SDG 1 appears in both "Only in Operations" and "Only in Curriculum" categories due to its association with "Sustainability in Administration" and "Environmental Education", respectively. As such, it is interpreted as being present in both domains. Similar overlaps occur with SDG 4 (Quality Education), SDG 5 (Gender Equality), and SDG 10 (Reduced Inequalities).

In contrast, SDG 16 (Peace, Justice, and Strong Institutions) is the only SDG classified solely under "Only in Operations," suggesting a particularly strong linkage to the university's administrative and governance structures.

On the other hand, SDGs 2, 3, 6, 7, 8, 9, 11, 12, and 14 are present in both "Only in Operations" and "Present in Both" categories, which may indicate a stronger emphasis on operational strategies within the university. SDG 13 (Climate Action) appears under "Only in Curriculum" and "Present in Both," implying a potentially greater influence from the academic dimension. Lastly, SDG 15 (Life on Land) is found only in "Present in Both", suggesting a balanced integration between the university's operations and curriculum for that goal.

4.3 Comparison with Previous Research

The findings align with a broader international picture, particularly across Latin America, which shows universities moving away from scattered "green" projects towards a more comprehensive integration of sustainability across operations, research, and teaching. A regional survey of 157 higher-education institutions revealed "campus operations" and "education and research" are the SDG dimensions most frequently addressed, yet they are typically approached in isolation rather than as part of a unified institutional strategy (Leal Filho et al. 2021). Likewise, bibliometric work by Bautista-Puig and Sanz-Casado (2021) confirmed Spanish universities (Brazilian ones might be added) still publish more on single environmental initiatives than on systemic transformation, underlining a need for the integrated governance model proposed in this study.

The energy- and climate-centered emphasis that emerged from S-SWOT (S5.1, O5.9) echoes several Brazilian case studies. At the University of Passo Fundo, targeted retrofits and on-site photovoltaic installations lowered energy consumption while advancing SDG 7 (Rebelatto et al., 2019). A similar gap analysis at UNICAMP highlighted a need for expanding solar power, sustainable mobility, and user engagement (de Melo et al. 2025), precisely the levers identified here (O4.8, O5.9, O7.5, O7.6). Experiments with hybrid photovoltaic–battery microgrids in Amazonian communities achieved cost savings with up to 97% renewable energy (Uruchi et al. 2025), suggesting universities of comparable scale can also deliver significant results (O5.8, O5.9). Combined with the European "reality check" by Ahonen et al. (2024) on the pathway to carbon–neutral campuses, those examples reinforce the present recommendation for pairing operational efficiency, governance, and education with long-term net-zero targets (T5.11).

The literature also underscores the importance of managing sustainability through robust metrics. Leal Filho et al. (2025) showed Brazil's top ten performers in the UI GreenMetric focus on energy, climate, and waste, yet still overlook water and transport, weaknesses likewise flagged in the present analysis (W1.4, W1.6, W1.7, W7.4). Budihardjo et al. (2021) argued embedding sustainability indicators in faculty development programs accelerates curricular reform, supporting the call for mandatory sustainability competencies in every degree program towards strengthening the curricular dimension (O4.6, O4.8). Complementarily, de Moraes et al. (2024) warned smart-campus dashboards must aggregate data across units for avoiding "indicator silos"—that is why centralized institutional storage was highlighted here as an operational priority (W10.5).

Finally, the study emphasized digital infrastructure as a cornerstone for making SDGs a campus reality. Zarpellon et al. (2024) demonstrated an open-source, campus-wide sensor network can provide real-time energy and water data (S10.2) and da Silva et al. (2018) showed the way such data streams feed an integrated energy-management system that halves distribution losses and trains 1350 "change agents"

(O4.6, S10.2). The cases illustrated the practical viability of the digital and capacity-building actions proposed and underscored the international relevance of the present results beyond the Brazilian context.

5 Conclusions and Implications

This article presented a theoretical analysis of the Thematic Environmental Policies established by the Superintendence for Environmental Management at the University of São Paulo, emphasizing the importance of aligning university operations and curriculum with Sustainable Development Goals through a Sustainability SWOT analysis. It also provided evidence an effective implementation of environmental policies at the university level can significantly contribute to advancing SDGs.

USP served as a case study demonstrating how sustainability can be institutionalized across multiple dimensions, including operations (e.g., water management, sustainability in administration) and curriculum (e.g., environmental education). The literature confirms universities adopting structured environmental policies can reduce environmental impacts and promote sustainable practices such as efficient waste management, rational use of natural resources, and integration of environmental criteria into academic and administrative processes, in the USP case.

Nonetheless, several challenges including lack of coordination among institutional sectors, financial constraints on sustainable initiatives, and resistance to changes within segments of the academic community still remain.

S-SWOT proved valuable for identifying internal and external factors influencing sustainability initiatives at USP, offering a systematic method for evaluations on how those initiatives align with global sustainability objectives. S-SWOT analysis identified 109 factors, classified as strengths, weaknesses, opportunities, or threats across the 11 TEPs and mapped them to specific SDGs. It revealed while some environmental policies are specific to operations (e.g., Water and Effluents) and others are curriculum-oriented (e.g., Environmental Education), many span both areas, which underscores the importance of adopting a holistic and integrated approach.

Figures 10 and 11 illustrate the relationships between SDGs and TEPs and the way SDGs are distributed across operations, curriculum, or both, which helps identify areas where USP can enhance its sustainability strategies.

Opportunities for future research include application of S-SWOT tool through interviews or surveys with key stakeholders at USP (e.g., staff, faculty, and students) who are actively engaged with TEPs towards comparisons between theoretical findings and the perspectives of domain experts.

Another direction involves exploring in detail the positive and negative influences of S-SWOT elements on each of the 17 SDGs, identifying specific advantages and barriers faced by USP for achieving sustainability goals through its current policy framework.

5.1 Theoretical Implications

This study contributes to the theoretical discourse on how SWOT analysis can be adapted to sustainability contexts in higher education, particularly by integrating the environmental, economic, and social dimensions of sustainable development. It also enhances the literature by empirically demonstrating how university environmental policies can be aligned with SDGs.

Moreover, it highlights the importance of interdisciplinary collaboration in addressing sustainability challenges and bridging the gap between operational management and academic curriculum and reinforces the theoretical premise that universities, viewed as "living laboratories", play a vital role in advancing sustainable development through research, education, and institutional practice.

5.2 Managerial Implications

The findings suggest university leaders should develop and implement structured environmental policies aligned with SDGs. USP's TEPs offer a compelling example of such an approach which includes establishment of sustainability committees, allocation of resources for green infrastructure, and promotion of active engagement of the entire academic community.

S-SWOT can also prioritize sustainability initiatives such as improvement in waste management (Policy 9), increase in energy efficiency (Policy 5), or guidance of more comprehensive efforts at other institutions. The study identified areas requiring further investments (e.g., infrastructure upgrades for sustainable buildings and cost-saving opportunities, including circular economy practices related to water use.

University managers should also expand environmental education programs and establish partnerships with external stakeholders (e.g., other universities, governments, and private organizations) towards maximizing the impact of sustainability initiatives.

5.3 Policy Implications

The study indicated stronger institutional policy frameworks can help standardize and consolidate sustainability practices and USP has emerged as a regional model through the development and implementation of its TEPs.

S-SWOT can serve as an effective tool for policymakers both within USP and at other institutions to align local initiatives with global sustainability targets and ensure measurable impact.

In summary, the study provides a strategic roadmap for integrating sustainability into the core functions of higher education institutions and a practical, evidence-based tool for decision-making and contributes meaningfully to advancing the global sustainability agenda.

Acknowledgements Funding: This research was partially financed by the Coordenação de Aperfeiçoamento de Pessoal de Nível Superior—Brazil (CAPES)—Finance Code 001 and partially supported by grants 312585/2021-7 and 404819/2023-0 from Conselho Nacional de Desenvolvimento Científico e Tecnológico (CNPq), Brazil.

References

Agyekum EB (2020) Energy poverty in energy rich Ghana: A SWOT analytical approach for the development of Ghana's renewable energy. Sustainable Energy Technologies and Assessments, 40, 100760. https://doi.org/10.1016/j.seta.2020.100760

Ahonen VL, Woszczek A, Baumeister S, Helimo UT, Jackson AK, Kopsakangas-Savolainen M, Kääriä J, Lehtonen T, Luoranen M, Pongrácz E, Soukka R, Vainio V, Geneidy SE (2024) Carbon neutral higher education institutions: a reality check, challenges and solutions. International Journal of Sustainability in Higher Education, 25(9), 293–315. https://doi.org/10.1108/ijshe-11-2023-0515

Alsharari NM (2020) Internationalization market and higher education field: institutional perspectives. International Journal of Educational Management, 34(2), 315–334. https://doi.org/10.1108/ijem-12-2018-0402

Alves LLM, Angelo H, Almeida AN, Silva GF, Matricardi EaT, Nunes A, Júnior CVNS (2024). Strategic analysis of the forest carbon market in Brazil. Sustainability, 16(16), 6898.https://doi.org/10.3390/su16166898

Association for the Advancement of Sustainability in Higher Education (AASHE) (2024) STARS Technical Manual. https://stars.aashe.org/wp-content/uploads/2024/05/STARS-Technical-Manual-v3.0.pdf

Awuzie B, Ngowi AB, Omotayo T, Obi L, Akotia J (2021) Facilitating successful smart Campus Transitions: A Systems Thinking-SWOT Analysis approach. Applied Sciences, 11(5), 2044. https://doi.org/10.3390/app11052044

Aznar Minguet P, Ull MA, Piñero A, Martínez-Agut MP (2014) La sostenibilidad en la formación universitaria: desafíos y oportunidades [Sustainability in university education: challenges and opportunities]. Educación XX1, 17(1). https://doi.org/10.5944/educxx1.17.1.10708

Bautista-Puig N, Sanz-Casado E (2021) Sustainability practices in Spanish higher education institutions: An overview of status and implementation. Journal of Cleaner Production, 295, 126320. https://doi.org/10.1016/j.jclepro.2021.126320

Benzaghta MA, Elwalda A, Mousa M, Erkan I, Rahman M (2021) SWOT analysis applications: An integrative literature review. Journal of Global Business Insights, 6(1), 55–73. https://doi.org/10.5038/2640-6489.6.1.1148

Budihardjo MA, Ramadan BS, Putri SA, Wahyuningrum IFS, Muhammad FI (2021) Towards Sustainability in Higher-Education Institutions: Analysis of contributing factors and appropriate Strategies. Sustainability, 13(12), 6562. https://doi.org/10.3390/su13126562

Camilleri MA (2019) Higher Education Marketing: Opportunities and Challenges in the Digital Era. Academia, 4–28. https://doi.org/10.26220/aca.3169

Cheng B, Huang J, Guo Z, Li J, Chen H (2023) Towards sustainable construction through better construction and demolition waste management practices: a SWOT analysis of Suzhou, China.

International Journal of Construction Management, 23(15), 2614–2624. https://doi.org/10.1080/15623599.2022.2081406

CNN Brasil (2025) USP é a única latino-americana no top 150 universidades com maior reputação [USP is the only Latin American university in the top 150 most reputable universities]. https://www.cnnbrasil.com.br/educacao/usp-e-a-unica-latino-americana-no-top-150-universidades-com-maior-reputacao

da Silva LCP, Villalva MG, de Almeida MC, Brittes JLP, Yasuoka J, Cypriano JGI, Dotta D, Pereira JTV, Salles MBC, Archilli GB, Campos JG (2018) Sustainable Campus model at the University of Campinas-Brazil: an integrated living lab for renewable generation, electric mobility, energy efficiency, monitoring and energy demand management. In World sustainability series (pp. 457–472). https://doi.org/10.1007/978-3-319-76885-4_30

de Melo MRS, Cavaliero CKN, Dalbelo T, Anholon R, de Souza Moraes B (2025) Driving energy sustainability and strategic climate action at a Brazilian public university: a case study of UNICAMP. International Journal of Sustainability in Higher Education. https://doi.org/10.1108/ijshe-10-2024-0732

de Moraes PA, Pisani F, Borin JF (2024) Smart University: A pathway for advancing Sustainable Development Goals. Internet of Things, 27, 101246. https://doi.org/10.1016/j.iot.2024.101246

Díaz NP, Gutiérrez LS, Gutiérrez JÁV, Guerra EML, Rodríguez EP (2019) Strategic actions of environmental education in the community La Majagua for its local development. Cooperativismo y Desarrollo, 7(3), 2310–340X. http://coodes.upr.edu.cu/index.php/coodes/article/view/273

Eifert A, Julmi C (2022) Challenges and how to overcome them in the formulation and implementation process of a Sustainability Balanced Scorecard (SBSC). Sustainability, 14(22), 14816. https://doi.org/10.3390/su142214816

Fahim A, Tan Q, Naz B, Ain QU, Bazai SU (2021) Sustainable Higher Education Reform Quality Assessment Using SWOT Analysis with Integration of AHP and Entropy Models: A Case Study of Morocco. Sustainability, 13(8), 4312. https://doi.org/10.3390/su13084312

Fissi S, Romolini A, Gori E, Contri M (2020) The path toward a sustainable green university: The case of the University of Florence. Journal of Cleaner Production, 279, 123655. https://doi.org/10.1016/j.jclepro.2020.123655

Flyvbjerg B (2006) Five misunderstandings about Case-Study research. Qualitative Inquiry, 12(2), 219–245. https://doi.org/10.1177/1077800405284363

Franco I, Saito O, Vaughter P, Whereat J, Kanie N, Takemoto K (2018). Higher education for sustainable development: actioning the global goals in policy, curriculum and practice. Sustainability Science, 14(6), 1621–1642. https://doi.org/10.1007/s11625-018-0628-4

Ghazinoory S, Abdi M, Azadegan-Mehr M (2011) SWOT methodology: a state-of-the-art review for the past, a framework for the future / SSGG metodologija: praeities ir ateities analizė. Journal of Business Economics and Management, 12(1), 24–48. https://doi.org/10.3846/16111699.2011.555358

Gill A (2009) Knowledge management initiatives at a small university. International Journal of Educational Management, 23(7), 604–616. https://doi.org/10.1108/09513540910990834

Hernández-Diaz PM, Polanco J, Escobar-Sierra M, Leal Filho W (2021) Holistic integration of sustainability at universities: Evidences from Colombia. Journal of Cleaner Production, 305, 127145. https://doi.org/10.1016/j.jclepro.2021.127145

Hitchens K (2024) Brazilian University Launches New Solar and Energy Storage Microgrid. Microgrid Knowledge. https://www.microgridknowledge.com/campus-microgrids/news/55247964/brazilian-university-launches-new-solar-and-energy-storage-microgrid

Jain S, Pant P (2010) Environmental management systems for educational institutions. International Journal of Sustainability in Higher Education, 11(3), 236–249. https://doi.org/10.1108/14676371011058532

Jornal da USP (2024) USP mantém a liderança em ranking de universidades latino-americanas [USP maintains its leadership in Latin American university rankings]. https://jornal.usp.br/institucional/usp-mantem-a-lideranca-em-ranking-de-universidades-latino-americanas/

Leal Filho W, Raath S, Lazzarini B, Vargas V, de Souza L, Anholon R, Quelhas O, Haddad R, Klavins M, Orlovic V (2018) The role of transformation in learning and education for sustainability. Journal of Cleaner Production, 199, 286–295. https://doi.org/10.1016/j.jclepro.2018.07.017

Leal Filho W, Amaro N, Avila LV, Brandli L, Damke LI, Vasconcelos CR, Hernandez-Diaz PM, Frankenberger F, Fritzen B, Velazquez L, Salvia A (2021) Mapping sustainability initiatives in higher education institutions in Latin America. Journal of Cleaner Production, 315, 128093. https://doi.org/10.1016/j.jclepro.2021.128093

Leal Filho W, Lopes G, Trevisan L, Bella RL, Rampasso I, Júnior ACB, Bichueti R, Quelhas O, Anholon R, Galleli B (2025) Green metrics: how are Brazilian universities doing? Discover Sustainability, 6(1). https://doi.org/10.1007/s43621-025-00864-z

Leiber T (2018) Impact evaluation of quality management in higher education: a contribution to sustainable quality development in knowledge societies. European Journal of Higher Education, 8(3), 235–248. https://doi.org/10.1080/21568235.2018.1474775

Leiber T, Stensaker B, Harvey LC (2018) Bridging theory and practice of impact evaluation of quality management in higher education institutions: a SWOT analysis. European Journal of Higher Education, 8(3), 351–365. https://doi.org/10.1080/21568235.2018.1474782

Livzeniece L, Pubule J, Blumberga D (2021) Sustainability Assessment of Wind Energy in Latvia: Sustainability SWOT and Multi-Criteria Analysis. Environmental and Climate Technologies, 25(1), 1253–1269. https://doi.org/10.2478/rtuect-2021-0095

Lozano R (2009) Diffusion of sustainable development in universities' curricula: an empirical example from Cardiff University. Journal of Cleaner Production, 18(7), 637–644. https://doi.org/10.1016/j.jclepro.2009.07.005

Lozano R, Lozano FJ, Mulder K, Huisingh D, Waas T (2013) Advancing Higher Education for Sustainable Development: international insights and critical reflections. Journal of Cleaner Production, 48, 3–9. https://doi.org/10.1016/j.jclepro.2013.03.034

Mian SH, Salah B, Ameen W, Moiduddin K, Alkhalefah H (2020) Adapting Universities for Sustainability Education in Industry 4.0: Channel of Challenges and opportunities. Sustainability, 12(15), 6100. https://doi.org/10.3390/su12156100

Neris A (2024) Brazil's largest microgrid goes online. Pv Magazine International. https://www.pv-magazine.com/2024/11/22/brazils-largest-microgrid-goes-online/

Nobrega JHC, Rampasso IS, Sanchez-Rodrigues V, Quelhas OLG, Leal Filho W, Serafim MP, Anholon R (2021) Logistics 4.0 in Brazil: Critical Analysis and Relationships with SDG 9 Targets. Sustainability, 13(23), 13012. https://doi.org/10.3390/su132313012

Pacheco RM, de Medeiros MM, Montalván RaV, de Carvalho Pinto, CRS (2019) Análise da sustentabilidade das operações de instituições federais de ensino superior com a ferramenta Stars: a experiência da Universidade Federal de Santa Catarina [Analysis of the sustainability of federal higher education institutions' operations using the Stars tool: the experience of the Federal University of Santa Catarina]. Revista de Gestão Ambiental e Sustentabilidade [Journal of Environmental Management and Sustainability], 8(2), 205–234. https://doi.org/10.5585/geas.v8i2.1018

Pesonen H, Horn S (2013) Evaluating the Sustainability SWOT as a streamlined tool for life cycle sustainability assessment. The International Journal of Life Cycle Assessment, 18(9), 1780–1792. https://doi.org/10.1007/s11367-012-0456-1

Pickton DW, Wright S (1998) What's swot in strategic analysis? Strategic Change, 7(2), 101–109. https://doi.org/10.1002/(SICI)1099-1697(199803/04)7:2<101::AID-JSC332>3.0.CO;2-6

Ramdan HB, Shams S, Imteaz MA, Ahsan A, Honda T (2023) Addressing sustainability through waste management: a perspective from higher education institutions in Southeast Asia. Journal of Material Cycles and Waste Management, 25(2), 873–885. https://doi.org/10.1007/s10163-022-01566-8

Rebelatto BG, Salvia AL, Reginatto G, Daneli RC, Brandli LL (2019) Energy efficiency actions at a Brazilian university and their contribution to sustainable development Goal 7. International Journal of Sustainability in Higher Education, 20(5), 842–855. https://doi.org/10.1108/ijshe-01-2019-0023

Roorda N, Rammel C, Waara S, Paleo UF (2009) AISHE 2.0 Manual - English Edition (2.0). Tilburg. https://niko.roorda.nu/management-methods/aishe/
Sá CMd, Parada LE, Strengari LAB, Hellmeister MS (2015) Smart campus: análise e proposta de aplicação de conceitos inteligentes na Cidade Universitária "Armando de Salles Oliveira" [Smart campus: analysis and proposal for the application of smart concepts at the "Armando de Salles Oliveira" University City]. Escola Politécnica, Universidade de São Paulo. https://bdta.abcd.usp.br/directbitstream/b2992b81-00df-4d67-9b93-3861bcc1d9fa/CAIO%20MURAKI%20DE%20S%C3%81%20TF-15.pdf
Saric J, Käser F, Lys J, Utzinger J, Breu T (2021) Synergising research and service activities at Swiss research institutions to accelerate sustainable development. Sustainability, 13(17), 9626. https://doi.org/10.3390/su13179626
Silva-Da-Nóbrega PI, Chim-Miki AF, Castillo-Palacio M (2022) A Smart Campus Framework: Challenges and opportunities for education based on the sustainable development goals. Sustainability, 14(15), 9640. https://doi.org/10.3390/su14159640
Smith G (2011) Developing a sustainability plan at a large U.S. College of Education. Journal of Teacher Education for Sustainability, 13(2), 5–16. https://doi.org/10.2478/v10099-011-0011-3
Soares CaB, de Nadae J, Nascimento DCD, Firmino PRA, Morioka SN (2023) Photovoltaic solar energy and sustainability in higher education institutions: a multiple case study. GEPROS. 18, e02939. https://doi.org/10.15675/gepros.v18i00.2939
Soutullo S, Ferrer JA, Seco O, López H, Sánchez MN, Vitale MJ, Reyes AL, Correa EN, de Diego L (2024) Sustainable transformation in the Latin American and Caribbean districts through the implementation of a qualitative methodology. Challenges and key aspects to be addressed. Journal of Cleaner Production, 472, 143336. https://doi.org/10.1016/j.jclepro.2024.143336
Stake RE (1995) The art of case study research. SAGE
Superintendência de Gestão Ambiental (SGA) [Superintendence for Environmental Management] (2025a) USP é a quinta universidade mais sustentável do mundo segundo ranking GreenMetric [USP is the fifth most sustainable university in the world according to GreenMetric ranking]. https://sga.usp.br/usp-e-a-quinta-universidade-mais-sustentavel-do-mundo-segundo-ranking-greenmetric/
Superintendência de Gestão Ambiental (SGA) [Superintendence for Environmental Management] (2025c) Política Ambiental na Universidade [Environmental Policy at the University]. https://sga.usp.br/politica-ambiental-na-universidade/
Superintendência de Gestão Ambiental (SGA) [Superintendence for Environmental Management] (2025b) Sobre a SGA [About the SGA]. https://sga.usp.br/sobre-a-sga/
Tasdemir C, Gazo R (2020) Integrating sustainability into higher education curriculum through a transdisciplinary perspective. Journal of Cleaner Production, 265, 121759. https://doi.org/10.1016/j.jclepro.2020.121759
Universidade de São Paulo (USP) [University of São Paulo] (2009) Portaria GR No 4448, de 29 de Setembro de 2009 [Ordinance GR No. 4448, dated September 29, 2009]. https://leginf.usp.br/?portaria=portaria-gr-no-4448-de-29-de-setembro-de-2009
Universidade de São Paulo (USP) [University of São Paulo] (2018) Resolução No 7465, de 11 de Janeiro de 2018 [Resolution No. 7465, dated January 11, 2018]. https://leginf.usp.br/?resolucao=resolucao-no-7465-de-11-de-janeiro-de-2018
Universidade de São Paulo (USP) [University of São Paulo] (2023) USP em números [USP in numbers]. https://dopar.usp.br/num/
University Leaders for a Sustainable Future (ULSF). (2009). Sustainability Assessment Questionnaire (SAQ) for Colleges and Universities. https://ulsf.org/sustainability-assessment-questionnaire/
Uruchi NT, Diaz VS, Torres NNS, Maciel JN, Ledesma JJG, Cavallari MR, Gazziro M, Lago TGSD, Junior OHA (2025) Techno-Economic optimization of an isolated solar microgrid: a case study in a Brazilian Amazon community. Eng—Advances in Engineering, 6(7), 133. https://doi.org/10.3390/eng6070133

Vlados C (2019) On a correlative and evolutionary SWOT analysis. Journal of Strategy and Management, 12(3), 347–363. https://doi.org/10.1108/jsma-02-2019-0026

Wang H, Wu J, Muhedaner M, Maihemuti S (2022) College students online education Evaluation through SWOT Analysis during COVID-19. IEEE Access, 10, 88371–88385. https://doi.org/10.1109/access.2022.3198082

World Resources Institute (WRI) (2012) sSWOT: A Sustainability SWOT. https://www.wri.org/sswot

Yin RK (2014) Estudo de Caso: Planejamento e Métodos [Case Study: Planning and Methods]. 5th ed. Bookman.

Zarpellon BO, Arenas LdO, Godoy EP, Marafão FP, Paredes HKM (2024) Design and implementation of a smart campus flexible Internet of Things architecture on a Brazilian university. IEEE Access, 12, 113705–113725. https://doi.org/10.1109/access.2024.3444471

Christian Eric Barrantes-Briceño is a PhD candidate in Production Engineering at the São Carlos School of Engineering, University of São Paulo (Brazil), where he investigates how organizations measure progress towards the UN Sustainable Development Goals. He holds a BSc in Industrial Engineering from Manuela Beltrán University (Colombia) and an MSc in Production Engineering from São Carlos School of Engineering. His work appears in *International Journal of Sustainability in Higher Education* and *Journal of Cleaner Production* and focuses on knowledge management, organizational excellence models, and multicriteria decision analysis for sustainable development.

Marcelo Seido Nagano is a professor in the undergraduate and graduate programs in Production Engineering at the São Carlos School of Engineering, University of São Paulo (Brazil). His research area integrates operations research, knowledge management, and innovation-driven entrepreneurship for improving production systems. As a CNPq Productivity Fellow, he has published in *Journal of the Operational Research Society*, *Computers and Industrial Engineering*, *International Journal of Advanced Manufacturing Technology*, and *Journal of Cleaner Production*, among other periodicals. He holds a BSc in Mechanical-Mechatronics Engineering and an MSc and PhD in Mechanical Engineering from the University of São Paulo and completed a postdoctoral fellowship in Mathematics and Applied Computing at the National Institute for Space Research.

Dropwise: A Game to Integrate Mathematics and Water Conservation

Isabela das Chagas Luiz, Elisa Henning, and Andreza Kalbusch

Abstract Water is one of the most important resources on Earth: there is no life, environment and economic production without it. No human activity takes place without water. As traditional school curricula do not always properly prepare individuals for the complex global reality, including limited water resources, critical environmental education becomes a necessity, a continuous and permanent process that must cover all school levels and stages of formal and informal education. Due to its importance, environmental education is transversely included in the documents that guide the curricular structure of Elementary School in Brazil. In this context, the purpose of this research work was to create a board game, aiming to contribute to the environmental training of students. The game, called "Dropwise", was designed to be applied in the sixth grade Mathematics classes of Elementary School. The game aims to enable the learning of practices to reduce water consumption, including recognizing places at home that can be equipped with water-saving devices, for example, dual-flush toilets and aerators on kitchen and bathroom faucets. The proposal meets the competencies of the National Common Core Curriculum for the sixth grade, working on problem-solving that uses research data on environmental contexts, enabling critical thinking. During the research, working with undergraduate students in Mathematics, who are, actually, prospective teachers, was proposed. The involvement of teachers in training is important for the inclusion of sustainability in the curricula of undergraduate programs. Preliminary results highlight the potential of the game in learning Mathematics contents and in the reflection on water conservation. The integration

I. das Chagas Luiz (✉) · E. Henning
College of Technological Science, Mathematics Department, Santa Catarina State University, Joinville, Brazil
e-mail: 11882734939@edu.udesc.br

E. Henning
e-mail: elisa.henning@udesc.br

A. Kalbusch
College of Technological Science, Civil Engineering Department, Santa Catarina State University, Joinville, Brazil
e-mail: andreza.kalbusch@udesc.br

W. Leal Filho et al. (eds.), *Sustainability in the Curriculum and Operations of Universities*, World Sustainability Series,
https://doi.org/10.1007/978-3-032-12678-8_12

of games into education is important, as it represents a playful approach to learning. By playing, children develop an understanding of the social world around them.

Keywords Environmental education · Mathematics · Prospective teachers · Water conservation · Sustainability

1 Introduction

As a vital natural resource, essential for ecosystem balance, human society, and economic growth, water plays a crucial role in sustaining life. (Yang et al. 2015). Despite its importance for the survival of the human race, data from 2022 by United Nations Children's Fund and World Health Organization (2024) show that 2.2 billion people had no access to safely managed drinkable water. From that perspective, "education is increasingly and globally seen to be a powerful tool for preparing individuals to enter the labor market as well as to create a peaceful and sustainable society" (UNESCO 2011; Lauder et al. 2006, as cited in Cars and West 2015). Thus, Education for Sustainable Development (ESD) becomes a necessity, covering all school levels, including Higher Education, especially in the training of future teachers.

At an international level, the United Nations Development Programme (UNDP) has developed seventeen "Sustainable Development Goals (SDGs)" as a universal call to action, aiming for all people to enjoy peace and prosperity by 2030 (UNDP 2015). The fourth SDG directly relates to the access to quality education, more specifically with goal 4.7: "by 2030, ensure that all learners acquire the knowledge and skills needed to promote sustainable development, including, among others, through education for sustainable development and sustainable lifestyles" (UNDP 2015). This brings to light the importance of education concerned with sustainable development.

The United Nations Educational, Scientific and Cultural Organization (UNESCO) sees Education for Sustainable Development as a field that encompasses Environmental Education (UNESCO 2005). Many conceptions of Environmental Education already include economical and social dimensions in line with the sustainable development perspective, in such a manner that speaking about Environmental Education is not different from speaking about Education for Sustainable Development (Barbieri and Silva 2011).

In Brazil, the environmental debate took place one year after the Stockholm Conference, in 1973, when the Special Secretariat for the Environment was created. Among other attributions, it aimed at controlling pollution and promoting formal and informal Environmental Education (EE). Furthermore, many achievements took place, such as environmental laws integrating education and the environment. In 1981, Federal Law 6938/81 of the National Environmental Policy (Brasil 1981) was established. It was the first law to establish Environmental Education as an instrument to search for solutions to environmental problems and to determine that it should be offered at all levels of education.

For the training of teachers, the inclusion of Environmental Education started legally in 2012, when the National Education Council established the National Curricular Guidelines for Environmental Education in Brazil, a document guiding the curricular structure using Environmental Education (Brasil 2012), reaffirming its presence in all levels of education and determining that it should also be present in the teachers' initial training.

According to a study by Corpuz et al. (2022), studies indicated the need for strengthening the integration of Environmental Education in teacher training programs. Thus, understanding that questions involving the environment can be addressed in university curricula is important, especially in Bachelor's degree programs (Arana and Bertoli 2021).

Thus, with these policies, teachers are expected to be better prepared to integrate Environmental Education into their classes, contributing to the environmental citizenship of the students. Teachers are responsible for carrying out Environmental Education, which is accomplished through their knowledge, perceptions, and intent concerning education and the environmental issues (Torales 2013). Specifically, the Mathematics teacher, in addition to playing the social role of educator, must be able to insert their students into different contexts and realities; to do so, they need to understand contemporary issues and carry out comprehensive and indispensable education to impact solutions found in a global and social context (Brasil 2001).

In the Mathematics for Elementary School course, the practice of Environmental Education is also explicit in the National Curricular Parameters, a set of educational guidelines from the Ministry of Education in Brazil, which recommends working with transversal themes within Mathematics, such as Environmental Education (Brasil 1997). Therefore, mathematical contents should be taught in a way that promotes reflection about socio-environmental issues, contributing to a change in behavior, respecting the environment and promoting common well-being and sustainable development.

Including Environmental Education in the educational system, interlinked with courses like Mathematics, calls for a new teacher profile, responsible for the development of a critical and transforming approach. Thus, Higher Education institutions must be ready to contribute to such training. However, Guerra and Figueiredo (2014) warn that schools and Higher Education institutions face major hurdles when implementing the inclusion of Environmental Education according to the guidelines proposed by the National Environmental Policy and by the National Curricular Guidelines for Environmental Education.

Universities play a crucial role in guiding society toward a sustainable future, as they function within a larger social framework and have the capacity to drive responsible development (Cortese 2003, as cited in Mohanty 2021). In the case of the Santa Catarina State University, Rosa (2020) suggested the inclusion of Environmental Education into the program's pedagogical project. The pedagogical project of the

2005 program brings a curricular component called Mathematics Teaching Laboratory III, in the fourth semester, whose syllabus includes teaching projects that contextualize mathematical contents, the development of interdisciplinary projects using manipulative materials and the creation of manipulative materials for the teaching of Mathematics, such as board games. Furthermore, the curricular component specifically aims to understand the possible relation between Mathematics Education and Environmental Education. It is a living laboratory for the future Mathematics teacher to experience practices and teaching activities that will, later, be applied in the classroom. From this perspective, a serious game called Dropwise was developed, which offers an opportunity to work with Environmental and Mathematics Education in courses such as the Mathematics Teaching Laboratory.

In this scenario, this research work's main goal was to create a board game (Dropwise), aiming to contribute to the environmental training of students. In this paper we seek to answer the following research question: how can the Dropwise game contribute to the training of students and of future Mathematics teachers, integrating Mathematical Education with water conservation practices in the context of sustainable development? Our proposal is innovative as it describes a serious game that contributes to student education by aligning Environmental Education, Mathematics' contents and water conservation, including the knowledge of water-saving devices.

2 Literature Review

Educational games are a powerful tool for engaging students by addressing complex problems in an enjoyable and pedagogically effective manner (Garcia et al. 2016). In addition to promoting learning, educational games add the elements of fun, competition, and engagement to the intended purpose of promoting environmentally sustainable behaviors (Tan and Nurul-Asna 2023).

Ho et al. (2022), in their paper, state that many studies sustain that the application of games aids students in understanding the necessary interdisciplinary concepts for the sustainable development practice, while contributing to changing behaviors. More recently, according to Bilancini et al. (2023), several works have emerged in the literature exploring the effectiveness of game-based learning for promoting sustainability and environmental education. Bilancini et al. (2023), in addition to a literature review on this topic, provide, in their study, a proposal for a game-based educational program which turned out as a potentially effective action for promoting sustainable water use practices, both for the students and for their families.

In this context, a category of games named "serious games" was proven to support the learning of abstract concepts, including water-related issues (Poděbradská et al. 2020). Serious games are designed for educational purposes and are often used alongside traditional teaching methods to help achieve various curriculum objectives (Coakley and Garvey 2015), as they allow for the integration of teaching and entertainment, creating an environment of engaging learning experiences, contributing to environmental awareness (Garcia et al. 2025).

Ahmadov et al. (2025) performed an extensive literature review on the application of serious games for sustainable environmental education, showing an increase in the number of research papers in the field which, in turn, reveals the recognition of games related to the topic. Furthermore, for Ahmadov et al. (2025), "expanding the focus of sustainability education beyond environmental aspects to include social sustainability dimensions empowers students to address multifaced societal challenges".

In this sense, serious games can be highlighted as a valuable tool in Environmental Education, as they have great potential to transform perceptions related to environmental conservation and to encourage sustainable behaviors (Tan and Nurul-Asna 2023). Furthermore, serious games can impact environmental knowledge (Juan and Chao 2015) and create a sense of conservation awareness (Terton and White 2014), especially when linked to real contexts and close to the students' experiences.

In the context of water consumption, engaging gameplay can serve as an effective means to illustrate the complex nature of water resource management (Douven et al. 2012; Rajabu 2007). Thus, serious games, especially those focused on water issues, constitute a promising strategy for the dissemination of sustainable behaviors, aligned with the Sustainable Development Goals (SDG).

3 The Dropwise Game

The Dropwise game (Fig. 1) was designed for application in Mathematics classes in the sixth grade of Elementary School by a Mathematics teacher. It aims to help with the learning of practices to reduce water consumption, including the knowledge of places in a house that can be equipped with water-saving devices, for example, dual-flush toilets and aerators on kitchen and bathroom taps. This proposal meets the competencies of the National Common Core Curriculum (Brasil 2018), the document that rules Brazil's school curriculum, for the sixth grade, working with problem solving with research data on environmental contexts, enabling critical thinking.

The use of games in education is productive, once it is a playful manner of learning. According to Ortega Zirawaga (2017), incorporating games into education plays an important role in engaging students by encouraging a practical approach. It is important, therefore, that the future Mathematics teacher should have elements in their training that can help in this process.

For the Dropwise game, players receive a model of a house and a water consumption card, whose value is changed as the game progresses (Fig. 1). The house has spaces to place tokens representing water consumption reduction equipment or practices, which are earned while playing (Fig. 2). The board is subdivided into spaces, with questions regarding consumption reduction practices and water-supply related terms for buildings, such as "what is a water meter?" There are also actions to promote player engagement and make the game more dynamic, such as losing a turn or interacting with another player's resources.

In addition to the learning of water conservation, the game covers some mathematical content. Hussein et al. (2022) observed that learning approaches based on games

Fig. 1 Parts of the dropwise game: house model, consumption card, board, pawns, cards, tokens and digital rulebook

are useful to improve learning performance of students in Mathematical Education. The use of Mathematics was linked to the water consumption theme, for example, with units' conversion from liters to cubic meters, using mathematical competencies expected for the target audience, sixth graders from Elementary School.

Thus, using games "can create engaging and impactful learning experiences, fostering collaborative skills and a deep understanding of science concepts" (Othman and Ching 2024), which are necessary for the learning of new water conservation ways, be it through new practices or by knowing consumption reduction equipment, which are necessary for an effective promotion of water conservation.

4 Methods

This research was exploratory with a qualitative approach (Creswell and Poth 2016). To answer the research question, the game was played by a group of Mathematics Education undergraduate students who are, in fact, future teachers (Fig. 3). The involvement of teachers in training is important for the inclusion of sustainability in undergraduate programs' curricula. The application of the game took place in November 2024, during a one-hour class of the Mathematics Teaching Laboratory III course from the Santa Catarina State University, with three Mathematics Education undergraduate students. After the application, a semi-structured interview was

Fig. 2 House model has spaces to place tokens representing water-saving devices or practices

conducted to capture the students' opinions and perceptions about the activity. The selection of the course and class was made based on a non-probabilistic sampling by judgment. The semester and the course matter because, at that moment of the program, students are starting the practical teaching courses. It is important to highlight that Environmental Education and the use of games for teaching are goals of the Mathematics Teaching Laboratory III course.

The semi-structured interview questions were open-ended and divided into three categories: experience while playing Dropwise; pedagogical applicability of the game; and contribution to the teacher's training. The answers were collected with a voice recorder and later transcribed verbatim. The analysis of the response data was carried out using content analysis from the perspective of Bardin (2016). A pre-analysis was carried out, extracting relevant excerpts from the interviews and categorizing the answers based on common factors, ending with the interpretation of the answers. The participating students were identified by code names (Student 1, Student 2 and Student 3). All participants authorized the use of their data by signing an Informed Consent Form approved by the Santa Catarina State University's Ethics Committee (CAAE no. 558483622.6.0000.0118). The categories and respective interview questions can be seen in Table 1.

Fig. 3 The dropwise game being played by a group of undergraduate students in mathematics education

5 Results and Discussion

During the application of the game, the students were observed to actively interact, debating the game, water conservation and the relationship with mathematics, showing engagement and involvement with the proposed themes. In the first category, regarding playing experience, Student 1 reported:

> I like that we have to think about which way to go, to do the math. To think if I already have this token, if it's worth going after it, because, at a certain point, I went to a space (on the board) that had this little piece I already had. (Student 1).

In this answer, the student is referring to a house improvement token as "this little piece". She says what she liked the most when playing was the decision making when deciding what path to take aiming to fully equip the house with the water-saving equipment. Another student added that she liked the game's dynamics of reducing consumption by correctly answering questions about practices and equipment that reduce water consumption or about mathematics. This type of engagement reflects more than simple participation, it involves reflection, analysis and choice. Such cognitive processes are aligned with what Gros (2007) describes as a key benefit of serious environmental games: by simulating real-world environmental challenges, the games encourage critical thinking, particularly in the formulation of problem-solving strategies.

When the question was asked about the game being engaging for sixth graders, the age group the game is intended to, Student 2 remarked that, for the game's target group, classes with different proposals than the traditional ones are engaging. Student 2 presented a positive aspect of using playful tools such as a board game: stimulating

Table 1 Questions script for the interviews carried out with the students

Category	Question
Experience when playing dropwise	Did you enjoy the game?
	What part did you like the most?
	What suggestions would you have to improve the game?
	What did you think of the game's approach to working on topics such as conscious water consumption and sustainability?
	Do you find the game engaging and suitable for the 6th grade age group? Why?
Pedagogical applicability	How do you think a game like this can be used in the classroom?
	How can the game facilitate learning interdisciplinary topics, such as Mathematics and the environment?
	Did you notice any moments in the game that encouraged interaction or collaboration between players?
	How can this be useful in pedagogical practice?
	What competencies, in addition to the curricular content, do you believe the game helps to develop in students?
	From zero to ten, how much did the game stimulate the teaching–learning process in a playful and environmental way?
	What do you imagine would be the main difficulty in using a game like this in the context of a real school?
Contribution to teacher training	Have you ever had contact with this type of material to teach mathematics integrating the environment in the undergraduate program context?
	While playing Dropwise, did you feel better prepared to develop and adapt playful materials for teaching?
	Did the game bring up ideas or strategies that you would apply in your own classroom?
	In your opinion, does the game help develop the ability to mediate discussions about sustainability and responsible consumption in the classroom?

knowledge in an engaging manner that is different from a traditional class. A study by Zainuddin et al. (2020) indicates that students who learn with gamified learning tools in a traditional classroom may be more engaged and improve their learning performance. Student 1 also said:

> I think they are very competitive at this age, and bringing this competitiveness into the classroom in a beneficial way, can be very useful, especially in terms of sustainability, which is something that needs to be worked on. (Student 1)

According to Student 1, the game presents a certain competitiveness between players. Furthermore, according to her, competitiveness provokes more student engagement in the discussion about sustainability, as the comments by the students during the application of the game show.

> Yours and hers only need two more (tokens to complete the house). Right now, if she (Student 3) ends (the game), I win. (Student 1)
>
> Yes. Yes. (Student 2)
>
> That's why she will choose to go to the question rather than land on the space with the token (Student 1).

In this situation, Student 1 realized that Student 3 needed to choose the space with a question instead of a token because, due to the level of her consumption meter, even if she was the first to fully equip her house, she wouldn't have the lowest consumption, as Student 1 herself answered more questions correctly during the game, earning her the lowest water consumption. Regarding the pedagogical applicability matter, the students provided the following remarks:

> I believe it can be used jointly, not just by one teacher, but in a joint effort, maybe the mathematics teacher with the mathematics questions, along with the science or geography teacher. (Student 3)
>
> So, maybe a 40-minute class is not enough (to apply the game). So, if another teacher who is also working this topic joins in, I believe it would be nicer. (Student 3)

These remarks show that the game can have a multidisciplinary perspective, which allows it to be applied along with other disciplines with themes related to sustainable development education. The students were also asked if the game allowed them to learn terms related to the environment and sustainability and the answers were:

> I think I can see the application of the things we hear about many times, right? Because they will think: "Look, I must reduce consumption of my little house, I must be more mindful when answering the questions. (Student 1).

This highlights how the game enables learning and visualizing concepts. Van Eck (2006) stated that educational games take place in a meaningful context and, for that reason, they have positive effects on the students' learning performance. Hence, using a board game allows us to integrate the water conservation theme into this teaching and learning process. The students were also asked about how much the game stimulated the teaching–learning process in a playful and environmental way, and one of the answers was:

> Me, I would rate it a 10. I have no complaints about the game. It is playful and teaches what it proposes really well. (Student 2)

The students highlighted that they did not know the water-saving devices, such as the dual-flush valves and the aerator, before the game, which shows the game's importance in the training and building of knowledge. The students answered:

> Me, as an 11 year old child, would definitely not know what those two buttons were for. I still get confused many times. (Student 1)
>
> I had no idea. I just found out about it. (Student 2)
>
> I didn't know what it (the aerator) was supposed to do. I thought it was just to make the tap look good (Student 3).

About the game's contribution to the training of teachers, the students were asked whether they had any contact with similar material to teach Mathematics integrated with water conservation during the Mathematics Education undergraduate program, to which they replied negatively. Still on this topic, they were asked if, during the game, they felt prepared to design or adapt playful materials for teaching or if the game had brought any ideas they could apply in class. The students unanimously stated that the game is inspiring and allowed them to observe the use of games for Environmental Education integrating Mathematics. Finally, they were asked if the game helped to develop skills and mediate discussions about sustainability and responsible water consumption in class. Student 2 answered:

> I believe that, no matter what happens in the classroom, there will be discussions about the consumption, about who won. (Student 2)

The answer was similar to what happened when the game was applied to the students. At the end of the activity, they had a debate regarding the fairness of the results. During the discussion, the students considered the number of correct answers per player and the completion of the house model with the water-saving tokens. Student 2 conquered the victory for having her house model fully equipped with the necessary tokens, even though she had correctly answered fewer questions than the other student. This interaction illustrates how games can stimulate meaningful engagement between participants, leading to reflective discussions on complex conservation-related issues (Neset et al. 2020). Furthermore, the results agree with Ahmadov et al. (2025), as they expand the knowledge beyond the environmental aspects, contributing to a more comprehensive education. This shows that the Dropwise game has potential as a pedagogical practice in the training of Mathematics teachers and, later, for application in the sixth grade of Elementary School. The perceptions of the students helped to clarify those aspects and offered the possibility to think about applications which will integrate more than one discipline.

The game clearly motivated the engagement of the students who will soon become teachers. This is important because, according to Ho et al. (2022), in the context of SDGs, "teachers should develop an instructional method to promote engagement as well as focused and spontaneous participation in learning". It also agrees with the ideas of Garcia et al. (2025), since it revealed itself as a teaching proposal which inspires student engagement.

The Dropwise game can contribute to students' Environmental Education by promoting the dissemination of knowledge about water conservation in homes, with practical application in the children's daily lives. Children, by learning about water conservation practices at school, can become agents of dissemination of such knowledge, sharing it with their families. These are expected results of the Dropwise game, with a contribution to the dissemination of Environmental Education through society. In addition, the game contributes to the expansion of this field of study in the literature, with an approach that integrates the teaching of mathematics and the subject of water conservation in the built environment.

6 Conclusions

Environmental education, especially regarding sustainable development, must be planned and part of the guiding documents of educational institutions, in all levels, including Higher Education, especially in Education undergraduate programs, which are the foundation of an Elementary School teacher's training. Games, especially serious ones, have been proven effective actions in Environmental Education, as well as in water conservation topics. Including games in the teaching and learning process allows to address this topic's interdisciplinary aspects, to support the understanding of abstract concepts and to promote student engagement.

This paper presented an application of the Dropwise game with three Mathematics Education undergraduate students, which showed the benefits of using a board game. The students were not familiar with water-saving equipment, which enabled them to learn about devices that reduce water consumption. The game also provides better training for future teachers who will be able to educate students for sustainable development.

Answering the question that drove this study, the Dropwise game can aid sustainable development education in the training of teachers, once the use of games is part of the syllabus and program contents in the Mathematics Education undergraduate programs. The game has potential, since it inspired the students, as observed in their statements, to use board games integrating Environmental Education and Mathematics. Furthermore, it yielded engagement and discussion on the topic. Thus, using the Dropwise game in a course with a focus on training future teachers on water-saving practices and devices while integrating with the Mathematics contents is a possible activity with positive results.

The Dropwise game is able to integrate Mathematics Education and water conservation, as the students highlighted it was engaging, playful and it allowed for a positive approach. The students also pointed out the possibility of working on the game from a multidisciplinary perspective. Besides that, the undergraduate students remarked that the game provided inspiration to integrate environmental concepts with Mathematics, contributing to their training, as well as to the skills traditionally addressed in the Laboratory course.

This work's limitation lies in the fact that it has only been applied to one group of students. Thus, the results might not be extended into other contexts. Regarding the continuation of this research, we expect to apply this board game to more undergraduate student classes, as they will become teachers in the future, since one limitation found during our research was the number of students in the class. The continuity of this research work also includes the game's validation from its application at a sixth grade class in Elementary School, with the participation of researchers from the different fields involved in the proposal.

Funding This research was funded by Fundação de Amparo à Pesquisa e Inovação do Estado de Santa Catarina (FAPESC, grant number 2023TR000329) and Conselho Nacional de Desenvolvimento Científico e Tecnológico (CNPq, grant number 423090/2021–6).

References

Ahmadov, T., Karimov, A., Durst, S., Saarela, M., Gerstlberger, W., Wahl, M. F., Karkkainen, T. (2025). A two-phase systematic literature review on the use of serious games for sustainable environmental education. Interactive Learning Environments, 33(3), 1945-1966. https://doi.org/10.1080/10494820.2024.2414429

Arana, A. R. A., Bertoli, S. C. (2021). Educação ambiental no currículo de uma instituição de ensino superior: o processo de ambientalização curricular [Environmental education in the curriculum of a higher education institution: the curricular environmentalization process]. Revista de Educação Pública, 30. https://doi.org/10.29286/rep.v30ijan/dez.7191

Barbieri, J. C., Silva, D. D. (2011). Desenvolvimento sustentável e educação ambiental: uma trajetória comum com muitos desafios [Sustainable development and environmental education: a common trend with many challenges]. RAM. Revista de Administração Mackenzie, 12, 51-82. https://doi.org/10.1590/S1678-69712011000300004

Bardin, L. (2016). Content analysis. Edições 70, 279.

Bilancini, E., Boncinelli, L., Di Paolo, R. (2023). Game-based education promotes practices supporting sustainable water use. Ecological Economics, 208, 107801. https://doi.org/10.1016/j.ecolecon.2023.107801

Brasil. Conselho Nacional de Educação. (2012). Resolução CNE/CP nº 1. Diretrizes Curriculares Nacionais para a Educação Ambiental. Diário Oficial da União, Brasília, DF. http://portal.mec.gov.br/dmdocuments/rcp002_12.pdf [Guideline 1: National Curricular Guidelines for Environmental Education]

Brasil. Ministério da Educação. Conselho Nacional de Educação. (2001). Parecer n. 1.302, de 6 de novembro de 2001. Diretrizes Curriculares Nacionais para os cursos de Matemática, bacharelado e licenciatura. Diário Oficial da União, Brasília, DF. http://portal.mec.gov.br/cne/arquivos/pdf/CES13022.pdf [Guideline 1302: National Curricular Guidelines for the Mathematics Programs]

Brasil. Ministério da Educação. (1997). Parâmetros Curriculares Nacionais: Matemática. Brasília, DF: MEC/SEF. http://portal.mec.gov.br/seb/arquivos/pdf/livro03.pdf [National Curricular Parameters: Mathematics]

Brasil. Ministério da Educação. (2018). Base Nacional Comum Curricular: Versão final. MEC/Secretaria de Educação Básica. https://www.gov.br/mec/pt-br/escola-em-tempo-integral/BNCC_EI_EF_110518_versaofinal.pdf [National Common Core Curriculum: Final version]

Brasil. (1981). Lei n.º 6.938, de 31 de agosto de 1981. Dispõe sobre a Política Nacional do Meio Ambiente, seus fins e mecanismos de formulação e aplicação e dá outras providências. Diário Oficial da União, Brasília, DF, 2 set. 1981. https://www.planalto.gov.br/ccivil_03/leis/l6938.htm [Law no. 6938 on National Policies for the Environment, its purposes,formulation mechanism and applications, and other determinations]

Cars, M., West, E.E. (2015). Education for sustainable society: attainments and good practices in Sweden during the United Nations Decade for Education for Sustainable Development (UNDESD). Environ Dev Sustain 17, 1–21 (2015). https://doi.org/10.1007/s10668-014-9537-6

Coakley, D., Garvey, R. (2015). The great and the green: Sustainable development in serious games. In ECGBL2015–9th European Conference on Games Based Learning: ECGBL2015 (Vol. 135). Academic Conferences and publishing limited.

Corpuz, A. M., San Andres, T. C., Lagasca, J. M. (2022). Integration of Environmental Education (EE) in Teacher Education Programs: Toward Sustainable Curriculum Greening. Problems of Education in the 21st Century, 80(1), 119.

Cortese, A. (2003). The critical role of higher education in creating a sustainable future. Planning for Higher Education, 31(3), 15–22.

Creswell, J. W., Poth, C. N. (2016). Qualitative inquiry and research design: Choosing among five approaches. Sage publications.

Douven, W., Mul, M. L., Fernández-Alvarez, B., Lam Hung, S., Bakker, N., Radosevich, G., & Van Der Zaag, P. (2012). Enhancing capacities of riparian professionals to address and resolve transboundary issues in international river basins: experiences from the Lower Mekong River

Basin. Hydrology and Earth System Sciences, v. 16, n. 9, p. 3183-3197, https://doi.org/10.5194/hess-16-3183-2012.

Garcia, X., Domene, E., Goenaga, X., Rodriguez-Benitez, A., Satorras, M., Acuña, V., Martínez-Ruiz, A., Boada, I., Corominas, L. (2025). Evaluating the effectiveness of a serious game to educate children about the urban water cycle. British Journal of Educational Technology, 00, 1–20. https://doi.org/10.1111/bjet.13578

Garcia, C., Dray, A., Waeber, P. (2016). Learning begins when the game is over: using games to embrace complexity in natural resources management. GAIA-Ecological Perspectives for Science and Society, v. 25, n. 4, p. 289–291, 2016. https://doi.org/10.14512/gaia.25.4.13

Gros, B. (2007). Digital games in education: The design of games-based learning environments. Journal of research on technology in education, v. 40, n. 1, p. 23-38, https://doi.org/10.1080/15391523.2007.10782494

Guerra, A. F. S., Figueiredo, M. L. (2014). Ambientalização curricular na Educação Superior: desafios e perspectivas [Curricular greening in Higher Education: challenges and perspectives]. Educar em Revista, 109-126.https://doi.org/10.1590/0104-4060.38110

Ho, S. J., Hsu, Y. S., Lai, C. H., Chen, F. H., Yang, M. H. (2022). Applying Game-Based Experiential Learning to Comprehensive Sustainable Development-Based Education. Sustainability, 14(3), 1172. https://doi.org/10.3390/su14031172

Hussein, M. H., Ow, S. H., Elaish, M. M., Jensen, E. O. (2022). Digital game-based learning in K-12 mathematics education: a systematic literature review. Education and Information Technologies, 27, 2859–2891 (2022). https://doi.org/10.1007/s10639-021-10721-x

Juan, Y. K., Chao, T. W. (2015). Game-based learning for green building education. Sustainability, 7(5), 5592-5608. https://doi.org/10.3390/su7055592

Lauder, H., Brown, P., Dillabough, J. A., Halsey, A. H. (Eds.) (2006). Education, Globalization, and Social Change. Oxford University Press.

Mohanty, A. (2021). Education for Sustainable Development: A Framework for Global Partnership and Sustainability in India. In: Leal Filho, W., Marisa Azul, A., Brandli, L., Lange Salvia, A., Wall, T. (eds) Partnerships for the Goals. Encyclopedia of the UN Sustainable Development Goals. Springer, Cham. https://doi.org/10.1007/978-3-319-95963-4_132

Neset, T. S., Andersson, L., Uhrqvist, O., Navarra, C. (2020). Serious gaming for climate adaptation—assessing the potential and challenges of a digital serious game for urban climate adaptation. Sustainability, 12(5), 1789. https://doi.org/10.3390/su12051789

Othman, M.K., Ching, S.K. Gamifying science education: How board games enhances engagement, motivate and develop social interaction, and learning. Educ Inf Technol **29**, 24525–24561 (2024). https://doi.org/10.1007/s10639-024-12818-5

Poděbradská, M., Noel, M., Bathke, D. J., Haigh, T. R., Hayes, M. J. (2020). Ready for Drought? A Community Resilience Role-Playing Game. Water, 12(9), 2490. https://doi.org/10.3390/w12092490

Rajabu, K. R. (2007). Use and impacts of the river basin game in implementing integrated water resources management in Mkoji sub-catchment in Tanzania. Agricultural Water Management, 94(1-3), 63-72. https://doi.org/10.1016/j.agwat.2007.08.010

Rosa, R. S. da. (2020). Educação ambiental nos cursos de licenciatura em matemática de Santa Catarina [Environmental education in mathematics undergraduate courses in Santa Catarina]. [Tese de Doutoramento, Universidade Tuiuti do Paraná]. https://tede.utp.br/jspui/handle/tede/1801

Tan, C. K., Nurul-Asna, H. (2023). Serious games for environmental education. Integrative Conservation, 2(1), 19-42. https://doi.org/10.1002/inc3.18

Terton, U., White, I. (2014). A computer-based educational adventure challenging children to interact with the natural environment through physical exploration and experimentation. In International Conference on Computer Supported Education (Vol. 2, pp. 93–98).

Torales S, M. A. (2013). A inserção da educação ambiental nos currículos escolares e o papel dos professores: da ação escolar à ação educativo-comunitária como compromisso politico pedagógico [The inclusion of environmental education in school curricula and the role of

teachers: from school action to educational-community action as a political-pedagogical commitment.]. Revista Eletrônica do Mestrado em Educação Ambiental, Rio Grande/RS, v. especial, p. 1–17, mar. 2013. https://repositorio.furg.br/handle/1/3948
UNDP. (2015). United Nations Development Programme. Sustainable development goals. https://www.undp.org/sustainable-development-goals
UNESCO. (2005). United Nations Decade of Education for Sustainable Development, 2005–2014: Draft International Implementation Scheme. UNESCO. https://unesdoc.unesco.org/ark:/48223/pf0000139937
UNESCO (2011). Education for sustainable development (ESD). UNESCO. https://www.unesco.org/en/sustainable-development/education
United Nations Children's Fund, World Health Organization. (2024). Progress on household drinking water, sanitation and hygiene 2000–2022: special focus on gender. World Health Organization. https://www.who.int/publications/m/item/progress-on-household-drinking-water--sanitation-and-hygiene-2000-2022---special-focus-on-gender
Van Eck, R. (2006). Digital game-based learning: It's not just the digital natives who are restless. EDUCAUSE review, 41(2), 16. https://commons.und.edu/tlpp-fac/25/
Yang, J., Lei, K., Khu, S., Meng, W. (2015). Assessment of Water Resources Carrying Capacity for Sustainable Development Based on a System Dynamics Model: A Case Study of Tieling City, China. Water Resour Management, 29, 885–899. https://doi.org/10.1007/s11269-014-0849-y
Zainuddin, Z., Shujahat, M., Haruna, H., Chu, S. K. W. (2020). The role of gamified e-quizzes on student learning and engagement: An interactive gamification solution for a formative assessment system. Computers & education, 145, 103729. https://doi.org/10.1016/j.compedu.2019.103729
Zirawaga, V. S., Olusanya, A. I., Maduku, T. (2017). Gaming in education: Using games as a support tool to teach history. Journal of Education and Practice, 8(15), 55-64.

Isabela das Chagas Luiz holds a Mathematics degree from Universidade do Estado de Santa Catarina (2026). She was an undergraduate research grant by CNPq at the Laboratory for Urban Water Study (Leau). She develops research in the field of Statistical Methods and investigation of factors related to water consumption in the built environment.

Elisa Henning has a Civil Engineering degree from Universidade do Estado de Santa Catarina (1992), a Master's degree in Environmental Engineering from Universidade Federal de Santa Catarina (1998), a specialization in Mathematics and Statistics (Universidade Federal de Lavras), a Master's degree in Statistics from Universidade Aberta de Portugal (2014) and a Doctor's degree in Production Engineering from Universidade Federal de Santa Catarina (2010). She is currently a tenured professor at Universidade do Estado de Santa Catarina. She develops research in the field of Statistics applied to Engineering, namely to Sustainable Development in Civil Construction and Statistical Methods for Research in Education.

Andreza Kalbusch is a Doctor in Civil Engineering from Federal University of Santa Catarina. She is currently a Professor in the Civil Engineering Department at Santa Catarina State University, located in Joinville, Southern Brazil. Dr. Andreza Kalbusch is the leader of the Laboratory for Urban Water Study (Leau) and holds a Research Productivity Grant from CNPq since 2020. Her research topics are water conservation, plumbing fixtures performance and water consumption in the built environment.

The Role of Student Leadership in the Adoption of Circular Economy Practices in Universities

Júlia Santos Humberto and Marcelo Seido Nagano

Abstract Since Circular Economy (CE) has emerged as a transformative framework for sustainability, the role of higher education institutions (HEIs) in fostering CE principles through education, research, and student engagement has become increasingly relevant. However, whereas institutional strategies and curriculum-based approaches to CE have received growing attention, the contribution of student leadership in advancing CE practices remains underexplored. Student leadership plays an important role in institutional transformation by bridging the gap between CE principles and practical implementation. This study provides a systematic and exploratory analysis of how student leadership promotes and implements CE initiatives within HEIs. It was guided by four objectives, namely, examination of the literature on student-led CE initiatives, analysis of the role of student leadership in CE implementation, identification of institutional barriers that hinder leadership development, and assessment of the perceived benefits of student engagement in CE transitions. The methodology included a systematic literature review of Scopus and Web of Science databases, which identified 15 relevant academic publications. Bibliometric and content analyses explored thematic patterns and initiative types and an additional exploratory literature review extended the analysis to leadership roles, challenges, and benefits. According to the findings, student leadership plays an important, yet often under recognized, role in driving circular transformations through activism, entrepreneurship, and institutional advocacy. However, the efforts are frequently constrained by limited support structures and integration into formal governance. The study contributes to the literature by framing student leadership as a strategic force for CE adoption in academia and calls for greater institutional commitment to empowering students as sustainability change agents.

Keywords Student leadership · Circular economy · Higher education institutions

J. S. Humberto (✉) · M. S. Nagano
Department of Production Engineering, São Carlos School of Engineering, University of São Paulo, São Carlos, SP, Brazil
e-mail: julia.s.humberto@usp.br

M. S. Nagano
e-mail: drnagano@usp.br

W. Leal Filho et al. (eds.), *Sustainability in the Curriculum and Operations of Universities*, World Sustainability Series,
https://doi.org/10.1007/978-3-032-12678-8_13

1 Introduction

Higher education institutions (HEIs) play an important role in advancing circular economy (CE) principles by cultivating student competencies, values, and behaviors aligned with sustainability goals (Venugopal and Kour 2020; Del Vecchio et al. 2021; Kostadinova et al. 2025). Through education, research, and campus-based initiatives, universities are positioned as influential agents in driving sustainability transitions (Venugopal and Kour 2020; Kostadinova et al. 2025).

Within such an evolving landscape, student leadership has emerged as a powerful force for institutional transformation, effectively bridging the gap between CE ideals and practical implementation. Acting as catalysts for change (Syed Azhar et al. 2020), student leaders mobilize their peers, shape university policies, and co-create initiatives that integrate circularity into everyday campus life, taking diverse forms, including activism, entrepreneurship, collaborative research, and curricular innovation, and collectively contributing to the development of more circular, sustainable campus cultures (Bonoli et al. 2018; Santacruz Espinoza et al. 2020; Mittal and Bansal 2024).

Nevertheless, the transformative potential of student leadership is not always matched by institutional support. Persistent barriers such as limited curricular integration of CE content, insufficient leadership training, and lack of structural resources impede students' ability to amplify their impact (Todorova et al. 2024). Although many students show interest and enthusiasm for CE practices, that is often not accompanied by the necessary professional confidence or technical competencies for the implementation of effective actions (Todorova et al. 2024).

Towards overcoming those barriers, HEIs must establish enabling environments that actively support and nurture student leadership in CE, which involves implementation of project-based learning methodologies, promotion of interdisciplinary collaboration, creation of innovation incubators, and supply of structured leadership development programs (Del Vecchio et al. 2021; Faludi et al. 2023). Such frameworks are essential for translating student motivation into measurable and scalable outcomes. Moreover, fostering effective student leadership in the CE context requires cultivation of specific core competencies, including systems thinking, regulatory literacy, strategic communication, and collaborative problem-solving (Del Vecchio et al. 2021). CE education must, therefore, extend to beyond the transmission of technical knowledge, empowering students with the skills to advocate, influence, and mobilize others towards systemic changes (Faludi et al. 2023).

Despite the increasing recognition of student leadership as a key enabler of sustainability within HEIs, the literature lacks a systematic examination of its specific role in advancing CE practices. To date and to the best of the authors' knowledge, no study has comprehensively analyzed the way student leaders actively promote and implement CE initiatives within academic settings, leaving a critical gap on the understanding of their contributions to institutional circularity transitions. Existing research has addressed student contributions to sustainability in general terms, often without distinguishing the challenges and the strategic actions associated with CE and, thus limiting the understanding of the transformative potential of student-led efforts in facilitating CE transitions within HEIs.

This study systematically investigated the role of student leadership in promoting and implementing CE initiatives in academic settings and the current state of research on student-led actions related to CE within HEIs, identifying key barriers and perceived benefits associated with those efforts. A two-part methodology that combined a systematic literature review (SLR) with bibliometric and content analyses, followed by an exploratory review focused on student leadership roles, associated barriers, and perceived benefits was adopted. The study was guided by the following research questions (RQs):

RQ1: What research on student-led action has been developed for promoting CE practices in HEIs?

RQ2: What is the role of student leadership in promoting and implementing CE practices within HEIs?

RQ3: What are the key barriers within HEIs to developing student leadership for CE education?

RQ4: What are the perceived benefits of student leadership in promoting CE initiatives within HEIs?

By addressing the questions and motivated by the recognition students are key stakeholders within higher education and the next generation of leaders and change-makers, the study expands the body of knowledge on student leadership and its role in driving CE transitions within HEIs. Investigating the contributions of student leadership to the implementation of CE practices reveals an untapped potential for catalyzing institutional transformation from the bottom up. Despite growing interest in sustainability across campuses, student-driven initiatives often remain informal, under-recognized, or disconnected from institutional frameworks. By elucidating the mechanisms through which student leaders promote and implement CE practices, the study, structured into five sections, informs on and enhances existing student-led initiatives and raises awareness between students and HEIs on effective strategies and practices that can advance circularity within university contexts.

The paper is structured into five sections. Section 2 presents the theoretical background with an emphasis on existing literature on student leadership and its role in advancing CE practices within HEIs; Sect. 3 outlines the methodological approach adopted; Sect. 4 details the key findings related to student leadership roles, implementation challenges, and perceived benefits of student-led CE initiatives; finally, Sect. 5 provides the conclusions, summarizing key insights, implications, and recommendations for future research.

2 Background

CE has emerged as a critical framework for addressing global sustainability challenges, offering an alternative to the traditional linear model of consumption and production by promoting resource efficiency, waste reduction, and regenerative systems (Venugopal and Kour 2020). Within such a transition, HEIs play a key role as catalysts for sustainability through research, innovation, and education of future change agents (Kostadinova et al. 2025) and, by integrating CE principles into their curricula and operations, can potentially shape student competencies, values, and behaviors aligned with sustainable development goals (Kostadinova et al. 2025; Del Vecchio et al. 2021).

Although institutional commitments to sustainability are on the rise, studies have shown CE content is often restricted to elective courses or peripheral disciplines, which limits its impact on broader educational pathways (Obrecht et al. 2022). In this context, developing student leadership becomes essential for a successful advancement of CE in HEIs. Student-led CE initiatives vary widely in form, ranging from activism and peer education to entrepreneurship, collaborative research, and curricular innovation (Bonoli et al. 2018; Mittal and Bansal 2024). Those efforts often bridge the gap between theory and practice, enabling students to translate CE principles into real-world campus applications.

Despite the growing body of research on student involvement in sustainability, no scholarly attention has been directed specifically towards student-led CE initiatives within HEIs, since studies tend to approach the topic from a broad sustainability perspective. Mittal and Bansal (2024) provided a broader examination of student-led environmental initiatives within HEIs, highlighting the role of student organizations and grassroots movements in advancing environmental awareness and implementing campus-based sustainability practices, such as composting, hydroponics, recycling systems, and green campus campaigns. They also presented a broad overview of student engagement in sustainability within HEIs, emphasizing environmental awareness, sustainable behaviors, and institutional efforts (e.g., Green Campus Initiatives) without focusing specifically on CE frameworks. The discussion includes sustainability education, experiential learning, media strategies, and barriers to student engagement in environmental initiatives. Although the authors contribute with valuable descriptive and practice-oriented insights into student engagement in sustainability, they do not advance the field through a systematic, leadership-focused, and CE-centered analysis of how students function as change agents within HEIs.

Siddique et al. (2021) introduced a structured, curriculum-based model for fostering youth leadership in CE through STEAM Platform at King Mongkut's University of Technology Thonburi (KMUTT), in Thailand. Despite demonstrating the value of formal CE education, the study adopted a top-down approach and did not explore student-driven initiatives, or the broader structural barriers, benefits, and leadership dynamics involved in CE implementation across higher education. Therefore, the literature faces a clear gap on the role of student leadership in advancing

CE practices and in the understanding of associated barriers and perceived benefits within HEIs.

3 Methodology

An SLR following PRISMA (Preferred Reporting Items for Systematic Reviews and Meta-Analyses) guidelines and using Scopus and Web of Science (WoS) was conducted towards answering RQ1. PRISMA is a widely acknowledged approach that streamlines the selection of relevant publications through three key stages, namely, (1) identification, (2) screening, and (3) inclusion (Moher et al. 2009). 107 articles were retrieved (Scopus: 57, WoS: 50) according to the following search string: (student OR young OR "student-led" OR "student leadership" OR "student leader") AND (initiative* OR action OR activit*) AND ("circular economy") AND (universit* OR "higher education" OR "educational institutions" OR "academic institutions" OR campus).

The inclusion criteria established CE initiatives should be student-led, i.e., students should play a central role in both implementation and leadership, while faculty and staff might participate in supporting roles. Only peer-reviewed journal articles published in English and available up to March 2025 were considered.

Additionally, a content analysis based on Bardin (2016) and consisting of the following three main phases was conducted:

1. Pre-analysis: organization of the material, floating reading, and definition of corpus and objectives of the analysis.
2. Exploration of the material: codification of texts and identification of categories, themes, and patterns that emerge from the data.
3. Treatment of results, inference, and interpretation: interpretation of the categorized content and generation of meaningful insights according to the research objectives.

Still in the first phase, a Bibliometric analysis was conducted with the use of Bibliometrix package in R, a comprehensive tool for quantitative research in bibliometrics (Aria and Cuccurullo 2017), and identified annual distribution of publications, geographical distribution of scientific production, and affiliations of authors contributing to the theme.

Figure 1 displays the research design adopted for the analysis of student-led CE initiatives in HEI.

An SLR was conducted in Scopus and WoS, complemented by an exploratory review in Google Scholar, for ensuring a broader coverage of relevant academic sources and answering RQs 2, 3, and 4. The following search string was applied for identifying literature focused on student leadership in the context of CE within higher education: (student OR young) AND (universit* OR "higher education" OR "educational institutions" OR "academic institutions" OR campus) AND ("circular economy") AND (leadership OR leader*).

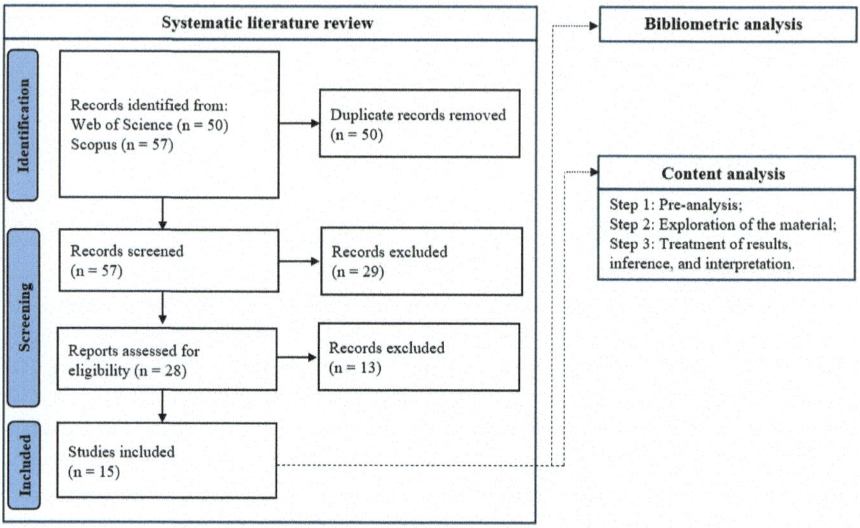

Fig. 1 Research design

A content analysis of the selected documents then identified recurring themes and patterns related to the roles of student leaders in advancing CE practices, barriers hindering the development of student leadership, and benefits and added value of student leadership in fostering sustainability transitions on campus.

4 Results and Discussion

4.1 Research on Student-Led Actions for Promoting CE Practices in HEI

An SLR combined with content and bibliometric analysis methods examined existing research on student-led action for promoting CE practices in HEIs focusing on four key dimensions, namely, (i) annual distribution of publications, (ii) scientific production per country, (iii) author affiliations, and (iv) types of student-led CE initiatives documented in the literature. 15 articles were selected through the SLR and the content analysis and covered the 2017 to 2024 period, indicating this is a relatively recent and emerging area of research. Figure 2 displays the annual distribution of publications related to student-led CE initiatives in HEIs.

Research on student-led CE initiatives in HEIs is still emerging, with low and inconsistent publication rates from 2017 to 2023 and typically with only 1 to 3 articles per year. However, a notable increase in 2024 signals growing interest and a possible shift towards more structured and institutionalized research.

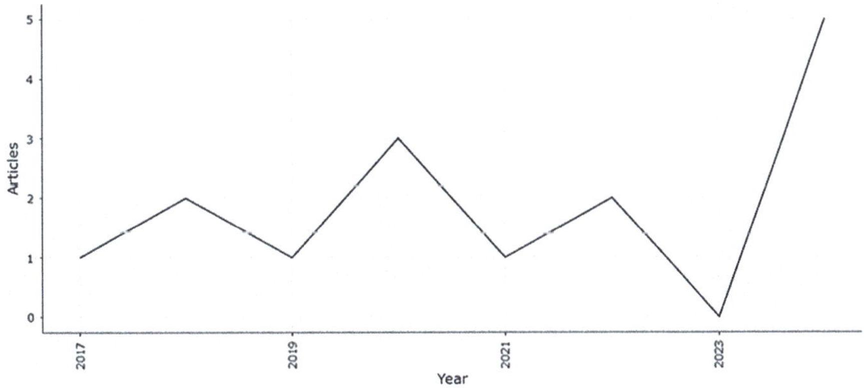

Fig. 2 Annual distribution of publications on student-led CE initiatives in HEIs

Geographically, contributions come from countries such as Italy, the UK, Spain, the Netherlands, China, South Korea, Brazil, Mexico, Colombia, and Bolivia. A world map (Fig. 3) visually represents it—shades of blue indicate publication volume and darker shades denote higher output.

Italy emerges as the leading contributor with 11 publications reflecting national and institutional backing, particularly through EU frameworks and supportive policies. The UK follows with seven publications, demonstrating academic interest, although student-led CE remains a niche area. Latin American countries (Mexico, Bolivia, and Brazil) show notable engagement, driven by local environmental challenges and inclusive circularity agendas, despite facing structural barriers to publication. In Asia, contributions from China and South Korea suggest growing alignment with national green growth strategies and higher education reforms. However, regions

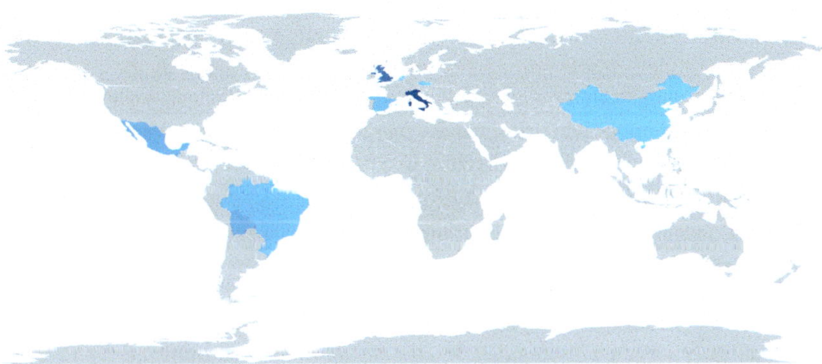

Fig. 3 Geographical distribution of scientific production on the theme

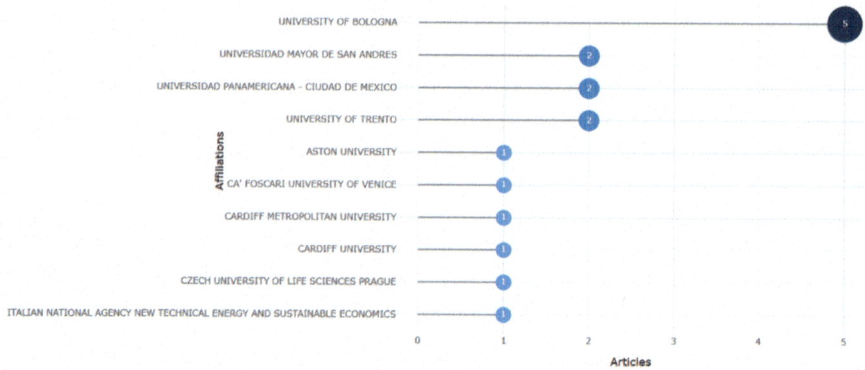

Fig. 4 Institutional affiliations of authors contributing to the theme

including Africa, South Asia, and Southeast Asia are largely absent from the peer-reviewed literature, likely due to limited research visibility and access rather than to inactivity.

Institutionally, the University of Bologna stands out with five publications, underscoring Italy's leadership, whereas presence of other Italian universities (e.g., Trento), along with Latin American institutions offers important Global South insights into the academic conversation. Figure 4 displays the affiliations of authors contributing to research on the theme.

Regarding types of initiatives, a growing body of literature highlights cases in which students take on leadership roles in conceptualization, development, and implementation of CE actions across academic, community, and industry settings. 15 student-led CE initiatives, discussed in what follows, were identified in the peer-reviewed literature and grouped into five main types, according to distinct forms of student engagement and pedagogical orientation.

1. **Campus-Based Sustainability Projects**: Within student-led CE practices in HEIs, three initiatives have emerged as campus-based sustainability projects driven by student leadership. In all of them, students acted as the main agents of change, identifying environmental issues and implementing sustainable solutions within their universities, while faculty and staff played supportive roles. As described by Bugallo-Rodríguez and Vega-Marcote (2020), students at the University of A Coruña, in Spain, participated in experiential learning activities for confronting issues, including waste, energy, water use, and planned obsolescence. They designed and executed solutions such as improvements in waste separation systems and promotion of repair and collaborative consumption, emphasizing student agency and behavioral change. Similarly, P-UMSA-Recycle project at Bolivia's Universidad Mayor de San Andrés, analyzed by Ferronato et al. (2020), showcased the work of student volunteers, the "young waste innovators," who led the creation of a selective recycling system, organized educational campaigns, and fostered recycling habits through social media outreach. Their

efforts were recognized formally and significantly influenced the university's waste management. At Cardiff University in the UK, the initiative examined by Hobson and O'Byrne (2024) involved students conceptualizing a Repair Café and Library of Things, engaging peers through surveys and workshops towards encouraging a culture of sharing and reuse. Although in its early stages, the project exemplified student-led activism and a commitment to embedding CE within everyday campus life.

2. **Industry Collaboration Projects**: Among the 15 student-led CE initiatives analyzed, four stood out as industry collaboration projects, characterized by active partnerships between HEIs and private companies. They enabled students to apply CE principles directly within real-world industrial settings, where they co-developed solutions addressing waste management, resource inefficiency, and sustainable design. Although companies and faculty were involved, students remained at the center of ideation, execution, and presentation, with external actors playing advisory and facilitative roles. In the initiative documented by Whitehill et al. (2022), 29 Surface Pattern Design students at Cardiff Metropolitan University collaborated with Orangebox Ltd., repurposing post-industrial textile waste into creative, sustainable prototypes using felting, overprinting, and embellishment, all under student direction. Similarly, Salinas-Navarro et al. (2024a) described how students from Tecnológico de Monterrey worked with small and medium enterprises (SMEs) for applying Kolb's experiential learning cycle, diagnosing inefficiencies and proposing CE strategies that were later validated with stakeholders. Another initiative, outlined by Williams et al. (2018), involved students from the University of Southampton working with Southern Water, conducting independent research and site analyses that resulted in actionable CE strategies for resource recovery. Finally, as reported by Salinas-Navarro et al. (2024b), the Social Lab for Sustainable Logistics (SLSL) featured collaborations between students and companies operating in logistics-intensive retail sectors, such as a wine and spirits distributor in Mexico, where learners led every stage, from problem identification to stakeholder engagement. Guided by faculty, students retained full creative and operational control throughout the experiential learning cycle.
3. **Living Lab/Civic/Community Innovation Initiatives**: A significant subset of student-led CE initiatives in HEIs includes living labs and community-based projects, as exemplified by five initiatives identified in this analysis. The projects thrive in transdisciplinary settings, where students collaborate with local communities, nonprofits, and diverse stakeholders to co-creating sustainable solutions. The living lab framework enables real-world experimentation, iterative learning, and strong community integration, with students leading project development, system design, and implementation, while faculty and institutional partners provide guidance and logistical support. Bonoli et al. (2018) described rAEEuse Lab, developed by engineering students at the University of Bologna, as part of the "Terracini in Transizione" living lab, focusing on electronics restoration and urban mining through feasibility study, workflow design, and implementation of reuse practices, driven by student leadership. Similarly, Rizzo et al.

(2017) detailed a hydroponic agriculture initiative that used recycled industrial waste as substrate, also under the same framework, and in which students led co-design process and testing phases with faculty support. Kumble (2019) documented a composting project initiated by graduate students from the University of Massachusetts Amherst in a Guatemalan neighborhood, culminating in the nonprofit AbonOrgániCo, with students independently creating composting protocols and a business model, exemplifying community-engaged learning and student-led implementation. As reported by Colucci et al. (2020), PINECUBE project emerged from UrbanFarm2019 challenge, according to which students from the University of Bologna transformed an abandoned school into a hydroponic medicinal plant farm, integrating technical design, stakeholder engagement, and a circular business model. Finally, Boeri et al. (2024) presented EN-ACTION project at the University of Bologna's Cesena campus, where students, acting as "transition ambassadors," led civic engagement efforts to promote energy citizenship and co-create a manifesto for a Renewable Energy Community, demonstrating how student leadership can connect academic institutions to local energy transition processes.

4. **Simulation and Design Studio Projects**: Two initiatives exemplify a distinctive form of student-led CE engagement in HEIs, where structured pedagogical environments are used to foster experiential, creative, and system-based learning. The initiatives simulate real-world complexity through design-based or gamified models, granting students full responsibility for decision-making, execution, and evaluation, while faculty act as facilitators. Gomes et al. (2022) described a long-running design studio at the University of Campinas (UNICAMP), in Brazil, where architecture students led projects focused on life cycle assessment (LCA) and circular design between 2017 and 2020. They acted as architects and sustainability consultants, developing design proposals, constructing LCA inventories, and calculating circularity indicators, with faculty providing methodological support. Such a hands-on approach integrated circular thinking into architectural education and emphasized student autonomy. In the initiative reported by Fraccascia et al. (2021), students at the University of Twente participated in an industrial symbiosis simulation game as part of a course on Circular Sustainable Business Development. Organized into teams representing virtual companies, they managed circular business relationships in response to dynamic market conditions, negotiated contracts, and strategized resource flows. Although the simulation framework was instructor-designed, students maintained full control over gameplay, promoting leadership, systemic thinking, and CE application in business contexts.
5. **DIY Product Design and Waste Transformation**: The Do-It-Yourself (DIY) product design and waste transformation category highlights a student-led CE initiative that fosters experiential learning, material innovation, and creative waste repurposing and is centered on transforming single-use plastics, particularly polypropylene (PP), into functional products through accessible and open-source tools. According to Nagatomo (2024), undergraduate students from the Department of Design at National Taiwan Normal University, in Taiwan, took

the lead in identifying the problem of plastic waste, documenting their consumption patterns and researching into PP's recyclability. They utilized resources such as Precious Plastic Universe database to guide their exploration of sustainable transformation methods and, through collaborative efforts, developed modular lighting fixtures via iterative prototyping, embodying a design-based research approach. Faculty acted mainly as mentors, while students maintained full control over the project's direction, technical development, and execution, showcasing leadership in sustainable design and hands-on circular practices.

Across all initiatives, students played central roles in shaping and implementing the projects, from problem identification and data collection to solution design, stakeholder engagement, and policy advocacy. The initiatives were globally distributed across Europe, Latin America, North America, and Asia and spanned diverse disciplines such as design, engineering, operations management, urban planning, logistics, and architecture, highlighting both interdisciplinary nature of CE and versatility of student involvement.

Student-led initiatives consistently delivered educational and societal benefits, enhancing skills such as systems thinking, project management, stakeholder engagement, and sustainability assessment. They also led to concrete outcomes, including product innovation, institutional change, and new community practices. The 15 cases illustrate empowering students as change agents fosters meaningful learning while driving institutional and societal transformation, contingent on institutional support, autonomy, and integration of real-world challenges.

4.2 *The Role of Student Leadership in Advancing CE in HEIs*

Student leadership has become a key force in advancing CE transitions within HEIs, serving as both a catalyst for institutional change and a bridge between sustainability ideals and actionable practices. Although universities are recognized as key players in promoting sustainability through education, research, and innovation (Kostadinova et al. 2025; Fonseca et al. 2018), the integration of CE into academic curricula remains sparse and disjointed (Obrecht et al. 2022). In this context, student leaders have taken on proactive roles as change agents (Syed Azhar et al. 2020), rallying peers, influencing institutional agendas, and co-creating initiatives that address real-world sustainability challenges. Obrecht et al. (2022) highlighted the marginal presence of CE within Slovenian higher education, where it is often confined to electives or treated as a peripheral topic. Such an institutional inertia has opened space for grassroots student activism, allowing student leaders to operate outside traditional hierarchies and employ advocacy, peer mobilization, and collaborative strategies for embedding CE into both academic and operational university structures (Helferty and Clarke 2009). The disconnection between academic theory and practical application was further underscored by UNESCO HEDGES baseline survey (Otieno et al. 2021), which points to a limited grasp of CE principles among students. The gap is particularly problematic in business and management education, where the need for CE literacy is urgent. Institutional responses, such as Erasmus + -funded Leaders of the Green Economy project, have attempted to bridge that divide by enhancing academia-industry collaboration and incorporating digital innovation towards supporting CE education in business schools. Yet, despite important, those top-down efforts often overlook the unique and transformative role that student-led initiatives play in driving systemic change from the ground up.

Student leadership in CE transcends advocacy, encompassing active embodiment and experimentation. Across various institutional settings, students translate CE principles into concrete actions through campus-based projects, entrepreneurial initiatives, and collaborative sustainability efforts. At the University of Bologna, rAEEuse Lab, emerging from the "Terracini in Transizione" initiative, demonstrates how students can innovate by creating systems for collecting, restoring, and reusing outdated IT equipment, thereby extending product lifespans, reducing e-waste, and embedding experiential learning in academia (Bonoli et al. 2018). Similarly, students at Hermilio Valdizán National University fostered daily CE practices that include recycling, minimizing consumption, and advocating for shared transportation, highlighting how student leadership can integrate sustainability into campus culture (Santacruz Espinoza et al. 2020). Such leadership also expands into research and policy development, as seen in the Mission to Mars initiative, in which students, in collaboration with the Dutch horticulture sector, engaged in interdisciplinary research on energy and waste management in greenhouses, shaping academic agendas with societal relevance (Vermeulen et al. 2020). On the other hand, despite their capacity for impact, many student leaders operate without sufficient institutional backing or pedagogical structures. Research by Todorova et al. (2024) revealed although eager to engage with CE, students in Bulgaria and North Macedonia often feel unprepared for its practical application. Addressing that gap requires integration of project-based learning, leadership training, interdisciplinary collaboration, and innovation support systems for converting student enthusiasm into lasting, sustainable change (Del Vecchio et al. 2021; Faludi et al. 2023).

An effective student leadership in CE requires a holistic mix of technical, strategic, and interpersonal competencies. Del Vecchio et al. (2021) emphasized the need for systems thinking, regulatory literacy, and strong communication skills so that the complexities of sustainability could be navigated. Faludi et al. (2023) added technical abilities in design for reuse and systemic circularity should be paired with soft skills such as collaboration, persuasion, and strategic leadership, empowering students to overcome institutional inertia, form interdisciplinary collaborations, and initiate cultural shifts. Kostadinova et al. (2025) further highlighted education's role in shaping sustainability mindsets, showing students across four European countries strongly supported key CE principles, including resource efficiency and renewable energy, signaling their readiness to lead. Towards unlocking such potential, HEIs must foster enabling environments that provide funding, institutional recognition, and platforms for student-led CE actions. A concrete example of that institutional approach is STEAM Platform at KMUTT, in Thailand, where GEN352 module equips students with knowledge in CE domains, such as LCA and sustainable financing, alongside leadership and communication training. As shown by Siddique et al. (2021), that experiential, interdisciplinary approach supports innovation and entrepreneurship, demonstrating how formal integration of CE leadership into academic curricula can drive meaningful behavioral and structural changes.

Student-led organizations are actively putting CE values into practice across HEIs worldwide through Green Campus Initiatives (GCIs), which encompass efforts such as energy conservation, recycling, and waste management (Mittal and Bansal 2024). Case studies that include Local Organic Gardening Initiative (LOGIC) at Southern Illinois University and MH Vatavaran Society at the University of Delhi illustrate the contribution of students to tangible sustainability outcomes through initiatives such as composting, hydroponics, and closed-loop systems. Kampus Sejahtera at Universiti Sains Malaysia exemplifies how institutional support for student leadership can strengthen environmental stewardship within campus culture. Similarly, Waste Management and Circular Economy Drive (WMCED) at Christ University and Parivarthana Student Engagement Program (PSEP) promote the 5Rs (Reduce, Reuse, Recycle, Research, and Restore) through student-driven mobilization and awareness campaigns (Mittal and Bansal 2024), fostering environmental impact reduction and development of leadership, civic responsibility, and systems thinking. O'Sullivan's (2017) longitudinal study at a private university in Dubai further emphasized the role of student leadership in sustainability transitions, showing how participation in environmental campaigns and community service equips students with essential competencies such as teamwork and communication while highlighting the importance of embedding students in decision-making processes for enhancing their role as co-creators of sustainable and circular systems. Altogether, those examples demonstrate student leadership is a powerful driver of CE transformation in HEIs, bridging institutional gaps and accelerating sustainability transitions through activism, innovation, and curriculum engagement. Towards fully leveraging that potential, universities must cultivate supportive environments that empower students through funding, recognition, mentorship, and decision-making inclusion.

Figure 5 illustrates a comprehensive categorization of student leadership roles in advancing CE within HEIs.

4.3 Barriers Within HEIs for the Development of Student Leadership for CE Education

The development of student leadership in CE within HEIs is constrained by a range of structural, curricular, and institutional barriers that limit universities' capacity to serve as drivers of sustainability transitions. A central issue is the limited integration of CE across academic programs, as highlighted by Obrecht et al. (2022), who

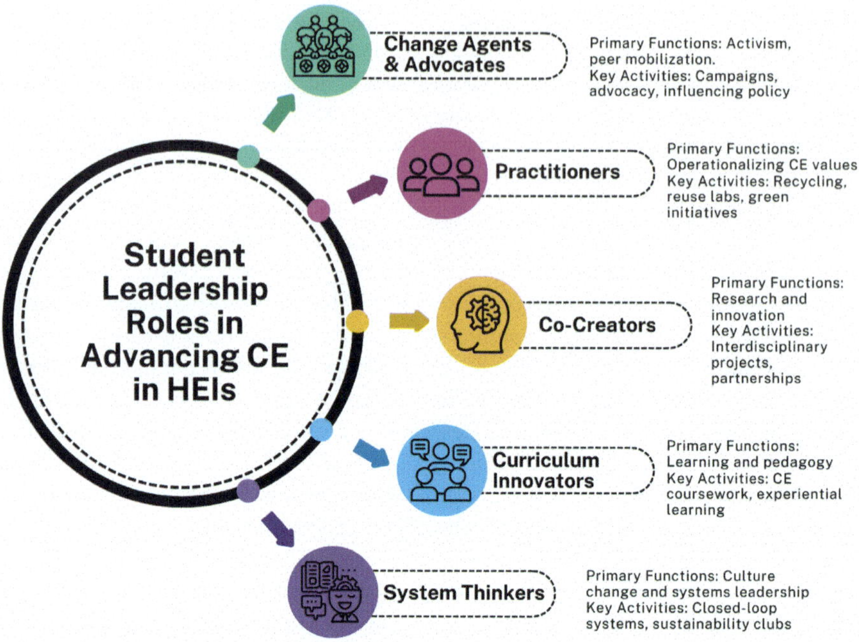

Fig. 5 Student leadership roles in advancing CE within HEIs

claimed CE is among the least represented sustainability topics in Slovenian university curricula, especially when compared to more traditional environmental themes such as ecology and environmental protection.

That curricular gap restricts students' exposure to critical CE concepts, hindering the cultivation of systems thinking, interdisciplinary collaboration, and entrepreneurial mindsets essential for sustainability leadership. Moreover, an inadequate inclusion of themes (e.g., social responsibility and civic engagement) reduces opportunities for students to evolve as inclusive and regenerative leaders. Institutional neglect also plays a significant role. Faludi et al. (2023) observed CE and sustainability are often marginalized within design education, typically confined to elective courses or superficial mentions, with similar patterns evident across other disciplines. The inconsistent and peripheral nature of CE content weakens the development of coherent and practical CE literacy among students. Compounding those issues is a widespread lack of faculty expertise and support in CE-related topics. As emphasized by the authors, educators who lack training in sustainability principles are less capable of mentoring students on circular initiatives, a challenge intensified by academic cultures that favor traditional disciplinary norms over transformative, sustainability-focused approaches.

Student-led CE initiatives in HEIs face significant operational and systemic barriers. Institutionally, students often encounter limited decision-making authority, minimal faculty involvement, and a lack of formal recognition or supportive frameworks, which restrict their capacity to lead effectively (O'Sullivan 2017). Those constraints are compounded by the challenge of juggling academic demands with leadership responsibilities and by peer-related social pressures that can discourage broader engagement. On a systemic level, the lack of coordinated strategies among academia, government, and industry hinders opportunities for experiential, project-based learning essential to cultivating CE leadership (Kostadinova et al. 2025). Additionally, inadequate funding for CE education and infrastructure limits both student's and faculty's abilities to innovate or expand CE initiatives. Traditional pedagogical models that prioritize theoretical instruction over practical skill-building further restrict the development of student leaders and their role as institutional change agents. Together, those issues point to a need for structural transformation in the way CE is integrated into curricula, supported through institutional ecosystems and fostered through inclusive, experiential, and collaborative learning cultures.

Figure 6 illustrates the barriers within HEIs for the development of student leadership for CE education.

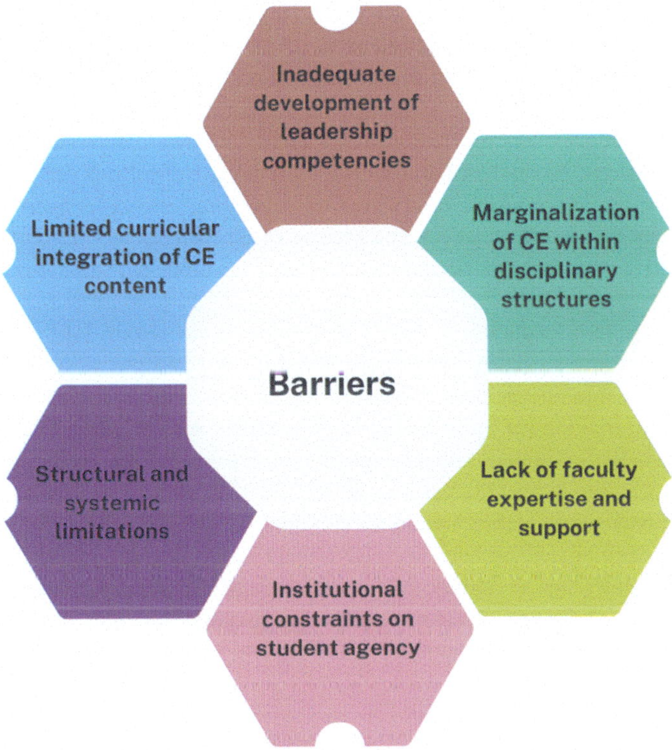

Fig. 6 Barriers within HEIs for the development of student leadership for CE education

4.4 *Benefits of Student Leadership for Promoting CE Initiatives Within HEIs*

In a variety of institutional settings, students who actively engage as leaders contribute significantly to educational, institutional, and societal transformations that support a broader transition towards sustainable development. A key benefit consistently highlighted is the development of personal and professional competencies that align closely with CE principles, including communication, collaboration, problem-solving, initiative, and critical thinking (O'Sullivan 2017; Kostadinova et al. 2025). Through involvement in CE-related projects, student leaders gain experiential knowledge by navigating complex systems, advocating for sustainability, engaging with diverse university stakeholders, and translating academic knowledge into practical solutions. Such an experience not only enhances academic motivation, but also equips students for future leadership roles in business, politics, and civil society (O'Sullivan 2017; Kostadinova et al. 2025).

Student leadership plays an important role in fostering institutional innovation, as illustrated by rAEEuse lab at the University of Bologna, where students act as co-creators of sustainable solutions by promoting the reuse of electronic waste through interdisciplinary collaboration (Bonoli et al. 2018). Their involvement extended to beyond technical skills in restoration and resource recovery for supporting the university's third mission of knowledge transfer and community engagement, thus reinforcing the societal value of student-led circular initiatives. Beyond the campus, students also drive broader cultural shifts, as seen at Hermilio Valdizán National University, where student-led campaigns on recycling, energy conservation, and sustainable mobility reflect a deep integration of CE principles into daily life and contribute to generational sustainability and civic awareness (Santacruz Espinoza et al. 2020).

Building student leadership capacity in CE initiatives within HEIs depends on supportive conditions such as mentorship from faculty and peers, structured leadership training, interdisciplinary collaboration, and student participation in institutional decision-making (O'Sullivan 2017). When those enabling factors are present, students are better equipped with strategic and creative skills for questioning norms and driving CE initiatives (Kostadinova et al. 2025). In this context, student leadership offers extensive benefits, including personal growth, skill development, institutional innovation, and broader societal impact. By encouraging student ownership, responsibility, and creativity, HEIs can integrate CE principles into their operations, teaching, and outreach efforts, advancing sustainability and preparing future professionals to lead circular transformations across various sectors.

5 Conclusions

This study has offered a systematic analysis of the role of student leadership in advancing CE practices within HEIs addressing four core dimensions, namely, current state of research on student-led CE initiatives, specific roles played by student leaders in promoting and implementing CE practices, key institutional barriers that hinder the development of student leadership, and perceived benefits associated with student engagement in CE within academic settings.

According to the findings, student leadership serves as a key driver of bottom-up change, mobilizing peer engagement, co-creating sustainability pathways, and influencing institutional strategies towards circularity. Unlike prior work, which focused primarily on environmental engagement or curriculum-based models of CE education, this study has provided a student leadership-focused and CE-specific contribution, emphasizing the strategic role of students as active agents in institutional transformation within higher education. It fills an important gap by framing student leadership not merely as a support mechanism, but also as a strategic force in embedding circularity within academic institutions and contributing to broader societal transitions towards sustainability. Moreover, it emphasizes the need for HEIs to fundamentally rethink the way they perceive and support student leadership in CE transitions.

Rather than viewing students as peripheral actors or passive recipients of sustainability education, HEIs should recognize them as key change agents capable of co-driving institutional transformation. The limited number of academic studies identified reveals a significant research gap concerning the nexus of student leadership and CE practices, calling for further empirical investigations into how students initiate change, co-create sustainability strategies, and contribute to embedding circularity within university systems. The findings point to student leadership as a strategically underleveraged force in CE transitions, advocating for institutional policies that actively involve students in sustainability governance, including decision-making processes on curriculum, waste systems, and campus operations. Yet, students face persistent structural barriers such as inadequate access to resources, minimal leadership training, and lack of institutional recognition. Overcoming those challenges requires HEIs to invest in enabling environments, curricular integration, interdisciplinary hubs, project-based learning, mentorship programs, and student-led innovation spaces that cultivate leadership capacity aligned with CE goals. Furthermore, the study has underscored the value of student leadership in enhancing peer engagement, enabling practical applications of CE principles, and driving cultural shifts within institutions. By actively engaging in CE-related projects, students not only develop general leadership skills, but also acquire practical experience and domain-specific knowledge in CE principles, thus being positioned as future professionals equipped to lead circular transitions in diverse organizational settings. That is important, for it ensures such professionals are sustainability-minded and provided with practical competencies and systemic understanding necessary for implementation and scaling of CE strategies across sectors.

The study offers key theoretical and practical contributions. Theoretically, it advances the understanding of student leadership as a strategic force in implementing CE practices within HEIs, filling a critical gap by focusing on student-led CE initiatives and providing a foundation for future cross-cultural and longitudinal research. Practically, it urges HEIs to recognize students as active change agents, recommending institutional reforms such as integrating students into sustainability governance, investing in leadership training, and embedding CE and leadership development into core curricula towards enhancing both educational quality and sustainability outcomes.

Moreover, it opens several avenues for future research at the intersection of student leadership and CE practices in HEIs. Given the limited number of academic publications, future studies should expand empirical investigations into student-led CE initiatives across diverse geographic, cultural, and institutional contexts. Finally, longitudinal research should examine the long-term impacts of student leadership on institutional changes and sustainability performance and assess the way student-led initiatives evolve over time and leadership experiences influence students' professional trajectories, and whether the efforts result in sustained circular practices within the institution.

References

Aria M, Cuccurullo C (2017) Bibliometrix: An R-tool for comprehensive science mapping analysis. J Informetr 11(4):959–975. https://doi.org/10.1016/j.joi.2017.08.007

Bardin L (2016) Análise de conteúdo. Rev. e atual. ed. Edições 70, Lisboa.

Boeri A, Turillazzi B, Sabatini F, Sassenou L-N, Manella G (2024) The threefold role of the university in fostering the energy transition: The case of Bologna and its Cesena campus. Energies 17:3751. https://doi.org/10.3390/en17153751

Bonoli A, Dolci N, Foschi E, Lalli F, Prandstraller D, Zanni S (2018) End of service scenario for universities' informatic equipment: Recovery and repair as educational and research tool for circular economy and urban mining. Detritus 4:90–97. https://doi.org/10.31025/2611-4135/2018.13747

Bugallo-Rodríguez A, Vega-Marcote P (2020) Circular economy, sustainability and teacher training in a higher education institution. Int J Sustain High Educ 21(7):1351–1366. https://doi.org/10.1108/IJSHE-02-2020-0049

Colucci N, Dall'Agnol N, De Biasi P, Orsini F, Tagliaferri N, Tonet E (2020) PINECUBE: Technologies for sustainable plant production and urban renewal in Belluno (Italy). Acta Hortic 1298:135–144. https://doi.org/10.17660/ActaHortic.2020.1298.20

Del Vecchio P, Secundo G, Mele G, Passiante G (2021) Sustainable entrepreneurship education for circular economy: Emerging perspectives in Europe. Int J Entrepreneurial Behav Res 27(8):2096–2124. https://doi.org/10.1108/IJEBR-03-2021-0210

Faludi J, Acaroglu L, Gardien P, Rapela A, Sumter D, Cooper C (2023) Sustainability in the future of design education. She Ji 9(2):157–178. https://doi.org/10.1016/j.sheji.2023.04.004

Ferronato N, Guisbert Lizarazu EG, Velasco Tudela JM, Blanco Callisaya JK, Preziosi G, Torretta V (2020) Selective collection of recyclable waste in universities of low-middle income countries: Lessons learned in Bolivia. Waste Manag 105:198–210. https://doi.org/10.1016/j.wasman.2020.02.014

Fonseca LM, Portela AR, Duarte B, Queirós J, Paiva L (2018) Mapping higher education for sustainable development in Portugal. Manag Mark 13(3):1064–1075. https://doi.org/10.2478/mmcks-2018-0023
Fraccascia L, Sabato A, Yazan DM (2021) An industrial symbiosis simulation game: Evidence from the circular sustainable business development class. J Ind Ecol 25(6):1688–1706. https://doi.org/10.1111/jiec.13183
Gomes V, da Silva MG, Kowaltowski DCCK (2022) Long-term experience of teaching life cycle assessment and circular design to future architects: A learning by doing approach in a design studio setting. Sustainability 14(12):7355. https://doi.org/10.3390/su14127355
Helferty A, Clarke A (2009) Student-led campus climate change initiatives in Canada. Int J Sustain High Educ 10(3):287–300. https://doi.org/10.1108/14676370910972594
Hobson K, O'Byrne M (2024) Sharing and repairing at university: On student practices, and the future of the circular campus. Circ Econ Sustain 4:2579–2598. https://doi.org/10.1007/s43615-024-00403-4
Kostadinova I, Todorova A, Ruskova S, Tomovska Misoska A, Letonja M, Tomljanović M (2025) Sustainability and the circular economy through the eyes of students: A comparative analysis among Bulgaria, North Macedonia, Slovenia and Croatia. Eur J Sustain Dev 14(1):1. https://doi.org/10.14207/ejsd.2025.v14n1p1
Kumble PA (2019) Reflections on service learning for a circular economy project in a Guatemalan neighborhood, Central America. Sustainability 11(17):4776. https://doi.org/10.3390/su11174776
Mittal P, Bansal R (2024) Effective strategies for engaging students in sustainability initiatives. In: Community engagement for sustainable practices in higher education. Palgrave Macmillan, Cham. https://doi.org/10.1007/978-3-031-63981-4_2
Moher D, Liberati A, Tetzlaff J, Altman DG (2009) Preferred reporting items for systematic reviews and meta-analyses: The PRISMA statement. The PRISMA Group. Journal of Clinical Epidemiology, 62(10), 1006–1012. https://doi.org/10.1016/j.jclinepi.2009.06.005
Nagatomo D (2024) Research on education for sustainable development with design-based research by employing Industry 4.0 technologies for the issue of single-use plastic waste in Taiwan. Sustainability 16(22):9832. https://doi.org/10.3390/su16229832
O'Sullivan K (2017) Student leadership in sustainable development in a private university in the UAE—A case study. In: Leal Filho W et al (eds) Handbook of theory and practice of sustainable development in higher education. World Sustainability Series. Springer. https://doi.org/10.1007/978-3-319-47868-5_13
Obrecht M, Feodorova Z, Rosi M (2022) Assessment of environmental sustainability integration into higher education for future experts and leaders. J Environ Manag 316:115223. https://doi.org/10.1016/j.jenvman.2022.115223
Otieno DB, Ngare I, Ogutu EA, Omwami D, Atima A (2021) Baseline report on green economy and sustainability: UNESCO Chair on Higher Education Development for a Green Economy and Sustainability, Kenyatta University. Kenyatta University. https://ir-library.ku.ac.ke/server/api/core/bitstreams/9a144843-e5e3-4320-a7dc-0d43c0cde8c5/content
Rizzo S, Cappellaro F, Accorsi M, Orsini F, Gianquinto G, Bonoli A (2017) Co-design for a circular approach in green technologies: Adaptation of reused building material as growing substrate for soilless cultivation of lettuce (Lactuca sativa var. capitata). Environ Eng Manag J 16(8):1775–1780. https://doi.org/10.30638/eemj.2017.193
Salinas-Navarro DE, Da Silva-Ovando AC, Palma-Mendoza JA (2024b) Experiential learning labs for the post COVID-19 pandemic era. Educ Sci 14(7):707. https://doi.org/10.3390/educsci14070707
Salinas-Navarro DE, Palma-Mendoza JA, Vilalta Perdomo E (2024a) Integrating circular supply chains into experiential learning: Enhancing learning experiences in higher education. Sustainability 16(23):10789. https://doi.org/10.3390/su162310789

Santacruz Espinoza A, Montenegro Muguerza H, Pizarro Alejandro A, Estacio Flores H (2020) Liderazgo transformacional y desarrollo sostenible ambiental verde en docentes de la Universidad Nacional Herminio Valdizán: Factores de liderazgo transformacional [Transformational Leadership and Green Environmental Sustainable Development among Faculty Members at the National University Herminio Valdizán: Factors of Transformational Leadership]. Rev Comun SEECI (53):135–151. https://doi.org/10.15198/seeci.2020.53.135-151

Siddique A, Nakseemok P, Liu L (2021) Youth leadership in a circular economy: Education enabled by STEAM platform. In: Liu L, Ramakrishna S (eds) An introduction to circular economy. Springer, Singapore. https://doi.org/10.1007/978-981-15-8510-4_29

Syed Azhar SNF, Mohammed Akib NA, Khelghat-Doost H, Sibly S (2020) Students as catalysts in achieving sustainable development in higher education: Initiatives and program highlights. Sustainability 13(6):271–281. https://doi.org/10.1089/sus.2019.0038

Todorova A, Kostadinova I, Stevcevska Srbinoska D (2024) Student attitudes towards the circular economy: A comparison between Bulgaria and North Macedonia. In: Proceedings of the International Conference on Economics and Social Sciences. The International Conference on Economics and Social Sciences. Editura ASE. https://doi.org/10.24818/icess/2024/039

Venugopal P, Kour H (2020) Integrating the circular economy into engineering programs in India: A study of students' familiarity with the concept. Ind High Educ 35(3):264–269. https://doi.org/10.1177/0950422220967542

Vermeulen ACJ, Hubers C, de Vries L, Brazier F (2020) What horticulture and space exploration can learn from each other: The Mission to Mars initiative in the Netherlands. Acta Astronaut 177:421–424. https://doi.org/10.1016/j.actaastro.2020.05.015

Whitehill S, Hayles CS, Jenkins S, Taylour J (2022) Engagement with higher education surface pattern design students as a catalyst for circular economy action. Sustainability 14(3):1146. https://doi.org/10.3390/su14031146

Williams ID, Roberts KP, Shaw PJ, Cleasby B (2018) Applying circular economy thinking to industry by integrating education and research activities. Detritus 1:134–143. https://doi.org/10.26403/detritus/2018.11

Júlia Santos Humberto is a PhD candidate in Production Engineering at the University of São Paulo, São Carlos campus (EESC-USP), Brazil, where she researches into application of Circular Economy principles, Blockchain, and Internet of Things (IoT) technologies in agriculture. She holds a Master's degree in Production Engineering from the University of Pernambuco (2020), where her research focused on risk management, particularly multi-criteria decision-making and climate-related disaster risks. She also holds a specialization in Occupational Safety Engineering (2020), as well as dual undergraduate degrees in Production Engineering, from Centro Universitário CESMAC, and Environmental and Sanitary Engineering, from the Federal University of Alagoas (2017).

Marcelo Seido Nagano is an Associate Professor at Department of Production Engineering at the São Carlos School of Engineering, University of São Paulo (EESC-USP), where he has served since 2004 under an exclusive dedication regime. He coordinates the Research Group in Operations Research and is recognized for his contributions to decision support systems, production process optimization, innovation management, and sustainability, particularly in the areas of Circular Economy and Sustainable Development Goals (SDGs).

He earned his BSc, MSc, and PhD in Mechanical Engineering from the University of São Paulo, with a specialization in Mechatronics, and later completed a postdoctoral fellowship at the National Institute for Space Research (INPE), focusing on evolutionary algorithms applied to production scheduling. In addition to his academic leadership, he is an Editor-in-Chief and editorial board member for several international journals in engineering and operations research.

SDG 12 in Focus: Geographic Approaches to Sustainable Consumption and Production at the Application School of the Federal University of Uberlândia

Elisângela de Azevedo Silva Rodrigues, Lidiane Aparecida Alves, and Suely Aparecida Gomes

Abstract A sustainable development model, as outlined in the "2030 Agenda for Sustainable Development" and the 17 Sustainable Development Goals (SDGs) of the United Nations, is urgent, necessary, and requires multiple approaches due to its complex and interdisciplinary nature. From the Geographical perspective, as a school subject offered in Basic Education, sustainability can be addressed within different curricular contents, considering teaching, research, and extension activities. This article presents possibilities for addressing themes related to SDG 12: Sustainable Production and Consumption, within the context of 5th-grade classes, from an investigative perspective committed to sustainability. The research-action methodology was employed, characterized by being participatory, dynamic, flexible, and interactive with playful didactic sequences that, based on real-world problems, facilitate active learning and promote meaningful, contextualized teaching aligned with contemporary needs for civic education. This article aims to contribute to discussions about methodological activities designed to reflect on the importance of promoting an integrated education for sustainability, in a practical and meaningful way, within the curriculum and the functioning of institutions such as the Application Schools of universities in Brazil.

Keywords Environmental education · Pedagogical practice · Methods · Waste

---

E. de Azevedo Silva Rodrigues (✉) · L. Aparecida Alves · S. Aparecida Gomes
Geography, Universidade Federal de Uberlândia, Uberlândia, Brazil
e-mail: elisangelarodrigues@ufu.br

L. Aparecida Alves
e-mail: lidianeaa@ufu.br

S. Aparecida Gomes
e-mail: suelygomes@ufu.br

W. Leal Filho et al. (eds.), *Sustainability in the Curriculum and Operations of Universities*, World Sustainability Series,
https://doi.org/10.1007/978-3-032-12678-8_14

1 Introduction

Currently, the world is experiencing an environmental crisis caused by the transformation of the geographical space, due to the increasingly unsustainable production and consumption patterns of today's society (Moretti et al. 2024). Geographical space is, therefore, approached from the understanding of the space inhabited by humans and modified throughout history by different cultural groups, through the use of techniques and technologies (Santos 1988).

Many of these transformations caused by human activity in geographical space have brought benefits to humans, making life more comfortable. However, on the other hand, they have also caused problems, such as the case of solid waste produced by modern society, which intensifies the depletion of environmental resources on the planet we live on and generates challenges like those related to climate change. From this, the importance of addressing this theme in educational institutions is justified (Andrade et al. 2021).

The "Solid Waste" matter has been included in the curricula of universities and schools of Basic Education in Brazil, surrounding discussions on solid waste management and sustainability in their educational programs. This theme is linked to environmental issues and is suggested by national curriculum guidelines, specifically from the National Curriculum Parameters (PCNs) (Brazil 1997, 1998), the National Common Curricular Base (BNCC) (Brazil 2018), the National Policy on Environmental Education (Brazil 1999), and the National Solid Waste Policy (Brazil 2010).

These documents guide that the theme "Environment" should be addressed from a transversal perspective, especially in the curricula of Geography and Science subjects, and should also be present in other subjects in an articulated/interdisciplinary manner, across all levels and modalities of education (Brazil 1999).

It is also important to add the 2030 Agenda and the 17 Sustainable Development Goals (SDGs) of the United Nations, which can be addressed, under different approaches, in teaching and learning practices focused on sustainability. These practices are committed to developing knowledge, skills, values, and attitudes essential to the preservation of environmental, social, and economic well-being, in compliance with the Education for Sustainable Development (ESD) of the United Nations Educational, Scientific and Cultural Organization (UNESCO 2021).

Among the various curricular components for students, although Environmental Education (EE) and Sustainability are not the exclusive responsibility of Geography as a school subject, this area of knowledge offers several possibilities for addressing these issues within different curricular contents, considering teaching, research, and extension activities.

Geography can contribute to EE and Sustainability because it is a science that studies geographical space, which is of great importance for civic education, as the concepts and themes addressed by this field of knowledge are directly related to the reality or daily life of the students. In this sense, it is the challenge of teachers who teach this subject to contribute to the formation of citizens committed to the society they belong to, perceiving themselves as transformative agents of the environment in which they live, since Geography studies geographical space from the perspective of the society/environment and society/society relationships (Callai 2015).

For the study of geographical space, other categories such as landscape, territory, region, and place are pivotal. In this sense, according to Andrade et al. (2021, p. 128), "the study of the geographical concepts of Landscape and Place provides conditions for teaching about solid waste, as they are present in the landscapes that make up the place." Moreover, according to Layrargues (2011, p. 185):

> The issue of waste has been pointed out by environmentalists as one of the most serious urban environmental problems of today. However, for students to understand this situation, it is necessary for them to see how solid waste affects their place and the world. The environmental and social problems caused by solid waste do not only affect the place of residence but also other places and their landscapes.

According to Andrade et al. (2021, p. 128), "the study of Landscape can contribute to the formation of more responsible citizens regarding the issue of waste disposal." For the author, the concepts of Landscape and Place are related to each other and work together to provide a better understanding of space by analyzing human actions on the environment and the impacts of solid waste caused in the surroundings.

In light of this, this article presents possibilities for addressing themes related to SDG 12: Sustainable Production and Consumption, within the context of three 5th-grade classes at the Application School of the Federal University of Uberlândia (Cap-UFU) during the year 2024, from an investigative perspective aimed at awakening individual and collective responsibility for the environment.

In the pilot model of environmental education or education for sustainable development presented below, solid waste plays a prominent role during the study of the geographical concepts of Landscape and Place, allowing students to understand the waste produced in their homes and at school. This facilitates a reflection on the impacts of their behaviors and consumption habits within the context of sustainability in the place where they live, due to the improper disposal of solid waste in geographical space.

1.1 Concept, Classification and Legislation of Solid Waste in Brazil

In Brazil, Law No. 9.795 (April 27, 1999), endorsed by the Ministry of Education (MEC) and the Ministry of the Environment (MMA), which deals with environmental education and establishes the National Policy on Environmental Education (PNEA). The law aims to promote the construction of values, knowledge, and skills necessary for environmental conservation. The main objective of PNEA is to encourage individual and collective actions that aim for sustainability, proposing the inclusion of the environmental dimension in the training of professionals across all fields of knowledge. The law also sets guidelines and parameters to incorporate environmental themes at various levels and modalities of education, as well as in government programs and informal educational activities.

PNEA understands Environmental Education (EE) as "the processes through which individuals and collectives build social values, knowledge, skills, attitudes, and competencies aimed at the conservation of the environment, a common good of the people, essential for healthy quality of life and its sustainability" (Brazil 1999, p. 1).

The Brazilian Standard (NBR) 10004 presents the classification of solid waste according to its potential risks to the environment in three classes, as follows: Class I hazardous waste, Class II non-inert or non-hazardous waste, and Class III inert waste (ABNT 2004). Table 1 presents the classification of waste and its characteristics.

Law No. 12,305/10, which establishes the National Solid Waste Policy (PNRS) and amends Law No. 9,605 of February 12, 1998, incorporates modern concepts of solid waste management and introduces new tools into Brazilian environmental legislation (Brazil 2010). Among the principles of this law, it is stated that solid waste is reusable and recyclable as an economic asset and has social value, generating work and income while promoting citizenship. On the other hand, waste is solid material that, after all possibilities of treatment and recovery through available and economically viable technological processes have been exhausted, has no other option but to be disposed of in an environmentally adequate final manner (Brazil 2010).

The National Solid Waste Policy, Law 12,305, in its article 13, also classifies waste according to its origin and hazardous nature. In this context, non-hazardous waste (Class II: non-inert) is considered, including those originating from domestic

Table 1 Classification of waste and its characteristics

Waste	Characteristics
Class I: hazardous	These are solid wastes or mixtures of wastes that, due to their characteristics of flammability, corrosiveness, reactivity, toxicity, and pathogenicity, may pose a risk to public health, causing or contributing to increased mortality or incidence of diseases and/or have adverse effects on the environment when improperly handled or disposed of
Class II: non-inert	These are solid wastes that do not fit into Class I or Class II-B. These wastes may have properties such as combustibility, biodegradability, or water solubility. They are basically waste with characteristics similar to household waste
Class III: inert	These are solid wastes or mixtures of solid wastes that, when sampled representatively according to NBR 10007 (ABNT 2004), and subjected to the solubility test according to NBR 10006 (ABNT 2004), do not have any of their constituents solubilized at concentrations exceeding drinking water standards, except for appearance, color, turbidity, hardness, and taste. These are wastes that do not degrade or decompose when disposed of in soil, such as construction and demolition debris, soils and rocks from excavations, glass, and certain plastics and rubbers that are not easily decomposed

Source ABNT-NBR 10004 (2004, p. 3)

activities in households, as well as those generated in public service establishments (such as papers, plastics, and various packaging) (Brazil 2010).

2 Methods Used

The basic premise of the methodology is action research, defined as "a line of research associated with various forms of collective action, guided by the resolution of problems or objectives of transformation" (Thiollent 2022, p. 7). This research methodology is characterized by being participatory, dynamic, flexible, and interactive. The context of its application involved playful didactic sequences, which, based on real-life issues, facilitate active learning and promote meaningful, contextualized teaching aligned with contemporary needs for citizenship education. Therefore, according to Gil (2021), it is a research approach that not only contributes to scientific production but also leads to social action.

For the development of activities in the pilot model, three 5th-grade classes were selected, comprising a total of 60 elementary school students. The study was conducted between October and December 2024, adopting various pedagogical tools that enable active and experiential learning, as well as changes in daily habits to be consolidated in the long term.

The theme of Solid Waste was discussed in the Geography curriculum component, in accordance with the Curriculum Parameters of the CAp-UFU in the Geography area and with the National Common Curricular Base (BNCC), focusing on the following skill (EF03GE08):

> Relate the production of household or school waste to the problems caused by excessive consumption and build proposals for conscious consumption, considering the expansion of habits of reduction, reuse, and recycling/disposal of materials consumed at home, at school, and/or in the surrounding area. (Brazil 2018, p. 371)

The methodology used was that of the Eco-Schools Program from the Litter Less Campaign (LLC), a project in partnership with the Mars Wrigley Foundation and the Foundation for Environmental Education (FEE). This initiative has been implemented since 2011 in various countries around the world, reaching the participation of approximately 3.5 million students (Eco-Schools Program 2022).

In Brazil, the project began in 2021 through the national operator, Environments Network Institute—*Instituto Ambientes em Rede* (IAR). The campaign goal is to engage and educate children and young people about the waste issue, with the aim of producing and expanding their knowledge on the topic, thereby changing behavior regarding waste management. The campaign methodology includes: (a) registering the school on the Litter Less Campaign website (https://www.ecoescolas.org.br/campanha-menos-lixo-2024/); (b) reading the campaign's prospectus documents, which provide guidelines for monitoring waste at school, the explanatory material about Education for the Sustainable Development Goals (SDGs) from the

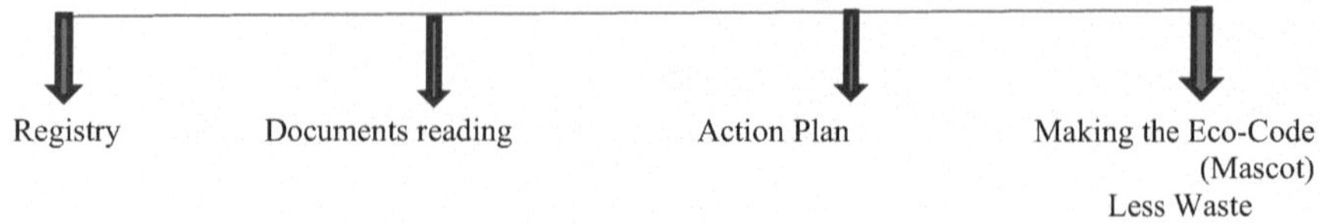

Fig. 1 Litter Less Campaign application method. *Source* Eco-Schools Program (2022)

United Nations Educational, Scientific and Cultural Organization (UNESCO), and the implementation of the Action Plan (Fig. 1).

The Action Plan spreadsheet consists of a table that outlines the actions to be developed throughout the Campaign (Table 2).

3 Results and Discussion

The different pedagogical approaches applied in this research allowed us to address complex and relevant issues both theoretically and empirically, in a contextualized manner. In other words, starting from the local issues experienced by the students, the aim was to reach the cognitive, socio-emotional, and behavioral domains concerning solid waste.

As shown in Table 2, in Theme 1, "Understanding Solid Waste," two lessons were developed, totaling one hundred minutes (100 min), during which students had the opportunity to engage in a discussion with the teacher about their perception of what solid waste is. Based on their responses, the concept of waste was introduced, with students highlighting that what might be considered waste for one person may not necessarily be for another. After the discussion, the teacher presented the concept of Solid Waste as defined by the Brazilian Association of Technical Standards NBR 10004 (ABNT 2004).

Following the program, a questionnaire with open-ended questions was applied for students to make a list of everything they consumed the previous day. During this activity, the students reflected on the idea that when we no longer find a "use" for something specific, it tends to become "trash." Afterward, the 5R principles (Reject, Rethink, Reduce, Reuse, and Recycle) were introduced, based on the list the students had created. They were then asked to categorize and tabulate the items that could be reused again or that were discarded, according to the 5R's (Table 3).

Among the students' responses, it was observed that the items can be classified into five main categories, in order of response: A—School Materials (Reuse), B—Food Packaging (Recycle), C—Clothing (Reuse), D—Electronic Devices (Reuse), E—Disposable Items (Recycle).

From theme 1, the students were able to reflect on what they consumed and the waste they produced, with the aim of reconsidering the true necessity of everything they consume daily. The first lesson plan aimed to provide a hands-on experience for the students regarding the classification of the waste they generated.

Table 2 "Litter Less Campaign" Action Plan spreadsheet

Theme	Objectives	Action	Costs/resources	Deadline
			What is the cost?	*When? For how long?*
1. Understanding waste	• Explain what constitutes waste • Identify ways to manage waste • Understand the importance of managing waste	The teacher asked the students to make a list of everything they used the day before, including toys, food, paper napkins, plates, notebooks, clothes, pencils, etc.	No cost materials for writing, such as notebooks and pens • Copy of the lesson plan—material 1—understanding waste management principles	Session 1 in class: 100 min for students to classify the items they used and introduce the concept of waste. Also, introduce the 5 Rs: Reject, Reuse, Rethink, Reduce, Recycle
2. School waste audit (lesson plan 2)	• Perform an audit to collect data on different types of waste generated at school • Record, analyze, interpret, and represent the data	The audit was conducted as a diagnostic tool to identify the current situation that needs correction and/or improvement, collecting and photographing different types of waste generated at school	No cost—materials for writing, such as notebooks and pens • Copy of the lesson plan 2—material 1—understanding waste management principles	Session 2 in class: 200 min to analyze the collected data

(continued)

Table 2 (continued)

Theme	Objectives	Action	Costs/resources	Deadline
			What is the cost?	*When? For how long?*
3. Research on household and school waste	• Conduct surveys to gather data on waste management and disposal practices at home and at school	Two questionnaires were applied 1. Household waste survey 2. School waste survey	No cost—materials for writing, such as notebooks and pens • Copy of the lesson plan 3—material 1—understanding waste management principles	Session 3 in class: 200 min to analyze the collected data
4. Mascot creation	• Draw a mascot representing waste management for the class	The students created drawings or a solid waste holder in pairs. Each class selected an eco-code about waste that represented their 5th-grade class (A, B, and C)	No cost—pencils, paper, colored pencils, recyclable materials	Session 4—2 weeks

Source Eco-Schools Program (2022)

Table 3 Items used by students and the 5Rs

No.	Things used	Can it be used again? (Yes/No)	Was it discarded? (Yes/No)	Reevaluate if discarded, and classify it under one of the 5R options				
				Reject	Rethink	Reduce	Reuse	Recycle
1								
2								

Source Eco-Schools Program (2022)

The learning processes included practical involvement, sharing answers through classroom interaction, group work, and analyzing the responses.

The results achieved in this first stage were the students' understanding of what solid waste is, what the 5Rs consist of, and how to identify and manage their own solid waste.

The plan for theme 2 aimed to carry out a "School Waste Audit," as it serves as a local diagnostic tool to identify the current situation and assess whether there is a need (or not) to correct or improve certain aspects. A waste audit, as a pedagogical proposal, can help students understand and determine the amount and types of waste generated daily, assisting them in identifying recyclable materials.

To conduct the waste audit at the school, the students worked in groups. As part of the audit, they noted the types of waste produced in the classroom, as well as visiting areas such as the cafeteria, courtyard, and sports courts, where large amounts of waste are generated. The research information was recorded by the groups in their notebooks.

After the school waste audit, a discussion was held to understand the types of waste found in different areas of the school (which were answered in Questionnaire 2: "Survey on School Waste") and to explore possible solutions to address the issue.

From the school waste audit, the students were able to identify the types of waste generated at the school and assess whether they are being disposed of correctly.

The lesson plan 3 encouraged the students to conduct research on household and school waste and carry out surveys to collect data regarding waste management and disposal practices both at home and at school. In this stage, two questionnaires were applied. The first was about "Household Waste," consisting of nineteen multiple-choice and open-ended questions, of which only fifteen were answered, as questions 2, 5, 12, and 13, related to monthly expenses on waste disposal, were not answered due to a lack of knowledge for formulating responses. It is worth noting that only 45 students participated in this research and submitted the completed questionnaires. Based on the answers, a graph (Graph 1) was created.

As seen in Graph 1, in "question 1," which asked, "If waste is not disposed of properly, can it pollute the environment?" all 45 students agreed that it can. In "question 3," the question was "Who disposes of the trash in your house?" 37 students responded that it is the responsibility of one of the family members. In "question 4," which asked, "Where is household waste discarded?" 37 students answered that it is in the trash bin. In "question 6," about the level of satisfaction with the current municipal waste collection system, 15 students considered it to be "good."; in "question 7,"

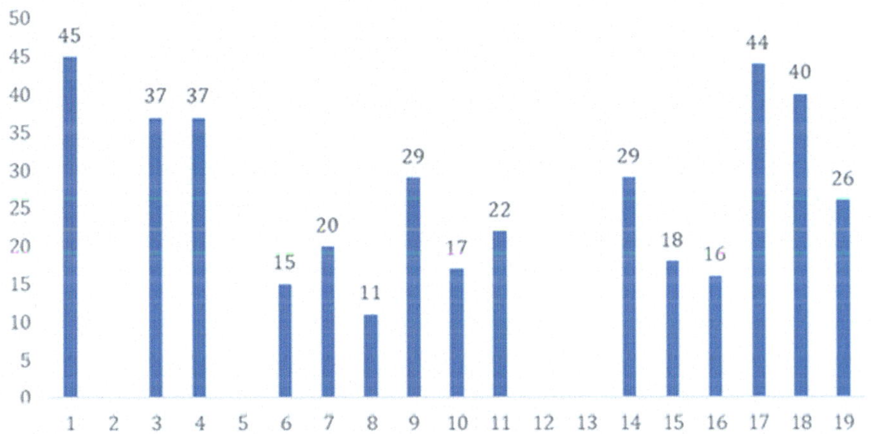

Graph 1 Research on household waste. *Source* Author's elaboration

which asked, "How often is your household waste disposed?" 20 students answered that it is "every two days." In "question 8," asking, "What do you use to dispose of your household waste?" 11 students answered that they use "any container." In "question 9," about "When do you dispose of your waste (time of day)?" 29 students answered that there is no set time for waste disposal. In "question 10," asking, "How often is waste collected?" 17 students indicated that it is "once every two days."; in "question 11," 22 students answered that they dispose of the waste in the trash can themselves. In "question 14," which asked, "Would you like to dispose of your household waste daily?" the majority (29 students) answered "no." In "question 15," when asked, "What types of waste are you willing to store for a few days?" 18 students responded "batteries." In "question 16," 16 students acknowledged that it is their responsibility to separate the household waste; in "question 17," almost all of the respondents (44 students) admitted that they are not separating any type of solid waste. In "question 18," it became clear that 40 out of the 45 students interviewed do not know where the household waste is taken. In "question 19," when asked, "Would you be willing to participate in an exercise to assess the volume of waste produced by different categories?" 26 students agreed.

By answering the questionnaire about household waste, the students were able to reflect on their waste disposal habits, considering the days of the week when waste collection occurs, the main types of waste produced, and unsustainable consumption habits/behaviors. In general, the survey also highlighted the students' lack of knowledge about the final destination of the waste produced in their households and the importance of their participation in the responsibility of sorting waste according to its types. Proper sorting, followed by recycling and the sale of recyclables, can even serve as a source of income for many families.

In the second questionnaire with 10 questions about "School Waste Survey," in question 1, when asked "I'm not at home and need to throw something away, so I…," 32.7% of the respondents said they throw it in the regular trash. In question 2, "Is waste disposal a concern for you?", 62% responded that yes, it is a concern. Question 3, which is a follow-up to question 2, asked "Why is improper waste disposal a concern for you?", 53.3% of respondents replied that it is because it makes the environment look dirty and they consider it harmful to health. In question 4, "Why do you think people dispose of waste improperly?", 35.5% of respondents believed it was due to bad habits, followed by 33.3% who answered "someone else will pick it up.", which suggests that the students understood that there are people who rely on school cleaning jobs and feel that they don't need to worry much about keeping the classroom clean, as it ensures employment for the school's cleaning staff. In question 5, when asked, "Where does the school waste go?", 53.3% believed it ends up in the river, indicating that most students are unaware of the proper destination for waste. In question 6, "In your opinion, what is the most common type of waste improperly discarded at your school?", the majority (53.3%) responded that it is packaging from chocolate, snacks, and juice boxes. This shows a clear awareness of the types of waste generated at school, though a lack of knowledge about their proper disposal. In question 7, when asked, "In your opinion, where is the most improperly discarded waste at school?", 57% of the students responded that it is in the classroom.

In question 8, "If there were enough trash bins, would you still throw trash in other places?", the students indicated that yes, they would, but they also highlighted the need for more trash bins to be installed at the school and suggested using signs to indicate the correct disposal areas. In question 9, when asked, "Do you think you can do something to reduce the improper disposal of trash at your school?", since this question was open-ended to express their opinion, the responses were quite similar to those in the previous question. The majority of the answers indicated that it would be the school's responsibility to install trash bins and provide clear signage for them. In question 10, when asked, "In your opinion, what is the most effective way to reduce/control waste?", 38.5% of respondents answered that it would be if waste collection were carried out by volunteers, 28.5% considered it would be through educational campaigns and awareness-raising, and 21.4% of respondents believe that fines or other penalties should be applied.

After the application of the questionnaires, an analysis/debate was conducted regarding the responses. The results showed ambiguity in the students' answers, as they stated they could dispose of waste if there were regular trash bins, even though the majority of students indicated that proper waste disposal is a concern for them. It is important to highlight that their responses demonstrated that they care about the issue of waste, primarily because it affects the aesthetics of the environment and could bring potential health problems. The students recognized that the areas with the most improperly discarded waste are the classrooms, where trash is often seen thrown on the floor or even inside the desks, such as scraps of paper or packaging from candies, gum, lollipops, chocolates, and snacks. At the same time, the students reflected on the need for increased awareness regarding proper waste disposal. They held the school administration accountable for not providing enough trash bins for proper disposal and suggested that more projects or campaigns related to solid waste should be developed within the school. Despite this, they also considered that poor habits could be corrected through punitive measures or other penalties outlined in the school's regulations.

After the discussion of the responses, some doubts were clarified regarding the location where waste generated in Uberlândia-MG is disposed of. In addition, there was a greater awareness about how to keep the classroom, cafeteria, playground, sports courts, and bathrooms clean.

The fourth stage involved creating a mascot, where students could choose between two options: a drawing or a type of solid waste container, which could be a battery collection container for the disposal of electronic waste. They chose the second option (Fig. 2).

For the creation of the mascot, the classroom was divided into groups of 2–3 students. One of the criteria was the use of recyclable materials, without the teacher's interference in the students' creative process. After the creation, the preferred mascot was chosen to represent their 5th grade class (A, B, and C).

The use of the four pedagogical moments facilitated work focused on Environmental Education. It was noticed that the lessons were well understood, as there was active participation from the majority of students in all the classes throughout the activities. It is essential to highlight that the goal of the proposal was to spread

Fig. 2 Mascotes de cada sala produzidos e escolhidos pelos/pelas estudantes. *Source* Author's elaboration

the concept of solid waste, present the 5Rs, and encourage a critical reflection on consumption habits and waste disposal in the daily lives of the students, both at school, at home, or in other places. Also, at the end of the implementation of this pilot model, the students achieved positive results. In addition to understanding the conceptual issues, they presented alternatives to the school for reducing consumption and ensuring the correct disposal of waste, such as expanding the installation of sufficient trash bins for solid waste disposal and the need to develop more projects or campaigns related to the theme "solid waste" at school. Many of them recognized the need for individual and collective changes regarding consumption habits and were creative in producing their mascots, keeping engaged throughout the lessons.

The results obtained reflect the awareness raised among the students, which mobilizes not only knowledge but also the capacity for action towards social transformation, through the consolidation of daily habits aligned with sustainability and environmental preservation.

4 Conclusions

The sustainable development model applied, in line with the theme related to SDG 12: Sustainable Production and Consumption, presented in this essay, was quite diverse and aimed to contribute to a more critical and reflective education for students on the issues of solid waste management and disposal. Through different pedagogical tools, the aim was to raise students' awareness about the topic, in order to contribute to the formation of more critical citizens committed to environmental preservation.

This pilot model encompasses different pedagogical approaches: it is based on the issue of solid waste, reflective and critical thinking, and active/creative learning, with students at the center of the teaching–learning process. Furthermore, it could be reapplied in other school years with the necessary curricular adjustments and age appropriateness, depending on the deepening of the learning process. Thus, beyond the conceptual domain, its potential for behavior changes stands out through the adoption of more sustainable practices, possibly extending beyond the classroom and the school.

This article aimed to contribute to discussions on methodological activities that promote education for sustainability, integrated in a practical and meaningful way into the curriculum and functioning of institutions, such as the University Application Schools in Brazil. In this context, it is important to highlight the relevance of supporting teaching practices based on methodologies like those discussed earlier, which stem from local issues and extend to a global scale, intrinsic to the Sustainable Development Goals (SDGs). Furthermore, strategic partnerships that provide technical resources and continuous support can expand the reach of such initiatives within the school and the community.

References

ABNT (Associação Brasileira de Normas Técnicas). (2004). *Resíduos sólidos – classificação: NBR-10004*. Rio de Janeiro. 63p. https://www.suape.pe.gov.br/images/publicacoes/normas/ABNT_NBR_n_10004_2004.pdf.

Andrade, A. S., Souza, T. F. M. S., Guimarães W. N. R. (2021). Gincana sobre resíduos sólidos na aula de Geografia. *Revista Brasileira de Educação Ambiental (RevBEA)*, 16 (6): 125–138.

Brasil. (1997). *Parâmetros Curriculares Nacionais*. Secretaria de Educação Fundamental-Brasília: MEC/SEF, 126p. http://portal.mec.gov.br/seb/arquivos/pdf/livro01.pdf.

Brasil. (1998). *Parâmetros curriculares nacionais: terceiro e quarto ciclos do ensino fundamental: temas transversais*. Secretaria de Educação Fundamental-Brasília, 436p. http://portal.mec.gov.br/seb/arquivos/pdf/ttransversais.pdf.

Brasil. (2018). Ministério da Educação. *Base Nacional Comum Curricular*. Brasília. Secretaria de Educação Fundamental, 600p.

Callai, H. C. (2015). A questão da cidadania nas séries iniciais. In: Callai H. C.; Toso C. E. I. (Org.) *Diálogos com professores*: cidadania e práticas educativas. Ijuí: Ed. Unijuí, pp. 23–40.

Gil, A. C. (2021). Como fazer pesquisa qualitativa. 1°. Ed. Barueri: Atlas. 190p.

https://www.gov.br/mec/pt-br/escola-em-tempo-integral/BNCC_EI_EF_110518_versaofinal.pdf.

Layrargues, P. P. (2011). O cinismo da reciclagem: o significado ideológico da reciclagem da lata de alumínio e suas implicações para a Educação Ambiental. In: Loureiro C. F. B, Layrargues, P. P, Castro, R. S. (orgs.). Educação Ambiental: repensando o espaço da cidadania. Cortez (5):185–225.

Lei nº 12.035 de 2 de agosto de 2010. (2010). Institui a Política Nacional de Resíduos Sólidos; altera a Lei no 9.605, de 12 de fevereiro de 1998; e dá outras providências. Presidência da República. https://www.planalto.gov.br/ccivil_03/_ato2007-2010/2010/lei/l12305.htm.

Lei nº 9597 de 27 de abril de 1999. (1999). Dispõe sobre a educação ambiental, institui a Política Nacional de Educação Ambiental e dá outras providências. Política Nacional de Educação Ambiental. Presidência da República. https://www.camara.leg.br/proposicoesWeb/prop_mostrarintegra?codteor=634068#:~:text=Entendem%2Dse%20por%20educa%C3%A7%C3%A3o%20ambiental,Art.

Moretti, E. C., Campos, B. F., Mader, A. M., P. (2024). Mundo da sustentabilidade e a produção da natureza no Brasil. Revista Territorial (ISSN 2317 0360), 3(2): 66–95. https://www.revista.ueg.br/index.php/territorial/article/view/16012.

Programa Eco-escolas. (2022). Campanha menos lixo 2022. https://www.ecoescolas.org.br/lle-menos-lixo-2022.

Santos, M. (1988). Metamorfoses do espaço habitado: fundamentos teóricos e metodológicos da geografia. 6ª Ed. São Paulo: Edusp. 136p.
Thiollent, M. (2022). Metodologia da pesquisa-ação. 19ª Ed. São Paulo: Cortez. 136 p.
UNESCO (Organização das Nações Unidas para a Educação, a Ciência e a Cultura). (2021). *Educação para o desenvolvimento sustentável*. https://www.unesco.org/en/sustainable-development/education#:~:text=Education%20for%20Sustainable%20Development%20empowers,a%20better%20future%20for%20everyone.

Artificial Intelligence and Higher Education: A Bibliometric Study

Abner Silveira Facina, Igor Silveira Facina, Jaluza Maria Lima Silva Borsatto, and Ananda Silva Singh De Carli

Abstract Integrating artificial intelligence (AI) into higher education has become a transformative force, offering innovations such as personalized learning, automated administrative tasks, and advanced data analytics. However, its rapid adoption raises critical questions about ethics, equity, and long-term sustainability in academic institutions. This bibliometric study examines the evolving discourse on AI in higher education, focusing on its implications for sustainable development, particularly in alignment with the United Nations Sustainable Development Goals (SDGs), including quality education (SDG 4), reduced inequalities (SDG 10), and global partnerships (SDG 17). The study analyzed 1803 peer-reviewed publications from the Web of Science and Scopus databases (2018–2024), using Bibliometrix® to map trends, key themes, and geographic contributions. The findings reveal a dramatic surge in research output, from fewer than 100 publications in 2018 to over 600 by 2024, reflecting heightened academic and institutional interest. Leading applications of AI in higher education include adaptive learning systems, predictive analytics for student performance, and AI-driven chatbots for student support. These tools promise to enhance educational accessibility and efficiency. Nevertheless, they also introduce challenges such as algorithmic bias, data privacy concerns, and risks to academic integrity issues prominently highlighted in highly cited works by Rudolph et al. (J Appl Learn Teach 6(1):342–363, 2023a; J Appl Learn Teach 6(1), 2023b) and Cotton et al. (Innov Educ Teach Int 61:228–239, 2024). Geographically, research production is uneven, with China (412 publications) and the United

A. S. Facina (✉) · I. S. Facina
Postgraduate Program in Administration, Federal University of Uberlândia, Uberlândia, Brazil
e-mail: abnerfacina@hotmail.com

I. S. Facina
e-mail: facinaigor@hotmail.com

J. M. L. S. Borsatto · A. S. S. De Carli
School of Management and Business, Federal University of Uberlândia, Uberlândia, Brazil
e-mail: jaluza.silva@ufu.br

A. S. S. De Carli
e-mail: ananda.singh@ufu.br

W. Leal Filho et al. (eds.), *Sustainability in the Curriculum and Operations of Universities*, World Sustainability Series,
https://doi.org/10.1007/978-3-032-12678-8_15

States (277) dominating the field, while contributions from the Global South remain limited. This disparity underscores the need for inclusive collaboration to ensure AI benefits are equitably distributed. The Astrophysical Journal and Education Sciences emerged as key sources, illustrating the interdisciplinary nature of AI research, spanning education, technology, and ethics. From a sustainability perspective, AI holds significant potential to advance SDG 4 by personalizing education and improving learning outcomes, particularly for underserved populations. If we prioritize ethical and equitable deployment, it also supports SDG 10 by bridging gaps in educational access. However, realizing these benefits requires robust frameworks for teacher training, transparent AI governance, and policies that address digital divides. This study contributes to the literature by synthesizing global research trends and highlighting actionable insights for universities. It calls for a balanced approach to AI integration that harnesses its transformative potential while mitigating risks through ethical guidelines, cross-border partnerships, and inclusive innovation. By doing so, higher education institutions can position AI as a catalyst for sustainable development, ensuring that technological progress aligns with the broader goals of equity and social responsibility.

Keywords Higher education · HE · Artificial intelligence · AI

1 Introduction

The evolution of technology has enabled the development of things once unimaginable, such as the creation of artificial intelligence (AI) and its variations (Saghiri et al. 2022). AI has become intrinsic to contemporary life, being present in areas such as entertainment and industry and increasingly in higher education (HE) (Lee et al. 2024).

AI has brought advancements in the academic field, such as the advancement of scientific research (Jumper et al. 2021), analysis of large volumes of data (Shallue and Vanderburg 2018), personalization and adaptation of teaching and learning (Holmes et al. 2022), automation of scientific review and publication (Bornmann and Mutz 2021), and the discovery of new materials with specific properties (Stokes et al. 2020). These advancements have expanded the boundaries of research and experimentation, making AI a powerful potential tool in teaching and reshaping the academic environment. However, AI's full capacity and applications have yet to be explored. Despite the relevance of these tools in restructuring educational institutions and transforming the process of teaching and learning, their feasibility of application is still questioned due to their potential adverse and collateral effects (Kumar et al. 2024; Alshahrani et al. 2024).

Teachers, technicians, and students must undergo training and meticulously analyze how to apply AI in higher education (Flores-Velásquez et al. 2024). Furthermore, there is growing concern about the legal, social, and ethical aspects of AI and uncertainty about how safe its use is (Alshahrani et al. 2024). Flores-Velásquez et al.

(2024) also discusses the importance of studying the relationship between higher education and artificial intelligence, especially in an era where the world has transitioned from in-person teaching to remote learning and back to in-person, making technology adoption crucial.

A study conducted by Zawacki-Richter et al. (2019) mapped the scientific production on AI in education, identifying four primary areas of application: personalized learning, automation of administrative tasks, assessment tools, and support for academic research. The study revealed increased publications starting in 2010, mentioning intelligent tutoring systems and adaptive platforms, allowing personalized learning tailored to students' needs. Additionally, the analysis indicated that the United States, China, and Europe lead scientific production in this area, reflecting global investment in educational technologies.

The bibliometric study by Chen et al. (2020) focused on the impact of AI in higher education. The authors identified that the main applications include predictive analysis of student performance, chatbots for student support, and plagiarism detection tools. The research also highlighted challenges, such as the need for teacher training to integrate AI into the classroom and ethical concerns related to the use of student data.

Luckin et al. (2022) explored the potential of AI to transform both higher education and basic education. The authors identified that AI is increasingly being used to automate assessments, provide personalized feedback, and create immersive learning environments, such as those based on virtual and augmented reality. However, the study also warned of significant ethical challenges, such as algorithmic bias, which can perpetuate existing inequalities, and the lack of transparency in AI decision-making systems.

Given the above, analyzing how AI in higher education is being discussed in literature is relevant, as highlighting the main challenges, applications, and ways in which artificial intelligence is being used in higher education can contribute to identifying theoretical gaps and offering real-world advice to universities. In this context, this work aims to analyze how the use of artificial intelligence in higher education is being discussed, highlighting the necessity to include, study, and invest not only in the relationship between AI students and teachers but also in how to include technicians and the university as a whole institution.

Regarding the Sustainable Development Goals (SDGs), this work's content falls into the 4th, 10th, 16th, and 17th goals: fourth quality education, 10th reduced inequalities, 16th peace, justice, and strong institutions, and 17th partnerships for the goals. 4th because helps educators and organizations to navigate what is being done in higher education about and with artificial intelligence and how is being done, 10th because helps countries levels theirs's educations standards on AI grounds, On the 16th, we aim to start building a bridge that enables professors, students, and other authorities involved in these organizations to use AI in a way that includes everyone and makes the knowledge generated in this environment accessible to all. On the 17th, we will help and guide these organizations in collaborating on the best ways to use and explore these new tools.

2 Literature Review

Since the mid-1940s, when the first works emerged, there has been discussion and research on artificial intelligence (AI). The discussion gained momentum in recent years, with a significant increase in publications in 2019 (Flores-Velásquez et al. 2024) and has become increasingly relevant due to its intrinsic role in human life. AI is a computational system that mimics human reasoning, meaning machines are capable of developing logical reasoning comparable to humans (Alves et al. 2023). AI applications include virtual assistants (chatbots), image recognition, machine learning, and more (Morais et al. 2020; Alves et al. 2023).

The daily use of technology, driven by its advancement, has fostered integration between all areas of life and technology, making living more convenient while facilitating and speeding up decision-making and information retrieval (Maphosa and Maphosa 2023). The use of technologies, particularly AI, has been widely adopted in education, including in fields such as linguistics, engineering, and medicine, becoming increasingly prevalent in higher education institutions (Crompton and Burke 2023).

Specifically, in the academic context, AI can be useful for identifying early signs of student dropout. Much research has focused on retention in higher education, aiming to understand, prevent, and predict student attrition. With this goal in mind, Gray and Perkins (2019) used machine learning techniques to map and identify early signs of students intending to leave their courses. The model developed by these authors identified opportunities for intervention in the early stages of dropout. As a result, the descriptive metric, the Bangor Engagement Metric, was created, which retains high predictive power.

Student performance is important to educational institutions, and it is often used to indicate institutional success. Early identification of at-risk students—those showing signs commonly associated with dropout, such as low grades, disengagement, or complaints about lack of institutional support—along with the implementation of preventive measures, can significantly improve their performance. Recently, machine learning techniques have gained popularity for prediction purposes. Unfortunately, most educators with knowledge of computing or artificial intelligence can only use these techniques (Gray and Perkins 2019).

Educational institutions have widely used data mining methods in addition to machine learning, as the two approaches complement each other. Effective use of these tools requires the determination of various variables, from defining student success to selecting relevant attributes and choosing the most suitable machine learning method for the problem at hand. Alyahyan and Düştegör (2020) propose a step-by-step guide for educators interested in using data mining techniques to predict student success, systematically compiling existing literature to address possible choices and parameters comprehensively and justifiably. This work aims to facilitate educators' access to data mining.

Information technologies, such as artificial intelligence and machine learning, are present in how people travel, manage their time, and acquire information. Incorporating such technologies in institutions has been a "game-changer," including in educational institutions. As a result, learning platforms are more aligned with students' needs and knowledge, making the learning process more efficient. The work of Kuleto et al. (2021) explores these technologies' potential and application areas in higher education, based on secondary data analysis (literature review) and surveys. The results demonstrated that these technologies significantly improved students' learning performance and provided an accessible and collaborative research environment. Immediate and interactive support in higher education institutions can be challenging due to low faculty-to-student ratios. Chatbots, designed to simulate human conversation, are increasingly used in various fields, including customer service, workplace productivity, and healthcare. They can also be applied in the academic context, helping educators better respond to students' needs. However, few empirical studies (Rudolph et al. 2023a, b; Cotton et al. 2023; Sullivan et al. 2023a) have examined the pedagogical effectiveness of chatbots in higher education, and even fewer have discussed the potential challenges and disadvantages of their use.

The study by Chen et al. (2023) addresses this gap in literature, exploring the opportunities, challenges, effectiveness, and ethical concerns of using chatbots as educational tools. They conducted two experiments with 215 students, using chatbot-guided interviews to understand the potential benefits of chatbots as intelligent student assistants and students' attitudes toward tasks. The results demonstrated the potential of chatbots to help students learn core content in a responsive, interactive, and confidential manner. The findings also provided insights into students' learning needs, which were used to design and develop an experimental chatbot assistant to teach basic AI concepts to 195 students. The second study suggests that chatbots have the potential to be engaging and responsive conversational learning tools for teaching basic concepts and providing educational resources (Chen et al. 2023).

Based on the previous information, the use of artificial intelligence in education is an important topic within the field of science. One of the AI tools commonly used is ChatGPT. It offers various benefits, including increased student engagement, collaboration, and accessibility. However, it also raises concerns about academic integrity and plagiarism. The study by Cotton et al. (2024) explores the opportunities and challenges of using ChatGPT in higher education and discusses such tools' potential risks and benefits. The research also addresses the difficulties of identifying and preventing academic fraud and suggests strategies universities can adopt to ensure these tools' ethical and responsible use. These strategies include developing usage policies and procedures, providing training and support, and employing various methods to detect and prevent fraud. The authors concluded that while AI in higher education presents significant opportunities and equally challenging issues, universities can effectively address these concerns by adopting a proactive and ethical approach to using these tools (Cotton et al. 2024).

From the above, it is evident that AI can be applied for various purposes in the academic context. It can predict dropout rates, assist students in developing academic work, and promote autonomous learning. However, its use is still questioned in terms of effectiveness, reliability, and credibility (Sullivan et al. 2023b; Rudolph et al. 2023a, b; Cotton et al. 2023).

3 Methods

The study has a qualitative nature, as it does not use mathematical data and because the data collection is designed as a documentary review (Raupp and Beuren 2004; Creswell 2007). With this state-of-the-art review on the proposed topic and its results, the present study can be considered descriptive in exploring already reviewed knowledge, primarily available in a foreign language. The data were collected from the Web of Science (WoS) and Scopus databases, which are peer-reviewed through a blind review process. The following keywords were used: ((“artificial intelligence”) OR (“AI”) OR (“inteligência artificial”) OR (“IA”)) AND ((“higher education”) OR (“HE”) OR (“ensino superior”) OR (“ES”)), being inteligência artificial and IA the Portuguese direct translation to artificial intelligence and AI, and ensino superior and ES the Portuguese direct translation to higher education and HE. Studies published between 2018 and 2024 were searched. Access to the databases was provided through CAFe, the CAPES (Coordenação de Aperfeiçoamento de Pessoal de Nível Superior) journal portal (Derviş 2019; Amiri et al. 2023; Groff et al. 2020). Bibliometrix®, a package within the RStudio® software, was used for data collection and analysis. This tool is capable of extracting and analyzing the collected data simultaneously.

The first round of data collection was conducted in the Web of Science database, with the criterion that the results would be extracted from the abstract of the texts. This phase resulted in 4729 results. The results were filtered according to the publication period (2018–2024), yielding 3117 texts. Next, the texts were restricted to articles, resulting in 2302 records. Then, the search was further restricted to articles in English and Portuguese, yielding 2140 results. Since WoS does not have an option to exclude articles that are not finalized, two options under the open access section were selected: “Green Accepted” and “Green Submitted,” as they are not yet finalized. All articles in this category were excluded, reducing the results to 1876. The final step was to filter for Open Access, completing the process with 1003 results. Then the same steps were followed on Scopus. Table 1 illustrates the steps for both databases.

Once these steps were finished, the results were exported in BibTeX format, compatible with Bibliometrix®, and in.csv format, compatible with Excel®. All the data were checked and organized in Excel®. Once these steps were finalized, a script with all the information was executed to combine data from both bases into a single document in RStudio®. The data was then processed again to exclude the duplicate documents, excluding 500 documents.

The tools above process the cleaned database, analyze it, and create all the required images and graphs.

Table 1 Inclusion and exclusion criteria Web of Science and Scopus

Selection criteria	Web of Science		Scopus	
	Excluded	Included	Excluded	Included
Research date: 10/03/2025	–	–	–	–
Research keywords: ((“artificial intelligence”) OR (“AI”) OR (“inteligência artificial”) OR (“IA”)) AND ((“higher education”) OR (“HE”) OR (“ensino superior”) OR (“ES”))	–	4729	–	8053
Publication period: 2018–2024	1612	3117	2885	5198
Publication type: article	815	2302	2651	2547
Language: English and Portuguese	162	2140	298	2249
Publication stage: finalized	264	1876	131	2118
Access: open access	873	1003	817	1301

4 Results

As illustrated in Fig. 1, the number of publications where their subjects navigate AI and education starts growing in 2019 (Flores-Velásquez et al. 2024), going from less than 100 in 2018 to more than 600 in 2024, when the productions have hit their peak so far. This situation underscores that the topic still needs to be addressed to test the best ways to apply this technology inside classes and how to use it to upgrade the workflow of an educational environment. Similarly, AI-assisted outputs from reviews pose dual challenges: preventing cheating and defining ethical use in education.

The Astrophysical Journal is the most relevant source since it has 44 publications, followed by Education Sciences, 37 publications; Computers and Education: artificial

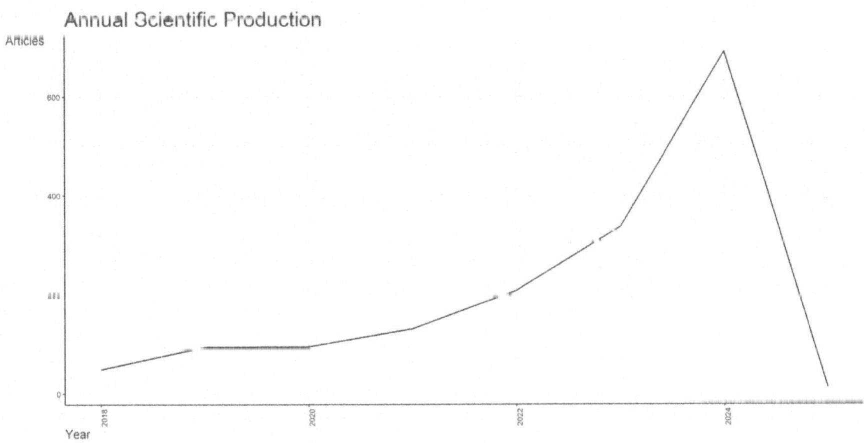

Fig. 1 Annual scientific production

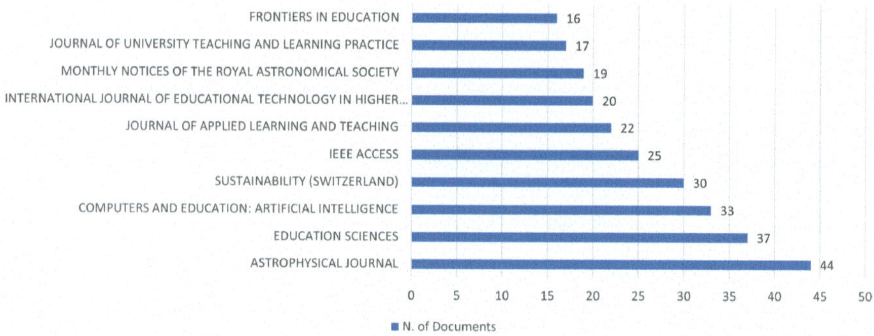

Fig. 2 Most relevant sources

intelligence, 33 publications; Sustainability (Switzerland), 30 publications; and IEEE Access, 25 publications, in fifth place, as shown in Fig. 2.

As the most cited documents, Rudolph et al. (2023a, b) emerge in first place, discussing ChatGPT, its origins and evolution, and the people behind it. Subsequently, an explanation about ChatGPT's work and principles is provided. They also test a few commands to analyze and compare the answers. After this process, the authors discuss the applications of AI and ChatGPT in education, from the students' and teachers' perspectives. Cotton et al. (2024), the second most referenced, pondered the ethical aspects of using ChatGPT to help the teaching and learning processes. It also addressed the challenges of using these tools and identifying their usage in in-class activities. The third most cited article is the work from Tao et al. (2019). The authors discuss using AI tools in the 4.0 industry, comparing cyber-physical systems and cyber-twins. The fourth place, a work from Byrne et al (2017), implemented an "artificial intelligence (AI) model for real-time assessment of endoscopic video images of colorectal polyps". And, at fifth place, Chan and Hu (2023) analyzed students' opinions about AI tools, like ChatGPT, and its usage in a teaching and learning scenario, to better understand how these tools can be applied in class and how they can create policies to regulate them. Figure 3 illustrates the citations of these articles.

Figure 4 shows a word cloud based on the abstracts, showing that the most referenced words are "AI", "education", "learning", "students", "study", and "research".

Figure 5 was based on the author's keywords; the most frequent ones are "artificial intelligence" in first place and "higher education" in second place.

These results illustrate the use and investigation of AI in higher education. These articles go beyond the usage of ChatGPT and explore the best ways to use these tools to improve the quality of life in higher education.

The United States of America and China lead the global publication market. Figure 6 shows these countries in darker blue to emphasize the concentration of publications.

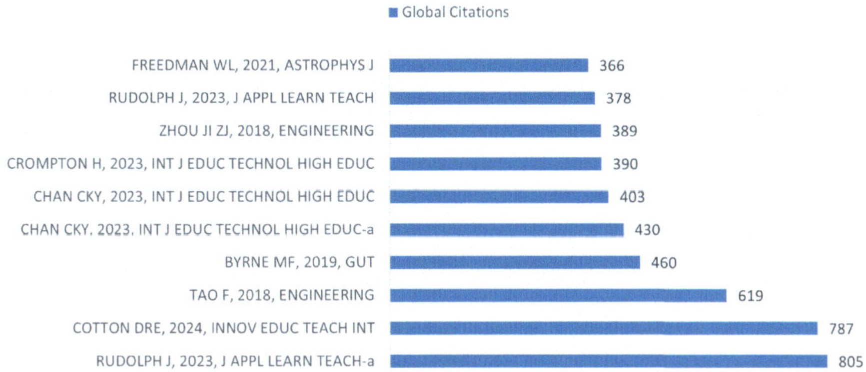

Fig. 3 Most global cited documents

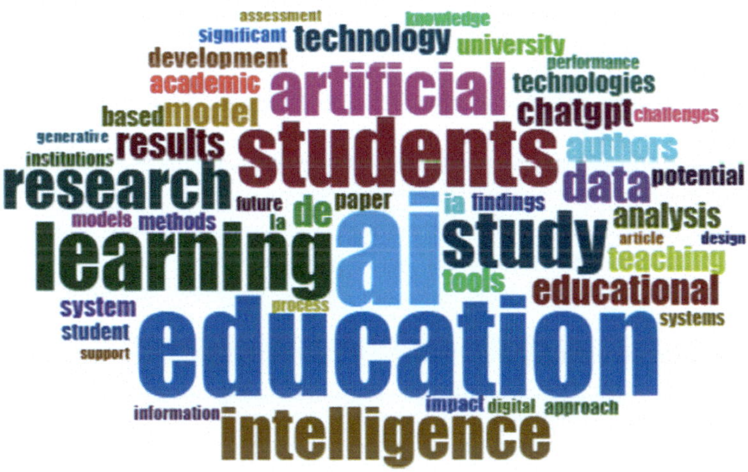

Fig. 4 Word cloud based on abstracts

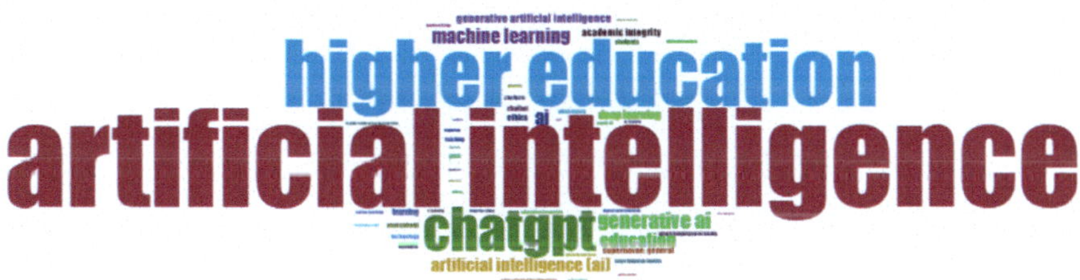

Fig. 5 Word cloud based on the author's keywords

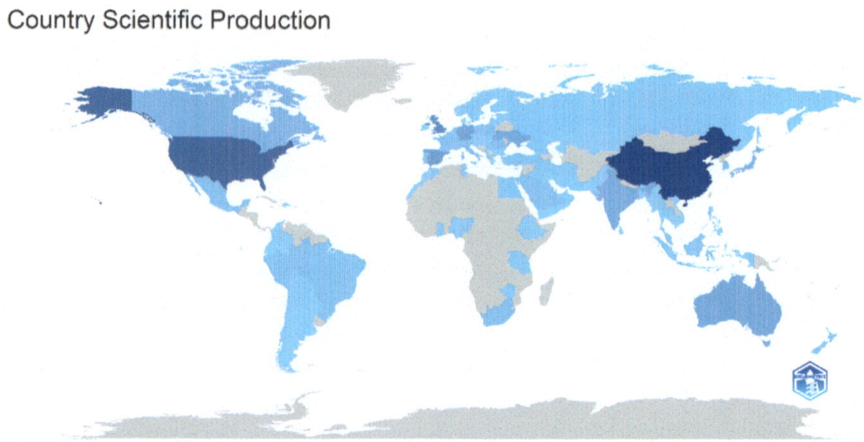

Fig. 6 Country scientific production

The top 4 countries, holding over 100 publications, are China with 412, the USA with 277, the United Kingdom with 166, and Spain with 115. The remaining countries score less than 100. Table 2 shows the top 25 countries in scientific production.

The production in the northern hemisphere is far more advanced than in the rest of the world, especially in third-world countries. These countries first appear in 18th place, Mexico; then 20th place, Brazil; and 22nd place, Peru.

The co-occurrence network demonstrates how the keywords and common themes establish relations, as illustrated in Fig. 7.

The strongest connection is visible between education and other domains, particularly AI, research, student learning, and artificial intelligence studies, highlighting their profound interrelationship.

5 Discussion

The sharp rise in publications, jumping from fewer than 100 in 2018 to over 600 by 2024, shows the growing scholarly interest in this connection. This expanding body of research highlights the transformative potential and many ways AI can impact teaching and learning methods, curriculum design, and administrative tasks within academic institutions. The sheer volume of recent articles suggests a key moment where AI moves from a theoretical idea to a practical tool in education. To ensure its responsible and beneficial integration, we must carefully study its effectiveness, ethical concerns, and teaching implications during this shift. The increasing number of articles also points to a global research effort, reflecting a widespread recognition of AI's disruptive and innovative capacity to reshape educational landscapes.

A closer look at the most important sources contributing to this growing knowledge base reveals a diverse range of academic journals. The Astrophysical Journal,

Table 2 Country publication

Country	Frequency
China	412
USA	277
UK	166
Spain	115
Germany	98
Australia	93
Ukraine	89
India	82
Japan	79
Canada	61
Romania	41
Netherlands	38
South Africa	38
Italy	36
New Zealand	36
South Korea	36
France	35
Mexico	35
Poland	35
Brazil	34
Saudi Arabia	34
Peru	33
Malaysia	28
Indonesia	27
Portugal	25

with 44 publications, stands out, possibly showing the interdisciplinary nature of AI research, where computational advancements in one field can inspire uses in others, and indicating studies on how AI is used in astrophysics education. Close behind, Education Sciences (37 publications) and Computers and Education: Artificial Intelligence (33 publications) directly focus on AI's teaching and technology aspects in education, indicating dedicated places for academic exchange in this field. These journals are key platforms for sharing research on AI's impact on learning processes, instructional design, and educational technology. Sustainability (30 publications) in the top five suggests a rising interest in the long-term societal and environmental effects of integrating AI into education, exploring how AI can help create sustainable educational practices or how ethical AI development in education aligns with broader sustainability goals. IEEE Access (25 publications), a broad open-access journal,

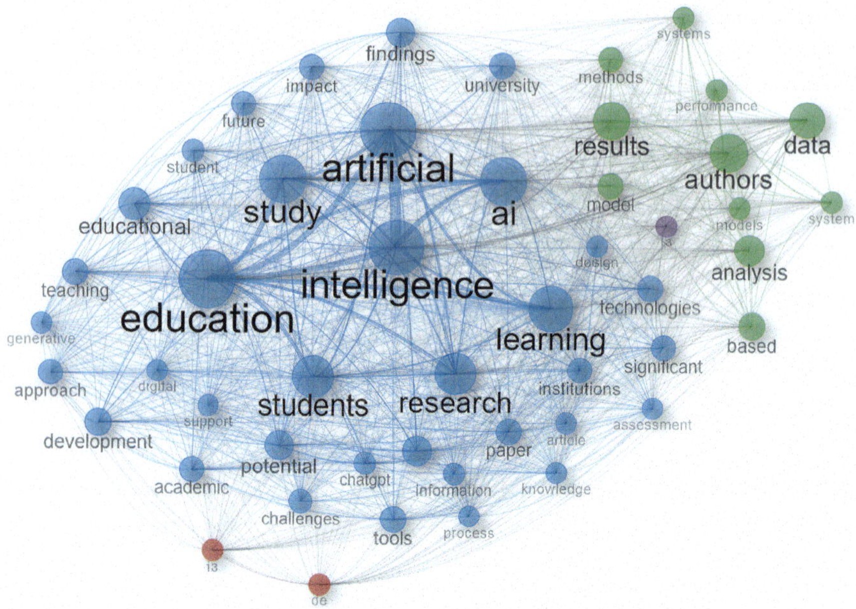

Fig. 7 Co-occurrence network

further shows the wide reach of AI research, covering various engineering and technology viewpoints relevant to educational uses. The prominence of these journals paints a picture of a strong and expanding research field, where different disciplines come together to explore the complex relationship between AI and education. The varied focus of these high-contributing journals also implies that AI in education isn't an isolated field but one that benefits from insights and methods drawn from computer science, educational psychology, ethics, and environmental studies.

The most cited articles offer insights into the most influential and foundational contributions shaping current discussions. Rudolph et al.'s (2023a, b) work on using ChatGPT in education is a key reference. This isn't surprising given the recent emergence and widespread attention received by large language models (LLMs) like ChatGPT, which have sparked intense debates about their potential to revolutionize teaching, learning, and assessment. The high citation count for this article indicates an urgent need for research to understand the teaching implications, ethical considerations, and practical uses of generative AI in educational settings. Similarly, Cotton et al. (2024) focus on the ethical aspects of AI in education, pointing to a critical and growing area of concern. Addressing issues like data privacy, algorithmic bias, academic honesty, and fair access to AI tools becomes crucial as AI systems become more common. This article's prominence signals a growing awareness within the academic community of the need to develop ethical frameworks and guidelines for the responsible deployment of AI in education. Finally, Tao et al. (2019) explore AI in Industry 4.0, while not directly about education, likely provides important context for understanding the skills and abilities needed for future workforces. This article's influence suggests that education researchers are considering how AI integration in higher education can prepare students for an increasingly automated and AI-driven industrial landscape. The emphasis on these three highly cited works indicates that current research priorities revolve around understanding the practical uses of cutting-edge AI technologies, navigating the ethical dilemmas they present, and aligning educational outcomes with the demands of an evolving global economy shaped by AI. These foundational papers will likely continue to inform future research paths, influencing theoretical ideas and practical actions in the field.

Further text analysis through keyword frequency reveals this research area's core themes. Not surprisingly, "AI" (Artificial Intelligence) and "education" are the most common terms, showing the central focus of the review. The consistent appearance of "learning" and "students" emphasizes a focus on AI's impact on learning processes and student experiences. This indicates that a large part of the research is student-focused, investigating how AI can improve, personalize, or otherwise transform the educational journey for students. These keywords together define the thematic landscape of AI in education, highlighting the exploration of intelligent systems in teaching contexts to improve learning outcomes and student engagement. The prevalence of these terms suggests that researchers are actively investigating how AI tools can be designed, implemented, and evaluated to support various aspects of learning, from adaptive tutoring systems to innovative assessment tools, and how these tools affect students' cognitive development, motivation, and overall academic performance. The recurring emphasis on "students" also points to a focus on user-centered design and considering student needs and preferences in developing and using AI in educational settings.

A global view of AI in education research shows an uneven distribution of academic output across different nations. With an impressive 412 publications, China leads, demonstrating a strong national commitment to advancing AI research, particularly within the education sector. This significant output reflects substantial government investment in AI development and its strategic integration into various parts of society, including education. The United States ranks second with 277 publications, which indicates a strong and active research community, although its publication volume is lower than that of China. The United Kingdom (166 publications) and Spain (115 publications) also show important contributions, reflecting well-established research infrastructures and academic interest in the field. The leading positions of these countries suggest coordinated efforts in both technological development and educational reform that use AI capabilities.

In contrast, Brazil's appearance in 20th place indicates a comparatively newer or less productive research landscape in this area. This difference in research output could be due to various factors, including funding levels, research infrastructure, policy priorities, and the maturity of AI development within each country. The global distribution highlights the need for increased international collaboration and knowledge sharing, especially for countries with developing research capacities, to foster a more balanced and complete understanding of AI's potential in education. Closing these research gaps could help adopt the best practices and develop contextually relevant AI solutions in education worldwide. The current scenario suggests that while some nations are rapidly moving forward, others are still building their foundations, highlighting opportunities and challenges for global collaborative research.

6 Conclusion

The main goal of this detailed literature review was to synthesize existing knowledge about how artificial intelligence is being used in higher education while also trying to make this complex information more accessible, especially for Portuguese speakers. The study successfully showed the rapid growth in academic publications dedicated to this interdisciplinary field, highlighting a significant and increasing scholarly engagement with AI's transformative potential in educational settings. By analyzing publication trends, identifying key journals, recognizing highly cited articles, and mapping the geographical distribution of research, this review has provided a broad view of the current state of AI in education research. The findings underscore the critical importance of understanding AI's wide-ranging impact on teaching practices, student learning outcomes, and the evolving landscape of academic institutions. The quick development of AI technologies, like the emergence of large language models, demands continuous and thorough investigation into their ethical implications, effective integration strategies, and potential to create more personalized and engaging learning environments. This review adds to the ongoing discussion by consolidating a substantial amount of research, offering a valuable resource for academics, policymakers, and practitioners looking to navigate the complexities and harness the opportunities presented by AI in higher education. Organizing this information aims to empower stakeholders with the knowledge needed to make informed decisions about the strategic implementation and responsible management of AI technologies within educational frameworks. Identifying key themes and leading research contributions provides a foundation for future inquiries, encouraging a deeper exploration of the detailed interactions between AI and various aspects of the educational process. Furthermore, by making this knowledge more accessible, especially to a broader linguistic audience, the review helps to foster a more inclusive global conversation about the future of education in the age of AI. The insights gained from this review serve as a benchmark for understanding the current research direction and identifying areas ripe for further investigation, ultimately contributing to advancing knowledge in this dynamic and fast-changing field.

Despite its valuable contributions, this review has limitations that should be acknowledged and considered for future research. First, the study's scope was naturally limited by its reliance on only two specific academic databases. While these databases are widely recognized for their thorough coverage of scholarly literature, excluding other potentially relevant sources might have unintentionally omitted important publications. This selective sample could have introduced a bias into the overall results, possibly overlooking significant contributions from journals or conferences not included in the chosen database. Consequently, the generalizability of the findings might be partly limited, as the insights gained mostly reflect the literature available within the selected sources. Second, the broad aim of the research, which was to cover the extensive field of AI in higher education, naturally carries the risk of introducing some gaps in the literature review. While efforts were made to capture the most important themes and influential works, the sheer volume and diverse nature of publications in this rapidly evolving field make it challenging to ensure exhaustive coverage of every subtopic or niche area. While advantageous in providing an overall perspective, this broad approach might have prevented a more in-depth exploration of specific applications, challenges, or theoretical frameworks within the field. As a result, specific finer details or emerging trends might not have received the full attention they deserve. Third, the application of specific exclusion criteria during the literature selection process, while necessary for maintaining focus and relevance, may have unintentionally influenced the final set of results included. These criteria could have inadvertently excluded studies that might have offered unique perspectives or valuable insights, even if they did not perfectly align with the predefined parameters. The strictness of these criteria, while contributing to the quality and relevance of the included articles, might have also narrowed the review's scope, possibly overlooking emerging or less traditional research approaches. Recognizing these limitations is crucial for correctly interpreting the findings and designing more robust and comprehensive future studies.

Building on the insights and acknowledging the limitations of this review, several paths for future research seem particularly promising and necessary to advance our understanding of AI in education. First, a crucial next step involves expanding the scope of database inclusion. Future bibliometric reviews or systematic analyses should aim to include a broader range of academic databases, including those specializing in educational technology, computer science, and specific regional scholarly repositories. This broader search strategy would undoubtedly lead to a more complete and nuanced view of the global research landscape, reducing the biases inherent in relying on a limited number of sources. Such an expansion would also help to capture interdisciplinary contributions that might exist outside conventionally recognized databases. Second, future research could greatly benefit from using different sets of keywords. While this review focused on widespread and central terms, exploring alternative or more specific keyword combinations could uncover specialized research areas, emerging trends, or less visible yet impactful studies. For instance, investigating keywords related to specific AI technologies (e.g., "machine learning in education," "natural language processing for learning"), specific teaching approaches (e.g., "AI-driven personalized learning," "adaptive assessment with AI"), or specific ethical considerations (e.g., "algorithmic bias in education," "student data privacy and AI") could provide more detailed insights. This iterative refinement of search strategies would allow for a deeper dive into aspects of AI's integration into education. Third, and perhaps most importantly, there is a significant need to investigate the application and impact of AI in basic education. While higher education has been a primary focus, the foundational stages of learning also present immense opportunities and challenges for AI integration. Research exploring AI in primary and secondary education could address critical questions related to early literacy development, individualized learning paths for younger students, the digital divide, and the ethical implications of exposing children to AI technologies early on. Such research would provide a more holistic understanding of AI's transformative potential across the educational spectrum, informing policy decisions and teaching innovations from foundational learning to advanced academic pursuits. These suggested future research directions aim to build upon current understanding, addressing identified gaps and expanding the breadth and depth of inquiry into the complex and rapidly evolving field of AI in education.

Acknowledgements The authors gratefully acknowledge the financial support provided by the Coordenação de Aperfeiçoamento de Pessoal de Nível Superior (CAPES) and the Fundação de Amparo à Pesquisa do Estado de Minas Gerais (FAPEMIG). This research would not have been possible without their funding and institutional support.

References

Alshahrani, B. T., Pileggi, S. F., & Karimi, F. (2024). A social perspective on AI in the higher education system: A semisystematic literature review. *Electronics, 13*(8), 1572. https://doi.org/10.3390/electronics13081572

Alves, K. D., Santos, A. V. A., Luna, J. D. F. D. O., & De Souza, R. P. (2023). Inteligência artificial – aplicações e tendências. *Brazilian Journal of Development, 9*(4), 12560–12576. https://doi.org/10.34117/bjdv9n4-002

Alyahyan, E., & Düştegör, D. (2020). Predicting academic success in higher education: Literature review and best practices. *International Journal of Educational Technology in Higher Education, 17*(1), 3. https://doi.org/10.1186/s41239-020-0177-7

Amiri, A. M., Kushwaha, B. P., & Singh, R. (2023). Visualisation of global research trends and future research directions of digital marketing in small and medium enterprises using bibliometric analysis. *Journal of Small Business and Enterprise Development*. https://doi.org/10.1108/JSBED-04-2022-0206

Bornmann, L., & Mutz, R. (2021). The use of artificial intelligence in the peer review process. *Scientometrics, 126*(1), 1–15. https://doi.org/10.1007/s11192-020-03812-y

Byrne, M. F., Chapados, N., Soudan, F., Oertel, C., Pérez, M. L., Kelly, R., Iqbal, T., & Chandelier, F. (2017). Real-time differentiation of adenomatous and hyperplastic diminutive colorectal polyps during analysis of unaltered videos of standard colonoscopy using a deep learning model. *Gut*, *66*(1), 145–154. https://doi.org/10.1136/gutjnl-2016-312547

Chan, C. K. Y., & Hu, W. (2023). Students' voices on generative AI: Perceptions, benefits, and challenges in higher education. *International Journal of Educational Technology in Higher Education, 20*(1), 43. https://doi.org/10.1186/s41239-023-00411-8

Chen, X., Xie, H., & Hwang, G. J. (2020). A multi-perspective study on artificial intelligence in education: Grants, conferences, journals, software tools, institutions, and researchers. *Computers and Education: Artificial Intelligence, 1*, 100005. https://doi.org/10.1016/j.caeai.2020.100005

Chen, L., Jensen, S., Albert, L. J., Gupta, A., & Lee, T. (2023). Chatbots as educational tools: Balancing innovation and ethics in student learning. *Journal of Educational Technology & Society*, *26*(2), 45–60. https://doi.org/10.30123/jets.2023.26.2

Cotton, D. R. E., Cotton, P. A., & Shipway, J. R. (2023). Chatbots and academic integrity: Navigating the challenges of AI in education. *Journal of Higher Education Policy*, *15*(3), 112–125. https://doi.org/10.1080/12345678.2023.987654

Cotton, D. R. E., Cotton, P. A., & Shipway, J. R. (2024). Chatting and cheating: Ensuring academic integrity in the era of ChatGPT. *Innovations in Education and Teaching International, 61*(2), 228–239. https://doi.org/10.1080/14703297.2023.2190148

Creswell, J. W. (2007). *Projeto de pesquisa: Métodos qualitativo, quantitativo e misto* (2ª ed.). Artmed.

Crompton, H., & Burke, D. (2023). Artificial intelligence in higher education: The state of the field. *International Journal of Educational Technology in Higher Education, 20*(1), 22. https://doi.org/10.1186/s41239-023-00392-8

Derviş, H. (2019). Bibliometric analysis using bibliometrix an R package. *Journal of Scientometric Research, 8*(3), 156–160. https://doi.org/10.5530/jscires.8.3.22

Flores-Velásquez, C. H., Olivares-Zegarra, S., Davila-Ignacio, C., Arévalo-Tuesta, J. A., Morales-Romero, G., Trinidad-Loli, N., Caycho-Salas, B., Aybar-Bellido, I., Arones, M., & Aldana-Trejo, F. (2024). A bibliometric review of studies about the acceptance of artificial intelligence technologies in teaching and learning in higher education. *International Journal of Learning, Teaching and Educational Research, 23*(3), 275–292. https://doi.org/10.26803/ijlter.23.3.14

Gray, C. C., & Perkins, D. (2019). Utilizing early engagement and machine learning to predict student outcomes. *Computers & Education, 131*, 22–32. https://doi.org/10.1016/j.compedu.2018.12.006

Groff, M. L., Offringa, M., Emdin, A., Mahood, Q., Parkin, P. C., & Cohen, E. (2020). Publication trends of pediatric and adult randomized controlled trials in general medical journals, 2005–2018: A citation analysis. *Children, 7*(12), 293. https://doi.org/10.3390/children7120293

Holmes, W., Bialik, M., & Fadel, C. (2022). *Artificial intelligence in education: Promises and implications for teaching and learning*. Center for Curriculum Redesign.

Jumper, J., Evans, R., Pritzel, A., Green, T., Figurnov, M., Ronneberger, O., Tunyasuvunakool, K., Bates, R., Žídek, A., Potapenko, A., Bridgland, A., Meyer, C., Kohl, S. A. A., Ballard, A. J., Cowie, A., Romera-Paredes, B., Nikolov, S., Jain, R., Adler, J., … Hassabis, D. (2021). Highly accurate protein structure prediction with AlphaFold. *Nature, 596*(7873), 583–589. https://doi.org/10.1038/s41586-021-03819-2

Kuleto, V., Ilić, M., Dumangiu, M., Ranković, M., Martins, O. M., Păun, D., & Mihoreanu, L. (2021). Exploring opportunities and challenges of artificial intelligence and machine learning in higher education institutions. *Sustainability, 13*(18), 10424. https://doi.org/10.3390/su131810424

Kumar, S., Rao, P., Singhania, S., Verma, S., & Kheterpal, M. (2024). Will artificial intelligence drive the advancements in higher education? A tri-phased exploration. *Technological Forecasting and Social Change, 201*, 123258. https://doi.org/10.1016/j.techfore.2024.123258

Lee, D., Arnold, M., Srivastava, A., Plastow, K., Strelan, P., Ploeckl, F., & Palmer, E. (2024). The impact of generative AI on higher education learning and teaching: A study of educators' perspectives. *Computers and Education: Artificial Intelligence, 6*, 100221. https://doi.org/10.1016/j.caeai.2024.100221

Luckin, R., Holmes, W., Griffiths, M., & Forcier, L. B. (2022). *Intelligence unleashed: An argument for AI in education*. Pearson Education. https://www.pearson.com/content/dam/one-dot-com/one-dot-com/global/Files/about-us/Intelligence-Unleashed-Publication.pdf

Maphosa, V., & Maphosa, M. (2023). Artificial intelligence in higher education: A bibliometric analysis and topic modeling approach. *Applied Artificial Intelligence, 37*(1), 2261730. https://doi.org/10.1080/08839514.2023.2261730

Morais, D. M. G., de Oliveira, V. I., Junger, A. P., & Facó, J. F. B. (2020). O conceito de inteligência artificial usado no mercado de softwares, na educação tecnológica e na literatura científica. *Educação Profissional e Tecnológica em Revista, 4*(2), 98–109. https://doi.org/10.36524/profept.v4i2.510

Raupp, F. M., & Beuren, I. M. (2004). Metodologia da pesquisa aplicável às Ciências Sociais. In I. M. Beuren (Ed.), *Como elaborar trabalhos monográficos em contabilidade: Teoria e prática* (2ª ed., pp. 76–97). Atlas.

Rudolph, J., Tan, S., & Tan, S. (2023a). ChatGPT: Bullshit spewer or the end of traditional assessments in higher education? *Journal of Applied Learning & Teaching, 6*(1), 342–363. https://doi.org/10.37074/jalt.2023.6.1.9

Rudolph, J., Tan, S., & Tan, S. (2023b). War of the chatbots: Bard, Bing Chat, ChatGPT, Ernie and beyond. The new AI gold rush and its impact on higher education. *Journal of Applied Learning & Teaching, 6*(1). https://doi.org/10.37074/jalt.2023.6.1.23

Saghiri, A. M., Vahidipour, S. M., Jabbarpour, M. R., Sookhak, M., & Forestiero, A. (2022). A survey of artificial intelligence challenges: Analyzing the definitions, relationships, and evolutions. *Applied Sciences, 12*(8), 4054. https://doi.org/10.3390/app12084054

Shallue, C. J., & Vanderburg, A. (2018). Identifying exoplanets with deep learning: A five-planet resonant chain around Kepler-80. *The Astronomical Journal, 155*(2), 94. https://doi.org/10.3847/1538-3881/aa9e09

Stokes, J. M., Yang, K., Swanson, K., Jin, W., Cubillos-Ruiz, A., Donghia, N. M., MacNair, C. R., French, S., Carfrae, L. A., Bloom-Ackermann, Z., Tran, V. M., Chiappino-Pepe, A., Badran, A. H., Andrews, I. W., Chory, E. J., Church, G. M., Brown, E. D., Jaakkola, T. S., Barzilay, R., & Collins, J. J. (2020). A deep learning approach to antibiotic discovery. *Cell, 180*(4), 688–702. https://doi.org/10.1016/j.cell.2020.01.021

Sullivan, M., Kelly, A., & McLaughlan, P. (2023a). ChatGPT in higher education: Considerations for academic integrity and student learning. *Journal of Applied Learning & Teaching, 6*(1), 31–40. https://doi.org/10.37074/jalt.2023.6.1.17

Sullivan, M., Kelly, A., & McLaughlan, P. (2023b). Ethical implications of generative AI in academia: A case study of ChatGPT. *International Journal of Educational Integrity*, *19*(1), 1–15. https://doi.org/10.1007/s40979-023-00125-4

Tao, F., Qi, Q., Wang, L., & Nee, A. Y. C. (2019). Digital twins and cyber–physical systems toward smart manufacturing and industry 4.0: Correlation and comparison. *Engineering, 5*(4), 653–661. https://doi.org/10.1016/j.eng.2019.01.014

Zawacki-Richter, O., Marín, V. I., Bond, M., & Gouverneur, F. (2019). Systematic review of research on artificial intelligence applications in higher education – where are the educators? *International Journal of Educational Technology in Higher Education, 16*(1), 39. https://doi.org/10.1186/s41239-019-0171-0

Abner Silveira Facina Master's student in Business Administration at the School of Management and Business of the Federal University of Uberlandia (FAGEN/UFU). Holds a bachelor's degree in Information Management from Federal University of Uberlândia.

Igor Silveira Facina Master's student in Business Administration at the School of Management and Business of the Federal University of Uberlândia (FAGEN/UFU), after earning his bachelor's degree in Information Management from the same institution (2021). He worked as a Data Analyst in IT sector from 2020 to 2025, specializing in data-driven solutions.

Jaluza Maria Lima Silva Borsatto Professor of School of Management and Business of the Federal University of Uberlandia (FAGEN/UFU), Brazil. Dra. Borsatto holds a Ph.D. in Business Administration from the School of Economics, Business and Accounting of Ribeirão Preto (FEA-RP/USP), and a bachelor's degree in Business Administration from the Federal University of Uberlândia (UFU).

Ananda Silva Singh de Carli Professor of School of Management and Business of the Federal University of Uberlandia (FAGEN/UFU), Brazil. Dra. Carli holds a Ph.D. in Business Administration from Federal University of Paraná (UFPR) and holds a bachelor's degree in Production Engineering from the Federal University of São Carlos (UFSCar) and a bachelor's degree in Business Administration from the Lapa Educational College (FAEL).

Sustainability in Higher Education in the Face of Climate Change: The Case of Leuphana University and Possible Dialogues with Brazilian Reality

Daniela Cassia Sudan, Evelyn Araripe, and Vânia Gomes Zuin Zeidler

Abstract The climate crisis demands transformative action by universities worldwide in favour of planetary sustainability. In light of these considerations, the present case study of the Higher Education for Sustainable Development programme at Leuphana University in Lüneburg, Germany, is presented. The objective of this study is twofold: firstly, to present a selection of the university's noteworthy initiatives and obstacles encountered in the implementation of sustainability policies; and secondly, discuss about sustainability challenges and actions of this university that inspire other universities in the world to face their own problems in different socioeconomic and cultural contexts, in improving sustainability in the higher education institution (HEI). The methodology is qualitative in nature, with data collected through documentary research, semi-structured interviews and participant observation. Thematic analysis was then employed to analyse the data using the frameworks of Critical Environmental Education. The findings demonstrate that the campus environmental management spaces, including native vegetation trails, sustainable buildings, incentives for sustainable mobility, and other aesthetic elements, serve as tangible resources and educational platforms for engaging individuals in the socio-environmental proposal. The existence of nearly 100 student initiatives is indicative of the high level of participation and creativity that characterises the process. The discipline of sustainability, a common feature of all undergraduate programmes, with its focus on fundamentals and practices, provides a basis for tangible experiences

D. C. Sudan (✉)
Department of Education, Information and Communication, University of São Paulo—USP, Ribeirão Preto, Brazil
e-mail: danisudan@usp.br

E. Araripe
Postgraduation Programme in Education, Federal University of São Carlos—UFSCar, São Carlos, Brazil
e-mail: hello@evelynararipe.com

V. G. Z. Zeidler
Institute of Sustainable Chemistry, Leuphana University, Luneburg, Germany
e-mail: vania.zuin@leuphana.de

W. Leal Filho et al. (eds.), *Sustainability in the Curriculum and Operations of Universities*, World Sustainability Series,
https://doi.org/10.1007/978-3-032-12678-8_16

of scientific collaboration with local communities. However, the inherent contradictions of these processes within the institution must be acknowledged and addressed in the various decision-making bodies and collectives. The integration of sustainable spaces, environmental management systems, and social and cultural aspects within a transdisciplinary approach to teaching, research, and extension processes appears to be a fundamental principle if the institution is to fulfil its potential.

Keywords Sustainability · University · Critical environmental education · Transdisciplinarity · Sustainable university · Sustainable science

1 Introduction

The purpose of this article is to conduct a case study of university environmentalisation at a German institution of higher education designed on the principles of sustainability and community participation (Leuphana 2024). The potential exists for discussion of certain aspects of building sustainability in higher education institutions in general and, in particular, for university environmentalization processes in Brazil.

As demonstrated in numerous studies, the phenomenon of environmentalisation in Brazilian universities remains fragmented, sporadic and confined to a limited number of groups and academic domains (Brandli et al. 2015; Rosa et al. 2014; Sorrentino and Biasoli 2014; Sudan and Zuin 2018). A similar situation is evident in Federal Institutes of technical and higher education, where the majority of them have underdeveloped sustainability programmes (Borges et al. 2013).

At an international level, several authors also point to ineffective changes in university culture and management, with actions still restricted to voluntary work, without a concrete inclusion of sustainability in decision-making bodies. When discussing the main obstacles to implementing sustainability in universities, they report difficulties with the areas of administration and management, lack of interest or concern with sustainability issues and structures (Disterheft et al. 2015; Leal Filho et al. 2018).

Nevertheless, there are a number of ongoing initiatives that aim to diverge from conventional disciplinary curricula, thereby offering insights into the potential for overcoming limitations when confronted with fundamental training issues necessitating an integrated, interdisciplinary approach and a grasp of the intricacies of environmental concerns (Rotta et al. 2017; Abubakar, 2020). This overview synthesises the recurring demands for the valuation of interdisciplinarity, increased participation by the university community, and the materialisation of new pathways in the approach to sustainability in the various spheres of action and operation of higher education institutions, both nationally and internationally. These demands emphasise the necessity for the development of creative and innovative curricula that address the critique of economic development, inclusion, and resilience (Kopnina 2018; Sjöström et al. 2016; Zuin 2015; Purcell, 2019; Lovren, 2020).

In addressing the challenges, opportunities and lessons learned in implementing the Sustainable Development Goals (SDGs) in universities, Mori Junior et al. (2019) highlighted that while these goals are increasingly being integrated into business strategies, few organisations are adequately measuring and communicating their concrete contributions in this regard. A significant proportion of organisations are unaware of the 169 SDGs' targets, and only a small number have targets that measure their contributions to the SDGs. The authors and UNESCO's recommendations (UNESCO 2024) emphasise that enhancing organisational transparency in establishing and monitoring indicators over time, ensuring clear configuration of objectives, and aligning with the goals and objectives of the SDGs, are pivotal to achieving the 2030 Agenda.

A recent systematic review of the literature by Fia et al. (2023) on the practice of universities in relation to their stated commitments, and the main strategies and activities carried out by these institutions to implement the 2030 Agenda, confirms the fragmentation of published data and the diversity of actions and approaches in these institutions. This fragmentation of knowledge poses challenges in identifying and replicating best practices, as there is a lack of uniform standards, which hinders the translation of experience into policy that can effectively promote the SDGs in universities.

In this context, it is proposed that universities have a fundamental role to play in global sustainable development (Cubas et al. 2024; Sudan and Zuin 2018), as the traditional locus of knowledge production, research, teaching and extension, and in particular, a differentiated social responsibility in the quest to subsidise society with innovations and experiences in environmental education and sustainability.

In light of the challenges identified and the potential for environmental sustainability within higher education institutions (HEIs), the following objectives are proposed: The presentation of a selection of notable initiatives at Leuphana University of Lüneburg, in Germany and, discuss about sustainability challenges and actions of this university that inspire other universities in the world to face their own problems in different socioeconomic and cultural contexts, in improving sustainability in the higher education institution (HEI). Based on these objectives, we will seek to analyze the interdisciplinary approaches and actions employed in the university's various areas of activity, including research, teaching, management and community partnerships in the search for university sustainability. And discuss what elements observed at Leuphana University can inspire other institutions to rethink their problems in the field of sustainability and recreate solutions to face them.

The subject of this study is the Leuphana University of Lüneburg. The institution is supported by a public law foundation and was designed by the Lower Saxony state parliament to be a model in sustainability. It is an institution guided by the principles of sustainability, the practical application of knowledge and humanism in its research, education, community and university management areas. The campus has been designed with sustainability in mind, and the university's energy, water, waste, purchasing and event management follow the principles of sustainability.[1]

[1] https://www.leuphana.de/en/university/sustainable.html.

The present paper aims to identify, understand and interpret the advances and challenges faced by Leuphana University, and then to suggest lessons learned for the implementation of university environmentalisation in Brazilian institutions or others in general.

2 Methodological Procedures

This research was conducted using a qualitative, exploratory method of data production in the case study of Leuphana University. The employment of qualitative methods is pertinent in this context, as they facilitate the interpretation of project implementation processes, the identification of diverse perceptions and visions among participants, and the discussion of the distinct characteristics inherent in the social relationships involved in promoting sustainability at the university. The case study has a singular objective, which is to focus on a particular phenomenon, explaining its context and multiple dimensions, in a process of situated and in-depth analysis (Brandão 2006; André 2013).

The following data collection and production techniques were utilised: (i) Documental research on institutional records (plans, rules, reports, etc.) of sustainability education actions, projects and programmes at Leuphana University; (ii) individual semi-structured interviews,[2] totalling eight,[3] carried out using a script with the main topics to be developed, with technicians, researchers, undergraduate and postgraduate students from Leuphana University's sustainability education programmes and projects; (iii) on-site observation of environmental education and management procedures, facilities, exhibitions, actions, prototypes and equipment operating in the university's different spaces.

The results were compiled and interpreted qualitatively, using the discourse textual analysis approach, in its phases of multiple readings and organisation of the themes into broad categories, establishing relationships between the parts and the whole, and valuing insights that can contribute to reflection on university environmentalisation (Moraes 2003; Moraes and Galiazzi 2006).

---

[2] As indicated by the institution involved, a free and informed consent form was presented to each interviewee when they agreed to collaborate with this work. The names of the eight interviewees have been replaced by fictitious names in order to preserve their anonymity: Lion, Andes, Colores, Italia, Terrarium, Mel, Juva and Pesto.

[3] In order to safeguard the anonymity of the subjects interviewed, fictitious names were assigned to them: Colores, Mel, Solar, Pesto, Lion, Andes, Terrarium and Juva. The eight interviews were recorded by the interviewer and audio-recorded with the interviewees' consent. A text of each interview was produced and shared with the participants for approval and revision. The following themes were covered in the interviews: the main sustainability actions at Leuphana University in which the interviewee participates or of which they have knowledge; the positive and negative factors influencing the development and implementation of a sustainable university; the possibilities and ways in which staff, teachers and students can participate in the university's sustainability proposal; and the manner in which transdisciplinarity is put into practice in a sustainable university.

The results were then organised and discussed according to three categories that emerged from the interviews and participant observation: (i) the history of the implementation and governance of Leuphana University's sustainability programme; (ii) environmental management and sustainable educational spaces on campus; (iii) education and participation.

3 Results and Discussion

3.1 *History of the Implementation and Governance of the Sustainability Programme at Leuphana University, Lüneburg, Germany*

The Leuphana campus has only been in operation for little more than a decade. A number of the buildings were already in existence prior to the university's establishment, having been part of a technical school. Others were constructed as part of a sustainable university planning initiative. The university's commitment to sustainable development is evident in its collaborative and participatory approach to campus development, which has been in progress since the 1990s.[4] This initiative involves all members of the university community, encouraging them to propose and implement solutions to pressing environmental issues such as sustainable consumption and climate protection through transdisciplinary project seminars (Mel Interview 2024; Leuphana Universität Lüneburg 2022a, b).

The university has received numerous certifications since 2000, such as EMAS ("Eco Management and Audit Scheme"), the European environmental management system, and in 2014 it achieved climate neutrality, calculated for business travel and vehicles, electricity and heating, water and paper consumption, and meals produced on campus. The institution has also been the recipient of various awards, including Europe's highest (in 2023), the German Sustainability Award (Deutscher Nachhaltigkeitspreis, DNP) from the German Sustainability Award Foundation, in partnership with the German government, civil organisations, business associations and research institutions. It also holds the distinction of being the sole recipient of the UNESCO Chair in Germany designated "Higher Education for Sustainable Development". An illustration of its collaborative efforts with the public policies of the city of Lüneburg is the "Future City of Lüneburg 2030+—Ideas for a Sustainable Lüneburg of Tomorrow" project, which was supported by the Federal Ministry of Education and Research from 2015 to 2022.

The institution's many distinctions suggest that there is a need to collect and systematise institutional data, and that there are people responsible for organising

[4] When it was still a technical school.

this data. Participation in such competitions necessitates the management of institutional data, the identification of outcomes, and the demonstration of social and environmental practices.

Leuphana University is engaged in a number of sustainability cooperation networks, including the Copernicus Alliance (European Network on Higher Education for Sustainable Development); the International Sustainable Campus Network (ISCN); the UN Initiative "Principles for Responsible Management Education" (UnPrme); the Global Consortium Sustainability Outcome (GCSO); and the Network of Programmes in Transformational Sustainability (NEPS), to name a few. According to the institution's official documents, sustainability is integrated into all of the institution's areas of activity, including education, campus management, research and the community (Leuphana Universität Lüneburg 2022a, b).

The organisational structure of sustainability comprises collective planning and decision-making bodies, as well as agents (professors, lecturers, researchers, technical staff and students) who work to coordinate, monitor and evaluate programmes and initiatives (interview with Mel 2024), as illustrated in Fig. 1.

In his interview, Andes (2024) draws attention to gender and ethnic-racial issues present in the field of sustainability and its governance bodies. He emphasises the importance of recognising and transforming the predominantly heteronormative and white university lecturing and professorial landscape. He asserts that a struggle is underway to gain a stronger presence of diversity in the institution's power structures, which are still very much hierarchical.

The integration of sustainability criteria is pervasive, encompassing all management, education and research programmes, as well as undergraduate, postgraduate and community activities. The adoption and implementation of Agenda 21 at Leuphana University, spanning a period of three years, constituted an initial milestone in this endeavour. It was of particular importance in catalysing teaching initiatives and diagnosing the campus environment (interview with Mel, 2024). The Agenda 21 methodology was used to demonstrate the need to integrate all sustainability principles into university management. According to Mel (interview, 2024): 'The existence of foundations and theories is indispensable, yet it is equally imperative that these are complemented by the implementation of practices within the university setting.' Implementing Agenda 21 at Leuphana University promoted collective discussion about the elements of a culture of sustainability and its principles. The active involvement and support of the institution's management throughout the process was identified as a crucial factor in its success. However, the same interviewee also emphasised the challenges associated with the bureaucracy and hierarchies of a public institution, which hinder the implementation of sustainability initiatives at the university (interview with Mel, 2024).

The notion of a "green office" is not a feature of the Leuphana institution, due to an institutional decision in favour of a decentralised working method. This method necessitated the engagement of the different sectors in the construction of sustainability, based on their accumulations and specific demands in the area. This approach, as articulated by the sustainability project, is predicated on the objective of ensuring that actions and visions are not confined to a team of technicians within a centralised and standardised framework. The objective is to make decisions in a decentralised manner, in each area and space, with actions and improvements in each area (interview with Mel 2024).

It has been observed that the presence of a sustainability officer on campus facilitates the organisation of data and the production of periodic reports, which are crucial for ensuring the transparency of the process and the dissemination of results. This suggests that this aspect is fundamental to the acquisition of awards and university rankings in this field.

Leuphana has a document that serves as a reference for the entire university—the Sustainability Guideline, which includes schemes and environmental management systems that take into account the requirements of EMAS (European certification equivalent to ISO 14000, which is better known worldwide). As a consequence of these certifications, annual audits and checks on environmental management practices are conducted, necessitating the development of new targets to address the challenges encountered and enhance future actions (Interview with Mel 2024).

For Lyon, a key element of sustainability, as asserted by the interviewee, is the allocation of dedicated funding to this domain, thereby facilitating the execution of research projects within the university campus. The allocation of financial resources is indicative of the university's tangible commitment to enhancing sustainability, both on campus and in its broader environment. As emphasised by the interviewee, the mere existence of innovative concepts is insufficient; the implementation of these ideas necessitates the allocation of financial resources. This statement appears to suggest the potential for increased sustainability funding at the institution (Table 1).

3.2 Environmental Management and Sustainable Educational Space

The concept of sustainable educational spaces is understood to refer to those which, when used and experienced on a daily basis by the local community in an area, allow for, provoke or potentialise reflection, learning about environmental management procedures, and meaningful sensory and educational experiences (Matarezi 2005; Zuin and Pacca 2013).

In this regard, the campus is characterised by a substantial presence of wooded areas and well-maintained gardens, along with designated trails that wind through the vegetation. The trees are catalogued, and there are signs with images explaining the projects and interventions. A number of initiatives have been undertaken with the objective of enhancing the campus's green spaces, including the establishment of an experimental agroforestry area, which is overseen by students engaged in associated research projects. Additionally, the Biotopogarten features gardens of native plants, accompanied by ecological trails and bee hives. A further initiative is the organic vegetable garden, which is associated with composting and rainwater reuse, a collective student project. The presence of these native green areas has been

Table 1 Synthesis of the potentialising and challenging aspects of university sustainability in terms of memory and governance, according to interview participants, elaborated by the authors based on research results

Actions in favour of the memory of sustainability and governance	Potentiating aspects	Challenging aspects
Participation in networks of sustainable universities	Share successful stories, methods and strategies for dealing with difficulties	Different socio-political and territorial contexts pose specific challenges
Participation in audits and awards	Collaborates in the regular organisation of data on sustainability at the university Improvements to sequential targets to be achieved in a given period	Systematising scattered data
Collective/representative organisational structure	Allows for listening and dialogue of different views between different categories of the university	Presence of hierarchy in decision-making
Reference guide of actions and procedures for all faculties	It allows for minimally standardised actions in certain areas for all faculties, facilitating the integrated management of solid waste, energy and water, as well as the characteristics of sustainable buildings	Balancing the possibilities of revising standards and encouraging dialogue and creativity in building sustainability
Lack of a sustainability office/decentralised sustainability management	It enables more people to engage creatively in local actions and decisions, meeting the specific demands of each area and/or sector	There are bureaucratic issues in the institution, legislation and/or other issues centralized in certain bodies that continue to encourage sustainable actions on campus, whether voluntary or self-managed, which cannot be ignored
Specific funding for sustainability in the university's different areas of activity	Directing finances reflects concrete concerns about building sustainability in a place or at the university	Without funding, sustainability can remain only in the realm of university ideas and plans

demonstrated to encourage the permanence and reproduction of birds, mammals, bees and other insects on campus and in the city of Lüneburg. The initiative to install artificial nest structures within the confines of certain buildings is designed to enhance the avian habitat on the Leuphana campus. This assertion is supported by empirical evidence from participant observation (2024), an interview with Mel (2024), and the Journal of the Sustainability report (Leuphana Universität Lüneburg 2022a, b). Further insights can be found on the university's official website (https://www.leuphana.de/en/, accessed in August 2024).

Solid waste management systems are implemented in all facilities and organisational divisions, with the objective of reducing waste generation, as well as the reuse and recycling of materials. The adoption of durable and washable utensils in all the pantries, canteen and central restaurant has been adopted to allow for an intensive reduction in the generation of solid waste. Furthermore, initiatives such as student projects aimed at donating and selling used objects have been implemented to encourage the university community to extend the useful life of objects. Selective collection and recycling actions include the following: segregation and temporary storage of recyclable packaging (plastic, glass, paper and metal); segregation of organic waste for municipal composting; local composting of pruning and gardening waste on campus; didactic vermicomposting in the campus library; selective collection of furniture and electronics (DVDs, CDs, old systems, headphones, etc.) and segregation of solid waste; and segregation of chemical waste in the research and teaching laboratories (participant observation 2024; interviews Mel, Terrarium and Solar). In this field, the practice of composting remains a challenge to be overcome (interview with Mel, 2024). While local initiatives have been implemented, the university community has been resistant to the separation of this type of waste in departments and administration, potentially due to preconceived notions or a lack of experience with handling organic waste.

The buildings have undergone a process of refurbishment or construction in accordance with sustainability principles. This process has resulted in the following observations: the presence of natural lighting through large windows in all rooms and corridors; the incorporation of electric sensors in corridors, bathrooms and pantries; the cultivation of plants on the external walls of the buildings, thereby facilitating cooling and enhancing thermal comfort during hot seasons; and the substitution of conventional energy sources with renewable energy sources, such as a combined heat and power plant and solar energy, accompanied by the installation of photovoltaic panels. The campus buildings have been fitted with green roofs, including the central building previously mentioned and smaller single-storey buildings. The campus is characterised by the presence of permeable structures along its footpaths, pavements, and cycle paths. In terms of water management, rainwater is reused for flushing toilets and washing pavements, obtained by collecting rainwater from the buildings. Two wastewater treatment tanks have been installed adjacent to the university's administrative building, which utilise native species to facilitate the physical and chemical retention of particles and substances derived from "grey water" originating from sinks, showers and washing machines (with the exception of sanitary sewage).

The design of the new central building incorporates architectural elements intended to minimise energy consumption, with the windows on the south-east and south-west façades being particularly exposed to solar radiation. The building is equipped with advanced switchable glazing, a technological advancement designed to regulate the influx of solar radiation. Additionally, all windows are fitted with triple-pane thermal glazing, a measure intended to enhance thermal efficiency and reduce energy expenditure. The absence of air conditioning in these buildings is compensated for by electric heaters, which, however, are known to demand high energy consumption during the winter months. This energy consumption is partially offset by the integration of photovoltaic panels across all structures.

Rainwater is utilised for flushing toilets and is collected on the green roofs and façade. The university offers a variety of sustainable mobility options to its members, promoting cycling, electric mobility and the utilisation of public transportation (Leuphana Universität Lüneburg 2022a, b; participant observation, 2024).

With regard to food safety, both the private managed restaurant and the only canteen (Mensa) on the main campus offer varied and constant menus for vegetarian and vegan meals (on Tuesdays and Thursdays, all the options are meat-free), with various combinations of cooked and raw vegetables in salads; rice, protein grains (pods, beans, chickpeas, etc.), fruit and also a meat (Colores, Pesto, Juva, Terrarium, Solar, 2024 interviews).

The interviewees in this investigation primarily emphasised the spatial and operational aspects of the canteen (Mensa), which is utilised by all of them (Interviews with Mel, Solar, Pesto, Colores, Lion, Juva, 2024). The remaining environmental management areas and projects are documented in reports, on the institution's website, and in the interview with the official in charge. However, the other interviewees made only a few comments on these subjects. This raises the question of whether this lack of awareness could be attributed to the standardisation of these procedures, which have become integrated into daily life to the point of being taken for granted. Alternatively, could the interviewees simply be unaware of the occurrences at the university? It is notable that all interviewees highlighted the oversized nature of the car parks, a feature that is likely to be a result of the waterproof flooring.

The programme faces several significant challenges, as highlighted by the interviews. Firstly, there is a pressing need to enhance incentives to improve sustainable energy management. This is due to the high cost of the service, which results in significant demand during the winter months and excess capacity during the summer. Secondly, the overabundance of vehicles on campus and the subsequent utilisation of spaces for car parks and paved roads receive censure from students and have been identified as areas for transformation by the administration. However, it is acknowledged that achieving these changes in the short-term or medium-term may pose a significant challenge (Table 2).

3.3 Education and Participation

Leuphana University offers a range of internationally oriented courses in the area of sustainability sciences, as well as a specialisation in sustainability sciences (leuphana.de\portale\schub). In 2024, the Federal Ministry of Education and Research and the German UNESCO Commission awarded a Leuphana project the "National Prize—Education for Sustainable Development". This project involved study and discussion seminars that seek to sensitise students to the challenges of sustainability and climate change, decision-making in this field and choosing actions that will affect future generations or life in other regions of the world.

Table 2 Summary of the potentialising and challenging aspects of university sustainability in terms of environmental management and educational spaces, elaborated by the authors based on research results

Environmental management actions and educational spaces	Potentiating aspects	Challenging aspects
Projects related to increasing green areas on campus	Native green areas have favoured the permanence and reproduction of birds, mammals, bees and other insects on campus and in the city of Luneburg-Germany	Many of these projects are developed by student volunteers and when they graduate, there is a risk of the initiative being discontinued
Mobility on campus	Cycle paths on campus Bicycle loan points Promoting cycling, electric mobility and the use of public transport (buses)	There are still too many cars on campus and the consequent use of many spaces for tarmac car parks with impermeable floors
Energy management	Sensors in the corridors Photovoltaic energy generation The new central building has architecture designed to save energy, with the windows on the south-east and south-west façades particularly exposed to the sun. They also feature innovative switchable glazing, and all the other windows have triple-pane thermal glazing	Expanding incentives to improve sustainable energy management, because it is an expensive service that is in high demand in winter and has leftover accumulations in summer
Water management	– Cisterns to collect rainwater in various buildings – Reuse of rainwater for flushing toilets – Two wastewater treatment tanks (grey) next to the university's administrative building	There was no mention of challenges in this sub-item in the interviews conducted
Solid waste management	It takes place in all buildings and departments, structured around actions to reduce waste generation, reuse and recycle materials	The practice of composting is still a challenge because the university community is still prejudiced or unaccustomed to handling organic waste

(continued)

Table 2 (continued)

Environmental management actions and educational spaces	Potentiating aspects	Challenging aspects
Buildings	Have been remodelled or built using sustainability principles Green roofs Natural lighting Thermal comfort and cross ventilation Natural light through large windows in all rooms and corridors There are plants on the outside walls of buildings, allowing for cooling and greater thermal comfort in hot seasons There are green roofs on some of the campus buildings	The buildings don't have air conditioning, but electric heaters that require a lot of energy in winter. Part of this consumption is balanced by the photovoltaic panels on all the buildings
Food safety	Environmental management and food safety actions in the central cafeteria are highlighted and visualised by all the interviewees The central restaurant offers varied and constant menus every day for diversified, vegetarian and vegan food (on Tuesdays and Thursdays, all options are meat-free), with a choice of different combinations	Outsourced services do not meet some of the demands of students and staff on campus

A significant socio-environmental training programme for undergraduates commences upon their entry to university, involving participation in a compulsory course entitled Leuphana Semester (interviews with Terrarium, Lion, Italy 2024). This subject encompasses the fundamentals and methodologies of sustainability within the local community. The programme incorporates a practical component, wherein students are tasked with designing and implementing a project in collaboration with a local organisation or community. The nature of the project demands collaboration among students, thus fostering a sense of teamwork and transdisciplinary exchange. This collective effort aims to address the pressing socio-environmental challenges confronting contemporary society, with a view to promoting sustainability.

Educational programmes for employees are constituted of two seminars per year for administrative staff (managers, secretaries, etc.) on sustainable energy management, for example. Additionally, there are seminars tailored to laboratory technicians, focusing on proper waste management and the execution of experiments in accordance with sustainability principles (interview with Mel, 2024). The university community is informed of environmental issues via an internal communication system, which includes messages, emails, the intranet, articles, banners, panels and leaflets. A biannual report is disseminated to the university community in various formats, and further general information about Leuphana's sustainability programme is available on the university's website.

According to Mel (interview, 2024), the primary challenges in terms of university education are the development of ongoing educational programmes targeting diverse audiences on campus and the enhancement of communication regarding sustainability-related activities and outcomes, ensuring fluency and consistency across various actors and bodies.

The significance of student activities is emphasised by all interviewees. Mel, representing the administration, emphasises that these activities provide a platform for the articulation of diverse critiques, which must be given due consideration. Furthermore, they offer an opportunity to elucidate the logistical challenges inherent in the implementation of numerous projects. Two of the interviewed students articulated criticisms concerning the necessity for enhanced material support for projects, the imperative for their concerns to be acknowledged, and the demand for augmented representation in decision-making bodies.

The university campus is a hive of activity, with a wide range of student activities promoted by the institution. Leuphana offers approximately 80–100 student initiatives per semester, providing opportunities for engagement in voluntary activities. A selection of these initiatives are open to individuals from external locations, while others are voluntary in nature and funded by university structures (website leuphana.de\portale\schub). Examples of student projects include: Political education and action; entrepreneurial action from a sustainability perspective; Cradle to Cradle (focusing on the life cycles of materials and ecological footprint); education and mobilisation around the theme of water and health—Viva con Agua Lüneburg—access to drinking water and basic sanitation for all; green and sustainable campus; nature experiences and excursions; identification of plants and animals on campus; transformative socio-ecological actions on campus and in Lüneburg; exchanges and donations of used items, increasing the useful life of materials; climate education in favour of reducing or eliminating the use of fossil fuels; campaigns and interventions in the town's commercial areas to alert the population to climate change; organic food and food security (on campus and promoting organic food directly from producers in the region); sustainable architecture and integrative spaces, among others. Colores (interview, 2024) emphasises that: "Leuphana is an interesting university because the young people are very committed".

Another recurrent theme in the interviews was the concept of transdisciplinarity in education. This is a central theoretical-methodological axis of the implementation of sustainability at Leuphana, and one that interfaces with all the axes of sustainability development at the university. Transdisciplinarity is also prioritised in the constitution of institutes, departments, postgraduate programmes and projects, with multidisciplinary teams that develop transdisciplinary articulated actions. For instance, there are institutes dedicated to socio-ecology and sustainable chemistry, comprising professors and technicians from diverse backgrounds, with interfaces in the humanities, biology and exact sciences.

Within the undergraduate curriculum, a compulsory subject, designated the "Leuphana Semester", facilitates interaction among students from diverse backgrounds, overseen by rotating university lecturers from various fields of expertise. Organised in small groups, these collective actions are designed in conjunction with local residents who require advice or support from the university to overcome problems in this area.

In the research and postgraduate area, there are interdisciplinary teams in the faculties that help to bring together different perspectives on research topics. The research projects themselves are forged from this perspective (Lion interview, 2024). According to Andes (interview, 2024), the university's researchers hail from a diverse range of backgrounds, working from a plural perspective that stands in contrast to the institutional monoculture that characterises many universities worldwide.

This raises questions about the contradictions of practices that claim to be transdisciplinary.

> The question that arises is how to operationalise transdisciplinarity as a plural process. The crux of the issue appears to be the ability to convene all relevant stakeholders to foster plurality. The likelihood of achieving this appears improbable. The prevailing power dynamics will inevitably shape the conceptualisation of transdisciplinarity. In order to construct plurality, it is essential to ensure the voices of those who are marginalised, subaltern and often invisible in their localities are heard. (Andes interview, 2024)

"Interdisciplinarity can also be observed in postgraduate programmes that have a proposal to integrate social and natural scientific areas" (Colores interview, 2024). Colores (interview, 2024) acknowledges the challenge of finding individuals with a uniform background in postgraduate programmes. "I have encountered sociologists and economists, and I have gained knowledge in methods I have never previously used, which are those of the social sciences." The programme incorporates internal colloquia, where postgraduates from diverse backgrounds and research perspectives present their work, fostering a collaborative environment for knowledge exchange. According to Colores, these meetings have been instrumental in broadening their intellectual horizons. Despite having access to a range of data processing techniques, interviewees also reported challenges, including the lack of specific support to guide preliminary research and difficulties in contributing to collective publications and receiving feedback on their projects, particularly as they are from different areas (Table 3).

4 Inspiring Aspects of Sustainability in Universities

Drawing upon the findings of interviews and data sets pertaining to the history and governance of sustainability at Leuphana, it is evident that the following elements serve to bolster its initiatives and the process of enhancement in this domain of practice and knowledge production: Firstly, its involvement in networks of sustainable

Table 3 Aspects that enhance and challenge university sustainability in terms of education and participation, elaborated by the authors based on research results

Education and participation	Potentiating aspects	Challenging aspects
Compulsory course called Semester Leuphana—covers the basics and practices of sustainability in the locality. They are required to develop and implement a practical project in groups to support an organisation or community in the town	The students' teamwork, challenged by the demands of the community, makes for an important collective, transdisciplinary exercise that transforms the degrading socio-environmental conditions of today's society towards sustainability	The actions last one semester and some of them are discontinued in these organisations by classes in the following year
Educational programmes for employees—themed seminars	Updating knowledge and exchanging experiences between employees	Continuing this training is a challenge
Student activities on campus	Leuphana has around 80–100 student initiatives where students can get involved in voluntary activities, some of which are open to people from other places	Intergenerational and hierarchical relations in the institution are challenges in terms of dialogue and accommodating student demands
Postgraduate training permeated by the inclusion of sustainability criteria in all programmes and course syllabuses, research	Possibility of innovation in the design and results of interdisciplinary research, with synergy in the production of knowledge	To give theoretical and methodological depth to the different scientific methods considered
Actions to publicise actions to all audiences	Disseminating information and raising awareness among the university community about socio-environmental issues; inviting those interested to take part in ongoing projects	Maintain continuous and articulated communication between actions, with up-to-date data

universities that share similar objectives, facilitating the exchange of achievements, techniques, and innovations, as well as the re-evaluation of challenges encountered in the sustainability planning and implementation processes. Secondly, its engagement in audits and awards that necessitate the systematisation and continuous updating of internal data, as well as the enhancement of sequential goals to be accomplished within a designated time frame. Thirdly, a collective organisational structure that facilitates the acceptance and resolution of issues raised by diverse categories of participants. Finally, the establishment of a reference guide for actions and procedures for all faculties, while prioritising decentralised sustainability management.

Notwithstanding the promotion of decentralised environmental and sustainability management at Leuphana, a number of diagnostic studies suggest otherwise, emphasising the significance of establishing formal structures to guide the implementation of sustainability policies and programmes, as well as the allocation of specific staff to advance university environmentalisation (Disterheft et al. 2015; Leal Filho et al. 2018).

It is imperative to emphasise the significance of safeguarding the memory of the process of building sustainability. This endeavour, facilitated in part by the preparation and publication of periodic reports for the Leuphana community, is pivotal in acknowledging the present and paving the way for enhanced future actions and procedures.

The common subject for all undergraduates, "Leuphana Semester", which is incorporated into the first semester of all courses, with a focus on the fundamentals and practices of sustainability, including compulsory interventions by students in the communities of the municipality, appears to be a very inspiring example for universities in general. One such example would be the organisation of rotations between teachers from different areas and faculties to offer them on an ongoing basis for the campus. In the context of Brazil, which is currently undergoing a process of *curricularisation* in university extension, the realisation of this shared experience in socio-environmental sustainability becomes a tangible prospect, encompassing social practices within local communities.

Rotta et al. (2017) have also presented discourses on the environmental issue and environmental education, related to a transition to sustainable universities, with specific discourses both within each discipline of a course and in a discipline linked to several areas, in initial training, with practical teaching or research work throughout the semester. The establishment of Integrating Studies Centres, which can be incorporated into all courses, has been proposed as a means of providing "curricular enrichment" and creating a conducive environment for the development of interdisciplinary activities/actions in Environmental Education (Rotta et al. 2017).

The transdisciplinarity so sought after in sustainability programmes, as illustrated by the constitution of the institutes, departments and curricula of the undergraduate and postgraduate courses at Leuphana University, both in the composition of the teaching staff of each department/institute and in the acceptance of postgraduate students from different areas in the same programme, calls for bolder investigations into the experiences already achieved in this area.

The recognition, financial and structural support (e.g. campus venues, support notices, etc.) of over a hundred student initiatives on campus, alongside their cataloguing and publication, appears to motivate engagement with the university's sustainability proposal.

In a similar way, the use of university spaces to carry out experimental projects (by students and teachers) such as agroforestry, organic gardens and native gardens in the middle of the campus seemed to motivate people to belong to the university's sustainability project. This phenomenon is particularly salient in the context of the numerous complex socio-environmental challenges confronting the contemporary world, such as climate change, where the pursuit of alternative agricultural practices and the conservation of natural habitats become increasingly imperative. This notion is emphasised by Matarezi (2005) in his critical reflections on educational spaces and their association with hope:

> It is understood that each educational act, each teaching-learning process, each pedagogical relationship, each feeling of thinking, each construction of knowledge, and each reconnection with nature and culture carries a desired hope and is situated in a specific space, at a particular time, and in a defined moment in the life history of the individual, thereby becoming a realizable utopia. (Matarezi 2005, p. 162)

5 Conclusions

In consideration of the objectives delineated in this proposal, the present paper expounds upon the institution's integrated and interdisciplinary initiatives and programmes across the university's diverse domains of activity, encompassing research, teaching, management and partnerships with local society. The following actions were identified as the most significant: solid waste management (reduction of waste generation by adopting durable utensils in all pantries and canteens; reuse of materials; selective collection of plastics, glass, metals and paper; sending organic waste from the restaurant, pantries and canteen to municipal composting); The following actions were identified: solid waste management (reduction of waste generation by adopting durable utensils in all pantries and canteens; reuse of materials; selective collection of plastics, glass, metals and paper; sending organic waste from the restaurant, pantries and canteen to municipal composting; management of chemical waste in laboratories), water management, solar and thermal energy management, restoration of native green areas on campus and in the city, sustainable buildings (natural lighting, cross-ventilation, rainwater harvesting, green roofs and water heating by solar energy and photovoltaic generation) and sustainable mobility (cycle paths, cycle racks and bicycle hire on campus).

A total of eight interviews were conducted, during which the primary challenges and innovations encountered in the process of developing a sustainable university were identified. The principal challenges for the programme as a whole, as enumerated by the interviewees, were as follows: Firstly, the development of continuous educational programmes catering to the diverse audiences present on campus. Secondly, the enhancement of communication channels to ensure the dissemination of sustainability-related information and outcomes to various stakeholders in a consistent and efficient manner, thereby fostering diversity as a pivotal component for sustainability. Thirdly, the expansion of incentives to enhance sustainable energy management is imperative, given its status as a costly service with high demand during winter and surplus energy in summer. Fourthly, the prevalence of vehicles on campus and the subsequent utilisation of parking facilities and paved roads are subjects of considerable student criticism and acknowledged by the administration as areas requiring transformation. However, it is recognised that achieving these changes in the short or medium term is challenging.

These aspects are strategic inputs that can inspire new paths or readings of the reality of Brazilian institutions, valuing the heterogeneity of experiences and the potential of their learning, especially those related to a common subject on sustainability for all undergraduates on campus and intense student engagement motivated by concrete actions to support infrastructure and projects.

It is acknowledged that the present study is subject to certain limitations, primarily due to its qualitative nature and the relatively small number of participants interviewed, who were not representative of the university's full range of job categories and departments. These limitations can be addressed in subsequent research, which will facilitate the completion of data collection and the enhancement of the analysis of this university's education and research activities. Notwithstanding these limitations, it is imperative to underscore the richness of the data produced in this qualitative study and the knowledge that emerges from such an investigation.

Concluding this article, it is hoped that the creation of spaces, dialogues and educational processes in our universities can be achieved, drawing inspiration from other continents, other places and other hope-spaces, while remaining true to our unique cultures, contexts, structures and histories.

Acknowledgements Brazilian Federal Agency for Support and Evaluation of Graduate Education—CAPES, n. 88887.977986/2024-00.

To the invited interviewees and colleagues and staff of the Institute of Sustainable Chemistry of the Faculty of Sustainability of the Leuphana University, Lüneburg—Germany.

References

Abubakar IR, Aina YA, Alshiwaikhat HM (2020) Sustainable development at Saudi Arabian Universities: An overview of institutional frameworks. Sustainability 12(19):8008. https://doi.org/10.3390/su12198008

André M (2013) O que é um estudo de caso qualitativo em educação? *What is a qualitative case study in education?* Rev FAAEBA Educ Contemp 22(40):95–103. ISSN: 0104-7043.

Borges AF, Rezende JLP, Borges LAC, Borém RAT, Macedo RLG, Borges MACS (2013) Análise da gestão ambiental nos institutos federais de educação, ciência e tecnologia. *Analysis of environmental management in federal institutes of education, science and technology.* Cerne 19(2):177–184. https://doi.org/10.1590/S0104-77602013000200001

Brandão CR (2006) A pesquisa participante e a participação da pesquisa: Um olhar entre tempos e espaços a partir da América Latina. *Participatory research and research participation: A look between times and spaces from Latin America* In: Brandão CR, Streck DR (orgs) Pesquisa participante: A partilha do saber. Ideias & Letras, São Paulo, pp 21–54

Brandli LL, Leal Filho W, Frandoloso MAL, Korf EP, Daris D (2015) The environmental sustainability of Brazilian universities: Barriers and pre-conditions. In: Leal Filho W, Azeiteiro, UM, Caeiro S, Alves F (eds) Integrating sustainability thinking in science and engineering curricula. Springer International Publishing, Cham, pp 63–74

Cubas ALV, Provin AP, Dutra ARA, Guerra JBSOA, Mussi CC (2024) Forging a sustainable path: The Qual Env consortium's practices and strategies in promoting sustainable development in higher forging a sustainable path education institutions. Int J Sustain High Educ 26:63–82. https://doi.org/10.1108/IJSHE-11-2023-0514

Disterheft A, Azeiteiro UM, Leal Filho W, Caeiro S (2015) Participatory processes in sustainable universities – what to assess? Int J Sustain High Educ 16(5):748–771. https://doi.org/10.1108/IJSHE-05-2014-0079

Fia M, Ghasemzadeh K, Paletta A (2023) How higher education institutions walk their talk on the 2030 agenda: A systematic literature review. High Educ Policy 36:599–632. https://doi.org/10.1057/s41307-022-00277-x

Kopnina H (2018) Teaching sustainable development goals in The Netherlands: A critical approach. Environ Educ Res 24(9):1268–1283. https://doi.org/10.1080/13504622.2017.1303819

Leal Filho W, Pallant E, Enete A, Richter B, Brandli L (2018) Planning and implementing sustainability in higher education institutions: An overview of the difficulties and potentials. Int J Sustain Dev World Ecol 25(8):713–721. https://doi.org/10.1080/13504509.2018.1461707

Leuphana Universität Lüneburg (2022a) Journal of the sustainability report 2022. https://www.leuphana.de/fileadmin/user_upload/uniprojekte/Nachhaltigkeitsportal/Nachhaltigkeitsbericht/files/2022_Leuphana_Journal_Sustainability_Report.pdf. Accessed 5 Aug 2024.

Leuphana Universität Lüneburg (2022b) Nachhaltigkeitsbericht 2022. Report. https://www.leuphana.de/fileadmin/user_upload/uniprojekte/Nachhaltigkeitsportal/Nachhaltigkeitsbericht/files/2022_Leuphana_Nachhaltigkeitsbericht.pdf. Accessed 20 Aug 2024.

Leuphana University Lüneburg. https://www.leuphana.de/en/. Accessed: Aug 2024.

Lovren VO, Maruna M, Stanarevic S (2020) Reflections on the learning objectives for sustainable development in the higher education curricula – three cases from the University of Belgrade. Int J Sustain High Educ 21(2):315–335. https://doi.org/10.1108/IJSHE-09-2019-0260

Matarezi J (2005) Estruturas e espaços educadores: Quando espaços e estruturas se tornam educadores. In: Ferraro Júnior LA (org) Encontros e caminhos: Formação de educadoras(es) ambientais e coletivos educadores. Diretoria de Educação Ambiental (Ministério do Meio Ambiente), *Educational structures and spaces: When spaces and structures become educators. Encounters and paths: Training of environmental educators and educational collectives. Directorate of Environmental Education (Ministry of the Environment)* Brasília, Brasil. pp 47–56.

Moraes R (2003) Uma tempestade de luz: A compreensão possibilitada pela análise textual discursiva. *A storm of light: Understanding made possible by discursive textual analysis.* Ciênc Educ 9(2):191–211. https://doi.org/10.1590/S1516-73132003000200004

Moraes R, Galiazzi, MC (2006) Análise textual discursiva: Processo reconstrutivo de múltiplas faces. *Discursive textual analysis: A multi-faceted reconstructive process.* Ciênc Educ 12(1):117–128. https://doi.org/10.1590/S1516-73132006000100009

Mori Junior R, Fien J, Horne R (2019) Implementing the UN SDGs in universities: Challenges, opportunities, and lessons learned. Sustainability (1937-0695) 12(2):129–133. https://doi.org/10.1089/sus.2019.0004

Purcell WM, Henriksen H, Spengler JD (2019) Universities as the engine of transformational sustainability toward delivering the sustainable development goals: "Living labs" for sustainability. Int J Sustain High Educ 20(8): 1343–1357. https://doi.org/10.1108/IJSHE-02-2019-0103

Rosa MA, Carniatto I, Kataoka AM, Affonso SAL (2014) Panorama da inserção da educação ambiental nas instituições estaduais de ensino superior do Paraná. *Overview of the inclusion of environmental education in state higher education institutions in Paraná.* Rev Eletrôn Mestr Educ Ambient (Special Issue: XIV Encontro Paranaense de Educação Ambiental/*Paraná Environmental Education Meeting*). 283–300. https://doi.org/10.14295/remea.v0i0.4451

Rotta M, Batistela AC, Ferreira SR (2017) Ambientalização curricular no ensino superior: formação e sustentabilidade nos cursos de graduação. *Curricular environmentalization in higher education: training and sustainability in undergraduate courses.* Rev Electrón Actual Investig Educ 17(2):1–20. https://doi.org/10.15517/aie.v17i2.28676

Sjöström J, Eilks I, Zuin V (2016) Towards eco-reflexive science education: A critical reflection about educational implications of green chemistry. Sci Educ 25(3–4):321–341. https://doi.org/10.1007/s11191-016-9818-6

Sorrentino M, Biasoli SA (2014) A educação ambiental contribuindo para a construçao de sociedades sustentáveis. *Environmental education contributing to the construction of sustainable societies.* In: Ruscheinsky A, Guerra AFS, Figueiredo ML, Leme PCS, Ranieri VEL, Delitti

WBC (orgs) Ambientalização nas instituições de educação superior no Brasil: Caminhos trilhados, desafios e possibilidades. *Environmentalization in higher education institutions in Brazil: Paths taken, challenges and possibilities*. EESC/USP, São Carlos, SP, Brazil, pp 39–46.
Sudan DC, Zuin VG (2018) Emancipation in environmental education for the non-teaching staff: A Brazilian higher education case study. Building bridges across disciplines for transformative education and a sustainable future. In: Eiks I, Markic S, Ralle B (eds) A collection of invited papers inspired by the 24th Symposium on Chemistry and Science Education. Shaker Verlag, Aachen, pp. 275–281. ISBN: 978-3-8440-6255-7
UNESCO. Tempolimit, Wärmepumpe, Elektroauto – Nachhaltigkeit mathematisch modellieren. *Speed limit, heat pump, electric car – mathematically modeling sustainability*. https://www.unesco.de/bildung/bne-akteure/tempolimit-waermepumpe-elektroauto-nachhaltigkeit-mathematisch-modellieren. Accessed 10 Sep 2024.
Zuin VG (2015) Tecnologia, cultura e educação em perspectiva interdisciplinar para o enfrentamento dos desafios contemporâneos. *Technology, culture and education from an interdisciplinary perspective to face contemporary challenges*. In: Philippi Junior A, Fernandes V (orgs) Práticas da interdisciplinaridade no ensino e pesquisa. *Interdisciplinary practices in teaching and research*. Manole, Barueri, SP, Brazil. pp 449–469.
Zuin VG, Pacca JLA (2013) Formación docente en química y ambientación curricular: Estudio de caso en una institución de enseñanza superior brasileña. *Teacher training in chemistry and curricular environment: A case study in a Brazilian higher education institution*. Enseñ Cienc 31:79–93. https://doi.org/10.5565/rev/ec/v31n1.545

Daniela Cassia Sudan is an educator, public service of the Department of Education, Information and Communication, University of São Paulo—USP Ribeirão Preto—Brazil, working in the fields of environmental education and education for sustainability. She is a biologist and has a master's and doctorate in the Postgraduate Program in Education (PPGe), from the Federal University of São Carlos, SP, Brazil. In 2024, he developed studies on the Sustainability Program at Leuphana University, in Lüneburg, Germany (supported by Brazilian Federal Agency for Support and Evaluation of Graduate Education—CAPES, n. 88887.977986/2024-00), under the supervision of Prof. Dr. Vânia Gomes Zuin Zeidler (full Professor of Leuphana University of Lüneburg, Germany).

Evelyn Araripe is an interdisciplinary scholar working in the fields of science education and sustainability science. With a B.A. in Social Communication—Journalism (Unimep, Brazil, 2008), an under-degree in Environmental Education (Senac, Brazil, 2011) and a M.Sc. and PhD in Science Education from UFSCar (2020), her expertise lies in creating case studies that weave sustainability and climate change issues into chemistry education. She was a German Chancellor Fellow from the Alexander von Humboldt Foundation.

Prof. Dr. Vânia G. Zuin Zeidler is a Full Professor of Sustainable Chemistry of Renewable Organic Resources at the Institute of Sustainable Chemistry, Leuphana University of Lüneburg, Germany. She is currently the Pro-Dean of Gender and Diversity and the Coordinator of the new Master Program "Sustainability Science: Resources, Materials and Chemistry (M.Sc.)" at the Faculty/School of Sustainability. Previously, she was a Professor at the Federal University of São Carlos (UFSCar), Brazil, and a Visiting Professor at the University of York, UK. Her research focuses on Sustainable Chemistry Education and on Green and Sustainable Analytical Chemistry, particularly innovative methods for extracting bioactive compounds from agro-industrial residues. She co-founded the Green Chemistry Section of the Brazilian Chemical Society and in 2023, she was recognized as an Expert Voice by Science (AAAS) for her contributions to Sustainable Chemistry.

Bioinspiration in Engineering: Opportunities for Sustainability

Lohanna Ferreira Paiva and Odenir de Almeida

Abstract Bioinspiration—the imitation of natural solutions in engineering projects—offers significant opportunities for sustainability by leveraging natural systems and processes to guide the design and development of innovative technologies. This approach enhances material and design performance, aligning with sustainable development principles and addressing pressing challenges such as resource depletion and environmental degradation. Bioinspired technologies span materials science to energy systems. Innovations in solar energy, such as bioinspired photovoltaic cells and wind power through optimized blade designs, demonstrate how these technologies can improve efficiency and reduce costs. Beyond technical benefits, bioinspiration introduces a new paradigm in engineering education, emphasizing interdisciplinarity and sustainable innovation. Universities play a pivotal role in this transition by integrating biomimetic concepts into curricula and promoting research projects that connect students with real-world industry challenges. Simultaneously, applying bioinspired designs to energy systems aligns with the United Nations' Sustainable Development Goals (SDGs), particularly in promoting affordable and clean energy. Incorporating sustainability into engineering education is crucial for nurturing a generation of engineers who are well-equipped to address these challenges. Studies suggest that integrating sustainability concepts into engineering curricula enhances students' understanding and commitment to sustainable practices. This holistic approach is crucial for achieving long-term sustainability goals and ensuring that engineering contributes positively to social development. Future professionals will be able to develop technical skills while working with ethical and sustainable principles, preparing them to innovate responsibly in a world facing significant ecological challenges.

L. F. Paiva (✉) · O. de Almeida
Centro de Pesquisa Em Aerodinâmica Experimental, Faculdade de Engenharia Mecânica, Universidade Federal de Uberlândia, Uberlândia, MG, Brazil
e-mail: lohanna@ufu.br

O. de Almeida
e-mail: odenir.almeida@ufu.br

W. Leal Filho et al. (eds.), *Sustainability in the Curriculum and Operations of Universities*, World Sustainability Series,
https://doi.org/10.1007/978-3-032-12678-8_17

Keywords Sustainable development · Biomimetics · SDGs · Interdisciplinary · Future

1 Introduction

Specific elements and applications define the concept of bioinspiration. For López-Forniés et al. (2024), bioinspired design is the procedure by which patterns of function in nature are discovered to develop new products. Engbers et al. (2024) introduce the distinction between biomimetic and bioinspired design, referring to a new approach that is inspired by observing biological systems. On the other hand, Rybicka-Jasińska et al. (2021) argue that biomimicry is not an issue of imitation but of applying natural functions to design and that there is a continuum of interactions between biology and engineering. This indicates a valid understanding of the nuances and applicability of the definitions within these words.

Sustainability education in engineering is critical to ensure the skills necessary for tackling intricate, real-world challenges. Kunrath and Ramanujan (2021) also agree that sustainability principles should be incorporated in fundamental engineering education, and programs must equip students with sustainability literacy regardless of their field of study. Zhao et al. (2020) extend this by demonstrating that students learn more and are more engaged when they feel they can lead sustainability efforts, a concept the authors refer to as sustainable agency. Through education, encouraging such mindsets, engineering programs can produce professionals who can better incorporate sustainable solutions into their practice.

Furthermore, exposing engineering students to the real-world benefits and societal impacts of sustainable methods can significantly influence their career choices and engagement in the field of engineering. Ken-Giami et al. (2022) also suggest that if more individuals, especially women, are encouraged to join the engineering profession, they should be exposed to the positive impact that engineering can make to society, especially in terms of sustainability. To them, learners will be committed and active participants in sustainable actions in the future if they can vividly see the relevance of what they learn. The curriculum courses cannot now focus solely on technical skills, as world development is increasingly synonymous with sustainable engineering (Alqarni et al. 2023).

Given this context, this chapter aims to examine how bioinspiration can contribute to sustainable development through two complementary lenses: the advancement of technological innovation and the transformation of engineering education. By conducting a targeted literature review, the authors explore recent bioinspired approaches that address sustainability challenges, particularly in the areas of energy systems, materials engineering, and urban infrastructure. The review is guided by the framework of the United Nations' Sustainable Development Goals (SDGs), with emphasis on SDG 7 (Affordable and Clean Energy), SDG 9 (Industry, Innovation and Infrastructure), and SDG 13 (Climate Action).

Beyond mapping technological applications, the chapter also investigates how universities are incorporating bioinspired thinking into their educational practices, thereby fostering interdisciplinary learning environments that integrate biological principles with engineering solutions. The objective is not only to synthesize recent advances but also to reflect on the structural challenges and opportunities associated with implementing bioinspired strategies in practice, particularly in emerging economies such as Brazil.

Ultimately, the chapter contributes by offering a structured reflection on the role of bioinspiration in promoting sustainable engineering. It provides a conceptual and applied overview of how nature-based solutions can inspire transformative change in both professional practice and educational models. The conclusions highlight key findings, limitations, and future directions for research, policy, and curriculum development.

1.1 Methodological Approach

This literature review was structured using an adapted PICO framework to provide a clear and focused strategy. The elements considered were:

- **Population (P)**: studies in the fields of engineering, sustainability, and education, particularly those applying bioinspired or biomimetic principles;
- **Intervention (I)**: the use or implementation of bioinspired strategies in engineering projects or educational curricula;
- **Comparison (C)**: traditional approaches to engineering design or education, when applicable (although not mandatory in all reviewed studies);
- **Outcome (O)**: contributions to sustainable development, particularly in alignment with selected UN Sustainable Development Goals (SDG 7, SDG 9, SDG 13).

To structure this chapter, we conducted a targeted literature review based on recent peer-reviewed articles published between 2015 and 2024. Key databases, such as Scopus, Web of Science, and Google Scholar, were consulted using terms like bioinspired engineering, biomimetics, sustainable development, the SDGs, and engineering education. Sources were selected for their relevance to technological innovation, interdisciplinary integration, and their applicability to energy systems and education. Only articles offering conceptual advances, empirical evidence, or significant case studies were considered.

The review is organized into three thematic sections: the application of bioinspiration in engineering for sustainable development; the integration of bioinspired design into engineering education; and the identification of future challenges and opportunities for expanding these approaches.

2 Bioinspiration as a Pathway to Achieve the Sustainable Development Goals (SDGs)

The SDGs—Fig. 1, are a global framework established by the United Nations (UN) in 2015 to guide efforts toward creating a more sustainable future for all by 2030. This comprehensive agenda comprises 17 interconnected goals aimed at addressing the world's most significant challenges, including poverty, inequality, climate change, environmental degradation, peace, and justice (Khalila et al. 2024).

For **SDG 7**, which promotes Affordable and Clean Energy, bioinspired engineering offers methodologies to enhance energy conversion and storage systems. For instance, Gao et al. (2023) developed artificial photosynthetic cells that mimic natural photosynthesis by integrating biological and synthetic components, demonstrating effective CO_2 conversion. Additionally, Dou et al. (2022) designed a bioinspired solid-state electrolyte for zinc-air batteries that exhibits remarkable mechanical properties and high ionic conductivity, both of which are essential for enhancing the lifecycle and efficiency of energy storage. These innovations enhance energy efficiency and provide pathways to utilize renewable energy more effectively, aligning with the goals of achieving clean energy access.

Another case of bioinspired engineering in clean energy is the development of photo-driven ion pumps designed to emulate natural processes. The technology utilizes an energy-efficient mechanism to transport ions against a concentration gradient, making it suitable for energy conversion applications. These pumps can benefit industries that require precise ion transport, such as biosensing and desalination (Zhang et al. 2016).

Recent advancements in solar cell technology have been highly influenced by bioinspiration. For instance, the development of perovskite solar cells has drawn inspiration from natural processes to enhance energy conversion efficiencies

Fig. 1 Sustainable development goals (United Nations 2024)

(Ramanathan et al. 2023). These cells utilize structures that mimic the efficiency of natural photosynthesis, offering a more cost-effective and efficient method for converting solar energy.

Hydrokinetic turbines, inspired by natural water flow patterns observed in fish and aquatic environments, are another remarkable bioinspired solution that promotes sustainable energy (Galdo et al. 2024). These turbines can harness kinetic energy from flowing water bodies, such as rivers and ocean currents, converting it into electricity.

Wind Lens technology is a notable advancement in wind turbine design, which incorporates a shroud around the turbine blades to enhance aerodynamic efficiency. This technology has been demonstrated to significantly improve the power output of wind turbines by concentrating and accelerating wind flow, much like a magnifying glass focuses light (Ohya and Karasudani, 2010; Griffiths et al. 2011). The Wind Lens design maximizes energy capture while minimizing acoustic noise, making it suitable for urban environments where noise pollution from traditional turbines can be a significant issue.

Innovative research conducted by Gu et al. (2011) explored the concept of bionic or bioinspired airfoils, specifically utilizing the design characteristics of the seagull wing, which enable efficient interaction with varying wind conditions. Their analysis suggested that using this biomimetic design could enhance the overall aerodynamic performance of wind turbines, resulting in increased energy output and improved reliability.

Research has also investigated bionic designs in solar thermal energy systems. Although the cited reference discusses energy-efficient building designs and biomimicry, it provides limited direct evidence linking bioinspired structures to improved thermal performance in solar absorbers (Varshabi et al. 2022).

Supporting Sustainable Development Goal 9 (**SDG 9**)—which focuses on resilient infrastructure, sustainable industrialization, and innovation—requires coordinated policy actions and investment. One successful application is the development of bioinspired thermal energy regulation systems. An et al. (2023) report on engineered thermal armor structures that draw inspiration from the hierarchical structures found in plant surfaces. These materials can effectively manage and dissipate heat while maintaining stability at high temperatures (over 1000 °C). Developing self-repairing materials, influenced by natural processes seen in biological organisms, is another promising branch of bioinspired engineering. Speck and Speck (2019) provided an overview of these materials, which mimic the mechanisms found in biological systems for healing and restoration.

Regarding Sustainable Cities (**SDG 11**), bioinspired technologies have demonstrated significant potential for urban applications. The innovative development of bright, functional surfaces, inspired by natural adaptations, can lead to enhanced urban infrastructure. Specifically, 3D micropatterned surfaces have been demonstrated to reduce drag and to improve material performance in wet conditions (Tricinci et al. 2023). Moreover, technologies harnessing bioinspired principles in urban design facilitate more resilient city planning to withstand climate impacts, addressing urban sustainability holistically.

Research by Sandak and Butina (2023) highlights the use of bioinspired materials that can mitigate the impacts of climate change, such as carbon sequestration and flood management through natural adaptations. Furthermore, integrating bioinspired surface designs for self-cleaning buildings, inspired by the lotus leaf, helps reduce maintenance costs and the environmental footprint in urban environments (Liu et al. 2015).

Climate Action (**SDG 13**) is significantly bolstered by bioinspired solutions to reduce environmental footprints and enhance resilience against climate change. Innovations, such as bioinspired materials that mimic natural systems for efficient energy capture, contribute to renewable energy applications that minimize reliance on fossil fuels (Wan et al. 2016). Furthermore, the development of bioinspired nanocomposites has demonstrated advancements in energy storage and conversion systems, targeting reductions in greenhouse gas emissions while promoting sustainable practices (Mei et al. 2021; Qu et al. 2024). By utilizing natural principles, such innovations facilitate effective responses to the challenges posed by climate change, encouraging practices that enhance carbon neutrality.

Another example is the "Wooden Armor" developed by Chen et al. (2022), which utilized micronano lignocellulose in a Bouligand structure to create lightweight yet durable materials to reduce damage from impacts. This aligns with developing resilient infrastructures that can withstand climate-induced extremes. Moreover, bioinspired nanomaterials have been designed to enhance energy storage capacities, thereby contributing to lower carbon footprints when incorporated into renewable energy technologies (Wu et al. 2022).

3 Interdisciplinary Integration in Engineering Education

Applying interdisciplinary projects in education encourages a holistic understanding of various subject matters. Samaniego et al. (2024) emphasize the importance of combining knowledge from diverse disciplines, where students tackle issues from multiple viewpoints, leading to the development of improved cognitive and creative capacities. This opinion is supported by Wanieck et al. (2020), who theorize that experiential learning exercises, such as interdisciplinary education, are more effective than traditional pedagogies because they foster the active application of newly acquired skills across diverse settings. Furthermore, Wang et al. (2020) disclose that effective interdisciplinary instruction enables educators to not only impart subject content but also to build key problem-solving skills.

Bioinspiration is indeed a promising way of knowledge synthesis between technical and environmental fields. For instance, the article by Zhang and Ferrand (2022) illustrates how sustainable material design can be informed by biological principles, thus integrating ecological science and engineering practice. By doing so, it can develop new, technologically advanced, and ecologically appropriate solutions. Moreover, Casarosa et al.'s (2020) study demonstrates that encouraging an interdisciplinary environment enables students to develop a well-rounded skill set that incorporates creativity, environmental awareness, and technical skills, which are in high demand in today's employment sector.

Social elements also contribute to the effectiveness of interdisciplinary education. For instance, Chen et al. (2019) have discussed how project-based learning in the community can significantly enhance students' understanding of issues in urban development while consolidating social responsibility. The same sentiment is echoed in Zhou and Bakhir's research (2014), where game-based digital learning facilitates an interactive experience that transcends disciplines, enhancing both technical and social learning outcomes for students. Furthermore, transitioning from traditional STEM (Science, Technology, Engineering, and Mathematics) education to STEAM, which incorporates the Arts, broadens students' creative capabilities and fosters a deeper understanding of complex systems within society.

One exemplary integration of bioinspiration in engineering education is highlighted in the work of Nagel and Pidaparti (2016), who discuss the significance of incorporating bio-inspired design courses into undergraduate engineering curricula. Their work emphasizes that such courses generate more economically viable and sustainable solutions than traditional engineering design methods. The efforts at various institutions to incorporate bioinspired design into curricula reflect a broader trend of teaching students how to draw concepts from nature to solve engineering issues successfully. In addition, the course "Biomimicry: from life to nanotechnological innovations" introduced at the Mines Nancy Engineering School demonstrates practical approaches to teaching biomimicry and applying it to engineering students (Cleymand et al. 2015).

Furthermore, Cayetano-Jiménez et al. (2024) describe how bioinspired robotics can facilitate interdisciplinary learning. Their project-based learning model yielded feasible robotic prototypes that addressed local environmental issues, thereby strengthening both science and engineering competencies through hands-on experience. This model highlights the importance of experiential learning, where students can apply bioinspired concepts to create real-life applications, thereby strengthening both their technical expertise and environmental awareness.

On the research front, advancements in bioinspired engineering have been facilitated through cooperation in interdisciplinary academic environments. Raman and Bashir (2017) chronicled the evolution of biological design, emphasizing the opportunities for biomimicry in developing new materials and engineering solutions. This follows growing recognition that an interdisciplinary methodology—integrating engineering, biology, and design—can significantly enhance learning and research in bioinspired environments.

Internationally, programs like "Eyes Toward Tomorrow" outlined by Full et al. (2021), promote interdisciplinary design principles that draw on biological discoveries, fostering interactions between students from various disciplines. The program demonstrates the potential for utilizing bioinspired design to improve learning outcomes and boost student confidence through collaborative efforts. However, while the University of California is highly reputed for its experiential programs with bioinspired principles, concrete examples are required to validate this point, as the basis of this assertion has not been comprehensively verified (O'Rourke and Seepersad 2015).

At the national level, the integration of bioinspired or biomimetic principles in engineering curricula has grown significantly in Brazilian universities, with several outstanding efforts being highlighted.

The Federal University of Minas Gerais (UFMG) hosts the Laboratory of Design and Biomimetics (LDBio), which creates technology and research solutions inspired by nature's design, engineering, and architecture. The lab aims to understand and apply nature's principles to develop innovative and sustainable projects (UFMG 2024).

Another example is the Biomimicry Laboratory of the Superior School of Industrial Design (ESDI) of Rio de Janeiro State University (UERJ). This lab focuses on biomimicry research and implementation, encouraging its application in design projects and related areas. Its interdisciplinarity aims to construct theoretical and practical foundations from biomimetic approaches (UERJ 2024).

Biosystems Engineering, offered by institutions such as the University of São Paulo (USP) and São Paulo State University (Unesp), also incorporates biomimetic principles in an interdisciplinary educational approach. These courses integrate biology and engineering, applying the knowledge of biological systems to develop sustainable technological solutions for various areas, including agriculture and environmental management.

The Federal University of Santa Maria (UFSM) also extensively researched applied biomimetic applications in civil engineering. An example is the research "Biomimicry Applied as a Concept in the Field of Construction," which evaluates sustainable alternatives in civil construction through nature-inspired concepts (UFSM 2024).

In addition, the Mauá Institute of Technology students demonstrated the real-world feasibility of biomimicry in the Mycena project—a mushroom-based decorative coating inspired by the structural arrangement of mushroom colonies. The project won the 5th Odebrecht Braskem Design Challenge, reflecting biomimicry's potential for driving sustainable innovation in design and engineering (Mongabay 2020).

In the field of experimental aerodynamics, the Federal University of Uberlândia (UFU) has made significant contributions through the Experimental Aerodynamics Laboratory (LAEX), part of the Experimental Aerodynamic Research Center (CPAERO). Among the research efforts carried out at UFU is the study titled "*Experimental Analysis of Biomimetic Blades for Small Wind Turbines,*" which explored biomimetic design principles to improve the efficiency of wind turbine blades. This work highlights the university's active engagement in biomimetic innovations that aim to advance renewable energy technologies (Paiva 2021).

Aside from academic courses and extracurricular and business-driven efforts, such as Amazu Biomimicry, biomimicry education in Brazil has also been significantly influenced by initiatives like these. Amazu has offered training, educational classes, and consultancy on biomimicry for over a decade, helping businesses and individuals convert nature's solutions into sustainable business innovations (Amazu 2024).

One of the most significant challenges is integrating sustainability into existing curricula. According to Zanitt et al. (2021), a harmonious integration of sustainability aspects is crucial for preparing future engineers with the competence to work with sustainable development. However, most engineering courses in Brazil are based on traditional paradigms that do not prioritize these aspects. This inertia often results in a lack of innovative pedagogical methods and real-world learning experiences necessary to develop a culture of sustainability among students.

Moreover, the lack of robust monitoring and evaluation (M&E) mechanisms significantly hampers the success of curriculum reforms. As Shawa and Botma (2020) contend, effective M&E is essential for identifying challenges that arise during the implementation of curricular innovations. Without such mechanisms, institutions and educators are unable to adequately recognize and address weaknesses within their programs, leading to poorly executed reforms and missed opportunities for meaningful improvement.

Another significant factor is resistance to change in educational organizations. The literature synthesized by Penrabel et al. (2022) indicates that profound cultural shifts are required to adopt competency-based curricula in sustainability. Educators must face substantial resistance in attempting to transform the dominant educational culture to one that embraces sustainability and interdisciplinarity. This often entails transforming teaching and assessment practices, which have been embedded for decades, and may be met with vehement resistance from faculty and administration.

Furthermore, aligning curricula with global sustainability goals is another challenge. According to Mulà et al. (2017), although there has been an international call for universities to adopt sustainability-based initiatives in higher education, some universities in Brazil face difficulties in implementing these initiatives within their respective contexts. The process of aligning curricula to include Sustainable Development Goals (SDGs) not only requires structural change but also an in-depth understanding on the part of teachers of how to implement these in a meaningful and effective manner.

Additionally, institutional resource differences can hinder the integration of bioinspiration and sustainability in engineering education. As the study of Cavalcante et al. (2024) illustrates, variations in investment in resources, infrastructure, and training between universities influence their ability to cultivate innovation and sustainability. Universities with limited funds may lack the proper equipment and staff to design effective programs with sustainable strategies, resulting in an uneven playing field of educational opportunities.

Lastly, assessing students' attitudes toward sustainability in engineering education reveals a significant disconnect. Colón-Flores et al. (2023) found that students often possess only a superficial understanding of sustainability and its relevance to engineering practice, indicating the need for a more in-depth exploration of the subject matter. This gap highlights the importance of pedagogical approaches that foster critical thinking and interdisciplinary perspectives, enabling students to develop holistic and sustainable engineering solutions.

4 Challenges and Opportunities for the Future

The implementation of bioinspired engineering in Brazilian industry and academia faces numerous technical, political and economic, and cultural challenges that hinder its widespread adoption. These challenges stem from diverse structural and educational barriers that must be addressed to enable its effective integration into conventional engineering practices.

Technical Barriers: The insufficient understanding of biological principles and the complexity of applying these principles to engineering solutions are significant technical issues. For example, despite the increased use of bioinspired design, engineers' knowledge of biological systems remains inadequate, and designs are not optimal (Broeckhoven and Plessis, 2022). Furthermore, interdisciplinary collaboration between engineers and biologists is weak, which is essential to release the potential of bioinspired methods fully (Chellapurath et al. 2023; Cayetano-Jiménez et al. 2024). This lack of synergy can lead to a disconnect between the available biological knowledge and its application in design, ultimately affecting the quality and feasibility of bioinspired projects (Fayemi et al. 2017).

Financial Barriers: Financial constraints are a significant challenge, particularly in an academic environment with limited research funds. The budgetary expense of cutting-edge research and development required for bioinspired initiatives can be unaffordable, and the probable uncertainty of commercial viability of these innovations deters investment from both public and private agencies (Zhang et al. 2023). Moreover, the novelty of bioinspired engineering suggests that firms may be hesitant to invest heavily in technologies that have yet to be tested through practical applications (Martín-Palma and Lakhtakia 2013). Therefore, a lack of early-stage venture capital for ventures in Brazil can significantly limit the development of bioinspired technology.

Political Barriers: The Brazilian government plays a crucial role in implementing bioinspired engineering in university curricula. Through its authority over national curriculum guidelines, research funding agencies, and innovation policies, the government can provide the necessary structure, incentives, and resources to support the inclusion of this interdisciplinary field. Without governmental support—such as updates to curricular directives, public research funding, and encouragement of university–industry collaboration—bioinspired engineering is unlikely to gain widespread adoption in academia. Moreover, as a strategic area aligned with sustainability and innovation, government promotion is essential to position Brazil as a leader in this emerging field, leveraging its rich biodiversity and scientific potential.

Cultural Barriers: On an artistic level, there can be resistance to cross-disciplinary practice in the educational context, where silos of education hinder the collaboration between disciplines. Reforms of the curriculum, which have suggested greater integration of science, engineering, and biology, have been encouraging but are not extensive (Cayetano-Jiménez et al. 2024). Students and professionals must be inspired to adopt an interdisciplinary approach, which is necessary for creating innovative bioinspired solutions (Ponnusamy et al. 2019). The public's perception of bioengineering and its implications can also vary significantly, leading to hesitation in adopting bioinspired practices, particularly when ecological concerns and ethical considerations are involved (Speck and Speck 2019).

The application of bioinspired engineering in diverse domains has the potential to transform sustainability, mainly through new application areas such as bioinspired design for circular economies, smart city technology development, and green transport solutions. With the expansion of this discipline, some of the future trends and promising research domains can be identified.

Bioinspired Circular Economy: The circular economy principle centers on the sustainability of resource utilization and waste minimization. Bioinspired engineering takes this up by developing materials and processes that mimic nature's cycles. A case in point is bioinspired silica research, which identifies sustainable production processes that ensure the optimal use of resources and minimize environmental impact (Baba et al. 2022). Bioinspired polymer innovations provide biodegradable material substitutes for conventional materials, thereby minimizing pollution and enhancing material lifecycle management (Omidian and Babanejad, 2024). Additionally, innovative approaches, such as synthesizing bioinspired materials, hold great potential for improving sustainability without compromising industrial requirements (Zhang et al. 2019).

Smart Cities and Sustainable Technologies: Bioinspired engineering has the potential to make a significant contribution to the development of smart cities by incorporating natural principles into urban planning and infrastructure. Intelligent systems emulating biological processes can maximize resource utilization, reduce urban waste, and promote sustainable living. For example, bioinspired robotics can drive ecologically responsible practices and improve conservation by collecting and utilizing data on ecological sustainability (Chellapurath et al. 2023). Second, incorporating biosolutions and innovative technologies into urban planning can foster public participation and ensure that sustainability becomes an overarching theme in urban planning and governance (Khalil et al. 2024; Turečková and Nevima 2020). The integration of these principles can lead to a data-driven approach for maximizing living standards and resource utilization in urban areas.

Sustainable Mobility Solutions: The transition to sustainable mobility requires creative engineering approaches that reduce reliance on fossil fuels and minimize environmental footprints. Bioinspired solutions are emerging as a feasible alternative in this regard. For example, bioinspired research on mechanical properties aims to design new materials that enhance the performance of vehicles and urban transportation networks, thereby reducing weight and energy consumption (Wijerathne et al. 2022). Additionally, bioinspired solutions for energy, such as utilizing natural materials in energy-efficient transport equipment, can lead to significant advancements in sustainable mobility initiatives. This includes studies on natural photonic structures that optimize energy absorption for vehicles to utilize (Mouchet 2025).

5 Conclusions

Bioinspiration presents itself not just as a technological tool but as a disruptive approach capable of transforming the face of engineering towards a more sustainable and ethical future. Through this review, it was possible to identify that nature-based solutions can enhance the performance of materials, energy systems, and infrastructure while simultaneously aligning with the SDGs, particularly SDG 7, SDG 9, and SDG 13, highlighting their potential for real-world impact. The examples covered—wind turbines mimicking the form of seeds to nanomaterials inspired by nature's designs—demonstrate how bioinspired innovations can provide scalable, affordable, and environmentally friendly technologies.

In parallel, bioinspiration is increasingly present in engineering education, particularly in Brazil, through university labs, interdisciplinary courses, and research projects connecting biology, design, and sustainability. UFMG, UERJ, UFSM, USP, Unesp, and UFU have led the way for integrating bioinspired ideas in teaching, research, and extension. From experimental aerodynamics on wind turbines to new materials and civil construction methods inspired by nature, these activities signal the enhanced applicability of biomimicry to Brazilian engineering education. These activities offer practical and pedagogical benefits: they promote systemic thinking, encourage innovation, and contribute to the training of professionals better equipped to face ecological and technological challenges.

However, despite these advances, structural and cultural barriers persist. Traditional curricula, funding gaps, and institutional resistance persist in hindering the widespread adoption of bioinspiration and sustainability as part of engineering education. These challenges must be addressed through increased interdisciplinarity, investment in pedagogical innovation, and cultural transformation, with sustainability at the core of engineering pedagogy.

Future developments should prioritize five key areas:

- Deeper integration of sustainability in engineering education and operations at the institutional and curricular levels;
- Expansion of interdisciplinary research, especially between engineering and life sciences;
- Development of finance and policy frameworks to incentivize investment in bioinspired innovation;
- Fostering public understanding and acceptance of biomimicry;
- Practical applications and measurable outcomes linking academic innovation to societal impact.

Ultimately, it is essential to enhance the relevance of sustainable engineering education to real-world issues. This includes promoting inter- and transdisciplinary learning, enabling future engineers to be both technically proficient and ethically and environmentally conscious. If implemented fully, bioinspiration has the potential to transform us into a more sustainable, innovative, and equitable future.

References

Alqarni AMZ, Afy-Shararah M, Salonitis K (2023) Integrating sustainability into engineering curricula: challenges, framework, and impact on student competencies in Saudi Arabian higher education. Adv Transdiscipl Eng. https://doi.org/10.3233/atde230921

Amazu Biomimicry (2024) Sobre a Amazu Biomimicry. https://amazu.bio/.

An S, Shi B, Jiang M, Fu B, Song C, Tao P, Deng T (2023) Biological and bioinspired thermal energy regulation and utilization. Chem Rev 123(11):7081–7118. https://doi.org/10.1021/acs.chemrev.3c00136

Baba YD, Chiacchia M, Patwardhan SV (2022) A novel method for understanding the mixing mechanisms to enable sustainable manufacturing of bioinspired silica. ACS Eng Au 3(1):17–27. https://doi.org/10.1021/acsengineeringau.2c00028

Broeckhoven C, Plessis A (2022) Escaping the labyrinth of bioinspiration: biodiversity as key to successful product innovation. Adv Funct Mater 32(18). https://doi.org/10.1002/adfm.202110235

Cavalcante DM, Ávila LV, Anholon R, Júnior ÁLN, Martins VWB (2024) Recovering sustainability in the Brazilian Amazon: an analysis of engineering courses from the perspective of innovation and sustainability. Int J Sustain High Educ 26(2):257–277. https://doi.org/10.1108/ijshe-09-2023-0436

Casarosa V, Ruggieri S, Salvatori E, Simi M, Turbanti S (2020) Educational ecosystems for information science: the case of the University of Pisa. Educ Inf 36(2):119–138. https://doi.org/10.3233/efi-190330

Casa Vogue. (2017). Desafio de design Odebrecht Braskem revela seus vencedores [Odebrecht Braskem design challenge reveals its winners]. Casa Vogue – Globo. https://casavogue.globo.com/Design/Gente/noticia/2017/12/desafio-de-design-odebrecht-braskem-revela-seus-vencedores.html

Cayetano-Jiménez IU, Bustamante-Bello R, Ramírez-Cadena M (2024) Bridging the gap: bioinspired robotics as catalyst for interdisciplinary education. Front Educ 9. https://doi.org/10.3389/feduc.2024.1375487

Chellapurath M, Khandelwal PC, Schulz A (2023) Bioinspired robots can foster nature conservation. Front Robot AI 10. https://doi.org/10.3389/frobt.2023.1145798

Chen Y, Daamen T, Heurkens E, Verheul WJ (2019). Interdisciplinary and experiential learning in urban development management education. International Journal of Technology and Design Education, 30(5), 919–936. https://doi.org/10.1007/s10798-019-09541-5

Chen Y, Dang B, Fu J, Zhang J, Liang H, Sun Q, Li H (2022) Bioinspired construction of micronano lignocellulose into an impact resistance "wooden armor" with bouligand structure. ACS Nano 16(5):7525–7534. https://doi.org/10.1021/acsnano.1c10725

Cleymand F, Rousseau M, Mano JF (2015) Introducing biomimetic approaches to materials development and product design for engineering students. Bioinsp Biomim Nanobiomater 4(4):207–212. https://doi.org/10.1680/jbibn.15.00004

Colón-Flores NDJ, Vargas-Martínez MR, Henríquez JCT, Domínguez-Valerio CM (2023) Environmental, social and economic attitudes and sustainable knowledge on the sustainable behaviour of engineering students: an analysis based on attitudes towards teachers. Sustain 15(18):13537. https://doi.org/10.3390/su151813537

Dou H, Xu M, Zheng Y, Li Z, Wen G, Zhang Z, Chen Z (2022) Bioinspired tough solid-state electrolyte for flexible ultralong-life zinc–air battery. Adv Mater 34(18). https://doi.org/10.1002/adma.202110585

Engbers S, Langevelde PHv, Hetterscheid DGH, Klein JEMN (2024) Discussing the terms biomimetic and bioinspired within bioinorganic chemistry. Inorg Chem 63(43):20057–20067. https://doi.org/10.1021/acs.inorgchem.4c01070

Fayemi P, Wanieck K, Zollfrank C, Maranzana N, Aoussat A (2017) Biomimetics: process, tools and practice. Bioinspir Biomim 12(1):011002. https://doi.org/10.1088/1748-3190/12/1/011002

Full RJ, Bhatti HA, Jennings P, Ruopp R, Jafar T, Matsui J, Estrada M (2021) Eyes toward tomorrow program enhancing collaboration, connections, and community using bioinspired design. Integr Comp Biol 61(5):1966–1980. https://doi.org/10.1093/icb/icab187

Galdo MIL, Rodríguez JD, Casanova AC, Damota JB, Caccia CG, Lorenzo JMR, Miranda JT (2024) Enhanced performance of a hydrokinetic turbine through a biomimetic design. J Mar Sci Eng 12(8):1312. https://doi.org/10.3390/jmse12081312

Gao F, Liu G, Chen A, Hu Y, Wang H, Pan J, Xiong Y (2023) Artificial photosynthetic cells with biotic–abiotic hybrid energy modules for customized CO2 conversion. Nat Commun 14(1). https://doi.org/10.1038/s41467-023-42591-x

Griffiths H, Ohya Y, Balleri A, Tong K, Al-armaghany A, Matsuura T, Karasudani T (2011) Measurement and analysis of the radar signature of a new type of wind turbine. Proc IEEE CIE Int Conf Radar. https://doi.org/10.1109/cie-radar.2011.6159671

Gu R, Xu J, Yang YB (2011) The investigation of the small bionic wind turbine based on the seagull airfoil. Adv Mater Res 347–353:3533–3539. https://doi.org/10.4028/www.scientific.net/amr.347-353.3533

Khalil ZF, Muazzinah M, Zamharira C, Zalikha SN, Dian W, Safitri S, … & Wulandari D (2024). Governance analysis in driving smart city policy in aceh. KnE Social Sciences. https://doi.org/10.18502/kss.v9i7.15470

Khalila NRABM, Sagar SALS, Basar MF (2024) An overarching summary of the sustainable development goals (SDGs). Int J Res Publ Rev 5(1):1234–1240. https://doi.org/10.55248/gengpi.5.0124.0209

Ken-Giami I, Simandjuntak S, Yang Y, Coats A (2022) A grounded theory approach to uncovering the process of how sustainability topics influence women engineers' career choice and engagement. Sustain 14(9):5407. https://doi.org/10.3390/su14095407

Kunrath K, Ramanujan D (2021) Fostering sustainable mindsets in engineering education. Proc Des Soc 1:1597–1606. https://doi.org/10.1017/pds.2021.421

Liu Y, Wang X, Fei B, Hu H, Lai C, Xin JH (2015) Bioinspired, stimuli-responsive, multifunctional superhydrophobic surface with directional wetting, adhesion, and transport of water. Adv Funct Mater 25(31):5047–5056. https://doi.org/10.1002/adfm.201501705

López-Forniés I, Asión-Suñer L, Sarvisé-Biec A (2024) Bioinspired prosumer. Interaction of bioinspired design methods in the prosumer scope. https://doi.org/10.20944/preprints202407.0188.v1

Martín-Palma RJ, Lakhtakia A (2013) Engineered biomimicry for harvesting solar energy: a bird's eye view. Int J Smart Nano Mater 4(2):83–90. https://doi.org/10.1080/19475411.2012.663812

Mei J, Liao T, Peng H, Sun Z (2021) Bioinspired materials for energy storage. Small Methods 6(2). https://doi.org/10.1002/smtd.202101076

Mongabay Brasil (2020) Biomimética: tecnologia inspirada na natureza avança no Brasil. https://brasil.mongabay.com/2020/03/biomimetica-tecnologia-inspirada-na-natureza-avanca-no-brasil/.

Mouchet SR (2025) Infrared absorbers inspired by nature. J R Soc Interface 22(223). https://doi.org/10.1098/rsif.2024.0284

Mulà I, Tilbury D, Ryan A, Mader M, Dlouhá J, Mader C, Alba D (2017) Catalysing change in higher education for sustainable development. Int J Sustain High Educ 18(5):798–820. https://doi.org/10.1108/ijshe-03-2017-0043

Nagel J, Pidaparti RM (2016) Significance, prevalence and implications for bio-inspired design courses in the undergraduate engineering curriculum. Proc ASME Int Conf Des Educ. https://doi.org/10.1115/detc2016-59661

Ohya Y, Karasudani T (2010) A shrouded wind turbine generating high output power with wind-lens technology. Energies 3(4):634–649. https://doi.org/10.3390/en3040634

Omidian H, Babanejad N (2024) Bioinspired polymers: bridging nature's ingenuity with synthetic innovation. Macromol 4(2):190–212. https://doi.org/10.3390/macromol4020010

O'Rourke JM, Seepersad CC (2015) Toward a methodology for systematically generating energy- and materials-efficient concepts using biological analogies. J Mech Des 137(9). https://doi.org/10.1115/1.4030877

Paiva LF (2021) Experimental analysis of biomimetic blades for small wind turbines. Undergraduate thesis, Federal University of Uberlândia (UFU), Uberlândia.
Penrabel R, Bastos PO, Biberg-Salum TG (2022) The perspectives and challenges of the competency-based curriculum in medical education: a literature review. Creat Educ 13(10):3191–3203. https://doi.org/10.4236/ce.2022.1310203
Ponnusamy PG, Nagel J, Rose C, Pidaparti RM (2019) Investigation of c-k theory based approach for innovative solutions in bioinspired design. Designs 3(3):39. https://doi.org/10.3390/designs3030039
Qu L, Gou Q, Deng J, Zheng Y, Li M (2024) A perspective of bioinspired interfaces applied in renewable energy storage and conversion devices. Langmuir 40(13):6601–6611. https://doi.org/10.1021/acs.langmuir.3c03679
Raman, R. and Bashir, R. (2017). Biomimicry, biofabrication, and biohybrid systems: the emergence and evolution of biological design. Advanced Healthcare Materials, 6(20). https://doi.org/10.1002/adhm.201700496
Ramanathan R, Mallik RC, Zinigrad M (2023) Effective approaches for perovskite solar cells: recent advances and perspectives. Phys Status Solidi A 220(14). https://doi.org/10.1002/pssa.202300091
Rybicka-Jasińska K, Derr JB, Vullev VI (2021). What defines biomimetic and bioinspired science and engineering?. Pure and Applied Chemistry, 93(11), 1275–1292. https://doi.org/10.1515/pac-2021-0323
Samaniego M, Usca N, Rosero JRS, Quevedo W (2024) Creative thinking in art and design education: a systematic review. Educ Sci 14(2):192. https://doi.org/10.3390/educsci14020192
Sandak A, Butina K (2023) Bioinspired building materials—lessons from nature. Front Mater 10. https://doi.org/10.3389/fmats.2023.1283163
Shawa M, Botma Y (2020) Practice guidelines for peer support among educators during a curriculum innovation. Afr J Health Prof Educ 12(4):186. https://doi.org/10.7196/ajhpe.2020.v12i4.1388
Speck O, Speck T (2019) An overview of bioinspired and biomimetic self-repairing materials. Biomimetics 4(1):26. https://doi.org/10.3390/biomimetics4010026
Tricinci O, Pignatelli F, Mattoli V (2023) 3D micropatterned functional surface inspired by Salvinia molesta via direct laser lithography for air retention and drag reduction. Adv Funct Mater 33(39). https://doi.org/10.1002/adfm.202206946
Turečková K, Nevima J (2020) The cost benefit analysis for the concept of a smart city: how to measure the efficiency of smart solutions? Sustain 12(7):2663. https://doi.org/10.3390/su12072663
UERJ - Universidade do Estado do Rio de Janeiro (2024) Laboratório de Biomimética – ESDI/UERJ. https://esdi.uerj.br/extensao/4964/laboratorio-de-biomimetica.
UFMG - Universidade Federal de Minas Gerais (2024) Laboratório de Design e Biomimética (LDBio). https://sites.arq.ufmg.br/ldbio/.
UFSM - Universidade Federal de Santa Maria (2024) Biomimética aplicada como um conceito na área de construção. Repositório Institucional UFSM. https://repositorio.ufsm.br/handle/1/22865.
United Nations (2025) Sustainable Development Goals. https://www.un.org/pt/teach/SDGs.
Varshabi N, Selçuk SA, Avinç GM (2022) Biomimicry for energy-efficient building design: a bibliometric analysis. Biomimetics 7(1):21. https://doi.org/10.3390/biomimetics7010021
Wan S, Peng J, Jiang L, Cheng Q (2016). Bioinspired graphene-based nanocomposites and their application in flexible energy devices. Advanced Materials, 28(36), 7862–7898. https://doi.org/10.1002/adma.201601934
Wang H, Charoenmuang M, Knobloch NA, Tormoehlen R (2020) Defining interdisciplinary collaboration based on high school teachers' beliefs and practices of STEM integration using a complex designed system. Int J STEM Educ 7(1). https://doi.org/10.1186/s40594-019-0201-4
Wanieck K, Ritzinger D, Zollfrank C, Jacobs S (2020) Biomimetics: teaching the tools of the trade. FEBS Open Bio 10(11):2250–2267. https://doi.org/10.1002/2211-5463.12963

Wijerathne B, Liao T, Ostrikov K, Sun Z (2022) Bioinspired robust mechanical properties for advanced materials. Small Struct 3(9). https://doi.org/10.1002/sstr.202100228

Wu G, Hui X, Hu L, Bai Y, Rahaman A, Yang X, Chen C (2022) Recent advancement of bioinspired nanomaterials and their applications: a review. Front Bioeng Biotechnol 10. https://doi.org/10.3389/fbioe.2022.952523

Zanitt JF, Rampasso IS, Quelhas OLG, Serafim MP, Filho WL, Anholon R (2021) Analysis of sustainability insertion in materials selection courses of engineering undergraduate programmes. Int J Sustain High Educ 23(5):1192–1207. https://doi.org/10.1108/ijshe-04-2021-0134

Zhang X, Chen G, Zhang H, Shang L, Zhao Y (2023) Bioinspired oral delivery devices. Nat Rev Bioeng 1(3):208–225. https://doi.org/10.1038/s44222-022-00006-4

Zhang Y, Ferrand HL (2022) Bioinspired self-shaping clay composites for sustainable development. Biomimetics 7(1):13. https://doi.org/10.3390/biomimetics7010013

Zhang Y, Mei J, Yan C, Liao T, Bell J, Sun Z (2019) Bioinspired 2D nanomaterials for sustainable applications. Adv Mater 32(18). https://doi.org/10.1002/adma.201902806

Zhang Z, Kong X, Xie G, Li P, Xiao K, Wen L, Jiang L (2016) "Uphill" cation transport: a bioinspired photo-driven ion pump. Sci Adv 2(10). https://doi.org/10.1126/sciadv.1600689

Zhao Q, Liu X, Ma Y, Zheng X, Yu M, Wu D (2020) Application of the modified college impact model to understand Chinese engineering undergraduates' sustainability consciousness. Sustain 12(7):2614. https://doi.org/10.3390/su12072614

Zhou S, Bakhir NM (2024) Interdisciplinary art learning through artistic digital game-based learning (DGBL): evaluating learning outcomes and processes among science and engineering students. Empir Stud Arts 43(1):584–617. https://doi.org/10.1177/02762374241246948

Lohanna Ferreira Paiva holds a Bachelor's degree in Aeronautical Engineering and a Master's degree in Mechanical Engineering from the Federal University of Uberlândia (UFU), Brazil. Her research focuses on applied aerodynamics, with an emphasis on biomimicry, flow visualization, and wind tunnel experimentation. She has professional experience in mechanical design projects across various industrial sectors, including the aeronautical, food, and electrical industries.

Odenir de Almeida is an Associate Professor in the Aeronautical Engineering Program at the Federal University of Uberlândia (UFU). He earned his B.Sc. (1998) and M.Sc. (2001) in Mechanical Engineering from São Paulo State University (UNESP – Ilha Solteira) and a Ph.D. from the Aeronautics Institute of Technology (ITA) in 2009, including a research period at the Institute of Sound and Vibration Research, University of Southampton, UK (2008). He has worked as a product development engineer at Embraer, focusing on propulsion and advanced technology. His research interests include turbulent flow simulation, fluid–structure interaction, computational aeroacoustics, and propulsion systems.

Mandatory SDG Classes as a Solution to Accelerate Sustainability Teaching at Higher Education Institutions?—A Case Study from Japan

Jana Walbaum

Abstract Higher Education Institutions (HEIs) worldwide have stated their commitment to sustainability and aim to successfully integrate Education for Sustainable Development into their teaching. However, approaches, concepts, and execution vary significantly across institutions. Research has shown the potential benefits (e.g., the increased likeliness of goal achievement) of incorporating the Sustainable Development Goals (SDGs) in HEI curricula. Yet, the presence of the SDGs in teaching mainly relies on individual initiatives. Only a small number of universities have taken the step to systematically introduce a mandatory SDG class for all of their students. One of these universities is the Toyohashi University of Technology, located in Japan. This paper presents qualitative research conducted in 2024 that consists of reviewing teaching materials and interviewing professors and administrative staff at the Toyohashi University of Technology. It shows how the SDG class is structured and implemented, what teaching methods are used, which training educators received, and what role the SDGs play outside the mandatory class. Particular emphasis is given to the problems perceived by the professors after the university's president introduced the mandatory class. The aim of this paper is to critically examine how a top-down approach to mandatory sustainability teaching functions in practice and to explore the practical and pedagogical challenges involved. The results will contribute to the understanding of top-down approaches and are relevant to educators, researchers, and other university actors aiming for an accelerated integration of sustainability into university teaching.

Keywords Education for sustainable development · Mandatory class · Higher education institution · Sustainability

J. Walbaum (✉)
TUM School of Social Science and Technology, Technical University of Munich, München, Germany
e-mail: jana.walbaum@tum.de

W. Leal Filho et al. (eds.), *Sustainability in the Curriculum and Operations of Universities*, World Sustainability Series,
https://doi.org/10.1007/978-3-032-12678-8_18

1 Introduction

The idea that universities have the task of contributing to a sustainable and climate-friendly future emerged in the 1960s (Bautista-Puig and Sanz-Casado 2021). What started as the concept of a green university based on its environmental impact transformed into a call for organizational changes at HEIs that include academic priorities and financial support for sustainability (Ryan et al. 2010). Researchers even extend the role of universities from contributors to leaders, mandating them to prevent global ecological collapse (Moore 2005). Educating students about sustainability is done, among other things, through knowledge transfer in the context of teaching (Hernández-Diaz et al. 2021). Next to the field of teaching, sustainability is also a central theme in campus operations, research, and governance at HEIs.

In Japan, climate change mitigation and sustainable development are key focus points. Since the earthquakes and tsunami in 2011 led to the nuclear disaster of Fukushima, the Japanese government accelerated its sustainability efforts (Nakai 2024). The concept of Society 5.0 was then developed and presented in a Basic Plan by the Council for Science, Technology and Innovation (Whittaker 2024). This concept focuses on a human-centered society solving challenges with innovation, including the need for green transformation. The same year, the Sustainable Development Goals (SDGs) were established by the United Nations (UN). The 17 goals aim to secure sustainable development on an economic, social, and ecological level and are aimed to be achieved by 2030 (UN 2015b). Since then, the SDGs have become a popular sustainability framework in Japan and part of Society 5.0. The government especially sees the potential and responsibility in higher education to achieve sustainable development (Edwards and Ashida 2021).

One important concept when considering sustainability in teaching is Education for Sustainable Development (ESD). According to the UN, ESD should help learners to confront global concerns such as climate change and unsustainable use of resources (UNESCO 2022). It equips people with the necessary knowledge, skills, and values to make informed choices and to enhance society's livelihood. In 2020, a new ESD program was created called "ESD for SDGs" (Federal Ministry of Education and Research 2025) which is tailored to achieve the SDGs by 2030. HEIs educate a huge proportion of society's future decision-makers. Incorporating the SDGs into teaching will, accordingly, increase the likelihood of achieving the goals (Zhou et al. 2020).

Current research has, therefore, focused on drivers and barriers of ESD at HEIs as well as SDGs implementation. Lazzarini and Pérez-Foguet (2018) emphasize the importance of qualitative research in sustainable development at HEIs. However, existing studies rarely incorporate the experiences and perceptions of educators (Christie et al. 2015), and the inclusion of different university actors and their understanding of the SDGs is missing (Cuesta-Claros et al. 2023). Acosta Castellanos and Queiruga-Dios (2022) highlight a lack of research regarding the integration of ESD into engineering curricula. They emphasize the need for universities and engineering faculties to conduct more experimentation in this area and publish their findings to build a validated knowledge base that can contribute to improving engineering programs.

While research has investigated the different ways in which universities incorporate the SDGs into their curricula and teaching, case studies that focus on the instrument of a mandatory SDG class are rare. This is due to the small number of universities that have taken the step to systematically introduce such a class for all their students. One of these universities is the Toyohashi University of Technology (TUT), located in Japan. The university is focused on engineering education and consists of 2,000 students. In 2024, the university management introduced a mandatory SDG class for all Japanese students. International students who understand Japanese can participate voluntarily.

This case study is part of a more extensive comparative study of different countries guided by the research question: How do mandatory SDG classes work as an instrument to accelerate sustainability teaching at HEIs? This paper investigates the implementation and design of the mandatory SDG class at TUT by interviewing the SDG office and different TUT educators. Emphasis is given to the perceptions of educators and the applied teaching methods, while relating to the broader integration of sustainability in teaching. The goal is to assess whether such a top-down initiative can accelerate sustainability education and to identify challenges and success factors that can guide other institutions in designing SDG classes. By providing an in-depth case study, the paper contributes to the ongoing discussion on effective strategies for embedding ESD at the institutional level.

2 Theory

The SDGs call for an increase of ESD at HEIs in their 4th goal (UN 2015a). At the same time, ESD aims to reach the SDGs, which also means including them in teaching practices (Fiselier et al. 2018). This interconnectivity of both policies, especially in the sphere of HEIs, is also found in the literature. For example, Leal Filho (2018) describes the SDGs as a chance to overcome barriers of previous ESD implementation at HEIs.

It is, therefore, crucial to look at the drivers and challenges of implementing ESD and SDGs teaching at HEIs. Etzkorn et al. (2019) identify the following drivers for promoting and distributing ESD in universities:

- Networks
- Engagement in student initiatives
- Political agenda setting
- A Sustainability Office and round tables at the university
- Professors and lecturers.

An individual lecturer who promotes the diffusion of ESD in universities is referred to as a change agent (Kapitulčinová et al. 2018). This promotion can happen in exchange environments such as the above-mentioned formats of networks or round tables. The mere existence of such offerings, as well as the agenda setting, can, in return, encourage new individuals to become change agents.

While these drivers positively impact sustainability in teaching, several challenges remain. Leal Filho et al. (2021) identified the following impediments by investigating HEIs in South Africa, Nigeria, the United States of America, Brazil, and Germany:

- Lack of institutional support
- Limited resources
- Lack of trained staff
- Limited emphasis on approaches.

Especially the German case study conducted at the Hamburg University of Applied Sciences points out the time constraints inherent for professors. As a result, it is challenging to motivate staff from different faculties to engage or improve sustainability in teaching. The issue of resource scarcity is also found in other publications. Ramakrishna (2021) argues that professors are overburdened with their general teaching and research commitment as well as administrative duties, leaving little room for engaging in sustainability at universities.

As another critical point, the study of Leal Filho et al. (2021) argues that the SDGs are not put to full use in teaching at HEIs. Elaboration on this statement is found in a viewpoint of the primary author in 2020 (Leal Filho 2020). Although the academic community played a significant role in raising awareness of the SDGs by informing students about their existence, acceleration is needed in order to fulfill them by 2030. According to Leal Filho (2020), a crucial point is integrating the SDGs into teaching programs. This can be achieved by linking issues in all scientific disciplines to the SDGs and making them a common theme in seminars and lectures. Leal Filho's view (2020) is congruent with that of other leading researchers. For example, (El-Jardali et al. 2018) state that the implementation of SDGs at universities is still in its early stages despite the great potential of universities to contribute to achieving these goals.

Apart from the above-mentioned challenges of distributing ESD, the study of Christie et al. (2015) sees teachers themselves as a challenging factor. According to their survey and interview findings, most professors are unaware of sustainability concepts. They would focus on the environmental dimension only and lack knowledge of which tools are needed for ESD. From other educational areas, studies have also concluded that if teachers fail to properly understand ESD, the learning of students suffers (Laurie et al. 2016). Consequently, educators should receive training that enables them to perform pedagogy in the transformative nature that is required for ESD (Pavlova 2013). The goal of training teachers at HEIs goes beyond improving individual teaching practices in the classroom and plays a crucial role in driving institutional transformation and shaping the broader discourse on teaching and learning in higher education (Mulà et al. 2017).

Furthermore, the process of ESD implementation can be started with a top-down or bottom-up approach (Mazon et al. 2020). Lu and Zhang (2014) compared the methods of British and Chinese universities. The bottom-up approach leads to a better inclusion of staff, which in turn makes them feel supported and empowered. The top-down approach displays institutional drive and commitment. The authors argue that to achieve significant changes in the promotion of ESD, a combination and proper balance of both approaches is needed.

The specific design of ESD classes has been a subject of research in the last few years. For example, Araneo (2024) focused on identifying recommendations for ESD course designers. Based on student interviews, they advise ESD courses to be science-based, locally relevant, practical, empowering, and requiring a critical thinking process. They also conclude that a specific language should be used: "*The voice of ESD—the language, content, and delivery—strive to be infectiously passionate, empowering, hopeful, inspiring, inclusive, validating, truthful, informative, emotionally engaging, interesting, relational, cocreated and transforming reflecting the urgency of the field.*" (Araneo 2024).

The University of New Haven introduced an interdisciplinary, university-wide introductory course on sustainability for first-year students, first as a voluntary one but with the idea to make it mandatory in the future. Aktas et al. (2015) analyzed the implementation and structure of the class. The course was team-taught by professors from various disciplines to provide a broad, interdisciplinary perspective on sustainability topics. The course structure included large-group lectures as well as smaller seminar sessions designed to facilitate discussions and in-depth analysis. However, the diverse academic backgrounds of the instructors led to differing expectations and initial inconsistencies in the course's execution. Additionally, many students had insufficient prior knowledge, making it difficult for them to actively engage in discussions. To other universities that want to implement a similar class, the authors recommend a long preparation period of at least one year for developing course materials and coordination among instructors.

In Japan, the Ministry of Education, Culture, Sports, Science, and Technology has cooperated with the Ministry of the Environment since 2007 to include sustainable development at HEIs in their policies (Nomura and Abe 2010). This caused Japanese HEIs to introduce several sustainability initiatives, often correlating with internationalization efforts of the HEIs (Edwards and Ashida 2021). In addition, Japanese universities are showcasing their achievements in sustainability, for example, by publishing their scores in rankings that include the SDG framework, such as the Times Higher Education Impact Ranking. (Urata et al. 2023). Liu et al. (2022) researched the integration of sustainability into teaching at Tohoku University, Japan. They conclude that inclusion of sustainability is especially relevant in non-environmentally related degrees and emphasize the benefits of interdisciplinary teaching for ESD. In total, the Japanese government and Japanese HEIs are increasing their efforts to accelerate ESD. However, mandatory university-wide classes have not become standard measures at Japanese universities so far.

3 Methods

Given the complex and context-dependent nature of sustainability education, a qualitative case study design was chosen for this research to gain an in-depth understanding of the perspectives and experiences of educators. The study is based on semi-structured interviews. The interview guide has a set of nine key questions that cover different thematic blocks relevant to answering the research question. Each key question contains a set of potential follow-up questions which can specify the content of the key question or lead in a different direction. They depend on the interviewee's answers, provide flexibility to explore, and allow participants to elaborate on their experiences and insights. The research follows established qualitative research principles, ensuring rigor through careful data selection, systematic collection, and thorough analysis.

Data Selection and Collection

TUT was chosen as a case study because of its unique mandatory SDG class combined with Japan's background in aiming to accelerate sustainable development. The SDG office of TUT was then contacted in January 2024 to organize a research stay to allow in-person interviews in July 2024. The SDG office provided a list of the lecturers of the mandatory SDG class. In addition, I used the course search to find subjects that include teaching sustainability. Therefore, keywords such as "sustainability", "environmental" and "sustainable" were applied. From these two lists, lecturers were selected randomly and based on availability for the interviews.

In total, ten interviews were conducted in person in the offices of the respective interviewees. A translator was present during two interviews as the interviewees communicated in Japanese. One interviewee was a member of the SDG office without a lecturer position. In this case, the interview guide was not used. Instead, tailored questions about the implementation process of the mandatory SDG class and the general work of the SDG office were asked. To support the findings of the interviews, supplementary documents such as the teaching materials (e.g., PowerPoint presentations) were reviewed. In addition, I observed two lessons in person. One of these lessons was part of the mandatory SDG class; one was a different class that focused on teaching international students about sustainability.

Data Analysis

The data analysis is based on Eisenhardt's (1989) three-step procedure. First, a within-case analysis with a detailed description of each interview is conducted. Second, cross-data patterns are examined by creating different comparison categories. Lastly, theory and data are compared. After transcribing the interviews with NoScribe (Dröge 2024), MAXQDA (VERBI Software 2024) was used for the qualitative analysis.

Table 1 shows which codes were used to categorize the practices and perceptions of the interviewees. Teaching methods, sustainability understanding, training of teachers, and SDGs application serve as comparison categories to (1) compare between different educators and (2) to compare between the mandatory SDG class and other classes. The code "Mandatory SDG class" was solely used to filter answers

Table 1 Coding categories

Code	Definition	Example
Teaching methods	Any means that are used to deliver teaching content to the students, such as discussions, group work, individual work, lectures, videos, audios, e-learning features	"Right, we are discussing a lot. We have a lot of questions." (Interview 2, Pos. 44)
Application of SDGs	Interviewee mentions the SDGs, for example, what role the SDGs play in their teaching, which specific ones are important, how they perceive them	"So sometimes it is very difficult to include the SDGs content." (Interview 4, Pos. 13)
Mandatory SDG class	Interviewee mentions the mandatory SDG class at TUT, for example, the process, the structure, the content, or the implementation	"But for SDG classes, we don't have so much time" (Interview 2, Pos. 42)
Training of teachers	Interviewee answers whether they have received any training, gives feedback, or explains expectations for a potential training	"Don't think I have participated in training for SDGs. But I have participated in training for engineering ethics" (Interview 3, Pos. 33)
Sustainability understanding	Interviewee mentions sustainability concepts other than the SDGs or explains their perception of sustainability	"That's why I'm saying we don't need to think about the sustainability" (Interview 2, Pos. 13)

that relate to the implementation and design of that class and, therefore, only relevant in the interviews with educators who teach the mandatory class. The code "Sustainability Understanding" was mainly used in the analysis of interviews from classes other than the mandatory SDG one, as there, sustainability can be applied generally, and a focus on SDGs is not necessarily guaranteed.

4 Results

The interview results can be divided into belonging to the mandatory SDG class and to regular TUT classes that are part of specific degrees but still teach sustainability-related content.

4.1 Mandatory SDG Class

The president of TUT has decided to introduce a mandatory class to accelerate sustainability teaching at the university. This follows the direction of the Japanese government which is promoting sustainable development at HEIs. However, the government does not impose rules regarding sustainable teaching, so the HEIs remain independent in their decision-making and the scope of designing teaching practices. The class at TUT is focused on the SDGs and has the title "Introduction to SDGs". It was created using a top-down approach. Initiated by the president of TUT, the task of creating and implementing a specific course design was then given to the SDG office.

The slide in Fig. 1 was created by the SDG office and it shows the goals as well as the structure of the mandatory SDG class. Each 90-min lecture is given by a different professor. The professors did not participate in the creation process of the class; however, they were required to design the content of their lessons themselves. The SDG office plans on having a meeting with all the lecturers to discuss the class content and make improvements (Interview 10).

The individual classes are based on the research background of the educators. This leads to an environmental focus and a lack of inclusion of social aspects. The SDG office reported that they are aware of the problem and want to include the social level of sustainability in the future but are missing the expertise from their educators due to the university's engineering focus.

During the interviews, a reluctance to teach the SDG class was shown by two interviewees.

> "But depending on the social situation, including the SDGs, we receive many pressure to teach some content." (Interview 4, Pos. 23)

> "Okay, I don't want to do that. But the president asked us to give a talk about the SDGs. So we produced a class" (Interview 2, Pos. 24)

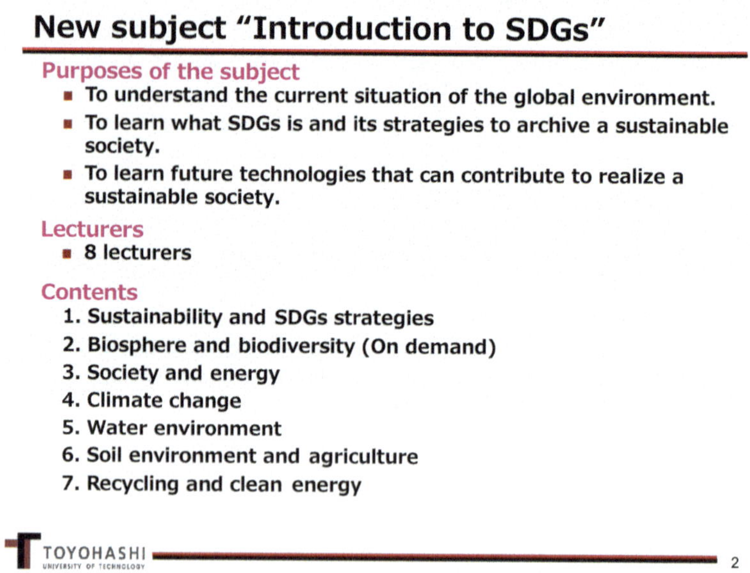

Fig. 1 Presentation slide from the mandatory SDG class

For one of these interviewees, the reluctance stemmed from a critical view of the SDGs. *"Okay, now we have to think about post-SDGs. We don't need to think about SDGs, especially for the young people" (Interview 2, Pos. 28).*

In general, the teachers experienced a knowledge gap among the students regarding the SDGs. "*And also, the students don't have much interest for the SDGs. Actually, they know the word SDGs but they don't understand what does it mean the SDGs. They don't know how to act for the SDGs.*" *(Interview 1)*. This is backed by in-class observation of students stating hearing about the SDGs for the first time

Interviewees who teach lectures for the SDG class were then asked how they apply the SDG content and which teaching methods they use.

The majority of the lecturers for the mandatory class base their content on their own research background and relate the SDGs to specific examples in the field of environmental sustainability, such as water security, energy efficiency, or climate change protection.

> "So maybe my target is the ocean and water. That's my target about the SDGs. But unfortunately, I maybe I don't pay attention to all the SDGs." (Interview 4, Pos. 11-12)

Two interviewees reported teaching about the basic concept of the SDGs.

> "I think it's important to teach the basic concepts of these areas. That's the basic core concept. And another thing I teach here is...For example, in SDGs 1 to 17, this is the goal. And there is a target for the goal. And the most important thing in the target is the fact. The fact is the target and the goal. This is the relationship. So, teaching the SDGs concept." (Interview 3, Pos. 9)"

> "So, during my SDG classes, I'm always saying, we have to think about it. No one leaves behind" (Interview 2, Pos. 54)

The teaching methods in the mandatory class are similar for each lecturer. The content is taught in a lecture-style manner, with a few questions targeted at the students.

> "But for the SDG classes, [...] I couldn't take a long time for the discussion." (Interview 2, Pos. 42)

This problem was also seen during the class observation. Two short questions were posted to the students, but the rest of the time was dedicated to a lecture by the professor.

One specific method used is taking articles to show students current news about the SDGs.

> "Sometimes I distribute some material about the SDGs topics in the news. Or some internet news or some newspaper. Recently, the students don't read the newspaper. So I copy the topics" (Interview 4, Post. 45)

Another theme that was used in the class is relating the SDGs to a public figure. A popular baseball player in Japan is known by the majority of the students, but the fact that he is promoting the SDGs is not as well known.

"Ohtani Shohei. He's a very popular baseball player. All over the United States, people know about him. Right, and the Japanese people. Very popular. And he is really interesting in SDG. So I always ask students Do you know why? Then students can identify, oh, we have that. There is a 17 goal. Because Ohtani Shohei, Ohtani Shohei, is in the SDG. He's always showing his back number" (Interview 2, Position 46).

4.2 *Outside of the Mandatory SDG Class*

To better understand the nature of the mandatory SDG class and allow comparison, the application of the SDGs and the teaching methods in other classes were also the subject of the interviews.

Three interviewees reported that the SDGs do not play a role in their teaching (Interviews 1, 6, and 9). One educator tries to incorporate some content about the SDGs although it is challenging in addition to teaching the general engineering content. This claim is backed up by another interviewee, stating that they relate one SDG that fits to their general teaching content but experience difficulties teaching about the other SDGs (Interview 4, Pos. 11, 13).

"We already have the curriculum. So if the government says that universities have to teach the SDGs content much more, we already have the original curriculum. So we try to include some parts" (Interview 4, Pos. 21)

One interviewee initiated a sustainability lesson themselves. They have to teach one lesson in the course "Life in Japan", which is a series of classes tailored to international students. The interviewee decided to focus the class on sustainability due to their interest in the topic and the importance of sustainability teaching. The SDGs are a core theme of this lesson. The students are informed about what the SDGs are and which targets they contain. Next, the students are required to interact and rate the goal achievement of their own countries based on their own perceptions. Furthermore, a section of the documentary "A Life on Our Planet" by David Attenborough is shown to the students. The idea is to visually display the urgency of climate change and the need to shift towards sustainable development to the students.

"I showed this video to them and they got a shock. They got a shock. They are really awake from that" (Interview 5, Pos. 76)

In addition, students have to come up with ideas on what they can do to achieve sustainable development and relate it to their areas of study.

"I put it in Google Classroom. So they'll have to try to find out what they can contribute in their field of study. What ideas do we have for example, if they're doing architecture, what can they do in their field to promote new developments that can contribute to sustainable or new materials that will be helpful, things like that. Then they'll give a lot of ideas" (Interview 5, Pos. 48)

4.3 Training of Educators

All interviewees reported that they had not received any training in teaching about sustainability. The interest in training for teaching about sustainability is higher for the educators who have to teach the mandatory class. They see the need for support in creating the lessons.

> "If there are some teaching classes for the teachers, or some material, it's very useful for us. Because in our university, we receive the information to please teach the SDGs, but we don't have any material or any information. So to teach the SDGs content, we have to start to learn what is the SDGs by ourselves before the teaching" (Interview 4, Pos. 25).

Time constraints remain a limiting factor in the interviewees' interest in participating in any training.

> "little bit interest but no time" (Interview 1, Pos. 37)

> "He always says, we have to discuss about the SDGs, the members who are giving the course. We have no time." (Interview 2)

One of the educators voluntarily participated in a general teaching training that offers guidance on teaching engineering classes, considers engineering ethics, and which competencies should be achieved. The interviewee emphasizes the importance of such training for all professors: "Training was helpful. Without that, it's impossible" (Interview 3, Post. 41).

5 Discussion

Researchers have called for better integration of the SDGs in teaching at HEIs by making them part of the curriculum. With the mandatory SDG class, TUT ensures that all students know what the SDGs are, get a basic understanding of the nature of the goals, and can be motivated for further research and action in their later professional lives. Creating this special class provides room for sustainability content and reduces pressure to fit in detailed SDGs content in other engineering classes, which was reported to be challenging.

The described phenomenon of change agents (Kapitulčinová et al. 2018) can also be seen in form of the professor at TUT who initiated a sustainability lesson outside the mandatory SDG class. This lesson contains student-centered activities and innovative learning strategies. As advised by Lu and Zhang (2014). experienced and motivated staff should be identified by the universities and included in the creation of a mandatory class.

In contrast, the teaching methods of the mandatory SDG class are mainly teacher based and lack student-centered activities (Table 2). Interviewees mentioned the time limitations of 90 min in each lesson as the reason for that. This bears the danger of creating passive learners. As Lu and Zhang (2014) described, active, creative, and empowered students are more likely to succeed in sustainability learning. A possible solution could be a restructuring of the class, moving away from conveying the content in a lecture-based chronological style towards a more innovative format that focuses on learning through interaction, initiative, and collaboration of the students.

Other impediments to proper SDGs integration and sustainability teaching that have been addressed in previous studies are time constraints and a lack of knowledge of sustainability concepts (Christie et al. 2015). These are also present at TUT. For instance, interviewees reported they feel overwhelmed and would need more guidance and material for teaching the mandatory SDG class. In addition, the environmental dimension of the SDGs is overrepresented within the TUT class as the educators themselves have an engineering background. In general, training to teach the SDG class or ESD content would be beneficial for TUT. This observation is consistent with the findings of other researchers (Pavlova 2013; Mulà et al. 2017). It has to be considered, though, that the training itself also requires available resources and time.

TUT's mandatory SDG class was implemented using a top-down approach. This shows commitment at the highest university level and sets it on the agenda for the whole university. While mandatory SDG classes could also be initiated as bottom-up by individual change agents, institutional support and implementation are needed. A participatory approach, where staff and educators are included in the top-down decision-making, is seen as leading to better commitment, quality, and empowerment (Disterheft et al. 2015). However, as seen from the New Haven case study, creating a mandatory sustainability class using a participatory approach takes time. Aktas et al. (2015) advise a preparation phase of at least one year to implement a university-wide sustainability class in a participatory approach. Against this suggestion stands the urgency of the climate crisis. Universities need to act now and cannot further delay far-reaching sustainability teaching. To accelerate the creation of mandatory SDG classes, HEIs can benefit from external sources, such as other universities that have

Table 2 Juxtaposition of the mandatory SDG class and regular classes at TUT

	Mandatory SDG class	Other classes
SDGs application	• SDGs are the main focus • The basic concept (Leave no one behind) • The structure (goal—target—indicator) • Examples from the research background that affect SDGs	• SDGs are not part of many classes • Individual change agents focus on them • Specific goals are picked that are related to the class subject
Teaching methods	• Teacher-centered • Mainly lecture style with a few questions	• Larger variety and more student-centered • Discussions, group work, videos
Training of educators (sustainability teaching)	• None received • Interest high as guidance is wanted	• None received • Interest is lower as necessity is not seen

already established classes. However, in sustainability teaching, there is no one-fits-all solution, and integrating one's own ideas, approaches, and teaching methods into existing classes remains important. In addition, mandatory SDG classes, such as the one at TUT, can still be changed and improved after their implementation. HEIs should, therefore, establish a continuous learning and feedback process.

Lastly, it should be considered that the SDGs are one part of ESD. A refocus from a specific SDG class towards a general ESD class can be beneficial. That would enable a focus on SDGs while they are present and urgent, e.g., until 2030, and allow a flexible shift in the teaching content afterward. In addition, other ESD practices can be easily integrated into the class.

6 Conclusions

From the case study of TUT, the following recommendations can be derived for universities that want to introduce a mandatory sustainability class:

1. **Participatory top-down approach**: Support and guide at the highest university level, involve educators and potential change agents in class design and implementation.
2. **Training of educators**: Provide material on ESD and teach educators how to activate critical and creative thinking in students.
3. **Variety of content**: Include the environmental, social, and political dimensions of sustainability.
4. **Teaching methods**: Make use of student-centered activities such as discussions and group work, include video and audio material, and relate to public figures and current events.

Limitations of this study lie in the fact that TUT is an engineering-focused institution with a small body of students and lecturers. This influences the nature of the introduced mandatory SDG class. While decision processes are shorter, and educating all students in person is more feasible, sustainability expertise and resources are scarcer. In addition, my study only focused on the educator's side and their perceptions and teaching methods. Future research should include the student perspective and analyze what effect a mandatory sustainability class has on their motivation and learning outcomes.

Due to the lack of available case studies, no clear conclusion can be drawn yet on whether mandatory SDG classes are a better instrument than voluntary SDG classes. However, this paper helps to understand the implementation of a mandatory class, its effect on educators, and impediments to consider. 2030, the year the SDGs are supposed to be achieved, is approaching rapidly, and the climate change is already affecting the lives of thousands of people. Therefore, action is needed, and HEIs should not wait too long until they have designed the supposedly perfect sustainability class but rather balance top-down and bottom-up approaches to create sustainability classes that can still be optimized during their operation phase.

References

Acosta Castellanos PM, Queiruga-Dios A (2022) From environmental education to education for sustainable development in higher education: a systematic review. International Journal of Sustainability in Higher Education 23:622–644. https://doi.org/10.1108/IJSHE-04-2021-0167

Aktas CB, Whelan R, Stoffer H, Todd E, Kern CL (2015) Developing a university-wide course on sustainability: a critical evaluation of planning and implementation. Journal of Cleaner Production 106:216–221. https://doi.org/10.1016/j.jclepro.2014.11.037

Araneo P (2024) Exploring education for sustainable development (ESD) course content in higher education; a multiple case study including what students say they like. Environmental Education Research 30:631–660. https://doi.org/10.1080/13504622.2023.2280438

Bautista-Puig N, Sanz-Casado E (2021) Sustainability practices in Spanish higher education institutions: An overview of status and implementation. Journal of Cleaner Production 295:126320. https://doi.org/10.1016/j.jclepro.2021.126320

Christie BA, Miller KK, Cooke R, White JG (2015) Environmental sustainability in higher education: What do academics think? Environmental Education Research 21:655–686. https://doi.org/10.1080/13504622.2013.879697

Cuesta-Claros A, Malekpour S, Raven R, Kestin T (2023) Are the sustainable development goals transforming universities? – An analysis of steering effects and depth of change. Earth System Governance 17:100186. https://doi.org/10.1016/j.esg.2023.100186

Disterheft A, Caeiro S, Azeiteiro UM, Filho WL (2015) Sustainable universities – a study of critical success factors for participatory approaches. Journal of Cleaner Production 106:11–21. https://doi.org/10.1016/j.jclepro.2014.01.030

Dröge K (2024) noScribe: AI-powered Audio Transcription. https://github.com/kaixxx/noScribe

Edwards S, Ashida A (2021) Higher education in Japan: internationalization, the Sustainable Development Goals and survivability. IJCED 23:104–119. https://doi.org/10.1108/IJCED-09-2020-0061

Eisenhardt K (1989) Building Theories from Case Study Research. The Academy of Management Review:532–550

El-Jardali F, Ataya N, Fadlallah R (2018) Changing roles of universities in the era of SDGs: rising up to the global challenge through institutionalising partnerships with governments and communities. Health Res Policy Syst 16:38. https://doi.org/10.1186/s12961-018-0318-9

Etzkorn N, Grapentin-Rimek T, Singer-Brodowski M (2019) Pfade der Transformation: Die Verbreitung von Bildung für nachhaltige Entwicklung im deutschen Bildungssystem, [Paths of Transformation: The Dissemination of Education for Sustainable Development in the German Education System] 1st edn. in the series "Ökologie und Erziehungswissenschaft" [Ecology and Educational Science] of the Commission on Education for Sustainable Development of the German Educational Society. Barbara Budrich, Leverkusen

Federal Ministry of Education and Research (2025) ESD for 2030. https://www.bne-portal.de/bne/en/esd-for-2030/esd-for-2030_node.html

Fiselier ES, Longhurst JW, Gough GK (2018) Exploring the current position of ESD in UK higher education institutions. IJSHE 19:393–412. https://doi.org/10.1108/IJSHE-06-2017-0084

Hernández-Diaz PM, Polanco J-A, Escobar-Sierra M, Leal Filho W (2021) Holistic integration of sustainability at universities: Evidences from Colombia. Journal of Cleaner Production 305:127145. https://doi.org/10.1016/j.jclepro.2021.127145

Kapitulčinová D, AtKisson A, Perdue J, Will M (2018) Towards integrated sustainability in higher education – Mapping the use of the Accelerator toolset in all dimensions of university practice. Journal of Cleaner Production 172:4367–4382. https://doi.org/10.1016/j.jclepro.2017.05.050

Laurie R, Nonoyama-Tarumi Y, Mckeown R, Hopkins C (2016) Contributions of Education for Sustainable Development (ESD) to Quality Education: A Synthesis of Research. Journal of Education for Sustainable Development 10:226–242. https://doi.org/10.1177/0973408216661442

Lazzarini B, Pérez-Foguet A (2018) Profiling research of the engineering academics who successfully promote education in Sustainable Human Development. Journal of Cleaner Production 172:4239–4253. https://doi.org/10.1016/j.jclepro.2017.08.234

Leal Filho W (ed) (2018) Nachhaltigkeit in der Lehre. Theorie und Praxis der Nachhaltigkeit. [Sustainability in Teaching. Theory and Practice of Sustainability.] Springer Berlin Heidelberg, Berlin, Heidelberg. https://doi.org/10.1007/978-3-662-56386-1

Leal Filho W (2020) Viewpoint: accelerating the implementation of the SDGs. IJSHE 21:507–511. https://doi.org/10.1108/IJSHE-01-2020-0011

Leal Filho W, Kovaleva M, Fritzen Gomes B, Fudjumdjum H, Emblen-Perry K, Platje J, Tuladhar L, Vasconcelos CR, LeVasseur TJ, Minhas A, Farinha CS, Buil-Fabregá M, Novo-Corti I, Țîrcă D-M, da Cunha DA (2021) Sustainability practices at private universities: a state-of-the-art assessment. International Journal of Sustainable Development & World Ecology 28:402–416. https://doi.org/10.1080/13504509.2020.1848940

Liu J, Watabe Y, Goto T (2022) Integrating sustainability themes for enhancing interdisciplinarity: a case study of a comprehensive research university in Japan. Asia Pacific Educ. Rev. 23:695–710. https://doi.org/10.1007/s12564-022-09788-z

Lu S, Zhang H (2014) A comparative study of education for sustainable development in one British university and one Chinese university. International Journal of Sustainability in Higher Education 15:48–62. https://doi.org/10.1108/IJSHE-11-2012-0098

Mazon G, Pereira Ribeiro JM, Montenegro de Lima CR, Castro BCG, Guerra, José Baltazar Salgueirinho Osório de Andrade (2020) The promotion of sustainable development in higher education institutions: top-down bottom-up or neither? IJSHE 21:1429–1450. https://doi.org/10.1108/IJSHE-02-2020-0061

Moore J (2005) Seven recommendations for creating sustainability education at the university level. IJSHE 6:326–339. https://doi.org/10.1108/14676370510623829

Mulà I, Tilbury D, Ryan A, Mader M, Dlouhá J, Mader C, Benayas J, Dlouhý J, Alba D (2017) Catalysing Change in Higher Education for Sustainable Development. IJSHE 18:798–820. https://doi.org/10.1108/IJSHE-03-2017-0043

Nakai T (2024) Japan's triple sustainability challenge. Asia Pacific Business Review 30:499–513. https://doi.org/10.1080/13602381.2024.2320541

Nomura K, Abe O (2010) Higher education for sustainable development in Japan: policy and progress. IJSHE 11:120–129. https://doi.org/10.1108/14676371011031847

Pavlova M (2013) Teaching and learning for sustainable development: ESD research in technology education. Int J Technol Des Educ 23:733–748. https://doi.org/10.1007/s10798-012-9213-9

Ramakrishna S (2021) Incorporating sustainability into the university curriculum. Drying Technology 39:985–988. https://doi.org/10.1080/07373937.2021.1908806

Ryan A, Tilbury D, Blaze Corcoran P, Abe O, Nomura K (2010) Sustainability in higher education in the Asia-Pacific: developments, challenges, and prospects. IJSHE 11:106–119. https://doi.org/10.1108/14676371011031838

UN (2015a) SDG 4. https://sdgs.un.org/goals/goal4

UN (2015b) SDGs. https://sdgs.un.org/

UNESCO (2022) What you need to know about education for sustainable development. https://www.unesco.org/en/education/sustainable-development/need-know

Urata S, Kuroda K, Tonegawa Y (2023) Sustainable Development Disciplines for Humanity. Springer Nature Singapore, Singapore. https://doi.org/10.1007/978-981-19-4859-6

VERBI Software (2024) MAXQDA, Berlin, Germany. maxqda.com

Whittaker DH (2024) Building a New Economy: Japan's digital and green transformation. Oxford scholarship online Economics and Finance. Oxford University PressOxford, Oxford. https://doi.org/10.1093/oso/9780198893394.001.0001

Zhou L, Rudhumbu N, Shumba J, Olumide A (2020) Role of Higher Education Institutions in the Implementation of Sustainable Development Goals. In: Nhamo G, Mjimba V (eds) Sustainable Development Goals and Institutions of Higher Education. Springer International Publishing, Cham, pp 87–96. https://doi.org/10.1007/978-3-030-26157-3_7

Jana Walbaum is a researcher and Ph.D. student at the Technical University of Munich. Her work focuses on education for sustainable development, sustainability integration at higher education institutions, and comparative international education. She has conducted fieldwork in Japan, Germany, and the United States, and is affiliated with several interdisciplinary sustainability projects. She secured her own research funding with a scholarship from the Foundation of German Business.

Potential for University-Government Collaboration: A Case of the Environmental Engineering Undergraduate Program and Brazil's First Smart Sustainable City

Adriano Bressane

Abstract Municipal governments often face challenges in addressing sustainability issues due to a lack of specialized expertise, while universities seek opportunities to apply academic knowledge to practical problems, enriching both teaching and student training. This scenario highlights the potential for mutually advantageous partnerships. Despite an increasing focus on sustainability, the processes required to establish effective collaborations between universities and governments remain insufficiently defined. Existing models often fall short of leveraging the combined benefits of academic insights for governance and applied problem-solving for educational institutions. This study aims to identify barriers and opportunities for university-government collaboration, focusing on the partnership between the Environmental Engineering program at UNESP and the local government of São José dos Campos, the first smart sustainable Brazilian city. By examining these dynamics, the study proposes a replicable framework for fostering effective partnerships. Key findings emphasize the transformative potential of integrating academic expertise into municipal governance, demonstrating how real-world problem-solving not only enhances student learning but also addresses urban sustainability challenges. The study concludes with actionable insights into aligning academic goals with municipal needs, showcasing how these partnerships can drive sustainability initiatives and contribute to broader educational and governance excellence.

Keywords Sustainability · University-government collaboration · Applied learning · Intelligent nature-based solutions · Urban development · Environmental engineering

A. Bressane (✉)
Research at Institute of Science and Technology, São Paulo State University, São José dos Campos, Brazil
e-mail: adriano.bressane@unesp.br

W. Leal Filho et al. (eds.), *Sustainability in the Curriculum and Operations of Universities*, World Sustainability Series,
https://doi.org/10.1007/978-3-032-12678-8_19

1 Introduction

Addressing sustainability in urban governance has become a global imperative as cities navigate increasingly complex environmental, social, and economic challenges. Municipal governments often lack the technical expertise needed to effectively manage these multifaceted problems. Meanwhile, universities face growing expectations to align educational content with real-world needs, ensuring students acquire practical skills for applied problem-solving. These circumstances present significant opportunities for collaboration between academia and local governments, where both parties stand to benefit: municipalities gain access to academic expertise, and universities offer students valuable hands-on experiences tackling sustainability issues. However, such partnerships remain underutilized, with limited research into the mechanisms needed to establish and sustain them effectively (Borges et al. 2020).

Integrating academic knowledge into municipal governance has proven to be a powerful approach to addressing sustainability challenges. Universities serve as hubs for innovation and research, with significant potential to support sustainable urban development when partnered with local governments. Theoretical insights into these collaborations reveal their importance and potential impact, with research indicating that cooperative frameworks improve both urban sustainability outcomes and the quality of education through experiential learning (Tolentino-Zondervan and van Winden 2024; Song and Anderson 2024).

The "triple helix" model, which highlights the dynamic interplay between universities, industry, and government, offers a valuable framework for fostering innovation and development (Etzkowitz and Leydesdorff 2000; Leydesdorff 2020). Applied to sustainability, this model emphasizes the co-creation of solutions that are both research-driven and practical (Lita 2022). Nonetheless, significant barriers persist, including policy misalignments, resource limitations, and ineffective communication, which hinder the success of these collaborations (Alhassan 2024).

Engaging students in applied sustainability projects enriches their academic experiences while positively impacting communities. Incorporating sustainability into curricula through practical projects has been shown to enhance education for sustainable development and advance the Sustainable Development Goals (Annan-Diab and Molinari 2017). By working on real-world urban challenges, students develop essential skills and foster civic responsibility, equipping them to become effective agents of change in their future careers (Song and Anderson 2024; Borges et al. 2020).

This chapter aims to bridge this gap by examining the mechanisms and impacts of university-government collaborations, focusing specifically on the partnership between the Environmental Engineering program at UNESP and the local government of São José dos Campos, Brazil's first smart sustainable city. The study's primary objectives are to (i) identify barriers and opportunities for collaboration; (ii) demonstrate how real-world problem-solving in municipal contexts enriches student learning and professional training; and (iii) develop a replicable framework that integrates academic expertise into governance improvements and urban sustainability initiatives.

The findings of this study provide actionable insights for stakeholders, showcasing the transformative potential of aligning academic goals with municipal needs. By leveraging municipal challenges as educational tools, this chapter highlights how such collaborations can enhance urban governance practices while fostering academic excellence. Ultimately, the study concludes with a framework for fostering synergistic partnerships that address sustainability goals, offering a scalable model for broader adoption.

2 Actions by Local Government for Sustainability

São José dos Campos, recognized as Brazil's first smart and sustainable city, has implemented innovative initiatives aligned with the principles of Intelligent Nature-Based Solutions, integrating advanced technology with ecological solutions (Bressane et al. 2024). These pioneering efforts transcend traditional urban practices, promoting urban sustainability and resilience.

The Urban Afforestation Program employs QR code and GIS technology to map and manage the urban forest. This initiative enhances green coverage, improves air quality, and promotes biodiversity. QR codes on trees provide information about species, ecological contributions, and care instructions, fostering environmental awareness and simplifying green area maintenance.

The "Observa" program utilizes satellite technology to monitor land use changes, including deforestation and irregular constructions. This system provides up-to-date, detailed data, enabling quick and efficient responses to urban challenges such as deforestation and environmental management (Resende 2022).

The city has integrated electric vehicles into its public transportation fleet, reducing greenhouse gas emissions and enhancing urban air quality. These vehicles, equipped with artificial intelligence technologies, provide energy efficiency and improved passenger experiences (Rodrigues et al. 2023).

Among the city's renewable energy projects are a photovoltaic solar plant for powering municipal buildings and a biogas project converting landfill gas into electricity. These initiatives reduce reliance on non-renewable energy sources and advance energy sustainability (Garcia et al. 2023).

New regulations mandate sustainable construction practices, including the use of eco-friendly materials and energy-efficient designs. These policies minimize environmental impact and ensure that urban growth aligns with conservation objectives (Silva et al. 2022).

The "Solar São José" program incentivizes the adoption of solar energy systems in homes and businesses. Additionally, urban agriculture initiatives improve food security and expand green spaces, encouraging sustainable practices among residents (Bressane et al. 2024).

These initiatives highlight São José dos Campos' commitment to sustainability, serving as a model for other cities facing similar urban challenges. The integration of natural and technological solutions demonstrates how local governments can lead innovative efforts to foster a more resilient and sustainable urban environment.

3 Undergraduate Curriculum in Environmental Engineering

The Environmental Engineering undergraduate program at UNESP, São José dos Campos, is strategically designed to prepare professionals capable of addressing environmental challenges using an integrated, innovative, and transdisciplinary approach. Established in response to the region's industrial and ecological demands, the program aligns its objectives with sustainability and development goals, producing professionals equipped to contribute effectively to society and the environment.

The curriculum is rooted in a balance between technical innovation, holistic understanding, and interdisciplinary collaboration. It incorporates foundational engineering principles while embedding sustainability across its courses. Courses such as Environmental Management, Sanitation, and Geoprocessing provide technical expertise in managing resources sustainably. Students are trained in advanced tools such as Geographic Information Systems (GIS) and modeling software for environmental monitoring (UNESP 2024).

The curriculum bridges gaps between disciplines by including elements from social sciences, natural sciences, and engineering, fostering a transdisciplinary approach essential for solving complex environmental problems. A strong emphasis on ethics and critical thinking helps students evaluate the broader implications of their work. Courses in environmental legislation and sustainable development ensure that students understand the socio-political frameworks influencing ecological policies.

The program aims to produce professionals capable of balancing ecological preservation with economic and societal needs. According to Bressane (2024), the role of an environmental engineer is defined by three critical attributes:

- Transdisciplinary Nature: Professionals are trained to synthesize knowledge from diverse fields, such as ecology, chemistry, and economics, creating comprehensive solutions for environmental challenges. For example, they may integrate ecological principles with engineering techniques to design effective waste management systems.
- Holistic and Systemic Thinking: Engineers are taught to evaluate the interconnectedness of natural and human systems. This systems-thinking approach is pivotal in understanding how local actions can have global implications, enabling effective intervention strategies.
- Adaptability and Innovation: Given the rapid evolution of environmental challenges, the curriculum instills adaptability, encouraging graduates to embrace lifelong learning and technological advancements.

Environmental engineers play key roles in industries and agribusinesses, managing waste and pollution control systems while ensuring compliance with environmental standards. Their contributions are critical in balancing industrial productivity with ecological preservation.

Consultants provide specialized services in environmental auditing, licensing, and risk assessment. This sector demands professionals adept at crafting solutions that meet legal and ecological requirements, often working closely with clients to align practices with sustainability goals.

Engineers in this sector contribute to the development and enforcement of environmental policies. Their expertise is instrumental in executing government programs focused on resource conservation, biodiversity protection, and climate change mitigation.

Opportunities in academia involve teaching, conducting cutting-edge research, and shaping the next generation of environmental professionals. Research activities often focus on developing innovative technologies and methods to address pressing ecological issues.

Non-Governmental Organizations (NGOs) leverage the skills of environmental engineers to implement grassroots initiatives, advocate for policy reforms, and educate communities about sustainable practices. This sector provides a platform for impactful work in conservation and sustainability advocacy.

The rising demand for green technologies has paved the way for entrepreneurial opportunities. Engineers can establish startups focusing on renewable energy solutions, sustainable product design, or environmental consultancy services, leveraging their technical expertise to create innovative market-driven solutions.

The inclusion of practical experiences through extension activities and internships bridges the gap between theoretical learning and real-world application:

- Articulated Extension Projects (PAEX): Students engage in community projects, such as participatory diagnoses and the development of socio-environmental solutions.
- These projects enhance technical skills while fostering civic responsibility and collaborative problem-solving (UNESP 2024);
- Internship Opportunities: Partnerships with industries and governmental bodies provide hands-on experiences, preparing students for the complexities of the professional environment.

Environmental engineering, while offering significant career potential, comes with challenges. Professionals must navigate the complexities of balancing economic growth with environmental preservation and contend with evolving regulatory landscapes.

The demand for environmental engineers has steadily risen, reflecting the global shift towards sustainable practices. Brazil's industrial and urban sectors, in particular, seek professionals capable of implementing efficient and ecologically sound practices.

Environmental engineers are uniquely positioned to lead multidisciplinary teams, integrating insights from biologists, geologists, and social scientists to create comprehensive solutions.

The Environmental Engineering curriculum at UNESP equips students with the technical acumen, ethical grounding, and interdisciplinary outlook required to navigate and address contemporary environmental challenges. By fostering a robust academic foundation, coupled with practical experiences and community engagement, the program not only prepares students for a dynamic job market but also positions them as pivotal contributors to a sustainable future.

4 Potential for University-Government Collaboration

The collaboration between São José dos Campos' local government and UNESP's Environmental Engineering program highlights a synergistic model for addressing sustainability challenges. This partnership leverages the city's innovative sustainability initiatives and the university's robust academic foundation, creating a dynamic interplay that benefits both governance and education. This section explores how these entities can collaborate effectively to drive urban sustainability while enriching academic and professional development.

São José dos Campos has established itself as a leader in urban sustainability through its implementation of Intelligent Nature-Based Solutions (INbS), such as the Urban Afforestation Program and the "Solar São José" project (Bressane et al. 2024). These initiatives demonstrate the city's commitment to innovative and eco-friendly solutions. Meanwhile, the Environmental Engineering program at UNESP prepares professionals adept at developing and implementing such initiatives, blending technical, ethical, and interdisciplinary training (UNESP 2024).

This alignment creates opportunities for mutual benefits:

- The city gains access to cutting-edge research and a pipeline of skilled professionals trained to address complex environmental challenges.
- The university can integrate real-world municipal challenges into its curriculum, offering students hands-on learning experiences through internships and extension projects.

Through a collaborative framework, São José dos Campos and UNESP can amplify the impact of existing sustainability initiatives:

- Urban Afforestation and Environmental Monitoring: Students and faculty from UNESP can contribute to the enhancement of the Urban Afforestation Program by developing advanced GIS analyses and proposing biodiversity-friendly urban planning strategies. This partnership also provides valuable data and case studies for academic research.
- Renewable Energy Projects: Collaboration on renewable energy initiatives, such as the biogas and photovoltaic solar projects, can foster innovative approaches

to energy management. Engineering students could conduct feasibility studies and optimize project designs, creating a feedback loop of improvement and implementation (Garcia et al. 2023).
- Policy Development and Community Engagement: UNESP's expertise in sustainability legislation and socio-environmental frameworks can support the city in refining its regulations, such as those on sustainable construction practices. Extension projects like PAEX can engage communities in sustainability education, aligning public understanding with municipal goals (UNESP 2024; Silva et al. 2022).

Collaboration with São José dos Campos offers Environmental Engineering students invaluable opportunities for applied learning:

- Hands-On Problem Solving: Internships with the city provide exposure to practical environmental management scenarios, such as monitoring land use changes through the "Observa" satellite program. This experience equips students with technical and managerial skills critical for professional success (Rodrigues et al. 2023).
- Research Opportunities: The city's sustainability programs create a living laboratory for research. Students and faculty can investigate the effectiveness of INbS initiatives, propose enhancements, and publish findings, contributing to the global body of knowledge on urban sustainability.

The partnership between São José dos Campos and UNESP can serve as a model for other cities and universities seeking to integrate sustainability into governance and education:

- Knowledge Dissemination: By documenting and sharing their collaborative efforts, these institutions can inspire similar initiatives worldwide. For example, the structured incorporation of extension projects and municipal internships into academic programs offers a replicable framework for bridging academia and local governance.
- Cross-Sector Innovation: The collaboration demonstrates the potential of the "triple helix" model, where academia, government, and industry work synergistically to drive sustainable development. By aligning objectives and pooling resources, this model can catalyze transformative change (Leydesdorff 2020).

While promising, university-government collaborations must navigate challenges such as resource limitations, regulatory misalignments, and communication gaps. Mechanisms for fostering mutual trust and clear communication are essential for success. Regular joint workshops and shared spaces for analysis and planning can mitigate these issues, ensuring alignment of objectives and efficient project execution (Borges et al. 2020).

The partnership between São José dos Campos and UNESP exemplifies the transformative potential of university-government collaborations. By leveraging their complementary strengths, these institutions can address urban sustainability challenges while enriching the educational experience. Such collaborations not only advance local development but also provide scalable models for fostering sustainable practices globally.

5 Barriers and Opportunities

Collaboration between universities and local governments, as exemplified by São José dos Campos and UNESP, offers significant potential to address urban sustainability challenges. However, realizing the full potential of such partnerships requires navigating a complex landscape of barriers while leveraging emerging opportunities. This section explores these dynamics, identifying critical challenges and the enabling factors for successful collaboration.

The disconnect between academic research timelines and government operational cycles presents a significant challenge. While universities focus on long-term studies and data-driven solutions, municipalities often operate under short-term political mandates, limiting the ability to implement and sustain collaborative projects (Borges et al. 2020). Moreover, bureaucratic hurdles and insufficient alignment of institutional regulations can delay project approvals and hinder the integration of academic expertise into municipal policies.

Local governments and universities often face budgetary constraints that restrict their ability to fund collaborative initiatives. Municipalities may lack the financial resources to adopt cutting-edge technologies proposed by academic research, while universities may struggle with limited funding for extension projects and internships (Rodrigues et al. 2023).

Effective collaboration requires clear and consistent communication between stakeholders. Differences in language, priorities, and expectations between academia and governance can lead to misunderstandings and misaligned goals. For instance, while local governments may prioritize immediate solutions, universities may focus on research with broader, long-term implications (Silva et al. 2022).

The lack of active engagement from key stakeholders, including community members, industries, and NGOs, can undermine the success of collaborative projects. Without comprehensive participation, initiatives may face resistance or fail to address community-specific needs (Bressane et al. 2024).

On the other hand, both universities and local governments share a vested interest in promoting sustainability, providing a strong foundation for collaboration. São José dos Campos' initiatives, such as renewable energy projects and the Urban Afforestation Program, align closely with UNESP's focus on environmental management and sustainable development (UNESP 2024; Garcia et al. 2023). This shared vision fosters a common purpose, ensuring that collaborative efforts are mutually beneficial.

Local governments offer real-world challenges that can enrich academic curricula and provide students with valuable hands-on experiences. Initiatives such as the "Observa" satellite monitoring program allow students to apply advanced GIS and remote sensing techniques, bridging the gap between theory and practice (Resende

2022). These opportunities enhance educational outcomes while simultaneously contributing to municipal objectives.

Universities serve as reservoirs of expertise and innovation, offering local governments access to cutting-edge research and technical solutions. Faculty and students can contribute to designing, implementing, and evaluating urban sustainability projects, such as solar energy initiatives and sustainable construction practices (Bressane et al. 2024; Silva et al. 2022). In return, municipalities provide a platform for testing and scaling academic innovations.

Collaboration opens doors to external funding opportunities, such as grants from international organizations, NGOs, and private sector partnerships. These resources can help overcome financial limitations and support large-scale projects, such as biogas energy systems and community-driven urban agriculture programs (Garcia et al. 2023).

Collaborative efforts that actively involve community stakeholders can amplify social impact and ensure the sustainability of initiatives. Extension projects, like UNESP's PAEX, serve as models for integrating community needs with municipal and academic objectives, fostering trust and shared ownership of projects (UNESP 2024).

5.1 Strategic Pathways to Foster Collaboration

Creating formal agreements that outline roles, responsibilities, and expectations can help align institutional priorities. Regular joint meetings and the establishment of advisory boards comprising representatives from academia, government, and the community can improve coordination and accountability (Borges et al. 2020).

Jointly developed action plans can ensure that academic research aligns with municipal needs. For instance, co-designed studies on urban resilience can inform policy changes while addressing real-time urban challenges (Silva et al. 2022).

Training programs and workshops for municipal staff and academic participants can improve understanding and foster collaboration. Capacity-building initiatives can focus on areas like data interpretation, sustainable urban planning, and stakeholder engagement (Rodrigues et al. 2023).

Transparent communication channels and inclusive decision-making processes can build trust among stakeholders. Ensuring that diverse voices, including marginalized communities, are heard can lead to more equitable and impactful outcomes (Bressane et al. 2024).

While barriers such as regulatory misalignments and resource limitations present challenges, opportunities for collaboration between universities and local governments are abundant. By addressing these barriers strategically and leveraging shared goals, practical learning opportunities, and external resources, institutions like UNESP and São José dos Campos can create robust frameworks for sustainable urban development. These partnerships not only advance local sustainability initiatives but also set an example for replicable and scalable collaborations worldwide.

6 Enriching Student Learning and Professional Training

Integrating real-world problem-solving into the academic framework of Environmental Engineering programs provides transformative educational and professional development opportunities. By engaging students in municipal sustainability projects, universities not only enhance academic learning but also cultivate the skills and competencies necessary for impactful professional careers. This section explores how collaborations between UNESP and São José dos Campos serve as a model for enriching student learning and fostering professional growth.

6.1 Bridging Theory and Practice

The collaboration between São José dos Campos and UNESP allows students to apply theoretical knowledge in practical scenarios, reinforcing their learning and developing critical problem-solving skills:

- Urban Sustainability Initiatives: Students participate in programs like the "Observa" satellite monitoring system and Urban Afforestation Program. These initiatives provide a hands-on understanding of geoprocessing and ecological management, bridging the gap between classroom instruction and field application (Resende 2022; UNESP 2024).
- Renewable Energy Projects: Engagement in renewable energy initiatives, such as biogas and photovoltaic systems, enables students to explore energy systems' technical and economic dimensions, equipping them with interdisciplinary expertise essential for tackling complex sustainability challenges (Garcia et al. 2023).

6.2 Developing Technical and Professional Skills

Collaborative projects in municipal contexts contribute significantly to the development of technical and professional skills:

- Technical Proficiency: Students gain proficiency in using advanced tools like Geographic Information Systems (GIS), environmental modeling software, and energy management systems, fostering their ability to address diverse environmental challenges (Bressane et al. 2024).
- Project Management: Participation in municipal programs enhances students' capabilities in project planning, execution, and evaluation. This hands-on experience is invaluable in preparing students for professional environments where multidisciplinary collaboration is crucial (UNESP 2024).

6.3 Enhancing Civic Responsibility and Ethical Awareness

Engaging with real-world sustainability issues promotes civic responsibility and ethical decision-making among students:

- Community-Centered Projects: Extension projects, such as participatory diagnoses in urban and rural settings, emphasize collaboration with community stakeholders. This exposure instills a sense of accountability and fosters an understanding of how engineering solutions impact societal well-being (Silva et al. 2022).
- Ethical Practices: Addressing challenges such as land use conflicts and resource management in partnership with municipal bodies provides students with a framework for ethical problem-solving, critical in aligning engineering solutions with societal and environmental goals (Rodrigues et al. 2023).

6.4 Fostering Interdisciplinary Collaboration

Municipal contexts provide a fertile ground for interdisciplinary collaboration, a cornerstone of Environmental Engineering education:

- Integration Across Disciplines: Projects like sustainable construction initiatives involve inputs from engineering, architecture, and urban planning, giving students firsthand experience in integrating diverse perspectives to create holistic solutions (Bressane et al. 2024).
- Collaboration with Stakeholders: Working alongside policymakers, NGOs, and community representatives enhances students' ability to communicate effectively across sectors, a skill critical for professional success (Borges et al. 2020).

6.5 Preparing Students for a Dynamic Job Market

Real-world problem-solving prepares students for the complexities of the job market by aligning academic training with professional demands:

- Internships and Practical Training: Internship programs with São José dos Campos offer exposure to ongoing municipal projects, providing insights into real-time challenges and equipping students with the skills to navigate regulatory and operational constraints (UNESP 2024).
- Career Readiness: The combination of technical, managerial, and ethical training ensures that students are well-prepared to assume roles in industries, government agencies, and consulting firms, addressing critical gaps in the sustainability workforce (Bressane et al. 2024).

6.6 Amplifying Research and Innovation

Collaborative municipal projects serve as platforms for advancing research and innovation:

- Living Laboratory: São José dos Campos' sustainability initiatives, such as urban agriculture and energy efficiency programs, act as living laboratories where students and faculty can test and refine innovative solutions (Garcia et al. 2023).
- Knowledge Creation: Research conducted in these contexts contributes to the global discourse on urban sustainability, with students co-authoring publications and presenting findings at academic and professional forums (Silva et al. 2022).

By embedding real-world problem-solving into the Environmental Engineering curriculum, UNESP and São José dos Campos demonstrate the transformative potential of university-government collaborations. These initiatives not only enrich academic learning but also equip students with the technical expertise, ethical grounding, and interdisciplinary skills necessary to excel in professional settings. This model of experiential education serves as a replicable framework for universities and municipalities worldwide, advancing both educational excellence and urban sustainability.

7 Integrating Academic Expertise into Governance for Urban Sustainability

The collaboration between São José dos Campos and UNESP exemplifies the transformative potential of university-government partnerships in advancing urban sustainability. By leveraging academic expertise to inform governance practices, these partnerships can address pressing environmental challenges while fostering innovation and education. Building on the lessons learned from this collaboration, this section outlines a replicable framework for establishing and sustaining effective partnerships that integrate academic resources into urban governance for sustainability.

7.1 Foundation: Aligning Goals and Vision

Effective partnerships require a shared vision and clearly aligned goals:

- Defining Common Objectives: Establish a joint commitment to sustainability that integrates governance needs with academic expertise. São José dos Campos' focus on INbS aligns well with UNESP's curriculum emphasis on environmental management and innovation, demonstrating the importance of shared priorities (Bressane et al. 2024).

- Early Stakeholder Involvement: Involve municipal agencies, academic institutions, community representatives, and private sector stakeholders in the planning process to ensure a comprehensive understanding of objectives and challenges (Borges et al. 2020).

7.2 *Collaborative Governance Structures*

Structured governance mechanisms ensure accountability and effective collaboration:

- Advisory Committees: Create advisory committees composed of representatives from academia, government, and key stakeholders. These committees oversee project design, implementation, and evaluation, ensuring transparency and alignment (Leydesdorff 2020).
- Memorandums of Understanding (MOUs): Formal agreements between universities and governments can delineate roles, responsibilities, and resource commitments, streamlining collaboration and reducing ambiguity (Silva et al. 2022).

7.3 *Joint Research and Development Initiatives*

Integrating academic research into municipal projects promotes innovation and knowledge transfer:

- Living Labs: Establish urban sustainability initiatives as "living laboratories" where students and researchers can test and refine solutions in real-world settings. For example, São José dos Campos' renewable energy and afforestation programs provide ideal platforms for experiential learning and innovation (Garcia et al. 2023).
- Data Sharing Agreements: Facilitate access to municipal data for academic research, enabling evidence-based decision-making and the development of practical solutions (Resende 2022).

7.4 *Capacity Building and Training Programs*

Building institutional capacity enhances the impact of partnerships:

- Workshops and Training: Offer regular training for municipal staff and students on topics such as GIS, energy management, and sustainability policy. These programs foster skill development and knowledge exchange (Rodrigues et al. 2023).
- Professional Development Opportunities: Encourage university faculty and municipal employees to participate in joint initiatives, promoting continuous learning and collaboration (UNESP 2024).

7.5 Community Engagement and Outreach

Community involvement ensures that initiatives are socially inclusive and address local needs:

- Participatory Planning: Engage residents and community organizations in the co-creation of sustainability initiatives. For instance, UNESP's extension projects emphasize participatory diagnoses to align solutions with community priorities (Silva et al. 2022).
- Public Awareness Campaigns: Use collaborative platforms to educate the public on sustainability goals and progress, fostering trust and active participation in governance efforts (Bressane et al. 2024).

7.6 Monitoring, Evaluation, and Scalability

Robust monitoring and evaluation frameworks are critical for assessing the success of initiatives and informing future strategies:

- Performance Metrics: Develop clear metrics to evaluate the impact of sustainability projects on environmental, social, and economic outcomes. Metrics can include reductions in greenhouse gas emissions, increases in green coverage, or improvements in community satisfaction (Borges et al. 2020).
- Documentation and Dissemination: Publish findings and lessons learned from collaborative projects in academic and professional forums to inform similar initiatives elsewhere. São José dos Campos' success stories offer valuable insights for global replication (Bressane et al. 2024).
- Scaling Successful Models: Identify components of initiatives that can be scaled to other municipalities or adapted to different contexts, such as the "Solar São José" program or the use of INbS for urban planning (Garcia et al. 2023).

This replicable framework provides a roadmap for fostering effective partnerships that integrate academic expertise into governance improvements. By aligning goals, building collaborative structures, and prioritizing capacity building and community engagement, municipalities and universities can address sustainability challenges more effectively. The experience of São José dos Campos and UNESP underscores the potential for such partnerships to transform urban governance, foster innovation, and advance educational excellence. This model offers valuable insights for global replication, empowering cities and universities to collaborate in achieving sustainable development goals.

8 Limitations and Future Prospects

While this chapter provides a detailed exploration of the potential for university-government collaboration to advance sustainability goals, several limitations constrain the scope and generalizability of its findings. Addressing these limitations offers directions for future research and practical advancements.

8.1 *Limitations*

The study centers on the partnership between UNESP's Environmental Engineering program and the local government of São José dos Campos. While this collaboration offers valuable insights, its findings may not fully apply to other cities or universities with different institutional structures, regulatory environments, or resource capacities.

The chapter identifies regulatory misalignments and political cycles as key barriers to collaboration. However, the analysis does not delve deeply into how specific legislative frameworks or shifting political priorities influence long-term sustainability partnerships. Limited financial and human resources in both academia and local governments are highlighted as challenges, but the study does not comprehensively address strategies to secure sustainable funding or mobilize additional resources for scaling these collaborations. Although the chapter emphasizes the potential benefits of university-government collaborations, it does not provide an in-depth evaluation of long-term outcomes, such as measurable improvements in governance practices or educational quality.

8.2 *Future Prospects*

Future research could explore similar partnerships in diverse socio-economic and cultural contexts to identify common barriers and replicable best practices. Comparative studies may uncover universal strategies that transcend local conditions.

Analyzing the impact of national and regional policies on the success of university-government collaborations can provide actionable recommendations for harmonizing regulatory frameworks and fostering supportive governance structures.

Research into public–private partnerships, international grants, or community-driven funding initiatives could uncover sustainable approaches to financing collaborative projects. Future studies should consider longitudinal assessments to measure the sustained impact of these collaborations on urban governance, student learning outcomes, and sustainability indicators over extended periods. Exploring the role of emerging technologies, such as artificial intelligence and advanced GIS tools, could enhance the operational efficiency and impact of university-government partnerships, particularly in areas like urban planning and resource management. Future work could focus on frameworks for engaging diverse stakeholders, including local communities, NGOs, and the private sector, to ensure that sustainability initiatives are inclusive and address localized needs.

8.3 Constraints of the Chapter

The chapter presents several constraints that define the scope and depth of its analysis. Firstly, the study primarily focuses on the partnership between UNESP and São José dos Campos, which limits its applicability to other regions or contexts. This narrow scope may restrict the generalizability of its findings. Additionally, some aspects of the collaboration rely on secondary data and anecdotal evidence, which can constrain the robustness and reliability of the conclusions drawn. Finally, the analysis reflects a temporal limitation, capturing a snapshot of ongoing initiatives rather than the evolution or long-term outcomes of the partnership. As such, the chapter provides insights within a specific timeframe, which may not fully account for future developments or broader implications.

9 Conclusion

This study highlights the transformative potential of university-government collaborations in advancing urban sustainability. Focusing on the partnership between São José dos Campos and UNESP's Environmental Engineering program, this research contextualizes the urgent need for such collaborations in addressing multifaceted urban sustainability challenges. By leveraging São José dos Campos' innovative Intelligent Nature-Based Solutions (INbS) initiatives and UNESP's robust academic curriculum, the study demonstrates how these partnerships can bridge gaps in expertise and resources while delivering tangible environmental and educational benefits.

The methodological approach centered on analyzing the dynamics of this collaboration, emphasizing both practical municipal applications and the academic framework at UNESP. Through a qualitative exploration of governance initiatives, curricular integration, and stakeholder engagement, the study identifies barriers and opportunities inherent in these interactions. Key findings include the critical role of aligning shared sustainability goals, utilizing real-world challenges for experiential learning, and fostering interdisciplinary collaboration. São José dos Campos' INbS initiatives, such as the Urban Afforestation Program and renewable energy projects, serve as living laboratories that provide students with hands-on experiences, enriching their technical skills and professional training. Simultaneously, the university contributes to governance improvements through research, innovation, and the development of actionable sustainability frameworks.

The implications of these findings are far-reaching. For municipal governments, partnerships with universities provide access to cutting-edge research, innovative solutions, and a pipeline of skilled professionals to address pressing sustainability challenges. For universities, such collaborations offer opportunities to align education with real-world applications, preparing students to become effective change agents in their future careers. These partnerships also foster a sense of civic responsibility and ethical decision-making among students.

While challenges such as regulatory misalignments, resource constraints, and communication gaps remain, the framework outlined in this study provides a replicable model for overcoming these barriers. By establishing clear governance structures, investing in capacity building, and promoting inclusive community engagement, universities and local governments can create mutually beneficial partnerships that drive sustainable development.

Nevertheless, the study acknowledges several limitations, including its focus on a single case study, the contextual specificity of São José dos Campos, and the lack of longitudinal data to assess long-term outcomes. Future research should expand on these findings by exploring diverse socio-economic and cultural contexts, analyzing the impact of national and regional policies, and developing sustainable funding models for collaborative initiatives. Additionally, longitudinal studies could offer deeper insights into the sustained impact of university-government collaborations on urban governance and educational outcomes.

This model not only advances local and regional sustainability goals but also offers valuable lessons for global replication. As the pressures of environmental, social, and economic complexities intensify worldwide, integrating academic expertise into governance emerges as a crucial strategy. Future work should explore the scalability of this model, incorporating emerging technologies and inclusive stakeholder frameworks to further enhance its impact. By fostering such partnerships, municipalities and universities can jointly contribute to a sustainable future, ensuring that education and governance evolve to meet the challenges of the twenty-first century.

References

Alhassan, A. Y. (2024). Uncovering the social fabric of urban governance: A social network analysis of actors in a road expansion project in Kristiansand (Doctoral dissertation). University of Agder.

Annan-Diab, F., & Molinari, C. (2017). Interdisciplinarity: Practical approach to advancing education for sustainability and the sustainable development goals. International Journal of Management Education, 15(2), 73–83.

Borges, G. A., de Abreu, S. R., de Souza, I. L., & Monteiro, L. S. (2020). Parceria entre governo local e universidades: Formação do Observatório da Cidade de Macaé no contexto de rede colaborativa interinstitucional e intersetorial [Partnership between local government and universities: Formation of the Macaé City Observatory in the context of an interinstitutional and intersectoral collaborative network]. Revista Vértices, 22(3), 412–428.

Bressane, A., Loureiro, A. I. S., Medeiros, L. C. C., Moruzzi, R., Formiga, J. K. S., Negri, R. G., & Saraiva, A. C. V. (2024). Intelligent nature-based solutions in the 1st smart sustainable Brazilian city: Insights and lessons learned. Nature-Based Solutions, 6, 100161. https://doi.org/10.1016/j.nbsj.2024.100161

Bressane, A. (2024). Introdução à Engenharia Ambiental: mercado e perfil profissional [Introduction to Environmental Engineering: Market and Professional Profile] (3rd ed.). Washington: KDP.

Etzkowitz, H., & Leydesdorff, L. (2000). The dynamics of innovation: From national systems and "Mode 2" to a triple helix of university–industry–government relations. Research Policy, 29(2), 109–123.

Garcia, L. B., Fiore, F. A., & Carvalho, F. L. (2023). Factors associated with the use of solar energy in urban households: Case study of São José dos Campos. International Journal of Energy Economics and Policy, 13(4), 522. https://doi.org/10.32479/ijeep.13474

Leydesdorff, L. (2020). Triple helix of university-industry-government relations. In E. G. Carayannis (Ed.), Encyclopedia of Creativity, Invention, Innovation and Entrepreneurship (pp. 2356–2364). Cham: Springer.

Lita, A. (2022). Reducing food safety hazards in China: A food system approach (Doctoral dissertation). Wageningen University and Research.

Resende, M. (2022). Satellite monitoring program. São José dos Campos City Hall. https://www.sjc.sp.gov.br/media/223966/comam-observa-160223_compressed.pdf

Rodrigues, G. S., Reis, J. G. M., Orynycz, O., Tucki, K., Machado, S. T., & Raymundo, H. (2023). A study on the viability of adopting battery electric vehicles in bus rapid transit in Brazil using the AHP method. Energies, 16(13), 4858. https://doi.org/10.3390/en16134858

Silva, C. C. P., Leite, B. A., Alves, A. C., Janson, I. J., Silva, L. C., & Matsukura, R. R. (2022). Manual of sustainable strategies. São José dos Campos City Hall. https://www.sjc.sp.gov.br/media/183656/manual-completo-publicacao.pdf

Song, L., & Anderson, T. F. (2024). Participatory planning and design research for the ARTery. Groundworks Journal. https://groundworks.io/journal/projects/115

Tolentino-Zondervan, F., & van Winden, W. (2024). Towards circular management of construction and demolition waste: Learning from innovative practices, barriers, and enablers for designing urban resource centers in European cities. In Sustainability in Practice: DIY Repair, Reuse, and Innovation.

UNESP (2024). Projeto Político Pedagógico – Engenharia Ambiental [Political-Pedagogical Project – Environmental Engineering]. Instituto de Ciência e Tecnologia, UNESP, São José dos Campos.

Adriano Bressane is a Professor of Environmental Engineering at the São José dos Campos campus of São Paulo State University (UNESP). He holds a Ph.D. in Environmental Sciences from UNESP Sorocaba and has completed postdoctoral studies in Urban Engineering at PUC-Campinas and Urban Geography at the University of Coimbra. His research focuses on nature-based solutions, fuzzy machine learning applications in environmental engineering, and actively participates in collaborative projects advancing sustainability in urban environments.

Practical Experiences on Education for Sustainable Development in Antonio Meneghetti Faculdade, a Higher Education Institution in the Rural Area of Brazil

Any Regina Rothmann, Clarissa Mazon Miranda, and Patricia Wazlawick

Abstract Although the implementation of sustainable development is a global movement in higher education institutions, it is rare to see the involvement of colleges located in rural areas. Nevertheless, the Antonio Meneghetti Faculdade is a higher education institution founded in 2007 in the rural area of Rio Grande do Sul, the southernmost state in Brazil, and has always been aligned with international movements regarding the teaching of sustainable development in its higher education courses. Among the five educational pillars of this institution, there are the pillars of internationalization and work, fostering students' practices in the work market. In alignment with these two pillars, this college constitutes an interesting case study. For example, it has taken part in the UNESCO Decade of Education for Sustainable Development; since 2011, it has been part of the PRME movement of the UN and of the UN Academic Impact movement; in 2018, the Antonio Meneghetti Foundation, the institution that supports and keeps the college, acquired special consultative status with the ECOSOC of the UN. This background has enabled students from this institution to be involved with the main discussions regarding sustainability promoted around the world. Because of this, the college and the Antonio Meneghetti Foundation promote 28 projects focused on the Sustainable Development Goals, which have students' leadership and benefit more than 4 thousand people from poor economic backgrounds who live in cities nearby the campus, free of charge. Also, a project from 2017 will be detailed here, in which students created SDG projects in the classroom as part of a special project. This paper may be useful to anyone interested in

A. R. Rothmann
Strategic Curator of the Antonio Meneghetti Foundation, Lugano, Switzerland

C. M. Miranda (✉)
Antonio Meneghetti Faculdade, Member of the International Affairs Office of Antonio Meneghetti Foundation, Lugano, Switzerland
e-mail: miranda.clarissa@gmail.com

P. Wazlawick
Antonio Meneghetti Faculdade and Coordinator of the Ontopsychology Department, Councilor of Antonio Meneghetti Foundation, Lugano, Switzerland

W. Leal Filho et al. (eds.), *Sustainability in the Curriculum and Operations of Universities*, World Sustainability Series,
https://doi.org/10.1007/978-3-032-12678-8_20

the different levels of emphasis given to sustainable development at higher education institutions in rural areas.

1 Introduction

The Antonio Meneghetti Faculdade (AMF) is a private higher education institution, founded in 2007 in the city of Restinga Sêca, in the center region of the southernmost state in Brazil, Rio Grande do Sul, being located in a rural area, commonly known as the 4th Colony of Italian Immigration of Rio Grande do Sul. It currently offers eight undergraduate programs and five graduate programs. Most of its students are young adults from 17 to 22 years old, who come from rural families in the region.

The college bases its educational activities on five pillars: study, science, high morality, internationality and work. These pillars foster the development of its higher education pedagogy and philosophy, which act towards sustainable development. This proposal is aligned with the statute of the Antonio Meneghetti Foundation, a foundation that maintains the college. These two entities follow the principles of Ontopsychology, as it established by Professor Academician Antonio Meneghetti, the founder of this science, and also support the initiatives of the United Nations (UN) towards sustainable development.

Aligned to that background, since its creation, the AMF has worked to be involved in the movements promoted by the UN and its agencies. In 2011, it became a member of the UN Academic Impact and Principles for Responsible Management Education (PRME). In 2012, a mission from the Antonio Meneghetti Faculdade and the Antonio Meneghetti Foundation took part in Rio + 20, the UN Conference on Sustainable Development (UNCSD). Until 2014, it was involved in and it followed the UNESCO Decade of Education for Sustainable Development principles. Since 2012, the AMF took part in the Youth Year of the UN and yearly celebrates the International Youth Day, always celebrated on August 12th. In 2018, the Antonio Meneghetti Foundation became an entity with special consultative status with the ECOSOC of the UN. Since then, every two years, the foundation, in partnership with the college, has promoted an international symposium about education in one of the UN offices or at UNESCO Paris. That was done in 2017 (UNESCO); in 2019 (UN Office Geneva); in 2022 (Peace Palace, The Hague); in 2025 (UNESCO). Also, courses were promoted in partnership with UNITAR for college students and the foundation took part in missions at the High-Level Political Forum and High-Level Segment of the ECOSOC yearly since 2019, in New York.

All of these actions are also ways to share with other higher education institutions and foundations the work that has been promoted locally by its 28 projects in our campus, in alignment with the SDGs. Moreover, these are occasions to learn from different entities. Considering the pillars that guide the teaching philosophy at the AMF, one of them, the internationalization pillar, is highly present in these initiatives. The present paper aims to discuss more on the specificity of some examples of how SDGs and sustainable development are worked in the classrooms of this college and how the 28 projects from the Antonio Meneghetti Foundation have impacted students' daily lives.

2 Methodology

This article uses the approach of case study to present how the SDGs can be integrated in a HEI in a rural area of Brazil. Data for the case study cover the connection between projects and the SDGs, report on observation from researchers/teams involved, and feedback from participating students.

From 2017 to 2019, through the creativity and determination of 76 young university students from undergraduate courses in Law and Administration, the AMF contributed to local improvements based on the UN SDGs. These projects were based on the questions posed to students in class: "What can you do?" (Wazlawick 2011). The institution used qualitative methodology, which involves conducting action and intervention research (Gil 2017) in locations identified as having problematic situations that require assistance for comprehensive development, fulfilling one or more SDGs.

It is noteworthy that the sources for recording and disseminating all the activities and interventions developed by the groups were: photographs, videos, recorded interviews, publication of articles in regional newspapers, the AMF website, and social media. The institutional proposal was to create and implement action projects and practices in the Region every academic semester, with new students, covering a greater number of entities and individuals, to gradually implement major changes in the educational, social, environmental, and economic aspects of our region, and to implement the best projects so as to leave no one behind.

This umbrella project operated for two and a half years, beginning its activities in February 2017 with the creation and execution of 22 applied projects, with actions and results in different areas of the social context of the Region where these young people were born, lived, worked, and studied. This text presents the proposal, the implementation methodology, a brief description of each project, as well as its results and possibilities for replication in other contexts.

3 Results and Discussion

In order to continue with all the work carried out with the implementation of the Millennium Development Goals (the 8 MDGs), which were in force during the period 2000–2014, in 2015, the UN approved a new international development agenda, the 2030 Agenda. This agenda is composed of 17 Sustainable Development Goals (SDGs) that involve 169 objectives to be achieved worldwide by 2030, with the participation of all signatory countries (Gadotti 2008; Malheiros et al. 2008; Buss Et Al. 2012; Nieto 2017).

According to the UN Development Program (UNDP) in Brazil, the SDGs, also known as the Global Goals, "are a universal call to action against poverty, to protect the planet, and to ensure that all people enjoy peace and prosperity" (UNDP 2018, n.d.; UNDP and IPEA 2019; Cabral and Campanatti 2019). The 17 goals include themes such as global climate change, economic inequality, innovation, sustainable consumption, peace and justice, among other priorities. The SDGs are interconnected, in the sense that the success of one SDG implies addressing issues that are associated with other goals.

It should be noted that all UNDP actions are aligned with the SDGs and its objective is to "leave no one behind" in the process of sustainable development. The UNDP also explains that the SDGs operate in a spirit of partnership and pragmatism so that people make the right choices to sustainably improve the quality of life for current and future generations. The SDGs offer clear guidelines and targets for all countries to adopt, according to their priorities and environmental challenges across the globe. Therefore, the SDGs are an extremely inclusive agenda (UNDP 2018).

Regarding the situation prior to the implementation of the umbrella project and its practices, it was conducted a diagnostic study at AMF, in which we had many students interested in carrying out actions to contribute to improving the social context in the municipalities of the Fourth Colony of Italian Immigration Region in Rio Grande do Sul (where the students came from); however, they did not know how to start and develop these actions.

This project was taken seriously due to two main aspects: (a) a large amount of young human capital interested in helping the context presented above, which is in academic, personal, and university professional training, and who, due to their interest, must be assisted in this process to train the human person in/for the social function (Meneghetti 2014, 2019), for the context in which they live and work: this, in our opinion, is one of the main roles of the university in contemporary times; (b) numerous entities, groups, communities, and a large percentage of the population of the municipalities from the Fourth Colony who found themselves in situations of vulnerability in all areas and who could and should be helped by the population, in a responsible manner, to learn a new life option.

3.1 The Antonio Meneghetti Faculdade: An Experience with SDG Projects in the Classroom

Therefore, there was a concrete opportunity to begin to change these situations and implement projects that would help achieve and attain the SDGs. Regarding the general objective of the overall project, it can be presented as follows: to carry out projects through the actions of young university students to contribute to the achievement of the SDGs internalized in the Recanto Maestro District and in the Fourth Colony of Italian Immigration Region in RS.

Based on this general objective, the specific objectives were summarized as follows:

(1) To find a problematic situation in their community/municipality, where an SDG project could be implemented to contribute to changes and improvements in the real problems identified;
(2) To develop the action plan to contribute to the SDGs in their specific context;
(3) To implement the action project, carry out the interventions, achieve concrete results to improve the identified situation, and document the process.

The target audience included various social entities, communities, and the general population of the Recanto Maestro district and the municipalities of Restinga Sêca, Agudo, Dona Francisca, São João do Polêsine, Faxinal do Soturno, and Nova Palma. In the sequence, we will present a brief description of the practical steps carried out in the implemented projects.

Based on the principles of action and intervention research (Gil 2017), the steps/actions followed to implement the practice, in chronological order for each academic semester, were:

(1) Preparation of the general project by the professors of the following courses: "Corporate Identity" (5th semester), "Business Training III" (3rd semester), from the AMF Law Course; and "Business Training and Leadership VIII" (8th semester), from the Business Administration Course at the same institution. All projects considered the principles of the UN PRME and focused on social responsibility. All projects were first presented to the College's Scientific Steering Committee and approved for implementation with students (from February 2017, February 2018, and March 2019);
(2) Presentation of the project to the students in the classroom and verification of agreement on group division (first half of March 2017, March 2018, and last half of March 2019);
(3) Elaboration of the groups' initial ideas, preparation of the preliminary project, analysis of the context and problematic situation for each group, and presentation in class (second half of March 2017, March 2018, and first half of April 2019);
(4) Preparation of the groups' introductory letter to the beneficiary entities (last week of March 2017, March 2018, and second half of April 2019);
(5) Initiation of group activities/interventions in the community (early April 2017, April 2018, and April 2019);
(6) Regular follow-up meetings with the teacher and each group (biweekly and/or monthly during the first half of 2017, the first half of 2018, and the first half of 2019);
(7) Implementation of group actions/interventions: April to June 2017 and the same period in 2018 and 2019;
(8) Verification of results and final presentation of the project to the teacher and class (for each group): first week of July 2017 and the same period in 2018 and 2019;

(9) Selection of projects that continued activities in the second half of 2017, 2018, and 2019 and their progress (second half of each of these years);

(10) In September 2017, the best projects, according to their evaluation, were presented through oral presentations and posters at the event held at UNESCO Headquarters in Paris, France, on the occasion of the International Symposium "Contemporary Pedagogy: Responsibility and Education of Youth for the Society of the Future" which celebrated the AMF's tenth anniversary at that event;

(11) In February 2018 and March 2019, the entire system and methodology for describing the practice stages continued and was carried out with the new classes of Law and Administration students taking those courses. The same steps were carried out in the new year, with new students and new projects, which would continue in the second half of 2018 and 2019. It is worth mentioning that this project went on for a year and a half, with 22 projects being carried out by young students from municipalities in the region of the Fourth Colony of Italian Immigration in Rio Grande do Sul.

When discussing and working with SDG projects, it is important to verify the feasibility and replicability of practices. In this sense, we emphasize that the practices adopted in the larger project, which were carried out in the classroom for the creation of student group projects, can be entirely replicated at other universities and also in local social contexts. Because these practices are simple and easy to replicate: you need people, an idea, a suitable context, power of will, creativity, a workforce, and perseverance in execution, creating the project, defining the SDGs to be participated in, defining the general objective, the specific objectives, the target public, the action plan and activities, the ways to measure the indicators (quantitative and/or qualitative), and work to achieve the objectives and results.

It is important to create a cover letter, present it to the public and/or private organization or entity to participate, and obtain their consent. The instructor must continuously monitor students' work as a means of supervising it. For example, the projects completed in 2017 were replicated, using the same methodology, in 2018 and 2019; and several schools in the Region received the projects and multiplied them. These are some measures for maintaining the practice: (a) continuation of the proposal in new semesters of the Course with new students; (b) financial support from the Antonio Meneghetti Foundation for the best expansion projects and continued implementation in other semesters.

Table 1 shows the projects implemented by the groups during the 2017–2019 period, after their creation, development, presentation, discussion, and approval in the classroom through dialogue between students/classes, groups, and professors.

In terms of results, the transformative action of responsible work by young university students produced changes, growth, improvements, and 24 new partnerships between public and private institutions. Everyone benefits: students, entities and locations in the community/municipalities, and the entire population receiving and involved in the services.

Table 1 SDG projects implemented in Recanto maestro and municipalities in the region

Name of the project	SDGs contemplated
1. Planting life	13, 14, 15, 17
2. City solare	7
3. *Quilombolas* in action: humanism and *quilombola* society	10, 16, 17
4. Sustainability and environmental education	3, 8, 11, 13
5. Quality education: study to be: a communication project	4
6. Self-composition in focus: a new paradigm in the resolution of conflicts	17
7. Interdisciplinary education as form of interaction	4
8. Sustainability in prisons: production of organic food	2, 3, 15, 17
9. Correct disposal: society awareness	11, 12
10. Feed a friend	15
11. PET—transformative education project: recycling is transforming	4, 11
12. Literature Corner	4
13. Painting and embroidery: inclusive education and equitable quality: it is our duty to promote learning opportunities for all	4
14. Support+	11, 13, 17
15. Small citizens project	4, 10
16. ISCAP—implementation of sustainable water control and associations	6, 12
17. Project happy grandmother	3
18. Training workshops: beauty professionals	5, 8
19. Teaching seminars: preparing children to be global citizens	4, 10, 16
20. Educating children to persevere through reading	4, 15
21. Training of youth: leadership and humanism	4, 16
22. Vegetable garden: self-sustenance	2, 3

Source Data from the research "Young students and society: practical application of the UN SDGs" (Antonio Meneghetti Faculdade—AMF)

Among the new partnerships established and confirmed, we highlight the following institutions: Municipality of Agudo, RS; Dom Érico Ferrari State School, RS; Municipality of Dona Francisca and its Infrastructure and Development Department, RS; São Miguel Velho Quilombola Community (Restinga Sêca, RS); Municipality of Restinga Sêca, RS; Francisco Giuliani Municipal School (Restinga Sêca, RS); Municipality of São João do Polêsine, RS; La Salle Municipal School (São João do Polêsine, RS); the Antonio Meneghetti Faculdade Foundation (Recanto Maestro-RS); Agudo-RS State Prison; Horto Florestal/Department of Forestry Sciences (Federal University of Santa Maria, UFSM); City of Santa Maria-RS; João Belém State Primary School (Santa Maria-RS); Santos Dumont Municipal Primary School (Agudo-RS); Relona/UFSM Project; João XXIII State Secondary School (São João do Polêsine-RS); Antonio Meneghetti Faculdade Student House; Lar das Vovozinhas

(Santa Maria-RS); Iara Nunes Beauty Studio in Restinga Sêca-RS; Paulo Freire Municipal Primary School (Faxinal do Soturno-RS); Mamãe Coruja Early Childhood School (Santa Maria, RS); Tiradentes State Elementary School (Nova Palma, RS).

Table 2 shows the main results achieved in the projects.

These were some of the results produced and achieved by the SDG projects carried out by the teams of young university students in 2017, 2018, and 2019. It is noteworthy that the associations and institutions received support for the implementation, but it was given them the freedom to carry out all the projects, both in technical and human aspects. The AMF and the Antonio Meneghetti Foundation provided financial support and space for the best projects to continue into the following semesters and to ensure their sustainability. After the implementation of the projects, several changes were visible and real, such as those shown in Table 2, as each group of students progressed in carrying out the projects.

Furthermore, regarding the results, it is important to address the participation of beneficiaries: social participation was active and productive, for both proponents and beneficiaries. The student proponents participated directly in the conception, planning, execution, and evaluation of the projects and co-involved the beneficiaries so that they also served as agents of transformation (interlocutors).

The beneficiaries' profiles include children, adolescents, young adults, adults, and the elderly from diverse social situations, with or without formal education, currently working professionally or not. The beneficiaries were instigated to action, challenged to take responsibility for the projects and for the continuity of the results. During the project period, we reached more than 600 people. The student proponents impacted reality and, through their work, impacted the beneficiaries, who were transformed and created a positive network of functional results in the social context. The students act as mediators and begin to develop maturity in relation to themselves, others, and society, as well as to face problems and implement solutions, and numerous partnerships are created in the process. This can be seen in some brief testimonies from students participating in the projects, who state:

> "We see the world through a different lens" (Tayná).
>
> "It is essential to be able to act practically, starting in our communities with actions that demonstrate the importance of a project as large as this one at the UN, the SDGs, being a conduit of information and an agent of change" (Déborah).
>
> "The greatest lesson is that change begins on its own and, with small actions, change will truly happen" (Ludmila).

The final considerations are based on the lessons learned during the implementation of these projects, contributing to the achievement of the SDGs, and based on an evaluation of the entire project and the young students' comments. In this sense, the first point to address is the main barriers encountered in the implementation of the internships for each of the projects. In a very practical way, we can list:

> "Since all the young members of the groups work, it was sometimes difficult to find time to meet. However, with the consent of each individual, it was possible to organize the activities successfully" (Raphaella).

Table 2 Results from the SDG projects—period 2017–2019

Young students	More mature, conscious, responsible, taking the initiative to propose actions to solve issues in their lives and in their social context, more autonomous, with greater willingness to act and resolve, establishing functional partnerships and relationships, and developing a more mature view of the context, of social issues and possible future actions
Entities that were served and beneficiaries	Creation of alliances for regional development in educational, social, environmental and economic aspects, new ideas for action and resolution from the contact with SDG projects of students in their places, empowerment of entities and beneficiaries, who began to see themselves as responsible agents of changes in their lives and contexts and with a more responsible vision in the training of future generations
Social Conditions	Contact with the *Quilombola* community, learning about their values and cultural diversity; Higher quality, innovations and creativity in educational contexts attended to, mainly in public schools; Academic support with classes on Portuguese, English, basic Libras knowledge, Braille, body language, culture, art, citizenship, and politics; Evidence that it is possible, even living in a poor community, to change and grow to have a future with better opportunities; Understanding mediation in the legal area; Learning about the three main stereotypes and concerns of young people: biology, critical idealism and consumerism (and how to change them); Creative classes for 30 children with the transformation of PET bottles into everyday objects and toys; Meetings among young people to read and produce texts and poems; Donation of useful school toys and material to foster creative classes with children
Environmental Conditions	Environmental awareness in children/adolescents (understanding the value and importance of preserving nature), allowing partners to consolidate and maintain the project in the municipality; Planting of more than 100 tree seedlings and seeds that help reforestation; Installation of solar panels and poles in the photovoltaic electrical system (access to clean and renewable energy); Fertilization without the use of chemical products and gaining quality food with less waste for the environment, less pollution; Less expenses for the prison to obtain fertilizers; Reuse of cooking oil in soap production, promoting awareness about proper disposal (avoiding pollution and contributing to a healthy environment); Collection of food and feed for donation to NGOs that care for vulnerable dogs; the production of 4 feeders for places with a large number of animals to feed; Collection of banners for the project Relona

(continued)

Table 2 (continued)

Young students	More mature, conscious, responsible, taking the initiative to propose actions to solve issues in their lives and in their social context, more autonomous, with greater willingness to act and resolve, establishing functional partnerships and relationships, and developing a more mature view of the context, of social issues and possible future actions
Economic Conditions	Creation of new job opportunities and social and economic actions in the Region; Training people in the community to improve techniques and efficiency in professional performance and in learning a new profession; Improvement of the municipal selective collection system: selective garbage collection and waste production and partnership for the development of the project with a cooperative organization of collectors; Cost reduction with waste collection, transportation and final destination; Increase in the number of members of the *Força no Braço* Association (selective waste collection); Income generation for professionals in the region; Creation of new individual and collective companies and initiatives
Institutional Conditions	Government support to carry out the projects, positively reaching the social body, with a positive performance in economic, social and environmental contexts, with responsibility and the creation of new partnerships; Support from the region's education and school departments for the implementation of educational, cultural and social projects that help in the education of children and adolescents and in the continued training of teachers
Convergence of practices with current public policies	All carried out practices are in line with the principles and guidelines of current public policies, as they aim at the well-being of the beneficiaries in the areas of education, environment, social assistance, economy, covering various aspects of citizenship and quality of life—and making contributions for the realization of public policies
Alignment to the SDGs and the 2030 Agenda	All practices involve people, promoting responsibility for the planet, prosperity, peace and associations, in alignment with the goals: 23; 3.5; 3.9; 4.1; 4.2; 4.3; 4.5; 4.6; 4.7; 7.2; 8.4; 10.2; 10.3; 11.a; 12.4; 12.5; 12.8; 13.3; 14.1; 15.1; 15.2; 15.3; 15.5; 15.a; 15.b; 16.3; 16.7; 16.b; 17.17. Through small actions, the results and the contributions of implemented projects help to achieve the highlighted goals

Source Data from "Young students and society: practical application of the Sustainable Development Goals (SDG) of the United Nations" (Antonio Meneghetti Faculdade—AMF)

"We often encounter a lack of willingness and disposition among many people to help or be part of a good cause for those most in need, considering that people are very selfish, profit-driven, and irresponsible. That is, if it involves doing something that isn't for their own benefit or profit, there's no willingness. We no longer have responsible people who care about the collective well-being, with a better place for everyone, respecting individual space and the rights of each species. And our group fought well against this and achieved concrete results!" (Eduarda).

> "I think one of our difficulties was being able to collect anything at the locations where we posted the signs. It's difficult to move society and try to help people do good. But we succeeded!" (Everson).
>
> "The difficulty in keeping the children's attention, as they were extremely hyperactive in class; the difficulty in organizing time for activities, as all the participants were interns and studied during the day" (Pedro).
>
> "Besides time, one of the barriers they faced was the children's behavior. They didn't stop, but over time, we got to know each other better. I'm not saying they became angels, but they improved a lot, because during the project they realized that the change in behavior was visible" (Fernanda).
>
> "In my opinion, the main challenge was managing the children. It feels uneasy and changes the entire course of a class because they didn't behave as we expected" (Rhuan).

Regarding the factors that contributed to the success of the projects' practices, as well as the aspects that can be addressed regarding overcoming barriers, it is possible to address: the persistence and skill that each group member had in finding ways to solve problems that arose or impeded the progress of the activity, without giving up; the combined knowledge of each member to make the project successful and ensure that the results materialize in practice.

> "For the success of the practice, we count on mutual support, as we have formed a good team since the beginning of university, composed of responsible and dedicated colleagues. The most important barrier to overcome was awakening in people a sense of duty and responsibility, helping those in need without relying on anyone's recognition, and doing on their own what they can best do for those who need it most at that moment" (Eduarda).
>
> "Changing the dynamics of the classes, trying a new one in each class that fits the profile of the children; the willpower of the group of participants, because everyone really wanted to be able, in some way, to contribute to the education of the children in our project" (Pedro).
>
> "Knowing the realities of each child and being able to collaborate for each one's growth in school; understanding the reality of teachers in the classroom, how difficult it is, and how much it requires a vocation to be an educator" (Rhuan).
>
> "I learned a lot, but the most surprising thing was getting to know myself better, understanding what I'm capable of learning to better manage my emotions, and being calmer. In addition to recognizing the importance of being a teacher, that responsibility, the concern they must have for us is something special and something we should take seriously. The sentence I'll take with me is 'I thought I was going to teach them, but I was the one who learned from the relationship with them'" (Débora).
>
> "I think the success of the project is due to the fact that the group remained united, from beginning to end, with a genuine desire to make everything happen and following a schedule we planned together. We decided, back in April, that the project would be something we would 'put our focus and strength into it', and we stuck with it until the end of the project (the end of the semester), without letting any of us get discouraged for any reason." (Raphaella)

It was possible to discover, therefore, that the implementation of all the internships in this project took 46 young college students out of their comfort zone and made them act on their social context and produce results for so many beneficiaries, voluntarily.

3.2 Projects Carried Out by the Antonio Meneghetti Foundation Involving Students from the Antonio Meneghetti Faculdade

All of the 28 projects supported by the Antonio Meneghetti Foundation[1] are guided by educational, cultural and social goals, having as bases the ontopsychological pedagogy and the SDGs. The teams that take the projects to practice are also composed by students and faculty members of the AMF. This way, they help the academic community to understand the SDGs, because they build their daily routines living close by to those projects every day.

The AMF and the Antonio Meneghetti Foundation have carried out SDG projects since the creation of the foundation in 2010. They are a tribute to the memory of Professor Academician Antonio Meneghetti, a man who always believed in the power of education to change the world. Nowadays, there are projects in the following areas: music education; environmental education; technological education; preservation of urban spaces; encouraging reading and writing; encouraging sports practice; scholarships for courses abroad; free-time employment for young people; and also supporting the sustainability of the region where the foundation is based, supporting the restoration and preservation of buildings, selective garbage collection, offering higher education scholarships to young people with limited financial resources, creating organic farms and flower cultivation areas, recovering native species of the campus, etc. This paper will address some of these projects more closely, especially those that are implemented in public primary and secondary schools. The demands addressed by the projects come from schools requesting support to offer students complementary courses, beyond what is provided by the school curriculum proposed by the government.

The contribution that the projects described below have created is to offer children and young people in the region the opportunity to complement their education through actions that provide citizenship training for children and young people in the region where the foundation is located. Annually, the projects' work is reported in terms of numbers and qualitative data in an activity report that is distributed to the community. The projects involve professors and educators from across the region to enable complementary education for children and young people, and contribute to raising citizenship awareness in these young people. This way, the projects developed have contributed to opening the market for educators and pedagogues, including pedagogy students in the higher education program offered by the AMF. This initiative reflects an international trend, which sees the third sector as a new avenue for professional activity for educators and pedagogues. Furthermore, the projects support the training of public primary school teachers so that they become familiarized with the SDGs and with ontopsychological pedagogy, and are also able to apply these principles in their classrooms.

[1] To know more: www.fundacaoam.org.br/projetos.

Although the practical methodology varies in each of the projects, the common thread among all the projects developed in this region by the aforementioned entities is the practical application of the UN SDGs and the application of ontopsychological pedagogy. The latter refers to a field of pedagogy developed, starting in the 1970s, by Professor Academician Antonio Meneghetti. Based on his studies, the ontopsychological pedagogy began to be applied in different educational institutions around the world. Its foundation lies in the principles applied in all the projects that the AMF and the Antonio Meneghetti Foundation maintain. These principles are: youth empowerment; helping students discover their vocation by valuing young people's abilities and choices; and understanding young people as individual citizens capable of motivating and leading economic and social change in their region. According to Meneghetti (2014), pedagogy is the art of how to help or involve a child in achieving. Therefore, "the practical scope is to educate the subjects to become and know themselves: to foster themselves as leaders in the world, to educate a logical-historical being with winning capabilities and behavior" (Meneghetti 2014, p. 14). To have an overview of how these principles become practice, below, some of them are presented.

(1) *Bola pra frente* (Ball Forward): a project aimed at children and young people in the city of Restinga Sêca. The project promotes various physical activities related to soccer, working on aspects of discipline, personal values, interaction between athletes, and it enriches academic performance, as good performance in education is one of the requirements for the continuity in the project. The classes involve a strong commitment among parents, teachers, students, and partner institutions to establish the comprehensive training provided by sports, incorporating values, character, discipline, behavior, physical-motor development, and good conduct in children. This learning contributes to their individual and social development as it is based on the educational premises of the classes conducted within the Project, understanding the meritocratic logic of rewards arising from effort. Other important aspects to consider are the improvement in young people's behavior and responsibility for their lives. The project primarily contributes to SDG 4, "Quality Education for All." The testimony of one participant in the project (Lucas Morais da Silva, 13) attests to the benefits it brings to its participants:

> For me, participating in the project is not just about playing soccer, but also about respecting differences, both teaching and learning from others, being more responsible, making new friends, having more experiences, and, most importantly, having fun together. Two years ago, I learned about the project and met so many new people; the project helped me so much. The project taught me to respect others and even taught me to enjoy studying, something I hated. The project is the only thing that makes me get up early. And this is a good thing; learning new things every week is incredible for me.

(2) *Desenvolvendo a Formação Inteligente por Meio da Leitura* (Awakening Intelligent Education through Reading): Understanding the power of knowledge and knowing that children's and young adult's literature is, for many children, the gateway to the world of knowledge and learning. Understanding that every message conveyed to children through books is of great importance for their beginnings in life, as they bring passages of growth and encouragement, always conveyed in a calm and pleasant manner. With each page read, readers awaken to a world completely different from

anything they have ever seen. Through these adventures, children begin to glimpse new possibilities and even become the authors of their own stories. Through books, they begin to realize that people must be the protagonists of their own lives and must be more responsible for the decisions they make, since, as in fairy tales, they always have consequences. Reading the various books in the project helps them understand who each person is and how they can better interact with their surroundings. On average, each child discovered and explored at least nine new books. The project primarily contributes to SDG 4 (Quality Education for All). The testimony of one of its participants clearly illustrates the results:

> One of the best things was meeting the new teachers. They teach various things. We learned to be better friends, to read more, to respect others, not to bully, and to respect and listen to teachers. Another great thing was discovering new books: Ugliness, Grandma Leda's Gift, and others. I really wanted the project to continue. (Edna Neves Ferreira, 10 years old)

(4) *Jovem e Tecnologia* (Youth and Technology)—More than 200 young people from public schools who, through classes, develop skills in very specific ways and can now launch themselves as strong candidates in the job market. It opens many doors, as it gives a practical understanding of how to work with technology and helps us understand how to use our full potential more responsibly. It also fosters entrepreneurship in various ways. Young people must maintain control over the dynamics of the countless information technologies and innovations that emerge at every moment, such as social media. Therefore, students are trained in everything essential to the world of technology and, therefore, in the future, they will know how to manage technology in the service of humankind, and not the other way around. In fact, AMF students themselves volunteer as instructors and teachers, enabling them to replicate classroom knowledge in a practical way with young people of all ages. The best of these young students are also recommended for internships in companies in the region, where they can begin their professional careers. However, the big difference, for everyone involved in the project, is the comprehensive training they receive: each participant is better prepared to be a solution in the labor market. The Youth and Technology project contributes mainly to SDG 4 (Quality Education for All), SDG 8 (Decent Work and Economic Growth), and SDG 5 (Gender Equality). The testimony of one of the students demonstrates the project's results.

> Participating in Youth and Technology is realizing that there are people who want to change the world with you. It's creating bonds and never breaking them. It's working as a team. It's very rewarding to see the results and the happiness of an entire group. Youth and Technology provided me with a great connection not only with my friends but also with myself. (Roberta Nunes, 14)

(5) OIKOS Project—Oikos is a project focused on the environment and fully associated with it. This opportunity highlights the responsibility of each student in preserving and caring for planet Earth. It is in it that one is born and grows, and therefore, one is also a part of it. In Greek, Oikos means "home," and it is caring for the space that the project involves and the person builds a lifestyle. To have a better world, we must first begin with education and individual responsibility. It is people themselves, with small individual actions, who transform a context, through the bias

of valuing human beings and the environment that surrounds them. This is how Oikos works: it teaches that everything is one and that there must be great harmony for tranquility in the coexistence between human beings and the environment. The results are explicit in the dimension of consciousness. When we talk about the need to preserve the environment so that, in the future, there will be an even better planet, or when we begin to help family members in their home gardens, it is clear that the seed planted by OIKOS is germinating and ready to bear beautiful fruit in the near future. It is the children and young people who will later form the group responsible for the development of the planet, maintaining consistent growth and respecting the environment. The OIKOS project reaches around 800 young people annually and contributes primarily to SDG 4 (Quality Education for All) and SDG 15 (Life on Earth). The testimony of one of the project's students demonstrates the results achieved.

> For me, the OIKOS project is a project that teaches many things, such as how to care for and preserve plants. It also ultimately teaches us how to work in groups, respect and listen to others, respect opinions, etc.—Eduarda Kleinpaul, 16.

These are just some examples, as we could not bring all of them to the limited space we have for this text. The important thing to underline might be that all this context allows the AMF and the Foundation to promote the SDGs locally, in the rural area of Brazil, and then to take their impact globally, by representing these projects in different initiatives and agencies of the United Nations' system.

4 Conclusions

In qualitative terms, the annual reports from the AMF and the Antonio Meneghetti Foundation present testimonies from young people who witness the changes that have taken place in their lives as a result of their free participation in the projects developed by these entities. The main result observed is that the projects have encouraged school teachers to meet to discuss the sustainable development goals in the classroom. Consequently, there has been an increase in civic awareness among both young people and these teachers.

Based on the experience with the projects carried out since the creation of the Antonio Meneghetti Foundation in 2010, in partnership with the AMF, it is possible to conclude that civil society organizations can and should act to improve education in their region. Experience has shown that simple projects, supported by professional educators available in the communities served to and are capable of bringing about changes in the habits and behavior of children and young people in public schools.

These projects enable these young people to: expand concentration, creativity, self-esteem, respect for others, the ability to work in groups, and communication skills; develop their own potential; discover their vocation and interests; expand their sense of citizenship; and expand their cultural awareness. The implementation of these projects also allows for the appreciation of the educational work carried out in schools, the expansion of the impact of teachers' work, and the enhancement of these professionals' educational awareness.

This case also shows that the future of education can go further with public–private partnerships. It is concluded that citizenship education projects are important steps toward achieving and building a society with greater opportunities for all. These projects can be replicated and shared across countries, as their common foundation involves an appreciation of local culture and universal principles, such as everyone's right to quality education.

References

Buss PM, Machado JM, Gallo E, Magalhães SD, Setti AF, Netto F, Buss DF (2012) Governança em saúde e ambiente para o desenvolvimento sustentável. In: Ciência & Saúde Coletiva, v. 17, n. 6: p. 1479–1491. https://doi.org/10.1590/S1413-81232012000600012

Cabral F, Campanatti M (2019) Imagina2030.

Gadotti M (2008) Educar para a sustentabilidade. Uma contribuição à Década da Educação para o Desenvolvimento Sustentável. São Paulo: Editora e Livraria Instituto Paulo Freire.

Gil A C (2017) Como elaborar projetos de pesquisa. (6th. ed) São Paulo: Atlas.

Malheiros T F, Philippi Jr A, Coutinho SMV (2008) Agenda 21 Nacional e Indicadores de Desenvolvimento Sustentável: contexto brasileiro. Saúde e Sociedade, v. 17, n. 1: p. 7-20.

Meneghetti A (2014) Pedagogia Ontopsicológica. (3rd. ed). Recanto Maestro (Brasil): Ontopsicológica Editora Universitária. (Originally published in Italian in 1979)

Meneghetti A (2019) Sistema e Personalidade. (3rd. ed.) Recanto Maestro: Ontopsicológica Editora Universitária.

Trejo Nieto A (2017) Crecimiento Económico e Industrialización en la Agenda 2030: Perspectivas para México. Problemas de Desarollo [online], vol.48, n.188 [citado 2025–08–04]: p.83–112. http://www.scielo.org.mx/scielo.php?script=sci_arttext&pid=S0301-703620170001 00083&lng=es&nrm=iso>. ISSN 0301–7036

Wazlawick P (2011) Declaração Universal dos Direitos do Homem: considerações à luz da dialética de responsabilidade e reciprocidade fundamentada na Ontopsicologia. Atos do Congresso Internacional Responsabilidade & Reciprocidade: p. 398–412.

Any Regina Rothmann Strategic Curator of the Antonio Meneghetti Foundation; visiting professor at AMF undergraduate and graduate courses. She holds a Bachelor's degree in Psychology from Saint Petersburg State University (2012). She is currently the Commercial Director of the Antônio Meneghetti College and a professor in the Bachelor's program in Ontopsychology at the Antonio Meneghetti College, AMF. She has also been a guest lecturer in the extension programs: Advanced Business Training and ISOMaster School at the same institution. She is the Strategic Curator of the Antonio Meneghetti Foundation for Humanistic, Cultural, and Educational Scientific Research, working on its management and project organization. She manages the Recanto Maestro Center for Humanistic Art and Culture. She has experience in Psychology, with an emphasis on Ontopsychology.

Clarissa Mazon Miranda Professor at Antonio Meneghetti Faculdade. Member of the International Affairs Office of Antonio Meneghetti Foundation. She holds a degree in Journalism from the Federal University of Santa Catarina (2005), a master's degree in Communication from the Federal University of Santa Maria (2012), and a PhD in Literature from the Federal University of

Santa Maria (2018). She also completed an MBA in Marketing from FGV (2010); a specialization in Knowledge Management and the Ontopsychological Paradigm from Antonio Meneghetti Faculty (2014), and a specialization in Ontopsychology from Saint Petersburg State University (2014). In 2023, she completed the Master's Degree in Audiovisual Screenwriting from The Core Entertainment School (Madrid, Spain) and the Certificate in Filmmaking from the National Film TV School (England) and the Certificate in Finding Your Voice: Screenwriting from the National Film TV School (Scotland). She is an undergraduate professor at Antonio Meneghetti Faculty in the Business Administration, Accounting, and Bachelor's Degree in Ontopsychology. She also works as an international relations and public relations advisor for the Antonio Meneghetti Foundation, a reporter for Performance Líder magazine, a project advisor for the OntoArte Association, and a screenwriter. She is one of the Coordinators of the Center for Excellence in Language (NEL). She is a member of the Screenwriting Research Network (and a full member of this research network's Executive Board), a member of the Association of Adaptation Studies, a member of the Brazilian Society for Cinema and Audiovisual Studies, and a member of the Brazilian Association of Screenwriters. At the AMF, she is a member of the Business Management, Humanistic Culture, and Intuition Research Line and a member of the Hard Cases Encounters between Administration and Ontopsychology research groups and the Ontopsychological Method: Theory and Practice research group.

Patricia Wazlawick Professor of Antonio Meneghetti Faculdade and Coordinator of the Ontopsychology Department. Councilor of Antonio Meneghetti Foundation. PhD in Psychology from the Federal University of Santa Catarina (UFSC 2010), in the Research Line Subject Constitution, Aesthetic Relations and Creative Processes. Post-doctoral fellow in Health Informatics (UFSC), developing studies on the theme of what soft skills are, their relationship to mental health, how to develop them in the academic-university context and the organization of an ontology of soft skills required for IT professionals in emerging contexts. Master's degree in Psychology from the Federal University of Paraná (UFPR 2004), Research Line Psychosocial Processes. Post-graduate specialist in Psychology with an Ontopsychological approach from Saint-Petersburg State University (SPbU 2012–2014), Post-graduate specialist in Knowledge Management and the Ontopsychological Paradigm (AMF 2011–2014). She holds a postgraduate degree in Positive Psychology, Science of Well-Being, and Self-Realization from the Pontifical Catholic University of Rio Grande do Sul (PUC-RS). She also holds a specialist degree in Transformative Education: Pedagogy, Foundations, and Practices from PUCRS. She holds a degree in Music Therapy from the Faculty of Arts of Paraná (2000). She is a professor at the Antonio Meneghetti Faculty in the Bachelor's Degrees in Ontopsychology, Information Systems, Accounting, and Bachelor's Degree in Pedagogy, as well as in the Lato Sensu Postgraduate Programs (MBA) in Corporate Identity, Specialization in Ontopsychology, and Specialization in Ontopsychological Pedagogy. She is the Coordinator of the Bachelor's Degree in Ontopsychology at AMF (Brazilian Academy of Social Sciences). She is a researcher and advisor for undergraduate and specialization thesis projects. She is a member of the AMF Graduate Program and a member of the School of Pedagogical Innovations Project. He has experience in the management and administration of higher education institutions, including participation in Institutional Accreditation and Reaccreditation Assessment Processes, Program Authorization, Recognition, and Renewal, CPA (Course Accreditation Process), personnel management, student support, and the internationalization of HEIs. He works at the AMF Interdisciplinary Research Center. He has experience and experience in academic and university internationalization activities, assisting in partnership processes for international student and faculty agreements. He is responsible for the Scientific, Academic, and Cultural Cooperation Agreement and international academic mobility activities with the University of Turin (UniTO) and the University for Foreigners of Perugia, Italy; and in Portugal with the Lusófona University of Porto and the Polytechnic Institute of Estoril. He conducts research in the areas of ontopsychology: Epistemology of Sciences and Epistemology of Ontopsychology; Ontopsychology, Quantum Physics, and the Critical Problem of Knowledge; Youth Education; Humanistic Culture, Ontopsychological Pedagogy and Education; Youth and the UN Sustainable Development Goals. She was a researcher in the areas of Music and Creative Activity; Music Education; Music Therapy; Teacher Training. Member of the Editorial Board of Saber Humano-Scientific Journal of the AMF (ISSN: 2446–6298) and of the Brazilian Journal of Ontopsychology. She was a member of the Research Center on Social Practices and Constitution of the Subject—Processes of Creation and Aesthetic Relationship (NUPRA) at UFSC, from 2006 to 2010, currently called the Research Center on Social Practices, Aesthetics and Politics. She is a member of the Research Group Psychology, Subjectivity and Phenomenology: Clinical-Social Theory and Practice, in the Research Line Contributions of Phenomenology to Psychology: Phenomenological Investigations of Psychic Phenomena, coordinated by Prof. Dr. Tommy Akira Goto, Federal University of Uberlândia. Member of the Brazilian Association of Ontopsychology and member of the Brazilian Society for the Advancement of Science (SBPC). Scientific articles on ResearchGate: https://www.researchgate.net/

The Perception of Federal University of Itajubá Staff Regarding Sustainability

Mileni Cristine das Chagas Azevedo, Maria Rita Raimundo e Almeida, Sara Ingrid da Silva Valim, Maria Tereza Alves Martins de Faria, and Frederico de Souza Nascimento

Abstract Sustainability and sustainable development are essential and interconnected concepts that require a comprehensive understanding. These topics have gained increasing prominence in contemporary discussions, engaging multiple sectors of society. As such, universities must adopt a proactive approach to socio-environmental challenges, fostering critical reflection and empowering individuals to become agents of transformation. In order to understand how these themes are perceived, this study aimed to analyze how staff at the Federal University of Itajubá (UNIFEI) perceive sustainability and sustainable development. We administered a questionnaire to both professors and administrative staff to assess their perceptions of the university's sustainability-related actions in teaching, research, outreach, and institutional management, as well as their level of engagement with these issues. The results indicate that, despite the high educational background of respondents, their understanding of sustainability remains superficial—primarily due to a perceived lack of institutional support and encouragement. Advancing sustainability within the university requires a collective commitment, with university leadership implementing effective management strategies and actively supporting staff, who, in turn, play a key role in transmitting sustainability values to students. Strengthening

M. C. das Chagas Azevedo · M. R. R. e Almeida (✉) · S. I. da Silva Valim ·
M. T. A. M. de Faria · F. de Souza Nascimento
Natural Resources Institute, Federal University of Itajubá, Itajubá, Brazil
e-mail: mrralmeida@unifei.edu.br

M. C. das Chagas Azevedo
e-mail: mileniazevedo@unifei.edu.br

S. I. da Silva Valim
e-mail: d2020020597@unifei.edu.br

M. T. A. M. de Faria
e-mail: d2020005812@unifei.edu.br

F. de Souza Nascimento
e-mail: d2020015891@unifei.edu.br

W. Leal Filho et al. (eds.), *Sustainability in the Curriculum and Operations of Universities*, World Sustainability Series,
https://doi.org/10.1007/978-3-032-12678-8_21

sustainability initiatives within the institution demands the integration of sustainability principles into both academic and operational practices, fostering a culture of environmental responsibility, and promoting of interdisciplinary collaboration.

Keyword Sustainability · Higher education · Environmental responsibility · Environmental awareness

1 Introduction

The characteristics of the current environmental crisis stem from the depletion of natural resources, driven by a consumption-centered lifestyle. According to Schneider (2015), increased production contributes to a higher overall level of wealth, which is essential for improving living conditions, especially the most disadvantaged populations. However, when this wealth is not equitably redistributed, it primarily benefits the wealthiest segments of society, thereby perpetuating social and economic inequalities. Income inequality, combined with socio-economic factors and environmental constraints, exacerbates the gap between wealthy and poor households, posing a significant challenge with far-reaching global consequences (Amar and Pratama 2020).

Humanity depends on natural resources not only to fulfill its basic needs but also to safeguard itself against threats such as epidemics and natural disasters. Contemporary economies are characterized by extreme inequality, and the mismanagement of resources represents a significant threat to the planet (Sachs 2017). Medeiros Júnior and Baracho (2016) examine how unbalanced economic development can lead to environmental degradation, thereby compromising human dignity and quality of life. They argue that the imbalance between economic growth and environmental concerns results in ethical and legal challenges, underscoring the urgent need for a new approach.

This new approach has emerged in postmodern society as an alternative paradigm, encompassing sustainability and sustainable development. These themes have gained prominence in contemporary discussions, encompassing diverse meanings and perspectives, and permeating various sectors of society in the pursuit of solutions to the environmental, social, and economic challenges faced globally.

In its logical sense, sustainability refers to the capacity to maintain and perpetuate itself (Mikhailova 2004). In practical terms, sustainability is a guiding principle that seeks balance through three pillars: social justice, economic viability, and environmental responsibility (Benites and Polo 2013). The concept of sustainability is closely linked to that of sustainable development, which are frequently conflated as synonyms (Souza et al. 2020). These conceptual ambiguities can hinder the effective implementation of both concepts (Souza and Armada 2017). According to Oliveira et al. (2009), sustainability can be understood as a value aimed at enhancing the current status quo, whereas sustainable development represents the outcome of actions derived from sustainability policies.

Thus, sustainability and sustainable development are two essential and interconnected dimensions that shape the present and future of our society. While sustainability can be understood as a broad concept aimed at harmonizing the economic, social, and environmental dimensions of human development, sustainable development focuses on achieving economic and social progress that improves the population's quality of life without harming or degrading the environment. A comprehensive understanding of sustainability enables the creation of a more resilient future, ensuring that human needs are met within the planet's limits (Oliveira et al. 2009). Although the challenge is complex, the pursuit of sustainable solutions remains both urgent and essential.

In this context, considering the urgency of an economic model that is socially fairer and more environmentally balanced, the United Nations (UN) formulated the Sustainable Development Goals (SDGs) in 2015. The SDGs aim to address these needs while mitigating negative impacts on the planet (Gomes and Ferreira 2018). Their implementation has enhanced representativeness and enabled all nations to participate in the decision-making process, including those that previously lacked such opportunities (Almeida Filho and Lauar 2021). Under these circumstances, society as a whole, along with its various actors, assumes responsibilities related to sustainability, including universities. As a Higher Education Institution (HEI), a university plays a crucial role in promoting sustainable regional development, in alignment with the objectives of the UN 2030 Agenda (Borsatto et al. 2024; Trevisan et al. 2024).

It is also important to emphasize that universities aim to educate professionals with critical thinking skills, capable of questioning and promoting change grounded in both personal and collective values. In this regard, they play a fundamental role in human development and education, as they are spaces where perspectives on various topics are formed and critical thinking is cultivated (Ribeiro et al. 2018). Thus, higher education should go beyond the mere transmission of knowledge, and assume an active role in addressing socio-environmental challenges, fostering critical reflection, and preparing students to act as agents of change. Therefore, universities should devote greater attention to social and environmental issues, acting within their capacities not only to inform, but also to educate individuals capable of transforming social reality (Tommasiello and Guimarães 2013).

Environmental education, in its various forms, enables the reassessment of social practices and promotes a deeper understanding of both global and local environments, the interconnections between challenges and solutions, and the importance of collective commitment to building a more just and ecologically sustainable society (Oliveira 2009). According to Grandisoli and Jacobi (2020), sustainability is most effectively integrated into education through the development of individuals who perceive reality in a more systemic, integrated, interdisciplinary, and transdisciplinary manner, grounded in collective values.

Therefore, it is crucial to ensure that, within universities, professors, administrative staff, and other education professionals recognize and understand the significance of these themes. This awareness enables them to influence and shape students' mindsets, attitudes, and practices through teaching, research, outreach, and innovation—ultimately preparing them to lead the socio-environmental transformations required for a more sustainable future.

According to Frizon and Eugénio (2022), based on a systematic review, universities in different countries are well represented in sustainability-related research, with most studies focusing on curricular content, professors, and students. Leal Filho et al. (2023) identify three emerging trends: despite the intrinsic value of sustainability research in higher education, the field is not as mature as one might expect; the range of topics addressed is broad and spans various areas; and individual efforts remain the most common mode of conducting research, while institution-level initiatives (at the university, department, or faculty level) are comparatively less frequent.

Research on the perceptions of actors within HEI has also been frequently conducted, including studies on the views of academic staff (Klein et al. 2022), students (Aleixo et al. 2021; Gomes et al. 2020), and managers—deans and heads of academic departments (Mousa et al. 2020). Trevisan et al. (2024) emphasize that future research should assess the effectiveness of existing strategies to support the transition to sustainability within HEIs, incorporate interviews with relevant stakeholders, and adopt diverse methodological approaches.

With this in mind, the present study surveyed staff at the Federal University of Itajubá (UNIFEI) to assess their perceptions of sustainability and related themes. Understanding these perceptions allows for an analysis of the current scenario and informs the development of more effective sustainability initiatives, aligned with the university's role in promoting environmental and social responsibility.

UNIFEI is a public HEI with campuses in the cities of Itajubá and Itabira, in the state of Minas Gerais, Brazil. The university offers a broad range of undergraduate and postgraduate programs across various fields of knowledge, with a strong emphasis on engineering and technology. Currently, it offers more than 30 undergraduate programs. The institution plays a significant role in the economic and technological development of southern Minas Gerais, serving as a hub for innovation and a training ground for skilled professionals. With regard to sustainability, UNIFEI carries out isolated initiatives but lacks a dedicated department or directorate specifically focused on these issues. It is also worth noting that the university offers undergraduate and postgraduate programs in environmental fields, such as Environmental Engineering and Water Resources Engineering, as well as a postgraduate program in Environment and Water Resources. These programs are involved in research and outreach projects that are aligned with sustainability principles.

2 Methodology

This study employed a questionnaire administered via Google Forms to collect and manage data, aiming to gain a deeper understanding of the university's sustainability initiatives. Based on Ferreira and Almeida (2024), the questionnaire was directed at professors and administrative staff at UNIFEI and was designed to capture their perceptions of key aspects related to sustainability in the areas of teaching, research, outreach, and management.

Overall, the questionnaire comprised both open-ended and predominantly closed-ended questions, addressing the following aspects:

- Personal information: gender, age, and educational level.
- Professional background: included questions about the respondent's *campus* (since UNIFEI operates in two locations—Itajubá and Itabira) and their role at the university (professors or administrative staff). Additionally, professors were asked about the undergraduate program(s) in which they teach.
- Teaching-related information: a section directed exclusively at professors, containing questions aimed at assessing the extent to which sustainability is integrated into their courses, how the topic is incorporated into their lectures, and their contribution to preparing students to become professionals aligned with sustainability principles.
- Environmental perception: this section included questions evaluating the respondents' knowledge of sustainability and sustainable development; the applicability of sustainability in their daily actions; their perception of UNIFEI's commitment to promoting awareness and encouraging sustainable practices; and the existence of sustainability-related initiatives within the university's teaching, research, outreach, and management.

We developed an action plan to disseminate the questionnaire to professors and administrative staff, using university email and social media platforms (Facebook and Instagram). The distribution of the questionnaire was strategically scheduled on days and at times with higher engagement rates over a four-week period, from April 13 to May 4, 2023.

The results were analyzed using descriptive statistics, and to enhance data visualization and interpretation, graphs were created in Excel. The findings were subsequently compared with those of other studies in the literature.

3 Results and Discussion

After four weeks of questionnaire distribution, 79 responses were received from professors and administrative staff at UNIFEI, covering both the Itajubá and Itabira campuses. At the time of the study, UNIFEI had 964 staff members, including professors and administrative staff (UNIFEI 2023). Thus, the resulting response rate was 8.2%. According to Batista et al. (2021) and Bastos et al. (2023), one of the primary challenges of online questionnaires is achieving high response rates, as participation is voluntary and respondents may choose not to participate. There is no consensus in the literature on an acceptable response rate for online questionnaire, but most studies suggest a range between 10 and 30% of potential respondents (Torini 2016). Given that this study is exploratory in nature and does not aim to provide a probabilistic sample, we consider the number of responses to be sufficient.

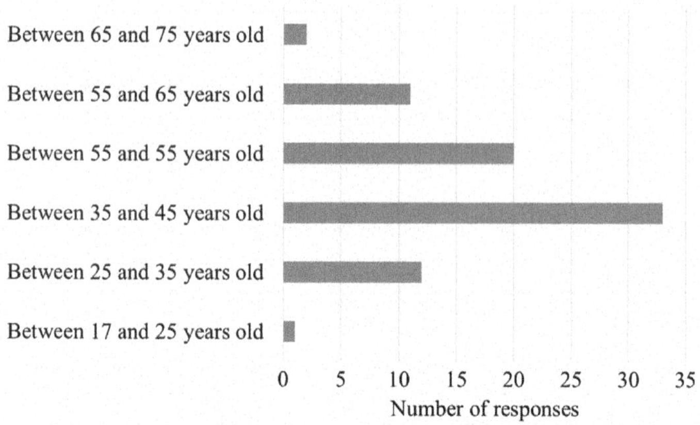

Fig. 1 Age distribution of respondents

Regarding personal information, the responses were balanced in terms of gender, with 40 male respondents (50.6%) and 38 female respondents (48.1%), while one respondent preferred not to disclose their gender. This slight male majority reflects the overall composition of the university staff, where 67% are male and 33% are female (UNIFEI 2023). The age group most represented in the sample was between 35 and 45 years old (41.8%) (Fig. 1). An analysis of the age distribution among all UNIFEI staff members (UNIFEI 2023) indicates that the predominant age group falls between 31 and 50 years. Therefore, the study sample is consistent with the demographic profile of the university staff.

In terms of educational level, 40.5% of the staff hold a Ph.D., 26.6% have a master's, 29.1% have completed undergraduate education, and 3.8% have incomplete higher education. The high level of education reflects the fact that the study involved professors. Therefore, it is evident that the staff members who responded to the survey possess a high level of education and expertise, which likely enhances their understanding of sustainability and sustainable development topics.

The questionnaire included staff members from both the Itajubá campus (86.1%—68 respondents) and the Itabira campus (13.9%—11 respondents), as well as administrative staff (58.2%—46 respondents) and professors (41.8%—33 respondents). According to university data, more staff members are allocated to the Itajubá campus, and the number of professors exceeds that of administrative staff (UNIFEI 2023). In this context, the distribution of respondents reflects the predominance of the Itajubá campus; however, it is noteworthy that the participation rate was higher among administrative staff.

Exclusively for the professors (33 respondents), a question was posed regarding the undergraduate programs in which they teach (Fig. 2). Multiple selections were allowed. In terms of the number of responses, the most represented programs were environmental engineering (11), production engineering (8), mechanical engineering

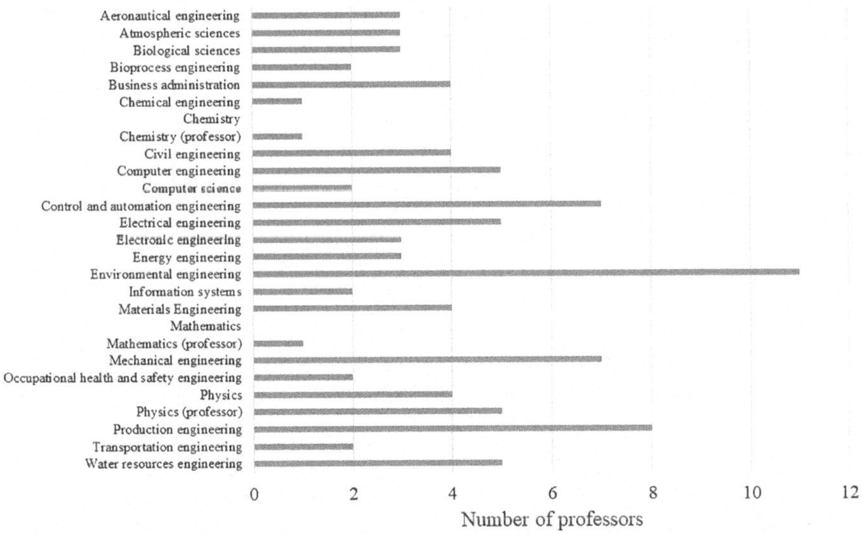

Fig. 2 Relationship between professors and the undergraduate programs they teach

(7), and control and automation engineering (7). In contrast, professors from mathematics and chemistry programs did not submit any responses. The alignment of the environmental engineering program with the theme of sustainability may explain the higher participation rate of its professors in the questionnaire.

The professors were also asked about the extent to which the concept of sustainability is incorporated into their lectures. The majority of respondents indicated that the topic is either minimally addressed (36.4%) or moderately covered (30.4%). A more in-depth approach, categorized as well-addressed and extensively covered, was reported by only 18.2% and 15.2% of the respondents, respectively.

Regarding their engagement with sustainability and sustainable development themes, among the 33 professors who participated, 17 (51.5%) had previously taught a course related to the topic, 16 (48.0%) had conducted research or outreach projects on sustainability, and 15 (approximately 45%) had supervised undergraduate theses on subject. However, when it came to assigning coursework related to sustainability or participating in sustainability initiatives at UNIFEI, 20 professors (60.6%) reported never assigning such tasks to students, and 21 (63.6%) had never participated in university-led sustainability initiatives.

Additionally, professors were asked to evaluate their contribution to preparing students to become responsible citizens and professionals with respect to sustainability and sustainable development. Abo-Khalil (2024) reflects on the essential role of interdisciplinary approaches and the critical need for active professors' involvement in fostering sustainability education. A five-point Likert scale was used, where 1 represented a low contribution and 5 indicated a high contribution (Fig. 3). The responses showed a predominance of moderate contributions—selecting scale 3

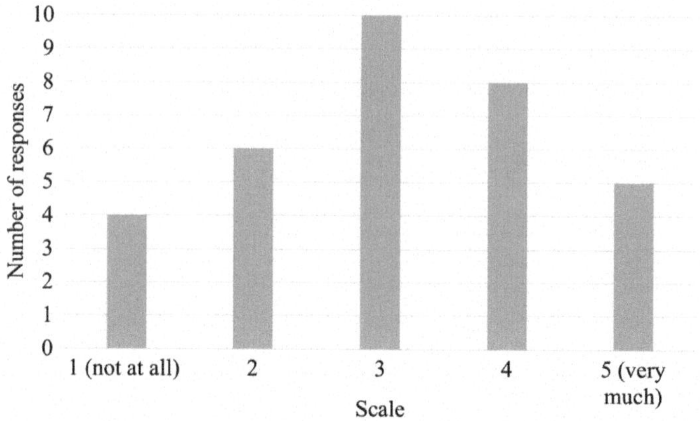

Fig. 3 Professors' contribution to students' education in sustainability

(30.3%, 10 responses), while only 15.2% (5 responses) perceived themselves as making a significant contribution—corresponding to scale 5.

In analyzing these findings, it is important to consider that one-third of the professors who participated in the survey (11 out of 33) teach in the environmental engineering program. Given the nature of this program, sustainability and sustainable development are inevitably central themes in its curriculum. Therefore, we believe that the higher level of engagement reflected in the responses is primarily influenced by this context. However, overall, based on the data collected from professors, sustainability education at UNIFEI remains limited. This limitation hinders the awareness and dissemination of information essential for student development—factors deemed indispensable by Jacobi (2003). In this context, the development of critical thinking, which universities should foster (Ribeiro et al. 2018), as well as a critical and innovative approach to environmental education (Jacobi 2003), are impeded when the complexity of sustainability and the necessary balance among its social, environmental, and economic dimensions are not adequately addressed.

The final section of the questionnaire, directed at both professors and administrative staff, aimed to assess the respondents' knowledge of sustainability and sustainable development, the applicability of sustainability in their daily actions, their perception of UNIFEI's commitment to promoting awareness and encouraging sustainable practices, and the existence of sustainability-related initiatives at the university. When asked about their knowledge of the term "sustainability", all respondents reported having heard of the concept in some capacity. Specifically, 2.6% had only heard of it, while 32.9% were familiar with the term but did not know much about it. The fact that some respondents were aware of sustainability but lacked deeper understanding supports the notion that, although widely discussed in theory, its meaning remains unclear for many. This lack of clarity hinders the implementation of practical and effective transformations (Claro et al. 2008). Conversely, 39.2% reported having studied the topic, and 25.3% considered themselves knowledgeable and actively engaged with sustainability. Most of these respondents were professors involved in environmental programs, such as environmental engineering, or administrative staff working in the university's environmental management sector. In other words, engagement with sustainability at UNIFEI remains largely limited to those who work directly with the topic.

When asked about sustainable development, the majority of respondents (53.2%) stated that it could be achieved if society changes its way of living. This response reflects an understanding of the multiple dimensions that must be considered in achieving sustainable development. It should not be viewed solely through an economic perspective, but should also encompass social, political, institutional, cognitive (including cultural, scientific, and technological), spatial-territorial, and temporal dimensions (Nascimento et al. 2013).

Furthermore, on a scale from 0% (none) to 100% (a lot), 13.9% of respondents reported applying sustainability in their daily lives at a low level (between 0–30%), 43% at a medium level (between 30 and 50%), 36.7% at a high level (between 50 and 70%), and 6.3% in a very high level (between 70 and 100%). To identify specific areas where sustainability is practiced, an open-ended question answered by 67 respondents revealed that the most frequently mentioned actions were waste separation (29 responses), energy conservation (27 responses), and water conservation (25 responses) as the most cited practices. These results suggest that knowledge of sustainable practices is limited, as respondents tend engage primarily in the most widely known and media-promoted actions.

Regarding staff evaluations of UNIFEI's sustainability actions in teaching, research, outreach, and management (Fig. 4), the most frequently selected rating was "fair". Among the four areas, management was perceived as performing the best—though still with limitations. In contrast, teaching received the lowest evaluation. This assessment is consistent with the feedback previously provided by professors.

Regarding UNIFEI's commitment to sustainable practices, on a scale from 1 (minimally committed) to 5 (highly committed), the majority of staff members

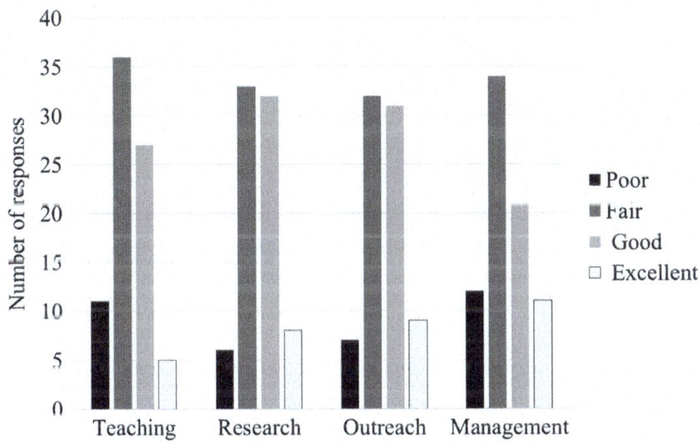

Fig. 4 Evaluation of UNIFEI's sustainability-related actions

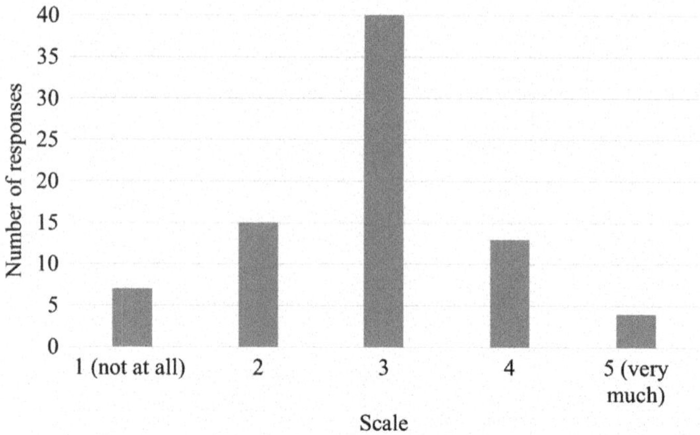

Fig. 5 Staff perception of UNIFEI's commitment to sustainable practices

(50.6%) perceived the university as only partially committed (Fig. 5). Furthermore, in response to a binary question (yes or no) about whether the university promotes awareness and training on sustainability, 67.1% of respondents stated that such initiatives do not exist. This perception of a partial or nonexistent institutional commitment is reflected in a similarly limited engagement among staff—and, by extension, among students.

According to Lara (2012), achieving effective outcomes in the pursuit of sustainability requires a strong and deliberate institutional commitment. Multiple studies indicate that while HEIs assign high importance to the integration of environmental sustainability at a strategic level, they generally lack dedicated policies for its integration into operational practices (Christou et al. 2024). Additionally, there is a mismatch between the widely acknowledged importance of education for sustainability and the actual implementation of its integration in HEIs, largely due to the multifaceted nature and complexity of the subject (Figueiró et al. 2022).

It is important to emphasize that UNIFEI may indeed engage in sustainability-related actions across teaching, research, outreach, and management. However, if such actions exist, staff, likely due to insufficient dissemination and communication, do not perceive them. Campos and Almeida (2019), in a study conducted with staff at the Federal University of Tocantins, found that employees had limited knowledge of the university's socio-environmental practices. The authors also highlighted that the university's sustainable management faces several constraints, particularly regarding communication with staff, as well as other barriers to the institutionalization of practices, such as a lack of alignment with or the enforcement of university regulations.

At the conclusion of the questionnaire, staff members were asked about their interest in joining a network of staff and students to discuss and implement sustainability initiatives at UNIFEI. A total of 25.3% indicated no interest, 32.9% expressed interest, and 41.8% responded "maybe". This level of interest could potentially increase with greater support and encouragement from the university administration.

Analyzing the questionnaire responses as a whole, we observed that despite the majority of respondents possessing a high educational level, UNIFEI's lack of commitment to promoting knowledge and raising awareness about sustainability and sustainable development directly impacts both staff and students, as they do not receive adequate training or encouragement from the university to incorporate these practices. This finding contrasts with Vaz et al. (2010), who asserts that universities are fundamental to sustainable development because they go beyond merely teaching and educating students; they play a vital role in the social context and bear the responsibility of shaping individuals conscious of the need to ensure a sustainable planet for future generations.

Given that universities have the responsibility to foster human development through critical thinking, the responses indicate that support for and the advancement of sustainability at UNIFEI remain limited. It is essential to emphasize the pivotal role of teaching in promoting sustainability and driving meaningful transformations. According to Romeiro (2001), environmental education should be designed to cultivate a sense of connection among individuals to inspire real change. Beyond teaching, it is equally important for the university to strengthen its sustainability efforts across research, outreach, and management.

In this regard, UNIFEI could implement improvements to encourage active participation from the academic community in integrating sustainability. Klein et al. (2022) highlight the relevance of aspects such as leadership, systemic vision to support staff activities, continuous improvement and waste elimination as daily practices, long-term thinking, and a focus on students as fundamental building blocks for the success of HEIs in promoting sustainability practices and fostering sustainability-oriented thinking. As Alves (2019) noted, achieving this objective requires fostering a new culture through seminars, integrative activities, and training courses to ensure that all actors understand and engage with the topic. Furthermore, increasing the accessibility of information regarding the university's sustainability practices is crucial to ensure that everyone benefits from this knowledge and feels motivated to adopt sustainable behaviors. According to Trevisan et al. (2024), despite the challenges, it is possible for higher education institutions (HEIs) to fully integrate sustainability into their core activities; however, it is essential to recognize that the transformation towards sustainability is an ongoing process that requires continuous learning and re-evaluation.

4 Conclusions

This study analyzed how staff, both professors and administrative staff, at the UNIFEI perceive sustainability and sustainable development, based on responses to a questionnaire. The findings reveal that the university faces significant challenges in promoting awareness and effectively integrating these themes into its academic environment.

A notable issue is the university's insufficient commitment to disseminating knowledge and raising awareness about these topics among its staff. The evident lack of training and institutional incentives directly affects staff development and, consequently, student learning. This institutional gap contradicts the fundamental role of HEIs in promoting sustainable development, as emphasized by several authors.

The lack of effective sustainability management within the institution also presents a significant barrier to transformative change. For meaningful transformation to take place, the university must establish a clear and comprehensible goal—one that is communicated and understood by all stakeholders, including staff and students. It is imperative that the institution actively contributes to shaping individuals who recognize the importance of sustainability for both present and future generations.

Additionally, staff members expressed considerable interest in seeing the university take a more active role in sustainability. Therefore, it is essential to strengthen the promotion and implementation of sustainability initiatives, including training courses, daily actions that encourage sustainable practices, and research and outreach projects related to the topic. It is also noteworthy that some staff members indicated interest in participating in a sustainability network within the university, if adequate institutional support and encouragement are in place.

In summary, the findings of this study underscore the urgent need for concrete actions by UNIFEI to promote awareness, provide training, and integrate sustainability and sustainable development into its academic environment. This effort requires not only changes in institutional management and policy but also the cultivation of a sustainability-oriented culture that actively engages the entire academic community. Only through such comprehensive commitment will the university be able to fulfill its role in shaping responsible citizens and professionals dedicated to building a sustainable future.

Although this research is relevant, it presents certain limitations, particularly regarding the number of responses obtained, which characterizes it as an exploratory study. Further studies should be conducted at UNIFEI to complement these findings and should also include the perspectives of other stakeholders, such as students, the external community, and managers—dean and heads of academic departments.

With respect to other universities, it is hoped that this study may serve as inspiration for assessing the perceptions of their academic communities and that UNIFEI's findings can contribute to broader discussions in similar contexts. As for the scientific contribution of this research, it aims to advance the debate on the challenges that HEI face in fulfilling their role in the context of sustainability and sustainable development.

References

Abo-Khalil, A. G. (2024). Integrating sustainability into higher education challenges and opportunities for universities worldwide. Heliyon, 10(9), e29946. https://www.cell.com/heliyon/fulltext/S2405-8440(24)05977-2

Aleixo, A. M., Leal, S., & Azeiteiro, U. M. (2021). Higher education students' perceptions of sustainable development in Portugal. Journal of Cleaner Production, 327, 129429. https://doi.org/10.1016/j.jclepro.2021.129429

Almeida Filho, N., & Lauar, A. S. S. (2021). Sustainable development and Sustainable Development Goals: distinct processes and convergent interests. Revista Brasileira de Políticas Públicas e Internacionais, 6(2), 3–29. https://doi.org/10.22478/ufpb.2525-5584.2021v6n2.57533

Alves, J. S. L. (2019). O papel da universidade na implementação dos objetivos de desenvolvimento sustentável das Nações Unidas: uma análise dos projetos de extensão da Universidade Federal da Paraíba [The role of the university in implementing the United Nations Sustainable Development Goals: an analysis of the extension projects at the Federal University of Paraíba]. (Under graduation project from Federal University of Paraíba). https://repositorio.ufpb.br/jspui/handle/123456789/16826

Amar, S., & Pratama, I. (2020). Exploring the link between income inequality, poverty reduction and economic growth: An ASEAN perspective. International Journal of Innovation, Creativity and Change, 11(2), 24–41. https://repo.uum.edu.my/id/eprint/27745/

Bastos, J. E. S., de Jesus Sousa, J. M., da Silva, P. M. N., & de Aquino, R. L. (2023). O uso do questionário como ferramenta metodológica: potencialidades e desafios [The questionnaire as a methodological tool: potentialities and challenges]. Brazilian Journal of Implantology and Health Sciences, 5(3), 623–636. https://doi.org/10.36557/2674-8169.2023v5n3p623-636

Batista, B., Rodrigues, D., Moreira, E., & Silva, F. (2021). Técnicas de recolha de dados em investigação: inquirir por questionário e/ou inquirir por entrevista [Data collection techniques in research: surveying by questionnaire and/or interviewing]. In Sá, P. et al. Reflexões em torno de metodologias de investigação: recolha de dados [Reflections on research methods: data gathering], 2, 13–36. https://doi.org/10.34624/ka02-fq42

Benites, L. L. L., & Polo, E. F. (2013). A sustentabilidade como ferramenta estratégica empresarial: governança corporativa e aplicação do Triple Bottom Line na Masisa [Sustainability as business strategic: the coroprate governance and application of the Triple Bottom Line in Masisa]. Revista de Administração da Universidade Federal de Santa Maria, 6, 827–841. https://www.redalyc.org/pdf/2734/273428928002.pdf

Borsatto, J. M. L. S., Marcolin, C. B., Abdalla, E. C., & Amaral, F. D. (2024). Aligning community outreach initiatives with SDGs in a higher education institution with artificial intelligence. Cleaner and Responsible Consumption, 12, 100160. https://doi.org/10.1016/j.clrc.2023.100160

Campos, M. C., & Almeida, L. A. (2019). Institucionalização da gestão sustentável em instituições de ensino superior: um estudo com servidores da Universidade Federal do Tocantins [Institutionalization of sustainable management in higher education institutions: a study with employees of the Federal University of Tocantins]. Revista Brasileira de Gestão Ambiental e Sustentabilidade, 6(14), 819–840. https://doi.org/10.21438/rbgas.061414

Claro, P. B. O., Claro, D. P., & Amâncio, R. (2008). Entendendo o conceito de sustentabilidade nas organizações [Understanding the sustainability concept in organizations]. Revista de Admnistração, 43(4), 289–300. http://www.redalyc.org/articulo.oa?id=223417504001

Christou, O., Manou, D. B., Armenia, S., Franco, E., Blouchoutzi, A., & Papathanasiou, J. (2024). Fostering a whole-institution approach to sustainability through systems thinking: an analysis of the state-of-the-art in sustainability integration in higher education institutions. Sustainability, 16(6), 2508. https://doi.org/10.3390/su16062508

Ferreira, P., & Almeida, M. R. R. (2024). Percepção dos docentes da Universidade Federal de Itajubá sobre os Objetivos de Desenvolvimento Sustentável [Professors' perception at Federal University of Itajubá on the Sustainable Development Goals]. Revista GUAL, 17(1), 160-180. https://doi.org/10.5007/1983-4535.2024.e95313

Figueiró, P. S., Neutzling, D. M., & Lessa, B. (2022). Education for sustainability in higher education institutions: A multi-perspective proposal with a focus on management education. Journal of Cleaner Production, 339, 130539. https://doi.org/10.1016/j.jclepro.2022.130539

Frizon, J. A., & Eugénio, T. (2022). Recent developments on research in sustainability in higher education management and accounting areas. The International Journal of Management Education, 20(3), 100709. https://doi.org/10.1016/j.ijme.2022.100709

Gomes, M. F., & Ferreira, L. J. (2018). Políticas públicas e os objetivos do desenvolvimento sustentável [Public policies and the objectives of sustainable development]. Direito e Desenvolvimento, 9(2), 155–178. https://doi.org/10.25246/direitoedesenvolvimento.v9i2.667

Gomes, S. F., Jorge, S., & Eugénio, T. P. (2020). Teaching sustainable development in business sciences degrees: evidence from Portugal. Sustainability Accounting, Management and Policy Journal. https://doi.org/10.1108/SAMPJ-10-2019-0365

Grandisoli, E., & Jacobi, P. R. (2020). O paradigma da sustentabilidade [The sustainability paradigma]. Instituto de Estudos Avançados da USP—IEA USP. http://www.iea.usp.br/pesquisa/projetos-institucionais/usp-cidades-globais/artigos-digitais/o-paradigma-da-sustentabilidade

Jacobi, P. (2003). Educação ambiental, cidadania e sustentabilidade [Environmental education, citizenship, and sustainability]. Cadernos de Pesquisa, (118), 189–205. http://educa.fcc.org.br/pdf/cp/n118/n118a08.pdf

Klein, L. L., Alves, A. C., Abreu, M. F., & Feltrin, T. S. (2022). Lean management and sustainable practices in Higher Education Institutions of Brazil and Portugal: A cross country perspective. Journal of Cleaner Production, 342, 130868. https://doi.org/10.1016/j.jclepro.2022.130868

Lara, P. T. R. (2012). Sustentabilidade em instituições de ensino superior [Sustainability in higher education institutions]. REMOA - Revista Monografias Ambientais, 7(7), 1646–1656. https://core.ac.uk/download/pdf/231170306.pdf

Leal Filho, W., Frankenberger Silva, F., Salvia, A., Shiel, C., Paço, A., Price, E.,... & Pretorius, R. W. (2023). An overview of research trends on sustainability in higher education–an exploratory study. International Journal of Sustainability in Higher Education, 24(5), 1161–1175. https://doi.org/10.1108/IJSHE-08-2022-0252

Medeiros Júnior, J. F., & Baracho, H. U. (2016). Desenvolvimento econômico, equilíbrio ambiental e ética: desafios jurídicos e filosóficos no (do) Séc. XXI [Economic development, environmental balance and ethics: juridical and philosophical challenges of the 21st century]. Revista de Direito Ambiental e Socioambientalismo, 2(2), 86–106. https://doi.org/10.26668/IndexLawJournals/2525-9628/2016.v2i2.1617

Mikhailova, I. (2004). Sustentabilidade: evolução dos conceitos teóricos e os problemas da mensuração prática [Sustainability: evolution of theoretical concepts and practical measurement problems]. Revista Economia e Desenvolvimento, 1(16), 22–41. https://core.ac.uk/download/pdf/231146376.pdf

Mousa, M., Massoud, H. K., Ayoubi, R. M., & Abdelgaffar, H. A. (2020). Should responsible management education become a priority? A qualitative study of academics in Egyptian public business schools. International Journal of Management Education, 18(1). https://doi.org/10.1016/j.ijme.2019.100326

Nascimento, E. P., Amazonas, M., & Vilhena, A. (2013). Sustainability and interdisciplinarity: innovations and challenges for postgraduate programmes on the Environment and Society. The case of the Centre for Sustainable Development at the University of Brasilia. Revista Brasileira de Pós-Graduação, 10(21), 661–689. https://doi.org/10.21713/2358-2332.2013.v10.417

Oliveira, M. (2009). Universidade e sustentabilidade: proposta de diretrizes e ações para uma universidade ambientalmente sustentável [University and sustainability: proposal of guidelines and actions for an environmentally sustainable university]. (Master's thesis from the Federal University of Juiz de Fora). http://www.repositorio.ufjf.br:8080/jspui/bitstream/ufjf/3929/1/marciodeoliveira.pdf

Oliveira, I. D., Montaño, M., & Souza, M. P. (2009). Avaliação ambiental estratégica [Strategic environmental assessment]. São Carlos: Suprema.

Ribeiro, M. M. C., Leite, R. C. M., Franco, S. C., & Max, C. Z. (2018). Práticas de divulgação, conscientização e capacitação para a sustentabilidade: uma proposta para as universidades federais brasileiras [Sustainable practices of disclosure, awareness and training: a proposal for Brazilian Federal Universities]. Revista de Administração IMED, 8(1), 146–168. https://doi.org/10.18256/2237-7956.2018.v8i1.2138

Romeiro, A. R. (2001). Economia ou economia política da sustentabilidade? [Economics or political economy of sustainability?]. Campinas: Instituto de Economia/UNICAMP.

Sachs, J. D. (2017). A era do desenvolvimento sustentável [The era of sustainable development]. Rio de Janeiro: Intrínseca.

Schneider, J. O. (2015). Cooperativismo e desenvolvimento sustentável [Cooperativism and sustainable growth]. Otra Economía, 9(16), 94–104. http://revistas.ungs.edu.ar/index.php/otraeconomia/article/view/otra.2015.916.07/720

Souza, L. L., Ferreira, L. V., Mendes, F. A. T., Borges, N. S., Costa, J. M., Ferreira, E. Y. de C. S., Aleixo, L. L. de S., & Silva, E. V. S. (2020). The debate around sustainability: sustainable rural development. Brazilian Journal of Development, 6(12), 96305–96322. https://doi.org/10.34117/bjdv6n12-209

Souza, M. C. S. A., & Armada, C. A. S. (2017). Desenvolvimento sustentável e sustentabilidade: evolução epistemológica na necessária diferenciação entre os conceitos [Sustainable development and sustainability: epistemological evolution in the necessary differentiation between concepts]. Revista de Direito e Sustentabilidade, 3(2), 17–35. https://doi.org/10.26668/IndexLawJournals/2525-9687/2017.v3i2.2437

Tommasiello, M. G. C., & Guimarães, S. S. M. (2013). Sustentabilidade e o papel da universidade: desenvolvimento sustentável ou sustentabilidade democrática? [Sustainability and the role of the university: sustainable development or democratic sustainability?]. Revista de Educação do COGEIME, (43), 11–26. https://doi.org/10.15599/0104-4834/cogeime.v22n43p11-26

Torini, D. (2016). Questionários on-line [Online Questionnaires]. In: Abdal, A., Oliveira, M. C. V., Ghezzi, D. R., & Santos Jr., J. Métodos de pesquisa em ciências sociais: Bloco Quantitativo [Research methods in social sciences: quantitative module], 52–75. 2016_E-BOOK_Sesc-Cebrap__Metodos_e_tecnicas_em_CS_-_Bloco_Quantitativo20190328-63448-1a2u85g-libre.pdf

Trevisan, L. V., Leal Filho, W., & Pedrozo, E. A. (2024). Transformative organisational learning for sustainability in higher education: A literature review and an international multi-case study. Journal of Cleaner Production, 447. https://doi.org/10.1016/j.jclepro.2024.141634

UNIFEI. Universidade Federal de Itajubá. (2023). UNIFEI em números [UNIFEI in numbers]. https://numeros.unifei.edu.br/pessoal/

Vaz, C. R., Fagundes, A. B., Oliveira, I. L., Kovaleski, J. L., & Selig, P. M. (2010). Sistema de gestão ambiental em instituições de ensino superior: uma revisão [Environmental management system in higher education institutions: a review]. Revista Gestão da Produção Operações e Sistemas, (3), 45–45. https://doi.org/10.15675/gepros.v0i3.327

Mileni Cristine das Chagas Azevedo Undergraduate student in Environmental Engineering at the Federal University of Itajubá.

Maria Rita Raimundo e Almeida Professor at the Federal University of Itajubá with a Ph.D. in Environmental Engineering Sciences.

Sara Ingrid da Silva Valim Undergraduate student in Environmental Engineering at the Federal University of Itajubá.

Maria Tereza Alves Martins de Faria Undergraduate student in Environmental Engineering at the Federal University of Itajubá.

Frederico de Souza Nascimento Undergraduate student in Environmental Engineering at the Federal University of Itajubá.

Sustainability in Higher Education in Brazil: An Overview of Contributors to the Literature

Walter Leal Filho, Jucelia Appio Frizon, Amanda Lange Salvia, Priscilla Cristina Cabral Ribeiro, and Gabriele da Cunha Lopes

Abstract Sustainability in higher education has emerged as a key research area globally, with Brazilian higher education institutions (HEIs) playing a significant role in advancing sustainable practices and education. This paper offers a comprehensive overview of key contributors to the literature on sustainability in higher education in Brazil, highlighting the substantial efforts and innovative initiatives undertaken by various universities to integrate sustainability into their curricula, campus operations, and community engagements. The study examines research on sustainability in higher education in Brazil from 2015 to 2024, utilizing a quantitative approach through a bibliometric review. A total of 158 articles were included for bibliometric analysis utilizing the VOSviewer software. Campus sustainability initiatives, community engagement, policy and governance, and interdisciplinary research. Key publications and journals featuring Brazilian contributions are also discussed. This paper underscores the transformative impact of these institutions in promoting sustainable practices and education, positioning Brazil as a leader in the

W. L. Filho
Department of Natural Sciences, Manchester Metropolitan University, Manchester, UK
e-mail: walter.leal2@haw-hamburg.de

W. L. Filho · A. L. Salvia
Hamburg University of Applied Sciences, Research and Transfer Centre "Sustainable Development and Climate Change Management", Ulmenliet Hamburg, Germany
e-mail: Amanda.Lange-Salvia@haw-hamburg.de

J. A. Frizon
Centre of Applied Research in Management and Economics, Polytechnic Institute of Leiria, Leiria, Portugal

School of Administration, Western Paraná State University, Francisco Beltrão, Paraná, Brazil

P. C. C. Ribeiro
Department of Industrial Engineering, Engineering School, Fluminense Federal University, Rio de Janeiro, Brazil
e-mail: priscillaribeiro@id.uff.br

G. da Cunha Lopes (✉)
School of Administration, Federal University of Paraná, Paraná, Brazil
e-mail: gabriele.clopes@gmail.com

W. Leal Filho et al. (eds.), *Sustainability in the Curriculum and Operations of Universities*, World Sustainability Series,
https://doi.org/10.1007/978-3-032-12678-8_22

global sustainability discourse within higher education. Through this examination, the paper aims to illuminate the critical role that higher education plays in driving sustainability initiatives and fostering a culture of environmental stewardship and social responsibility, thereby contributing to the broader objectives of sustainable development. It concludes with recommendations for enhancing sustainability practices in Brazilian universities and calls for continued research and collaboration to strengthen the impact of higher education in addressing the pressing challenges of sustainability in a rapidly changing world.

Keywords Sustainability · Higher education · Universities · Teaching · Research

1 Sustainability in the Higher Education Sector in Brazil

The urgency to address environmental issues puts universities around the globe in a leading position on the sustainable development agenda (Pires and Pereira 2023). For decades, sustainability researchers have insisted on transforming higher education institutions into more sustainable and inclusive campuses. Yet, as the 2030 Agenda seems unlikely, universities are struggling to meet the fourth Sustainable Development Goal (SDG 4) before the 2030 deadline (Gamage et al. 2022). Reducing the environmental impacts of campus operations is a key step in achieving sustainability at universities (Pires and Pereira 2023). Universities are the spaces—often referred to as "living laboratories"—where future professionals and decision-makers are trained to live and work in smart cities and to understand the current and future reality of urban spaces (Dalla Gasperina et al 2022).

Over the past two decades, recognition of universities' role in supporting the transformation of societies into sustainable societies has grown. Among the possible actions, the institutionalization of sustainability in higher education stands out, seeking the transition to a model that has been called a sustainable university (Mazon et al. 2020). Even though the significant achievements in sustainability are within sustainable development's environmental or economic dimensions, the social dimension is still a pending task (Gamage et al. 2022). Intending to study more about this dimension in these themes, curricula, campus operations, and community engagement should be discussed for a better understanding of the context in Brazilian universities.

The Latin America region, which is extended from Mexico in the North, to Argentina and southern Chile in the south, including southern Brazil, is characterized by a set of developing countries. The region has many environmental problems, which include degradation of ecosystems, exposure to contaminants, deforestation and increased pollution (Leal Filho et al. 2021a, b).

Brazil has more than 2500 higher education institutions (HEIs), around 310,000 teachers and around 10 million students (National Institute of Educational Studies and Research Anísio Teixeira 2023), where students spend 4–10 h a day learning and living. Although Brazilian universities have been implementing green practices and taking part in sustainability rankings, their performances on these platforms are poorly appreciated. Therefore, the Brazil is a fertile ground for universities to adopt the SD agenda, but there are few studies investigating how HEIs in the country engage in university sustainability rankings (Pires and Pereira 2023).

Related to community engagement (Bizerril et al. 2018), in their study about Portuguese-speaking countries, discovered weaknesses in the sample analysed in the HEI operations and institutional framework, including Brazilian universities. Frandoloso and Scorsatto (2021) published a paper that described the results of a project of the University of Passo Fundo (UPF) concerning the so-called "Beira Trilhos" territory. In these areas along the railway that cut the urban area of Passo Fundo, Brazil, a community composed of approximately 1500 families occupied this territory. This extension project aimed to contribute to the legal, urban, and social aspects, expanding its perspectives and interfaces. Besides, it seeks a solution that contributes to the construction of houses for populations regarding the University's Social Responsibility Policy. The project aimed to improve the quality of life and respect social, legal, cultural, and aesthetic needs, according to the Sustainable Development Goals in a global and integrated perspective.

Also at UPF, Dalla Gasperina et al. (2022) found smart practices that (mainly) reach quality education (SDG 4) and innovation and infrastructure (SDG 9), especially due to aspects of teaching technologies for the new classroom models and the optimization of campus infrastructure management. Related to campus operations, the main practices were waste reduction and reuse; awareness of mobility; the institution's computer system for better organization and records of orders, calls, and maintenance; smart app of real-time data; online guidance books; consumption, cost, and quality management; intelligent LED lighting; open data online; camera monitoring; and control and reduction of electricity consumption in lighting and air conditioning systems.

In Brazil, there is a lot of potential for improvement of campus sustainability, especially regarding different aspects of campus operations (Pires and Pereira 2023). The University of São Paulo's (USP) Superintendence of Environmental Management (SGA) adds appeal to faculty members, as their project ideas, input, and expertise can be incorporated into landmark campus retrofits and new builds, coordinating the environmental dimension of institutional strategic plans. USP has seven multi-city campuses, and according to the Environmental Master Plan, each campus establishes local priorities and pathways in compliance with minimum standards prescribed centrally. The general mission is to make the university's built environment operationally sustainable and to reach ambitious goals for solarization, "Zero Waste by 2025" strategy (Robinson et al. 2023).

Some papers discussed the themes of HEIs and sustainable actions, such as Mazon et al. (2020), where they analysed scientific articles to identify implementation approaches for sustainable development and the involvement of the technical staff in the process, the teachers, and the students. These efforts are focused on a 'smart' teaching process, curricular discussions, and changes for improvements, adding sustainability to classes' contents.

In Brazil, since 2004, the National System of Higher Education Evaluation has been in operation (National Institute of Educational Studies and Research Anísio Teixeira 2022). It evaluates the performance of graduates of undergraduate courses due to the programmatic contents in the courses' curricular guidelines and the development of skills and abilities to allow for the deepening of general and professional training, institutions, and student performance. It considers aspects such as teaching, research, extension, social responsibility, and management of the institution and faculty (National Institute of Educational Studies and Research Anísio Teixeira 2022; Palma et al. 2023).

According to Robinson et al. (2023), the University of São Paulo (USP) recognizes that many sustainability curriculum initiatives do not attract the wider student body. Because of this, the University offers a series of introductory undergraduate sustainability courses for all, which were developed and e-delivered in collaboration with universities in seven countries across the GreenMetric World University Rankings Network.

In a study with 90% of Business Administration undergraduate majors in Brazilian federal universities, Palma et al. (2023) found that most courses related to sustainability continued to require 60 h of class time (2% of the total course load in most majors). Compared with a previous study in 2011, at that time, most majors focused on sustainability at more advanced stages of the programs, while now this class is in intermediate semesters.

It is widely acknowledged that research and publications in peer-reviewed journals offer important metrics in describing the academic outputs of higher education institutions on the one hand and their societal impacts on the other. Peer review is a well-tested method for quality control and has been successfully deployed over many decades in academic journals worldwide (Leal Filho et al. 2021a, b). Despite increasing global interest in sustainability in higher education, there remains a lack of comprehensive and up-to-date analysis of the literature of publications in developing countries, particularly Latin America. Existing studies tend to focus on isolated case studies and specific approaches, without offering a broader synthesis of the topic, authors and emerging themes. This study aims to fill in this gap, and focus on Brazil given its extension, number of institutions and regional significance. The review is guided by the following research questions: (1) What is the current state of the scientific literature on sustainability in higher education in Brazil? (2) Who are the main authors, institutions, and sources contributing to this body of research? (3) What are the main themes, trends, and knowledge gaps identified in this literature?

By answering these questions, this paper aims to provide an overview of **key contributors to the literature on sustainability in higher education in Brazil**, highlighting the substantial efforts and innovative initiatives undertaken by various universities in integrating sustainability into their operations, including curriculum integration, campus sustainability initiatives, community engagement, policy and governance, and interdisciplinary research.

2 Methodology

This study covers research on **sustainability in higher education** from 2015 (to coincide with the launch of the 2030 Agenda) to 2024. It adopted a quantitative approach, which consisted of a bibliometric review (Abad-Segura and González-Zamar 2021). Responding to recommendations, a bibliometric analysis is based on the analyses of a series of research aspects, such as journals, the number of articles per year, fields of research, authors and co-authors, collaborations, institutions, and countries (Trevisan et al. 2024).

Scopus is the most widely used database for bibliometric analysis (Trevisan et al. 2024) and is considered the largest repository of peer-reviewed resources (Abad-Segura and González-Zamar 2021). Among the main advantages of Scopus is the variety of data provided by each selected article, its analysis, and the comparison between them (Abad-Segura and González-Zamar 2021). Several examples of bibliometric analyses on the subject of sustainability in higher education are available on this database (Abad-Segura and González-Zamar 2021; Dawodu et al. 2022; Leal Filho et al. 2024; Halmaghi et al. 2023; Meseguer-Sánchez et al. 2020; Razali et al. 2024; Shi and Wan 2024; Trevisan et al. 2024). Therefore, this research started by identifying search terms in the Scopus database. The search was held on May 6, 2024, and used the following combination query of keywords: (sustainability) and (sustainable AND in AND higher AND education) or (sustainable AND in AND higher AND education). This combination query should be present in the titles, abstracts, and/or keywords of the publications. The following criteria were also established and included: articles published between 2015 to May 2024; limited to Brazil as the country/territory of the authors; in English; and article as document type. Figure 1 outline the process mentioned.

A total of 162 articles were initially obtained from the search, but four were excluded due to missing DOIs. The remaining 158 articles underwent a bibliometric analysis, which covered several aspects: number of articles per year, research categories, journal, fields of research, authors, co-authors, co-authorship analysis, citations analysis, and the creation of bibliometric maps. Bibliometric mapping helps detect similarities between articles (Abad-Segura and González-Zamar 2021), and VOSviewer, a freely available program developed by van Eck and Waltman (2010), was used for this purpose.

3 Results

3.1 Year

Figure 2 illustrates the number of publications over the years, with a peak in 2022, reaching 32 publications. In contrast, 2016 saw the lowest number of publications on the topic. This trend is likely influenced by the adoption of the 2030 Agenda in 2015,

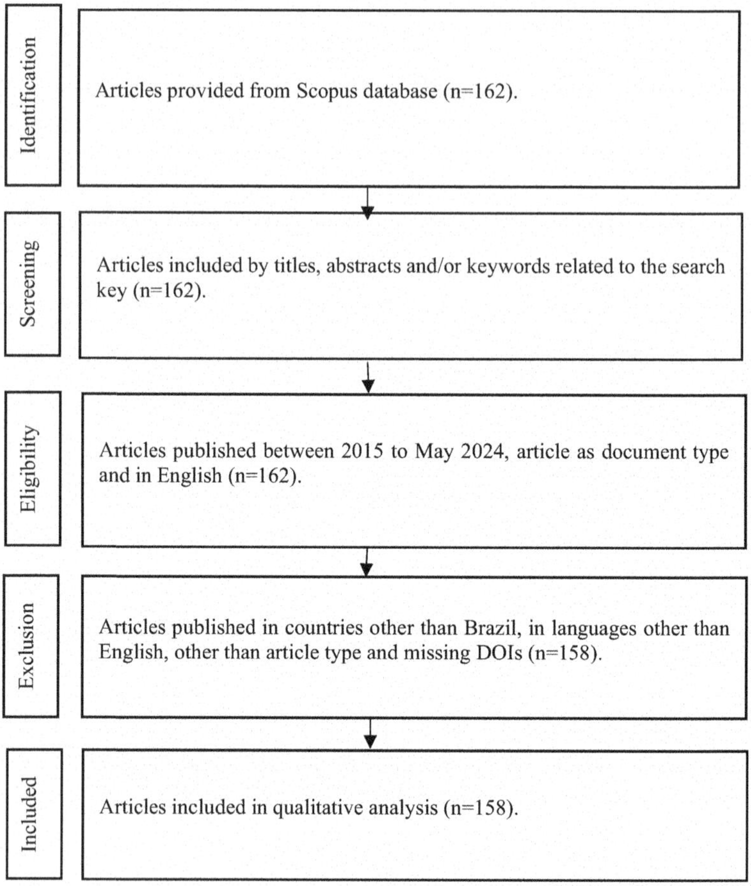

Fig. 1 Research approach

as interest in the subject gradually increased in the following years. The decrease in publications in 2020 may be attributed to disruptions caused by the COVID-19 pandemic, affecting global research activities (Leal Filho et al. 2022).

In the chart below, the years are listed, and the number of publications per year is shown. The visualization shows the number of publications published each year.

3.2 *Research Categories—Sustainable Development Goals*

After the filters, within 158 documents found in the Scopus search, 141 papers were selected aligned with the Sustainable Development Goals (SDGs), as shown in Fig. 3. A clear emphasis is placed on Quality Education (SDG 4), with 98 papers focusing on this goal, making it the most prominent topic among the publications. Responsible

Fig. 2 Growth trajectory of the thematic of Sustainability in Higher Education by Brazilian researchers. *Note* Scopus 2015–May 6, 2024 (n = 158)

Consumption and Production (SDG 12) follows with 13 papers, while Affordable and Clean Energy accounts (SDG 7) for 8 papers. Other SDGs, such as Sustainable Cities and Communities (SDG 11) and Partnerships for the Goals (SDG 17), each had five papers. Several other goals, including Zero Hunger (SDG 2), Climate Action (SDG 13), and Life on Land (SDG 15), had only a few papers, highlighting the varying degrees of attention these goals have received in the academic literature.

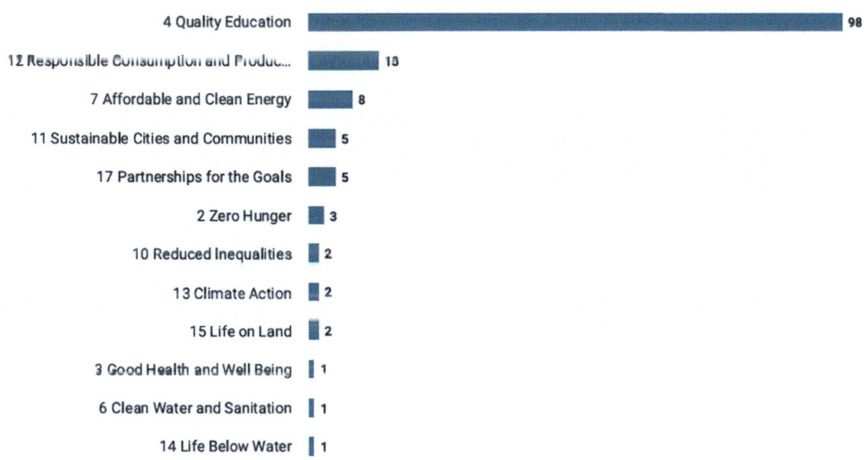

Fig. 3 Number of publications in each research category in the SDGs

3.3 Journal, Fields of Research, and Metrics

The bibliometric analysis reveals that the publications are distributed across 54 journals. Table 1 highlights the top 10 journals, including the number of publications, research fields, and impact factors (JJR, CiteScore, and impact factor). The most prolific journal is the International Journal of Sustainability in Higher Education, with 49 publications, followed by the Journal of Cleaner Production, with 31 publications, and Sustainability, with 17 publications. This distribution underscores the interdisciplinary nature of the research, covering fields such as sustainability in education, engineering, and the built environment. The VOSviewer results align with the broader scientific landscape, indicating that journals focused on sustainability in education (e.g., International Journal of Sustainability in Higher Education and Sustainability) and sustainable engineering practices (e.g., Journal of Cleaner Production) have prominently featured studies developed by Brazilian authors.

3.4 Authors', Co-Authors, and Organization for Country

The authors and co-authors represent a wide range of affiliations, as shown in Table 2. Among the most prolific researchers are Walter Leal Filho, with 46 papers and 2429 citations; Amanda Lange Salvia, with 29 papers and 999 citations; Luciana Londero Brandli, with 23 papers and 1741 citations; and Rosley Anholon, with 15 papers and 711 citations. The articles were selected based on the criterion that at least one author or co-author identified as Brazilian. The analysis highlights significant international collaboration, with contributions from researchers in countries such as Germany, Chile, Portugal, Malta, the United States, the United Kingdom, and Japan.

Collaborative structure indicators, which measure links between authors and countries, have been analysed using processing tools and network maps due to their reliability in bibliometric analysis (Abad-Segura and González-Zamar 2021). Networks of scientific collaboration have been systematically studied since 1960 by researchers from various disciplinary backgrounds. In general, it is agreed that institutional, national, and global research collaborations play a vital role in leading the research (Qasim 2017). In the studies, the importance of research synergies and collaborations are observed and are well acknowledged by several studies (Abad-Segura and González-Zamar 2021; Leal Filho et al. 2024; Meseguer-Sánchez et al. 2020; Razali et al. 2024; Trevisan et al. 2024).

3.5 Co-Authorship Analysis

In this study, we conducted a co-authorship analysis to assess the level of collaboration between researchers. A co-authorship network forms when two or more authors

Table 1 Top 10 journals per number of publications

Rank	Journal	NN	Fields of research	JJR 2022	CiteScore 2022	Impact factor (2023)
1	International Journal of Sustainability in Higher Education	49	Education; Education Systems; Education Policy, Sociology and Philosophy; Environmental Sciences; Environmental Management; Curriculum and Pedagogy; Pollution and Contamination	0.752	6.1	3
2	Journal of Cleaner Production	31	Engineering; Built Environment and Design; Manufacturing Engineering	0.981	18.5	9.7
3	Sustainability	17	Commerce, Management, Tourism and Services; Education; Education Policy; Sociology and Philosophy; Education Systems; Strategy, Management and Organisational Behaviour	0.664	5.8	3.3
4	Environment, Development and Sustainability	8	Agricultural, Veterinary and Food Sciences; Environmental Sciences; Human Society; Agriculture, Land and Farm Management; Environmental Management; Development Studies	0.835	4.90	4.7

(continued)

Table 1 (continued)

Rank	Journal	NN	Fields of research	JJR 2022	CiteScore 2022	Impact factor (2023)
5	International Journal of Sustainable Development & World Ecology	6	Environmental Sciences; Human Society; Environmental Management; Development Studies	0.922	9.2	6.5
6	The International Journal of Management Education	5	Commerce, Management, Tourism and Services; Education; Strategy, Management and Organisational Behaviour; Education Systems; Education Policy, Sociology and Philosophy	1.060	7.1	6.0
7	Sustainability Science	2	Education; Education Systems; Education Policy, Sociology and Philosophy	1.530	6.0	5.1
8	Gestão & Produção	2	Commerce, Management, Tourism and Services; Economics; Strategy, Management and Organisational Behaviour; Applied Economics	0.195	1.2	0.800

(continued)

Table 1 (continued)

Rank	Journal	NN	Fields of research	JJR 2022	CiteScore 2022	Impact factor (2023)
9	Environmental Quality Management	2	Information and Computing Sciences; Education; Environmental Sciences; Education Systems; Library and Information Studies	0.291	1.8	1.5
10	Sustainable Development	2	Education; Education Policy, Sociology and Philosophy; Education Systems	2.040	15.2	4.9

Note In the table only journals with more than one papers published

collaborate on a publication (Razali et al. 2024). In Fig. 4, the size of the nodes reflects the number of articles produced by each author, highlighting the intensity of research collaboration. Larger nodes indicate authors with a higher publication output, while the node colours represent different clusters, signifying the groups of researchers who tend to collaborate more frequently (van Eck and Waltman 2010). This visual representation underscores the dynamic and interconnected nature of the research community.

The map in this figure was constructed based on co-authorship data, with colors representing the clusters to which each author was assigned through a clustering technique. Walter Leal Filho from HAW Hamburg, Germany, stands out in the red cluster with 59 links across 46 published papers. The blue cluster includes Amanda Lange Salvia from the University of Passo Fundo, Brazil, who has 55 links and 29 papers, alongside Luciana Londero Brandli, also from the University of Passo Fundo, with 47 links and 23 papers. The yellow cluster comprises researchers primarily affiliated with universities in Santa Catarina, Brazil.

3.6 *Citations Analysis*

The co-citation map was created by considering only articles as document types. The authors selected the top 100 researchers with the highest number of citations, following van Eck and Waltman (2010) recommendations. Using the co-citation matrix, VOSviewer generated the co-citation map displayed in Fig. 5.

Table 2 Authors', co-authors and organization for country

Name	Organization, country	N	Citations	Citations mean
Walter Leal Filho	HAW Hamburg, Germany	46	2.429	52.80
Amanda Lange Salvia	University of Passo Fundo, Brazil	29	999	34.45
Luciana Londero Brandli	University of Passo Fundo, Brazil	23	1.741	75.70
Rosley Anholon	State University of Campinas, Brazil	15	711	47.40
Izabela Simon Rampasso	Catholic University of the North, Chile	12	308	25.67
Arminda Do Paço	University of Beira Interior, Portugal	12	689	57.42
Lucas Veiga Ávila	Federal University of Santa Maria, Brazil	11	1118	101.64
José Baltazar Salgueirinho Osório De Andrade Guerra	Universidade do Estado de Santa Catarina, Brazil	9	269	29.89
Osvaldo Luiz Gonçalves Quelhas	Fluminense Federal University, Brazil	9	527	58.56
Mark Christopher Mifsud	University of Malta, Malta	9	674	74.89
Laís Viera Trevisan	Federal University of Rio Grande do Sul, Brazil	8	18	2.25
Claudio Ruy Portela Vasconcelos	Federal University of Paraíba, Brazil	7	160	22.86
Fernanda Frankenberger	Pontifícia Universidade Católica do Paraná, Brazil	7	374	53.43
Ulisses Miranda Azeiteiro	University of Aveiro, Portugal	6	882	147.0
João Henrique Paulino Pires Eustachio	HAW Hamburg, Germany	6	163	27.17
Markus Will	HAW Hamburg, Germany	6	365	60.83
Todd Jared Levasseur	College of Charleston, United States	6	102	17.00
Valeria Ruiz Vargas	Manchester Metropolitan University, United Kingdom	5	993	198.60

(continued)

Table 2 (continued)

Name	Organization, country	N	Citations	Citations mean
Chris Shiel	Bournemouth University, United Kingdom	5	520	104.00
Aline Bento Ambrosio Avelar	Municipal University of São Caetano do Sul, Brazil	5	121	24.20
Ayyoob Sharifi	Hiroshima University, Japan	5	50	10.00
Tony Wall	Liverpool John Moores University, United Kingdom	5	222	44.4
Elizabeth A C Price	Manchester Metropolitan University, United Kingdom	5	259	51.80
Maria Alzira Pimenta Dinis	Fernando Pessoa University, Portugal	5	47	9.40
Carlos Rogério Montenegro De Lima	University of Southern Santa Catarina, Brazil	5	139	27.80

Note Shows authors and co-authors with more than 5 published articles

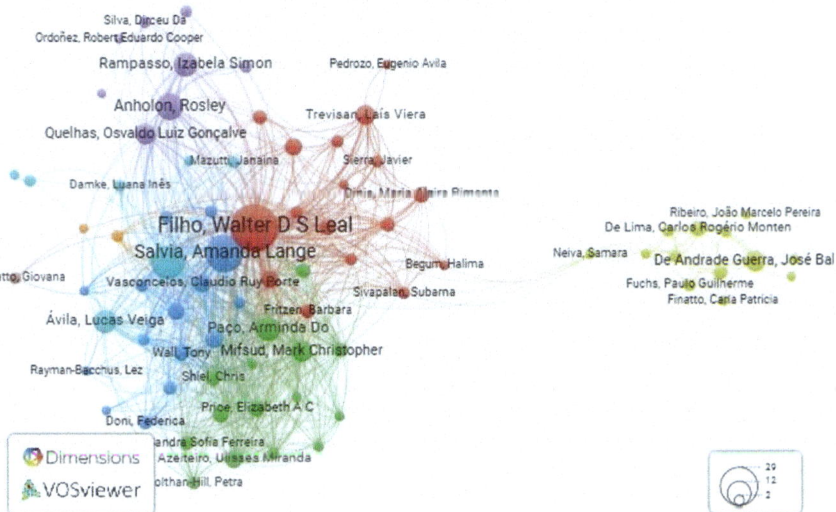

Fig. 4 Co-authorship analysis relationship between 100 researchers

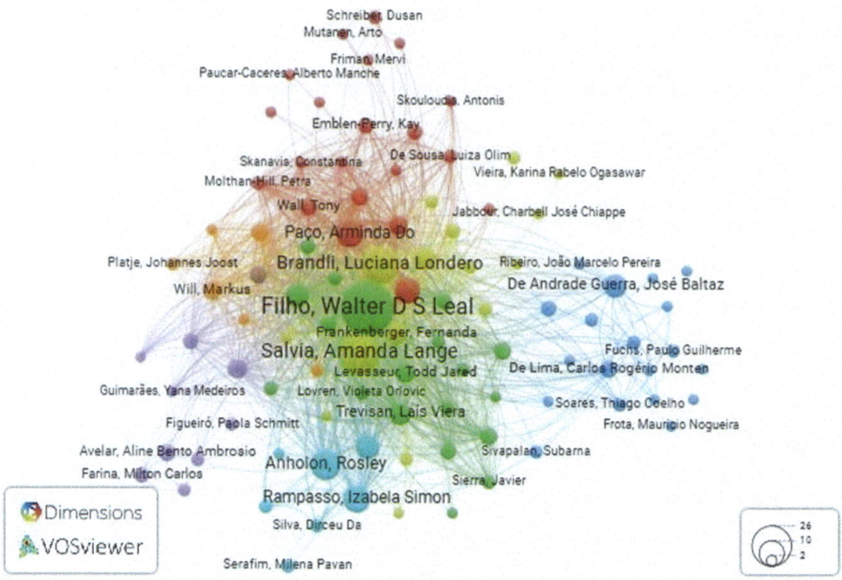

Fig. 5 Co-citation analysis relationship between 100 researchers

Our co-citation map provides a comprehensive overview of the relationships among authors in the field. Clusters of related researchers can be identified, with proximity in the map indicating closely related scientific groups. According to van Eck and Waltman (2010), clusters near one another reflect closely linked fields. The relatedness of researchers is determined by the frequency with which they cite each other (van Eck and Waltman 2018), often due to thematic affinities. The closer the authors are to the map, the more frequently they are co-cited. These graphic representations facilitate a clearer visual understanding of the knowledge connections among publications (Zhao et al. 2018).

The analysis indicates that authors such as Walter Leal Filho, with 1,685 citations; Luciana Londero Brandli, with 1075 citations; and Amanda Lange Salvia, with 1013 citations, highly represent the field. The lower citation counts for some authors may be attributed to their recent publications, which, while potentially significant, have yet to gain substantial recognition (Qasim 2017).

3.7 Opportunities and Challenges in Brazil and in Other Developing Countries Opportunities

According to the literature found in the search, the number of courses related to sustainability in Business Administration programs in Brazilian HEIs increased considerably in ten years. However, these courses still represent a low percentage of hours in the overall curriculum. Moreover, almost a third of Business Administration programs may produce graduates who have never addressed sustainability during their studies (Palma et al. 2023).

Leal Filho et al. (2021a, b) research shows that sustainability elements are being incorporated in more than 80% of the sampled universities, Latin American HEIs, and Brazilian institutions. "The curriculum innovation initiative showed that Latin-American HEIs are increasing their awareness of the importance of teaching sustainability in their academic programs" (Leal Filho et al. 2021a, b, p. 6).

According to Palma et al. (2023), the National Student Performance Exam results show that sustainability issues were increasingly addressed in the exams. Even though the questions require some interdisciplinary knowledge, they do not require critical thinking associated with challenges related to sustainability. In Brazil, since 2018, there has been an obligation to include extension practices in the disciplines of MEC. This is an opportunity for students to develop projects to improve results in small producers, to provide a better quality of life, and to learn by doing.

Some ordinary activities such as teaching and outreach provide HEIs the opportunity to support SD at a local level, while research engages the staff with national and international organizational change processes (Radinger-Peer and Pflitsch 2017). Besides, academic conferences include engagement with external stakeholders and lead the HEIs to integrate with these partners outside their environment, allowing exchanges and enlarging networks. Therefore, these conferences minimize external barriers supported by national policy frameworks, creating new opportunities for HEIs to achieve organizational change toward sustainable development (Vargas et al. 2019).

Leal Filho et al. (2023) mentioned that HEIs leadership can provide platforms for co-creation and collaboration among faculties and disciplines, which may have a greater impact on their initiatives. Besides, the HEIs' administrators can collaborate to create programs and initiatives with collective action within surrounding communities and take their sustainability training into regional jobs that require sustainability skills and competencies. These consequences of investments in SD in HEIs close the loop on best practice impacts of effective HEI governance for sustainability.

Challenges

One of the difficulties is related to the sustainability meanings. For professors and coordinators, it is necessary to understand it for curricula and course design. Another challenge is the understanding of the institutions' role in changes required by Sustainability (Palma et al. 2023).

Vargas et al. (2019) state that HEIs need an organizational change process to achieve sustainable development in their activities. The key new insight is that the existing networks identified in the policy frameworks may not support the effective integration of sustainable development in higher education. This occurs, first, because the stakeholders do not have sufficient interactions for the effective integration of sustainable development. Second, the policy frameworks are not sufficiently coordinated. Third, there is no clarity related to the financial model at the network level. Improvements in planned organizational change toward sustainable development in higher education could occur by increasing network density; establishing shared governance; and developing clear financial models for ensuring overall policy review updates.

Opportunities and Challenges in Developing Countries

There is a growing awareness that also teaching methods matter. The competencies and skills required in sustainability education stretch beyond what traditional education usually delivers. In South America, there is still a large inertia in educational methods in higher education (Coppens et al. 2020). For changes in this behaviour, HEIs need flexibility and agility. The flexibility in the education system supports different learning styles, learning pace, and circumstances, and agility in education extends beyond flexibility. It implies that the type of teaching and learning should demonstrate the implementation of transformative change, embracing creativity and innovation while breaking from conventional behaviours and modes of thinking (Rasli et al. 2024).

This flexibility concept confirms what Coppens et al. (2020) stated, related to Sustainable Development projects' implementation in HEIs, that is "project implementation had to be sensitive to national and local constraints and opportunities" in countries in the same category ("developing countries"), although in different geopolitical condition, culture, resources, ideology, governance and so on.

Another opportunity for HEIs is digital transformation and/or technological innovation, which is defined as a process of disruption and change in which institutions strategize to create value-added programs and activities (Rasli et al. 2024).

The unprecedented adoption of Information and Communication Technologies (ICT) devices by organizations and individuals' fuels e-waste, forcing governments, industry leaders, and regulators to battle for solutions to protect the environment and public health. Educational institutions invest in data centres to ensure continuous access to e-learning servers and support services for students and staff. Data centres are usually remotely located to provide a secure location for processing large amounts of data (Maphosa 2022). The author investigated the ICT adoption in HEI in Zimbabwe and confirmed that these technologies support green initiatives, reducing expenditures with travel to attend lectures, printing, and less electricity usage and cooling by limiting the usage of lecture rooms. On the other hand, the ICT usage increases e-waste. To solve this problem, Maphosa (2022) suggests more e-waste awareness programs to increase knowledge and influence the attitudes of ICT personnel when acquiring and maintaining ICT equipment. More institutions should be encouraged to adopt cloud computing technologies and reduce data centre storage to limit carbon emissions. The absence of elementary recycling in the country is worrying, and HEIs are losing valuable metals that are recoverable from e-waste.

In this context, blended learning (remote and in-person) has become the way forward for some institutions. This new trend implies different teaching tools that challenge educators, namely Big Data and Data Mining, Artificial Intelligence, Machine Learning, Data Science, STEM Education, Education 2.0, 4.0, 5.0; Neurotechnology, Bioinformatics, Hybrid Classrooms or Hy Flex, and Intelligent Tutoring (Deroncele-Acosta et al. 2023). Interestingly, while some educators accept these educational tools with open arms, others are reverting to the traditional option (Rasli et al. 2024). These new technologies for blended learning demand that professors must be trained for a new learning environment. These training courses should be supportive instead of demanding resources that universities cannot provide and skills that professors don't have.

Rasli et al. (2024) promoted a discussion of the use of ICTs and AI tools among five professors from Malaysia, the Netherlands, India, and Bangladesh. The professors from developing countries agreed that the HEIs need to be resilient and that digitalization must be moved forward. A professor from Malaysia answers a question, emphasizing the importance of preparedness to overcome the challenges after the pandemic time. Another professor from Malaysia detaches the interchanging benefits among students when they are provided with the opportunity to interact with peers from international universities in developing intercultural competencies and digital skills. In the end, there were other proposals, such as an educational reform (for all 5 interviewees), and from the professor of Bangladesh, who suggested a long-term educational sustainability that encompasses the integration of environmental considerations and the establishment of international networking.

In nine Pakistan public universities, the interviewees stated that there is a lack of governance for sustainability, which has three reasons, such as the absence of stakeholders' inclusivity; confusing governance that is a mixture of democratic and authoritarian styles; and a weak decision-making process that hampers PPUs from achieving institutional targets and objectives. Hence, the process is not yet mature enough to support the system, and sustainability is affected (Shah Bukhari et al. 2022). This issue causes three other problems:

- lack of institutional change—it has two major shortcomings, lack of visionary leadership and planning;
- lack of value, training, and commitment—hinders them from having an in-depth knowledge of leadership techniques, strategic planning, management, and decision-making for effective academic administration which may cause the missing of development, the decline of quality of education, and creates a barrier for sustainability;
- lack of resources—it depends on the other 2 because the training failure in securing, monitoring, and auditing financial activities affects the infrastructure and the development of universities. It is also linked with governance. If governance is strong, it creates opportunities for capacity building in areas of financial development (Shah Bukhari et al. 2022).

In Brazil, (Moura et al. 2019) discovered that 28 of 92 institutions offer courses and lectures in the area as their only form of sustainable practice. Sustainability appeared in majors such as engineering, architecture, and business. Narrowing the analysis to the 8 HEI with higher evaluations by MEC the main sustainable practices were behaviour, teaching, and curriculum.

It is also identified that the factors required to enable universities to undertake and include a sustainable change in all areas of the institution consist of a strong environmental policy, obtaining strategic resources, the importance of sustainability for future generations, implementing more pedagogical practices, developing a sense of community, inspire people to change behaviour and social policies, build a scientific understanding, connect local environmental issues and redefine values. In addition, leadership, incentives, and demand are needed to move forward in improving sustainable practices. Despite many good points, the authors emphasized that the implementation of sustainable practices requires a lot of dedication and financial assistance.

Rebelatto et al. (2019) analysed actions of energy efficiency implemented in a Brazilian university, such as replacing fluorescent lamps with LED lamps, constructing photovoltaic solar power generation, and opting for the free energy market. The results were double the global rate of improvement in energy efficiency; universal access to affordable, reliable, and modern energy services; and increase renewable energy in the global energy mix, respectively. Using SWOT analysis, the authors found an opportunity, to reach a certification to label the HEI or a building according to energy efficiency parameters or levels. The participation of the HEI in a free energy market can be analysed also as an economic approach considering the Brazilian electrical power bills, promoting financial savings and therefore investing in other issues. The threats were an absence of energy efficiency funding and the free energy market legislation changes.

Drahein et al. (2019) after analysing seven case studies in the South of Brazil, concluded that these organizations are adopting sustainable practices, but they lack coordination efforts in terms of governance. Other challenges were the lack of governmental incentives and the difficulty for managers to associate the curriculum with service operations. A solution for these HEIs that are in the technological education system, is to participate in the international efforts for sustainable development.

Leal Filho et al. (2018) in their multiple case studies selected one in Brazil. The results of this HEI showed that it is necessary to provide a new vision to engineers about the sustainable management models developed by companies. The integration of sustainability principles in technical education necessarily requires students not only knowledge but also skills and competencies related to sustainability, such as the development of critical thinking, future envisioning, systemic thinking, as well as the questioning of one's own values.

Contributions to the Literature

The papers contributed to the literature by detaching the opportunities for HEIs to implement SD practices, such as interdisciplinary knowledge and extension practices in undergraduate courses and investing in governance. The staff must engage with national and international organizational change processes, connecting with external stakeholders, and co-creating and collaborating among faculties and disciplines.

Related to the challenges, the papers suggested overcoming them to increase network density; establish shared governance and develop clear financial models ensuring overall policy review updates. Besides, HEIs need to be more agile and flexible, searching for new pedagogical practices, technology innovations (including ICTs and AIs tools) for classes (in person and remote) and campus operations, changes in behaviours for waste reduction, and use of training from the teaching activities for applications with the community.

4 Discussions

The integration of sustainability in higher education is a vital endeavour that addresses global challenges related to environmental stewardship, social responsibility, and economic viability. The literature reflects a robust engagement with these themes across various contexts, particularly in the context of the SDGs. In Brazil, HEIs are increasingly recognizing the significance of embedding sustainability into their curricula and operational practices, yet they face unique challenges and opportunities. The metrics derived from the VOSviewer and Dimensions platforms encompass various aspects, including growth trajectories, the number of publications by SDG, the distribution of journals by publication count, and information on authorship, co-authorship, and co-citation analysis. We analysed the ten most cited articles to gain deeper insights into the gaps, barriers, and opportunities identified in the literature. Table 3 presents the results of this analysis.

One of the primary themes in sustainability literature is the incorporation of SDGs into academic programs. This integration enhances the educational experience and fosters student engagement with sustainability concepts. However, gaps exist in terms of resources, training, and institutional support, which can hinder effective implementation. Brazilian universities often grapple with these challenges, including a lack of structured frameworks for SDG integration. Other barriers include a lack of resources, materials, time, effective planning, focus, an active environmental committee, applicability and continuity, resistance to behavioral change, absence of formal policies, inadequate coordination, lack of interest from administration or the academic community, weak partnerships, insufficiently trained staff, and limited institutional support (Avelar et al. 2019; Leal Filho et al. 2019a, b, c). Notably, some gaps, such as the absence of formal policies and partnerships, significantly contribute to these barriers. These obstacles create voids, hinder progress, and restrict access to further opportunities.

Effective leadership is crucial for advancing sustainability initiatives in higher education. Research indicates that sustainability leaders within HEIs play a pivotal role in driving change and fostering a culture of sustainability (Leal Filho et al. 2020; Leal Filho et al. 2019a, b, c). However, many Latin American institutions encounter a lack of interest or support from the administration and faculty, which can hinder progress (Leal Filho et al. 2021a, b). Establishing formal policies prioritizing sustainability is essential in creating an enabling environment for these initiatives, and Brazilian universities have the opportunity to develop comprehensive sustainability policies that align with international standards and address local needs and contexts (Galleli et al. 2022).

Regarding the gaps in the actions of HEIs, the articles indicate that formal structures are critical for the effective implementation of sustainability initiatives within

Table 3 Gaps, barriers and opportunities provided from the most cited articles

Title	Gaps	Barriers	Opportunities	References
Sustainable development goals and sustainability teaching at universities: falling behind or getting ahead of the pack?	Integrate SDG in learning process at curriculum, informal way in lectures, research and extension activities, conferences, courses and capacity, living labs, isolated actions, master and dictate program, etc.	Contribute to SDG 14 and 1; lack of opportunity, materials, time, among others	Combine curricula with SDG them, use learning methods and transformative approaches, engage with students	https://doi.org/10.1016/j.jclepro.2019.05.309
The role of transformation in learning and education for sustainability	Researching and rethinking how appropriate and successful educational practices	Bureaucratic systems; resistance; conservative mentality; lack of funding; limited capacity to integrate sustainability concepts in the curricula	Collaborative approaches; redesign disciplines; include a multicultural vision of sustainability	https://doi.org/10.1016/j.jclepro.2018.07.017
Barriers to innovation and sustainability at universities around the world	Relationship between university and external partners	Lack of planning and focus, environmental committee, applicability and continuity, resistance to changes in behavior, among others	Building new technology	https://doi.org/10.1016/j.jclepro.2017.07.025
Identifying and overcoming obstacles to the implementation of sustainable development at universities	Establish individual implications	Lack of support from management, appropriate technology, awareness and concern, among other	Improve campus operation, research and teaching. Formal structures to guide SD implementation	https://doi.org/10.1080/1943815X.2017.1362007

(continued)

Table 3 (continued)

Title	Gaps	Barriers	Opportunities	References
Sustainable development policies as indicators and pre-conditions for sustainability efforts at universities	Polices and practice regarding SD	Lack of formal policy	Engagement of SD in practice	https://doi.org/10.1108/IJSHE-01-2017-0002
A framework for the implementation of the sustainable development goals in university programs	Contribution for SDG two	Implement in university structures	Include SDG holistically in the programs	https://doi.org/10.1016/j.jclepro.2021.126915
The role of higher education institutions in sustainability initiatives at the local level	Offer information and training programs on local partnerships to staff at universities	Lack of resources/incentives and lack of coordination	Cooperation; local communities'	https://doi.org/10.1016/j.jclepro.2019.06.059
Sustainability leadership in higher education institutions: an overview of challenges	Investments in education for sustainable development (ESD), sustainable procurement, and reporting	Lack of interest by the university administration and among some members of the academic community, as well as lack of expertise and materials or resources	Institutional programs to incentive SDG adoption	https://doi.org/10.3390/su12093761
Education for advancing the implementation of the sustainable development goals: a systematic approach	Train responsible managers for SDG issues	Partnership with corporations to extend knowledge	Implementation of polices	https://doi.org/10.1016/j.ijme.2019.100322

(continued)

Table 3 (continued)

Title	Gaps	Barriers	Opportunities	References
Planning and implementing sustainability in higher education institutions: an overview of the difficulties and potentials	Adhere to SDGs and another's international initiatives	Limited resources, lack of trained staff, institutional support, planning and limited emphasis on approaches, such as problem-based learning	Adhere UN Declaration "The World we Want"	https://doi.org/10.1080/13504509.2018.1461707

HEIs (Leal Filho et al. 2017). The lack of institutional policies that integrate sustainability into all processes and activities results in reduced adherence to the SDGs within higher education (Leal Filho et al. 2019a). This is crucial because, without formal structures, promoting sustainable practices may be inconsistent or neglect important sustainable issues. Consequently, education, research, and extension activities may suffer, evidenced by the barriers identified (Leal Filho et al. 2018). Several studies indicate that investments are necessary to enhance sustainability initiatives (Leal Filho et al. 2019a, b, c; Leal Filho et al. 2019a, b, c). Leal Filho et al. (2019a, b, c) encourage HEIs to invest in learning processes across various platforms, including curriculum development, informal lectures, research and extension activities, conferences, courses, capacity-building initiatives, and living labs. Furthermore, Leal Filho et al. (2019a, b, c) advocate for investment in reporting progress to underscore the importance of sustainability to society. Finally, we highlight the gap in partnerships (Avelar et al. 2019; Leal Filho et al. 2019a, b, c). Engaging with external actors, such as private organizations and civil society, is crucial to foster more success in implementing sustainability initiatives.

Nevertheless, there are opportunities for Brazilian institutions to adopt innovative pedagogical approaches, such as experiential learning and interdisciplinary collaboration, to create more dynamic and engaging sustainability curricula. The literature suggests the potential for holistically incorporating the SDGs into educational programs, combining curricula with sustainable development topics, and employing transformative learning methods (Leal Filho et al. 2019a, b, c). Additionally, the authors emphasize the importance of creating policies and adhering to international declarations to foster a supportive institutional environment (Avelar et al. 2019). Such institutional approaches can provide better opportunities for faculty and students to engage with sustainability topics. Another critical aspect of sustainability in higher education is community engagement. Literature suggests that partnerships between universities and local communities can enhance the relevance and effectiveness of sustainability initiatives (Leal Filho et al. 2019a, b, c). Brazilian HEIs have a key opportunity to leverage their resources and expertise to address pressing local challenges, enriching the educational experience while contributing to societal goals (Amorim et al. 2020).

In summary, the literature on sustainability in higher education underscores the importance of integrating sustainability concepts into curricula, addressing systemic barriers, promoting effective leadership, and engaging with local communities. As they navigate the challenges and opportunities presented by sustainability, Brazilian HEIs have the potential to make substantial contributions to both academic advancement and societal progress.

5 Conclusions

Sustainability in higher education is a rapidly evolving field, reflecting global concerns about environmental stewardship, social responsibility, and economic viability. This field has garnered significant attention in Brazil, with various higher education institutions making substantial contributions to the literature and practical advancements in sustainability.

As this paper has shown, the presence of Brazilian universities in the literature is quite visible. Several themes have emerged from the literature on sustainability in higher education in Brazil. Curriculum integration is a central theme, with universities embedding sustainability concepts across various disciplines and developing specialized programs. Campus sustainability initiatives are also prominent, focusing on implementing green technologies and promoting sustainable living practices. Community engagement is another vital theme, with universities actively collaborating with local communities to advance sustainability. This engagement benefits the communities and enriches academic experience by providing real-world contexts for sustainability studies.

Policy and governance studies reveal how institutional policies and governance models can support sustainability efforts. These studies provide insights into the frameworks that facilitate sustainable practices within HEIs. Many Brazilian researchers have established interdisciplinary research centres and projects, fostering collaboration and innovation in sustainability.

Key publications such as the International Journal of Sustainability in Higher Education and the Journal of Cleaner Production and Sustainability feature numerous contributions from Brazilian researchers, reflecting the country's active role in global sustainability discourse.

In conclusion, Brazilian higher education institutions have made significant contributions to the literature and practice of sustainability. Through innovative teaching programs, interdisciplinary research, and active community engagement, these institutions are advancing sustainability within their campuses and contributing to broader societal goals. Their work exemplifies the transformative potential of higher education in promoting a sustainable future, positioning Brazil as a leader in this critical field.

References

Abad-Segura, E, González-Zamar, M (2021) Sustainable economic development in higher education institutions: A global analysis within the SDGs framework. Journal of Cleaner Production, 294. https://doi.org/10.1016/jjclepro2021126133

Amorim, W, Neiva, S, Castro, B, Deggau, A (2020) Higher Education Institutions as Drivers of Sustainable Communities: A Case Study of the University of Southern Santa Catarina Empowering the Community. In: Leal Filho, W, Tortato, U, Frankenberger, F (eds) Universities and Sustainable Communities: Meeting the Goals of the Agenda 2030. https://doi.org/10.1007/978-3-030-30306-8_50

Avelar, A, Silva-Oliveira, K, Pereira (2019) Education for advancing the implementation of the Sustainable Development Goals: A systematic approach. International Journal of Management Education, 17:3. https://doi.org/10.1016/jijme2019100322

Bizerril, M, Rosa, M J, Carvalho, T, Pedrosa, J (2018) Sustainability in higher education: A review of contributions from Portuguese Speaking Countries. Journal of Cleaner Production, 171. https://doi.org/10.1016/jjclepro201710048

Coppens, T, Felipe, A, Pineda, V, Henao, K, Rybels, S, Jonghe, N De Camacho, H (2020) Innovating Education for Sustainable Urban Development through Problem Based Learning in Latin America: Lessons from the CITYLAB Experience Journal of Problem Based Learning in Higher Education, 8:11–18.

Dalla Gasperina, L, Mazutti, J, Londero Brandli, L, dos Santos Rabello, R (2022) Smart practices in HEIs and the contribution to the SDGs: implementation in Brazilian university. International Journal of Sustainability in Higher Education, 23:356–378. https://doi.org/10.1108/IJSHE-12-2020-0480

Dawodu, A, Dai, H, Zou, T, Zhou, H, Lian, W, Oladejo, J, Osebor, F (2022) Campus sustainability research: indicators and dimensions to consider for the design and assessment of a sustainable campus. Heliyon, 8:12. https://doi.org/10.1016/jheliyon2022e11864

Deroncele-Acosta, A, Palacios-Núñez, M, Toribio-López, A (2023) Digital Transformation and Technological Innovation on Higher Education Post-COVID-19. Sustainability (Switzerland), 15:3. https://doi.org/10.3390/su15032466

Drahein, A, De Lima, E, Da Costa, S (2019) Sustainability assessment of the service operations at seven higher education institutions in Brazil; Journal of Cleaner Production 212:527–536. https://doi.org/10.1016/jjclepro201811293

Frandoloso, M, Scorsatto, E (2021) Beira Rails: The Implementation of the Right to Housing and the City by an Extension Project in Southern Brazil. In: Leal Filho, W, Tortato, U, Frankenberger, F (eds) (Integratin, p 393) Springer Nature Switzerland AG. https://linkspringercom/chapter/101007/978-3-030-59975-1_26

Galleli, B, Teles, N, Freitas-Martins, M, Semprebon, E, Hourneaux Junior, F (2022) Sustainability in management undergraduate courses: mapping the Brazilian higher education institutions. International Journal of Sustainability in Higher Education, 23:1628–1647. https://doi.org/10.1108/IJSHE-03-2021-0109

Gamage, K, Munguia, N, Velazquez, L (2022) Happy Sustainability: A Future Quest for More Sustainable Universities. Social Sciences 11(1). https://doi.org/10.3390/socsci11010024

Halmaghi, E, Ranf, D, Badea, D (2023) Interdisciplinary Exploration between Organizational Culture and Sustainable Development Management Applied to the Romanian Higher Education Environment. Sustainability (Switzerland) 15(13). https://doi.org/10.3390/su151310688

Leal Filho, W, Wu, Y, Brandli, L, Avila, L, Azeiteiro, U, Caeiro, S, Madruga, L (2017) Identifying and overcoming obstacles to the implementation of sustainable development at universities Journal of Integrative. Environmental Sciences, 14:93–108. https://doi.org/10.1080/1943815X2017136 2007

Leal Filho, Walter, Brandli, L, Becker, D, Skanavis, C, Kounani, A, Sardi, C, Papaioannidou, D, Paço, A, Azeiteiro, U, de Sousa, L O, Raath, S, Pretorius, R W, Shiel, C, Vargas, V, Trencher, G, Marans, R W (2018) Sustainable development policies as indicators and pre-conditions for sustainability efforts at universities: Fact or fiction? International Journal of Sustainability in Higher Education 19:85–113 https://doi.org/10.1108/IJSHE-01-2017-0002

Leal Filho, W, Raath, S, Lazzarini, B, Vargas, V, de Souza, L, Anholon, R, Quelhas, O L, Haddad, R, Klavins, M, Orlovic, V (2018) The role of transformation in learning and education for sustainability. Journal of Cleaner Production 199: 286–295 https://doi.org/10.1016/jjclepro201807017

Leal Filho, W, Shiel, C, Paço, A, Mifsud, M, Ávila, L, Brandli, L, Molthan-Hill, P, Pace, P, Azeiteiro, U, Vargas, V, Caeiro, S (2019a) Sustainable Development Goals and sustainability teaching at universities: Falling behind or getting ahead of the pack? Journal of Cleaner Production 232:285–294. https://doi.org/10.1016/jjclepro201905309

Leal Filho, W, Vargas, V, Salvia, A, Brandli, L, Pallant, E, Klavins, M, Ray, S, Moggi, S, Maruna, M, Conticelli, E, Ayanore, M, Radovic, V, Gupta, B, Sen, S, Paço, A, Michalopoulou, E, Saikim, F, Koh, H, Frankenberger, F, … Vaccari, M (2019b) The role of higher education institutions in sustainability initiatives at the local level Journal of Cleaner Production, 233: 1004–1015. https://doi.org/10.1016/jjclepro201906059

Leal Filho, W, Will, M, Salvia, A, Adomßent, M, Grahl, A, Spira, F (2019c) The role of green and Sustainability Offices in fostering sustainability efforts at higher education institutions. Journal of Cleaner Production 232:1394–1401 https://doi.org/10.1016/jjclepro201905273

Leal Filho, Walter, Eustachio, J, Caldana, A, Will, M, Salvia, A, Rampasso, I, Anholon, R, Platje, J, Kovaleva, M (2020) Sustainability leadership in higher education institutions: An overview of challenges. Sustainability (Switzerland) 12:9. https://doi.org/10.3390/su12093761

Leal Filho, W, Wall, T, Salvia, A L, Frankenberger, F, Hindley, A, Mifsud, M, Brandli, L, Will, M (2021a) Trends in scientific publishing on sustainability in higher education Journal of Cleaner Production 296:126569. https://doi.org/10.1016/jjclepro2021126569

Leal Filho, W, Salvia, A, Amaro, N, Avila, L, Brandli, L, Damke, L, Vasconcelos, C, Hernandez-Diaz, P M, Frankenberger, F, Fritzen, B, Velazquez, L, Salvia, A (2021b) Mapping sustainability initiatives in higher education institutions in Latin America Journal of Cleaner Production. https://doi.org/10.1016/jjclepro2021128093

Leal Filho, W, Lange Salvia, A, Abubakar, I, Mifsud, M, Azadi, H, Sharifi, A, LeVasseur, T, Luetz, J, Velazquez, L, Singh, P, Pretorius, R, Akib, N, Savelyeva, T, Brandli, L, Muthu, N, Lombardi, P (2022) Impacts of the COVID-19 Pandemic on Routines of Higher Education Institutions: A Global Perspective. Sustainability (Switzerland) 14:1–19. https://doi.org/10.3390/su142114105

Leal Filho, W, Abubakar, I, Mifsud, M, Eustachio, J, Albrecht, C, Dinis, M, Borsari, B, Sharifi, A, Levesque, V, Ribeiro, P, LeVasseur, T, Pace, P, Trevisan, L, Dibbern, T A (2023) Governance in the implementation of the UN sustainable development goals in higher education: global trends. Environment, Development and Sustainability. https://doi.org/10.1007/s10668-023-03278-x

Leal Filho, Walter, Trevisan, L, Dinis, M Ulmer, N, Paço, A, Borsari, B, Sierra, J, Salvia, A (2024) Fostering students' participation in the implementation of the sustainable development goals at higher education institutions. Discover Sustainability, 5(1). https://doi.org/10.1007/s43621-024-00204-7

Maphosa, V (2022) Sustainable E-waste Management at Higher Education Institutions' Data Centres in Zimbabwe. International Journal of Information Engineering and Electronic Business 14:15–23. https://doi.org/10.5815/ijieeb20220502

Mazon, G, Pereira Ribeiro, J, Montenegro de Lima, C, Castro, B, Andrade Guerra, J (2020) The promotion of sustainable development in higher education institutions: top-down bottom-up or neither? International Journal of Sustainability in Higher Education 21:1429–1450. https://doi.org/10.1108/IJSHE-02-2020-0061

Meseguer-Sánchez, V, Abad-Segura, E, Belmonte-Ureña, L, Molina-Moreno, V (2020) Examining the research evolution on the socio-economic and environmental dimensions on university social responsibility. International Journal of Environmental Research and Public Health 17:1–30. https://doi.org/10.3390/ijerph17134729

Moura, M, Frankenberger, F, Tortato, U (2019) Sustainability in Brazilian HEI: practices overview. International Journal of Sustainability in Higher Education 20:832–841. https://doi.org/10.1108/IJSHE-01-2019-0021

National Institute of Educational Studies and Research Anísio Teixeira (2022) Higher Education Census. https://wwwgovbr/inep/pt-br/areas-de-atuacao/pesquisas-estatisticas-e-indicadores/censo-da-educacao-superior/resultados

National Institute of Educational Studies and Research Anísio Teixeira (2023) Higher Education Census 2022: statistical notes https://wwwgovbr/inep/pt-br/centrais-de-conteudo/acervo-linha-editorial/publicacoes-institucionais/estatisticas-e-indicadores-educacionais/censo-da-educacao-superior-2022-notas-estatisticas

National Institute of Educational Studies and Research Anísio Teixeira (2024) National Student Performance Exam. https://wwwgovbr/inep/pt-br/areas-de-atuacao/avaliacao-e-exames-educacionais/enade

Palma, L, de Oliveira, L, Alves, N, Figueiró, P (2023) Sustainability in Business Administration programs in Brazil: what curricula changes have taken place in the past ten years? International Journal of Sustainability in Higher Education 24:1347–1363. https://doi.org/10.1108/IJSHE-11-2021-0482

Pires, L, Pereira, B (2023) Are Brazilian universities greening? An analysis of their performances in an international sustainability ranking. Environment, Development and Sustainability. https://doi.org/10.1007/s10668-023-04172-2

Qasim, M (2017) Sustainability and Wellbeing: a Scient metric and Bibliometric Review of the Literature. Journal of Economic Surveys 31:1035–1061. https://doi.org/10.1111/joes12183

Radinger-Peer, V, Pflitsch, G (2017) The role of higher education institutions in regional transition paths towards sustainability: The case of Linz (Austria). Review of Regional Research 37:161–187. https://doi.org/10.1007/s10037-017-0116-9

Rasli, A, Fei, Z, Abas, I, Tee, M, Prestianawati, S, Lajuma, S, Yusof, R (2024) A framework for higher education institutions sustainability: A multi-method study. Journal of Infrastructure, Policy and Development 8:1–19 https://doi.org/10.24294/jipdv8i53566

Razali, R, Arifin, M, Shafie, L, Azizan, F, Ishak, M (2024) Mapping the landscape of talent management research in higher education: a bibliometric analysis. Cogent Business and Management. https://doi.org/10.1080/2331197520232298300

Rebelatto, B, Lange Salvia, A, Reginatto, G, Daneli, R, Brandli, L (2019) Energy efficiency actions at a Brazilian university and their contribution to sustainable development Goal 7. International Journal of Sustainability in Higher Education. https://doi.org/10.1108/IJSHE-01-2019-0023

Robinson, J, Alhakim, A, Ma, G, Alam, M, Brando, F, Braune, M, Brown, M, Côté, N, Espinosa, D, Garza, A, Gorman, D, Hajer, M, Madden, J, Melnick, R, Metras, J, Newman, J, Patel, R, Raven, R, Sergienko, K, … Wiek, A (2023) Odd couples: reconciling academic and operational cultures for whole-institution sustainability governance at universities. International Journal of Sustainability in Higher Education. https://doi.org/10.1108/IJSHE-07-2022-0244

Shah Bukhari, S, Said, H, Gul, R, Ibna Seraj, P (2022) Barriers to sustainability at Pakistan public universities and the way forward. International Journal of Sustainability in Higher Education. https://doi.org/10.1108/IJSHE-09-2020-0352

Shi, R, Wan, X (2024) A bibliometric analysis of knowledge mapping in Chinese education digitalization research from 2012 to 2022. Humanities and Social Sciences Communications 11(1). https://doi.org/10.1057/s41599-024-03010-8

Trevisan, L, Eustachio, J, Dias, B, Filho, W, Pedrozo, E (2024) Digital transformation towards sustainability in higher education: state-of-the-art and future research insights, Environment, Development and Sustainability. https://doi.org/10.1007/s10668-022-02874-7

van Eck, N, Waltman, L (2010) Software survey: VOSviewer, a computer program for bibliometric mapping. Scientometrics. https://doi.org/10.1007/s11192-009-0146-3

van Eck, N, Waltman, L (2018) VOSviewer Manual - version 168. http://wwwvosviewercom/documentation/Manual_VOSviewer_154pdf
Vargas, V, Lawthom, R, Prowse, A, Randles, S, Tzoulas, K (2019) Sustainable development stakeholder networks for organisational change in higher education institutions: A case study from the UK. Journal of Cleaner Production 208:470–478. https://doi.org/10.1016/jjclepro201810078
Zhao, H, Zhang, F, Kwon, J (2018) Corporate social responsibility research in international business journals: An author co-citation analysis. International Business Review. https://doi.org/10.1016/jibusrev201709006

Walter Leal Filho Professor Walter Leal Filho (BSc, Ph.D., DSc, DPhil, DTech, DL, DLitt, FSB, FLS, FRGS) holds Professorial positions at universities in Germany (Hamburg University of Applied Sciences), UK (Manchester Metropolitan University) and Poland (WSB Merito University). He directs the European School of Sustainability Science and Research (ESSSR) and the Inter-University Sustainable Development Research Programme, the largest network of universities specifically focusing on research on matters related to sustainable development. Professor Walter Leal Filho has over 30 years experience on sustainable development research and in excess of 800 publications to his credit.

Jucelia Appio Frizon is currently a postdoctoral fellow in Management at the Polytechnic Institute of Leiria, Portugal. She has a Ph.D. in Administration (Universidade Positivo, 2012) and was a visiting Ph.D. student at the Institut Supérieur de Gestion, Paris, France and a Master's in Administration from Fundação Universidade Regional de Blumenau (2010); and a degree in Foreign Trade Administration (Honorable Mention as the best student in the course) from Universidade Paranaense (2005). She is currently a Assistant Professor at the State University of Western Paraná; member of the research group in organizational studies; working mainly on the following topics: People Management, Higher Education, Social Responsibility, Ecological Awareness, Sustainability. She acts as a reviewer in Brazilian journals and conferences, including EnANPAD, Perspectivas Contemporâneas journal, EBAPE, Modern Regulations and Practices for Social and Environmental Accounting.

Dr. Amanda Lange Salvia is a Project Manager and Researcher at the Hamburg University of Applied Sciences in Germany, where she serves as Director of the FTZ-NK Sustainable Development Networks. She is also a Deputy Editor of the Encyclopaedia of the Sustainable Development Goals and the International Journal of Sustainability in Higher Education. With a background in Environmental Engineering, her work focuses on the role of universities in advancing sustainability, the impacts of climate change, and the implementation of the Sustainable Development Goals.

Priscilla Cristina Cabral Ribeiro is an Associate Professor of the Production Engineering Department at Universidade Federal Fluminense (UFF), Bachelor's degree in Economics (1995) from Universidade Federal Rural do Rio de Janeiro, and Doctorate's degree in Industrial Engineering (2009) at Universidade Federal de São Carlos. During her doctoral studies, from 2009 to 2009, she was a visiting scholar at Utah State University. She has a range of publications in internationally indexed journals. Her interests are related to Supply Chain Management, Information Technology and Artificial Intelligence, and Sustainable Development in Higher Education Institutions.

Gabriele Lopes is a Ph.D. candidate at the Postgraduate Program in Administration at the Federal University of Paraná, focusing on Innovation and Technology. Currently peer reviewer for five academic journals, an online MBA tutor, part of the Social Impact team for UFPRs Postgraduate

Programme in Administration and research associate at Research and Transfer Center Sustainable Development and Climate Change Management (HAW Hamburg). Open to working on projects involving Organisational Sustainability, Corporate Social Responsibility, Education for Sustainability—especially Responsible Management, and Competencies for Sustainability.

Sustainability Teaching and Online Learning in Higher Education: The Experiences of the DITTS Project

Luciana Londero Brandli, Amanda Lange Salvia, Bianca Gasparetto Rebelatto, Gabriela Rodrigues, Rudi W. Pretorius, Melanie D. Nicolau, João Henrique Paulino Pires Eustachio, Jasmin Röseler, and Walter Leal Filho

Abstract With less than six years for 2030 and to achieve the Sustainable Development Goals (SDGs), initiatives to support their implementation need to be urgently accelerated. Including the goals into higher education teaching is one of these initiatives, considering the role played by universities in promoting the SDGs, offering learning opportunities and training future generations. The aim of this study is to present the international project "Digital introduction of the Sustainable Development Goals into Higher Education Teaching" (DITTS)—a partnership between University of Passo Fundo (Brazil), University of South Africa (South Africa) and Hamburg University of Applied Sciences (Germany), and the materials prepared. The project fosters the creation and dissemination of open educational materials for higher education institutions as a strategy to introduce the SDGs to students from all areas. The collaboration among the partner institutions and countries leverages the expertise and regional insights of each team, that enrich each teaching module. The paper will present the main lessons from the teaching modules prepared by the research team, and recommendations to support further South-South and North–South collaboration and sustainability initiatives in higher education institutions.

Keywords Global goals · Sustainability agenda · International collaboration · Open educational resources (OER)

L. L. Brandli · A. L. Salvia (✉) · B. G. Rebelatto · G. Rodrigues
Graduate Progamme in Civil and Environmental Engineering, University of Passo Fundo, Passo Fundo, Brazil
e-mail: amandasalvia@gmail.com

R. W. Pretorius · M. D. Nicolau
Department of Geography, University of South Africa, Pretoria, South Africa

J. H. P. P. Eustachio · J. Röseler · W. Leal Filho
Research and Transfer Centre Sustainable Development and Climate Change Management (FTZ-NK), Hamburg University of Applied Sciences, Hamburg, Germany

W. Leal Filho et al. (eds.), *Sustainability in the Curriculum and Operations of Universities*, World Sustainability Series,
https://doi.org/10.1007/978-3-032-12678-8_23

1 Introduction

In 2015, the United Nations launched the Sustainable Development Goals (SDGs), a set of 17 interconnected global objectives aimed at addressing pressing challenges such as poverty, inequality, climate change, and environmental degradation by 2030. These goals provide a universal call for action to ensure that no one is left behind. As we approach the halfway mark to this ambitious target, urgent action is needed to accelerate the implementation of the SDGs, especially within the realm of higher education. Universities, as hubs of knowledge, innovation, and leadership, have a unique responsibility to drive forward this global agenda (Altbach 2021). However, despite the increasing recognition of the importance of sustainability, research has shown that the integration of SDGs into academic curricula is often fragmented or insufficiently implemented in many institutions worldwide (Jöns 2022). There is a growing need for concrete and coordinated efforts to ensure that higher education systems fully embrace and contribute to the achievement of the SDGs.

Higher education institutions are at the forefront of not only fostering societal transformation but also in preparing the next generation of local, national, and global decision-makers. By incorporating sustainability into academic programs, universities can cultivate the competencies, attitudes, and skills necessary to tackle complex global challenges (Leal Filho 2021). The role of universities goes beyond delivering graduates that can drive meaningful transformation; these institutions are also key players in generating cutting-edge research that addresses local, national and global sustainability challenges. Through interdisciplinary research, higher education institutions can generate more holistic and practical solutions that can contribute to achieving the SDGs and to provide students with the necessary tools to apply their knowledge and skills within real-world contexts (Sachs 2022). The dual role of education and research position universities as indispensable contributors in the global push for sustainable development.

Despite these capabilities, there remains a significant gap between the awareness of the SDGs and their practical integration into the various tiers of academic structures. The lack of structured and accessible educational resources further exacerbates the challenge of effectively incorporating sustainability into higher education. As a result, there is a critical need for innovative educational approaches that can bridge the gap between sustainability education and actionable solutions that can be provided from within universities, and these could include OERS to transform education by making sustainable education more inclusive (Jha et al. 2020).

In response to this need, the project The digital introduction of the Sustainable Development Goals into Higher Education Teaching—regional aspects in implementing the SDGs from Brazil and South Africa (DITTS) was created with the goal of developing and disseminating open educational resources (OERs) to support the integration of sustainability and the SDGs into higher education curricula. The DITTS project represents a key initiative aimed at empowering universities with high-quality, accessible teaching materials grounded in sustainability principles. By promoting the use of OERs, this project strives to ensure that educational resources are not only widely available but also equitable and adaptable to diverse academic contexts around the world. The development of these resources is vital for fostering a global learning community committed to achieving the SDGs and equipping future generations to confront the challenges of a rapidly changing world (UNESCO 2020a).

This study explores the outcomes of the DITTS project on higher education, focusing on how the creation and dissemination of open educational materials can enhance sustainability education and contribute to the realization of the SDGs. By examining the project's approach, this paper aims to provide valuable insights into the role that higher education can play in shaping a more sustainable future. Additionally, analyse how innovative initiatives like DITTS are transforming the way sustainability is taught and learned in universities around the world, thereby advancing the global push for sustainability.

2 Methodology

The DITTS project, which run from 2022 to 2025, is an international collaboration aimed at fostering sustainable partnerships in higher education. Led by Hamburg University of Applied Sciences (HAW) in Germany, the initiative brought together the University of Passo Fundo (UPF) in Brazil and the University of South Africa (Unisa). The primary objective was to develop joint educational resources that enable Faculty in higher education institutions to introduce undergraduate students to the SDGs while highlighting their relevance to the broader Agenda 2030 (Leal Filho et al. 2022). The project is part of the SDGs Partnerships Programme of the German Academic Exchange Service (DAAD), which is funded by Germany's Federal Ministry for Economic Cooperation and Development (BMZ). The DAAD programme emphasises the need for international collaboration to address global challenges through interdisciplinary approaches and knowledge exchange, fostering partnerships between German universities and institutions in the Global South (DAAD 2022, 2023, 2024).

The development of teaching materials and academic exchanges within the DITTS project is guided by three core principles, grounded in frameworks for Education for Sustainable Development (UNESCO 2020b) and for integrating the SDGs in higher education (Hessen and Schmelkes 2022). Firstly, the initiative considers regional challenges, differences, and commonalities. Secondly, it incorporates best practices for implementing the Sustainable Development Goals (SDGs). Lastly, it fosters an enriching learning environment that enables undergraduate students to engage with the SDGs. The teaching materials were designed to emphasise both the relevance and practical application of these resources, ensuring they promote mutual learning and greater awareness of the SDGs.

In addition to curriculum development, the project seeks to build upon and further strengthen formal and informal collaborations among researchers and academics from the three partner institutions. This is achieved through joint research activities, engagement with external stakeholders, scholarly publications, reports on sustainability in higher education, and conference organisation and participation. Furthermore, these efforts have been reinforced through South-South academic exchanges between Unisa and UPF, as well as South-North meetings with HAW. Figure 1 illustrates the workflow of the project.

HAW, the lead partner, had already developed OERs for eight of the 17 SDGs, specifically SDGs 3, 4, 5, 6, 7, 12, 13, and 14, in a previous initiative (https://dl4sd.org/my/). In response, the DITTS project focused on developing teaching handbooks for the remaining nine SDGs—namely, SDGs 1, 2, 8, 9, 10, 11, 15, 16, and 17. The handbooks adopt a problem-based learning (PBL) approach, which plays a crucial role in preparing students for sustainability-oriented careers. Nguyen et al. (2024) highlights the significance of PBL, while Silviariza and Handoyo (2021) extend this concept to spatial problem-based learning (SPBL), which enhances students' spatial thinking abilities. Given the complexity of implementing and achieving the SDGs,

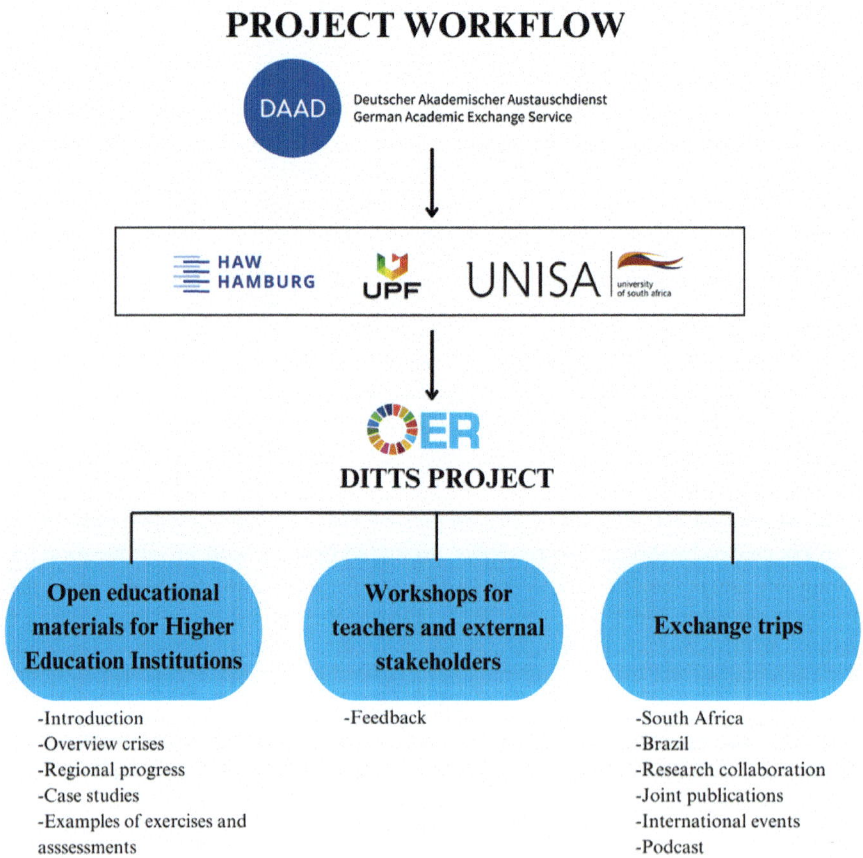

Fig. 1 The project's workflow. *Source* the authors

both PBL and SPBL—reflected in the regional focus of each handbook, are considered by the authors as essential competencies required for addressing sustainability challenges effectively. The use of PBL and SPBL is highlighted by UNESCO (as quoted by Ichinose 2024) as critical in curriculum within sustainability education as the exposure to real-life situations may help to encourage learners to embrace more transformative actions towards sustainability within their own communities. Ichinose (2024) further emphasise that PBL can help the learner to make connections to the realities at hand in their own. Morais et al. (2024) argues further that PBL is a critical tool to foster critical thinking, problem-solving, and interdisciplinary collaboration among learners and this results in more active student participation in the learning process, as they are required to reflect on the problem at hand and then need to consider potential techniques, processes, and tools to find a solution to the problem. SPBL on the other hand encourages learners can encourage students to think in space (geography) in which they live. The integration of PBL and SPBL within a regional approach in the OERS developed for DITTS creates a more innovative learning model that could help to address the spatial phenomena or problems (Silviariza and Handoyo 2021).

Additionally, the teaching materials were developed through a collaborative process that brought together diverse perspectives from the project partners. Monthly meetings were held with the involved researchers to discuss updates, challenges, and ideas for each module. Furthermore, yearly partner meetings were organised to deepen these discussions before finalising the modules, allowing for revisions and the integration of feedback from all team members. The combination of different regional contexts, academic backgrounds, and educational priorities represented a challenge, but it played a central role in shaping the content and approach of the modules. This valuable exchange supported the creation of materials that are both locally and globally relevant, exploring sustainability challenges and opportunities in each region.

Each teacher's handbook is dedicated to a specific SDG and is structured into six sections. The first section provides a general introduction to the SDGs, while the second section delves into the selected SDG, exploring its significance, interconnections with other goals, and the opportunities and challenges associated with its implementation. The third section examines the regional (Africa, Europe, and Latin America) impact of three major global crises (climate change, COVID-19, and conflicts) on the achievement of the specific SDG. Section four then outlines the progress made toward achieving the goal within each region. In section five, three best practices per region are presented, following both PBL and SPBL methodologies. Lastly, section six provides assessment questions for each section, along with exercises designed to help learners integrate sustainable practices into their daily lives. Throughout the handbook, references to additional reading materials and links to OERs relevant to the SDG are included.

In addition, each teaching handbook is supplemented with a PowerPoint presentation, which educators can use or adapt to integrate into their teaching materials or classroom activities. Story Maps were created for two case studies per SDG (one from

Africa and one from Latin America) offering an innovative approach to engaging students with regional perspectives.

The handbooks were initially developed in English and later professionally translated into Portuguese and two indigenous South African languages, isiZulu, and Sesotho sa Leboa (Northern Sotho). This translation effort aligns with research (e.g., Mweri 2020) highlighting the importance of indigenous languages in preserving regional cultural heritage. Tahiri (2025) argues that mother tongue teaching is a key component in the development of sustainable and inclusive education (SDG4) and as such the translation into local languages is not only in line with the overall principles of the United Nations SDGs, but it also enhances communication skills; national identity; and fosters the acquisition of cultural and human values. By making the materials available in multiple local languages, DiTTS supports the inclusion of indigenous knowledge, practices, and values as key components of sustainability, particularly in the Global South.

3 Results and Discussion

The nine teaching handbooks developed as part of the project are available as Open Educational Resources (OERs) and can be accessed through the corporate websites of the three participating universities (Table 1). These resources are published under the Creative Commons Attribution-ShareAlike 4.0 (CC-BY-SA 4.0) license, which permits users to share and adapt the materials, provided that proper attribution is given to the original creators. This open access approach ensures that the handbooks can be freely utilized, modified, and redistributed, contributing to the wider dissemination and adaptation of sustainable development content in higher education.

Table 1 Electronic availability of the teachers' handbooks

Name of the partner university	Link to study material
Hamburg University of Applied Sciences	https://dl4sd.org/course/view.php?id=4
University of Passo Fundo	https://www.upf.br/greenoffice/modulos-ods
University of South Africa	https://www.unisa.ac.za/sites/corporate/default/Colleges/Agriculture-&-Environmental-Sciences/Schools,-departments,-centres-&-units/School-of-Ecological-and-Human-Sustainability/Department-of-Geography/The-Digital-Introduction-to-the-SDGs-into-Higher-Education-Teaching

3.1 Discussion About SDGs 1, 2, 8, 9 and 10

The eradication of poverty in all its forms remains a fundamental and urgent global challenge, as underscored by SDG 1. Despite some progress, over 700 million people still live below the poverty line, and crises such as climate change, the COVID-19 pandemic, and armed conflicts have significantly hindered global and regional advances toward achieving this goal. A key lesson from the implementation efforts is that poverty is multidimensional, affecting access to education, health, housing, employment, and political participation (Alkire et al. 2020). Moreover, SDG 1 is deeply interconnected with other goals, particularly SDGs 2 (Zero Hunger), 3 (Good Health), 4 (Quality Education), and 10 (Reduced Inequalities), indicating that poverty eradication requires integrated and multisectoral approaches, along with localised strategies (Gupta et al. 2015). One of the most effective strategies has been the expansion of social protection systems and resilience-building policies, which buffer vulnerable populations from economic and environmental shocks. However, challenges such as unequal access to resources, weak institutional frameworks, and the need for more inclusive policy design remain persistent. It is therefore recommended that countries prioritise targeted investments in basic services, adopt gender-sensitive and pro-poor policy frameworks, strengthen international cooperation, and enhance local capacities to build long-term resilience against future crises (UNDP 2022). PBL forms the focus of the suggested exercises that include the suggestion that learners hold a 'hunger banquet' that will encourage them to develop a deeper understanding of world food distribution and poverty. By incorporating a methodology that exposes students to diverse contexts and food items from various socioeconomic groups (wealthy, middle class and low-income) and regions (Global North and Global South), the approach fosters a deeper awareness of poverty and hunger worldwide.

In the SDG 2 module, the manual underscored the importance of food security targets (2.1 and 2.2) and their links to sustainable agricultural practices (2.3, 2.4, and 2.5), with a focus on balancing production, sustainability, and equity (FAO 2022; Allen et al. 2018; Fanzo et al. 2020). It also addressed the challenges of achieving these targets, particularly the tensions between environmental sustainability and food production. Futhermore, the manual highlighted the need for transformative changes in the global food system, emphasizing investment in sustainable agriculture, shifts in consumption patterns, and the reduction of food waste. One of the suggested student exercises is titled "Food Tales", which involves a problem-based learning approach. Students are tasked with researching of the typical journey of a food item—from where it is grown to how it ends up on their plate. This exercise promotes an understanding of how daily personal choices have on impact on climate change and how everyday food choices contribute to carbon footprint and affect the environment. SPBL is particularly relevant to this exercise as the movement of food across regions form a major component of the activity. Exercises such as this are critical to assist students in understanding how they can contribute towards achieving the targets of SDG 2 and thus contribute to the broader objectives of the 2030 Agenda.

SDG 8 underscores the importance of promoting sustained, inclusive, and sustainable economic growth, full and productive employment, and decent work for all. The treatment of SDG 8 in the DITTS project converges with the International Labour Organization's (ILO 2023) emphasis on inclusive labour markets and sustainable economic models. While acknowledging persistent inequalities in employment and income distribution, the materials highlight structural vulnerabilities exacerbated by crises such as COVID-19, supporting observations made by the OECD (2021). PBL is incorporated into this manual by advocating that students create a business plan for sustainable start-ups that focus on sustainable products or services. In such an exercise they would have to consider factors such as environmental impact, social responsibility, and long-term viability to promote sustainable economic growth. By promoting the development of green and digital skills, and supporting micro and small enterprises, the project's handbooks offer concrete strategies that reflect recommendations in the work of Jackson and Victor (2019), who advocate for rethinking growth considering environmental and social limits.

SDG 9 aims to build resilient infrastructure, promote inclusive and sustainable industrialisation, and foster innovation. The teacher's manual highlights that while progress has been made, several global crises—including climate change, the COVID-19 pandemic, and armed conflicts—have significantly undermined the pace of implementation, particularly in developing and least developed countries. A key lesson learned is the critical interdependence of SDG 9 with other goals such as SDG 1 (No Poverty), SDG 7 (Affordable and Clean Energy), and SDG 13 (Climate Action), as infrastructure and industrialisation influence virtually all dimensions of sustainable development. Notably, digital, and technological gaps, unequal R&D investments, and underdeveloped infrastructure continue to exacerbate disparities between regions. The goal's successful implementation demands multisectoral approaches, increased investments in sustainable infrastructure and green technologies, and the empowerment of small-scale enterprises through financial and policy support. As such, the manual recommends coordinated policy action, capacity-building at local and national levels, and fostering global partnerships to promote innovation, facilitate equitable access to resources, and build climate-resilient infrastructure systems. Emphasising inclusive growth and sustainability, SDG 9 is essential for shaping a more equitable, innovative, and environmentally sound future.

The DITTS module on SDG 9 provides a nuanced view of the role of infrastructure and innovation in sustainable development, echoing concerns raised by Mazzucato (2018) about the need for mission-oriented innovation policies. The handbook stresses disparities in infrastructure investment and technological readiness, especially between developed and developing nations—issues also highlighted in the Global Innovation Index (WIPO 2023). The focus on empowering small-scale enterprises and fostering green technologies aligns with the views of Schot and Steinmueller (2018), who stress the importance of inclusive innovation pathways. The regional case studies from Africa and Latin America illustrate how localised industrial strategies can enhance resilience and sustainability, offering a grounded perspective for learners. SPBL is encouraged in this manual through a suggested exercise that requires students to select a case study from a 'good practices' platform (for example https://sustainabledevelopment.un.org/partnerships/goodpractices). Students are then requested to present the case study, discuss it with their peers. In their discussion they need to reflect on the lessons learnt, and how these can be replicated in other contexts, for example by comparing and discussing the economic growth in Latin America, Africa and Europe, or between the global north and the global south. Sustainable Development Goal 10 addresses one of the most pressing and multifaceted challenges of our time: the reduction of inequality within and among countries. The teacher's manual highlights that despite decades of global growth, disparities in income, wealth, access to opportunities, and social inclusion remain alarmingly high and, in many regions, are increasing. A key lesson learned is that inequality is not only an economic issue but also a social, legal, and political concern that undermines human rights, social cohesion, and sustainable development. The COVID-19 pandemic, climate change, and global conflicts have disproportionately affected already vulnerable groups, exacerbating structural inequalities. SDG 10's broad scope—encompassing income distribution, social inclusion, representation in global institutions, and migration—makes it a complex yet foundational goal, with strong interdependencies with SDGs 1 (No Poverty), 4 (Quality Education), 5 (Gender Equality), 8 (Decent Work), and 16 (Peace, Justice, and Strong Institutions). The manual also reveals that vague language in targets and the lack of institutional accountability hinder effective implementation. Therefore, it is recommended that governments adopt precise and rights-based policy frameworks, enhance transparency and accountability mechanisms, and foster inclusive participation in decision-making. International cooperation and solidarity, along with stronger commitments from high-income nations, are also essential to narrow global inequalities and deliver on the transformative promise of Agenda 2030. To familiarise students to the context of SDG 10, they could be encouraged to participate in a Community Engagement Project, where students are able to engage in a hands-on project that addresses a specific inequality issue within their local community. It could involve organizing a fundraising event, volunteering at a local organization focused on reducing inequalities, or conducting surveys and interviews to understand the experiences and needs of marginalized groups. Through such a project, students could have the opportunity to make a positive impact and develop empathy and understanding towards the challenges faced by others, and better understanding the goal and target for SDG10.

Inequality remains a central challenge in sustainable development, and the DITTS project addresses this with a multidimensional approach to SDG 10. This aligns with the global analysis presented by Piketty (2020) and UN DESA (2022), who argue that reducing inequality requires systemic reform in governance, taxation, education, and access to opportunity. The handbook's attention to the political, legal, and economic dimensions of inequality provides a comprehensive teaching tool that goes beyond income measures. Furthermore, the critique of vague policy targets and weak accountability is consistent with critiques from the academic community (Fukuda-Parr 2019), highlighting the need for clearer metrics and stronger international commitments. The emphasis on rights-based frameworks and inclusive decision-making mirrors calls from the Human Rights Council and reinforces the transformative ambition of the 2030 Agenda.

3.2 Lessons Learnt from the Development of the OERs

The development of the teachers' handbooks within the DITTS project revealed several key lessons. First, the flexibility of the handbooks allowed faculty members to tailor the material to their specific teaching needs. Faculty could adapt the content or use it verbatim, incorporating it into existing curricula or creating new materials. This adaptability also provided faculty with the opportunity to deepen their own understanding of SDGs, which they could subsequently impart to students. The regional approach adopted in the handbooks encouraged faculty to contextualise the SDGs within their own regions while being mindful of their global impact. Additionally, the case studies, exercises, and assessment questions provided practical examples of best practices, which faculty could integrate into their teaching through PBL and SPBL that will enhance the awareness of, the relevance too and the application of sustainability topics within real world scenarios.

The international collaboration between the partner universities was crucial in enriching the teaching modules. By incorporating regional perspectives from Africa, Latin America, and Europe, the project fostered a deeper understanding of how the SDGs are locally impacted by global crises, such as climate change, COVID-19, and conflict. This diversity of viewpoints allowed for the development of teaching materials that were not only contextually relevant but also reflective of the unique challenges and opportunities each region faces. The collaborative nature of the project also allowed faculty and students to engage in cross-regional learning, enriching their perspectives and fostering a better understanding of the global and local dimensions of sustainability.

For future collaborations focused on sustainability initiatives in higher education, several recommendations emerge from the DITTS project. First, it is essential to maintain and expand South-South and North–South collaborations, as these provide rich opportunities for mutual learning and the exchange of knowledge. The inclusion of multiple regional perspectives enhances the relevance and applicability of educational materials, ensuring they address the unique challenges faced by different regions. Second, fostering a hybrid approach to digital collaboration, which combines online interaction with in-person exchanges, can strengthen partnerships. While digital platforms are sustainable and convenient, visiting partner institutions and experiencing the local context firsthand greatly enriches the collaborative process. Lastly, continuing to make teaching materials available as Open Educational Resources (OERs) is critical to ensuring their accessibility and adaptability. This approach supports the widespread adoption and contextualisation of SDG content, empowering educators globally to integrate sustainability into their curricula more effectively.

4 Conclusion

The DITTS project has demonstrated the transformative potential of collaborative, open-access, and regionally contextualised approaches to sustainability education in higher education institutions. By developing teaching handbooks focused on nine SDGs and integrating pedagogical strategies such as PBL and SPB in the form of Story Maps, the initiative not only produced valuable educational materials but also encouraged meaningful engagement with sustainability challenges from diverse regional perspectives. Universities are uniquely positioned to lead efforts towards achieving the SDGs, both by educating future generations and by acting as hubs for research, innovation, and societal transformation. The DITTS project reinforces the idea that integrating the SDGs into curricula requires more than awareness; it demands deliberate pedagogical design, interdisciplinary collaboration, and accessibility of teaching resources across linguistic and cultural boundaries.

As the 2030 deadline for the SDGs approaches, accelerating initiatives such as DITTS becomes increasingly urgent. The experience gained through this project highlights the importance of hybrid and inclusive educational strategies that support both digital innovation and intercultural academic exchange. The continued expansion and dissemination of open educational resources, rooted in regional relevance and global interconnectedness, will be key to enabling universities worldwide to equip students with the knowledge, skills, and values necessary to build a more sustainable and equitable future.

Despite the positive outcomes of the DITTS project, some limitations should be acknowledged. First, the implementation and testing of the handbooks have been limited to a small number of institutions and academic contexts, potentially affecting the generalisability of results. Second, while the translation of materials into Portuguese and indigenous South African languages enhances accessibility, additional languages and cultural contexts could further broaden the impact. Third, the evaluation of student learning outcomes and the long-term integration of the materials into institutional curricula remain ongoing and are beyond the scope of this initial phase.

Future initiatives should prioritise broader piloting and feedback to assess the effectiveness and adaptability of the teaching handbooks across different cultural and disciplinary settings. Strengthening mechanisms to monitor and evaluate the impact of these materials on student learning outcomes would provide valuable evidence to guide curriculum development. Moreover, continued investment in multilingual translation, inclusive content development, and both virtual and in-person academic exchanges will be essential to enhance global reach and relevance. Finally, integrating the SDGs into formal institutional policies and faculty development programmes can help ensure long-term commitment and impact.

References

Altbach, P. G. (2021). The role of higher education in achieving the Sustainable Development Goals. Global Education Review, 8(4), 1–14.

Allen, T., de Brauw, A., & Gelli, A. (2018). Evaluating the promise of multi-sectoral programs to improve child nutrition: A case study of Ethiopia's Productive Safety Net Programme. *World Development, 110*, 105–122. https://doi.org/10.1016/j.worlddev.2018.05.026

Alkire, S., Kanagaratnam, U., & Suppa, N. (2020). *The Global Multidimensional Poverty Index (MPI) 2020*. UNDP and OPHI. https://ophi.org.uk/multidimensional-poverty-index/

DAAD. (2022a). *Virtual Launch Conference for New Projects in the DAAD SDG – Partnerships Programme, 25–26 April*. DAAD, Bonn.

DAAD. (2023). *Project Leader Conference: Higher Education Partnerships as a Key to Sustainable Development, 24–26 April 2023, Costa Rica*. DAAD, Bonn.

DAAD. (2024). *Project Leader Conference: Educating tomorrow's leaders to pioneer sustainable development – The role of academia, 22–24 July, Dar Es Salaam*. DAAD, Bonn.

DiTTS. (2022b). *Project planning summary for submission to the DAAD* [Documento não publicado]. HAW, Hamburg.

DiTTS. (2025). *The digital introduction of the Sustainable Development Goals into Higher Education Teaching*. https://www.haw-hamburg.de/forschung/forschungsprojekte-detail/project/project/show/ditts/

Fanzo, J., Haddad, L., McLaren, R., & Marshall, Q. (2020). *Food systems are failing: The status of food systems around the world*. Global Panel on Agriculture and Food Systems for Nutrition.

FAO – Food and Agriculture Organization. (2022). *The state of food security and nutrition in the world 2022*. https://www.fao.org/publications/sofi/2022/en/

Fukuda-Parr, S. (2019). Keeping out extreme inequality from the SDG agenda: The politics of indicators. *Global Policy, 10*(S1), 61–69. https://doi.org/10.1111/1758-5899.12601

Gupta, J., Vegelin, C., & Höhne, N. (2015). The role of theory and practice in achieving the SDGs: Challenges for development studies. *Development Studies Research, 2*(1), 5–22. https://doi.org/10.1080/21665095.2015.1020187

Hessen, D. and Schmelkes, S., 2022, Higher Education and the SDGs. A synthesis based of the Report of the Global Independent Expert Group on the Universities and the 2030 Agenda (EGU2030). Paper commissioned for the World Higher Education Conference 18–20 May 2022.

Ichinose, T. (2024). The Possibility of Problem-Based Learning Targeted Toward SDGs to Undertake Transformative Actions for Sustainability. In *The Routledge International Handbook of Life and Values Education in Asia* (pp. 460–475). Routledge.

International Labour Organization (ILO). (2023). *World employment and social outlook – Trends 2023*. https://www.ilo.org/global/research/global-reports/weso/2023/lang--en/index.htm

Jackson, T., & Victor, P. (2019). *Managing without growth: Slower by design, not disaster* (2nd ed.). Routledge.

Jha, R. K., Ganguly, S., & Mishra, S. (2020). Alignment of OER platforms with SDGs: An exploratory study. In *Handbook of Research on Emerging Trends and Technologies in Library and Information Science* (pp. 77–96). IGI Global Scientific Publishing.

Jöns, H. (2022). Sustainable Development Goals and higher education: Trends in curriculum integration. *International Journal of Sustainability in Higher Education, 23*(3), 395–410.

Leal Filho, W. (2021). *Sustainability in higher education: A global perspective*. Springer.

Leal Filho, W., Brandli, L. L., & Pretorius, R. (2022). *Completed project description form for submission to the DAAD* [Documento não publicado]. HAW, Hamburg.

Mazzucato, M. (2018). *The entrepreneurial state: Debunking public vs. private sector myths*. Penguin.

Morais, G., Martins, M., Cunha, D., Azevedo, M., & Gois, M. (2024). Sustainability in higher education: an experience report on the application of problem-based learning to address challenges related to sustainable development goals. In *EDULEARN 24 Proceedings* (pp. 7437–7442).

Mweri, J. (2020). Sustainable development goals: Reaching people through their mother tongue. *Linguistics and Literature Studies*, *8*(1), 14–25.
Nguyen, L. T. V., Cleveland, D., Nguyen, C. T. M., & Joyce, C. (2024). Problem-based learning and the integration of Sustainable Development Goals. *Journal of Work-Applied Management*, *16*(2), 218–234.
OECD – Organisation for Economic Co-operation and Development. (2021). *The impact of the COVID-19 crisis on sustainable development and financing in developing countries.* https://www.oecd.org/coronavirus/policy-responses/
Piketty, T. (2020). *Capital and ideology*. Harvard University Press.
Sachs, J. D. (2022). *The age of sustainable development*. Columbia University Press.
Schot, J., & Steinmueller, W. E. (2018). Three frames for innovation policy: R&D, systems of innovation and transformative change. *Research Policy*, *47*(9), 1554–1567. https://doi.org/10.1016/j.respol.2018.08.011
Silviariza, W. Y., & Handoyo, B. (2021). Improving critical thinking skills of geography students with spatial problem-based learning (SPBL). *International Journal of Instruction*, *14*(3), 133. https://doi.org/10.29333/iji.2021.1438a
Tahiri, A. (2025). Mother Tongue Teaching and Educational Policies for Sustainable Global Quality Education (SDG4). *Journal of Lifestyle and SDGs Review*, *5*(5), e5147–e5147.
UN DESA – United Nations Department of Economic and Social Affairs. (2022). *World Social Report 2022: Inequality in a rapidly changing world*. https://www.un.org/development/desa/dspd/world-social-report/2022-2.html
UNDP – United Nations Development Programme. (2022). *Human development report 2021/2022: Uncertain times, unsettled lives – Shaping our future in a transforming world*. https://hdr.undp.org/
UNESCO. (2020a). *Open Educational Resources (OER) and the SDGs: A global movement*. https://www.unesco.org/en/digital-learning/oer
UNESCO. (2020b) Education for Sustainable Development: a roadmap. Available at: https://unesdoc.unesco.org/ark:/48223/pf0000374802.locale=fr
United Nations. (2015). *Transforming our world: The 2030 Agenda for Sustainable Development*. https://sdgs.un.org/2030agenda
WIPO – World Intellectual Property Organization. (2023). *Global innovation index 2023*. https://www.wipo.int/global_innovation_index/en/

Developing and Defining Contextualised Education for Sustainable Development Related Graduate Attributes for Myanmar

Bo Bo Lwin, Andy Lane, and Rachel Slater

Abstract The literature contains many examples of the attributes that graduates are expected to have if they are to fulfil the ambitions of the Education for Sustainable Development (ESD) movement, including those set out in the Sustainable Development Goals (SDGs). These attributes are variously expressed as knowledge, skills, attitudes, behaviours, and competencies and are usually distilled from discussions and debates between researchers and practitioners from the fields of education and sustainable development. These abstracted lists are set out as guides but run the danger of being based on certain cultures and traditions and needing to be interpreted for different settings. A different approach is to develop and define the most relevant set of graduate attributes for a country through asking educators, students, and graduates what they think they are in relation to a desired sustainable future. This paper reports on one such study focussed on Myanmar. The study employed both qualitative and quantitative methods to iteratively develop and define a set of graduate attributes that were then priority ranked and that could be compared with published lists. Fourteen attributes were defined, all of which were highly ranked, thus providing a robust guide for Myanmar universities to adopt in future.

Keyword Education for sustainable development · Graduate attributes · Myanmar

B. B. Lwin (✉) · A. Lane · R. Slater
Faculty of Science, Technology, Engineering and Mathematics, School of Engineering and Innovation, The Open University, Walton Hall, Milton Keynes, UK
e-mail: bobolwin@gmail.com

A. Lane
e-mail: andy.lane@open.ac.uk

R. Slater
e-mail: rachel.slater@open.ac.uk

W. Leal Filho et al. (eds.), *Sustainability in the Curriculum and Operations of Universities*, World Sustainability Series,
https://doi.org/10.1007/978-3-032-12678-8_24

1 Introduction

There are tensions between the desire to establish a global agenda for change and with the implementation of such an agenda in specific contexts. The Sustainable Development Goals (SDGs) are one example of producing sets of interlinked goals and targets that all countries within the United Nations (UN) contributed to through a deliberative consultation process requiring collaboration and compromise. Once the SDGs were approved by all countries in 2015, they were then expected to take actions to meet those goals and targets by 2030. With 17 goals and 179 targets there has been much scope for developing different pathways to meeting them, depending both on the starting point of a particular country and how they are interpreted, and interventions enacted, in reaction to that context. This paper looks at one target within one goal dealing with Education for Sustainable Development (ESD) and focuses on one aspect of ESD—the expected graduate attributes that will support sustainable development. It does so by comparing the attributes or competencies set out through international discussions and deliberations with those defined through a research study into attributes seen by graduates themselves as being relevant to one country's sustainable future—Myanmar.

2 Education for Sustainable Development

While academic discourse notes the variety of terms and meanings around sustainable development (Jackson 2017; Merrill 2018), international political discourse has tended to be more focussed. The Brundtland report (Brundtland et al. 1987) was the first political milestone for defining sustainable development, raising worldwide concerns about the impact of human activities, and inspiring a series of international conferences and meetings. Then, the first UN Conference on Sustainable Development in Rio de Janeiro (Brazil) in 1992 agreed on 'Agenda 21' as the first in a series of international commitments on sustainable development as part of a global partnership. Subsequently, in 2000, world leaders made the Millennium Declaration as a new global partnership to implement the Millennium Development Goals (MDGs), which included eight targeted goals with a deadline of 2015. The MDGs mainly targeted developing countries and are claimed to have saved millions of lives and improved livelihoods for many people through global, regional, national, and local efforts (United Nations 2015a). MDG 2 focused on accessibility and the completion of primary education in all countries but did not mention the role of higher education and other education levels. MDG 7, targeting environmental sustainability, mainly focused on addressing deforestation, greenhouse gas emissions, water sanitation, and pollution issues. There was also no explicit link between the environmental sustainability goal and educational institutions' involvement in MDGs. Nevertheless, the "Experiences and evidence from the efforts to achieve the MDGs" through this global partnership on sustainable development provided insights for the next

15 year-strategic-plan of Sustainable Development Goals (SDGs) "for all" (United Nations 2015a, p. 3).

Despite this link not being present in the MDGs, the role of education in sustainable development was part of Agenda 21. Chapter 36 of Agenda 21, promoting education, public awareness, and training underpinned the Decade of Education for Sustainable Development 2005–2014 (UNESCO 2009). Moreover, the Action plan 36.1 of Chapter 36, in Agenda 21, clearly stated "Education, raising of public awareness and training are linked to virtually all areas in Agenda 21" while 36.2 also recommends that education programmes should be oriented towards sustainable development. Action Plan 36.2 also recognised that all types of education be involved "including formal education, public awareness and training' while Action plan 36.3 stated that "education is critical for promoting sustainable development and improving the capacity of the people to address environment and development issues" (United Nations 1992, p. 320).

At the UN General Assembly in 2015, all member states adopted the 17 Sustainable Development Goals (SDGs) to extend the global partnership for the next set of 15-year targets and the 2030 Agenda for Sustainable Development was launched on 1st January 2016. Out of 17 SDGs, the role of education is set out in SDG 4: "Ensure inclusive and equitable quality education and promote lifelong learning opportunities for all" (United Nations no date). Amongst ten different targets under SDG 4, sustainability specifically features in target 4.7, commonly known as Education for Sustainable Development (ESD), which states:

> By 2030, ensure all learners acquire knowledge and skills needed to promote sustainable development, including among others through education for sustainable development and sustainable lifestyles, human rights, gender equality, promotion of a culture of peace and non-violence, global citizenship, and appreciation of cultural diversity and of culture's contribution to sustainable development. (United Nations 2015b)

At the same time, this statement for target 4.7 still left questions as to what ESD covers—particularly what graduate attributes should be expected—and how ESD could be integrated into the activities of Higher Education Institutions (HEIs), amongst all other educational institutions. However, over time, many authors have set out their views on what ESD within higher education should cover, reflecting the argument of Landorf et al. (2008) that "ESD is a complex, contested and constantly evolving concept" (p. 222). For example, Calder and Clugston (2003) claimed that ESD should be broader than environmental education as it should include international development, cultural diversity, social and environmental equity. While this appears to emphasise disciplinary knowledge authors like Robley et al. (2005), Barnett (2012) and Donleavy (2012) support the idea that universities should also develop important employability skills, life skills, and behaviours. More specifically, other authors (Sterling 2004; Paten et al. 2005; Tilbury 2011; Shephard et al. 2015; Tierney 2016; Dlouha et al. 2018; Leal Filho et al. 2018; Gul et al. 2019; Vargas et al. 2019) feel that HEIs need to enable students to develop competencies that contribute to the prosperity of society by addressing important social and environmental issues. In other words, the graduates' competencies or attributes are expected to be able to contribute to the sustainable development agenda.

3 Graduate Attributes for ESD

Just as the concept of ESD itself has many dimensions, there have also been many attempts to specify what skills and behaviours are required of graduates in order to contribute to sustainable development. Some authors have suggested that sustainability literacy should be seen as an important attribute that HEIs need to inculcate in graduates (Sterling 2004; Stibbe and Luna 2009; Winter and Cotton 2012; Zizka and Varga 2021) whilst recognising that sustainability remains a contested term. El Ansari and Stibbe (2009) claimed that sustainability literacy has evolved from environmental literacy and ecological literacy while Stibbe and Luna (2009), noted that "building a more sustainable self and society requires far more than knowledge about sustainability—it requires sustainability literacy" (p. 10). Over time this idea of sustainability literacy has been commonly interpreted in terms of the competencies, knowledge, skills, values, and attitudes—the graduate attributes—required for sustainable development (Stibbe and Luna 2009; Sterling 2012). Indeed, based on Sterling's (2012) Future Fit Framework, sustainability literacy can be "... characterised by such qualities as resilience, resourcefulness, creativity, systemic and critical thinking, enterprise, and co-creative and caring outlook" (2012, p. 23).

Additionally, Mochizuki and Fadeeva (2010) argued that to achieve sustainability literacy is a challenging task as it requires working with more ambitious educational goals for preparing students to build a society that is resilient to economic, environmental, and social shocks and to anticipate and envision different futures. Most statements on ESD merely cover the competences believed to be able to contribute to sustainability, but do not include how to instil those competencies in graduates. While this is a significant issue, this paper focuses on what the graduate attributes could be and not how HEIs should teach or support them.

When setting out the attributes that graduates should achieve, it is important to note that there have been many ways in which these have been expressed within and across countries. Many HEIs use the language of Learning Outcomes in describing their students' development: "a statement of what a student is expected to know, understand and be able to do at the end of a module or qualification" (Edwards 1996, p. 4). Alternatively, Svanstrom et al. (2008), talk about learning outcomes as a set of competencies which includes knowledge, skills, and attitudes beyond a specific qualification. Indeed, employability is the predominant perspective in the literature when discussing outcomes, attributes and/or competencies (see Edwards 1996; Amin and Amin 2003; Hager and Holland 2006; McCabe 2010; Spronken-Smith et al. 2015). Consequently, graduates' competences have gained more attention amongst governments, policy makers, employers, and professional bodies, and many of them have developed a set of general and specific competencies for which HEIs are expected to prepare the graduates through their qualifications (Lane 2017). These developments have also led to different ways in which to describe both generic and specific attributes or competencies, such as key, transferable, cognitive and/or practical skills, knowledge and understanding, attitudes, values, and behaviours (see Spronken-Smith et al. 2015; Hill et al. 2016).

The determination of a widely agreed set of graduate attributes for ESD has come about through various routes and developments. The national and international discussions happening about ESD and the inclusion of ESD as part of the SDGs has also led to UNESCO developing a set of eight ESD competencies for graduates that HEIs can use as a guide. These competencies were developed by authors from Europe and reviewed by scholars and experts from different countries (UNESCO 2017). They were reportedly informed by the concepts and models from the three articles: one which discussed 12 sub-competencies of design skills or shaping competencies (de Haan 2010), key sustainability competencies based on the literature (Wiek et al. 2011), and the key ESD competencies from the perspectives of senior members in higher education, NGOs and public administration (Rieckmann 2012). UNESCO's eight competencies are categorised into three domains: cognitive (for knowledge); socio-emotional (for attitudes and values); and behavioural (for skills) (UNESCO 2017). These UNESCO competencies for ESD have been slightly adapted by the UK's Quality Assurance Agency (QAA) and Advance HE (QAA and Advance HE 2021) by using different terminologies for the three domains—ways of thinking (cognitive), ways of being (socio-emotional) and ways of practicing (behavioural) (see Table 1).

The idea of three domains of learning has long been recognised as being a process rather than an event (Lynch et al2009; Hoque 2016; Smith 2018). This can be seen in the Transformative Sustainability Learning (TSL) framework introduced by Sipos et al. (2008) and shown in Fig. 1 where the three sets of attributes can be referred to by three H's: Head for cognitive domain (way of thinking); Heart for affective domain (way of being); and Hand for psychomotor domain (way of practicing).

The '3Hs' model is particularly relevant to the Myanmar context as it is widely used in Buddhist communities as well as by development practitioners in the NGO/CSO communities.

4 Higher Education and ESD in Myanmar

Myanmar, previously known as Burma, gained its independence from the British Empire in 1948 and encountered a chaotic political situation and civil unrest for 14 years in the post-colonial period. Then it was ruled by the military junta for 48 years: 26 years (1962–1988) under the Burma Socialist Programme Party and 22 years (1988–2010) under the State Peace and Development Council. A new political system, known as a quasi-democratic system, was started from 2010 through a 2008 constitution. The Union Solidarity and Development Party (USDP), which is backed by the military, won the first election in 2010 as the first term for a semi-civilian government in Myanmar and initiated some educational reforms. Then, the National League for Democracy (NLD) won the second term's election in 2015 and the education reform process was escalated. The subsequent election was won in a landslide by the NLD party in November 2020. However, on February 1, 2021,

Table 1 The UK QAA and Advance HE's adaptation of UNESCO's ESD competency framework

Subject knowledge and knowledge of all SDGs	Competency	Students who display this competency can	
	Systems thinking	Recognize and understand relationships; analyze complex situations; consider how systems are embedded within different domains and scales; and deal with uncertainty	Ways of thinking
	Anticipatory	Understand and evaluate multiple futures; create their own visions for the future; apply the precautionary principle; assess the consequences of actions; and deal with risks and changes	
	Critical thinking	Question norms, practices, and opinions; reflect on own one's values, perceptions, and actions; and take a position in the sustainability discourse	
	Strategic	Develop and implement innovative actions that further sustainable development at the local level and further afield	Ways of practicing
	Collaboration	Learn from others (including peers, and others from inside and outside of their institutions); understand and respect the needs, perspectives, and actions of others; deal with conflicts in a group; and facilitate collaborative and participatory problem solving	
	Integrated problem-solving	Apply different problem-solving frameworks to complex sustainability problems; develop viable, inclusive, and equitable solution options; and utilize appropriate competences to solve problems	
	Self-awareness	Reflect on their own values, perceptions, and actions; reflect on their own role in the local community and the global society; continually evaluate and further motivate their actions; and deal with their feelings and desires	Ways of being
	Normative	Understand and reflect on the norms and values that underlie one's actions; negotiate sustainability values, principles, goals, and targets, in a context of conflicts of interests and trade-offs, uncertain knowledge and contradictions	

Source QAA and Advance HE (2021, p. 20)

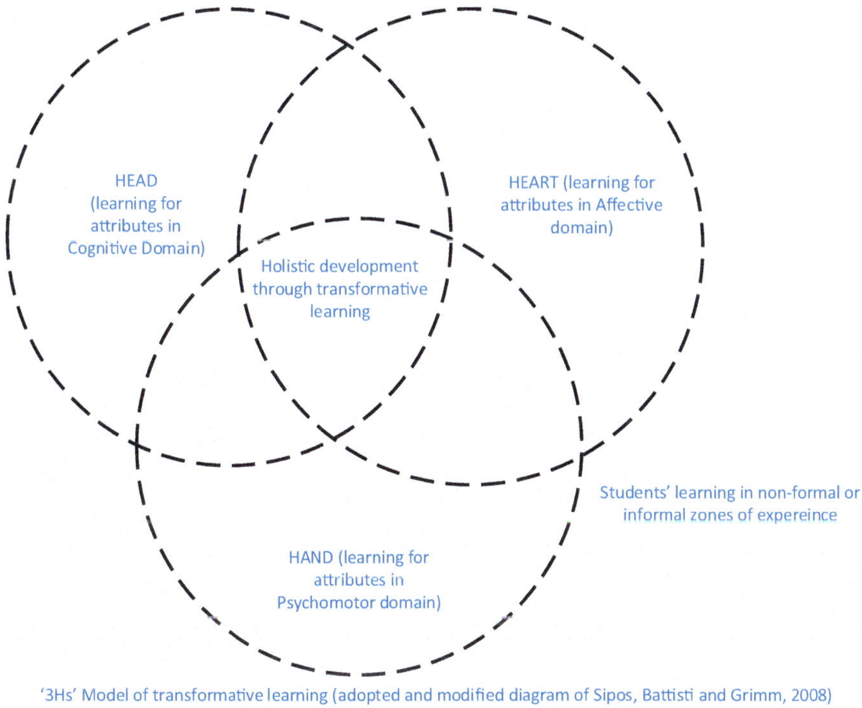

Fig. 1 Head, Heart, Hand Model of students' transformative learning [modified with simplified descriptions from Sipos et al. (2008, p. 75)]

the military seized power from the civilian government and thus both sustainable development and education reform processes have been disrupted since early 2021.

The USDP-led Government's Comprehensive Education Sector Review (CESR) (2012–2014) aimed to analyse the educational situation, shape new policies, and draft a comprehensive education plan by 2014. After winning the 2015 election, the NLD-led Government carried on some of these reform processes amongst which educational transformation was a high priority in their political agenda as well as being a key driver for social and economic development (Heslop 2019). As another important milestone of education reform in Myanmar, the 'National Education Strategic Plan (NESP): 2016–2021' was launched in 2016. In 2018, the NLD-led Government set out a 'Myanmar Sustainable Development Plan: 2018–2030 (MSDP)' which was heavily informed by the UN's SDGs, and was considered as a road map for the country's sustainable development. The MSDP provided a reference point for the conceptualisation of ESD in Myanmar higher education despite lacking guidance on the roles of HEIs and expected outcomes from higher education as a contribution towards sustainability. The NESP provided comprehensive strategic guidelines for almost all sectors of education reform promoting quality education which includes teaching, assessment and curriculum development, accessibility to education, development of alternative education (accommodating the needs of different groups of learners with restricted access to formal education), technical and vocational skills development and higher education sector development.

A significant feature of HEIs in Myanmar is the historic multi-ministry administration structure, with different universities under the administration of different ministries. There are reportedly 171 HEIs which come under the control of eight different ministries (Ministry of Education 2016) with weak or no cooperation amongst these ministries in how they manage their universities. Although the 2013 CESR report gave an overall picture of challenges and suggestions for higher education in Myanmar, the higher education sector was not widely mentioned in both NESP (2016–2021) and MSDP (2018–2030). Nevertheless, the role of higher education is acknowledged as an important sector for the country's socio-economic development in NESP and the commitment of the Ministry of Education on the transformational shift of HE to be achieved by 2021 was stated as:

> Students have equitable access to a world-class higher education system leading to better opportunities for employment and significant contributions to a knowledge-based economy. (Ministry of Education 2016, p. 189)

HEIs in Myanmar are generally beset by a lack of autonomy, curricula are centrally driven, and prescribed teaching content and methods are centrally incorporated (Lall 2016). A very common critique of the teaching and learning system is about rote learning, locally named 'parrot learning' (Comprehensive Education Sector Review 2013) and lacking effective quality assurance systems. Equally, the higher education system has a lower rate of student enrolment and has not prepared graduates with the skills required by the labour market compared with other neighbouring countries (OECD 2014).

Thus, there has been no explicit national drive to integrate ESD in higher education in Myanmar and there is a lack of evidence on how ESD is viewed in Myanmar. To fill this gap a doctoral study explored ESD in Myanmar higher education and in particular the concepts and models needed to influence the integration of ESD in, and transformation of, HEIs. One key aspect of this study was developing and defining the graduate attributes that HEIs need to inculcate to meet a sustainable future for Myanmar.

5 Developing and Defining Contextualised Graduate Attributes for Myanmar HEIs

There were two interrelated approaches used to develop and define possible lists of graduate attributes relevant to Myanmar's future sustainability. First, an initial list of graduate attributes emerged from the systematic analysis of qualitative data acquired through seven Key Informant Interviews (KIIs) with experienced professionals from different backgrounds and two Focus Group Discussions (FGDs) with university teachers and students from Myanmar. Both the KIIs and FGDs were recorded, transcribed, and verified with the participants concerned. The list of attributes produced by the systematic analysis of these transcripts were then refined through both a process of verification and iterative consultation with the various participants. The fourteen graduate attributes developed from this systematic and in-depth analysis were then tested using an online survey with a large number ($n = 417$) of Myanmar respondents to understand their priority rankings in the Myanmar context. These respondents were all Myanmar citizens, some based in Myanmar and others not, and all were graduates or university students themselves.

The online survey instrument included descriptors developed by the authors for each proposed attributes that explained in more depth what they covered, as an attempt to minimise the potential overlaps and variation in understanding of terms by survey respondents (see Table 2). These descriptors were themselves refined based on the research participants' comments and suggestions. Finally, the list was prepared in both Myanmar and English Languages to aid comprehension. The survey used a 'numerical scale rating' in which each attribute was to be rated based on a scale of minimum '0' (Zero as 'not at all important') to maximum '9' (Nine as the most important). Respondents were invited to rate the given graduate attributes based on their perceived importance in contributing towards sustainable development in Myanmar. It also included open questions for additional comments and suggestions. Following some small revisions during a pilot stage, the online survey was conducted through social media and direct email contact. The data from the 417 responses to the online survey were quantitatively analysed using SPSS software.

As well as a statistical analysis of the priority rankings of the 14 attributes raising from the systematic analysis of the KII and FGD data, a separate thematic analysis of the qualitative data was conducted using NVivo software. The qualitative data from the open questions of the online survey were used alongside the original qualitative data from the KIIs and FGDs. This alternative analysis resulted in a broader set of seven graduate attributes that can be compared with the more specific fourteen graduate attributes (see Table 3).

One difficulty that arose from these two separate analyses was naming an attribute for a particular group of qualities as some of the more specific skills and abilities can be shared across two or more broader attributes' names. For example, some qualities in social abilities or personal abilities can also be found in professionalism or leadership. Similarly, certain qualities of all other attributes can match employability skills because all these skills or attributes are commonly discussed as employability skills. Therefore, while it is not always easy to pair up these two sets of attributes as they have come from two different analyses, although the degree of relatedness is sufficient to believe that both represent a robust set of attributes arising from the responses of the different sets of participants.

Table 2 Graduate attributes and descriptors used in the online survey

Graduate attributes	Descriptors
Social abilities	Being reflexive and open-minded for feedback; Social dealing skills e.g., interpersonal communication, negotiation, conflict resolution, and being able to build social networks; Having teamwork skills e.g., being collaborative, adaptable, flexible, and confident to discuss, question or debate
Cultural sensitivity	Being knowledgeable about human rights, equality, social conflicts, different cultures/values/beliefs/traditions, psychology, and country's context; Having basic awareness on peace, respect rights/freedom of minority/marginalised identities (e.g., LGBT, minority ethnic/religious/disable groups); Ability to think of being inclusive, work in multi-cultural contexts, and adapt to new cultural contexts
Management skill	Being reliable; Having accountability and responsibility skills; Having problem solving skills and facilitation skills to address the issues; Being able to organise activities or projects and being competent in time management, resource management, project, and/or business management
Environmental literacy	Having basic environmental knowledge (e.g., ecology, biodiversity, symbiotic relationships, climate change, deforestation, etc.) and sustainability concepts; Understand basic principles of sustainability in environmental and resources management, and the interdependence of human beings and ecosystems; Having value, respect, and love for the natural environment, and Ability to think of future impact and act and apply environmentally sound management practices in their activities
Thinking proficiency	Being good in reasoning, logical thinking, analytical and critical thinking, and creative thinking, leading to wisdom
Language proficiency	Proficiency in English language skills (at least communicable), Myanmar language; Ability to use any ethnic or foreign languages depending on the situation and individuals' preferences
Work experience	Having work experience in the job environment, or research field or communities (such as internship or volunteer) during study programmes
Digital literacy	Ability to use a computer and commonly used software such as Office software (Microsoft Office: word, excel, PowerPoint etc.,); Ability to use internet, search data and apply communication software such as Skype/Zoom etc., including social media
Personal abilities	Having self-awareness, moral conduct, contentment, self-esteem, altruism, compassion, and sympathy; Being curious and open-minded to learn, being motivated to learn from work/experience, and able to learn independently; Being able to read and write well (e.g., proposals/reports/research papers)
Researching	Having fundamental research knowledge and skills
Entrepreneurship	Having skills and enthusiasm to create/initiate innovative projects/business
Teaching	Ability to teach or to transfer knowledge to others

(continued)

Table 2 (continued)

Graduate attributes	Descriptors
Professionalism	Being professionally ethical, having respect to one's own profession and others; Being punctual and honest; Knowing one's own rights as well as being responsible
Resilience	Having perseverance, grit, adaptability, and commitment; Having a good focus/attention on what one is studying or working on; Ability to overcome problems and hardship

Table 3 Comparison of two sets of graduate attributes

Seven graduate attributes from thematic analysis (broader view)	Fourteen graduate attributes from systematic, in-depth analysis (specific view)
Sustainability literacy and practice	Environmental literacy
Social abilities	Social abilities
	Cultural sensitivity
Professionalism	Professionalism
	Research knowledge and skills
Personal abilities	Personal abilities
	Resilience
	Thinking proficiency
Leadership and entrepreneurship skill	Entrepreneurship
	Management skill
General knowledge and life skill	Digital literacy
Employability skills	Work experience
	Teaching
	Language proficiency

6 Priority Ranking of Graduate Attributes by Myanmar Graduates and University Students

A variety of statistical tests were carried out on the survey data. A Friedman test revealed a statistically significant difference in the perceived priority values between the different graduate attributes. As the data were not normally distributed, median and mode were considered in the rank-based non-parametric tests and the graduate attributes were arranged in the order of mean rank generated by the Friedman test (see Table 4). Notably, all the 50th percentile median values of the fourteen graduate attributes fall between 7 and 9 showing that most respondents highly prioritized the given fourteen graduate attributes, even though there were some statistical differences. Therefore, it can be interpreted that majority of the respondents in the survey perceived the priority values for all the graduate attributes as important and above the mid-point scale, although there were a few outliers with a slightly lower priority.

Table 4 Perceived priority values of graduate attributes by mean rank and medians

Graduate attributes	N	Percentiles			
		25th	50th (Median)	75th	Mean rank
Professionalism	417	7	9	10	9.45
Thinking proficiency	417	6	9	10	8.66
Personal abilities	417	6	8	10	8.61
Social Abilities	417	6	8	10	8.56
Management	417	6	8	10	8.35
Digital literacy	417	6	8	10	8.09
Cultural sensitivity	417	6	8	9	7.86
Resilience	417	6	8	9	7.74
Environmental literacy	417	6	8	9	7.13
Teaching	417	6	8	9	6.93
Work experience	417	5	8	9	6.50
Researching	417	5	7	9	6.05
Language proficiency	417	5	7	9	5.93
Entrepreneurship	417	5	7	8	5.13

To further test this high level of perceived importance, another part of the study looked at these attributes from two dimensions: their degree of importance and impact over time using an online platform. A google Jam-board[1] was used and the FGD participants were invited to place the fourteen graduate attributes on a three-by-three matrix by considering three priority levels (Low, Medium, and High) from two dimensions—Y axis for the degree of importance and X axis for the degree of impact over time. Fifteen FGD participants responded with their ratings with the mean results shown in Fig. 2.

The data sets (degree of importance and impact over time) were again tested for median percentiles and mean rank using a Friedman test. The emerging order of the ranks was slightly different from the former analysis, for example environmental literacy and entrepreneurship were moved up to the higher priority rows in the dimension of potential impact over time. Nevertheless, apart from slight changes in ranking order, all the graduate attributes were still rated as important for their potential contribution to the sustainable development of Myanmar. Therefore, all these graduate attributes can be considered as an important part of how ESD can be conceptualised for implementation in Myanmar HEIs. To be specific, they can be named as generic graduate attributes which include, but are not limited to, desirable sets of skills and values regardless of their specialised degrees (Spronken-Smith et al. 2015). Moreover, these generic graduate attributes can be claimed as contextually relevant to Myanmar HEIs. Nevertheless, there are 5 attributes that score more the 2.5 on both scales—thinking proficiency, personal abilities, social abilities,

[1] https://workspace.google.com/intl/en_uk/products/jamboard/

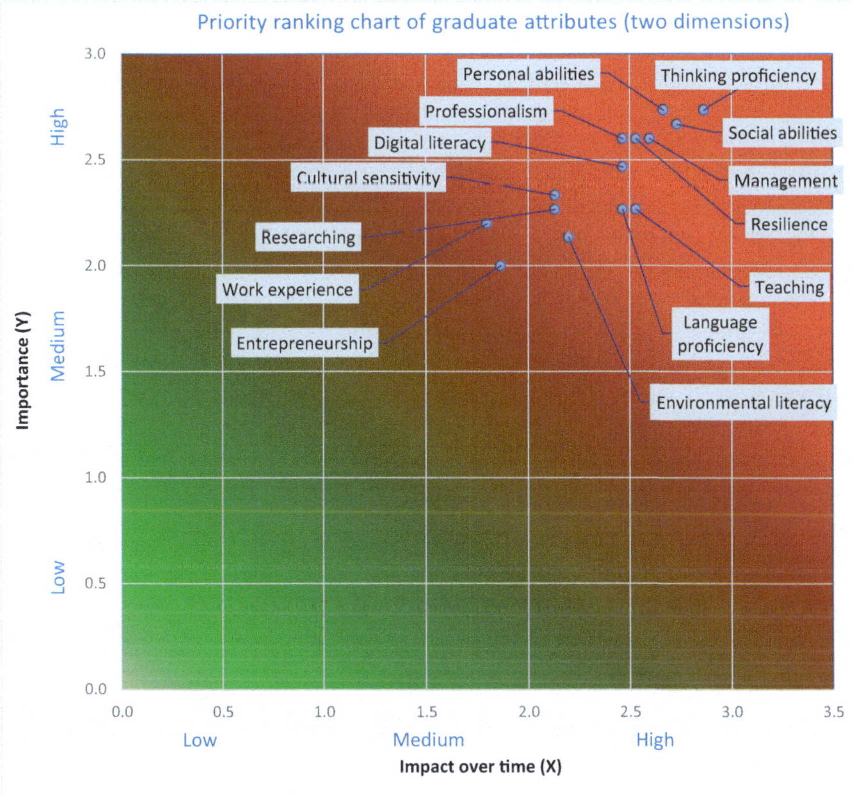

Fig. 2 Graduate attributes in priority ranking chart with two dimensions: importance and potential impact for the sustainable future of Myanmar [0–1 = Low; 2–3 = Medium; 2–3 = High]

management and professionalism, while environmental literacy is one of the lower ranking attributes. This may reflect a belief that sustainability should arise from or be embedded in these generic attributes.

7 Comparing the Contextualised List of ESD Graduate Attributes from This Myanmar Focused Study, with ESD Graduate Attributes from the Literature

Having developed and defined a set of contextually relevant graduate attributes or competencies for Myanmar, there remains a question about how these compare to the ESD attributes and competencies in the literature. Comparing the seven broader graduate attributes from the thematic analysis against the UNESCO ESD competencies (see Table 5), five out of seven attributes from this research can be paired up, whilst acknowledging that they were not always fully related to their counterparts, partly due to limited descriptors.

Two attributes from this study which are not as explicit in the UNESCO list are 'sustainability literacy and practice' and 'general knowledge and life skills.' These might be considered as being holistic because they can enfold all three types of qualities from the 3H's model: Head, Heart, and Hand. From the research participants' view on Myanmar ESD attributes, they expected graduates to be knowledgeable and ethically sound professionals as well as having the ability to lead or contribute to the betterment of society. In addition, the desired sustainability literacy and habitual practices are mostly related to environmental and social aspects while UNESCO's framework appears to rely upon this being covered by knowledge of the SDGs.

While the graduate attributes expected by Myanmar nationals were similar to ones in the literature, it can be claimed that the pattern and coverage of the attributes that emerged from this research are more diverse and holistic—covering many aspects of competencies, attitudes, and behaviours. This is not a new issue, for example, using the '3Hs' model Sandri's (2014) proposed sustainability learning outcomes can be presented as:

> HEAD: Sustainable literacy, Systemic and holistic thinking, Critical thinking and reflection, Foresighted, anticipatory and future thinking
> HEART: Implicit development of values or values of learning and reflection, empathy, compassion, self-motivation and sense of identity
> HAND: Capability and motivation or lifelong learning, Reflexivity, working with complexity and uncertainty, Interdisciplinary skills, and ability to work with stakeholders. (p. 38).

Table 5 Comparison of ESD graduate attributes from this research and UNESCO's recommended competencies

Seven categories of ESD graduate attributes emerged from the analysis of the research data	ESD graduate attributes adapted by QAA and Advance HE (UK) based on UNESCO's recommended competencies	
Personal abilities/professionalism	Self-awareness competency	HEART
Social abilities	Normative competency	
Employability skills/personal abilities	Collaboration competency	HAND
	Integrated problem-solving competency	
Leadership and entrepreneurship skill	Strategic competency	
	Systems thinking competency	HEAD
Professionalism/personal abilities	Anticipatory competency	
	Critical thinking competency	
Sustainability literacy and practice		Holistic (HHH)
General knowledge and life skills		

8 Concluding Remarks

The framing of concepts and their understanding by others is predicated on the social, cultural, educational, and spiritual background of the authors. While scientific endeavour attempts to be objective and account for biases this is not as easily done for contested topics that are subject to much political debate. Determining policy and guidelines therefore leads to compromise but such compromise can still favour certain viewpoints or ignore the tensions and differences in their interpretation. The UNESCO competencies for ESD are such an example, with more academic terminology and no explicit attempt to account for or reconcile the overlaps between the different competencies. Thus, the UNESCO competencies can be seen as an idealised set of competencies from which HEIs can adopt and adapt as they see fit, as exemplified by the QAA and Advance HE (2021) guidance document. Nevertheless, this list can also be seen as a 'top down' list drawn from a literature biased towards the research emanating from so-called western or developed country traditions.

This paper has reported on a 'bottom up' list of graduate attributes for one country that were developed and defined by an in-depth participatory process involving one key group of stakeholders from that country, namely graduates and higher education students themselves (albeit that some participants graduated many years earlier). This list was not directly derived from the UNESCO list and the latter only indirectly informed the process through the mediation of the primary researcher. As far as possible the two lists of attributes that emerged reflected the social, cultural, educational, and spiritual context of Myanmar. They therefore represent a primary guide to HEIs in Myanmar for when they implement ESD in their curricula, leaving the UNESCO framework as a secondary influence. Indeed, no such list is definitive and should be subject to ongoing review. Further research could, for example, include the perspectives of employers in Myanmar or of both national and international professional bodies.

As noted by Lane (2017), the diversity of frameworks is a feature of complex subjects such as the environment and sustainable development and that "the constructive alignment of such frameworks can best be achieved through constant and ongoing conversations between the actors so as to resolve any unnecessary differences but to enable requisite diversity to meet the many complexities of the real world" (p. 425). It is to be hoped that the contextualised framework for Myanmar HEIs to consider described in this paper will soon be able to inform the implementation of ESD in those same HEIs.

References

Amin, M.R. and Amin, N.A. (2003). Benchmarking learning outcomes of undergraduate business education, Benchmarking. An International Journal, 10(6), pp. 538–558.

Barnett, R. (2012). Learning for an unknown future. Higher Education Research and Development, 31(1), pp. 65–77.

Brundtland, G.H. et al. (1987). Our Common Future: Report of the World Commission on Environment and Development (UN Document A/42/427). Geneva.
Calder, W. and Clugston, R.M. (2003). International Efforts to Promote Higher Education for Sustainable Development. Planning for Higher Education, 31(3).
Comprehensive Education Sector Review (2013). Phase 1: Rapid Assessment: Technical Annex on the Higher Education Subsector.
de Haan, G. (2010). The development of ESD-related competencies in supportive institutional frameworks. International Review of Education, 56(2), pp. 315–328.
Dlouhá, J., Henderson, L., Kapitulčinová, D. and Mader, C. (2018). Sustainability-oriented higher education networks: Characteristics and achievements in the context of the UN DESD. Journal of Cleaner Production, 172, pp.4263-4276.
Donleavy, G. (2012). Proclaimed graduate attributes of Australian universities: Patterns, problems and prospects. Quality Assurance in Education, 20(4), pp. 341–356.
Edwards, C. (1996). Learning Outcomes in Higher Education. Studies in Higher Education, 21(1), pp. 93–108.
El Ansari, W. and Stibbe, A. (2009). Public health and the environment: What skills for sustainability literacy - and why? Sustainability, 1(3), pp. 425–440.
Gul, A. et al. (2019). Internationalization of Higher Education from the Perspective of Educational Leadership. Journal of Managerial Sciences, 14(4).
Hager, P. and Holland, S, (2006). Introduction. In P. Hager and S. Holland (Eds.), Graduate Attributes, Learning and Employability. Springer, pp. 1–15.
Heslop, L. (2019). Encountering internationalization: higher education and social justices in Myanmar [Unpublished Doctoral thesis]. University of Sussex.
Hill, J., Walkington, H. and France, D. (2016). Graduate attributes: implications for higher education practice and policy: Introduction. Journal of Geography in Higher Education, 40(2), pp. 155–163.
Hoque, M.E. (2016). Three Domains of Learning: Cognitive, Affective and Psychomotor. The Journal of EFL Education and Research, 2(2), pp. 2520–5897.
Jackson, L. (2017). Asian Perspectives on Education for Sustainable Development. Educational Philosophy and Theory, 49(5), pp. 473–479.
Lall, M. (2016). Education Challenges for Myanmar's New NLD Government. In Khin Mar Mar Kyi and Matthew J Walton (Eds.), Oxford-Myanmar Policy Brief Series, 1(1).
Landorf, H., Doscher, S. and Rocco, T. (2008). Education for sustainable human development: Towards a definition. Theory and Research in Education, 6(2), pp. 221–236.
Lane, A. (2017). The systemic implications of constructive alignment of higher education level learning outcomes and employer or professional body based competency frameworks. In The Online, Open and Flexible Higher Education Conference. The Open University (UK), pp. 411–426.
Leal Filho, W., Raath, S., Lazzarini, B., Vargas, V.R., de Souza, L., Anholon, R., Quelhas, O.L.G., Haddad, R., Klavins, M. and Orlovic, V.L., (2018). The role of transformation in learning and education for sustainability. Journal of Cleaner Production, 199, pp.286-295.
Lynch, D.R. et al. (2009). Beyond the Cognitive: The Affective Domain, Values, and the Achievement of the Vision. Journal of Professional Issues in Engineering Education and Practice.
McCabe, G. (2010). Graduate Attributes and Employability: helping universities and students prepare for the changing landscape. Interchange, Centre for Teaching, Learning and Assessment (TLA). The University of Edinburgh, pp. 1–3.
Merrill, M.Y. (2018). Introduction - Education for Sustainability in Asian contexts. In M.Y. Merrill et al. (Eds.), Education and Sustainability: Paradigms, Policies and Practices in Asia. Routledge, pp. 3–27.
Ministry of Education (2016). National Education Strategic Plan: 2016–2021.
Mochizuki, Y. and Fadeeva, Z. (2010). Competences for sustainable development and sustainability: Significance and challenges for ESD. International Journal of Sustainability in Higher Education, 11(4), pp. 391–403.

OECD (2014). Multi-dimensional Review of Myanmar. OECD (OECD Development Pathways). https://doi.org/10.1787/9789264220577-en.
Paten, C.J., Palousis, N., Hargroves, K. and Smith, M. (2005). Engineering sustainable solutions program: Critical literacies for engineers portfolio. International Journal of Sustainability in Higher Education, 6(3), pp.265-277.
QAA and Advance HE (2021). Education for Sustainable Development Guidance. QAA and Advance HE.
Rieckmann, M. (2012). Future-oriented higher education: Which key competencies should be fostered through university teaching and learning? Futures, 44(2), pp. 127–135.
Robley, W., Whittle, S. and Murdoch-Eaton, D. (2005). Mapping generic skills curricula: Outcomes and discussion. Journal of Further and Higher Education, 29(4), pp. 321–330.
Sandri, O.J. (2014). Good Practice Learning and Teaching for Sustainability in Higher Education. [Unpublished PhD Thesis]. School of Global, Urban and Social Studies, RMIT University. https://researchbank.rmit.edu.au/eserv/rmit:160976/Sandri.pdf.
Shephard, K., Harraway, J., Lovelock, B., Mirosa, M., Skeaff, S., Slooten, L., Strack, M., Furnari, M., Jowett, T. and Deaker, L. (2015). Seeking learning outcomes appropriate for 'education for sustainable development' and for higher education. Assessment & Evaluation in Higher Education, 40(6), pp.855-866.
Sipos, Y., Battisti, B. and Grimm, K. (2008). Achieving transformative sustainability learning: engaging head, hands and heart. International Journal of Sustainability in Higher Education, 9(1), pp. 68–86.
Smith, M. (2018). The Emotional Learner. Routledge.
Spronken-Smith, R., Bond, C., McLean, A., Frielick, S., Smith, N., Jenkins, M. and Marshall, S. (2015). Evaluating engagement with graduate outcomes across higher education institutions in Aotearoa/New Zealand. Higher Education Research & Development, 34(5), pp.1014-1030.
Sterling, S. (2004). Higher Education, Sustainability, and the role of Systemic Learning. In P.B. Corcoran and A.E.J. Wals (Eds.), Higher Education and the Challenge of Sustainability. Springer, pp. 49–70.
Sterling, S. (2012). The Future Fit Framework—An introductory guide to teaching and learning for sustainability in HE. Higher Education Academy.
Stibbe, A. and Luna, H. (2009). Introduction. In A. Stibbe (Ed.), The Handbook of Sustainability Literacy: Skill for a changing world. Green Books Ltd., pp. 9–16.
Svanstrom, M., Lozano, J. and Rowe, D. (2008). Learning Outcomes for Sustainable Development in Higher Education. International Journal of Sustainability in Higher Education, 9(3), pp. 339–351
Tierney, W.G. (2016). Portrait of higher education in the twenty-first century: John Henry Newman's "The idea of a university". International Journal of Leadership in Education, 19(1), pp. 5–16.
Tilbury, D. (2011). Higher Education for Sustainability: A Global Overview of Commitment and Progress. In Higher Education in the World 4 Report, Higher Education's Commitment to Sustainability: From Understanding to Action. Global University Network for Innovation, pp. 18–28.
UNESCO (2009). Review of contexts and structures for Education for Sustainable Development 2009: Key findings and ways forward. UNESCO Digital library. https://unesdoc.unesco.org/ark:/48223/pf0000187757
UNESCO (2017). Education for Sustainable Development Goals: Learning Objectives. Paris: UNESCO.
United Nations (1992). Agenda 21—United Nations Conference on Environment and Development. Rio de Janerio, Brazil, 3 to 14 June 1992. http://www.un.org/esa/sustdev/agenda21.htm.
United Nations (no date). Goal 4|Department of Economic and Social Affairs. https://sdgs.un.org/goals/goal4
United Nations (2015a). The Millennium Development Goals Report-2015. United Nations
United Nations (2015b). 70/1. Transforming our world: the 2030 Agenda for Sustainable Development. UN resolution [Preprint]. United Nations.

Vargas, V.R. et al. (2019). Sustainable development stakeholder networks for organisational change in higher education institutions: A case study from the UK. Journal of Cleaner Production, 208, pp. 470–478.

Wiek, A., Withycombe, L. and Redman, C.L. (2011). Key competencies in sustainability: A reference framework for academic program development. Sustainability Science, 6(2), pp. 203–218.

Winter, J. and Cotton, D. (2012). Making the hidden curriculum visible: Sustainability literacy in higher education. Environmental Education Research, 18(6), pp. 783–796.

Zizka, L. and Varga, P. (2021). Teaching Sustainability in Higher Education Institutions: Assessing Hospitality Students' Sustainability Literacy. Journal of Hospitality and Tourism Education, 33(4), pp. 242–257.

Leading Change for Sustainability—Business: Preliminary Evaluation of a Sustainability Simulation for Management Education

Philip Hallinger, Chatchai Chatpinyakoop, Parinya Showanasai, and Vien-Thong Nguyen

Abstract Simulations are among the active learning strategies advocates have claimed are well-suited for teaching and learning about sustainability challenges and solutions. Recent research reviews have, however, identified a need for simulations that address a broader range of sustainability issues. This is also the case in management education, where the predominant paradigm focuses on traditional shareholder value-based conceptions of corporate responsibility. This article presents the new sustainability-oriented, online computer simulation, Leading Change for Sustainability-Business. The authors present preliminary evaluation data from a study of 87 students who learned with the simulation in a Master of Management program in Thailand. Data analyses focus on students' engagement with the simulation and their perceptions of its efficacy as an online learning tool. The relevance of this research was enhanced due to the fact that the study was conducted during the COVID-19 pandemic, when all university courses were executed in a fully online learning mode. Initial results support the efficacy of the Leading Change for Sustainability-Business simulation as a helpful tool for learning how to apply knowledge of change management to the challenges of corporate transformation for sustainability. More broadly, the results affirm the viability of simulation-based learning when used in a fully online mode of instruction at the university level.

P. Hallinger · C. Chatpinyakoop (✉)
College of Management, Mahidol University, Bangkok, Thailand
e-mail: chatchai.cht@mahidol.ac.th

P. Hallinger
e-mail: hallinger@gmail.com

P. Showanasai
Faculty of Management Sciences, Prince of Songkla University, Songkla, Thailand
e-mail: tum.parinya@gmail.com

V.-T. Nguyen
Faculty of Education, University of Social Sciences and Humanities, Vietnam National University Ho Chi Minh City, Ho Chi Minh City, Vietnam

W. Leal Filho et al. (eds.), *Sustainability in the Curriculum and Operations of Universities*, World Sustainability Series,
https://doi.org/10.1007/978-3-032-12678-8_25

Keywords Leading change for sustainability · Simulation-based learning · COVID-19 · Online learning

1 Introduction

Over the past several decades, the quest for corporate sustainability has gained increasing currency in Asia. For example, 'human capitalism,' with its strong employee focus, has enabled Japan to reach an advanced stage of capitalism (Ozaki 1991). Another variation comes from Southeast Asia, where governments have assumed a more proactive role in creating, shaping, and guiding markets. This has required firms to take greater responsibility for the social welfare of their employees (Sachs 2012). Singapore is a successful example of this business model. In Thailand, the Sufficiency Economy Philosophy, which aims for the balanced development of a sustainable society, has been acclaimed as a viable approach to corporate sustainability (Avery and Bergsteiner 2011; Kantabutra 2012). However, even as research has begun to demonstrate the efficacy of these alternatives to prevailing corporate management models, managers continue to be mainly educated based on the prevailing paradigm of shareholder capitalism, which has contributed to the unsustainable exploitation of people and the planet (Avery and Bergsteiner 2011; Sachs 2012). Thus, current and future managers require education and training to successfully reorient companies toward the vision of sustainable progress in the United Nations' Sustainable Development Goals (SDGs).

An extensive body of literature has examined strategies used by managers to implement change successfully in organizations. Scholars conclude that leaders are 'facilitators' and 'gatekeepers' of change (Folke et al. 2002; Kotter 2007). As gatekeepers, managers—at every company level—can open or close the door to change. However, research also affirms that this does not guarantee successful change even when managers open the door. Facilitating changes adopted by the company—even those that managers accept as priorities—requires skillful managerial support to succeed. Thus, management education programs, whether at the corporate level or university degree programs, have made 'change management' a standard competency in the 21st-century management education curriculum (Benn and Dunphy 2009).

However, the change to 'sustainability' presents an even more significant challenge than the usual 'internal organizational changes' that managers face. The change to sustainability involves complex, systemic changes at the individual, organizational, and social system levels (Lozano 2012; Senge et al. 2015). Changes that fall under the rubric of sustainability are typically addressed through interaction with the organization's environment and under conditions of uncertainty (Avery and Bergsteiner 2011; Benn et al. 2014; Ferdig 2007; Grin et al. 2010; Smit and Pilifosova 2003). This suggests a need for new and more powerful tools to assist managers and their teams understand how to succeed under these challenging conditions.

One approach that has emerged in efforts to learn how to 'make the change to sustainability has been computer simulation (Bekebrede et al. 2017; Crookall 2013;

Moratis et al. 2006). Simulations have been developed on corporate social responsibility (Moratis et al. 2006), climate change (Crookall 2013; Ducrot et al. 2015), and sustainable urban development (Mayer et al. 2005). Simulations have emerged as a 'tool of choice' among educators when addressing sustainability issues due to their superior ability to capture the complexity of systemic problems (Crookall 2013; Gosen and Washbush 2004; Lovelace et al. 2016; Sitzmann 2011; Sterman 1994). Simulations not only place learners in the situation of tackling complex problems and making decisions based on incomplete information, but they also provide feedback on those decisions, thereby enhancing capacities for problem-solving. Simulation-based learning is ideal for the subject matter and problems that require the ability to model and anticipate future scenarios (Avolio et al. 1988; Gosen and Washbush 2004; Salas et al. 2009).

A recent review of the literature on sustainability simulations and games identified a need for more simulations that can be used in management education and empirical research that sheds light on how much they achieve their stated goals. This includes high learner engagement, practical learning application, holistic, analytical thinking development, and learning transfer to the world outside the classroom (Hallinger et al. 2020).

This article addresses several, though not all, gaps in the research on using simulations and games in education for sustainability. More specifically, the goals of this article are to:

1. Elaborate on the instructional process used with the simulation.
2. Present preliminary evaluation data based on student perceptions of learning with the Leading Change for Sustainability—Business simulation.

This article is devoted to the instructional use of the LCS-B simulation and a preliminary empirical study. The empirical research included data collected from 87 part-time graduate students studying in a Master of Management degree program in Thailand. The 'treatment' consisted of a three-week, fully online learning module designed around the LCS-B simulation. Data analyses included data on student engagement with the simulation, student perceptions of the efficacy of the learning module based on end-of-course evaluations, and qualitative feedback obtained from open-ended questions answered by students during and following the module.

The significance of this article lies in three domains. First, the detailed instructional use of the LCS-B simulation may be of practical significance for educators interested in developing or using simulations. Second, the research highlights student perceptions of the effectiveness of the simulation in creating an active and engaging learning environment, enhancing their understanding of core concepts. Finally, since this research was conducted during the COVID-19 pandemic, the LCS-B simulation was delivered in a fully online synchronous learning mode. Thus, the findings also offer empirical insights into the efficacy of simulations as a mode of active learning when used in a fully online learning environment.

2 Instructional Process

The design features incorporated into the LCS-B simulation drew upon prior literature on experiential learning (Kolb 2014), simulation-based learning (Britain 2004; Burgos et al. 2007), and problem-based learning (Bransford 1993; Hallinger and Bridges 2007). Thus, the LCS-B simulation design and instructional process emphasize the principle that 'learning should start with the problem,' a fundamental tenet of problem-based learning (Barrows and Tamblyn 1980). Therefore, while the simulation is grounded in several theories, these are not introduced to students beforehand. Instead, the theories are elaborated on by students in mini-lectures and readings as they engage in solving the problem. This contrasts with case teaching, in which instructors typically teach a theory and provide a case for the application. Thus, for example, the Adopter Type theory is introduced to the students when they struggle to engage resistant stakeholders. Students can then apply the theory and see how it can help 'solve the problem.'

As indicated in Table 1, in the LCS-B module, students study in class for three three-hour sessions (a total of nine instructional hours). During class each week, the instructor divides the session into three parts (see Table 1). In the first 45 min, the instructor delivers a 'mini-lecture' on change management and sustainability concepts. For the next 90 min, teams consisting of three students take the 'project team' role and play the simulation in separate breakout rooms. The instructor periodically visits the breakout rooms to answer questions and offer consultation on change strategies. In the last half-hour of class, the instructor brings the whole class back together and leads a debriefing of the change obstacles and strategies they employ. Periodic debriefing of their simulation strategies accelerates learning, enhances the development of higher-order thinking, and clarifies the mental models essential to solving complex systemic problems (Dreifuerst 2009; Fanning and Gaba 2007).

In addition to learning in class, students are expected to complete a reading assignment each week, participate in an Online Discussion Forum (on Moodle), and play the simulation several times from home. The last two assignments are critical to learning with the simulation. Each week at the end of class, the instructor emphasizes that learning via simulation requires learners to play numerous times to see patterns associated with successful strategies. That is, the 'learning from experience' process requires multiple cycles of planning, doing, reflecting, and adjusting (Kolb 2014).

Indeed, it would be impossible to obtain a passing grade on the simulation exam without playing multiple times at home. The amount of time spent playing the simulation in class is simply insufficient. In Week 1, when students were unfamiliar with the people, activities, and method of play, it could take up to two hours to play an entire three-year simulation session. However, students can complete the full three-year simulation in 30 to 45 min by the third or fourth iteration. Nonetheless, learning via simulation requires a substantial time commitment from the students. Playing the simulation at home is 'incentivized' by allocating 25% of the course grade to results on the individual simulation exam.

Table 1 The basic instructional sequence is used with the LCS-B simulation

Week	Activities in class	Individual Homework
1	Mini lecture: Sustainability concepts Resistance to change Introduction to LCS-B simulation Teams of 3 play simulation in breakout rooms Whole class debriefing	• Simulation practice • Post in Week 1 Online Discussion Forum discussion question • Reading assignment
2	Mini lecture: Sustainability concepts Stages in the change process Change Adopter Types Teams of 3 play simulation in breakout Whole class simulation debriefing	• Simulation practice • Post in Week 2 Online Discussion Forum Reading assignment
3	Mini lecture: Sustainability concepts Apply Kotter's 8-stage model of change Synthesize change models Teams of 3 play simulation in breakout Whole class simulation debriefing	• Simulation practice • Post in Week 3 Online Discussion Forum • Prepare for the Simulation Exam in Week 4 • Reading assignment
4	Individual simulation exam	• Post in Week 4 Online Discussion Forum

The Online Discussion Forum is designed to 'continue the conversations' started among students in class and to support individual problem-solving as they play the simulation at home. Each week, students are expected to post a response to the question in the Online Discussion Forum and also comment on each other's posts. For example, following the first class, three discussion questions are posted:

1. What was the most difficult problem that you encountered so far in implementing change for sustainability at the Best Company? How did you solve it?
2. What one piece of advice would you like to share with others?
3. What one problem would you like advice on from others?

Thus, the Online Discussion Forum is another form of 'student-centered learning' integrated into the module.

3 Method

The main objective of this study is to preliminarily evaluate the simulation based on descriptive data, including student engagement and perception data. Future research of this study is to use additional quantitative and qualitative tools, including inferential analysis, to analyze students' simulation performances and discussion responses.

Participants. The participants in this study consisted of 87 students studying in a Master of Management degree program at a university in Thailand. The students averaged 28 years of age and five years of work experience, mainly in the corporate sector. Sixty-five percent of the students were female, and 35% were male.

Instructional Treatment. The learning module employed a simulation as an elective course in which enrolment was capped at 25 students to stimulate interaction. The simulation module was designed to draw upon and extend their prior learning of management theory. Students did not, however, have any prior coursework related to sustainability concepts. Therefore, the data presented in this analysis were collected from four classes over three trimesters between May 2021 and February 2022. The same instructor taught all four classes using the version of the simulation and support materials (i.e., lectures, readings, and discussion questions). The instructional process closely followed the description provided above. However, due to the COVID-19 pandemic, it should be noted that all instruction was delivered in a fully online mode. The weekly class sessions were delivered using the Zoom platform, supplemented by Moodle for storing materials and a link to the simulation.

Data collection. The research reported in this article is based on three data sources. First, student engagement was tracked by monitoring the frequency of simulation sessions played by individual students. Each time a student played, his/her 'strategy' (i.e., sequence of activities), results, and level achieved were saved on a server. Thus, the researchers could determine who played, when, for how long, how many times, and with what results.

The primary data source for this report was an end-of-course evaluation used by the college. The course evaluation survey used by the college consisted of 16 questions answered on a five-point Likert scale (1 Not at All to 5 To a great extent). Students completed the survey online outside of class at their convenience. We extracted student responses to four questions deemed most relevant to our understanding of the impact of the LCS-B simulation. These were:

1. Rate how well this course developed your analytical ability.
2. Rate the instructor's ability to develop your understanding of the subject.
3. Rate the ability of the instructor to make the course content practical.
4. Rate the extent to which the course is actively involved in learning.

We chose to focus on these questions for several reasons. First, delivering the module in a fully online mode presented a challenge for engaging students actively in simulation-based learning. Second, prior research suggested that simulation-based learning is well-suited to developing analytical ability and higher-order thinking. Thus, we were interested in whether students perceived this as the case. Third, simulation-based learning is also an explicit goal to enhance students' ability to apply theory to practical problems.

Finally, these quantitative data were supplemented by qualitative data from student feedback in the Online Discussion Board and responses to open-ended questions on the end-of-course evaluations.

Data Analysis. Preliminary statistical analysis confirmed no significant differences in student characteristics across the four classes in terms of age, college study duration, grade point average, or prior work experience. Thus, we treated the 87 students as a single group. Descriptive statistical analysis was conducted at the item level. Therefore, no reliability analysis was conducted. Data analysis reported the mean response of 87 students and the grand mean of students on the same items for all courses taught in the college during the same terms.

4 Results

This section presents qualitative and quantitative results concerning the efficacy of the LCS-B simulation in the context of a course offered in a graduate school of management in Thailand.

Before introducing the results of this study, however, we wish to highlight several features of the learners and the learning environment. First, although the students had learned Kotter's change model in an introductory course, they had no experience applying this or other change theories in practice. Second, a knowledge pre-test given in the first week of the module (not tabled) confirmed that the students had little or no baseline knowledge of basic sustainability concepts (e.g., triple bottom line, social sustainability, waste management, life cycle assessment). Thus, despite their work experience and prior coursework, the students came to the LCS-B module as 'novices' regarding change management and sustainability. Finally, we wish to reemphasize that all students in the four classes included in this research were working full-time and taking three classes concurrently. Consequently, their time for study outside of class was an ever-present constraint.

4.1 *Impact of the Simulation on Student Engagement*

The first result concerns the frequency with which students played the simulation. Table 2 shows the descriptive analysis of student engagement with the LCS-B simulation. The mean level of 22 game sessions per student per class was both consistent and 'high.' Using a low-end estimate of 30 min per game session as a standard, students spent an average of 3.7 h per week playing the simulation at home. The data in Table 2 do, however, highlight considerable dispersion among students in their engagement with the simulation, as evidenced by the min/max counts in each class and the relatively high standard deviations.

Table 2 Frequency count of LCS-B simulation games played by students over the three weeks of the module (n = 87 students in total)

LCS-B class	Min	Max	Mean	SD
Class 1	1	73	20.8	18
Class 2	1	48	21.8	14.3
Class 3	4	49	19.9	13.7
Class 4	3	70	24.8	16.7
All four classes	1	73	22.2	15.8

When interpreting the data in Table 2, it is also useful to know that 82 of the 87 achieved the highest level, Change Master, on the individual simulation exam. Thus, the Min counts shown in Table 2 were truly outliers regarding the number of times they played the simulation and their achievement on the 'exam.' Moreover, it would be incorrect to interpret the high percentage of students achieving on the individual exam as indicating that the simulation was 'easy.' Because students can 'see' the level of their success at the end of each three-year simulation session, they are motivated to continue playing until they can succeed. This is consistent with the principles of mastery learning.

These frequency data, which we interpret as a proxy for student motivation, were elaborated by quantitative data gathered on student participation in the Online Discussion Forum. More specifically, we assessed the level of student participation in the Online Discussion Forum for each of the four weeks. Analysis of the four classes found that 82 of 87 students (94%) participated in all four Online Discussion Forum activities outside of class.

In addition, qualitative data were gathered from student responses in the Online Discussion Forum and open-ended questions in the course evaluation survey conducted at the conclusion of the module. Some relevant responses included the following.

- The simulation helps students enjoy playing together while learning how to make the company achieve sustainability.
- Too much homework, comments to post in the discussion forum, and simulation practice
- The simulation game is fun, challenging, and interesting.
- The class environment is very active, and students can participate a lot.
- Really love the simulation game. It is really fun to learn from the simulation, and it is helpful in real life as well.

4.2 Student Perceptions of the Efficacy of the LCS-B Simulation

Table 3 shows the average ratings on the four survey questions for each LCS-B class and all college courses assessed in the same trimester. When interpreting the data in Table 3, two contextual factors deserve comment. First, it should be noted that the mean evaluation scores for 'all college courses' were consistently 'high' on the five-point Likert scale during each of the terms during which data were collected for this study (i.e., M = 4.22; Min 4.13/Max 4.31). The college has long emphasized active learning methods, including implementing a problem-based learning track in the curriculum (Hallinger and Bridges 2007). Thus, the scores obtained for the LCS-B simulation in this study were measured against a 'high standard.'

Second, it should be reemphasized that data collection for this study began one year into the COVID-19 pandemic (i.e., May 2021). Prior to the onset of the pandemic a year earlier, no courses in the college had been offered in a fully online learning

Table 3 Comparison of student perceptions of the LCS-B module with all college courses

Course	Class 1	Class 2	Class 3	Class 4	Grand mean
Rate how well this course developed your analytical ability					
LCS-B module	4.58	4.43	4.55	4.39	4.49
All courses for the term	4.23	4.13	4.13	4.19	4.17
% difference	+ 7.0%	+ 6.0%	+ 8.4%	+ 8.4%	+ 7%
Rate the instructor's ability to develop your understanding of the subject					
LCS-B module	4.79	4.62	4.5	4.5	4.6
All courses for the term	4.26	4.2	4.2	4.16	4.21
% difference	+ 10.6%	+ 8.4%	+ 6.0%	+ 6.8%	+ 8%
Rate the ability of the instructor to make the course content practical					
LCS-B module	4.63	4.52	4.55	4.56	4.57
All courses for the term	4.29	4.23	4.23	4.17	4.23
% difference	+ 6.8%	+ 5.8%	+ 6.4%	+ 7.8%	+ 6.7%
Rate the instructor's ability to actively involve students in learning					
LCS-B module	4.74	4.71	4.6	4.5	4.64
All courses for the term	4.31	4.25	4.25	4.23	4.26
% difference	+ 8.6%	+ 9.2%	+ 7%	+ 5.4%	+ 7.5%
Grand Mean 4 Items					
LCS-B module	4.69	4.57	4.55	4.49	4.58
All college courses	4.27	4.20	4.20	4.19	4.22
% difference	+ 8.4%	+ 7.4%	+ 7%	+ 6%	+ 7.2%

Notes courses were rated on a 5-point Likert scale. The LCS-B modules averaged an 86% student response rate to the course evaluation survey. + indicates the advantage of the LCS-B mean compared with All College Courses mode. Indeed, no instructors had used a synchronous learning platform such as Zoom, and relatively few had even used Moodle extensively. This is relevant because all college courses, not just the LCS-B module, were being delivered in a fully online learning mode during the entire data collection period.

With these factors in mind, Table 3 shows very high grand scores for the LCS-B module on each of the four selected survey items (i.e., 4.49, 4.6, 4.57, 4.55). The grand mean across the four items was 4.55. These data provide perceptual evidence that learning through the LCS-B simulation developed student understanding of the subject (i.e., change management for sustainability), enhanced their analytical ability (i.e., change strategy), made the content practical, and provided an active and motivating learning environment.

The results of the course evaluation data provide clear, if somewhat limited, evidence that the LCS-B simulation provided an active, engaging, and challenging learning environment. The data also suggest that students experienced growth in their understanding of change management and sustainability concepts and increased interest in sustainability. Nonetheless, affirmation of these conclusions awaits further analysis of quasi-experimental data on actual learning gains and attitude change during the LCS-B module.

Comments from the Online Discussion Forum and evaluation survey elaborated on these quantitative perceptions. Students made the following observations concerning developing an understanding of the content and enhancing their analytical ability.

- I am very impressed with this simulation because it seems like you are in the organization. Actually, in real life, you cannot replay or redo everything, but in this simulation, you know that you could try to do it and redo it again. It inspires me to know exactly what the process of change is and the ways of embedding a sustainability attitude, instead of just knowing the concepts.
- For me at the beginning of this course, I had no knowledge about sustainability and questioned why sustainability can be beneficial to the company and industry. Now, I feel like there is a big surprise with many examples of sustainable practices that can be implemented. I don't feel like it is out of people's reach.
- I am surprised by the simulation. Sustainability is not only an environmental issue but also an impact on people at the individual, corporate, and societal levels. The simulation was challenging but fun and gave me the whole concept of sustainability.
- I learned that being a planned person but flexible to different circumstances is an excellent way to deal with problems. The simulation helped me learn how to think ahead and anticipate the consequences of my plan. I also learned to be prepared for unexpected situations in the future. In real life, we cannot expect everything to be predictable, and making any major change in the company is also like that.

Students also emphasized how the simulation made theories applicable in their corporate and personal lives.

- I was impressed by the simulation in class. The program helps to create a process of thinking about the concept of sustainability in ways that can be used in real life.
- The simulation is similar to real life in my workplace. When a company wants to implement a new policy or make a change, it always faces resistance at first. At that point, the strategy we use is critical. We must determine who will be the first group to change and how to influence more people to support the change as time passes.
- To me, I think the problem of sustainability is big enough that we should have a course on it at the university. The fact that we have this class in the Master's program reflects the importance of the issue. Even though right now we know just only the tip of the iceberg, it is really useful because eventually, we will have to use this knowledge no matter where we are working or what we are doing.

Analysis of the LCS-B results against 'all college courses' offers additional insight into the meaning of these results. On each of the four items shown in Table 3, the mean scores for the LCS-B module evidenced a positive advantage when compared with the mean scores for all college courses (i.e., an average difference of 7%). Based on these data, we conclude that students perceived the fully online delivery of the LCS-B module to be successful at creating an active learning environment that enhanced their ability to think strategically and practically about the transformation of companies for sustainability.

5 Discussion and Conclusion

This article has described a project undertaken to design an online computer simulation focusing on leading change for sustainability in the corporate sector. The relevance of this research was unexpectedly enhanced by the onset of the COVID-19 pandemic, which intensified the need for identifying active teaching and learning methods that could be employed in a fully online mode of instruction. In this final section of the article, we reflect on the research method's limitations and interpret the findings' meaning and implications.

5.1 Interpretation and Implications

This article has focused on the initial evaluation of a fully online computer simulation intended to educate prospective and practicing managers. The LCS-B simulation offers a potent illustration of 'student-centered learning' delivered in a fully online learning mode. For example, our data offered insights into the extent of student engagement during the module. Frequency counts on the number of game sessions played by students represented objective data confirming the substantial amount of time students invested in learning outside class. Many students used words like 'fun,' 'stimulating,' engaging,' challenging,' 'frustrating,' and 'exciting' to describe their experience with the simulation. Notably, a few students objected to their time learning with the simulation, reading, and participating in the Online Discussion Forum. Nonetheless, from our perspective, even this negative assessment can be viewed as an affirmation that students were at the 'center' of learning, doing the hard work needed to make meaning from their 'simulated experience' gained through the game sessions.

Several implications follow from the findings presented in this article. First, this theory-driven simulation resulted from a multi-year, funded research and development project to enhance higher education for sustainability. The potential utility of this simulation as a resource for learning how to 'make the change to sustainability' is quite broad. The simulation could be used with undergraduate or graduate education students in various disciplines. Similarly, it could be used as part of a company's effort to raise staff awareness as it embraces sustainable practices. Finally, because the simulation is hosted 'online,' it could also be accessed and played outside of formal learning settings. Indeed, we repeatedly emphasize the learning gained from playing the simulation.

A second implication lies in the perceived efficacy of learning with the simulation in a fully online mode. The COVID-19 pandemic challenged educators to seek effective pedagogical models and tools that could be adapted to meet the constraints of 'distance learning' (Lantis 2022; Singhal et al. 2021; Tan et al. 2020). The preliminary analysis presented in this article lends credence to the potential that lies in simulations and games as a pedagogical framework for online learning. Our results highlight the motivational advantages inherent in well-designed simulations and their potential for reducing the gap between theory and practice.

Anecdotally, the authors wish to reference the perceptions of the instructors and students in this regard. As instructors, we observed a few disadvantages to using the simulation in a fully online mode compared with prior terms when it was used in a blended learning mode (i.e., meeting with students in a physical classroom). To our surprise, this perception was mirrored in student responses. For example, as our college began transitioning back to face-to-face instruction, the instructor allowed students to hold the third and fourth class sessions online or face-to-face. Seventy-five percent of the class opted for online (i.e., via Zoom), and 25% expressed no preference. None opted explicitly for face-to-face instruction. Discussion with the students surfaced the perception that learning with the online simulation would be equally effective.

Third, positive student response to the LCS-B simulation has paved the way for the design of additional sustainability simulations. For example, the LCS-B simulation has been translated into Thai for a broader audience in Thailand. In addition, two new versions focusing on different organizational contexts for implementing sustainability are in the design and testing process: Leading Change for Sustainability—Schools (Nguyen and Hallinger 2022) and Leading Change for Sustainability—Higher Education.

5.2 *Limitations of the Study*

The key limitation of this research lies in the predominant reliance on student perception data to gain insights into the efficacy of the LCS-B simulation. In fact, the authors have been among those scholars who have criticized the use of 'weak research designs and methods in research on simulations and games in educating for sustainability' (Gosen and Washbush 2004; Hallinger et al. 2020). To be fair, this criticism also applies to the empirical research presented in this article, which consisted of a descriptive analysis of student perceptions.

Our rationale for proceeding with the presentation of these admittedly preliminary results is based on two factors. First, the main purpose of this article was to provide a detailed description of the design features underlying a new sustainability-oriented computer simulation. Space limitations would preclude the inclusion of such detailed information on the LCS-B simulation in an article whose primary purpose was to report the results of an empirical study of student learning outcomes.

Thus, while not ideal, the analysis of student perceptions presented in this paper offers useful insights concerning the motivational impact of the simulation and its efficacy from the learners' perspective. Indeed, student perceptions of online learning became paramount during the COVID-19 pandemic when universities were challenged to adapt to a fully online instructional environment (Baber 2020; Bączek et al. 2021). Thus, we suggest that student perceptions in this preliminary study can be justified.

5.3 *Further Studies*

Finally, we wish to emphasize that this report is just the first in a research program on the learning efficacy of the LCS-B simulation and its offshoots. As future reports will elaborate, automatic data collection on a central server greatly facilitates research on online simulations. This data can be analyzed to gain insights into the thinking and decision-making of learners (Showanasai et al. 2013).

References

Avery, G.C., Bergsteiner, H. (2011). How BMW successfully practices sustainable leadership principles. *Strategy & Leadership 39*(6), 11–8.

Avolio, B.J., Waldman, D.A., Einstein, W.O. (1998). Transformational leadership in a management game simulation: Impacting the bottom line. *Group & Organization Studies 13*(1), 59–80.

Baber, H. (2020). Determinants of students' perceived learning outcome and satisfaction in online learning during the pandemic of COVID-19. *Journal of Education and E-learning Research 7*(3), 285–92.

Bączek, M., Zagańczyk-Bączek, M., Szpringer, M., Jaroszyński, A., Wożakowska-Kapłon, B. (2021). Students' perception of online learning during the COVID-19 pandemic: a survey study of Polish medical students. *Medicine 100*(7), 1–6.
Barrows, H.S., Tamblyn, R.M. (1980). *Problem-based learning: An approach to medical education.* Springer: New York, NY.
Bekebrede, G., van Bueren, E., Wenzler, I., van Veen, L. (2017). Challenges in the transition towards a sustainable city: The Case of GO2Zero. In Simulation Gaming. Applications for Sustainable Cities and Smart Infrastructures: 48th International Simulation and Gaming Association Conference, ISAGA 2017, Delft, The Netherlands, July 10–14, 2017, Revised Selected Papers 48 2018 (pp. 67–74). Springer Cham
Benn, S., Dunphy, D. (2009). Action research as an approach to integrating sustainability into MBA programs: An exploratory study. *Journal of Management Education, 33*(3), 276–295.
Benn, S., Edwards, M., Williams, T. (2014). *Organizational change for corporate sustainability.* Routledge: London, UK.
Bransford, J. (1993). *Who ya gonna call? In Cognitive perspectives on educational leadership*; Hallinger, P., Leithwood, K., Murphy, J., Eds.; Teachers College Press: New York, NY.
Britain, S. (2004). *A review of learning design: concept, specifications, and tools. A report for the JISC E-learning Pedagogy Programme.*
Burgos, D., Tattersall, C., Koper, R. (2007). Re-purposing existing generic games and simulations for e-learning. *Computers in Human Behavior. 23*(6), 2656–2267.
Crookall, D. (2013). Climate change and simulation/gaming: Learning for survival. *Simulation and Gaming 44*(2–3), 195–228.
Dreifuerst, K.T. (2009). The essentials of debriefing in simulation learning: A concept analysis. *Nursing Education Perspective 30*(2), 109–114.
Ducrot, R., Van Paassen, A., Barban, V., Daré, W.S., Gramaglia, C. (2015) Learning integrative negotiation to manage complex environmental issues: Example of a gaming approach in the peri-urban catchment of São Paulo, Brazil. *Regional Environmental Change, 15,* 67–78.
Fanning, R.M., Gaba, D.M. (2007). The role of debriefing in simulation-based learning. *Simulation in Healthcare 2*(2),115–125.
Ferdig, M. A. (2007). Sustainability leadership: Co-creating a sustainable future. *Journal of Change Management, 7*(1), 25–35.
Folke, C., Carpenter, S., Elmqvist, T., Gunderson, L., Holling, C., Walker, B. (2002). Resilience and sustainable development: Building adaptive capacity in a world of transformations. *AMBIO: A Journal of the Human Environment 31*(5), 437–441.
Gosen, J., Washbush, J. (2004). A review of scholarship on assessing experiential learning effectiveness. *Simulation & Gaming 35*(2), 270–293.
Grin, J., Rotmans, J., Schot, J. (2010). Transitions to sustainable development: new directions in the study of long-term transformative change. Routledge: London, UK.
Hallinger, P., Bridges, E.M. (2007). A Problem-based Approach for Management Education: Preparing Managers for Action. Springer Science and Business Media: Berlin, Germany.
Hallinger, P., Wang, R., Chatpinyakoop, C., Nguyen, V.T., Nguyen, U.P. (2020). A bibliometric review of research on simulations and serious games used in education for sustainability, 1997–2019. *Journal of Cleaner Production 256,* 120358.
Kantabutra, S. (2012). Putting Rhineland principles into practice in Thailand: Sustainable leadership at Bathroom Design Company. *Global Business and Organizational Excellence, 31*(5), 6–19.
Kolb, D.A. (2014). *Experiential learning: Experience as the source of learning and development.* Prentice Hall: Englewood Cliffs, NJ.
Kotter, J.P. (2007). Leading Change. *Harvard Business Review 85,* 96–103.
Lantis, J.S. (2022). Introduction: active learning for a post-pandemic world. In Active learning in political science for a post-pandemic world; Lantis, J.S., Ed., Palgrave Macmillan, Cham. pp. 1–15.

Lovelace, K.J., Eggers, F., Dyck, L.R. (2016). I do and I understand: Assessing the utility of web-based management simulations to develop critical thinking skills. *Academy of Management Learning & Education 15*(1), 100–121.
Lozano, R. (2012). Orchestrating organisational changes for corporate sustainability: Overcoming barriers to change. *Greener Management International, 57*, 43-64.
Mayer, I.S., Van Bueren, E.M., Bots, P.W., Van Der Voort, H., Seijdel, R. (2005). Collaborative decision-making for sustainable urban renewal projects: a simulation gaming approach. *Environment and Planning B: planning and design, 32*(3), 403–423.
Moratis, L., Hoff, J., Reul, B. (2006). A dual challenge facing management education: Simulation-based learning and learning about CSR. *Journal of Management Development, 25*(3): 213–231.
Nguyen, V.T., Hallinger, P. (2022). Creating the Leading Change for Sustainability in Schools (Vietnam) Simulation: A Research and Development Project. In Developments in Business Simulation and Experiential Learning, Proceedings of the Annual ABSEL conference, Pittsburgh, PA.
Ozaki, R. (1991). Human capitalism: The Japanese model for market economies. Kodansha: New York, NY.
Sachs, J.D. (2012). From the millennium development goals to sustainable development goals. *Lancet, 379*:2206-2211.
Salas, E., Wildman, J.L., Piccolo, R.F. (2009). Using simulation-based training to enhance management education. *Academy of Management Learning and Education 8*(4): 559–573.
Senge, P., Hamilton, H., Kania, J. (2015). The dawn of system leadership. *Stanford Social Innovation Review 13*(1), 27–33.
Showanasai, P., Lu, J., Hallinger, P. (2013). Developing tools for research on school leadership devel opment: an illustrative case of a computer simulation. *Journal of Educational Administration 51*(1):72–91.
Singhal, R., Kumar, A., Singh, H., Fuller, S., Gill, S.S. (2021). Digital device-based active learning approach using virtual community classroom during the COVID-19 pandemic. *Computer Applications in Engineering Education, 29*(5):1007–1033.
Sitzmann, T. (2011). A meta-analytic examination of the instructional effectiveness of computer-based simulation games. *Personnel Psychology, 64*(2): 489–528.
Smit, B., Pilifosova, O. (2003). Adaptation to climate change in the context of sustainable development and equity. *Sustainable Development 8(*9), 9.
Sterman, J.D. (1994) Learning in and about complex systems. *System Dynamic Review, 10*, 291–330
Tan, H.R., Chng, W.H., Chonardo, C., Ng, M.T., Fung, F.M. (2020). How chemists achieve active learning online during the COVID-19 pandemic: using the Community of Inquiry (CoI) framework to support remote teaching. *Journal of Chemical Education, 97*(9), 2512–2518.

Defining the Scope and Framework of STEM4S for Design Education: Case Analysis of Design Course Topic with Single-Use Food Packaging Waste

Daisuke Nagatomo

Abstract Design education is increasingly essential in fostering sustainability by promoting critical thinking and problem-solving through design thinking and hands-on learning. STEM education for sustainability (STEM4S) extends traditional STEM education by addressing global challenges such as climate change and the SDGs through multidisciplinary approaches. This study investigates the integration of STEM4S into design education in Taiwan, focusing on single-use food packaging waste at National Taiwan Normal University. Focusing on the Asia-Pacific region, this research addresses the unique sustainability challenges of a rapidly developing area, emphasizing the role of education in driving innovative solutions. The study employs qualitative analysis of student design projects and quantitative analysis of pre- and post-course questionnaires to evaluate the effectiveness of STEM4S in cultivating sustainability competencies. The findings demonstrate how integrating STEM4S enhances students' ability to tackle real-world challenges, contributing to sustainable development and fostering critical thinking within the region.

Keywords STEM4S · Sustainability · SDGs · Design · PBL

1 Introduction

In the past two decades, STEM (Science, Technology, Engineering, and Mathematics) has emerged as a transformative educational movement, offering students cognitive, procedural, and attitudinal benefits through interdisciplinary activities (Martín-Páez et al. 2019). STEM education has demonstrated its potential to equip students with the skills and literacy needed to thrive as 21st-century citizens and

D. Nagatomo (✉)
Department of Design, National Taiwan Normal University, Taipei, Taiwan
e-mail: dn@ntnu.edu.tw

International Degree Program in Climate Change and Sustainable Development, National Taiwan University, Taipei, Taiwan

W. Leal Filho et al. (eds.), *Sustainability in the Curriculum and Operations of Universities*, World Sustainability Series,
https://doi.org/10.1007/978-3-032-12678-8_26

professionals (Ritz and Fan 2015). Defined as an interdisciplinary theoretical and instructional approach (Mohr-Schroeder et al. 2015), STEM emphasizes project-based and problem-based learning to encourage the application of knowledge in tackling real-world challenges (LaForce et al. 2017). Design thinking has become a core component of STEM, providing aesthetically innovative outcomes and practical solutions to complex problems (Li et al. 2019).

The Sustainable Development Goals (SDGs) introduced by the United Nations in 2015 redefine the responsibilities of various sectors, including business, manufacturing, land use, and education (Nations 2015). Higher education plays a crucial role in advancing these goals through the implementation of innovative pedagogical methods (Zamora-Polo and Sánchez-Martín 2019). However, the methods of incorporating SDGs into curricula remain in the development stage, highlighting the urgent need for research to bridge this gap (Gyasi et al. 2021). Studies exploring the integration of SDGs into education have demonstrated how the goals equip students with knowledge and skills to address sustainability challenges (Borges et al. 2017). Avelar et al. (2019) emphasize the importance of education in advancing SDG implementation, citing their review of 193 articles published from 2015 to 2018. Their findings underscore the critical need to train students through integrating SDGs into educational frameworks.

These educational frameworks have added significance in the Asia–Pacific region, where economic development and environmental challenges intersect. By aligning STEM with SDGs, higher education can prepare students to address region-specific sustainability challenges while fostering global prosperity and peace. Therefore, this study explored defining the scope and framework of STEM aligning with SDGs in higher education.

2 Research Background

The concept of STEM education for sustainability (STEM4S) broadens the traditional interdisciplinary scope of STEM education to emphasize multidisciplinary critical thinking and a sustainability mindset (Bascopé and Reiss 2021). By integrating sustainability into STEM curricula, STEM4S challenges students to consider solutions beyond the confines of their disciplines, addressing global issues like climate change and SDGs (Petrun Sayers et al. 2021). This approach equips the next generation of leaders with the tools to ensure sustainable development while fostering critical thinking and problem-solving skills (Campbell and Speldewinde 2022).

Critical thinking, a fundamental goal of higher education, is pivotal in expert decision-making and professional competence (Facione and Facione 2013). Students actively engage with real-world tasks through problem-based or project-based learning (PBL), integrating theory and practice to develop actionable knowledge (LaForce et al. 2017). STEM4S leverages PBL methodologies to nurture critical thinking.

Table 1 The ideal scope of STEM4S

Scope	Description
Scope A	Adopt STEM interdisciplinary knowledge and train problem-solving abilities via sustainability-related issues
Scope B	Create a multidisciplinary environment that enables students to work on the issues from a diverse perspective
Scope C	Enable the project assignment to build on social contexts to promote critical thinking
Scope D	Equip the instructor with multidisciplinary knowledge and skills for guiding students' literacy

2.1 Expanding STEM education for sustainability

Originally aimed at developing a workforce proficient in science and engineering to boost global competitiveness and national security (Gonzalez 2012), STEM education has evolved to address broader societal needs. STEM4S develops technical competence and emphasizes sustainability as a foundation for action in addressing global challenges. One of the aims is to cultivate the habit of using scientific evidence to justify sustainable practices and promote a deeper understanding of the societal value of STEM (Pahnke et al. 2019). The STEM4S approach also integrates social contexts, such as local knowledge, into education to make sustainability issues relatable and actionable for students (Bascopé and Reiss 2021). Multidisciplinary curricula encourage students to think critically about sustainability issues outside their immediate fields while enabling educators to bridge the gaps between traditional interdisciplinary STEM education and STEM4S (Petrun Sayers et al. 2021).

Table 1 describes the ideal scope of STEM4S. These dimensions underscore STEM4S's transformative potential to align educational objectives with sustainable development goals, fostering individual and societal progress.

2.2 Framework for innovation

"Design and Design Thinking are vital to creativity and innovation and has become increasingly important in the current movement of developing and implementing integrated STEM education" (Li et al. 2019). Incorporating design-based tasks could provide opportunities for students to explore their creativity (Hathcock et al. 2015). As Fortus et al. (2004) demonstrated, design can be a vehicle for scientific knowledge and real-world problem-solving skills. Banathy (1996) states that design manifests dynamic interaction between feedback and feedforward, reflection and creation, and divergence and convergence. The dynamics of divergence and convergence form a double-diamond relationship in the design process. The UK's Design Council later applied this double diamond model to define the framework for innovation (Fig. 1). The framework for innovation consists of four steps: discover, define, develop, and deliver. The first diamond includes *discover* and *define*. During the *discover* phase, people begin to understand rather than assume. After gathering insight in the *discover* phase, people need to define the challenge. The second diamond consists of *develop* and *deliver*. The development phase enables people to give different answers to clearly defined problems. *Deliver* involves testing solutions and evaluating outcomes to create innovative solutions (Council 2011). The framework is not necessarily a linear process, as it also brings the process back to the beginning to identify underlying problems.

This framework intends to provide a clear challenge in the student's assignment and create a guideline for their project development. The project assignment needs to ensure the student's knowledge and skill level match the degree of the challenge, and it should also provide them with tools to develop under the instructor's guidance. Setting clear expectations for the student's outcome is also necessary. The instructor must encourage students' creativity to avoid design fixation in their development process. This research utilizes the double-diamond framework to align the STEM design thinking process with students' design projects.

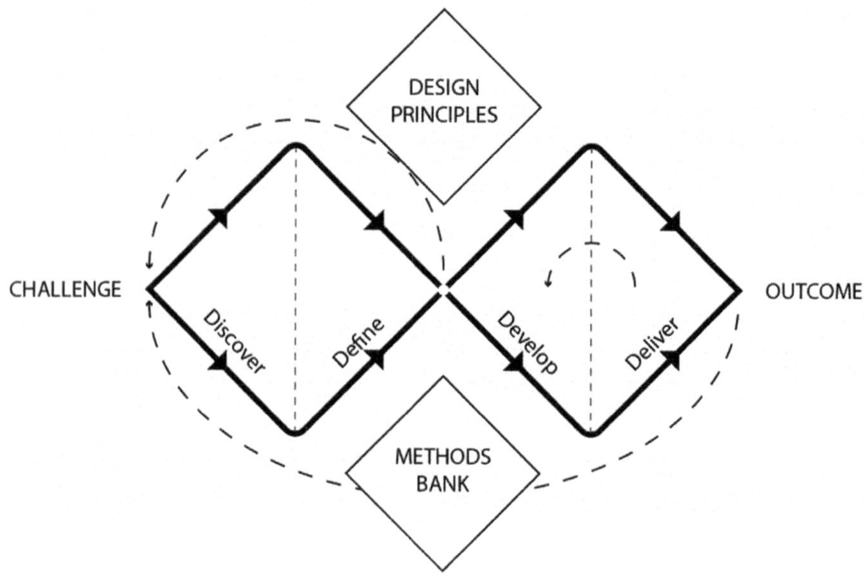

Fig. 1 The framework for innovation from the UK design council (recreated by the author)

3 Research Method

This research examines implementing the STEM4S framework in the undergraduate course curriculum by incorporating sustainable development goals and multidisciplinary knowledge-based activities to develop students' cognitive skills through design projects. The research data was collected from the Innovation Design II course at the NTNU Department of Design. This course is mandatory for undergraduate sophomore students in the product design group. The course aims to enhance project-based learning through intensive design research and problem-solving by following the double-diamond framework for innovation. Pre- and post-course questionnaires are conducted to observe how the STEM4S framework impacted students' awareness and enhanced their learning competency.

3.1 Pre-course questionnaire

The pre-course questionnaire was distributed to the students prior to the course introduction in the first week. Table 2 indicates the questions provided to the students. The questions mainly concern their awareness of the social issues related to sustainability, covering areas such as SDGs, climate change, recycling of trash, circular economy, and single-use plastic waste. The answering choices are indicated with one to five points based on whether the students agree with the question. The responses collected in Table 3 were used to analyze knowledge and awareness of social issues before the course, students' learning competency after the course, and the effectiveness of the STEM4S framework in cultivating their critical thinking on those issues.

3.2 Project assignments

The objective of STEM4S is to cultivate students' critical thinking skills by presenting sustainability challenges and training their problem-solving abilities by incorporating multidisciplinary knowledge and skills. Therefore, the topic for the project assignment involves considering how students can identify themselves as part of the problem. The students are tasked with addressing social issues related to food, which can directly relate to SDGs 11 (Sustainable City and Community), 12 (Responsible Consumption and Production), 13 (Climate Action), and 14 (Life Below Water). Furthermore, by incorporating STEM4S, the course aims to achieve SDG 4 (Quality Education).

Table 4 describes each project topic and the expectations for the student's project. The social issues related to food are complex rather than simple disciplinary solutions. Therefore, the development process must incorporate a multidisciplinary design thinking process. The two design projects are embedded into the double-diamond

Table 2 Pre-course questionnaire questions

No.	Question
1	Do you have a complete understanding of the United Nations' SDGs?
2	Have you taken any courses that focus on SDGs or any sustainability topics?
3	Do we need to take action to address climate change?
4	Is our responsibility in climate change clear to you?
5	Do you know the difference between the linear economy and the circular economy?
6	Are you aware of how much waste you generate?
7	Do you separate trash for recycling?
8	Do you know how much plastic goes to landfill?
9	Do you think sustainability can be promoted by design?
10	Are you familiar with the circular design approach?
11	Is plastic a hazardous material for the environment?
12	Can we live without using plastic products?
13	Do you think recycling plastic is a key to saving the Earth?
14	How can we reduce the usage of plastic products?
15	Can you define design innovation?

Table 3 Pre-course questionnaire result

No.	Strongly disagree (1)	Disagree (2)	Neutral (3)	Agree (4)	Strongly agree (5)	Mean	SD
1	5	5	3	3	3	2.68	1.45
2	7	8	3	0	1	1.95	1.03
3	0	0	1	6	12	4.58	0.61
4	1	0	5	8	5	3.84	1.01
5	6	2	1	7	3	2.95	1.58
6	1	3	9	2	4	3.26	1.15
7	0	0	1	10	8	4.37	0.60
8	4	4	9	0	2	2.58	1.17
9	0	0	2	8	9	4.37	0.68
10	4	4	7	3	1	2.63	1.16
11	0	0	5	7	7	4.11	0.81
12	3	8	4	3	1	2.53	1.12
13	1	0	1	7	10	4.32	1.00
14	2	2	5	8	2	3.32	1.16
15	1	6	7	4	1	2.89	0.99

Note n = 19

Table 4 Innovation design II assignment description

Assignment	Title	Description
(I)	Food packaging regenerative design	This assignment explores the potential of repurposing food packaging waste into a new lifecycle of design products by utilizing Precious Plastic methods. This assignment aims to understand food packaging waste by collecting daily food-related waste and processing it as a repurposable material form
(II)	Repack design challenge	This assignment aims to develop a new design prototype of a circular system on food packaging. Design is utilized to create a centered circular system, contributing to society's smooth shift into the circular mindset. In this assignment, a circular system is ideated by re-designing food and packaging

framework for innovation, and the course syllabus is constructed to cohere with the framework. Each project lasts eight weeks, and students present their outcomes at the end of the eighth week.

3.3 Methods bank and multidisciplinary implementation

The methods bank is a set of tools that helps students develop the project. The resources could be physical tools, such as equipment and computers, and observation tools, such as field trips and interviews. The purpose of assignment I is to repurpose food packaging waste into a new design product that elevates the value of the material. Therefore, the method requires physical hardware, such as a plastic shredder and heat press, for repurposing plastic flakes (Fig. 2a and b). Each student joined the tutorial session to handle the equipment and learn how to use it before working on their design project. Furthermore, the student group participated in a workshop at Trash Kitchen, operated by Miniwiz, Taiwanese entrepreneurs who recycle waste into innovative products. During the workshop, the students repurposed plastic waste into phone charging devices and gained insights into Taiwanese recycling practices (Fig. 3a and b).

Assignment II required students to explore the relationship between packaging and food and develop circular food systems to reduce associated waste. Initially, the students struggled with the challenge of replacing plastic wrapping, particularly for mass-produced food packaging. To address this issue, the course organized a field trip to U-Market, a traditional Taiwanese street market, where students interviewed a restaurant owner who collaborates closely with the market. The owner shared innovative practices, such as allowing customers to bring food purchased elsewhere

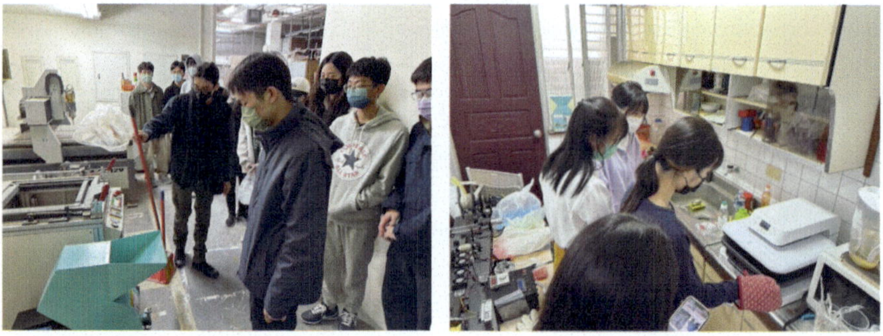

(a) (b)

Fig. 2 **a** Plastic shredder operation tutorial to repurpose plastic waste, **b** heat-press operation tutorial to repurpose plastic waste

(a) (b)

Fig. 3 **a** Learning about upcycling plastic waste at the Trash Kitchen workshop, **b** workshop at trash kitchen for upcycling plastic waste to phone charger

at the U-Market while ordering drinks and other menu items. This initiative promotes local culture and supports sustainable food distribution methods. Inspired by the field trip, several students shifted their focus to developing projects centered on locally distributed foods (Fig. 4a and b).

STEM4S aims to integrate multidisciplinary knowledge to address sustainability challenges. However, as the course is mandatory for the Department of Design, it exclusively comprises design students, limiting opportunities for collaboration with peers from other disciplines. To introduce diverse perspectives and facilitate a more multidisciplinary learning experience, professionals from outside the design industry were invited to share their insights on food packaging issues. These guest speakers included a circular economy consultant from Taiwan and a director from a Japanese circular economy research corporation (Fig. 5a and b).

(a) (b)

Fig. 4 **a** Interviewing the restaurant owner next to the traditional food market in Taipei city, **b** field trip to the traditional food market in Taipei city

(a) (b)

Fig. 5 **a** Assignment I review with a Taiwanese professional consultant working in the circular economy, **b** assignment II workshop regarding food and packaging with a Japanese circular economy research company

The guests' contributions enriched the curriculum by providing real-world insights and fostering critical discussions. By integrating these expert perspectives with course assignments, review systems, and activities, the double-diamond STEM4S framework was successfully implemented (Fig. 6).

3.4 Post-course questionnaire

The post-course questionnaire was collected after 16 weeks of coursework work to evaluate the effectiveness of the STEM4S framework. The questionnaire comprised 35 questions in seven sections:

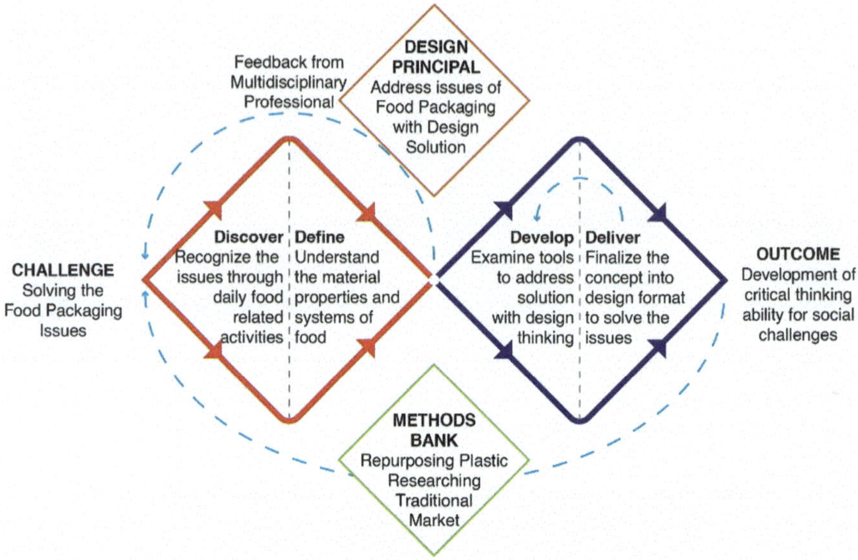

Fig. 6 The double-diamond STEM4S framework for innovation design II (illustrated by the author)

1. Sustainability knowledge (Sec. 1: Question numbers 1.1–1.5)
2. Course goal (Sec. 2: Question numbers 2.1–2.5)
3. Assignment topic (Sec. 3: Question numbers 3.1–3.5)
4. Course activities (Sec. 4: Question numbers 4.1–4.5)
5. Course works (Sec. 5: Question numbers 5.1–5.5)
6. Plastic-related (Sec. 6: Question numbers 6.1–6.5)
7. Food packaging-related (Sec. 7: Question numbers 7.1–7.5)

Each section had five questions to determine which elements most influenced students' learning competency. Sec. 1 questions focused on how students enhance and utilize sustainability knowledge in their project development. Sec. 2 and 3 questions asked students about the course goal and assignment topic and how they reacted to the course content. Sec. 4 and 5 enabled students to indicate whether the course activities helped them execute their coursework and whether they felt good about their work. The last two sections, Sec. 6 and 7, asked detailed questions about Assignments I and II. Table 5 lists the questionnaire questions, while Table 6 shows the students' responses.

Table 5 Post-course questionnaire questions

No.	Question
1.1	Did this course provide enough opportunities to expand your knowledge of sustainability issues?
1.2	Did this course provide enough guidance for you to explore sustainability issues?
1.3	Did this course provide you with new knowledge of sustainability issues?
1.4	Did this course boost your ability to think about global sustainability issues?
1.5	Did new knowledge in sustainability issues help you complete the course?
2.1	Was the course goal clear to you?
2.2	Did you find it easy to follow the course goal?
2.3	Were you interested in this course goal?
2.4	Do you think this course goal is helpful for your future career?
2.5	Do you think this course trains your problem-solving ability?
3.1	Was the assignment topic clear to you?
3.2	Did you find it easy to work on the assignments?
3.3	Was there enough guidance to develop the assignments?
3.4	Did the assignments motivate you to work on the course?
3.5	Did the assignments boost your ability to think about innovation?
4.1	Did the workshop at Trash Kitchen motivate you to participate in this course?
4.2	Did the workshop at Trash Kitchen expand your interest in sustainability?
4.3	Did the RE: Public guest lecture give you more idea of the circular economy?
4.4	Did the RE: Public workshop help you to think about the issue of food packaging?
4.5	Did the field trip to U-Market help you to develop the idea of circular packaging?
5.1	Were you satisfied with your coursework?
5.2	Did you have enough time to develop your coursework?
5.3	Did you have enough knowledge to complete the coursework?
5.4	Did the review enable you to improve your work?
5.5	Did you find inspiration from your classmates during the course?
6.1	Does repurposing plastic waste help achieve the circular economy?
6.2	Does design play a critical role in repurposing plastic?
6.3	Does repurposing plastic help your creativity?
6.4	Does repurposing plastic add a higher value to plastic waste?
6.5	Does repurposing plastic create an emotional connection to plastic waste?
7.1	Is it possible to change food packaging problems?
7.2	Do we need to replace plastic to achieve circular packaging?
7.3	Does the food distribution system need to change to become more sustainable?
7.4	Is focusing on local food culture more than mass-produced food the key to being circular?
7.5	Can circular food packaging solve the single-plastic problem?

Table 6 Post-course questionnaire result

No.	Strongly disagree (1)	Disagree (2)	Neutral (3)	Agree (4)	Strongly agree (5)	Mean	SD
1.1	0	0	2	15	2	4.00	0.47
1.2	0	1	6	9	3	3.74	0.81
1.3	0	0	2	14	3	4.05	0.52
1.4	0	0	2	10	7	4.26	0.65
1.5	0	1	4	10	4	3.89	0.81
2.1	1	1	7	8	2	3.47	0.96
2.2	1	8	9	0	1	2.58	0.84
2.3	0	1	6	11	1	3.63	0.68
2.4	1	1	3	10	4	3.79	1.03
2.5	1	0	2	12	4	3.95	0.91
3.1	0	3	7	7	2	3.42	0.90
3.2	2	12	5	0	0	2.16	0.60
3.3	0	6	6	6	1	3.11	0.94
3.4	0	2	8	8	1	3.42	0.77
3.5	1	0	4	13	1	3.68	0.82
4.1	1	0	2	11	5	4.00	0.94
4.2	1	1	1	9	7	4.05	1.08
4.3	0	0	6	10	3	3.84	0.69
4.4	0	0	7	9	3	3.79	0.71
4.5	0	0	3	12	4	4.05	0.62
5.1	1	2	9	6	1	3.21	0.92
5.2	2	6	7	3	1	2.74	1.05
5.3	1	9	5	4	0	2.63	0.90
5.4	0	1	4	13	1	3.74	0.65
5.5	1	1	2	14	1	3.68	0.89
6.1	0	3	4	10	2	3.58	0.90
6.2	0	1	6	8	4	3.79	0.85
6.3	1	0	6	10	2	3.63	0.90
6.4	1	1	3	10	4	3.79	1.03
6.5	0	2	3	11	3	3.79	0.85
7.1	0	2	5	9	3	3.68	0.89
7.2	0	1	9	8	1	3.47	0.70
7.3	0	0	3	11	5	4.11	0.66
7.4	0	6	2	6	5	3.53	1.22
7.5	0	1	6	8	4	3.79	0.85

Note n = 19

4 Results

4.1 Pre-course Questionnaire Results

Prior to the commencement of the course, a survey was conducted to assess students' sustainability-related knowledge. Their understanding of SDGs varied (Q1: Mean = 2.68, SD = 1.45), while the anticipation level of past sustainability-related education was low (Q2: Mean = 1.95, SD = 1.03). However, students expressed the need for action towards climate change (Q3: Mean = 4.58, SD = 0.61) and considered plastic an environmentally hazardous material (Q11: Mean = 4.11, SD = 0.81). Furthermore, the awareness level of the difference between linear and circular economies varied the most (Q: 5 Mean = 2.95, SD = 1.58), although the students were eager to recycle trash (Q7: Mean = 4.37, SD = 0.6). Before the course, students demonstrated uncertainty in addressing social issues, with low scores for understanding climate change (Q4: Mean = 3.84, SD = 1.01), waste management (Q6: Mean = 3.26, SD = 1.15; Q8: Mean = 2.58, SD = 1.17), the circular design approach (Q10: Mean = 2.63, SD = 1.16), and design innovation (Q15: Mean = 2.89, SD = 0.99). Regarding the issue of plastic waste and how to reduce its usage, the students' thoughts varied (Q12: Mean 2.53, SD = 1.12, and Q14: Mean = 3.32, SD = 1.16). They considered the significance of recycling plastic (Q13: Mean = 4.32, SD = 1.00) and thought design could promote sustainability (Q9: Mean = 4.37, SD = 0.68).

4.2 Student' Coursework

Assignment I, "Food Packaging Regenerative Design," required students to design a new product using food packaging waste. Students logged their daily food-related items and explored how much plastic waste they generated. Furthermore, the students were encouraged to appreciate the various types of plastic waste and repurpose them through the facilities in the school. Their design solutions ranged from practical products, such as a fastening device for bags, to more decorative products, like small bags and accessories (Fig. 7a, b, and c). The students initially faced difficulty repurposing plastic waste; however, they achieved creative designs by integrating a hands-on trial method.

Assignment II, Repack Design Challenge, asked students to develop a new design prototype for a food packaging circular system. Students tried to replace plastic wrapping for mass products at the beginning of the project; however, the attempts were unsuccessful since plastic was the most efficient material for wrapping food due to its waterproof ability and the distance most foods needed to travel after leaving the factory. Therefore, the students shifted their focus to local food supply, which did not require long-distance travel between production and consumption. The design solutions ranged from reducing packaging by incorporating cookable packaging for local Taiwanese tea eggs or using paper-made containers for local rice cakes to utilizing the traditional compressing tealeaf technique to make a tea block (Fig. 8a, b, and c).

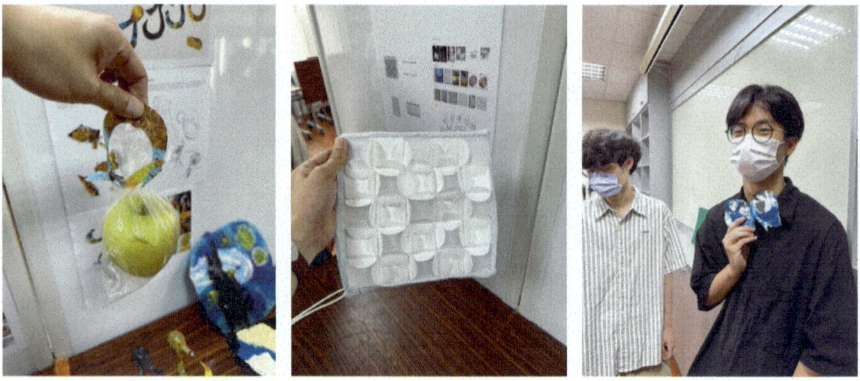

(a) (b) (c)

Fig. 7 **a** Utilizing repurposed plastic into a fastening device for the bag, **b** incorporating the bottoms of milk bottles in a bag design, **c** repurposing plastic into a bowtie by enhancing the color of the material

(a) (b) (c)

Fig. 8 **a** Cookable egg packaging for traditional Taiwanese tea-flavored egg, **b** paper-made container for Taiwanese local rice cake, **c** tea blocks utilizing a compressing method to reduce packaging

4.3 Post-course Questionnaire Results

The post-course questionnaire feedback was collected in seven sections comprising 35 questions. In section 1, Sustainability Knowledge, the feedback was mostly positive; most students said the course provided enough guidance and opportunity to expand their knowledge. Furthermore, the knowledge they obtained helped them complete the coursework. The course goal, Sect. 2, revealed that the course is challenging for the students (Q2.2: Mean = 2.58, SD = 0.84), even though the course goal was relatively straightforward (Q2.1: Mean = 3.74, SD = 0.96). There were different views regarding the purpose of a course that may help individuals achieve their desired career goals (Q2.4: Mean = 3.79, SD = 1.03). Additionally, the majority of students concurred that the course offers the potential to enhance their problem-solving abilities (Q2.5: Mean = 3.95, SD = 0.91).

In section 3, Course Assignment, most students encountered difficulties working on the assignment (Q3.2: Mean = 2.16, SD = 0.60), although the objective was explicit. However, the course provided sufficient support to develop their projects during the assignments while they were motivated by the challenging topics. In addition, most students developed innovative ideas while working on the assignment. The feedback from Section 4, Course Activities, shows that the implemented activities successfully impacted students' learning (Average Mean = 3.95, Average SD = 0.81). Among the course activities, outside-of-school events, such as joining workshops and field trips, gained most positive feedback from students.

Regarding Section 5, Course Works, only some students were satisfied with their project outcomes. Moreover, many noted that the time to develop should be increased and that they need further knowledge to complete the coursework. However, the review sessions provided opportunities for them to improve their developing project. Moreover, the students derived inspiration from their classmates during the course. Sections 6 and 7 explored more detailed feedback on the assignment topics, such as plastic and food packaging issues. Although students started the course feeling pessimistic about plastic issues, the coursework allowed them to revisit the issue and apply creative solutions to solve problems. Moreover, they concurred that the design could play a crucial role in addressing social issues. Finally, based on their feedback, this course successfully widened students' concerns about sustainability issues.

The pre- and post-course questionnaires used in this study were designed to assess students' knowledge, attitudes, and skills related to sustainability and design. The questions were categorized into six key themes to facilitate the analysis (Table 7). Each theme reflects a core competency or focus area of the STEM4S framework, enabling targeted evaluation of the course's impact on student learning. For example, questions addressing foundational knowledge, such as the SDGs or climate action, were grouped under "Sustainability Knowledge," questions focusing on hands-on activities, like field trips and workshops, were categorized as "Reflection and Engagement." This categorization allowed a structured analysis of the pre- and post-course data, providing insights into areas of improvement and those requiring further development.

Furthermore, the pre- and post-course mean ± SD values across six key categories were compared to illustrate the course's impact on students' understanding and skills. The visualization highlights improvements in sustainability knowledge, design innovation, and the practical application of circular economy principles. By providing a side-by-side comparison, the result emphasizes the effectiveness of the STEM4S framework in fostering critical thinking and problem-solving abilities related to sustainability challenges.

Table 7 Six key categories of the pre- and post-course questionnaires

Category	Pre-course question	Post-course question
Sustainability knowledge	1, 2, 3, 4	1.1, 1.2, 1.3, 3.3
Circular economy and waste management	5, 6, 7, 8	4.3, 5.1, 6.2, 7.4
Role of design in sustainability	9, 10, 15	3.5, 5.3, 6.4, 7.3
Plastic waste and usage reduction	11, 12, 16, 14	6.1, 7.2, 7.5, 7.6
Course goals and assignments	N/A	2.1, 2.2, 3.1, 3.4
Reflection and engagement	N/A	4.1, 4.2, 5.4, 5.5

Figure 9 reveals significant gains in sustainability knowledge, with the mean score rising from 3.26 ± 1.03 pre-course to 4.01 ± 0.61 post-course. This result demonstrates the course's effectiveness in broadening students' understanding of global sustainability issues while achieving its core educational objectives. This significant gain reflects the success of the course in expanding students' awareness of global sustainability issues and aligning with its educational objectives. Similarly, Circular Economy, Waste Management, and Role of Design exhibited substantial growth, with post-course means of 3.72 ± 0.89 and 3.70 ± 0.84 from pre-course 3.29 ± 1.19 and 3.30 ± 1.29. These findings indicate that students better understood how design and circular economy principles can address environmental challenges. Moreover, the reduced standard deviations in these categories suggest increased consistency in students' responses, pointing to a more uniform grasp of these concepts.

In contrast, Plastic Waste Reduction demonstrated a more modest improvement, with the mean score rising from 3.57 ± 1.04 pre-course to 3.63 ± 0.80 post-course.

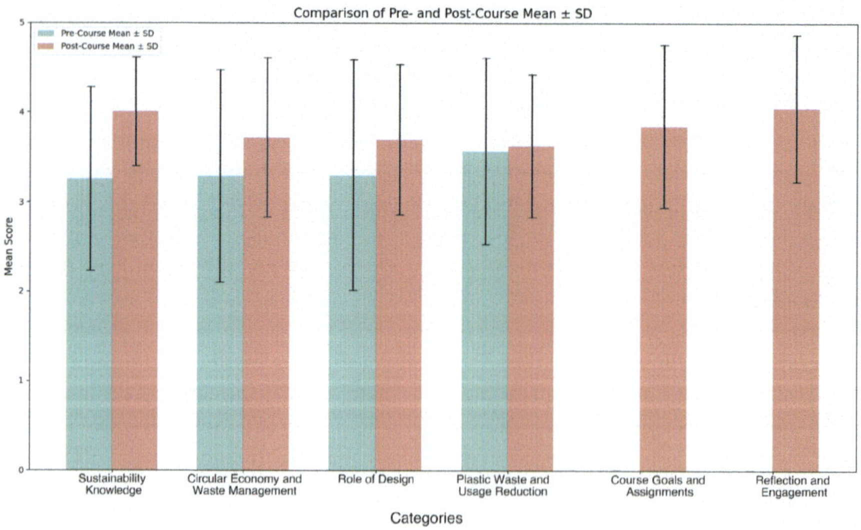

Fig. 9 Bar chart of comparison pre- and post-course questionnaire in six categories

This smaller gain may indicate the inherent complexity of reducing plastic waste in design contexts or potential gaps in the course's hands-on activities. Variability in student responses suggests that some participants found this topic more challenging to conceptualize or apply. To address this, future course iterations could allocate additional resources to practical workshops and collaborative exercises focused on reducing plastic waste.

Among the six evaluated categories, reflection and engagement received the highest post-course scores (Mean = 4.05, SD = 0.82), reflecting the value of activities such as the Trash Kitchen workshop and U-Market field trip. These hands-on experiences enhanced students' understanding of sustainability while encouraging creativity and collaboration. Peer-to-peer feedback further enriched the learning process, helping students refine their projects and broaden their perspectives.

While sustainability knowledge scored slightly higher (Mean = 4.01, SD = 0.61) than course goals and assignments (Mean = 3.85, SD = 0.91), the latter still demonstrated the effectiveness of clear objectives and well-aligned assignments in guiding students through complex sustainability topics. The slight difference in scores may indicate that some students found the course goals or assignments challenging to follow, potentially due to the multidisciplinary nature of the tasks. However, positive feedback underscores the importance of transparent communication and structured activities in enhancing student learning outcomes.

Overall, the bar chart underscores the course's success in equipping students with the knowledge and skills to tackle sustainability challenges through design. These results validate the effectiveness of the STEM4S framework and highlight its potential for broader application in higher education, particularly in the Asia-Pacific region. By aligning with the region's sustainability goals, this pedagogical approach demonstrates its relevance to fostering critical competencies in students, preparing them to address complex real-world issues.

5 Discussion

Clark and Button (2011) found that the mutual learning implicit in STEM allows participants to mutually expand their understanding of sustainability. STEM can be used as a teaching tool to promote shared learning, which can impact instructors and learners significantly more than the traditional educational model. Smith and Watson (2019) argue that the current STEM education trend focuses on modern technological-centered issues rather than critical and sustainability-focused real-world problems. Therefore, implementing sustainability concerns in the STEM curriculum should be a priority for higher educational institutions. Furthermore, Rogers et al. (2015) highlight the need for a multidisciplinary approach to solve twenty-first-century problems. Moreover, the multidisciplinary curriculum scope encourages students to work collaboratively and develop communication skills. The present research also reveals that most students find that outside-of-school activities involving interaction with other disciplinary professions promote their creativity and higher learning outcomes.

Regarding STEM as an interdisciplinary educational method for a single discipline, STEM4S provides students with a more collaborative environment for multidisciplinary knowledge by enhancing students' communication skills (Fig. 10). However, the current higher education system faces the challenge of crossing the disciplinary boundary. Additionally, it is vital to develop an easy-to-implement educational framework to help educators meet the sustainability challenges of the twenty-first century (Rogers et al. 2015). Since society expects the new generation to be equipped with twenty-first-century skills to face global challenges (Bridgstock 2017), higher education must consider how to enhance student's abilities to be leaders in facing challenges. This present research examines the framework and scope of STEM4S in the design course when handling food packaging-related sustainability issues. Although the most challenging part of framework and scope is creating a multidisciplinary collaborative environment in a typical higher education course format, the research shows that the STEM4S implemented course provides students with broader opportunities to explore real-world problems and promote their critical thinking.

The study findings have significant theoretical and practical implications for promoting sustainability in higher education in the Asia-Pacific region. Theoretically, integrating STEM4S into design curricula provides a framework for addressing complex, region-specific sustainability challenges. By emphasizing interdisciplinary collaboration and critical thinking, the STEM4S framework aligns with the broader educational goals of fostering sustainable development as a pathway to prosperity and peace. This approach highlights the importance of equipping students with the skills and mindsets necessary to navigate sustainability's interconnected economic, social, and environmental dimensions.

Practically, the course activities—such as repurposing single-use food packaging waste and designing circular systems—demonstrate how higher education can directly address sustainability issues relevant to the Asia-Pacific context. For example, the focus on local food packaging solutions aligns with regional goals for reducing plastic waste and promoting a circular economy. Moreover, incorporating field trips, workshops, and industry collaborations exemplifies how experiential learning can bridge the gap between theoretical knowledge and real-world application, fostering a sense of agency and innovation among students.

These findings underscore the role of higher education in advancing the United Nations' SDGs within the Asia-Pacific region. By embedding sustainability into curricula, universities can prepare students to address local challenges and contribute to global efforts to create equitable, prosperous, and peaceful societies.

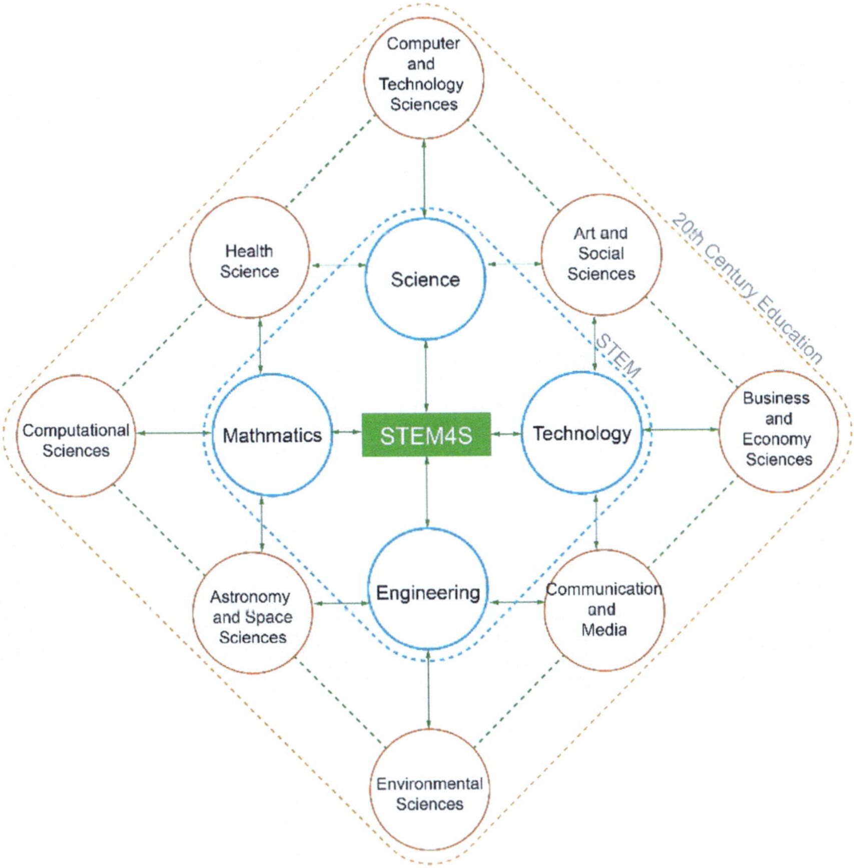

Fig. 10 Comparison of educational models (created by the author)

6 Conclusion

This study investigates the framework and scope of the STEM4S in a Design course. We analyzed the course outcome by collecting participants' pre- and post-course questionnaire feedback. Based on the pre-course questionnaire, students needed more knowledge and confidence in sustainability despite having concerns about related issues. The post-course questionnaire shows that the STEM4S implemented course allows students to gain confidence in sustainability-related issues while training their problem-solving abilities. Furthermore, students would be more capable of facing difficulties through enhanced instructor guidance and more knowledge provided during the course, even if the course assignment topic appeared challenging.

STEM4S is committed to providing students with a learning environment that integrates multiple disciplines and encourages diverse perspectives. In this case study, the STEM4S framework was implemented in a single-discipline design course that focused on extracurricular activities such as workshops, field trips, interviews, and guest critiques during reviews. These activities proved helpful in enabling students to develop their projects and explore related topics of interest. The issue of plastic waste associated with food packaging was a major topic of discussion throughout the course. Despite the negative attitudes toward plastic waste at the beginning of the course, students used the available tools and had their creativity challenged. Furthermore, students concurred that design can add value to plastic waste and build an emotional connection with users to extend the material's lifespan. Therefore, design plays a significant role in establishing a sustainable society.

Although the results indicate positive outcomes, there are limited opportunities for implementing a multidisciplinary learning environment. Additionally, the study collected a limited amount of feedback from questionnaires, and the effectiveness of STEM4S may vary depending on the number of students on the course. Future research should investigate other course types, such as seminars and lectures, to further evaluate the effectiveness of STEM4S.

Acknowledgements This research was founded by the Taiwan National Science and Technology Council (Project Number: 111-2621-M-003-004-'Investigation of alternation of recycling plastic-waste in Circular Economy strategies with digital and open-source technologies') and the Taiwan Ministry of Education (Project Number: PHA1110020' Research of STEM4S scope and framework for design education). The authors thank all the participants and supporting entities who assisted during the research process.

References

Avelar, A. B. A., Silva-Oliveira, K. D. d., & Pereira, R. d. S. (2019). Education for advancing the implementation of the Sustainable Development Goals: A systematic approach. *The International Journal of Management Education, 17*(3), 100322. https://doi.org/10.1016/j.ijme.2019.100322

Banathy, B. H. (1996). What Is Design? Why Do We Need It? In B. H. Banathy (Ed.), *Designing Social Systems in a Changing World* (pp. 11–47). Springer US. https://doi.org/10.1007/978-1-4757-9981-1_2

Bascopé, M., & Reiss, K. (2021). Place-Based STEM Education for Sustainability: A Path towards Socioecological Resilience. *Sustainability, 13*(15).

Borges, J. C., Ferreira, T. C., Borges de Oliveira, M. S., Macini, N., & Caldana, A. C. F. (2017). Hidden curriculum in student organizations: Learning, practice, socialization and responsible management in a business school. *The International Journal of Management Education, 15*(2, Part B), 153–161. https://doi.org/10.1016/j.ijme.2017.03.003

Bridgstock, R. (2017). The University and the Knowledge Network: A New Educational Model for Twenty-first Century Learning and Employability. In (pp. 339–358). Palgrave Macmillan UK. https://doi.org/10.1057/978-1-137-57168-7_16

Campbell, C., & Speldewinde, C. (2022). Early Childhood STEM Education for Sustainable Development. *Sustainability, 14*(6).

Clark, B., & Button, C. (2011). Sustainability transdisciplinary education model: interface of arts, science, and community (STEM). *International Journal of Sustainability in Higher Education, 12*(1), 41–54. https://doi.org/10.1108/14676371111098294

Council, D. (2011). Design for Innovation. London, UK (p. 14). Retrieved 21 June, 2016. In.

Facione, P., & Facione, N. (2013). Critical Thinking for Life. *Inquiry: Critical Thinking Across the Disciplines, 28*, 5–25. https://doi.org/10.5840/inquiryct20132812

Fortus, D., Dershimer, R. C., Krajcik, J., Marx, R. W., & Mamlok-Naaman, R. (2004). Design-based science and student learning. *Journal of Research in Science Teaching, 41*(10), 1081–1110. https://doi.org/10.1002/tea.20040

Gonzalez, H. B. (2012). *An analysis of STEM education funding at the NSF: Trends and policy discussion.* Congressional Research Service Washington, DC.

Gyasi, J. F., Zheng, L., & Zhou, Y. (2021). Perusing the Past to Propel the Future: A Systematic Review of STEM Learning Activity Based on Activity Theory. *Sustainability, 13*(16).

Hathcock, S. J., Dickerson, D. L., Eckhoff, A., & Katsioloudis, P. (2015). Scaffolding for Creative Product Possibilities in a Design Based STEM Activity. *Research in Science Education, 45*(5), 727–748. https://doi.org/10.1007/s11165-014-9437-7

LaForce, M., Noble, E., & Blackwell, C. (2017). Problem-Based Learning (PBL) and Student Interest in STEM Careers: The Roles of Motivation and Ability Beliefs. *Education Sciences, 7*(4).

Li, Y., Schoenfeld, A. H., diSessa, A. A., Graesser, A. C., Benson, L. C., English, L. D., & Duschl, R. A. (2019). Design and Design Thinking in STEM Education. *Journal for STEM Education Research, 2*(2), 93–104. https://doi.org/10.1007/s41979-019-00020-z

Martín-Páez, T., Aguilera, D., Perales-Palacios, F. J., & Vílchez-González, J. M. (2019). What are we talking about when we talk about STEM education? A review of literature. *Science Education, 103*(4), 799–822. https://doi.org/10.1002/sce.21522

Mohr-Schroeder, M. J., Cavalcanti, M., & Blyman, K. (2015). Stem Education: Understanding the Changing Landscape. In A. Sahin (Ed.), *A Practice-based Model of STEM Teaching: STEM Students on the Stage (SOS)TM* (pp. 3–14). SensePublishers. https://doi.org/10.1007/978-94-6300-019-2_1

Nations, U. (2015). The 17 Goals. In.

Pahnke, J., O'Donnell, C., & Bascopé, M. (2019). Using science to do social good: STEM education for sustainable development. *Proceedings of the Second International Dialogue on STEM Education (IDoS), Berlin, Germany*, 5–6.

Petrun Sayers, E. L., Craig, C. A., Skonicki, E., Gahlon, G., Gilbertz, S., & Feng, S. (2021). Evaluating STEM-Based Sustainability Understanding: A Cognitive Mapping Approach. *Sustainability, 13*(14).

Ritz, J. M., & Fan, S.-C. (2015). STEM and technology education: international state-of-the-art. *International Journal of Technology and Design Education, 25*(4), 429–451. https://doi.org/10.1007/s10798-014-9290-z

Rogers, M., Pfaff, T., Hamilton, J., & Erkan, A. (2015). Using sustainability themes and multidisciplinary approaches to enhance STEM education. *International Journal of Sustainability in Higher Education, 16*(4), 523–536. https://doi.org/10.1108/IJSHE-02-2013-0018

Smith, C., & Watson, J. (2019). Does the rise of STEM education mean the demise of sustainability education? *Australian Journal of Environmental Education, 35*(1), 1–11. https://doi.org/10.1017/aee.2018.51

Zamora-Polo, F., & Sánchez-Martín, J. (2019). Teaching for a Better World. Sustainability and Sustainable Development Goals in the Construction of a Change-Maker University. *Sustainability, 11*(15).

Daisuke Nagatomo is an Associate Professor of the Department of Design at the National Taiwan Normal University (NTNU) and a PhD researcher in the International Degree Program in Climate Change and Sustainable Development (IPCS) at National Taiwan University (NTU). He is also a design practitioner recognized internationally, including the iF award from Germany, the Muse Design Award from the USA, the A'Design Award from Italy, and the Golden Pin Award from Taiwan. Before taking his position at NTNU, he taught at Chung Yuan Christian University (2016–2020), National Chiao Tung University (2012–2014), and Tokyo Denki University in Japan (2011). His research interests include design thinking with virtual reality technology, sustainability in design education, circular economy and design, and bilingual education in design practice.

Listen Here! Sustainability Education from the Ground up

Fiona Harrisson and Judy Rogers

Abstract This paper discusses the development and delivery of an interdisciplinary, field trip based elective that has been offered at RMIT University for 25 years. Titled Farming the Future, during the course students are introduced to differing perspectives and approaches to sustainability in different contexts and localities. The paper explains how the learning and teaching approach has shifted and changed over time and also what has remained the same. It also explores the research question—how does being in the field contribute to and transform students' understandings of sustainability? It does so by focussing on two field trip location, the first a coastal location 150 km south east of Melbourne, Victoria's capital city and the second a peri urban region to the west of Melbourne. During the course students are asked to listen carefully to speakers and each other and to reflect on what they have heard in the field. They are also asked to consider how this iterative process challenges their predetermined assumptions. In doing so the idea of what sustainability is, or could be, is opened up for discussion, debate and reimagining. The paper argues that listening calls for empathy, critical reflection and (self) awareness enabling more respectful processes of engagement leading to the possibility of a more peaceful and prosperous future.

Keywords Listening · Sustainability · Sustainable development · Interdisciplinarity

---

F. Harrisson (✉)
School of Architecture and Urban Design, RMIT University, Melbourne, VIC, Australia
e-mail: fiona.harrisson@rmit.edu.au

J. Rogers
School of Property, Construction and Project Management, RMIT University, Melbourne, VIC, Australia
e-mail: judith.rogers@rmit.edu.au

W. Leal Filho et al. (eds.), *Sustainability in the Curriculum and Operations of Universities*, World Sustainability Series,
https://doi.org/10.1007/978-3-032-12678-8_27

1 Introduction

Farming the future is an interdisciplinary elective that has been offered at RMIT University for over 25 years. Originally developed by Dr Judy Rogers and the late Professor Jim Sinatra the focus has been, and continues to be, on providing insights into the complex issues facing and challenging the future of land management in Australia. Framed in terms of sustainability the course is based around fieldtrips through which students are introduced to differing perspectives and approaches to sustainability in different contexts. Fieldtrip locations have included the peri-urban fringe (Alvarez and Rogers 2006a), extensive and intensive farming districts (Alvarez and Rogers 2006b), and coastal areas (Rogers 2013, 2016). Listening to other points of view has always been central to the course and its early formulation was based on Sinatra and Murphy's publication 'Listen to the people Listen to the land' (Sinatra and Murphy 1999). In the book the authors highlight and celebrate the stories of both indigenous and non-indigenous Australians and their relationship to the landscape. It asks us to allow ourselves to listen and '...understand the land from different points of views, to observe the world differently from how we have been trained in our particular cultural environment ....in a way which many people have lost through busy, complicated, modern lifestyles' (Sinatra and Murphy 1999). Another key feature of the course is that on each field trip students and staff undertake some hands-on work—usually tree planting or weeding not only as a way to 'give back' but also as a way of building a sense of being part of a larger student group as students come from a range of different disciplines and few know each other before they begin the course. It is also important to emphasise here that the field trips are not 'one off'. Visits to different regions have taken place over many years as a way of building trust with the people we meet—whether government representatives, landholders, community protesters, community groups, indigenous caretakers. Since its early formulation, however, the course has evolved and it is this evolution that is one of the main focuses of this paper. Field trips have, however, always been central to the delivery of the course and the paper explores the research question—how does being in the field contribute to and transform students' understandings of sustainability?

The paper begins with an overview of the learning and teaching methods adopted on the course before moving to a brief discussion of what we understand as the difference between education 'about' and 'for' sustainability. The paper then highlights the importance of listening and what it means to listen. Drawing on examples from two field trip locations the paper highlights the importance of interdisciplinary field work-based learning as an alternative to learning 'about 'sustainability in a lecture theatre.

2 Learning and Teaching Approach

There are number of phases or key moments in the learning and teaching approach adopted in Farming the Future. The course begins with three pre field classes which aim to prepare students for the field trip experience. In these classes students discuss and explore two main questions: What is Sustainability? And What does it mean to listen? Most of the students who enrol into Farming the Future already have some education 'about' sustainability and most are also familiar with the sustainable development goals. However, education *about* sustainability generally focusses on content and techniques that need to be learnt and understood before implementation can occur. This content, which is generally disciplinary specific is then often used to evaluate the merits of practices in the 'real' world, or the world of lived experience. As a result, students are often quick to judge between 'right' and 'wrong', 'good' and 'bad' sustainability practices based on these predetermined understandings. In pre field trip classes we, therefore, ask students to read about and discuss what they understand as sustainability and then to consider the different ways it is spoken about. This involves carefully listening to each other, and rather than assume that this will happen what it means to listen also becomes a topic of conversation. The importance of listening will be returned to throughout the paper but it needs to be highlighted here that this is one key shift in the way the course has been delivered over time. Initially students were simply asked to listen and to take field notes as a way of demonstrating that they had listened. Over time listening has become much more central to the delivery of the course as has the focus on the disciplinary understandings of sustainability that students bring with them. Discussions focus not only on differences but also how they constrain students' ability to imagine possibilities and potentials for a peaceful, prosperous and sustainable future for all.

Before embarking on the field trip, students also engage in group work where each group investigates and presents on a specific issue or topic such as water, biodiversity, industry, infrastructure, land use, or population. This activity aims to familiarize the class with key field-related challenges and also to promote a sense of community. It is the field trip, however, that holds pivotal importance in the course curriculum. The benefits of field work based education have been widely discussed (Fedesco et al. 2020, Jones and Washiko 2022, Moscovici and Witt 2018, Ramachandiran 2016) and include problem-based learning, active learning, experiential learning and 'deep learning'. Jones and Washiko (2022) have identified four key benefits of field based learning—active learning, knowledge co creation, feedback and place-based values. They define active learning as 'an instructional practice where students complete activities, use higher-order thinking to consider what they are doing, and focus more on skills than information'. This requires a shift away from the traditional lecture/tutorial approach which largely involves the transmission of knowledge to a "learner-centered" approach that involves students "doing" things and reflecting on what they are doing (Misseyanni et al. 2018). Active learning is closely related to knowledge co creation (Robertson 2018) where students participate in the construction of knowledge, allowing them space to challenge mainstream thinking and opinions.

According to Misseyanni et al. (2018), 'this is an essential step in the development of informed, socially responsible, and creative individuals'. On field trips students, presenters, lecturers and the landscape itself become co-creators of knowledge for sustainability. Positioned in this way shifts the role of teacher from instructor to facilitator or Storberg-Walker and Yu (2021) argued

> The relationship between student and teacher also changes from a "sage on the stage" to a "guide on the side" in many ways. When a teacher introduces contemplative practices into the curriculum, she/he is inviting this new relationship – a relationship of mutuality and co-creation. Yes, the course content needs to be delivered and grades need to be submitted. However, moving the relationship from just head to head (e.g. cognitive relationship only) to a fully embodied relationship (e.g. head, heart and spirit) means that the commonality of being human can be more salient during the semester, both inside and outside of class. This admittedly is a new space for many instructors, but one that I believe needs to happen because teaching for sustainability/flourishing is not about techniques or skills necessarily – it is about a state of being in relationship to others and the planet in a profoundly deep way... It cannot be learned cognitively.

What this means is that we too as instructors also need to acknowledge our own disciplinary limitations and also that often we don't have the 'answers'. On all of the field trips and in pre field trip classes we therefore emphasise that our role is to facilitate a process of learning and that we are also learning from people in the field and from the students themselves. Knowledge co creation relies on a process of constant feedback between students and staff and also between peers as students reflect on and consider what they have heard in the field. Finally place based learning provides insights into 'real world' issues and perspectives. Place here refers to a location or site that has meaning mediated through human experience.

Throughout the course students are assessed not only on what they know but also how they have come to know it. In their daily journals and in their final reflection students are asked to reflect on their participation through four kinds of listening' (Brearley 2010) and to describe how the different ways in which their engagement with others (peers, speaker, etc.) informed their learning. They are also asked to reflect on their own practices and disciplinary assumptions in relation to understandings of sustainability.

3 Education About and for Sustainability

As noted above education about sustainability is generally involves the delivery of content and while this is important initial step, education for sustainability relies on a deeper level of engagement, providing opportunities for students to reflect on their own attitudes and behaviours and understand how they impact both planet and people. Such an education, according to Huckle and Sterling 'is essentially process-driven, is participative and empowering, is liberatory and continuous' (Huckle and Sterling 1996). Open dialogue, discussion, debate, and exchange of ideas are foregrounded rather than the transmission of content or the 'how to' of sustainability. This is the approach adopted in farming the future. What students learn on the field trips is that sustainability is much more than 'right' or 'wrong', 'good' or 'bad'—it is complex, contested, and messy. The aim is to encourage students to think beyond prescriptive and fixed ideas about sustainability, to grapple with the possibilities of what it could mean, to effectively turn the concept back upon itself and open it up for critique in dialogue with others. In the process understandings of sustainability are transformed. Sipos et al. (2008) have defined such an approach to education for sustainability as engaging with the head, the hands, and the heart. based on the premise that 'In identifying all sustainability education to have a common vision of perspective transformation, we may better encourage and enable all participants of education to challenge and be open to change in their own minds, beliefs and behaviours' (Sipos et al. 2008).

4 Case Study: Wonthaggi

One particular region that we visited for 10 years is Wonthaggi, a small coastal town, 150 km south east of Melbourne, Victoria's capital city. It is the site of a desalination plant that was intended to reduce water insecurity for Melbourne's population, (exacerbated by ten consecutive years of drought). Announced by the then Premier of Victoria in 2007 at a time when Melbourne's storage levels had dropped to 28.4% the project immediately attracted a great deal of criticism across the state and in the broader region where the plant would be sited. Construction of the plant commenced in 2009 and was completed in 2012, however, by then Melbourne's water supply had recovered and the plant was placed in 'stand-by' mode. The other 'sustainability' credentials of the plant are that it is covered by an extensive green roof using mainly indigenous vegetation is surrounded by an expansive 255 ha ecological reserve. Field trips to the area began in 2008 when the protests were at their height and field trips continued until 2018 post construction of the plant and then ceased due to the pandemic. The field trips included visits to both the site of the desalinisation plant and an adjacent wind farm along with visits to the community blockade of the project, an archaeological and cultural heritage site and a regenerative farm where students were involved in tree planting under the guidance of the landholder. Students also listened to representatives from local government, from the local action group, to residents and landholders, and to local experts who introduced them to the flora and fauna and the history of the area (Fig. 1).

Most of the students on the field trips were initially vocal in their support of the desalinisation plant as a sustainable solution to Melbourne's water, however, listening to a diversity of perspectives allowed them to understand that this was just one solution and encouraged them to discuss and consider alternatives. Described by some locals as a 'white elephant', a 'knee jerk reaction' for others it was an 'insurance policy' against future droughts. What was clear however was that the plant was not intended to not service the local area but rather the capital city of Victoria, Melbourne. Others in the community supported the development arguing

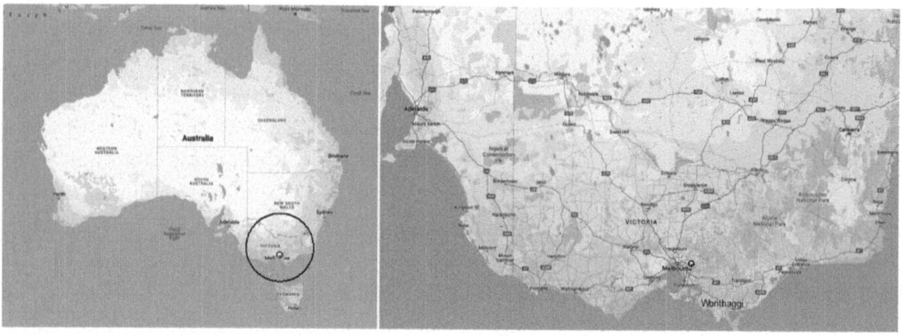

Fig. 1 Location of Wonthaggi, in Victoria, Australia

that the injection of funds represented a significant boost to the local economy. And the concerns were different too. While some of the speakers focussed on the loss of visual amenity in the region others were much more concerned about climate change and the impact on local flora and fauna.

Likewise, from afar a wind farm can and does appear as a sustainable solution, however, for some of the locals it was noisy, a blight on the landscape, a threat to native birds. For others it enhanced the landscape while also providing employment opportunities in the local area. It was these local embedded understandings, experiences, and knowledge that students were given access to and gained so much from during the field trips. Through interacting with a range of community members students were also given the chance to recognise that sustainability is largely about negotiating change, about developing strategies for the future, about identifying with the present and identifying how ideas of sustainability are embedded in communities in place. In this way students built an awareness that sustainability is not fixed, nor linear but contingent.

Students also realize that the issues involved cannot be resolved simply by reversing the situation, by privileging those who opposed the desalinisation plant or the wind farm at the expense of the broader Victorian public or members of the local community who welcomed the project, but also realised that the disquiet within the community could not simply be resolved using categories like right or wrong, good or bad, sustainable, or unsustainable. It was much more complex than this particularly as the human cost of development was revealed. Attitude values and beliefs were transformed (Figs. 2 and 3).

As a result, in their reflections students often highlighted the importance of local voices and how they shifted their understanding not only about sustainability but also tertiary education more generally. As one student put it, '*Getting an insight from a local (perspective) was really good, there was always two sides to a story and hearing from the locals is just something that cannot be taught in Uni*' (Bachelor of Engineering student).

Fig. 2 Wind farm on the Bass Coast (image: Fiona Harrisson)

Fig. 3 Local protest against the proposed desalinisation plant which also attracted opposition across the State of Victoria (Image: Fiona Harrisson)

Two other students also highlighted the way in which the experience of being out there in the field transformed their position on sustainability and their professional practice:

> Being in the construction industry, wastage and high demand for materials is on a large scale. Now that I have witnessed the effects of removing these resources I have attached a

> different worth to these materials we don't think about using. Also, the fact that it is so quick to remove but to replace is a slow process (Bachelor of Construction Management student).

And:

> Before this field trip, the definition of sustainability to me was mainly focussed on strategies organisations in the corporate world take to achieve sustainability for their own firms, but not the environment ... companies are constantly strategizing the term 'sustainability' to suit their needs and wants. The corporate world sells sustainability ... During the field trip the aspect that most contributed to my learning was hearing opinions from local people ...and different contradictory opinions ... hearing from both parties ...deepened my understanding of sustainability ... seeing the effort and passion, perseverance of the local people in Bass Coast has taught me that sustainability does not happen overnight, it takes time and maintenance, much like planting a tree (Bachelor of Business student).

In addition, as discussed above, as we became increasingly aware of the importance of listening as a practice we began to place less emphasis on definitions of sustainability and instead shifted our attention to the centrality of listening to self and to others as a way to consider and become open to the possibilities and potentials of sustainability and sustainable development.

5 Listening

In pre field trip classes as well as on field trip what is means to listen has therefore become a topic of conversation. According to Brearley (2010), 'Deep Listening',

> ...describes a way of learning, working and being together. It is informed by the concepts of community and reciprocity. It means listening with a sense of responsibility to stories that are told. It also means listening and observing the self as well.

Brearley introduces the work of both Miriam Rose Ungunmerr-Bauman (2002) and Scharmer (2018). Miriam Rose Ungunmerr-Bauman, indigenous woman and aboriginal elder, describes 'Dadirri' which translates as inner deep listening and quiet still awareness'. She describes dadirrii as, *'the gift that Australia is thirsting for'* (Ungunmerr-Bauman 2002). We are inspired by Miriam Rose Ungunmerr-Bauman, and strive to listen more deeply and to work with listening with our students, while also acknowledging this is not 'deep listening' but an initial step towards a lifelong practice. Brearley recognized a correlation between indigenous deep listening practices and listening coming through the Western lineage as defined by Scharmer (2018), who has identified four fields of listening. The first—downloading is listening that simply reconfirms what we already know. Factual listening means that we become aware of information that what we know and so requires an open mind and 'the capacity to suspend our habits of Judgment'. Empathic listening, on the other hand, requires opening our heart to tune into another person's view. Finally, generative listening allows, 'for the highest future possibility to show up while holding a space for something new to be born' Scharmer (2018). Such listening takes practice.

Through reading and thinking about the work of Brearley (2010), Ungunmerr-Bauman (2002) and Scharmer (2018) we understand listening as a human 'technology' that can be practiced in every moment. It is more than taking in sounds as words into one's ears but rather is a whole-body action, of being present in the present. Listening requires putting aside what one already knows to allow room for the unknown in a space between oneself and another(s), and other 'others'. Listening, therefore, provides a bridge – between people and worldviews and cannot be practiced without respect. Listening practiced in this way as part of 'dialogue' can allow new ways of thinking and acting to emerge. In the context of sustainability education, listening re-orientates the idea of the ''expert, we argue and it is this human capacity that in the ground for peace and prosperity.

6 Case Study: Grassy Plains

In their reflective journal on a 2023 field trip, students were not only able to identify the centrality of listening but also to understand it as a practice. On this particular field trip we investigated the future sustainability of endangered native grasslands to the west of Melbourne, Victoria's capital city. Prior to colonisation the grasslands were actively managed by indigenous people to ensure an on-going food supply. The use of fire in particular ensured that plants were replenished as food for not only people but also the animals that grazed there (Goss et al. 2015). Early explorers and settlers celebrated the abundant grassland and they were rapidly taken over by sheep (Williams and Morgan 2015) (Fig. 4).

The grasslands have been under increasing pressure since colonisation and those pressure have increased in recent times. As Williams and Marshall (2015) explain

> For most of the year native grasslands appear as drab paddocks, making them a 'hard sell' compared to more familiar conservation causes such as the tall forests that are, ironically, much better represented in Australia's national reserve system. It is only during spring and early summer that the diversity and small-scale beauty of native grasslands becomes apparent.

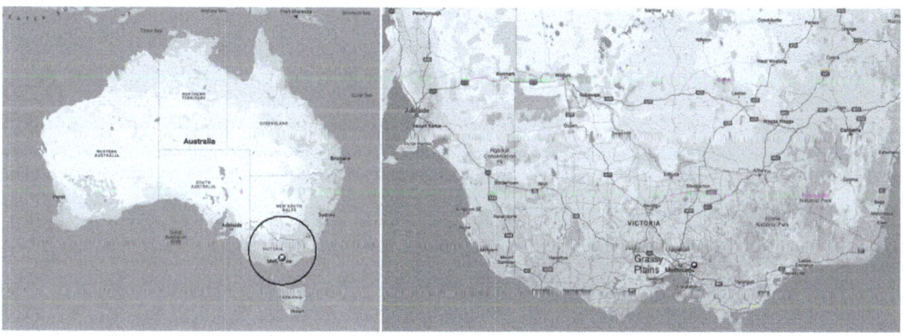

Fig. 4 Location of Grassy Plains

As a result urban development, industrialisation, and agriculture are all competing interests and while a few reserves have been established and remnants remain 'in lightly grazed paddocks, road and rail reserves, airport lands, cemeteries and travelling stock routes, all areas that generally have a history of not being fertilised, ploughed or heavily grazed (Williams and Morgan 2015) there are large areas of grassland that remain under threat even though less than 2% of the pre colonisation grasslands remain (Williams and Morgan 2015). The 'simple' sustainability response, of course, is to conserve the grasslands at all cost. But what does this mean for peace and prosperity for people living there? On this field trip we visited an intensive farm, a restorative farm, a regeneration project, a sewerage treatment plant that recycles water. Students were also involved in weeding a remnant grassland site in a railway reserve. We also listened to the voices of local government, state government authorities and NGO representatives and all of their stories challenged and shifted what the students understood as sustainability and they began to realise and acknowledge the limitations of their own disciplinary understanding of the concept. Issues that emerged included the need for affordable housing, the challenges posed for intensive farming by urban development and what that means, intergenerational farming and the desire of established farmers to pass on to future generations, water scarcity, food security and the possibilites and potentials of recycled water and then, of course, the fragility of the grasslands and how the remaining remnants could be managed and preserved (Fig. 5).

In their reflections during the field trip students were able to begin to articulate the importance of listening as a long-term learning practice, the different modes of listening and how this practice could be used as a way to realise change toward a sustainable, peaceful and prosperous future. As two students noted:

Fig. 5 Site visit to a regenerative farmer (image: Fiona Harrisson)

> At the beginning of the course, my viewpoint on sustainability was more about passively listening. As the course advanced and we embarked on the field trip, it shifted my perception from merely downloading information to actively engaging in empathetic listening – understanding the speakers' perspectives. Through generative listening, I'm allowing myself to consider other people's perspectives to achieve the same goal (Construction Management Student).

> The four kinds of listening I found to be a theory and a practice that I kept coming back to regardless of who it was that we met. Much of my time was spent objectively listening as the information being presented was entirely new… However, I would claim that I have tried to use empathetic listening and generative through any time we were in the field. The course has overall taught me to not only listen but to have the desire to listen, as that is the way we create the change (Bachelor of Business student).

The other key shift that has occurred is that while initially we focussed on definitions of sustainability and how it is contested and contestable out there in the field, students are now asked to consider the limitations of their own disciplinary understandings of sustainability. They begin to understand that sustainability is a process towards change not an outcome.

7 Interdisciplinarity

In pre field trip classes and on the field trip we asked students to not only reflect on what they had listened to but were also asked to consider how their own disciplinary understanding shifted as a result of listening to other students and people in the field. While it has become a truism that interdisciplinarity/multi-disciplinarity is central to the realisation of sustainability and the sustainable development goals what this means for teaching and learning for sustainability is often not articulated nor realised. As two students commented on the 2023 field trip.

> The act of engaging with different disciplines contributed most to my learning …It wasn't only the discussion moments, it was the general interactions and noticing how different disciplines communicated and engaged with the content and each other, during excursions, at the bus, and out to dinner. In Environment and Society we speak a lot on the idea of the 'definition' of sustainability and the "importance" of multi-disciplinary engagement. But, at the end of the day, in the classes that we can choose and the work that we do, we are in a bubble, an echo chamber and we aren't challenged or presented with anything new (Bachelor of Social Science student, 2023)

And

> In construction we are outcome driven in the sense that when we work on something we already know the outcome we expect… However, with this course, I understand it is a bit different from my discipline because it explores the possibility of more than outcome depending on how you are either approaching it, your perceptions, your values and what you believe is important according to your overall take on sustainability (Bachelor of Construction Management Student, 2023).

In the interdisciplinary space offered in the elective students begin to acknowledge not only the limitations of their own understanding of sustainability they also begin to identify the importance of being able to share different disciplinary perspectives rather than hold onto their pre-determined values and assumptions. They learn more from each other than they do from instructors, and often even people in the field. In group exercises and in informal conversations on the bus, out to dinner they, together, begin to discuss and imagine alternative ways of thinking through and acting upon what are seemingly intractable problems that they may never have thought of before. This allows them to let go of positions that are about 'good' and 'bad', 'right' and 'wrong' as they begin to build a critical reflective practice and consider how they might bring this learning to their own disciplinary practice. This shift opens possibilities for them to consider much more actively actions and measures *for* sustainability.

8 Limitations and Future Prospects

It should be highlighted here, however, that not all students who enrol in farming the future benefit equally. At RMIT University all undergraduate students are required to complete three electives. Farming the future often attracts students because it is offered in intensive mode and is therefore seen by some as a way to complete the elective requirement quickly. The requirement to complete electives is an administrative decision that could be overturned at any stage. In addition, as farming the future is offered only once a year with a class size of approximately 20 students its reach is very narrow. Small class sizes also threaten its future viability.

The other key limitation moving forward is that despite the many benefits of field trip-based education in higher education such opportunities are in decline (Jones and Washiko 2022). Issues of liability, workload, financial challenges and other institutional barriers have been identified as some of the issues that have led to this decline along with a shift towards virtual field trips.

9 Conclusion

In this paper we introduced a learning and teaching approach that has been developed over 25 years at RMIT University, Victoria, Australia. It asked the question *how does being in the field contribute to and transform students' understandings of sustainability?* The paper argues that education for sustainability should do more than simply equip students with content 'about' sustainability. Rather it should challenge, destabilise and present a range of different approaches and possibilities, allowing space to imagine alternative sustainable futures.

The paper also highlights what we see as the importance of field trip-based education for sustainability by providing students with opportunities to listen to, acknowledge and grapple with the diversity of voices and stories that exist in place. In doing so the idea of what sustainability is or could be is opened up for discussion, debate and reimagining. Pre-determined assumptions are transformed and in the process the role of 'expert' and of teacher shifts. In this interdisciplinary space students begin to reflect on the limitations of their own disciplinary understandings and become much more open to the potentials and possibilities of sustainability as they acknowledge the messiness, open-end-ness and the contradictions involved.

Over the 25 years the elective has been offered the learning and teaching approach has evolved. Part of this evolution has been to increase the focus on listening and disciplinary understandings of sustainability.

What the paper highlights is the importance of listening as a practice *for* sustainability. As students are introduced to this practice of listening to self and others, they begin to move from thinking about sustainability as black or white, right or wrong, good or bad and in doing so begin to develop an empathetic awareness that sustainability is a process of change towards a more peaceful and prosperous future.

The on-going delivery of the course is not, however, without its challenges. Field work-based education it has been noted is in decline and so there are a range of institutional barriers moving forward. These include financial constraints, liability and issues of workload.

References

Alvarez A, Rogers, J (2006a) Imaging sustainability on the urban fringe: A learning experience. In H. Lewis & C. Ryan (Eds.), Imaging sustainability (pp. 93–103). Melbourne, Australia: RMIT University Press

Alvarez A, Rogers, J. (2006b) "Going "out there": learning about sustainability in place", International Journal of Sustainability in Higher Education, No. 2, pp. 176–188

Brearley L (2010) Gulpa Ngawal Indigenous Deep Listening, RMIT University Press, Melbourne

Fedesco H.N, Cavin D Henaares, R (2020) Field-based Learning in Higher Education Journal of the Scholarship of Teaching and Learning 20 (1): 65–84

Gott B, Williams, N.S.G, Antos M. (2015) 'Humans and grasslands a social history' Chapter 2 in Land of Sweeping Plains: Managing and Restoring the Native Grasslands of South-Eastern Australia, edited by Nicholas S. G. Williams, et al., CSIRO Publishing

Huckle J, Sterling S (1996). Introduction. In J. Huckle & S. Sterling (Eds.) Education for Sustainability. Earthscan. London.

Jones J C, Washko S (2022) More than fun in the sun: The pedagogy of field trips improves student learning in higher education, Journal of Geoscience Education, 70:3, 292–305, https://doi.org/10.1080/10899995.2021.1984176

Misseyanni, A., Lytras, M.D., Papadopoulou, P. and Marouli, C. (2018), "Introduction", Misseyanni, A., Lytras, M.D., Papadopoulou, P. and Marouli, C. (Ed.) *Active Learning Strategies in Higher Education*, Emerald Publishing Limited, Leeds, pp. 1–13. https://doi.org/10.1108/978-1-78714-487-320181001

Moscovici D, Witt, E. (2018), "Active Learning Strategies: Stories and Lessons Learnt – Studying Environment in the Field", Misseyanni, A., Lytras, M.D., Papadopoulou, P. and Marouli, C. (Ed.) *Active Learning Strategies in Higher Education*, Emerald Publishing Limited, Leeds, pp. 133–150. https://doi.org/10.1108/978-1-78714-487-320181006

Ramachandiran, Dhanapal S (2016) Evaluation of the Effectiveness of Field Trips in the Teaching and Learning of Biosciences in Tang S F, Logonnathan L (eds.), Assessment for Learning

Within and Beyond the Classroom, Springer Science+Business Media Singapore https://doi.org/10.1007/978-981-10-0908-2_15
Robertson L (2018), "Toward an Epistemology of Active Learning in Higher Education and Its Promise", Misseyanni, A., Lytras, M.D., Papadopoulou, P. and Marouli, C. (Ed.) *Active Learning Strategies in Higher Education*, Emerald Publishing Limited, Leeds, pp. 17–44. https://doi.org/10.1108/978-1-78714-487-320181002
Rogers J (2013) 'Education for and Through sustainability: Toward Interdisciplinary Dialogue' in S. Caeiro et al. (eds.), Sustainability Assessment Tools in Higher Education Institutions, https://doi.org/10.1007/978-3-319-02375-5_23, Springer International Publishing Switzerland
Rogers J (2016) Sustainability and Performativity. In: Davim, J., Leal Filho, W. (eds) Challenges in Higher Education for Sustainability. Management and Industrial Engineering. Springer, Cham. https://doi.org/10.1007/978-3-319-23705-3_10
Scharmer O (2018) The essentials of theory u: Core principles and applications. Berrett-Koehler Publishers, Oakland, CA
Sinatra, J and Murphy, P (1999) Listen to the People, Listen to the Land Melbourne University Press, Melbourne
Sipos, Y, Battisti, B, Grimm K (2008) Achieving transformative sustainability learning: Engaging head, hands and heart. International Journal of Sustainability in Higher Education, 9(1), 68–86
Storberg-Walker, J, Yu D (2021) 'THE QUANTUM SHIFT Teaching sustainability leadership critically and mindfully' in Ivanova E, Rimanoczy I (eds) Revolutionizing Sustainability Education: Stories and Tools of Mindset Transformation, Taylor & Francis Group
Ungunmerr-Bauman M.R (2002) Dadirri: Inner deep listening and quiet still awareness, Emmaus Productions
Williams, N.S.G and Marshall, A (2015) 'Introduction' in in Nicholas S. G. Williams (eds) Land of Sweeping Plains: Managing and Restoring the Native Grasslands of South-Eastern Australia, CSIRO Publishing
Williams N. S.G, Morgan, J.W (2015) 'The native temperate grasslands of south-eastern Australia' Chapter 2 in Nicholas S. G. Williams (eds) Land of Sweeping Plains: Managing and Restoring the Native Grasslands of South-Eastern Australia, CSIRO Publishing

Fiona Harrisson is a practicing landscape architect, design consultant, contemplative teacher and passionate gardener. In her roles as a senior lecturer and creative practice research (recently completed PhD at RMIT) University, she explores relational practices to open pathways for situated ways of knowing to engage with the social and ecological complexities unfolding in our midst.

Judy Rogers is an Associate in the School of Property, Construction and Project Management at RMIT University. She has previously taught in the Landscape Architecture Program at RMIT and was Program Manager of the Master of Disaster, Design and Development. Judy has written extensively in the field of sustainability education.

Sustainable Entrepreneurship in Education for Sustainable Development—Case Method and Design Approach

Hoe Chin Goi and Kishore Kumar François

Abstract Understanding the approach to implementing Sustainable Entrepreneurship (SE) in Education for Sustainable Development (ESD) is paramount to empower potential sustainable entrepreneurs with the knowledge and skills to tackle economic, social, and environmental challenges. However, the current research streams have not explored enough the intersections of content, methodology, and context for teaching and applying the ideas within this theme. This exploratory study describes the process of employing the case method and design approach as the learning model, and applies the "mymizu 2022" case, a Japanese social startup, to generate a Business Model (BM) for sustainability by 70 international undergraduate students from Nagoya University of Commerce and Business. The research sheds light on a positive relationship between the learning approach and SE in ESD. The study shows three findings: the impact on the awareness of responsible consumption and scaling social impact of students through case and case method, impact on Entrepreneurial Intention (EI) of students through decision and design approach, and the new BM generation for sustainability. The study discusses the impact on EI for nurturing future sustainable entrepreneurs in ESD, however, adding the new dimension of Creating Shared Value (CSV). The study contributes to the theoretical fronts of ESD by highlighting the efficacy of the case method and design approach to impact EI as cognitive outcomes of SE and proposes the "Learner-centric and Action-oriented Approach for Sustainable Entrepreneurship" (LAASE) as the integrated learning model for ESD. The study provides implications for Higher Education Institutions (HEIs) to integrate the knowledge of SE into the content dimension, develop local cases, and employ the progressive pedagogical approach for the cultivation of entrepreneurs for sustainability, with the support of policymakers, in the Asia–Pacific region.

H. C. Goi (✉)
Faculty of Management, NUCB Business School, Nagoya University of Commerce and Business, Nagoya, Aichi, Japan
e-mail: goi_hc@gsm.nucba.ac.jp

K. K. François
College of Business and Law, RMIT University, Melbourne, VIC, Australia
e-mail: kishore.francois@student.rmit.edu.au

W. Leal Filho et al. (eds.), *Sustainability in the Curriculum and Operations of Universities*, World Sustainability Series,
https://doi.org/10.1007/978-3-032-12678-8_28

Keywords Sustainable entrepreneurship · Case method · Design approach · Education for sustainable development · Business model

1 Introduction

Education for Sustainable Development (ESD) was a concept first proposed by Japan at the 2002 World Summit on Sustainable Development. Under the support of UNESCO, it has led to the adoption of the "Decade of Education for Sustainable Development (UN DESD)" (2005–2014), then to the "Global Action Programme on Education for Sustainable Development (ESD) (GAP)" by the Japanese government since 2015 till 2019 (MEXT, 2017). During this period, the adoption of the Sustainable Development Goals (SDGs) by the United Nations in 2015 (United Nations, 2015) emphasized sustainability as the universal call to action for countries, organizations, and individuals around the world. In light of this, education was a catalyst for change necessary to drive sustainable development, positioning ESD as a key enabler of all 17 SDGs. It was also an integral element of SDG 4 on Education through Target 4.7, aiming to have ESD as one of the pillars that enables learners to acquire the knowledge and skills required to promote sustainable development. Hence, following the previous programmes, the "Education for Sustainable Development: Towards achieving the SDGs (ESD for 2030)" framework was adopted, and a roadmap was established to strengthen the ESD implementation and to increase the role of education in order to build a sustainable world.

Specifically, the roadmap emphasizes four priority areas: transforming learning environments, building the capacities of educators, empowering and mobilizing youth, and accelerating local-level actions. They focus on enhancing the educational environment, supporting educators, engaging learners, and fostering impactful initiatives. As such, there is a need for educators to actively integrate the vast concept of sustainability into the content and pedagogical approach to enable learners to become agents of change for driving sustainable development (Persson et al., 2023). It is increasingly important for students to acquire the relevant knowledge, competencies, and develop the attitudes to solve challenging issues for sustainability (MEXT, 2021; NIER, 2012). Thus, Higher Education Institutions (HEIs) undertake the role as part of the ecosystem to create knowledge and pedagogy to support the goals of a sustainable society. HEIs with ESD are claimed to make the greatest impact on sustainability by shaping the leaders for tomorrow by challenging and empowering them, and in the case of business schools, to also promote the benefits of sustainable business practices to shift away from business as usual (Cotgrave and Kokkarinen, 2011; Lourenço et al., 2012).

The Asia–Pacific context is primordial as the largest region, with more than half of world population, continues to undergo an important rapid economic growth and social change to be recognized as the most important booster of the global economy, albeit at the cost of higher consumption of resources, negative impact on the environment, and social inequality (Li et al., 2021; Naeem and Neal, 2012). Such developments have prompted HEIs to adopt and promote SDGs actively in line with ESD, while governments and companies incorporate new measures and improve their standards to align with global requirements (Liu et al., 2022). The region boasts 534 out of 1410 reported universities in the 2022 Times Higher Education (THE) impact ranking (THE, 2022). Therefore, further attention needs to be paid to the implementation of efforts in ESD within this context for the local sustainability issues.

Governments and corporations play a major role in policymaking and strive to create new markets. However, they often lack the capability to resolve and respond to the ever-changing needs of society and the environment. To this extent, Sustainable Entrepreneurship (SE) is relevant to respond to the rigidity of policies and the gaps in the new markets; hence, startups are undertaking important roles to generate new businesses that can address such socio-environmental issues in sustainability. As such, integrating sustainability and Business Models (BM) into ESD is crucial for preparing aspiring sustainable entrepreneurs to address economic, social, and environmental challenges. For this reason, HEIs play a major role in laying the foundational competencies for SE and enabling the SE ecosystem (Diepolder et al., 2021; Lans et al., 2014).

SE in ESD was traditionally seen as different components within HEIs: entrepreneurship education in business schools and sustainability grounded in policy, life science, and environmental education (Lans et al., 2014). However, the complexity of sustainability requires drawing from various foundational disciplines, leading HEIs to develop inter-, multi-, and trans-disciplinary courses and programs to teach sustainability to bridge the social, environmental, and economic systems (O'Byrne et al., 2015). Thus, SE in ESD is required to provide learners with the competencies to evaluate business opportunities considering environmental and societal needs (Diepolder et al., 2021) and effectively build their Entrepreneurial Intention (EI). Reaching this goal of competency development and growing the intentions also requires effective pedagogical approaches to impart the knowledge, skills, attitudes, values, and behaviors. The literature shows that the traditional lecturing way of teaching cannot solely accomplish this mission, prompting the teaching and learning to be learner-centric and action-oriented (e.g., Bacon et al., 2011; Hermann et al., 2020; Strachan, 2018; Vasconcelos et al., 2017; Woo et al., 2012, etc.). While methodologies such as project-based learning, problem-based learning, and case method are suggested, there is a need to bridge the gap in teaching materials and methodologies within each context to arrive at the learning outcomes.

This exploratory study aims to understand the effects of content and pedagogy as a learning approach to build awareness and generate new BM for sustainability.

Research Questions

- *RQ1: What (content) should students learn to increase their awareness of sustainability?*
- *RQ2: How (pedagogy) to make students exploit resources to generate a Business Model for sustainability?*

The exploratory study employs a single case analysis (Hancock et al., 2021; Yin, 2018) based on the context of the "mymizu 2022" case (case as course content; Christensen and Hansen, 1987; Rebeiz, 2011) used through the case method (pedagogical approach; Moskovitz, 1992; Servant-Miklos, 2019) in a "Sustainability Project" with approximately 70 international undergraduate students at the Nagoya University of Commerce and Business. Mymizu is a highly relevant case for the course content, as it is a social entrepreneurship case of a platform-based solution for plastic reduction and water refilling in Japan, with the goal to drive behavioral change for responsible consumption. Furthermore, the Asia–Pacific region is notorious for plastic pollution, accounting for 60.6% of global mismanaged plastic waste (Ritchie, 2018). The study employs a mixed-methods approach through project surveys conducted on a weekly basis, combining qualitative data (open-ended questions) and quantitative data (Likert scale questions) to gather insights on students' perceptions, learning outcomes, and overall experience.

The findings show the gain in awareness on responsible consumption, scaling social impact, and SE as an emerging opportunity for students. Secondly, there is a positive impact on EI through the decision and design approach by getting the students to act on real-life-based projects. Furthermore, the learning approach is also effective in resulting in a new BM generation for sustainability. The study discussed the imperative of broadening our understanding of social and environmental issues pertinent to influencing EI for SE. Furthermore, it emphasized the importance of integrating the dimension of Creating Shared Value (CSV) into EI, thereby broadening the scope and relevance of SE for sustainability. The study also proposed a learning approach, "Learner-centric and Action-oriented Approach for Sustainable Entrepreneurship" (LAASE), relevant for nurturing future entrepreneurs and business leaders in the field of SE and to help generate BM for sustainability in ESD. The study contributes to the theoretical fronts of ESD by highlighting the efficacy of the case method and design approach to impact EI as cognitive outcomes of SE with the proposed pedagogical framework. The study's implications also extend to the cultivation of future entrepreneurs and business leaders for sustainability, offering insights for the BM generation that are relevant for the context of HEIs in the Asia–Pacific region. Finally, the study will also provide HEI educators and policymakers with a tool to incorporate SE in ESD that can work in this regional context.

The outline of the study is as follows: Sect. 2 explores the literature review on ESD, SE, and the pedagogical approach, focusing on the interrelationships to study the pedagogy's impact on SE education in ESD. Sect. 3 details the research methodology employed, and Section 4 presents the findings of this exploratory study. We then discuss their implications and contribution in Sect. 5 to propose a learning approach effective for learners to test SE. Section 6 concludes the chapter with the limitations and lays out the steps to conduct further research.

2 Literature Review

This literature review begins with the overarching theme of ESD, its implementation by HEIs, and its goals, delves into the concept of SE, and finally focuses on the pedagogical approaches.

2.1 Education for Sustainable Development

Education for Sustainable Development in Higher Education Institutions

UNESCO posits that ESD comprises three major dimensions: economic, social, and environmental, with themes focused on globalization, trade and development, poverty alleviation, sustainable consumption and production, social equity, gender perspective, and the understanding of different cultures (Chen et al., 2022). And for any given dimension and theme, ESD contributes changes towards sustainability, and to achieve it, it contributes towards the development of change agents (Strachan, 2018). Therefore, HEIs are clearly among the primary forces driving the transformation of education for sustainability. In implementing ESD, HEIs adopted various approaches to gradually incorporate it: from developing courses across various disciplines to incorporating sustainability within existing programs and even developing separate programs and schools focused solely on sustainability (O'Byrne et al., 2015). This has led to the "whole-of-university" approaches, transforming the learning environment, and shifting the traditional educational practices from having learners as knowledge receivers to becoming able to act on it (Barth et al., 2014; Woo et al., 2012).

Broad and Specific Competencies

Through the "whole-of-university" approach, the values of the HEIs need to reflect the values being taught for sustainability to spark changes in attitudes, values, and behavior of the learners (Strachan, 2018; Woo et al., 2012). This is part of ESD's characteristics, which are to achieve transformation, envision change, and provide a holistic approach (Jellinek, 2018). In addition, due to sustainability's complexity, there is a need for core and general competencies that can empower learners to take responsible actions: systems-thinking competence, anticipatory competence, normative competence, strategic competence, and interpersonal competence (Bacon et al., 2011; Činčera and Aspengren, 2018). However, these competencies and characteristics may be too broad and do not necessarily fit the intention and motivation of learners for their specific domains and disciplines, such as nurturing the mindset and behavior in entrepreneurship in the context of this study.

Teaching and Learning Approaches

Whether it is to develop general or specific competencies, ESD recognizes the need for an innovative pedagogical approach. Indeed, the reality of HEIs is that teaching and learning approaches remain conventional for the majority, focusing on lectures to transmit knowledge from educators to learners as passive recipients, making it an individual learning (Bacon et al., 2011). But with the ongoing paradigm shift through ESD, curricula and pedagogical approaches are expected to be learner-centric, by activating and engaging learners with real-world problems, while educators don the role of facilitators rather than truth-holders (Činčera and Aspengren, 2018; Jellinek, 2018; Vasconcelos et al., 2017). The implementation of such a learner-centric approach may be carried out through various methods: case studies, simulations, project-based learning, problem-based learning, etc., all of which provide mutual learning, higher involvement in learning, and the opportunity to apply knowledge to specific and real-life situations (Vasconcelos et al., 2017; Woo et al., 2012). Even though the current literature suggests various teaching methods to accommodate the needs of different disciplines and learners, there is a gap in connections to be made between ESD and entrepreneurship fields. Hence, there is a need for understanding how new pedagogical approaches can influence the training of entrepreneurs who are inclined towards sustainability that focuses on improving the environment and/or society while increasing profits (Lourenço et al., 2012; Strachan, 2018; Woo et al., 2012).

Given the complexities of addressing the needs of ESD, it necessitates a curriculum and pedagogical approach distinct from traditional methods. This approach should empower learners to cultivate the competencies necessary to drive sustainable change (Bacon et al., 2011; Woo et al., 2012), harmonizing economic, social, and environmental imperatives. Therefore, there is a compelling case to integrate the principles and lens of SE into the discourse.

2.2 *Sustainable Entrepreneurship*

Scope of Sustainable Entrepreneurship

SE has become increasingly important alongside and for governments, businesses, universities, and international organizations due to its key role in the transformative process towards sustainability through its innovative market solutions for solving sustainability challenges (Diepolder et al., 2021; Sreenivasan and Suresh, 2023). Entrepreneurship has been traditionally related to economic development and wealth generation, but with the business opportunities arising for sustainable development, SE has become a pressing and practical need to bring forward sustainability innovations to the mass market while benefiting society and the environment (Rosário et al., 2022; Sarango-Lalangui et al., 2018; Schaltegger and Wagner, 2011). It lies in the intersection of three different entrepreneurship concepts: conventional entrepreneurship, social entrepreneurship, and environmental entrepreneurship; pursuing the corresponding triple bottom line of economic, social, and ecological goals, whereas the others focus on one or two goals only (Belz and Binder, 2017). This brings sustainable entrepreneurs to focus on a holistic meaning of value by integrating these three dimensions, whether as a start-up, or within existing Small- and Medium-sized Enterprises (SMEs) and large firms (Kummitha and Kummitha, 2021; Sreenivasan and Suresh, 2023). Rosário et al. (2022) suggested that positive attitudes are necessary towards both sustainability and entrepreneurship to gain key driving intentions for SE, and knowledge enhancement can bring positive attitude changes, giving ESD an important role to drive SE (Woo et al., 2012). However, there is a lack of empirical evidence on how this knowledge enhancement should be structured in HEIs to drive these changes towards SE and to drive intentions.

Competencies and Intentions

HEIs' contributions towards sustainability are partly done by building the foundational competencies for SE, providing support through an SE ecosystem to aspiring learners, as well as enhancing the self-efficacy of nascent entrepreneurs (Kummitha and Kummitha, 2021; Lans et al., 2014). Traditionally, sustainability and entrepreneurship were taught as separate fields to develop different competencies. However, advancements in ESD have merged these areas to cultivate competencies that encompass knowledge, skills, attitudes, values, and behaviors. This integration empowers individuals to responsibly assess business opportunities, considering environmental and social needs (Diepolder et al., 2021).

Although learners develop these competencies through experience and reflection facilitated by the pedagogy, they do not necessarily lead to EI, which is the deciding factor in whether one engages in entrepreneurial activities and performs entrepreneurial behaviors (Davidsson, 2023; Liñán et al., 2024). Therefore, HEIs need to provide an environment where the competencies and intentions can complement each other (Lyu et al., 2023). Building upon the Theory of Planned Behavior (TPB) of Ajzen (1991, 2011), Liñán (2004) proposed that EI can be explained by three dimensions: personal attitude towards start-up, subjective norm, and perceived behavioral control. Personal attitude towards start-up represents the positive or negative individual valuation of being an entrepreneur, subjective norms represent the perceived social expectations and support from others to carry out entrepreneurial behaviors, and perceived behavioral control represents the perception of how easy it is to become an entrepreneur and to overcome its challenges (Ajzen, 2020). However, EI is only discussed in the literature of conventional entrepreneurship education to generate wealth (Liñán et al., 2011), so there is a lack of understanding of how EI can be developed towards SE. Hence, there is a need to focus on ESD and how its pedagogical approaches can change the mindset of learners who focus on EI for sustainability (Nájera-Sánchez et al., 2023).

Education for Sustainable Entrepreneurship

Education for SE converges the field of sustainability and entrepreneurship to help aspiring learners and nascent entrepreneurs evaluate business opportunities to generate economic, social, and environmental values that benefit society and the environment; thus, being multi-disciplinary within ESD. The literature shows that ESD is transforming the teaching and learning approaches from being lecture-based and teacher-guided to becoming learner-centric and action-oriented through experiential and innovative methods to involve learners in developing solutions for real-world sustainability challenges (Diepolder et al., 2021; Halberstadt et al., 2019). Thus, there is a need for HEIs to focus on not only the content being taught but also the pedagogical approach on how the content is being delivered to the students.

This pedagogical approach promotes critical thinking, collaboration, reflection, inspiring and motivating learners, sparking transformative personal growth, and changes in attitudes and values, all of which are necessary for the paradigm shift (Seatter and Ceulemens, 2017; Strachan, 2018). In its implementation, a wide range of methods and tools are used by educators to involve the learners in sustainability challenges to solve issues that fit the local context and starting by raising their awareness of their unsustainable practices in society (Jellinek, 2018). Within ESD and SE, certain methods appear to be more recurrent than others, with a particular focus on case studies, simulations, project- and problem-based learning with the integration of real-life challenges, with which learners get to act and practice, sometimes in collaboration with stakeholders (Bacon et al., 2011; Brundiers and Wiek, 2013; Woo et al., 2012). This requires an in-depth understanding of these methods for the effective implementation of SE in ESD to nurture potential future sustainable entrepreneurs.

2.3 Case Method and Design for Education for Sustainable Development

Origin and Practice of the Case Method

The case method originated from the Harvard Law School by Christopher Columbus Langdell in 1870 and was then adopted by the Harvard Business School (HBS) in 1911, pioneering this method as the standard for graduate schools of business around the world (Moskovitz, 1992; Servant-Miklos, 2019). Given its popularity, HBS started developing and compiling cases (a factual description of human experience in a real organization facing specific problems) in the early 1920s to provide learners with many realistic business situations (Christensen and Hansen, 1987; Rebeiz, 2011). The case method works with the educator utilizing a case as the main course content to lead class discussions on analyzing and solving the said issues, which brings practical business learning to the classroom (Fisher, 1978; Forman and Rymer, 1999; Nohria, 2021). These case discussions fall within the rubric of active

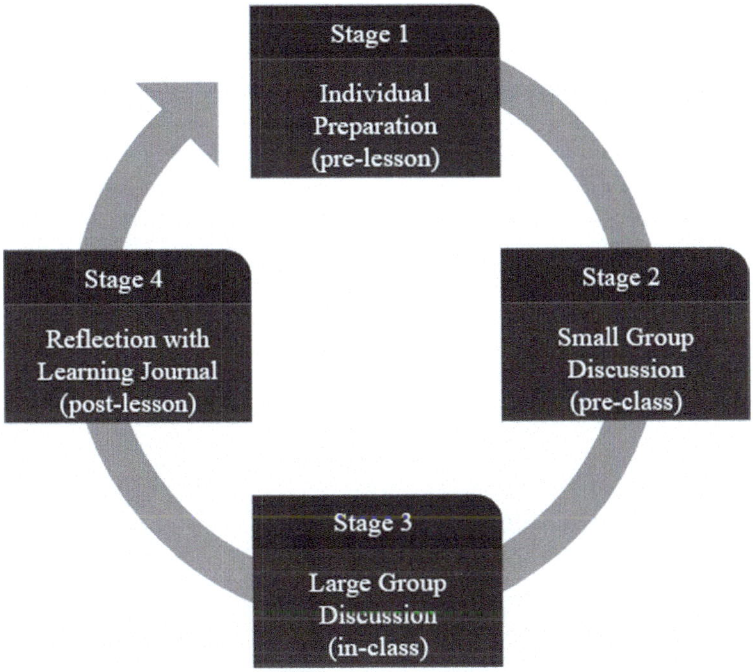

Fig. 1 Four-stage case method approach adapted from Harvard Business School. *Source* Created by the authors

learning as it "involves students in doing things and thinking about the things they are doing" (Bonwell and Eison, 1991). Through problem identification, analysis, and developing practical solutions, the learners become able to make crucial decisions (Schuster and Radel, 2018).

At present, HBS employs a case method with a four-stage process, as shown in Fig. 1. Firstly, the learners should read and analyze the case from the viewpoint of a real person (the protagonist) leading a real organization. Secondly, they discuss the case with a small team of classmates from diverse professional backgrounds. Thirdly, they engage in class discussion, which requires highly engaged conversation facilitated by the faculty member but driven by the classmates' comments and experiences. Finally, they reflect on the learnings and takeaways.

Going Beyond the Case Method for Sustainability: Learner-Centric and Action-Oriented

It is suggested that a deductive course architecture, by using cases, could be used to help the educators and learners test theories on the business scenarios. However, there is a lack of opportunity to act on the decisions to be taken (Musara, 2024;

Neck and Greene, 2011), marking the need for a prescriptive theory. Entrepreneurs need the ability to use the appropriate theory and/or predict the result based on the actions they plan to take, leveraging the knowledge, skills, attitudes, competencies, and intentions (Lyu et al., 2023). For this entrepreneurial development, experiential learning is put forward to allow students to experience, reflect, think, and act (Motta and Galina, 2023). Design involves both divergent and convergent processes that require skills through a combination of activities, such as observation, synthesis, feedback, visual representation, and value creation (Neck and Greene, 2011).

Consequently, there is a pressing need for pedagogy within ESD that not only imparts knowledge but also fosters practical decision-making and facilitates action through "Design-Based Learning" that focuses on BM generation. The cultivation of competencies and intentions, integral to SE, could serve as pivotal outcome indicators within this pedagogical framework.

3 Research Methodology

3.1 Mixed Methods Approach

The study employed a single case study with a mixed-methods approach, combining qualitative and quantitative methods to gather data on students' perceptions, learning outcomes, and overall experience to develop a better understanding of results, allowing for the emergence of new perspectives or new measurement tools (Dawadi et al., 2021; Dunning et al., 2008). The case study approach involves an in-depth investigation of a contemporary phenomenon within its real-world context to comprehend the case and understand how and why it functions (Hancock et al., 2021; Yin, 2018). Utilizing a case study was deemed fitting as our objective was to acquire a profound understanding of the "Sustainability Project" within the confines of a campus-based setting for ESD, from the perspective of the learners who may aspire to become entrepreneurs (Zucker, 2010). Given the complexity of the phenomenon, it becomes necessary to adopt several perspectives, which are enabled by both qualitative and quantitative methods (Shan, 2022). A qualitative analysis was employed to examine the students' awareness, actions, and output produced as outcomes within the educational environment. Our analysis would develop a conceptual understanding through an inductive reasoning process (Reid-Searl and Happell, 2012; Wagner et al., 2010).

3.2 Case Study: Sustainability Project

The Sustainability Project aimed to promote awareness of sustainability and to create sustainable innovation regarding the SDGs. The project provided opportunities for an in-depth study and discussion on the BM and social impact of Mymizu, a Japanese startup, and proposed a sustainability solution which considered the resources and constraints of the students' immediate environment and its alignment with the SDGs by the United Nations.

The project was executed in two phases: "case method" and "business model creation" phases. For the case method phase, students were given a 23-page case of "mymizu 2022" (François, 2022), a social startup established in 2019. Mymizu originated from a crowdfunding campaign and became a platform-based water refill application in Japan. It aims to achieve two key objectives: reducing PET bottle usage by encouraging the refilling of plastic bottles and promoting behavioral change for responsible consumption. The application features an open-source database with public water refill locations and participating companies, such as shopping malls and restaurants that offer free water refills. This initiative helps people avoid purchasing single-use plastic bottled drinks, as highlighted in Fig. 2.

In phase one of the project, students had undergone the four stages of the case method: pre-research on the case as a case assignment; a 20-min. small group discussion, a 90-min. classroom discussion, and an individual learning journal for reflection on the takeaways from the lesson. Subsequently, in phase two of the project, the students had to form groups to tackle sustainability issues within the campus environment by generating a new BM that could achieve not only economic outcomes, but also social and environmental impact. The process of generating the BM was executed over a seven-session course, with 90-min. sessions. It is divided into seven stages: identifying sustainability issues on campus, brainstorming ideas, creating the sustainability plan, integrating into BM, team consultation, project presentation, and individual reflection. The theme of each session and project activities are specified in Table 1.

3.3 *Data Collection*

Data was be collected before and after the case discussions to measure changes in students' knowledge and attitudes. Analysis involved comparing students' initial and post-case responses, identifying common themes in discussions, and evaluating the depth of critical thinking demonstrated during the case analysis.

This project was tailored for a total of 70 international students spanning two cohorts of the undergraduate course at Nagoya University of Commerce and Business (NUCB), who have completed the "Business Model Design" and "Leading the SDGs" modules in June–July 2023. The participating students represented a rich and diverse background from 25 different countries. The data sources included pre-course surveys, after-lesson journals, and a final report upon completion.

Fig. 2 The "mymizu 2022" case. *Source* Adapted from the "mymizu 2022" case by the authors

Table 1 Process of the Sustainability Project (with "mymizu 2022" case and project work)

Session	1	2	3	4	5	6	7
Theme	Policy	Lifestyle	Environment	Economy	Society	Stakeholders	Leadership
Task	Issues	Ideas	Plan	Business	Consultation	Presentation	Reflection

Source Created by the authors

3.4 *Content Analysis Data Code*

Content analysis was conducted on the learning journals during and at the end of the course, as shown in Table 2. This method serves as an effective tool to explore and describe students' knowledge and attitudes, create common themes, and draw analysis that can be integrated with other existing data, as well as the past studies (Hsieh and Shannon, 2005; Neuendorf and Kumar, 2016; Stemler, 2001). Furthermore, the content analysis allows educators to better understand their teaching and the students' learning, leading to improvement of educational practices (Graneheim et al., 2017; Kleinheksel et al., 2020). This table brings out three key themes: "awareness of sustainability", "entrepreneurship attributes related to entrepreneurial intention", and "generation of business model", which led to the analysis and discussion.

Table 2 Content analysis from the learning journals and reports by students

Code	Category	Theme
Not always money	Social purpose	Awareness of sustainability
Engage responsibly in society	Social responsibility	
Contribution to society	Social responsibility	
Refill the bottle and use fewer PET bottles	Sustainable action	
Encourage peers	Sustainable behavior	
CSR / ESG	Sustainable action	
Turn the old into new	Adaptation	Entrepreneurship (Entrepreneurial intention)
Need to act and execute	Action and execution	
Act on ideas	Action and execution	
Interest in entrepreneurship	Entrepreneurial intention	
Passion and dream	Ambitions	
Collaborate efficiently and openly in a safe space	Collaborative action	
Value proposition	Value proposition	Generation of business model
Company purpose and mission	Purpose and mission	
Economic and social balance	Balance	
Alternative revenue streams	Revenue streams	

Source Created by the authors

4 Findings

4.1 Awareness of Responsible Consumption and Scaling Social Impact for Students Through Case and Case Method

Students have increased their awareness of generating a new BM that can be effective for scaling social impact due to the role of responsible consumption from the case and the perspectives from different contexts in different countries. Through the discussion using the "mizmizu" case and the process of collaborating with group members to create a new BM for sustainability, students could simulate their role as responsible individuals as well as learners. This has impacted their values and behaviors to become good citizens in their attitude towards the SDGs.

> "I only knew that large corporations could make profit through sustainability. From the mymizu case and discussion from peers, I learned that startups can also achieve that through responsible consumption"". – by an exchange student from Germany.

Through the Sustainability Project, the findings show that by engaging the students in case discussions and BM creation activities, the project has successfully established a pivotal link between students and their immediate context, encompassing both societal and environmental dimensions. With the group of 70 students, 80% of them possess only limited knowledge about the generation of BM, while 50% are acquainted with SDGs. Additionally, 65% of the respondents expressed a tentative inclination, responding with a "maybe," regarding their EI for the future. Through the post-course survey discussion, about 92% (total of rating of relevance for levels 4 and 5) of the students indicated the relevance of the "mymizu 2022" case and case discussion for their learning about the mindset and behavior of startups, and the BM creation for sustainability, as shown in Fig. 3. This implies the inclination amongst students who may create an economically viable business if they receive the knowledge based on a real-life case in entrepreneurship for sustainability.

The interlink of learner-centric and action-oriented learning process through the project acts as a catalyst, facilitating the exchange of knowledge between real-world cases and learners, while also encouraging reflective engagement between end-users' actions and the learners themselves. These outcomes closely align with the social and environmental dimensions inherent to SE, as highlighted by Shepherd and Patzelt (2011), thus confirming the project's efficacy in changing attitudes and strengthening the subjective norms to form new "genre" systems (Forman and Rymer, 1999) supporting sustainability for entrepreneurs. Furthermore, the case discussions underscore the significance of using the relevant sustainability cases, such as "mymizu 2022", in conjunction with case methods. This approach enables learners to gain insights into the importance of driving consumer behavior change and scaling social impact through their BM (François and Goi, 2023).

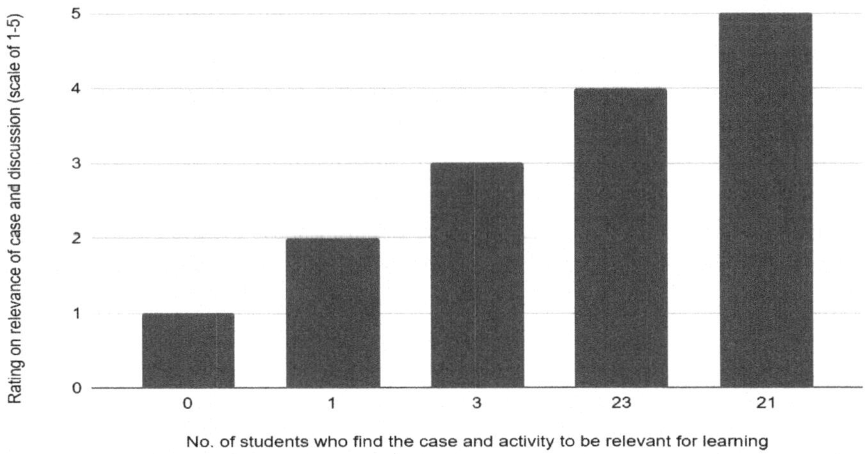

Fig. 3 Feedback on the relevance of the case and the discussion through the case method by students. *Source* Created by the authors

4.2 *Impact on Entrepreneurial Intention of Students Through Decision and Design Approach*

Students can take actions to design the new BM for sustainability based on the knowledge gained from case discussion ("mymizu 2022" case), which is an indication of EI on the dimension of Perceived Behavioral Control (PBC) (Fayolle et al., 2006).

> "The project also sparked my interest in an entrepreneurial pursuit that I never really thought of before, which in this case, is the business idea (the pick-up car service) my group had presented." – by a full-time student from the Philippines.

The student from the Philippines demonstrated a noticeable shift in his mindset regarding his capacity to generate innovative business ideas with his group and effectively present their proposal to the class, reflecting a heightened sense of PBC akin to that of an entrepreneur. This observation serves as compelling evidence of the effectiveness of the project's emphasis on learning-centric and action-oriented methodologies. Furthermore, it suggests the potential impact of such approaches on shaping students' EI, particularly through the lens of SE. Moreover, the transformative outcomes observed in EI serve to inspire students with career aspirations to pursue paths as social entrepreneurs or business leaders within the realm of sustainability in the future (Fayolle et al., 2006). Hence, such practices not only enrich the educational experience but also empower learners to grasp the intricacies of SE in a tangible and impactful manner. Hence, this reflects the importance of a combination of decision and design lenses as an integrated approach in ESD to fulfill the lack of knowledge, competencies, and opportunities amongst students who may be interested in undergoing a process of action in entrepreneurship.

4.3 New Business Model Generation for Sustainability by Students

Students could compare the current BM to generate a new BM for sustainability by forming new assumptions and creating a new prototype as evidence on the extension from a decision-making lens to a design lens.

One of the student groups created the "ReFlow" business proposal as a campus-based solution to reduce the usage of plastic bottles within the university as a partnership with mymizu. They created their logo, product design, digital QR code, and product launch video to execute the business, as shown in Fig. 4.

> "There should be a nice balance between social impact and profitability, so to increase profits, you could diversify the ways to earn." – by a full-time student from Japan, in the ReFlow group.

By integrating the case method, particularly with the "mymizu 2022" case, and expanding BM creation activities, the project has effectively equipped students with the means to apply their entrepreneurial skills to the business model canvas. Additionally, they have been empowered to develop video prototypes that visually articulate their designs for sustainable solutions. This approach not only fosters practical application of entrepreneurial competencies but also underscores the project's commitment to facilitating innovative problem-solving in sustainability. Moreover, the project's emphasis on generating a BM linked to sustainability highlights the profound impact of entrepreneurship education in addressing sustainability challenges while concurrently generating scalable and sustainable impact within a financially viable framework.

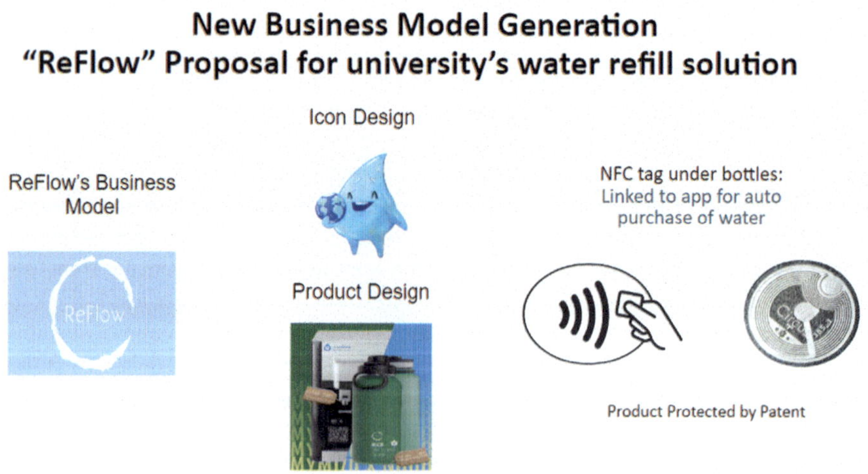

Fig. 4 Prototype of business model by ReFlow group. *Source* Adapted by the authors from the ReFlow proposal of the students

5 Discussion

5.1 New Scope and Dimension of Entrepreneurial Intention for Sustainable Entrepreneurs

This study delved into SE as an expanded framework within sustainability, adopting an entrepreneurial approach that not only seizes long-term scaling of social impact opportunities but also effectively balances awareness and efficacy in addressing economic, social, and environmental challenges. Our study underscores the significance of nurturing essential dimensions of EI pertinent to the promotion of SE in ESD. Firstly, our research emphasizes the necessity of broadening awareness and fostering responsible consumption behavior towards the environment and society among aspiring entrepreneurs and their peers through cases like "mymizu 2022" as they provide valuable lessons and insights (Sreenivasan and Suresh, 2023). This expansion is crucial for shaping Personal Attitude (PA) and Subjective Norms (SN) dimensions within EI, and there is a stark difference in this area between conventional entrepreneurship and SE. Secondly, our findings highlight the importance of instilling responsible action and values while expanding the scope of Perceived Behavioral Control (PBC) dimension within EI, particularly relevant for sustainable entrepreneurs. Within PBC, the ability to navigate entrepreneurship challenges should encompass addressing social and environmental concerns to cultivate self-efficacy (Bandura, 2000) and perceived feasibility (Shapero and Sokol, 1982). Thirdly, our study advocates for incorporating the dimension of CSV into EI, enabling sustainable entrepreneurs to leverage resources with stakeholders towards long-term goals rather than solely prioritizing short-term economic outcomes. Positioned as a unitary societal perspective, CSV (Porter and Kramer, 2011) underscores the value of fostering economic and social benefits relative to costs, thereby maximizing profit while simultaneously creating community value aligned with stakeholders' interests, as emphasized by SE and ESD on the triple bottom line (Tate and Bals, 2018). Consequently, the dimension of CSV will be a relevant component that triggers HEI to focus its resources on fostering partnerships with actual stakeholders who are dealing with the issue of sustainability in the context of the Asia–Pacific region.

Thus, the preceding discussion illuminated the first research question and answered the appropriate content that students should learn to increase their awareness of sustainability and, beyond that, impact the EI. Furthermore, the study emphasizes the importance of HEIs in incorporating knowledge of SE that addresses local issues in the Asia–Pacific region. This alignment is essential for the content dimension and for utilizing EI as supplementary entrepreneurship indicators, ultimately aiding in the development of the urgently needed entrepreneurs in the region.

5.2 *Proposed Framework: Learner-Centric and Action-Oriented Approach for Sustainable Entrepreneurship (LAASE)*

Our study delves into the content and pedagogical approach for integrating SE into ESD, expanding upon existing literature in this field. We posit that sustainable entrepreneurs adopt an entrepreneurial approach that not only capitalizes on long-term scaling social impact opportunities, but also effectively balances awareness and efficacy in addressing economic, social, and environmental challenges.

Building on our findings, we proposed LAASE as a learning model. LAASE emphasizes the significance of case method learning and BM creation in influencing EI, serving as cognitive outcomes for sustainable entrepreneurs in ESD (Fig. 3). This model incorporates a stage focused on the design and generation of BM, extending knowledge into cognitive actions that empower learners to devise solutions for sustainability and enact meaningful impact.

Following HBS' case method, the study employed a four-stage process: stage one for analyzing the case from the viewpoint of a real person (the protagonist) leading a real organization; stage two for discussing the case within a small team from diverse backgrounds; stage three for engagement in class discussion based on active participation by learners, and stage four for reflection on the learnings and takeaways. Extending the traditional case method process, we propose a fifth stage, "Design and Generation of Business Model", that requires the learners to extend knowledge into practice to resolve sustainability issues through generating a new BM for economic, social, and environmental outcomes to satisfy the purpose of a sustainable entrepreneur, as shown in Fig. 5. Hence, this five-stage process would seamlessly integrate the case studies, simulations, project- and problem-based learning with the integration of real-life challenges, with which learners get to act and practice, sometimes in collaboration with stakeholders as discussed in SE literature (Bacon et al., 2011; Brundiers and Wiek, 2013; Woo et al., 2012).

Thus, the above discussion answered the second research question based on the pedagogical approach that led to the outcome of generating a new BM for sustainability by students with the given resources and environment in HEI. Furthermore, the above suggests that case method faculty and practitioners in HEIs should develop and teach with more localized cases that address sustainability issues within the Asia–Pacific context. They should creatively employ a progressive and integrated learning approach that fosters both learning and action to achieve more unified learning goals.

6 Conclusion

Our study addresses the first research question regarding the effectiveness of using a case with content capturing a real-life practice of SE, aiming to achieve not only economic outcomes but also the scaling of social impact. Additionally, we answered

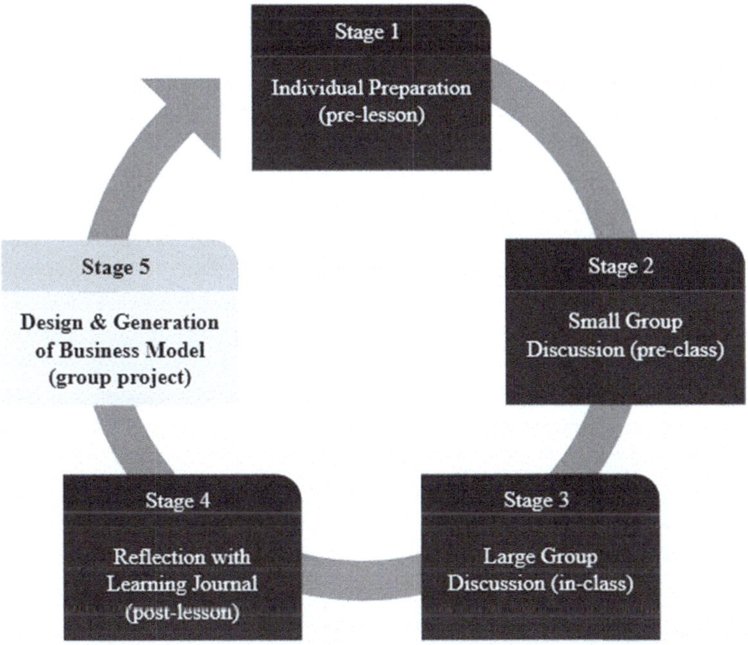

Fig. 5 Learner-centric and action-oriented approach for sustainable entrepreneurship. *Source* Created by the authors

the second research question, which focuses on the necessity of developing a learning approach that integrates learning-centric methods to enhance decision-making skills and action-oriented methods to leverage resources within the existing environment. The study also addressed the imperatives of SE relevant for ESD with several key findings. These findings include the impact on the awareness of responsible consumption and scaling social impact by students, the impact on the EI of students through the decision and design approach, and the new BM generation for sustainability.

This research contributes significantly to both theories in EI for SE and the pedagogical approach for ESD. We discuss the imperative of including social and environmental issues and the local context that are pertinent to influencing EI for SE. Furthermore, we emphasize the importance of integrating the dimension of CSV into EI, thereby creating a new scope and relevance of SE for sustainability. On the theoretical front for ESD, we proposed LAASE as a learning model that underscores the importance of case method learning and BM creation to impact EI as cognitive outcomes of sustainable entrepreneurs in ESD. In practice, the study advocates for the promotion of case development on SE and the utilization of the case method, coupled with a design approach in and beyond classroom settings for creating best practices in ESD. Both the theoretical and practical contributions are relevant to address the pedagogical gaps in pedagogical understanding, the application of the case method, and the design approach in ESD amongst HEIs in the Asia–Pacific region.

The study also highlights the implications for HEIs on the integration of knowledge of SE that addresses local issues in the Asia–Pacific region and inclusion of EI as additional entrepreneurship indicators to support the long-term outcomes in nurturing the entrepreneurs urgently needed in the region. Secondly, HEIs should develop and teach with more localized cases that address sustainability issues within the Asia–Pacific context and employ a progressive and integrated learning approach for learning and action. We would also point out to policymakers to provide resources and support for the development of ESD through infrastructure for case method development and teaching, as well as funding for student start-ups on their sustainability actions.

We acknowledge the limitations in this study, such as the restricted sample size and potential biases arising from the roles of the case author and the case method educator. There are also constraints in the control of educational settings to enable more activities to be simulated in the learning process. Environmental constraints would include the classroom settings, availability of cases, and the time lapse between the session for case method and project implementation for observation and data collection, which are typical amongst HEIs in the Asia–Pacific region. Contextual constraints would be cultural differences amongst the international students and their orientations towards the subject of sustainability, which may be mediating factors in pedagogical studies. For future research, we could dive into an expanded pedagogical approach that aims at bolstering the implementation of BM strategies and fostering the gestation of sustainable businesses for ESD. One way of achieving this would be to partner with relevant stakeholders to provide learners with the opportunity to interact with practitioners throughout the design process (Diepolder et al., 2021; Hermann et al., 2020; Woo et al., 2012). Second, prospective studies should investigate the correlations of sustainability orientations and cultural factors that may affect the EIs of international students. Such endeavors would empower the educators in HEI to craft course plans with a transformative potential and empower learners to implement their solutions, ensuring sustainable outcomes beyond the initial phases of knowledge acquisition and practical application.

References

Ajzen, I. (1991). The theory of planned behavior. *Organizational Behavior and Human Decision Processes, 50*(2), 179–211. https://doi.org/10.1016/0749-5978(91)90020-T

Ajzen, I. (2011). The theory of planned behaviour: Reactions and reflections. *Psychology & Health, 26*(9), 1113–1127. https://doi.org/10.1080/08870446.2011.613995

Ajzen, I. (2020). The theory of planned behavior: Frequently asked questions. *Human Behavior and Emerging Technologies, 2*, 314–324. https://doi.org/10.1002/hbe2.195

Bacon, C. M., Mulvaney, D., Ball, T. B., DuPuis, M. E., Gliessman, S. R., Lipschutz, R. D., and Shakouri, A. (2011). The creation of an integrated sustainability curriculum and student praxis projects. *International Journal of Sustainability in Higher Education, 12*(2), 193–208. https://doi.org/10.1108/14676371111118237

Bandura, A. (2000). Self-efficacy: The foundation of agency. In W. J. Perrig and A. Grob (Eds.), *Control of human behavior, mental processes, and consciousness: Essays in honor of the 60th birthday of August Flammer* (pp. 17–33). Lawrence Erlbaum Associates Publishers.

Barth, A. E., Tolar, T. D., Fletcher, J. M., and Francis, D. (2014). The effects of student and text characteristics on the oral reading fluency of middle-grade students. *Journal of Educational Psychology, 106*(1), 162–180. https://doi.org/10.1037/a0033826

Belz, F. M., and Binder, J. K. (2017). Sustainable entrepreneurship: A convergent process model. *Business Strategy and the Environment, 26*(1), 1–17. https://doi.org/10.1002/bse.1887

Bonwell, C. C., and Eison, J. A. (1991). Active learning: Creating excitement in the classroom (1991 ASHE-ERIC Higher Education Reports). ERIC Clearinghouse on Higher Education, The George Washington University. https://eric.ed.gov/?id=ED336049

Brundiers, K., and Wiek, A. (2013). Do we teach what we preach? An international comparison of problem-and project-based learning courses in sustainability. *Sustainability, 5*(4), 1725–1746. https://doi.org/10.3390/su5041725

Chen, C., An, Q., Zheng, L., and Guan, C. (2022). Sustainability literacy: Assessment of knowingness, attitude and behavior regarding sustainable development among students in China. *Sustainability, 14*(9), 4886. https://doi.org/10.3390/su14094886

Christensen, C. R., and Hansen, A. J. (1987). Teaching and the case method: Text, cases, and readings. Harvard Business School Press.

Činčera, J., and Aspengren, P. (2018). Sustainability and an educational perspective: Roots in environmental education. In K. Iwińska, M. Jones, and M. Kraszewska (Eds.), *Widening interdisciplinary sustainability education.* Collegium Civitas. https://civitas.edu.pl/wp-content/uploads/2015/03/Widening-Interdisciplinary-Sustainability-Education-light.pdf

Cotgrave, A. J., and Kokkarinen, N. (2011). Promoting sustainability literacy in construction students: Implementation and testing of a curriculum design model. *Structural Survey, 29*(3), 197–212. https://doi.org/10.1108/02630801111148185

Davidsson, P. (2023). Ditching discovery-creation for unified venture creation research. *Entrepreneurship Theory and Practice, 47*(2), 594–612. https://doi.org/10.1177/10422587211030870

Dawadi, S., Shrestha, S., and Giri, R. A. (2021). Mixed-methods research: A discussion on its types, challenges, and criticisms. *Journal of Practical Studies in Education, 2*(2), 25–36. https://doi.org/10.46809/jpse.v2i2.20

Diepolder, C. S., Weitzel, H., and Huwer, J. (2021). Competence frameworks of sustainable entrepreneurship: A systematic review. *Sustainability, 13*(24), 13734. https://doi.org/10.3390/su132413734

Dunning, H., Williams, A., Abonyi, S., and Crooks, V. (2008). A mixed method approach to quality of life research: A case study approach. *Social Indicators Research, 85*, 145–158. https://doi.org/10.1007/s11205-007-9131-5

Fayolle, A., Gailly, B., and Lassas-Clerc, N. (2006). Assessing the impact of entrepreneurship education programmes: A new methodology. *Journal of European Industrial Training, 30*(9), 701–720. https://doi.org/10.1108/03090590610715022

Fisher, C. M. (1978). Ataxic hemiparesis: A pathologic study. *Archives of Neurology, 35*(3), 126–128. https://doi.org/10.1001/archneur.1978.00500270008002

Forman, J., and Rymer, J. (1999). The genre system of the Harvard case method. *Journal of Business and Technical Communication, 13*(4), 373–400. https://doi.org/10.1177/105065199901300401

François, K. K. (2022). *Mymizu 2022* [Unpublished master's dissertation]. NUCB Business School, Nagoya University of Business and Commerce.

François, K. K., and Goi, H. C. (2023). Business model for scaling social impact towards sustainability by social entrepreneurs. *Sustainability, 15*(18), 14027. https://doi.org/10.3390/su151814027

Graneheim, U. H., Lindgren, B.-M., and Lundman, B. (2017). Methodological challenges in qualitative content analysis: A discussion paper. *Nurse Education Today, 56*, 29–34. https://doi.org/10.1016/j.nedt.2017.06.002

Halberstadt, J., Schank, C., Euler, M., and Harms, R. (2019). Learning sustainable entrepreneurship by doing: Providing a lecturer-oriented service learning framework. *Sustainability, 11*(5), 1217. https://doi.org/10.3390/su11051217

Hancock, D., Algozzine, B., and Lim, J. (2021). *Doing case study research: A practical guide for beginning researchers, fourth edition* (D. Hancock, B. Algozzine, and J. Lim, Eds.). Teachers College Press. https://www.tcpress.com/ng-case-study-research-9780807765852

Hermann, R. R., Bossle, M. B., and Amaral, M. (2020). Lenses on the post-oil economy: Integrating entrepreneurship into sustainability education through problem-based learning. *Educational Action Research, 30*(3), 480–506. https://doi.org/10.1080/09650792.2020.1823239

Hsieh, H.-F., and Shannon, S. E. (2005). Three approaches to qualitative content analysis. *Qualitative Health Research, 15*(9), 1277–1288. https://doi.org/10.1177/1049732305276687

Jellinek, N. (2018). Sustainability and the teaching context: The role of university educators. In K. Iwińska, M. Jones, and M. Kraszewska (Eds.), *Widening interdisciplinary sustainability education*. Collegium Civitas. https://civitas.edu.pl/wp-content/uploads/2015/03/Widening-Interdisciplinary-Sustainability-Education-light.pdf

Kleinheksel, A. J., Rockich-Winston, N., Tawfik, H., and Wyatt, T. R. (2020). Demystifying content analysis. *American Journal of Pharmaceutical Education, 84*(1), 7113. https://doi.org/10.5688/ajpe7113

Kummitha, H. R., and Kummitha, R. K. R. (2021). Sustainable entrepreneurship training: A study of motivational factors. *The International Journal of Management Education, 19*(1), 100449. https://doi.org/10.1016/j.ijme.2020.100449

Lans, T., Blok, V., and Wesselink, R. (2014). Learning apart and together: Towards an integrated competence framework for sustainable entrepreneurship in higher education. *Journal of Cleaner Production, 62*(1), 37–47. https://doi.org/10.1016/j.jclepro.2013.03.036

Li, M., Zhang, Y., Fan, Z., and Chen, H. (2021). Evaluation and research on the level of inclusive green growth in Asia-Pacific region. *Sustainability, 13*, 7482. https://doi.org/10.3390/su13137482

Liñán, F. (2004). Intention-based models of entrepreneurship education. *Small Business, 3*(1), 11–35.

Liñán, F., Rodríguez-Cohard, J. C., and Rueda-Cantuche, J. M. (2011). Factors affecting entrepreneurial intention levels: A role for education. *International Entrepreneurship and Management Journal, 7*, 195–218. https://doi.org/10.1007/s11365-010-0154-z

Liñán, F., Jaén, I., and Domínguez-Quintero, A. M. (2024). An action phase theory approach to the configuration of entrepreneurial goal and implementation intentions. *International Journal of Entrepreneurial Behavior & Research, 30*(11), 64–90. https://doi.org/10.1108/IJEBR-07-2023-0772

Liu, J., Kitamura, Y., and Savelyeva, T. (2022). Building an 'ecosystem' for transforming higher education teaching and learning for sustainability. *Asia Pacific Education Review, 23*, 539–542. https://doi.org/10.1007/s12564-022-09794-1

Lourenço, I. C., Branco, M. C., Curto, J. D., and Eugénio, T. (2012). How does the market value corporate sustainability performance? *Journal of Business Ethics, 108*, 417–428. https://doi.org/10.1007/s10551-011-1102-8

Lyu, J., Shepherd, D., and Lee, K. (2023). From intentional to nascent student entrepreneurs: The moderating role of university entrepreneurial offerings. *Journal of Innovation & Knowledge, 8*(1), 100305. https://doi.org/10.1016/j.jik.2023.100305

MEXT. (2017). *ESD (Education for Sustainable Development)*. Ministry of Education, Culture, Sports, Science and Technology. https://www.mext.go.jp/en/unesco/title04/detail04/sdetail04/1375695.htm

MEXT. (2021). *Implementation plan for education for sustainable development in Japan (the second ESD implementation plan).* Ministry of Education, Culture, Sports, Science and Technology. https://www.mext.go.jp/en/unesco/title04/detail04/sdetail04/mext_00030.html
Moskovitz, M. (1992). Beyond the case method: It's time to teach with problems. *Journal of Legal Education, 42*(2), 241–270. http://www.jstor.org/stable/42893216
Motta, V. F., and Galina, S. V. R. (2023). Experiential learning in entrepreneurship education: A systematic literature review. *Teaching and Teacher Education, 121,* 103919. https://doi.org/10.1016/j.tate.2022.103919
Musara, M. (2024). Using the case study method in undergraduate entrepreneurship education. *Administrative Sciences, 14*(4), 60. https://doi.org/10.3390/admsci14040060
Naeem, M., and Neal, M. (2012). Sustainability in business education in the Asia Pacific region: A snapshot of the situation. *International Journal of Sustainability in Higher Education, 13*(1), 60–71. https://doi.org/10.1108/14676371211190317
Nájera-Sánchez, J. J., Pérez-Pérez, C., and González-Torres, T. (2023). Exploring the knowledge structure of entrepreneurship education and entrepreneurial intention. *International Entrepreneurship and Management Journal, 19,* 563–597. https://doi.org/10.1007/s11365-022-00814-5
Neck, H. M., and Greene, P. G. (2011). Entrepreneurship education: Known worlds and new frontiers. *Journal of Small Business Management, 49*(1), 55–70. https://doi.org/10.1111/j.1540-627X.2010.00314.x
Neuendorf, K. A., and Kumar, A. (2016). Content analysis. In G. Mazzoleni (Ed.), *The international encyclopedia of political communication.* John Wiley & Sons. https://doi.org/10.1002/9781118541555.wbiepc065
NIER. (2012). *Gakkou ni okeru jizokukanouna hatten no tameno kyouiku (ESD) ni kansuru kenkyu (Saishu Houkokusho)* [Research on education for sustainable development (ESD) in the schools (final report)]. Tokyo: National Institute for Educational Policy Research (NIER).
Nohria, N. (2021). What the case study method really teaches. *Harvard Business Review.* https://hbr.org/2021/12/what-the-case-study-method-really-teaches
O'Byrne, D., Dripps, W., and Nicholas, K. A. (2015). Teaching and learning sustainability: An assessment of the curriculum content and structure of sustainability degree programs in higher education. *Sustainability Science, 10,* 43–59. https://doi.org/10.1007/s11625-014-0251-y
Persson, C., Einarson, D., and Melén, M. (2023). Educating the educators to be a driving force in higher education towards sustainable development. *International Journal of Sustainability in Higher Education, 24*(9), 197–212. https://doi.org/10.1108/IJSHE-10-2022-0332
Porter, M. E., and Kramer, M. R. (2011). Creating Shared Value. *Harvard Business Review.* https://hbr.org/2011/01/the-big-idea-creating-shared-value
Rebeiz, K. S. (2011). An insider perspective on implementing the Harvard case study method in business teaching. *US-China Education Review, 5,* 591–601. https://eric.ed.gov/?id=ED527670
Reid-Searl, K., and Happell, B. (2012). Supervising nursing students administering medication: A perspective from registered nurses. *Journal of Clinical Nursing, 21*(13-14), 1998–2005. https://doi.org/10.1111/j.1365-2702.2011.03976.x
Ritchie, H. (2018). *Plastic pollution.* Global Change Data Lab. https://ourworldindata.org/plastic-pollution
Rosário, A. T., Raimundo, R. J., and Cruz, S. P. (2022). Sustainable entrepreneurship: A literature review. *Sustainability, 14*(9), 5556. https://doi.org/10.3390/su14095556
Sarango-Lalangui, P., Santos, J. L. S., and Hormiga, E. (2018). The development of a sustainable entrepreneurship research field. *Sustainability, 10*(6), 2005. https://doi.org/10.3390/su10062005
Schaltegger, S., and Wagner, M. (2011). Sustainable entrepreneurship and sustainability innovation: Categories and interactions. *Business Strategy and the Environment, 20*(4), 222–237. https://doi.org/10.1002/bse.682
Schuster, R. J., and Radel, J. (2018). A reflection on the (Harvard) case method from a group dynamics perspective: Connecting transcendent knowledge with immanent phenomena. In H. Gölzner and P. Meyer (Eds.), *Emotionale Intelligenz in Organisationen* (pp. 279–314). Springer.

Seatter, C. S., and Ceulemans, K. (2017). Teaching sustainability in higher education: Pedagogical styles that make a difference. *Canadian Journal of Higher Education, 47*(2), 47–70. https://doi.org/10.47678/cjhe.v47i2.186284

Servant-Miklos, V. F. C. (2019). The Harvard connection: How the case method spawned problem-based learning at McMaster University. *Health Professions Education, 5*(3), 163–171. https://doi.org/10.1016/j.hpe.2018.07.004

Shan, Y. (2022). Philosophical foundations of mixed methods research. *Philosophy Compass, 17*(1), e12804. https://doi.org/10.1111/phc3.12804

Shapero, A., and Sokol, L. (1982). *The social dimensions of entrepreneurship.* University of Illinois at Urbana-Champaign's Academy for Entrepreneurial Leadership Historical Research Reference in Entrepreneurship.

Shepherd, D., and Patzelt, H. (2011). The new field of sustainable entrepreneurship: Studying entrepreneurial action linking "what is to be sustained" with "what is to be developed." *Entrepreneurship Theory and Practice, 35*(1), 137–163. https://doi.org/10.1111/j.1540-6520.2010.00426.x

Sreenivasan, A., and Suresh, M. (2023). Twenty years of entrepreneurship education: A bibliometric analysis. *Entrepreneurship Education, 6*(1), 45–68. https://doi.org/10.1007/s41959-023-00089-z

Stemler, S. (2001). An overview of content analysis. *Practical Assessment, Research, and Evaluation, 7*(17). https://doi.org/10.7275/z6fm-2e34

Strachan, G. (2018). Can education for sustainable development change entrepreneurship education to deliver a sustainable future? *Discourse and Communication for Sustainable Education, 9*(1), 36–49. https://doi.org/10.2478/dcse-2018-0003

Tate, W. L., and Bals, L. (2018). Achieving shared triple bottom line (TBL) value creation: Toward a social resource-based view (SRBV) of the firm. *Journal of Business Ethics, 152*, 803–826. https://doi.org/10.1007/s10551-016-3344-y

THE. (2022). *Times Higher Education Impact Ranking 2022.* Times Higher Education (THE). https://www.timeshighereducation.com/impactrankings

United Nations. (2015). *Transforming our world: The 2030 agenda for sustainable development.* United Nations. https://sdgs.un.org/2030agenda

Vasconcelos, C., Faria, J., and Cardoso, A. (2017). Sustainability and case-based methodology. In I. Katarzyna, J. Michael, and M. Kraszewska (Eds.), *WISE Handbook* (pp. 28–31). Erasmus+. https://core.ac.uk/download/pdf/162557622.pdf

Wagner, E., Alves, H., and Raposo, M. (2010). An exploratory research on the stakeholders of a university. *Journal of Management and Strategy, 1*(1). https://doi.org/10.5430/jms.v1n1p76

Woo, Y. L., Mokhtar, M., Komoo, I., and Azman, N. (2012). Education for sustainable development: A review of characteristics of sustainability curriculum. *OIDA International Journal of Sustainable Development, 3*(8), 33–44. https://ssrn.com/abstract=2031102

Yin, R. K. (2018). *Case study research and applications: Design and methods* (6th ed.). Sage Publications. https://uk.sagepub.com/en-gb/eur/case-study-research-and-applications/book250150

Zucker, D. (2010). How to do case study research. In M. Garner, C. Wagner, and B. Kawulich (Eds.), *Teaching research methods in the social sciences* (pp. 264). Routledge.

One Village One CEO as Social Innovation for Sustainability and the Curriculum: Teaching About and for Sustainable Development in Indonesia

Mohammad Iqbal Irfany, Handian Purwawangsa, Alim Setiawan Slamet, Bintoro Pujo Prawiro, and Daffa Aqomal Haq

Abstract Rapid economic growth is projected in developing countries across East Asia and the Pacific, with rates expected to rise from 3.5% in 2022 to 5.1% in 2023. To sustain this growth, it is essential for younger generations to understand sustainability while addressing both economic and nutritional challenges. The abundance of land in rural areas presents an opportunity to develop agricultural enterprises that can meet food demands. The *One Village One CEO* (OVOC) initiative serves as a social innovation program designed to build business ecosystems that empower youth through the utilization of local resources. This initiative integrates focus groups, interviews, and direct assistance as part of its implementation strategy. In Cianjur, West Java, the program increased the average income from IDR 2,200,000 in 2019 to IDR 3,325,000 in 2021, generated employment for 9000 individuals, and facilitated academic partnerships involving 343 university students. As of today, OVOC has reached 10 provinces, 35 regencies or municipalities, and 452 villages in Indonesia. It contributes directly to the achievement of multiple Sustainable Development Goals (SDGs), including: (1) No Poverty, (2) Zero Hunger, (8) Decent Work and Economic Growth, (9) Industry, Innovation, and Infrastructure, (12) Responsible Consumption and Production, (14) Life Below Water, and (17) Partnerships for the Goals.

Keywords Social innovation · Sustainable ecosystem business · Curriculum sustainability · One Village One CEO · Community development

---

M. I. Irfany (✉) · A. S. Slamet · D. A. Haq
Faculty of Economics and Management, IPB University, Bogor, Indonesia
e-mail: iqbal.irfany@apps.ipb.ac.id

H. Purwawangsa
Faculty of Forestry and Environment, IPB University, Bogor, Indonesia

B. P. Prawiro
Directorate of Agromaritime Community Development, IPB University, Bogor, Indonesia

W. Leal Filho et al. (eds.), *Sustainability in the Curriculum and Operations of Universities*, World Sustainability Series,
https://doi.org/10.1007/978-3-032-12678-8_29

1 Introduction

The Asia–Pacific region, including Indonesia, is experiencing rapid economic growth, with projections indicating an increase from 3.5% in 2022 to 5.1% in 2023 (World Bank, 2023). While promising, this growth presents a dual challenge: achieving economic progress while ensuring environmental sustainability and social equity. Education—particularly higher education—plays a critical role in addressing these challenges by equipping younger generations with the knowledge, skills, and mindset needed to drive sustainable development (SDSN Australia/Pacific, 2017). Universities are uniquely positioned to act as catalysts for change, bridging the gap between academic knowledge and practical solutions. As noted by Land & Herzog (2017), higher education institutions are key players in fostering sustainable practices, promoting research-driven innovations, and developing future leaders capable of addressing global sustainability challenges.

In the Indonesian context, the demographic bonus—characterized by a large proportion of the population being of productive age—presents both an opportunity and a challenge (Achmad et al., 2024). With this demographic window expected to peak between 2020 and 2035, it is imperative to leverage this potential workforce through quality education, skill development, and innovative approaches. However, despite this demographic advantage, rural areas in Indonesia continue to face significant socio-economic disparities, including limited access to education, technological infrastructure, and sustainable economic opportunities. According to Kamil et al. (2019), education and skills training are crucial in empowering rural youth, providing them with viable pathways to employment and entrepreneurship.

The Fourth Industrial Revolution has introduced transformative changes in global production systems, marked by digitalization, automation, and interconnectivity (Schlechtendahl et al., 2015). In Indonesia, however, these advancements have largely been concentrated in urban areas, leaving rural communities with limited access to emerging technologies. Consequently, rural development strategies must adopt innovative solutions to bridge this digital divide and strengthen rural economies. Village-based development thus emerges as a vital strategy for achieving the Sustainable Development Goals (SDGs), particularly in the areas of food security, economic growth, and environmental sustainability. OECD (2023) underscores that rural innovation ecosystems—supported by higher education and technology—can significantly boost productivity and community resilience.

As one of the universities committed to achieving the SDGs, IPB University aims to become a global techno-socio-entrepreneurial institution, particularly in the fields of agriculture, maritime affairs, and tropical biosciences. The techno-socio-entrepreneurial university is a vision that integrates research, innovation, and entrepreneurship, with outcomes that serve not only economic purposes but also social impact. To address these values and challenges, IPB University introduced a notable initiative called One Village One CEO (OVOC). OVOC is a social innovation model designed to build sustainable rural business ecosystems by integrating academic knowledge, student engagement, and multi-sector collaboration. As a result of this initiative, IPB University became the first institution to receive the 1st Best Award in Indonesia's SDGs Action in 2022–2023. The university's contributions to the SDGs focus on sustainable development at the village level.

Through OVOC, students are actively involved in real-world problem-solving, applying theoretical knowledge in practical settings to foster innovation and entrepreneurship in rural communities. Similar initiatives have succeeded in other regions, such as the One Village One Product program in Japan, which underscores the importance of localized innovation in achieving sustainable development (Schumann, 2016). However, unlike Japan's movement—which operates independently of higher education—OVOC integrates university resources and capacities to empower students and amplify impact through academic, entrepreneurial, and social innovation. This integration makes OVOC a unique and pioneering initiative in the field of sustainable development.

The OVOC program is also closely aligned with the Merdeka Belajar Kampus Merdeka (MBKM) curriculum—an educational policy introduced by Indonesia's Ministry of Education, Culture, Research, and Technology to enhance student competencies through experiential learning. The MBKM framework allows students to engage in community development projects, fostering a deeper understanding of sustainable development principles and their practical application in rural contexts. Research by Phan (2024) supports the integration of experiential learning into higher education curricula, demonstrating its effectiveness in building problem-solving skills and cultivating social responsibility.

This study is theoretically grounded in Experiential Learning Theory (Kolb, 1984), which emphasizes learning through experience, reflection, and active experimentation. This framework aligns closely with the OVOC approach, where students solve real-world problems and apply academic knowledge in community settings. Additionally, Collaborative Governance Theory (Ansell & Gash, 2008) offers a relevant lens for understanding the multi-sector collaboration within OVOC, highlighting the importance of inclusive decision making, trust building, and shared goals in achieving sustainable outcomes.

Despite the significant strides made by initiatives such as OVOC, a critical research gap remains regarding the integration of social innovation frameworks into higher education curricula across the Asia–Pacific region. Existing studies tend to focus on isolated aspects of sustainable development, often overlooking the role of universities in developing scalable, community-driven innovation models. This study seeks to address this gap by examining OVOC as a social innovation model embedded within a higher education curriculum, emphasizing its contributions to sustainable rural development, student capacity building, and broader socio-economic impacts.

The primary objectives of this study are threefold: (1) to analyze the role of the OVOC program as a social innovation initiative integrated into higher education curricula for sustainable development; (2) to explore the role of multi-sectoral collaboration in building a sustainable rural business ecosystem; and (3) to evaluate the impacts and achievements of the OVOC program in contributing to the SDGs. By addressing these objectives, this research aims to offer valuable insights for policymakers, educators, and stakeholders on how higher education institutions can drive sustainable development through innovative, community-centered initiatives.

Ultimately, this study contributes not only to the theoretical discourse on social innovation in higher education but also provides practical implications for designing and implementing sustainability-oriented curricula aligned with the socio-economic and environmental goals of the Asia–Pacific region. The findings are expected to inform future research, policy frameworks, and collaborative strategies to advance sustainability in higher education institutions across the region.

2 Literature Review

The integration of sustainable development into higher education curricula has become increasingly vital, particularly in the Asia–Pacific region, where diverse environmental, social, and economic challenges persist. Recent studies emphasize the critical role of Higher Education Institutions (HEIs) in promoting sustainability through innovative curricular approaches and active community engagement. Scholars highlight the importance of culturally relevant sustainability education that reflects the region's varied socio-economic contexts (Leal Filho et al., 2019; Žalėnienė & Pereira, 2021). However, a persistent gap remains between theoretical knowledge and practical application, underscoring the need for educational approaches that bridge academic learning with real-world impact.

Experiential Learning Theory, proposed by Kolb (1984), offers a valuable framework for understanding how students can engage with sustainability challenges through direct experience. This theory emphasizes the significance of concrete experiences in the learning process, enabling students to reflect, conceptualize, and apply their knowledge in practical contexts. In sustainability education, experiential learning fosters critical thinking and problem-solving skills, allowing students to connect theory with practice. Wiek et al. (2011) advocate for experiential learning approaches that immerse students in sustainability issues—an especially relevant strategy in the Asia–Pacific, where complex environmental and socio-economic problems require localized, context-specific solutions.

Social innovation also emerges as a pivotal component of sustainability education, offering creative solutions to complex societal challenges. Murray et al. (2010) describe social innovation as a process that generates new ideas to meet social needs, enhancing capabilities and optimizing the use of resources. These practices are increasingly leveraged to address local needs and strengthen community resilience. A study by Cobo-Gómez (2024) highlights the critical role of universities in fostering social innovation through partnerships with local communities, which promote sustainable entrepreneurship and enhance community well-being. The study underscores the potential of universities as agents of transformation, providing opportunities for students, faculty, and community members to engage in meaningful, impactful initiatives.

Similarly, Juliani et al. (2017), through a Brazilian case study, emphasize the importance of integrating social innovation into university missions to advance the often-overlooked social dimension of sustainable development. The study calls for overcoming institutional barriers to address social exclusion and promote community development, thereby supporting a more comprehensive approach to sustainability. Numerous documented case studies further demonstrate how social innovation—facilitated by collaborations among universities, governments, and communities—has led to sustainable community development.

However, within the Asia–Pacific context, there remains a notable lack of research on how social innovation is embedded in university curricula. This study seeks to address this gap by examining the One Village One CEO (OVOC) program as a model for integrating social innovation into higher education, thereby enhancing students' capacity to contribute to sustainable development.

Integrating sustainability into curricula involves more than simply including relevant content; it also requires adopting pedagogical approaches that foster critical thinking and problem-solving skills. Kolb's Experiential Learning Theory (1984) offers a foundational framework for this integration, emphasizing the value of learning through experience and reflection. This approach is particularly pertinent in the Asia–Pacific, where distinct environmental and socio-economic challenges demand context-specific solutions. Sterling (2013) and Wiek et al. (2011) advocate for experiential learning opportunities that connect students with real-world sustainability issues—an approach highly relevant in this region. While there is a growing trend toward interdisciplinary curricula that embed sustainability across various disciplines (Barth, 2014), the effectiveness of these programs in cultivating sustainable development competencies remains underexplored. By examining the One Village One CEO (OVOC) initiative, this study contributes to a better understanding of how curriculum integration can be effectively operationalized to advance sustainability goals in higher education.

Governance Theory—particularly the collaborative governance model proposed by Ansell and Gash (2008)—emphasizes the importance of multi-stakeholder collaboration in achieving sustainable outcomes. This framework is especially significant in sustainability education, where partnerships among universities, governments, and communities can amplify the impact of academic initiatives. The OVOC program exemplifies such collaboration, fostering cross-sector partnerships that support sustainable community development while enriching students' educational experiences.

Despite the growing body of literature, several research gaps remain. First, there is limited empirical evidence on the impact of sustainability education initiatives on community development in the Asia–Pacific. This study addresses that gap by evaluating the outcomes of the OVOC program in Indonesia, providing insights into its effectiveness as a model for sustainable community engagement. Second, although social innovation is increasingly acknowledged as a catalyst for sustainability, its integration into higher education curricula is still under-documented. This research investigates how social innovation can be systematically embedded within university programs to enhance students' practical competencies and societal contributions. Finally, the thematic intersection of higher education, sustainability, and peace in the Asia–Pacific remains insufficiently explored. By situating this study within that framework, the research underscores the potential of educational initiatives to contribute not only to sustainability but also to regional stability and prosperity.

3 Method

In implementing the One Village One CEO (OVOC) program, it is essential to gather comprehensive insights into the effects and outcomes of similar initiatives within Indonesia. This study presents case studies that highlight both the successes and challenges encountered by communities striving to create employment opportunities and foster entrepreneurial activities—efforts that contribute to economic growth and revitalization in areas previously facing economic hardship. The selected cases were chosen based on their notable achievements and the availability of detailed post-implementation analyses.

Field studies, literature reviews, and in-depth interviews were conducted with key stakeholders involved in the program, including local leaders, program administrators, and community volunteers. Additional data were collected from regional development agencies and local government reports to provide a thorough overview of the implementation process.

This study also draws on documentary analysis and historical narratives surrounding the development of the OVOC initiative across various regions in Indonesia. A comparative method was adopted to critically examine the implementation of the program in different contexts, allowing for the identification of success factors as well as areas in need of improvement. The findings from this study, along with recommendations for future implementation, are presented to provide valuable insights for other communities considering similar strategies to promote local economic development and sustainability.

4 Results and Discussion

4.1 One Village One CEO as Social Innovation for Sustainability and the Curriculum

Merdeka Belajar Kampus Merdeka (MBKM) is a policy launched by the Ministry of Education and Culture that aims to encourage students to acquire a diverse set of knowledge and skills beneficial for entering the workforce. Through the Merdeka Belajar—Kampus Merdeka framework, students are given the opportunity to spend one semester—or the equivalent of 20 credits—studying outside their study program within the same university. One of the programs under this policy focuses on village development by involving multiple sectors and concentrating on one or two projects aimed at improving community welfare and enhancing the entrepreneurial capacity of communities, MSMEs, or village-owned enterprises (BUMDes).

The high number of underdeveloped villages in Indonesia remains one of the country's pressing challenges. Based on data from the Ministry of Villages as reported in the 2022 Developing Village Index (Indeks Desa Membangun or IDM), there are 9,220 underdeveloped villages and 4,366 highly underdeveloped villages. This condition is driven by persistent obstacles to local economic development in rural areas, such as the difficulty in identifying leading products for development, low production efficiency, limited added value due to the lack of innovation and appropriate technology, and inadequate access to capital, markets, and marketing strategies.

In response, the Village Project initiated by the Ministry of Education, Culture, Research, and Technology serves as a strategic program that can be synergized with the MBKM initiative. According to Ministerial Regulation Number 3 of 2020 on National Higher Education Standards, this program facilitates the enrichment, deepening, and expansion of student insight and competence through eight off-campus learning activities, including village projects.

In line with this, IPB University launched a social innovation initiative called MBKM Sociopreneur: One Village One CEO (OVOC), designed to foster village business ecosystem development. This program aligns with IPB's vision and mission to become a techno-socio-entrepreneurial university at the forefront of strengthening national dignity through globally excellent higher education in agriculture, maritime affairs, and tropical biosciences. The OVOC program aims to build self-sufficient and prosperous communities by integrating innovation and appropriate technology through a techno-sociopreneur approach. It involves mentoring students in villages for six months to one year.

In developing a sustainable village business ecosystem, the One Village One CEO program divides its approach into three stages: One Village One Product, One Village One Innovation, and One Village One Exporter (Slamet et al., 2024). The implementation of each stage adapts to the existing conditions of the village or local business ecosystem. The stages of the OVOC program are described in the Fig. 1:

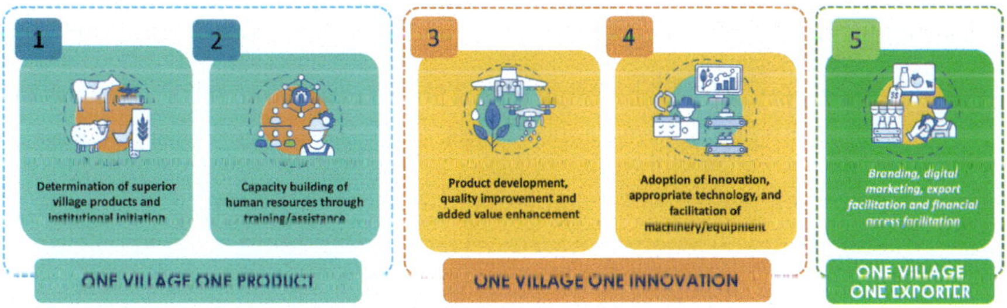

Fig. 1 Stages of the One Village One CEO program

a. One Village One Product

At this stage, each village identifies its superior products along with the local business units and stakeholders that will govern its economic ecosystem. All participating villages in this initiative have determined their stakeholders and leading products, facilitating the creation of a localized business ecosystem. The program's superior products are categorized into agriculture, animal husbandry, agricultural processing, fisheries, crafts, and carbon offset. By 2023, the program had supported 452 villages across Indonesia.

b. One Village One Innovation

This stage focuses on enhancing product quality and adding value through the application of appropriate technological innovations, equipment procurement, and product refinement based on higher predefined specifications. These technologies are introduced by experts, specifically lecturers from IPB University, as part of the institution's commitment to community engagement and knowledge dissemination. The innovation process must adopt a basic approach to ensure societal acceptance. The OVOC program requires that innovations be tailored to community needs to ensure accurate targeting. One such example is the coffee farming business model, which encompasses cultivation, harvesting, processing, and marketing. The direct application of such innovations has led to the development of multiple sustainability-oriented social and economic development curricula.

c. One Village One Exporter

This final stage focuses on digital branding and facilitating market access for products that have gained added value and long-term competitiveness in local, national, and international markets. One of the main challenges in this stage is integrating advanced IT systems with the unique characteristics and digital limitations of each village. Inequalities in digital access can be mitigated by forming community groups and holding regular meetings to ensure continuous information exchange, even in areas with limited connectivity. Despite these IT challenges, the OVOC program has successfully facilitated the export of several village-based products. Notable examples include Garut Indonesian Coffee (GIC), which has been exported to eight countries; Goathai, exported to eleven countries; as well as sorghum, its derivative products, and cowpeas. These successes indicate that food products remain in high demand and underscore Indonesia's potential as Asia's food basket, given its tropical climate and agricultural diversity.

OVOC represents a social innovation model aimed at developing village business ecosystems through a participatory approach and multi-sectoral collaboration. Participation is defined as an active process initiated by community members themselves, shaped by their own perspectives, and implemented through methods and mechanisms they can effectively control (Ansell & Gash, 2008). Multi-sectoral collaboration—often referred to as collaborative governance—is a governance strategy that involves multiple actors in formulating legal or regulatory frameworks and implementing public policy to achieve shared goals (Lima, 2021). In the context of OVOC,

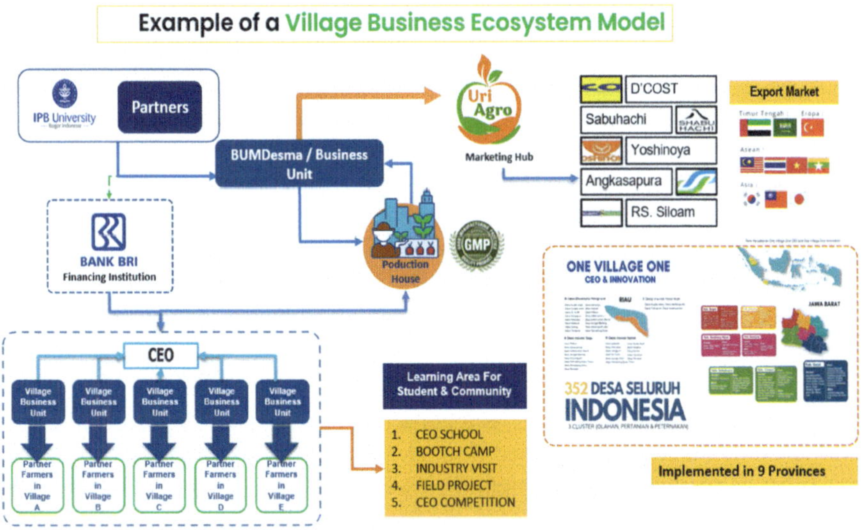

Fig. 2 Multi-sector Collaboration in Village Business Ecosystem Model

these sectors include expert lecturers, students, central and regional governments, industrial partners, and community-based organizations.

One example of a village business ecosystem model includes multiple assisted villages with similar superior commodities managed under a single CEO, supported by a centralized production house. This production house ensures product quality consistency by adhering to Good Manufacturing Practice (GMP) standards. The production house is managed by BUMDes (village-owned enterprises) as a formal business unit. Products are then connected to a marketing hub for distribution. Financing institutions provide essential business capital, while student involvement includes training and capacity-building to prepare them for the CEO role in nurturing the village business ecosystem (Fig. 2).

A more detailed overview of the roles of each multi-sector stakeholder in building a sustainable village business environment is provided below:

- Expert Lecturers: Deliver training and facilitate appropriate technology transfer to enhance product diversification and value addition.
- Students: Serve as CEOs who assist in managing the village business ecosystem and implementing context-relevant technologies.
- Central Government–Ministry: Supports the OVOC initiative through the Merdeka Belajar—Kampus Merdeka (MBKM) policy, provides funding via the Matching Fund scheme in collaboration with private sector partners, and offers data on potential partner villages.
- Regional Government: Endorses and facilitates program implementation through local funding and logistical support.

- Private sector: Acts as a program partner by offering financial support, supplying equipment, identifying partner villages, and serving as off-takers for high-quality village products.
- Export Financing Institution: Provide export training and facilitate the entry of superior products into international markets.
- Village-Level Business Units: Oversee the full cycle of business activities—aggregating superior commodities from farmers, processing them into value-added products, and marketing them.
- Financing Institutions: Offer credit and capital to businesses and partners in assisted villages.
- Community: Functions as the primary implementer of business activities across the value chain and the main beneficiary of the village business ecosystem.

The implementation of OVOC as a curriculum for sustainable development in rural communities can be assessed through the learning experiences of students who serve as program implementers (Wiek et al., 2011). The program serves as a learning medium that enables students to engage directly with societal issues. The syllabus is designed by aligning expected learning outcomes and instructional hours with the core characteristics of sociopreneurship. It refers to the Merdeka Belajar Kampus Merdeka (MBKM) curriculum launched by the Ministry of Education, Culture, Research, and Technology, as well as IPB University's 2020 curriculum (K2020), thereby ensuring its validity and recognition.

Student activities in the program include participant registration and selection, a grand launching event, CEO School, boot camps, industrial visits, field projects, reporting, and a final evaluation through the CEO Competition. Participation in these activities is equivalent to 20 academic credits.

a. CEO School

   The CEO School stage introduces the OVOC program and emphasizes the role of students in community empowerment, as well as the role of industry in strengthening villages and contributing to the SDGs. At this stage, students receive comprehensive training to build a village-based business ecosystem. Topics covered include business fundamentals, social mapping, business planning, technology adoption, business legality, negotiation, communication, BUMDes, SDGs, digital marketing, and export business development.

b. Bootcamp

   The MBKM One Village One CEO bootcamp enhances students' technical skills in entrepreneurship and community development. It also addresses social and economic goals by promoting local entrepreneurship and improving community capabilities. As part of practical preparation, students are equipped with business plans and the skills required to contribute meaningfully at the community level, making the bootcamp a vital step in nurturing future leaders and local change agents.

c. Industrial Visit

   Equipping students with knowledge and skills related to village-developed product commodities is essential to the success of the OVOC program. This is achieved through field visits to relevant agencies or institutions aligned with the commodities of the destination village. These industrial visits aim to provide students with practical, hands-on knowledge relevant to their assigned commodities.

d. Field Projects

During the Field Project stage, students are placed in partner villages to develop local business ecosystems. Activities include mapping local potentials, identifying stakeholders and community institutions, designing training and mentoring plans using design thinking, and implementing programs. Students receive support through mentoring by IPB experts, collaborative training, and ongoing guidance from both partners and the local community.

e. CEO Competition

The program concludes with the CEO Competition & Expo, which serves as both a final presentation and an evaluation platform. The competition includes presentations of field project results and a scientific poster session as a medium for student scientific communication. An innovation product expo is also held to showcase students' innovative products to the academic community and the general public.

4.2 *Impact of Implementation of the OVOC Program*

From 2019 to 2023, the OVOC program successfully engaged students, lecturers/experts, partners, and communities in numerous villages across several Indonesian provinces. In its implementation, the community acts not only as a beneficiary but also as a key driver in the development of each local business ecosystem. Starting from the village administration, the program empowers heads of households, housewives, and youth to participate actively in its execution. Table 1 presents the community's involvement and the progression of development programs over the years:

The data above illustrates the program's development from 2019 to 2023. The number of partners involved increased from 1 in 2019 to 5 in 2023. The program also expanded geographically, from 1 province in 2019 to 9 in 2022, although it

Table 1 OVOC's growth distribution from 2019 to 2023

Details	Years					Total
	2019	2020	2021	2022	2023	
Number of partners	1	1	2	5	5	5
Number of provinces	1	1	3	9	1	10
Number of villages	53	53	319	439	13	452
Number of MBKM students	0	0	103	279	50	329
Number of lecturers involved	5	5	10	59	19	78
Communities affected by the program	3.975	5.300	23.925	32.899	936	67.035

dropped back to 1 province in 2023. The number of participating villages grew from 53 in 2019 to 452 in 2023. Student participation in the MBKM program rose significantly from 103 in 2020 to 329 in 2023. With a total community outreach of 67,035 individuals, the data reflects the growing societal impact of the program.

OVOC's efforts to develop self-reliant and empowered villages through business ecosystems are guided by various concepts adapted to local needs and superior commodities (Irfany et al., 2024). In agriculture, key products include cantina papaya, low-glycemic rice, sacha inchi nuts, secondary crops, crystal guava, sweet potatoes and their derivatives, lemon, mangosteen, durian, avocado, and dragon fruit. In the livestock sector, the focus is on poultry, goats, cattle, and Kelulut honey. The fisheries sector includes catfish, tilapia, and gourami.

The development of business ecosystems extends beyond primary commodities to include derivative and supporting products, such as dried fruit chips, low-calorie salted eggs, husk charcoal planting media (Goathai), and sago noodles. All development efforts are tailored to the environmental, social, and economic conditions of each location. The following section provides an example of OVOC's successful implementation in developing a rural business ecosystem up to the export stage.

OVOC for Product Development of Cikajang Coffee as an Alternative Solution for Environmental Preservation

Cikajang, a sub-district in Garut Regency, frequently experiences flooding due to the conversion of forested areas into rice fields and residential zones—primarily driven by the economic needs of the local community. To address this issue, IPB University introduced the concept of agroforestry, promoting the use of economically valuable commodities that also contribute to ecological balance, such as coffee (Fig. 3).

OVOC contributes to the development of the coffee business ecosystem by engaging farmers, village-owned enterprises (BUMDes), and the surrounding community. Lecturers and experts deliver training and technology transfer to equip local residents with knowledge and skills in coffee cultivation, harvesting, post-harvest processing, and marketing—including export-oriented activities. Subsequently, students accompany the community in applying the delivered technology.

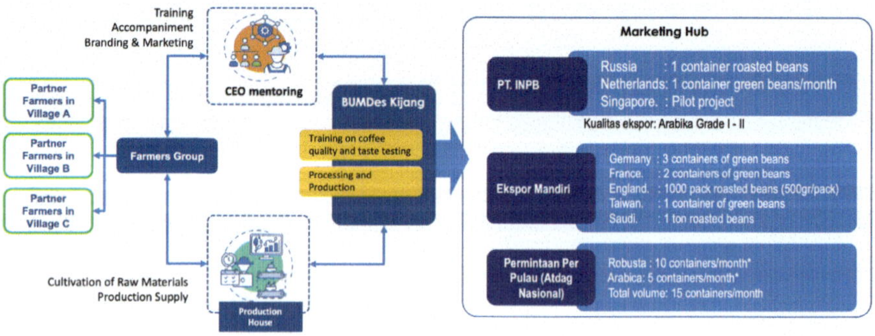

Fig. 3 Cikajang Coffee business ecosystem by OVOC (Slamet et al., 2024)

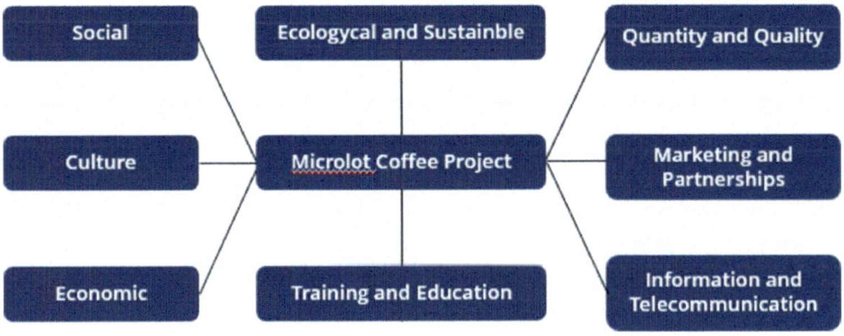

Fig. 4 Microlot Coffee Project

These training and mentoring initiatives are integrated into the Microlot Coffee Project (Fig. 4).

The Microlot Coffee Project is designed to enhance the quality and distinctiveness of coffee commodities while addressing social, cultural, economic, and environmental dimensions. Post-harvest training plays a pivotal role in ensuring coffee products meet market demands. The community receives training in coffee bean processing, wheel roasting, and flavor profiling. IPB University has also established a coffee school for local youth. In addition, the university collaborates with partner company PT Astra International to provide supporting equipment such as solar dryers, roasters, and other necessary tools.

Building partnerships and market networks is critical in the agroforestry business ecosystem, particularly within the coffee community. Cooperatives serve as intermediaries between local coffee producers and export markets, significantly enhancing the community's economic benefits. Through strategic partnerships, cooperatives help ensure market stability and compliance with export quality standards. Market networks also offer access to vital market information, technological advancements, and certification processes essential for entering global markets. The OVOC team supports quality testing and flavor profiling to obtain certification. Government and private sector involvement—such as that of PT Astra International Tbk—further expands global market access and provides financial support and production infrastructure.

These efforts have enabled Cikajang coffee to be exported to eight countries under three schemes: collaboration with PT Inovasi Produk Bangsa, independent exports, and Per Island Demand (National Atdag). This achievement is underpinned by the active participation of the Cikajang Coffee Community and close collaboration with academic experts.

The "One Village One CEO" program has had a positive socio-economic impact in the Cikajang Coffee Community. Monthly coffee production rose from 55 tons in 2020 to 70 tons in 2021, accompanied by an increase in the average monthly income from IDR 1,450,000 in 2019 to IDR 3,325,000 in 2021. The program also created 200 new jobs in 2020 and 250 in 2021, while community participation grew from 2,200 individuals in 2020 to 3,788 in 2021. Sustaining this progress requires robust strategies to ensure supply chain stability, maintain product quality, and safeguard farmers' interests amid future economic and environmental challenges.

Partnerships and Ecosystem Development in Airputih Village Through the OVOC Program

The implementation of the One Village One CEO (OVOC) program in Airputih Village, Bengkalis Island, serves as a compelling example of a successful, partnership-driven initiative fostering sustainable rural development. This collaboration brings together multiple stakeholders—IPB University, the Peatland and Mangrove Restoration Agency (BRGM), BUMDes Air Putih Jaya, and the local community—to address both socio-economic and environmental challenges.

The program centers on sago, a local and abundant commodity, as the foundation for a sustainable rural business ecosystem (Purwawangsa et al., 2024). Through OVOC, IPB University has played a central role in facilitating capacity-building activities, including technical training, innovation transfer, and mentoring programs involving faculty and students. One major achievement is the transition from producing wet sago noodles—with a shelf life of just three weeks—to instant sago noodles with a shelf life of up to three months. BRGM contributed significantly by providing vital production equipment, such as drying machines, which enhanced product durability and marketability—opening access to national and even international markets.

Key collaborative activities included establishing an efficient production process, introducing innovative packaging solutions, and leveraging digital marketing strategies to expand market reach. Farmers contributed raw materials by cultivating sago on peatlands, while BUMDes Air Putih Jaya handled production, branding, and marketing. BRGM's involvement ensured environmental sustainability by incorporating peatland rewetting and mangrove restoration into the business model, aligning with global climate action goals. These efforts not only preserved the local environment but also enriched the socio-economic fabric of the community by creating jobs and empowering farmers to become entrepreneurs. The OVOC program followed a structured approach through a three-phase development model: One Village One Product, One Village One Innovation, and One Village One Exporter. This model has proven effective in fostering local entrepreneurship and integrating rural businesses into global value chains. For instance, the production capacity of sago noodles increased tenfold, from 40 to 400 kg per day, while potential daily revenue rose from IDR 560,000 to IDR 16 million. This transformation has significantly improved the livelihoods of community members, demonstrating the program's success in linking education, innovation, and sustainability.

The implementation of One Village One CEO contributes to the achievement of several Sustainable Development Goals (SDGs). It supports efforts to eliminate poverty (SDG 1) by increasing community income through the development of the rural business ecosystem. In terms of food security, the program enhances people's ability and purchasing power for food (SDG 2) while ensuring sustainable food availability through community agricultural production. It also promotes decent work and economic growth (SDG 8) by creating new job opportunities in cultivation, processing, and marketing of local products. Furthermore, the program encourages innovation and infrastructure development (SDG 9) by facilitating the transfer of innovations from universities to communities, which in turn supports rural industrialization as both a producer and supplier of food raw materials. Responsible consumption and production are addressed through upstream-to-downstream program management (SDG 12), which helps reduce food loss in daily consumption. Environmental sustainability is promoted by integrating mangrove ecosystem approaches into economic development strategies that protect marine ecosystems (SDG 14). Lastly, the program strengthens global partnerships by fostering multi-sector collaboration among researchers, students, the private sector, communities, and media (SDG 17). These examples highlight the transformative role of partnerships and innovation in shaping a scalable and replicable model for sustainable rural development, offering valuable lessons for higher education institutions and policymakers across the Asia–Pacific region.

4.3 Achievements and Recognition of the OVOC Program

The OVOC program has achieved remarkable recognition, including winning 1st Place at the 2023 SDGs Action Award organized by the Ministry of National Development Planning of the Republic of Indonesia (Bappenas). In 2022, OVOC was selected as IPB University's delegate for the Bappenas SDGs Award. That same year, it was also named the winner of the KBA–DSA INNOVATION 2021 in the categories of DSA Coffee and Agriculture, Livestock, and Food Processing.

OVOC was recorded in the SDGs Accord in both 2021 and 2022 and marked a significant milestone in 2021 by exporting manure products from supported villages in Cianjur to 11 countries. In 2022, the program successfully exported coffee from supported villages in Garut to the Netherlands and several other European countries, with a transaction value reaching IDR 4 billion.

The program's sustainability is evident through active participation, knowledge transfer, and replication across multiple locations. Between 2018 and 2023, IPB University's various community empowerment programs have reached 4,579 villages (equivalent to 6.1% of all villages in Indonesia), with 452 villages directly engaged under the OVOC program. OVOC has successfully facilitated inclusive participation of rural business units and connected them with markets, local universities, local governments, and industries.

Developing a rural business ecosystem is crucial for ensuring the sustainability of these ventures. Business sustainability, in this context, refers to a condition in which businesses are able to maintain, develop, and safeguard their resources while meeting ongoing needs within their industries (Tutuba, 2024).

4.4 Implications for Higher Education Institutions, Policy, and Regional Relevance

The findings of this study highlight the essential role that higher education institutions (HEIs) play in advancing sustainable development through innovative, community-centered programs like One Village One CEO (OVOC). HEIs can serve as catalysts for change by incorporating experiential learning models into their curricula, equipping students with the skills needed to address complex socio-economic and environmental issues. The OVOC program exemplifies how academia can foster community-based solutions by bridging theoretical knowledge with practical applications in rural development.

For HEIs, integrating social innovation into the curriculum not only enhances students' critical thinking and problem-solving abilities but also nurtures a sense of global responsibility. This approach aligns with UNESCO's (2021) vision of educational institutions as hubs for sustainable development—institutions that generate knowledge while also creating direct, positive impacts in the communities they serve. The participatory model of OVOC demonstrates how universities can both develop leadership skills in students and empower local communities to meet Sustainable Development Goals (SDGs), particularly Goals 1 (No Poverty), 8 (Decent Work and Economic Growth), 9 (Industry, Innovation, and Infrastructure), and 17 (Partnerships for the Goals).

From a policy perspective, OVOC presents a replicable framework for promoting sustainable development through education. Its proven ability to address socio-economic and environmental challenges makes it a valuable reference for policymakers aiming to integrate sustainability and innovation into higher education. The Indonesian Ministry of National Development Planning has acknowledged the importance of such programs by supporting the establishment of 50 SDGs Centers across public and private universities. These centers reflect the government's commitment to embedding Science, Technology, and Innovation (STI) in the national agenda to meet the SDGs—especially those focused on quality education and innovation-driven growth.

Additionally, the study underscores the need for policymakers to invest in digital infrastructure and capacity-building programs in rural areas. Addressing digital inequality is crucial, as shown in the OVOC initiative, where digital tools and training enabled rural entrepreneurs to access wider markets. Embedding sustainable practices within rural business ecosystems—supported by clear, responsive policy frameworks—can foster long-term socio-economic resilience and environmental stewardship.

The OVOC model also reinforces the importance of community-driven innovation tailored to regional contexts. By adopting a multi-stakeholder approach, as advocated by Collaborative Governance Theory (Ansell & Gash, 2008), policymakers can create inclusive solutions that bridge sectors and scales. Replicating and adapting the OVOC approach to local realities can help address persistent disparities in rural development and accelerate national and regional progress toward the SDGs.

5 Conclusion

The One Village One CEO (OVOC) program represents a transformative model that integrates higher education, social innovation, and sustainable development. Grounded in experiential learning principles, the program equips students with practical skills while simultaneously addressing socio-economic and environmental challenges in rural Indonesia. Through its structured focus on product development, innovation, and market integration, OVOC has improved livelihoods, fostered entrepreneurship, and enhanced community resilience.

The findings underscore the program's significant contributions to several Sustainable Development Goals (SDGs), including (1) No Poverty, (2) Zero Hunger, (8) Decent Work and Economic Growth, (9) Industry, Innovation, and Infrastructure, (12) Responsible Consumption, (14) Life Below Water, and (17) Partnerships for the Goals. The program's multi-sectoral collaboration model illustrates the potential of aligning academic institutions, government agencies, private sectors, and local communities in pursuit of shared development objectives. For higher education institutions, OVOC highlights the value of embedding sustainability into curricula and encouraging student engagement with real-world challenges. For policymakers, it offers a replicable framework for advancing rural development through innovation and inclusive governance.

Future research could examine the long-term impacts of OVOC on rural economies and assess its scalability across diverse socio-economic contexts. Further studies may also quantify the environmental benefits of similar initiatives, particularly their role in mitigating climate change and promoting sustainable resource management. Despite its successes, the program faces ongoing challenges such as digital inequality and limited access to capital, which call for continuous adaptation and cross-sectoral support. Addressing these challenges can further amplify the program's impact, positioning OVOC as a model for sustainable development across the Asia–Pacific region and beyond.

References

Achmad, W., Nurwati, N., Fedryansyah, M., Sumadinata, R. W. S., & Sidiq, R. S. S. (2024). Taking Advantage of Indonesia's Demographic Bonus In 2024: Challenges And Opportunites Memanfaatkan Bonus Demografi Indonesia Di Tahun 2024: Tantangan Dan Peluang. *Management Studies and Entrepreneurship Journal*, *5*(2), 4425–4434. http://journal.yrpipku.com/index.php/msej

Ansell, C., & Gash, A. (2008). Collaborative Governance in Theory and Practice. *Journal of Public Administration Research and Theory*, *18*(4), 543–571. https://doi.org/10.1093/jopart/mum032

Barth, M. (2014). *Implementing Sustainability in Higher Education*. Routledge. https://doi.org/10.4324/9780203488355

Cobo-Gómez, J. C. (2024). Social innovation in university-community partnerships in Latin America: Exploring collaborative models. *Sustainable Technology and Entrepreneurship*, *3*(2), 100061. https://doi.org/10.1016/j.stae.2023.100061

Irfany, M. I., Purwawangsa, H., Slamet, A. S., Prawiro, B. P., & Haq, D. A. (2024). Sustainable Rural Business Ecosystem of Pineapple Commodity in Peatland and Mangrove Ecosystem with One Village One CEO (OVOC) Program. *IOP Conference Series: Earth and Environmental Science, 1359*(1), 012049. https://doi.org/10.1088/1755-1315/1359/1/012049

Juliani, D. P., Silva, A., Cunha, J., & Benneworth, P. (2017). Universities' Contributions to Sustainable Development's Social Challenge: A Case Study of a Social Innovation Practice. *International Journal of Social Ecology and Sustainable Development, 8*(3), 1–18. https://doi.org/10.4018/IJSESD.2017070101

Kamil, M., Lutfiansyach, D. Y., & Sukmana, C. (2019). Rural Youth Entrepreneurship Training Based on Local Potential. *Proceedings of the 2nd International Conference on Educational Sciences (ICES 2018).* https://doi.org/10.2991/ices-18.2019.30

Kolb, D. A. (1984). *Experiential Learning: Experience As The Source Of Learning And Development.* Prentice Hall. http://www.learningfromexperience.com/images/uploads/process-of-experiential-learning.pdf!

Land, H. van't, & Herzog, F. (2017). *HIGHER EDUCATION PAVING THE WAY TO SUSTAINABLE DEVELOPMENT: A GLOBAL PERSPECTIVE.*

Leal Filho, W., Vargas, V. R., Salvia, A. L., Brandli, L. L., Pallant, E., Klavins, M., Ray, S., Moggi, S., Maruna, M., Conticelli, E., Ayanore, M. A., Radovic, V., Gupta, B., Sen, S., Paço, A., Michalopoulou, E., Saikim, F. H., Koh, H. L., Frankenberger, F., … Vaccari, M. (2019). The role of higher education institutions in sustainability initiatives at the local level. *Journal of Cleaner Production, 233,* 1004–1015. https://doi.org/10.1016/j.jclepro.2019.06.059

Lima, V. (2021). *Collaborative Governance for Sustainable Development* (pp. 1–11). https://doi.org/10.1007/978-3-319-71066-2_2-1

Murray, R., Caulier-Grice, J., & Mulgan, G. (2010). *The open book of social innovation.* https://api.semanticscholar.org/CorpusID:204409614

OECD. (2023). *Enhancing Rural Innovation in the United States.* OECD. https://doi.org/10.1787/22a8261b-en

Phan, N. (2024). Bridging the Gap: A Framework for Experiential Learning in Higher Education. *Business and Professional Communication Quarterly.* https://doi.org/10.1177/23294906241292009

Purwawangsa, H., Irfany, M. I., Slamet, A. S., Rustiadi, E., Prawiro, B. P., & Haq, D. A. (2024). Sustainable Rural Business Ecosystem: Sago Noodles in Airputih Village, Bengkalis Island. *IOP Conference Series: Earth and Environmental Science, 1359*(1), 012054. https://doi.org/10.1088/1755-1315/1359/1/012054

Schlechtendahl, J., Keinert, M., Kretschmer, F., Lechler, A., & Verl, A. (2015). Making existing production systems Industry 4.0-ready. *Production Engineering, 9*(1), 143–148. https://doi.org/10.1007/s11740-014-0586-3

Schumann, F. R. (2016). *A Study of One Village One Product (OVOP) and Workforce Development: Lessons for Engaging Rural Communities around the World.*

SDSN Australia/Pacific. (2017). *Getting Started with The SDGs In Universities A Guide for Universities, Higher Education Institutions, and the Academic Sector. Australia, New Zealand and Pacific Edition.* Sustainable Development Solutions Network. www.acts.asn.au

Slamet, A. S., Purwawangsa, H., Prawiro, B. P., Isbayu, M., Irfany, M. I., & Haq, D. A. (2024). Community-based business ecosystem of coffee with the One Village One CEO Program at Cikajang Garut. *IOP Conference Series: Earth and Environmental Science, 1358*(1), 012040. https://doi.org/10.1088/1755-1315/1358/1/012040

Sterling, S. (2013). The Future Fit Framework: An Introductory Guide to Teaching and Learning for Sustainability in HE (Guide). *Journal of Education for Sustainable Development, 7*(1), 134–135. https://doi.org/10.1177/0973408213495614b

Tutuba, N. (2024). Perspective Chapter: Business Ecosystems – A Structure to Commercialize Value Chain of Rural Economies in Developing Areas. In *Rural Areas - Development and Transformations.* IntechOpen. https://doi.org/10.5772/intechopen.112259

Wiek, A., Withycombe, L., & Redman, C. L. (2011). Key competencies in sustainability: a reference framework for academic program development. *Sustainability Science, 6*(2), 203–218. https://doi.org/10.1007/s11625-011-0132-6

World Bank. (2023). *World Bank East Asia and Pacific Economic Update, April 2023: Reviving Growth*. The World Bank. https://doi.org/10.1596/978-1-4648-1983-4

Žalėnienė, I., & Pereira, P. (2021). Higher Education For Sustainability: A Global Perspective. *Geography and Sustainability, 2*(2), 99–106. https://doi.org/10.1016/j.geosus.2021.05.001

Mohammad Iqbal Irfany is a lecturer at the Faculty of Economics and Management, IPB University, and currently serves as the Assistant Director of Social Entrepreneurship at the Directorate of Agromaritime Community Development, IPB University. His research interests include development economics, finance, climate change, and rural industrialization. In 2023, he led the Matching Fund Program focused on applying integrated farming technologies to develop resilient and self-sustaining pesantren communities with sustainable economic and environmental outcomes.

Handian Purwawangsa is a lecturer at the Faculty of Forestry and Environment, IPB University, and serves as the Director of the Directorate of Agromaritime Community Development. He earned his Bachelor's degree in 2004, Master's degree in 2008, and Doctorate in 2022, all from the Faculty of Forestry, IPB University. He is one of the initiators of the "One Village One CEO" (OVOC) program. His expertise includes rural business ecosystems, forest policy, agroforestry, and carbon stock. Handian is actively involved in sustainable community development programs, including serving as Chair of the 2022 Matching Fund Program on business incubation and science and technology management in the peat and mangrove ecosystems of Riau, Kepulauan Riau, and South Sumatra provinces. In 2023, he led the Matching Fund Program on the adoption of appropriate technology to strengthen food system resilience and enhance community economic welfare. In 2024, he leads another Matching Fund Program focusing on innovation through value-added forest products to support sustainable village ecosystems.

Alim Setiawan Slamet is a lecturer at the Faculty of Economics and Management, IPB University, and currently serves as Rector of IPB University. He earned his Doctorate (S3) in Agribusiness and Agricultural Economics from Ehime University, Japan, in 2017, his Master's (S2) degree in Agricultural Industrial Technology from IPB University in 2009, and his Bachelor's (S1) degree in the same field in 2005. He is one of the initiators of the "One Village One CEO" (OVOC) program. His research interests include agricultural marketing, supply chain management, strategic management, operations research, and operations management. He has been actively involved in community development through his leadership and participation in several Kedaireka Matching Fund programs from 2021 to 2023. These programs have included collaborations with Astra International, the Peat and Mangrove Restoration Agency, and various universities to develop rural business ecosystems, support national food security, and promote green and blue economy initiatives.

Bintoro Pujo Prawiro is the Assistant Manager for Policy, Research, Innovation, and Monitoring at YBM BRILiaN. He earned his Bachelor's degree in Nutrition and Feed Technology from the Faculty of Animal Husbandry, IPB University, and his Master's degree in Rural Sociology from the Department of Communication and Community Development, Faculty of Human Ecology, IPB University. Previously, he served as a Community Development Officer at the Directorate of Agromaritime Community Development, IPB University. His experience includes serving as Faculty Advisor for IPB Outsco at the Future We Want Model United Nations in New York, USA, in 2020, and leading IPB's "One Village One CEO" program, which was recognized as a top public service innovation by the Ministry of Education and Culture in 2021. He has led various initiatives in rural business development and technological collaboration to promote sustainable community empowerment.

Daffa Aqomal Haq is a Master's student at the Department of Environmental and Resource Economics, Faculty of Economics and Management, IPB University. He earned his Bachelor's degree in Islamic Economics from the same faculty. He currently serves as a research assistant and community development officer at the Directorate of Agromaritime Community Development, IPB University. His academic and professional interests include econometric analysis, social return on investment, macroeconomics, resource economic valuation, and agribusiness.

Analysis of the Hiring Trends of Japanese Private Sector Companies Considering Gender and Educational Background in the SDG Era

Ludmila Soares Carneiro and Michael Henry

Abstract Sustainable Development is the focus of discussion in many aspects of society. The Sustainable Development Goals (SDGs) became effective in 2016, consisting of 17 goals to be achieved by 2030. This chapter explores hiring practices of Japanese private sector companies before and after the implementation of SDGs (2016 and 2021), to investigate the changes to the new global initiative. The hiring practices intersect with Sustainable Development Goal 8, aiming for inclusive economic growth and employment. However, in Japanese society gender disparities are also common in hiring process, with women often relegated to part-time roles due to familial responsibilities, also highlighting challenges in achieving SDG 5 on gender equality. Despite efforts, Japan faces significant challenges in meeting both SDGs. This chapter examines the hiring trends of young graduates in 2016 and 2022 to assess Japanese private sector hiring number related to gender and age, shedding light on progress towards these global goals of equal opportunities of employment. The results from the Japanese private companies investigated in this study shows that there is still a pattern of high percentage of males in the general and young workforce. The changes in percentage of young females hired from 2016 to 2022 were small when comparing both years (3.4%), which demonstrates low progress even after the implementation of the SDGs. The reasons could be pointed as a reflection of the corporate culture for hiring employees and a perception of women as the main carrier of household tasks.

Keywords Sustainable development goals · Hiring · Young employees · Private sector companies · Japan

L. S. Carneiro (✉) · M. Henry
Shibaura Institute of Technology, College of Engineering, Koto, Tokyo, Japan
e-mail: i046817@shibaura-it.ac.jp

W. Leal Filho et al. (eds.), *Sustainability in the Curriculum and Operations of Universities*, World Sustainability Series,
https://doi.org/10.1007/978-3-032-12678-8_30

1 Introduction

Education contributes to pave the way for advancing sustainable development (Berchin et al., 2021). The higher education institutions (HEIs) play a crucial role in shaping societies toward a more sustainable future by generating and sharing knowledge through research, education, and community engagement. The higher education institutions are responsible for ensuring academic and professional training, providing the human resources that contribute to support the society towards a more sustainable future (Berchin et al., 2021). Japan presents outstanding indicators for the percentage of the population that have received education, being considered a reference globally (OECD, 2024). In the Summit of the Future in 2024, Japan reinforced the commitment to contribute to international efforts for comprehensively achieve Sustainable Development Goals (SDGs) by the year 2030 (MFAJ, 2024–1). The country's leaders also promise to "invest in people," guided by the concept of human security. Prime Minister Kishida stated that empowering women, children and youth is a top priority (MFAJ, 2024–2). The empowerment of these groups includes providing means and opportunities for them to support themselves by increasing the chances of getting a job position.

In Japan, after concluding the compulsory education, students can choose the path of technical training or university, which contributes to prepare them to be a part of the workforce after graduation. Japan has developed a labor market for new graduates based on an institutionalized linkage between companies and schools, and the companies employ new graduates mostly from the beginning of April every year (Sugayama, 2011). This structure builds a continuous school-to-work transition (Fig. 1) that is considered a unique social custom in Japan (Sugayama, 2011). These peculiarities in transition from school to work are a reflection of the personnel management of post-war Japanese companies which also includes annual pay raises, regular staff reassignment, and seniority-based promotion (Tanaka, 1980). The same hiring culture applies to students who graduate from higher education institutions. The culture was actively started in 1870s by the large companies; however, it is still unclear when the recruitment style started and spread because, in the past, only a few graduates occupied positions in private companies due to the public sector presenting higher social status (Sugayama, 2011). The connection between educational institutions and companies has been established in Japan up to now as a system for the simultaneous hiring of new graduates from different educational backgrounds (Sugayama, 2011).

Japanese companies have also a conservative and strict selection process for hiring new employees. The selection of candidates involves long interview rounds that test the commitment of the candidate to the company, and the recruiters are actively looking for candidates who can be shaped by their policies and ideas to stay until retirement age. Nagano (2014) mentioned that the companies refer to young graduates as "white cloth," an analogy to exemplify that the newcomers are equivalent to white cloth that could be dyed any color, and are thus preferred by companies seeking to imprint their in-house standards and corporate culture. The young candidates

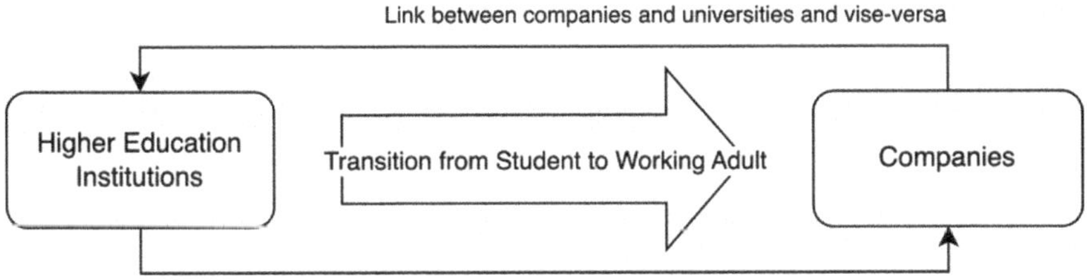

Fig. 1 Transition from higher education institutions to companies

are preferred by the companies, creating a stagnation of mid-career entrance of employees.

The hiring of new graduates or, young candidates, is related to SDG 8 which aims for inclusive and sustainable economic growth, full and productive employment, and decent work for all. From SDG 8, there are targets related to increasing productive employment and decent work for women and men, including younger people, and reducing the proportion of unemployed youth. Furthermore, the hiring opportunities for women and men in Japan are not the same, as a woman is more likely to work part-time because of the responsibility of taking care of the family (Brinton et al., 2008). The promotion of equality between women and men falls under SDG 5, which aims to reduce gender inequalities in society, including the work environment. The hiring trends of young employees from different educational backgrounds can thus indicate the current situation of Japanese private companies towards the achievement of SDG 5 and SDG 8. The global sustainable development agenda aims to enhance and secure decent job opportunities for all people. Gender and educational background can be considered important factors that influence the possibility of getting a job in Japanese society. However, in the SDGs progress report of 2022, Japan's achievement of SDG 5 was noted as "major challenges remain," and the achievement of SDG 8 was considered "significant challenges remain." There is also not enough information about the private sector performance in the SDGs era. Therefore, this research aims to better understand how Japanese private sector companies are hiring the young generation of employees.

The transition from student to employees can be influenced by many factors. Considering Japan, the path from schools to working force is influenced by local culture pattern. Japanese society have adopted the global initiative of SDGs so it is important to observe if the goals presented influence in the hiring of young graduates in 2016 and 2022. The progress can be tracked by analyzing the number of people hired by gender in the Japanese private sector companies. These years also correspond to the time immediately before and six years after implementation of SDGs internationally and, by observing trends in these years, it may be possible to gain insights if there were any significant change the hiring patterns in Japan in the SDGs era.

2 Bibliographic Review

2.1 Gender Roles and Corporate Culture in Japan

Education, at any level, must reinforce equity as well as the path after graduation, securing fair job opportunities for every individual. In Japan, aside from the good numbers on mandatory education level, there are some resistances to the engagement of woman in the higher education institutions and job opportunities. An article from The Guardian (2018) reported that one of Japan's most prestigious medical schools admitted deliberately changing the results of admission tests to restrict the number of female students and ensure more men become doctors. The news had international repercussion resulting in a public apology from the university and an investigation of the causes and those responsible. The proportion of females and males entering HEIs influences the posterior number of young graduates that will finish the courses and, consequently, the number of women that will compete for a job position in the same field of work.

Regarding job opportunities for females in Japan, Yamaguchi (2019) discusses in his work about gender segregation in the Japanese profession that a gender can reflects the hiring practices. It is because gender stereotypes create a lack of opportunity for women to go into professions other than those "suitable for women." The careers of women in Japanese society would be an extension of the traditional family roles such as children's education, nursing, and other supportive roles. The employers fail to recognize the workplace as not an extension of the home attributing supportive jobs to women. Yamaguchi (2019) also emphasizes that the workplace should be where individuals, independent of gender, can fulfill their potential and contribute to society instead of reflecting biased gender roles of society. There is also an effort of the Japanese government to promote equal salaries for workers, but it also have to be complemented by promoting occupational opportunities for men and women. The hiring process have been influencing the opportunities for women as also the work conditions. Japan's company culture is characterized by long working hours that are required from the employees. These conditions result in a poor of work-life balance. A better balance could enhance the number of employees and give better opportunities for women to join the workforce. Better flexibility in work would also encourage a change in society's culture that assumes child-care and home responsibilities as primarily woman roles (Yamaguchi, 2019).

According to OECD life index 2024, the percentage of employees in Japan that are working very long hours is 15.7%, higher than the OECD block average of 10% (OECD 2024). The ranking for time devoted to leisure and personal care, Japan occupies 37 between 41 countries. The balance between personal life and work also influences in the fertility rates because parents could find difficult to combine work and family commitments. The country presents s fertility rate of 1.42 children in 2014 and is considered to be among the "lowest-low" countries and that fact contributes to the decline in the Japanese population. Japan has important legal systems related to work-life balance including the child care and family care leave act, the Act on Advancement of Measures to Support Raising Next-Generation Children, and the Basic Act for Gender-Equal Society (Noda 2020). These acts help workers to achieve a better work life balance, however the act for gender equal society does not require directly that companies support their employees work life balance. It stills presenting a viewpoint that gender specific division of labor where men spend most time in the work and woman are responsible for the family (Noda 2020), reinforcing some gender bias. According to Noda (2020) mostly Female workers have taken advantage of the systems, however it does not stop the vicious cycle of work styles that reinforce gender-specific roles and the placing of the burden of family responsibility on women. The acts help to support the employees however, it is left to the individuals to decide using the benefits based on their personal preferences and culture.

2.2 *Japanese Commitment to Gender Equality and Decent Work for Youth in the SDGs Era*

The gender bias is still a part of the Japanese society being maintained historically, however the country has been continuously seeking to modernize. After a long period of being closed to other countries from the Meiji era, Japan has been aiming to modernize the traditional business practices and systems based in the western countries. The isolation of the country previously contributed to Japan creating its own way of doing business and interacting in society (Ueda 2015). This has influenced up to now in the Japanese corporate culture and many aspects of gender roles. However, over the years there have been some changes in the environment surrounding corporate culture in Japan. The main changes are related to the types of stakeholders within the Japanese market, and the new initiatives for Japanese markets to attract international investors being more diverse. This context also makes Japan very interested in global initiatives to adequate the Japanese corporative culture to international expectations such as Sustainable Development Goals (SDGs). This kind of initiative can decrease the isolation of the country, contributing to the attraction of international attention.

After the adoption of the SDGs by the UN Sustainable Development Summit in September 2015, Japanese government established in May 2016 the "SDGs Promotion Headquarters" (MOFA, 2016). The implementation of the goals was organized in a guiding that mention specific measures to achieve SDGs (MOFA, 2016). The document presented eight priority areas and policies, being gender equality and decent work for youth mentioned in the national plan in the priority area of empowerment of all people and third, for creating growth market, revitalization of rural areas, and promoting technical innovation.

The SDGs directly related to gender equality and hiring of young candidates are Goal 5 and 8. While SDG 8 mentions women's employment, there is also SDG 5 that mentions more specific indicators focusing on gender equality in general.

This goal complements the Japanese problem of the step forward of young women from education to the workforce. Specifically, Target 5.1 regards ending all forms of discrimination against all women and girls everywhere, that regard the issue of restricting the entrance of young females in HEIs or job opportunities based on gender criteria. Also, Target 5.4 states that it is important to recognize and value unpaid care and domestic work through the provision of public services, infrastructure, and social protection policies and the promotion of shared responsibility within the household and the family as nationally appropriate. The target agrees with the vision of a better work life balance, making possible to accommodate the day life routine with work for both genders, making use of support systems created by public institutions. In the SDGs progress report of 2022, Japan's achievement of SDG 5 was noted as "major challenges remain.

SDG 8 is related to promoting full and productive employment and decent work for all. Specifically, Target 8.5 states an objective by 2030 to achieve full and productive employment and decent work for all women and men, including for young people and people with disabilities, and equal pay for work of equal value. The report also mentions in the Target 8.6 that states to reduce the proportion of youth not in employment, education, or training (population from age of 15 to 29). This performance indicator is marked as already achieved by the report however, the result does not consider the gender of the young population observed. Japan shows in the SDGs report (former the SDG Index & Dashboards) from 2023 that, Goal 8 is moderately improving but still significant challenges remain. So, there are still gaps in the information needed to understand the progress of Japan towards SDGs 5 and 8.

3 Data Description and Methodology

The database used in this study was published by Toyo Keizai Inc., in the years 2016 and 2022, corresponding to surveys on CSR in the Japanese private sector conducted in 2015 and 2021, respectively. The major features of the database from Toyo Keizai are the extensive range of item fields (approximately 1000) and company coverage (approximately 1700), together with the rich historical data (Toyo Keizai, 2024). The data is used not only by business companies to analyze the initiatives of their competitors, but also for screening investment targets as well as extracting ESG factors for use in ESG investment. The use of this data for this research contribute to the understanding of the private sector companies in Japan due to the range of respondents.

The questionnaire survey used for the present study was carried with 1325 respondents' companies for 2016 and 1631 of 2022. Among the respondents, 981 companies answered both years of the survey, being than selected for this study. The questions that become indicators for the analysis were part of the Human Resources database. The indicators were divided into three categories for this study: (1) assessment of general profile of Japanese private sector companies regarding age and gender of employees by year of investigation (companies age overview), (2) the number of young employees (< 30 years old) being hired in the first trimester of the correspondent year by gender, and (3) number of people hired from educational background of university graduates and above only by gender (a) and in general (b). Due to data availability, gender is not discussed in the graduate students for the category 3(b).

Considering the indicators selected for the analysis:

1) The assessment of average age and gender of employees by year of investigation contribute to understand the general picture of the companies that are in the questionnaire survey. By knowing this data, it is possible to draw conclusions about the entrance of young graduates in the workforce using the changes in average age of employees by industry type. It can also be seen the sectors that are more attractive to young employees for both genders by analyzing if the average age increased or decreased over the time, meaning that more young people joined the company.
2) The second indicator of number of young employees (< 30 years old) being hired in the first trimester of the correspondent year by gender, correspond to a close look in the younger portion of the employees only, giving more specific analysis of the dataset.
3) The academic background from Higher Educational Institutions (HEIs) is presented in this indictor. The number of people hired by university graduates and above gives the information about how many employees were hired from HEIs for each industrial sector (a). The number of people hired from HEIs are so compared to the Junior College and High school graduates in section "b" to enhance the analysis of young employees that joined the companies of each industrial sector. Due to data availability, gender is not discussed in the graduate students for the category 3b.

The profile of the selected dataset of companies used for this study can be seen on Table 1. The table shows the specific number and percentage of companies' that answered both years of the questionnaire survey by industry type. The percentage of companies by industrial sector is presented as a heat map with gradient green color that represents the variation on relative to the total sum of companies (Grand Total = 981, or 100%).

The industrial sectors with larger number of companies are Wholesale_Trade (9.17%), Chemistry (9.38%), and Eletricl_Equi (9.79%). The smallest are Mining (0.20%), Air_Transp (0.31%) and Petro_Coal_Prod (0.31%). The size of the companies was not considered for this analysis. The results for age are shown in the next sections as average and the number of people in percentage related the total number of people hired. The data were evaluated by using cross-tabulation analysis observing the variations in information for each year by selected category and presented in tables and bar charts.

Table 1 Number of companies by industry type in alphabetical order

Nº	Industry Type	Number of Companies	%	Nº	Industry Type	Number of Companies	%
1	Air_Transp	3	0.31%	18	Other_FinancServ	11	1.12%
2	Banking	32	3.26%	19	Other_Prod	32	3.26%
3	Chemistry	92	9.38%	20	Petro_Coal_Prod	3	0.31%
4	Constr_IND	51	5.20%	21	Pharma	23	2.34%
5	Eletricl_Equi	96	9.79%	22	Precis_Mech_Equip	13	1.33%
6	Eletrici_Gas_IND	10	1.02%	23	Pulp_Paper	6	0.61%
7	Fiber_Prod	21	2.14%	24	Real_State	24	2.45%
8	Fish_Agro_Forestry	4	0.41%	25	Retail_Businn	65	6.63%
9	Glass_Clay_Prod	10	1.02%	26	Rubber_Prod	12	1.22%
10	Groceries	48	4.89%	27	Secur_Commodities	8	0.82%
11	Info_Commu_IND	65	6.63%	28	Serv_IND	68	6.93%
12	Insur_Businn	11	1.12%	29	Shipping_IND	4	0.41%
13	Land_Transp	19	1.94%	30	Steel	12	1.22%
14	Machine	60	6.12%	31	Transp_Equi	45	4.59%
15	Metal_Prod	19	1.94%	32	Warehou_Transp	9	0.92%
16	Mining	2	0.20%	33	Wholesale_Trade	90	9.17%
17	Non_FerreMetal	13	1.33%		**Grand Total**	**981**	**100.00%**

Color legend of heatmap: Green: higher values; White: lower values

4 Results and Discussion

4.1 Profile of Japanese Workforce for 2016 and 2022—General Results for Age/Gender

The first result of this study shows the average age of the employees of Japanese Private Sector companies by gender (Table 2). The results are separated in average age of "All employees," "Male," and "Female" by year. Also, the average age of employees for all industries can be seen in the last row. Observing the average age of employees in both 2016 and 2022 contributes to observe trends in the entrance of young workforce in the Japanese Private Sector Companies. From Table 2, the average age for all industries increased from 2016 to 2022 for all results (all employees, male and female)

.

The general range of average age for 2016 results for all industries are from 38.2 years to 43.2 years old. The industries with lower average age are Shipping_IND (38.6), Serv_IND (38.3) and Info_Commu_IND (38.2). The highest average ages for the same year are Insur_Businn (43.2), Constr_IND (42.7) and Eletrici_Gas_IND (42.6). However, for 2022 results the average age of employees by industry become higher, ranging from 38.5 to 43.9 years old. The industrial sectors with younger employees changed to Serv_IND (39.5), Steel (39.5) and Shipping_IND (38.5). The Steel Industry employees average age changed from 40.2 to 39.5 from 2016 to 2022. The result means that young employees have had joined the company in the time interval, moving the average age down. On the other hand, for 2022 the group of three higher average age also changed. Insur_Businn maintained the position as industrial type with older average age employees (43.9), but Non_FerreMetal (43.9) and Eletricl_Equi (43.1) substituted Constr_IND (43.0) and Eletric_Gas_IND (42.7)

Table 2 General average age distribution of industry type by gender for 2016 and 2022

Industry Type	2016			2022		
	All Employees	Male	Female	All Employees	Male	Female
Air_Transp	40.9	45.3	36.0	40.6	44.9	36.4
Banking	39.5	41.9	35.8	40.1	41.9	37.3
Chemistry	40.7	41.5	38.6	41.6	42.1	39.7
Constr_IND	42.7	43.1	38.6	43.0	43.7	39.2
Eletricl_Equi	41.9	42.5	39.6	43.1	43.7	41.2
Eletrici_Gas_IND	42.6	42.6	39.8	42.7	43.1	40.6
Fiber_Prod	41.0	41.6	39.9	42.0	42.8	41.2
Fish_Agro_Forestry	41.0	41.3	38.7	41.6	42.9	39.7
Glass_Clay_Prod	41.0	41.4	39.8	41.9	42.3	39.6
Groceries	40.8	42.0	37.5	41.1	42.4	38.5
Info_Commu_IND	38.2	39.3	35.4	40.0	41.4	36.6
Insur_Businn	43.2	43.8	41.9	43.9	45.3	42.9
Land_Transp	42.2	42.7	36.1	42.7	44.1	37.7
Machine	41.2	41.7	38.9	41.7	42.4	40.4
Metal_Prod	41.0	41.9	38.5	41.9	42.8	39.6
Mining	39.7	40.6	35.9	40.4	41.4	38.2
Non_FerreMetal	42.1	42.3	39.8	43.9	42.8	40.0
Other_FinancServ	39.3	41.5	36.2	41.6	43.1	39.4
Other_Prod	40.8	42.1	36.2	42.1	43.4	38.2
Petro_Coal_Prod	42.3	42.5	39.9	42.7	43.1	41.1
Pharma	41.4	42.3	38.6	42.4	43.2	38.0
Precis_Mech_Equip	41.0	42.2	38.2	42.5	42.9	39.4
Pulp_Paper	40.3	41.5	36.7	41.0	42.3	38.0
Real_State	40.0	41.2	35.7	40.9	41.8	37.1
Retail_Businn	38.8	41.2	36.4	41.5	42.9	37.5
Rubber_Prod	40.1	39.9	38.7	40.7	40.7	39.6
Secur_Commodities	41.1	41.8	37.7	42.9	42.6	39.7
Serv_IND	38.3	38.9	35.2	39.5	41.4	36.4
Shipping_IND	38.6	38.6	40.3	38.5	38.2	41.0
Steel	40.2	40.2	40.1	39.5	39.6	38.5
Transp_Equi	39.8	40.1	36.8	40.6	40.9	37.8
Warehou_Transp	38.7	42.0	36.2	39.6	42.2	38.8
Wholesale_Trade	40.6	41.4	36.3	41.6	43.1	37.8
All Industries	**40.5**	**41.5**	**37.6**	**41.5**	**42.6**	**38.8**

Legend: All values are in years; Heatmap color scale: Blue: lower values; Red: higher values; White: middle values.

as the sectors with older employees. The result can be influenced by different factors, such as more hiring of middle carrier professionals or decrease in the number of hiring of young employees. Both scenarios represent a decrease in young graduates entering in job positions in these industrial sectors. The results also emphasize the aging problem of the Japanese population indicating that the workforce is getting older and fewer young people are joining companies, increasing the average age of the employees as an effect.

Another noticeable result from Table 2 is the difference in average age between male and female employees. Female employees show average age lower than male for the majority of the industrial types for both years of the study. However, Shipping_IND is the only sector that showed opposite behavior for both years. The average age for this industry in 2016 is 40.3 for female and 38.6 for male. In 2022 the industrial sector presented 41.0 for female and 38.2 for male employees. Shipping industry is related to transportation of goods via sea routes, the sectors is critical for global trade being one of the most significant sectors in Japan. In the dataset, Shipping industry correspond to 4 companies being 0.41% of the dataset information, however the number of workers in the companies was not considered for this analysis. The high average age for woman means that young females are not interested to join the companies in this field, however young male are joining the companies in more number, dropping the average age of the male employees.

The other 32 industrial sectors showed the age of men higher than woman. The age of males varies between 38.6 to 45.3 for 2016, and 38.2 to 45.3 for 2022. The sectors with higher average age of males for both years of 2016 and 2022 are Air_Transp, Constr_IND, Insur_Businn, and Land_Transp. The industries with younger average age in the sample for 2016 are Info_Commu_IND (39.3), Serv_IND (38.9) and Shipping_IND (38.6). The result for 2022 also shows Shipping_IND (38.2) as lower average age however, together with Rubber_Pro (40.7) and Steel (39.6).

The range of average age of woman in the industries are from 35.2 to 41.9 years old in 2016, and 36.4 to 42.9 years old in 2022. The sectors with highest values for 2016 are Insur_Businn (41.9), Shipping_IND (40.3) and Steel (40.1). In 2022, Insur_Businn (42.9) still have the higher result however, now together with Fiber_Prod (41.2) and Eletricl_Equi (41.2). The lowest values for average age of females in 2016 are Real_State (35.7), Info_Commu_IND (35.4) and Serv_IND (35.2), and for 2022 are Info_Commu_IND (36.6), Serv_IND (36.4) and Air_Transp (36.4). The lowest value for Information and Communication Industry in both years indicates that a high number of young women may have enter in this field, decreasing the average age for female employees. The sector includes telecommunications, software, and IT services that are critical for digital transformation across other industries. The sector has been presenting a rapid grow and in Japan, Information and Communication Industry focusses on mobile technologies and 5G networks.

Table 3 shows as a heat map with the differences in average age by industry type between 2016 and 2022. The blue color indicates that the average age of the employees from 2022 is lower than from 2016 which results in a negative answer demonstrating that, on average, the group got younger. The red color indicates that the difference in average age from 2022 to 2016 is positive, so the number in the later year is higher than the earlier. The red color indicates that the group average age increased from 2016 to 2022 and the average age of the members became older

.

It can be noticed from the table that only a few industrial sectors experienced a decrease in the average employee age for all groups of this study. The "All Industries" average number for "All Employees" shows an increase in 1 year older from 2016 to 2022. The males average age increased by 1.1 years and the female by 1.2. Considering "All Employees" by industry type, only the industrial sectors of Shipping (−0.1), Air_Transp (−0.3), and Steel (−0.7) showed reduction in the average age. The Steel industry showed the highest negative variation with a decrease in the overall average age of all employees by 0.7 years. The other 30 industrial sectors presented results of average general age in 2022 as higher than in 2016. The industries that showed higher increase in average age were Retail_Businn (2.8) and Other_FinancServ (2.3).

The changes in average age of male employees by industry shows that 29 industries increased, being Serv_IND (2.5) and Info_Commu_IND (2.1) the highest values from

Table 3 Changes in the age of employees from 2016 to 2022

	Industry Type	All Employees	Male	Female
1	Air_Transp	-0.3	-0.5	0.3
2	Banking	0.6	0.0	1.4
3	Chemistry	0.9	0.6	1.1
4	Constr_IND	0.3	0.6	0.6
5	Eletricl_Equi	1.2	1.2	1.7
6	Eletricl_Gas_IND	0.2	0.5	0.8
7	Fiber_Prod	0.9	1.2	1.4
8	Fish_Agro_Forestry	0.6	1.6	1.0
9	Glass_Clay_Prod	1.0	0.9	-0.2
10	Groceries	0.3	0.5	0.9
11	Info_Commu_IND	1.8	2.1	1.2
12	Insur_Businn	0.7	1.5	1.0
13	Land_Transp	0.4	1.4	1.6
14	Machine	0.4	0.7	1.4
15	Metal_Prod	0.8	0.9	1.1
16	Mining	0.8	0.8	2.4
17	Non_FerreMetal	1.8	0.5	0.3
18	Other_FinancServ	2.3	1.6	3.2
19	Other_Prod	1.4	1.2	2.0
20	Petro_Coal_Prod	0.4	0.6	1.2
21	Pharma	1.0	0.9	-0.6
22	Precis_Mech_Equip	1.5	0.7	1.2
23	Pulp_Paper	0.7	0.7	1.3
24	Real_State	0.9	0.6	1.4
25	Retail_Businn	2.8	1.7	1.1
26	Rubber_Prod	0.6	0.7	0.9
27	Secur_Commodities	1.8	0.8	2.0
28	Serv_IND	1.2	2.5	1.2
29	Shipping_IND	-0.1	-0.4	0.7
30	Steel	-0.7	-0.6	-1.6
31	Transp_Equi	0.8	0.9	1.0
32	Warehou_Transp	0.9	0.2	2.6
33	Wholesale_Trade	0.9	1.7	1.5
	All Industries	**1.0**	**1.1**	**1.2**

Legend: All values are in years; Heatmap color scale: Blue: lower values; Red: higher values; White: middle values

2016 to 2022. Banking did not present variation of average age in the same period, and Shipping_IND (−0.1), Air_Transp (−0.3), and Steel (−0.7) presented negative values that means average age for male employees decreased.

The results for female employees show that in general the average age for All Industries increased by 1.2 years old. The individual results by industry reveals that Glass_Clay_Prod (−0.2), Pharma (−0.6), and Steel (−1.6) sectors decrease the average age of female employees. However, the other 30 industries presented average age increase. Among them, Other_FinancServ (3.2), Warehou_Transp (2.6) and Mining (2.4) showed the highest values.

After analyzing the average age of the employees, it is also possible to observe the proportion of employees in the Japanese Private Sector companies of the dataset by gender. The results can be seen on Table 4. The distribution of the heatmap can be interpreted as higher colored in dark blue, lower in red and the numbers close to 50% in white. In general, the All Industries results shows that the number of males in the companies is larger than female for both years of the research. The year of 2016 reveals that 74.3% are male and 25.7% of females, and for the year of 2022 males represent 70.6% and females are 29.4%. However, considering the results for industry types, some sectors demonstrate opposite behavior containing more females than males. The results for 2016 show that two for Air_Transp and Insur_Businn, females workers correspond to 55.9% and 78.7%, respectively. For the year of 2022, the number of industries with higher percentage of females increased to five. Other_FincServ (50.4%), Banking (50.8%), Serv_IND (52.2%) are also included together with Air_Transp (58.6%) and Insur_Businn (81.2%) as having more female than male. However, the overall results for All Industries, indicate that around one-fourth of the percentage of total employees were women in the 981 companies in this study for both years

.

In order to better understand the changes in number of female and male workers, Fig. 2 shows a closer view of the changes from 2015 to 2022, focusing in the percentage of female employees. The percentage of female employees increased for 32 industry types with the highest value being the Fishing, Agriculture, and Forest Industry (27.8%), decreasing only for the Land and Transportation industry by 8.6% when comparing the year 2022 to 2016. The overall result shows an increase of 3.7%

Table 4 Percentage of employees by gender

Industry Type	2016		2022	
	Male	Female	Male	Female
Air_Transp	44.1%	55.9%	41.4%	58.6%
Banking	54.9%	45.1%	49.2%	50.8%
Chemistry	81.8%	18.2%	81.8%	18.2%
Constr_IND	87.3%	12.7%	84.8%	15.2%
Eletricl_Equi	84.7%	15.3%	82.2%	17.8%
Eletrici_Gas_IND	89.8%	10.2%	87.6%	12.4%
Fiber_Prod	67.5%	32.5%	61.8%	38.2%
Fish_Agro_Forestry	86.0%	14.0%	58.2%	41.8%
Glass_Clay_Prod	84.2%	15.8%	78.0%	22.0%
Groceries	78.5%	21.5%	77.1%	22.9%
Info_Commu_IND	79.6%	20.4%	74.2%	25.8%
Insur_Businn	21.3%	78.7%	18.8%	81.2%
Land_Transp	76.1%	23.9%	84.7%	15.3%
Machine	88.9%	11.1%	87.8%	12.2%
Metal_Prod	80.0%	20.0%	79.2%	20.8%
Mining	83.8%	16.2%	80.9%	19.1%
Non_FerreMetal	88.6%	11.4%	86.6%	13.4%
Other_FinancServ	57.5%	42.5%	49.6%	50.4%
Other_Prod	79.3%	20.7%	71.4%	28.6%
Petro_Coal_Prod	90.5%	9.5%	86.0%	14.0%
Pharma	77.4%	22.6%	75.3%	24.7%
Precis_Mech_Equip	85.2%	14.8%	79.1%	20.9%
Pulp_Paper	91.9%	8.1%	89.8%	10.2%
Real_State	72.1%	27.9%	65.2%	34.8%
Retail_Businn	68.3%	31.7%	59.7%	40.3%
Rubber_Prod	91.5%	8.5%	89.9%	10.1%
Secur_Commodities	60.7%	39.3%	57.7%	42.3%
Serv_IND	62.4%	37.6%	47.8%	52.2%
Shipping_IND	76.5%	23.5%	76.4%	23.6%
Steel	92.7%	7.3%	90.1%	9.9%
Transp_Equi	90.6%	9.4%	89.1%	10.9%
Warehou_Transp	70.0%	30.0%	65.7%	34.3%
Wholesale_Trade	76.8%	23.2%	73.1%	26.9%
All Industries	**74.3%**	**25.7%**	**70.6%**	**29.4%**

Legend: Heatmap color scale: Blue: higher values; Red: lower values; White: middle values (50%)

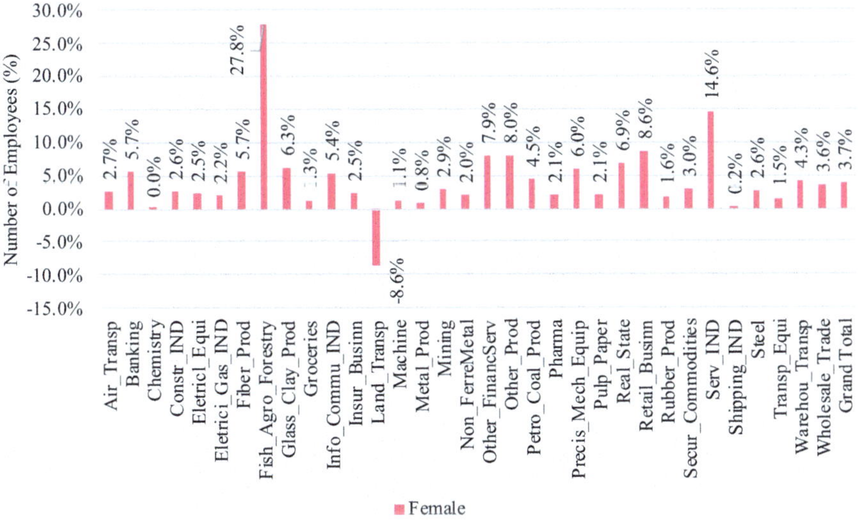

Fig. 2 Difference in the average number of female employees from 2016 to 2022

in percentage of women in the Japanese private sector companies investigated in this study.

As this part of the study considered the age and percentage for the total number of employees by gender, it is possible to visualize the overall image of the 981 private sector companies. From age perspective, the male group is older than females, also females represent around 25% of the employees in 2016 and 30% in 2022. The females are in low percentage meaning that there is a gender unbalance in the overall results for All Industries. Also, the age issue contributes to the continuation of the patterns in hiring process and genders bias, repeating old patters. As young and diverse the group become, higher the possibility that the innovative ideas and concepts are going to be part in the corporate culture of Japan.

4.2 *Profile of Young Employees in Japanese Workforce by Gender*

After observing the general profile of the industrial sectors in the previous section, a specific analysis of the employees under 30 years old by gender is also presented in this study. The distribution of percentage of workers between females and males under 30 years old by industry can be seen in Table 5. The color scale indicates the highest numbers as dark blue and the lowest as red.

The results for employees under 30 years old shows a similar pattern of the previous results of general workers (Table 4). The All Industries results for employees under 30 years old show on Table 5 that there are more male than female employees

Table 5 Percentage of employees under 30 years old by gender

Industry Type	2016		2022	
	Male	Female	Male	Female
Air_Transp	13.6%	86.4%	14.3%	85.7%
Banking	37.7%	62.3%	38.9%	61.1%
Chemistry	76.7%	23.3%	79.1%	20.9%
Constr_IND	81.6%	18.4%	78.8%	21.2%
Eletricl_Equi	80.7%	19.3%	78.0%	22.0%
Eletrici_Gas_IND	86.8%	13.2%	83.5%	16.5%
Fiber_Prod	64.4%	35.6%	57.0%	43.0%
Fish_Agro_Forestry	71.2%	28.8%	68.0%	32.0%
Glass_Clay_Prod	83.0%	17.0%	74.9%	25.1%
Groceries	68.7%	31.3%	66.1%	33.9%
Info_Commu_IND	67.2%	32.8%	64.5%	35.5%
Insur_Businn	18.2%	81.8%	17.0%	83.0%
Land_Transp	78.8%	21.2%	71.3%	28.7%
Machine	87.3%	12.7%	83.8%	16.2%
Metal_Prod	72.7%	27.3%	77.0%	23.0%
Mining	79.7%	20.3%	76.4%	23.6%
Non_FerreMetal	87.2%	12.8%	81.4%	18.6%
Other_FinancServ	42.4%	57.6%	41.8%	58.2%
Other_Prod	63.0%	37.0%	54.7%	45.3%
Petro_Coal_Prod	91.3%	8.7%	89.2%	10.8%
Pharma	66.0%	34.0%	60.4%	39.6%
Precis_Mech_Equip	76.7%	23.3%	68.7%	31.3%
Pulp_Paper	90.7%	9.3%	85.6%	14.4%
Real_State	57.3%	42.7%	52.1%	47.9%
Retail_Businn	55.4%	44.6%	46.1%	53.9%
Rubber_Prod	90.5%	9.5%	88.7%	11.3%
Secur_Commodities	50.5%	49.5%	53.8%	46.2%
Serv_IND	48.2%	51.8%	37.2%	62.8%
Shipping_IND	81.1%	18.9%	80.8%	19.2%
Steel	92.9%	7.1%	85.3%	14.7%
Transp_Equi	85.1%	14.9%	83.8%	16.2%
Warehou_Transp	61.6%	38.4%	48.6%	51.4%
Wholesale_Trade	64.2%	35.8%	60.5%	39.5%
All Industries	**67.3%**	**32.7%**	**63.7%**	**36.3%**

Legend: Heatmap color scale: Blue: higher values; Red: lower values; White: middle values (50%)

in general. However, among the young, the percentage of female are larger when compared to the overall of all employees in the sectors of Table 4. The result of Table 5 for 2016 indicates that there are 67.3% of males and 31.7% of female. In 2022 the number increased to 36.3% of female. Also, for 2016 five industrial sectors show the percentage of young female as higher than males (Serv_IND, Other_FinancServ, Banking, Insur_Businn and Air_Transp), increasing to seven on 2022 with the inclusion of Warehou_Transp and Retail_Businn. The Air transportation industry showed a high percentage of women with 86.4% in 2016 and 85.7% in 2022 that could be influenced by females' members in the cabin crew as the sector includes the passengers and cargo services. The industrial sectors with the smallest percentage of females are Steel (7.1%) in 2016 and Petro_Coal_Prod (10.8%) in 2022.

The difference between 2016 and 2021 percentage of female workers under 30 years old by industry can be seen in Fig. 3. Among the industry groups, Air transportation, Banking, Chemistry, Metal Production, Security, and commodities presented a decrease in the number of young women in the workforce even with some of them showing high numbers in the previous results. The other groups presented a

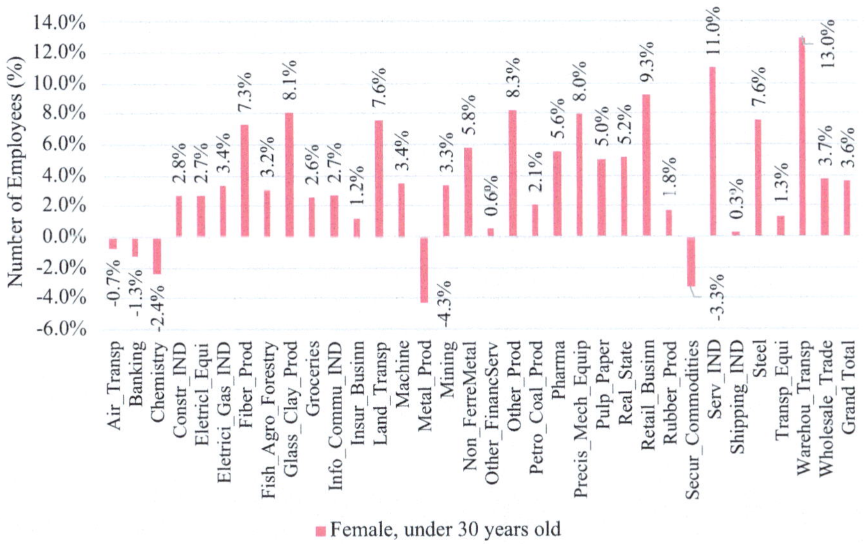

Fig. 3 Difference in percentage of female employees under 30 years old from 2016 to 2022

tendency to increase the number of female workers from 2016 to 2022. The average of the total result shows an increase of 3.6% for females under 30 years old.

The hiring of young women reflects a concern from the Japanese private sector to welcoming this category into their workforce providing policies that incentivize the numbers to grow. The cultural understanding is that young women could be inclined to get in charge of the family duties and raise the children, making it a consideration that discourages the investment of companies in young women hiring in many sectors. The gap between the percentage of female and male employees are shown in the overall.

4.3 *Employment of Workers by Education Background*

(a) University Graduate or Above by Gender

The first group considered for educational background are the university graduates and above by gender and the results can be seen in Table 6. The heatmap indicates the highest numbers in dark blue and the lowest in red. The university graduates demonstrate the majority of males hired by the private companies in the dataset, increasing from 62.8% in 2016 to 63.4% in 2022. The number of females was 37.2% in 2016 and decreased to 36.6% in 2022.

The industrial types with the largest percentage of female hired from university and above were four in 2016, the sectors are Retail_Businn (50.3%), Serv_IND (53.6%), Banking (61.5%), and Insur_Businn (76.3%), however the number decreased to

Table 6 Percentage of employees from university graduate and above by gender

Industry Type	2016		2022	
	Male	Female	Male	Female
Air_Transp	69.1%	30.9%	60.9%	39.1%
Banking	38.5%	61.5%	54.9%	45.1%
Chemistry	70.7%	29.3%	71.3%	28.7%
Constr_IND	77.3%	22.7%	77.6%	22.4%
Eletricl_Equi	79.1%	20.9%	76.7%	23.3%
Eletrici_Gas_IND	82.2%	17.8%	77.3%	22.7%
Fiber_Prod	56.6%	43.4%	65.0%	35.0%
Fish_Agro_Forestry	66.0%	34.0%	62.8%	37.2%
Glass_Clay_Prod	76.6%	23.4%	78.6%	21.4%
Groceries	64.3%	35.7%	58.5%	41.5%
Info_Commu_IND	68.5%	31.5%	62.5%	37.5%
Insur_Businn	23.7%	76.3%	31.7%	68.3%
Land_Transp	66.4%	33.6%	67.3%	32.7%
Machine	83.3%	16.7%	80.9%	19.1%
Metal_Prod	74.1%	25.9%	69.1%	30.9%
Mining	73.7%	26.3%	67.4%	32.6%
Non_FerreMetal	83.3%	16.7%	75.4%	24.6%
Other_FinancServ	50.4%	49.6%	52.3%	47.7%
Other_Prod	64.4%	35.6%	59.3%	40.7%
Petro_Coal_Prod	77.5%	22.5%	74.6%	25.4%
Pharma	62.5%	37.5%	53.6%	46.4%
Precis_Mech_Equip	70.7%	29.3%	67.0%	33.0%
Pulp_Paper	74.7%	25.3%	72.0%	28.0%
Real_State	53.0%	47.0%	55.2%	44.8%
Retail_Businn	49.7%	50.3%	41.2%	58.8%
Rubber_Prod	73.0%	27.0%	77.0%	23.0%
Secur_Commodities	54.7%	45.3%	57.7%	42.3%
Serv_IND	46.4%	53.6%	43.9%	56.1%
Shipping_IND	84.3%	15.7%	73.0%	27.0%
Steel	84.6%	15.4%	83.7%	16.3%
Transp_Equi	83.4%	16.6%	81.5%	18.5%
Warehou_Transp	54.0%	46.0%	55.9%	44.1%
Wholesale_Trade	67.5%	32.5%	60.9%	39.1%
All Industries	**62.8%**	**37.2%**	**63.4%**	**36.6%**

Legend: Heatmap color scale: Blue: higher values; Red: lower values; White: middle values (50%)

three in 2022 because Banking hired only 45.1% of females. The differences in percentage from 2016 to 2021 for female hires with educational background of university graduate or above can be seen in Fig. 4.

The percentage changes results from Fig. 4 indicate that 12 groups of industries had reduced the percentage of female hired from university or above background. Banking, Chemistry, Construction, Fiber production, Glass and clay production, insurance business, Land and transportation, Other financial services, Real estate, Rubber production, Security and commodities, and Warehouse trade reduced the number of female employees from 2016 to 2022.

(b) Miscellaneous Educational Background

The content of Table 7 indicates the percentage of employees' different educational backgrounds hired by the companies in this study not considering gender. The general number in the database of the employees from university graduates and above decreased by 0.5% when comparing 2016 to 2022. The Junior college and high school students had a variation of 0.4% and 0.1% between the years, respectively. The highest variation in university graduated or above can be seen as a 53.9% increase

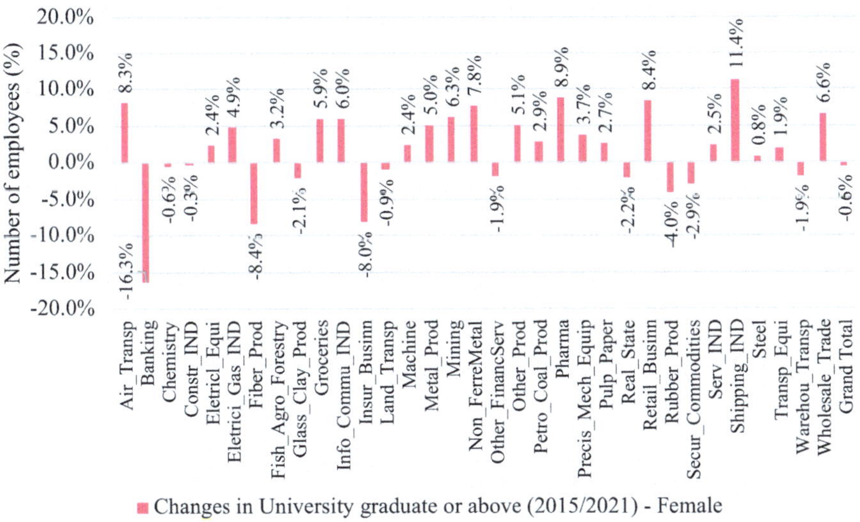

Fig. 4 Changes in the percentage of Females hired from university graduates and above

from Other Financial Industries, a 10.9% increase for Land and Transportation of Junior College, and finally, a 29.8% increase in High School graduate students hired in the Petroleum and Coal Production. The lowest numbers are presented by the Petroleum and coal industry (−26.6%) in university graduates and Other financial services for Junior college and high school background employees (24.6 and 29.2%).

5 Conclusions

Japan adopted the global initiative of the SDGs in 2016, and the changes in the hiring of young graduates in 2016 and 2022 can be seen in the results of this research for private sector companies. The study observed the progress by analyzing the number of people hired by gender. So, the trends that were observed can be indicated as:

(1) The general results for average of employees and gender showed that Steel industry increased the number of young professionals in their sector. However, the percentage of woman employees for this industry still low (7.3%). Considering age, the lowest value for females in 2016 were for Serv_IND (35.2), and in 2022 are Serv_IND and Air_Transp, both with 36.4 years old. The general result shows that the percentage of female employees increased for 32 industry types with the highest value being the Fishing, Agriculture, and Forest Industry (27.8%), decreasing only for the Land and Transportation industry by 8.6% when comparing the year 2022–2016. The overall result shows an increase of 3.7% in percentage of women in the Japanese private sector companies investigated in this study.

Table 7 Changes in hired employees of different backgrounds and education in percentage from 2016 to 2022

	Industry Type	University graduates and above	Junior college specialty	High school graduates, etc.
1	Air_Transp	-6.1%	6.1%	0.0%
2	Banking	-0.2%	-0.7%	0.9%
3	Chemistry	-4.0%	1.3%	2.7%
4	Constr_IND	-3.5%	1.1%	2.4%
5	Eletricl_Equi	0.3%	-1.1%	0.8%
6	Eletrici_Gas_IND	8.9%	-4.8%	-4.1%
7	Fiber_Prod	-3.6%	-4.7%	8.3%
8	Fish_Agro_Forestry	-1.4%	-3.6%	5.0%
9	Glass_Clay_Prod	7.7%	-3.0%	-4.7%
10	Groceries	-4.0%	-2.1%	6.1%
11	Info_Commu_IND	4.2%	-2.7%	-1.4%
12	Insur_Businn	0.4%	-0.4%	0.0%
13	Land_Transp	-0.6%	10.9%	-10.3%
14	Machine	5.5%	-1.5%	-3.9%
15	Metal_Prod	-14.5%	1.2%	13.2%
16	Mining	1.3%	-0.3%	-1.0%
17	Non_FerreMetal	-4.9%	-0.1%	5.0%
18	Other_FinancServ	53.9%	-24.6%	-29.2%
19	Other_Prod	1.1%	0.0%	-1.1%
20	Petro_Coal_Prod	-26.6%	-3.2%	29.8%
21	Pharma	-5.7%	-0.3%	6.0%
22	Precis_Mech_Equip	-14.3%	4.1%	10.2%
23	Pulp_Paper	0.9%	1.4%	-2.3%
24	Real_State	8.6%	-8.6%	0.0%
25	Retail_Businn	9.0%	3.1%	-12.2%
26	Rubber_Prod	-10.5%	-1.1%	11.5%
27	Secur_Commodities	-0.5%	0.5%	0.0%
28	Serv_IND	-0.8%	3.0%	-2.2%
29	Shipping_IND	0.2%	-0.2%	0.0%
30	Steel	2.2%	-0.5%	-1.6%
31	Transp_Equi	-8.2%	-0.7%	8.9%
32	Warehou_Transp	-1.4%	0.8%	0.7%
33	Wholesale_Trade	-3.6%	0.7%	2.9%
	Average Result	**-0.5%**	**0.4%**	**0.1%**

Color legend: Green: lower values; Red: higher values; White: middle values

(2) Among the employees under 30 years old, the percentage of woman is higher than in the general results for employees of all ages. The result of 2016 indicates that there are 31.7% of female in All Industries under 30 years old. In 2022 the number increased to 36.3%. The result for employees under 30 years old and female is higher than the percentage of females of all ages. Also, for 2016 five industrial sectors show the percentage of young female as higher than males (Serv_IND, Other_FinancServ, Banking, Insur_Businn and Air_Transp), increasing to seven on 2022 with the inclusion of Warehou_Transp and Retail_Businn. This shows a positive outcome for gender equality (SDG 5) in the private sector of Japanese companies.

(3) The educational background results shows that the percentage of female hired decreased by 0.6% from 2016 to 2022. In the results, 12 groups of industries out of the 33 had reduced the percentage of female hired from university or above background (Banking, Chemistry, Construction, Fiber production, Glass and clay production, insurance business, Land and transportation, Other financial

services, Real estate, Rubber production, Security and commodities, and Warehouse trade). The result from academic background shows different results from the employees under 30 years old that indicates an increase in female employees by 3.6%. A possible scenario is that the female employees under 30 are coming from a different educational background than university or above, however it is not possible to confirm from the current database.

This work compared the variation in the percentage of young graduates hired by gender in Japanese private sector companies for 2016 and 2022. The time-lapse corresponds to a period before and after the implementation of SDGs internationally, observing these years, it made it possible to gain insights if there were any significant changes in the hiring practices in Japan for the period. During this time interval, there was an increase in international pressures to promote gender equality and work-life balance, consequently, the entry and permanence of women in the Japanese workforce. However, the results showed that there are still issues related to the cultural understanding of women's roles in Japanese society creating bias in their progress from education as the research shows low overall percentage of females in the workplace.

There is also a huge unbalance between the percentage of males and females in the companies of this study. The results showed that around one-fourth of the general workforce will be female in 2016 and 2022. Female university graduates and above comprise one-third of the new hires for both years. These factors indicate that low progress was made in the direction of a more diverse work environment and there are still issues that have to be addressed to incentivize and maintain the female workers. The results show a stagnation of the progress toward women being part of the workforce over the six years of implementation of SDGs.

Often, Japanese workplace practices make it difficult for parents to match personal life with work. The result reflects factors such as low birth rates and culminates in an economic problem for the future. A solution would include increasing the opportunities for young women to join the workforce in a higher percentage. Financial stability combined with proper distribution of household responsibilities would increase the work-life balance for men and women to be secure enough to start a family.

Acknowledgments This research was made possible by a scholarship from the Ministry of Education, Culture, Sports, Science, and Technology of Japan (MEXT).

References

Berchin, I. I., de Aguiar Dutra, A. R., & Guerra, J. B. S. O. D. A. (2021). How do higher education institutions promote sustainable development? A literature review. Sustainable Development, 29(6), 1204–1222.

Brinton, M. C., & Mun, E. (2008). Gender and part-time work in Japan. ResearchGate. Retrieved January 21, 2025, from https://www.researchgate.net/publication/29458429_Gender_and_part-time_work_in_Japan

Ministry of Foreign Affairs of Japan. (2024–1). Japan's foreign policy in 2024: Overview [PDF]. Retrieved December 17, 2024, from https://www.mofa.go.jp/policy/other/bluebook/2024/pdf/pdfs/1a.pdf

Ministry of Foreign Affairs of Japan. (2024–2). Japan's initiatives in the United Nations: Disarmament and non-proliferation. Retrieved December 17, 2024, from https://www.mofa.go.jp/fp/unp_a/pageite_000001_00569.html

MOFA—Ministry of Foreign Affairs of Japan. (2016). *Efforts toward sustainable development goals (SDGs)*. Retrieved January 7, 2025, from https://www.mofa.go.jp/policy/oda/sdgs/effort/index.html

Nagano, T. (2014). Trends in corporate hiring of recent graduates. Japan Labor Review, 11(2), 24–36. Retrieved January 21, 2025, from https://www.jil.go.jp/english/JLR/documents/2014/JLR42_nagano.pdf

Noda, H. (2020). Work–life balance and life satisfaction in OECD countries: A cross-sectional analysis. Journal of Happiness Studies, 21(4), 1325–1348. doi:https://doi.org/10.1007/s10902-019-00131-9

OECD, 2024. Organization for Economic Co-operation and Development. Country profile: Japan. OECD Education GPS. Retrieved January 21, 2025, from https://gpseducation.oecd.org/CountryProfile?primaryCountry=JPN&topic=EO&treshold=10

OECD—Organization for Economic Co-operation and Development. (2024). *Work-life balance*. Retrieved January 7, 2025, from https://www.oecdbetterlifeindex.org/topics/work-life-balance/

Sugayama, S. (2011). The Evolution of Regular Hiring of New Graduates in Modern Japan. Japanese Research in Business History, 28, 61–83. https://doi.org/https://doi.org/10.5029/jrbh.28.61

Tanaka, H. (1980). Low Growth and Aging Labor Force. Japan Quarterly, 27(1), 39. Nagano, H. (2014). Trends in corporate hiring of recent graduates: focus on developments since the global financial crisis. Japan Labor Review, 11(2), 23–36.

The Guardian. (2018). *Tokyo medical school admits changing results to exclude women*. The Guardian. Retrieved January 8, 2025, from https://www.theguardian.com/world/2018/aug/08/tokyo-medical-school-admits-changing-results-to-exclude-women

Toyo Keizai Inc. (2024, April 5). Toyo Keizai Data Services | The Best Database of Japanese Companies. Retrieved from https://biz.toyokeizai.net/en/data/

Ueda, R. (2015). How is corporate governance in Japan changing? [Journal article]. OECD. https://doi.org/10.1787/5jrw7j3s37hh-en

United Nations. (2024). High-level week 2024. Retrieved January 20, 2025, from https://www.un.org/en/high-level-week-2024

Yamaguchi, K. (2019). Japan's Gender Gap. Finance & Development, 56(1), A009. doi:https://doi.org/10.5089/9781484398784.022.A009

Accounting and Sustainability—A Pedagogical Experience in a Multicultural Context

Teresa Eugénio, Jucélia Appio Frizon, and Ricardo Manso

Abstract Teaching methods must be innovative, promote forward thinking and enable the acquisition of higher order thinking skills, interdisciplinarity and the linking of local and global issues. Learning methods that emphasise interdisciplinarity and values and promote critical thinking, participatory action and medium and long-term planning are essential. Multilateral co-operation between schools and local stakeholders and authorities, scientific communities, the private sector and NGOs enables richer educational experiences that are closer to economic reality. This study aims to provide educational experiences in accounting and sustainability in a multicultural context. It reports on the experiences of five European higher education institutions (Portugal, Belgium, Finland, Spain and France) as part of a project that began in 2012 and is carried out annually. Each year, around 80 students and 10 teachers take part in this project to analyse the sustainability strategy of real companies in each country and compare best practises. Topics such as the Sustainable Development Goals, circular economy, non-financial reporting, auditing of sustainability reports and others are covered. The aim of this study is to present to the academic world a real example of an educational innovation project that has borne real and concrete fruit. Apart from the fact that this project also includes European countries, we believe that this study could inspire new similar projects around the world, especially in the Asia-Pacific region, and contribute to prosperity and peace.

T. Eugénio (✉) · J. A. Frizon
CARME–Centre of Applied Research in Management and Economics, Polytechnic Institute of Leiria, Leiria, Portugal
e-mail: teresa.eugenio@ipleiria.pt

J. A. Frizon
e-mail: jucelia.frizon@unioeste.br

J. A. Frizon
State University of Western Paraná (UNIOESTE), Cascavel, Brazil

T. Eugénio · R. Manso
School of Technology and Management, Polytechnic Institute of Leiria, Leiria, Portugal
e-mail: ricardo.manso@ipleiria.pt

W. Leal Filho et al. (eds.), *Sustainability in the Curriculum and Operations of Universities*, World Sustainability Series,
https://doi.org/10.1007/978-3-032-12678-8_31

Keywords Education, pedagogical experiences · Sustainable development · Accounting and sustainability

1 Introduction

Rethinking education is of fundamental importance, especially at a time when Brundtland's (1993, p. 22) statement is emphasised: "We live in an age where complicated issues are presented in 30-second sound bites. This tends to confuse people's willingness to concentrate on in-depth thinking about the future". This debate has also taken place at international level. One example is the report of the United Nations Conference on Sustainable Development (Rio + 20), which emphasises the need to improve the capacity of education systems to prepare societies for sustainable development (Leal Filho et al., 2015).

This study presents a pedagogical experience carried out at a Portuguese higher education institution (HEI) as a partner of four other European HEIs for students of accounting, finance and management programmes. The topic is accounting and sustainability. There is much debate about the teaching of accounting. Some comments claim that accounting courses are outdated, with too much technical emphasis and very analytical approaches that do not allow students to develop a more critical approach and perspective on the socio-political context (Brown & Dillard, 2019). These authors also point out that traditional accounting was a discipline that focussed more on standards, measurement and disclosure of financial issues and ignored many issues facing today's society. Corporate responsibility, environmental sustainability and social justice issues are cited as examples: "Environmentalists, trade unions, ethical investors, indigenous peoples and social movements all have concerns - about climate change, poverty, social injustice, economic power and globalisation - that are ignored or glossed over in accounting" (Brown & Dillard, 2019, p. 26).

Other scholars, focusing their studies in the area of social and environmental accounting, have come to similar conclusions (e.g. Milne & Grey, 2013; Bebbington et al., 2007; Bebbington & Unerman, 2018; Grey et al., 2014), arguing that the narrow focus of conventional accounting is insufficient for professionals' goals of serving the public interest due to a lack of deep awareness of the social, economic, political and environmental impacts of corporate activities. Gray et al., (2014) call for critical thinking and state that in the fields of management, economics, accounting and finance, a balance between the skills taught to students and critical thinking is essential. Gray et al., (2014) also argue that social and environmental accounting offers enormous potential for the exploration of both theoretical and practical components in conjunction with a critical and innovative framework: "Education is inevitably a combination of the teaching of skills and knowledge and the development of the thinking muscle - the development of analysis and critical thinking" (Gray et al., 2014, p. 326).

There is an urgent need to reflect on education for sustainable development, with higher education institutions playing a fundamental and leading role in developing skills and knowledge and driving prosperity and peace. In particular, by including the topic of social and environmental accounting in the curriculum of courses in the business area and in their various programmes, because as Leal et al. (2023) note, few teachers integrate sustainable development comprehensively into their curricular units.

We therefore believe that the pedagogical experience reported in this study will help to close this gap and contribute to an accounting education that better prepares future professionals for the current challenges in organisations. It can provide students with skills in topics that combine sustainability and accounting. The project, that started in 2012, is called "Accounting for Sustainability" and involves five European HEI from Portugal, Belgium, Finland, Spain, and France. Around 80 students and 10 teachers take part of this international project, each year. The aim is to analyse the sustainability strategy of real companies in each country and compare good reporting practices. Topics covered include the Sustainable Development Goals (SDGs), the circular economy, non-financial reporting, and the audit of sustainability reports. This study aims to present to the academic world a real case of pedagogical innovation that has borne real and concrete fruit and could inspire other higher education institutions to undertake similar projects. We believe this example of a pedagogical experience in a multicultural context could be a significant contribution to the ssustainable development in Higher Education in the Asia-Pacific Region.

The present study is structured in 5 sections. Following the introduction, in Sect. 2, the authors present a literature review focused on the teaching accounting for sustainability in business schools, education for Sustainable Development and the SDGs, and learning activities for sustainability. Section 3 describes the methodology adopted and in Sect. 4, the authors describe the Pedagogical experience—Accounting for Sustainability. Finally, the authors present the conclusions, and future research clues.

2 Literature Review

2.1 The Teaching Accounting for Sustainability in Business Schools

Universities have a prominent role to play, which they can realise through their actions in four categories: (1) providing a model of sustainable practices for society; (2) equipping students with skills and competences in integration, synthesis and systems thinking and in tackling complex problems needed to address sustainability challenges; (3) conducting practice-based research; and (4) fostering and strengthening engagement between individuals and institutions within and outside universities to shape universities as transdisciplinary, highly integrated and engaged actors with other societal institutions (Gomes et al., 2018, cited in Zilahy & Huisingh, 2009 and based on).

As business schools are responsible for the future leaders of organisations, this topic should be reflected in their curricula and integrated into their mission. This notion is reinforced by Swaim et al. (2014) and Eugenio et al. (2022), who point out that many organisations are already placing greater emphasis on environmental, social and economic goals and seeking to integrate them into their strategy, so business schools need to prepare future leaders to strongly defend corporate sustainability initiatives.

Studies such as Bebbington et al., (2017); Cooper et al., (2005); Thomson & Bebbington (2005); Bebbington & Unerman, (2018) and Frizon & Eugénio (2022) attempt to develop ideas to operationalise this visibility and competence in students and future professionals. In the study by Bebbington et al., (2017), the authors share their individual learning experiences in a logic of reflections and suggestions on accounting and sustainable development; Cooper et al., (2005) explore alternative ways to make social injustice visible; Thomson & Bebbington (2005) show how accounting can support sustainable development. The article by Bebbington & Unerman (2018) provides evidence on how accounting research can be a tool to implement the SDGs. This article provides a synthesis of interdisciplinary perspectives on sustainable development and their integration with the accounting for sustainability literature. Frizon & Eugénio (2022) also examined recent developments in sustainability in higher education research, specifically seeking to understand what is happening in the fields of management and accounting. By determining the production of scientific research for the period 2014–2020 in journals classified in the Scimago Journal Rank (SJR), they concluded that although some steps have been taken, there is still a long way to go. Further scientific studies are therefore needed.

2.2 *Education for Sustainable Development and the SDGs*

Education is explicitly recognised in several of the SDGs, and universities have a direct role to play in making them a reality. However, the contribution of universities to the SDGs is much broader, as they can support the implementation of each of the SDGs. Some of the key areas of contribution include strengthening public engagement and participation in addressing the SDGs, initiating and facilitating dialogue with the academic community, and ensuring representation of higher education (Sustainable Development Solutions Network (SDSN)—Australia/Pacific, 2017).

The Sustainable Development Goals (SDGs) recognise the importance of education for sustainable development (ESD) through SDG 4, which calls for 'ensuring inclusive and equitable quality education and promoting lifelong learning opportunities for all', and SDG 11, which calls for 'making cities and human settlements inclusive, safe, resilient and sustainable' (United Nations, 2015). The SDGs are a universal set of goals, targets and indicators adopted by the UN in 2015 to end poverty in all its forms by 2030 and to balance the three dimensions of sustainable development: economic, social and environmental. All countries are called upon to integrate the ambitious 17 goals and 169 targets into their agendas and policies, and to work towards their realisation. For the SDGs to be successful at the global level, the goals must be operationalised in a management tool that helps countries develop implementation strategies, measure their progress and ensure accountability of all stakeholders (Leal Filho et al., 2019).

Universities have a fundamental role to play in contributing to the realisation of the SDGs. Quality education brings significant benefits to individuals, communities, and countries in terms of the SDGs. It is also an important means of supporting and accelerating the global capacity to realise the SDGs. The Sustainable Development Solutions Network (SDSN) was launched in 2012 by United Nations Secretary-General Ban Ki-moon to mobilise scientific and technical expertise from academia, civil society, and the private sector to support problem-solving for sustainable development at local, national, and global levels. The SDSN operates national and regional networks of knowledge institutions and thematic networks focused on solutions and contributes to the achievement of the SDGs (Sustainable Development Solutions Network (SDSN)—Australia/Pacific, 2017).

According to the SDSN, the role of universities in implementing the SDGs relates to the core functions of learning and teaching, research, organisational governance, culture and operations, and external leadership. The SDGs provide great opportunities to create, strengthen and communicate the links between these areas and therefore a university-level approach to addressing the SDGs is favoured (Sustainable Development Solutions Network (SDSN)—Australia/Pacific, 2017).

ESD is a fundamental part of UNESCO's mission and a central aspect of Agenda 21 (United Nations Conference on Environment and Development [UNCED], 1992). ESD refers to educational programmes and experiences that aim to empower people to acquire the knowledge, skills and values needed to shape a sustainable future (Biasutti & Frate, 2017; Tilbury, 2011).

The role of education is present in Agenda 21, with a particular focus on Chap. 36. Chapter 36 of Agenda 21 was one of the first calls for action on education for sustainability and formed the basis for the development of ESD networks, which address the following three objectives: Refocusing education on sustainable development, raising public awareness and promoting education (United Nations Conference on Environment and Development [UNCED], 1992). Several other initiatives have been implemented by UNESCO, such as the inclusion of ESD indicators for the monitoring and evaluation of education and its progress, the monitoring of literacy indicators, and the measurability and evaluation of education policies, programmes and systems (UNESCO, 2014).

Scoullos (2013) outlined the following characteristics of ESD learning methods: (a) interdisciplinary and holistic, student-centred and participatory; (b) value-centred, promoting critical thinking; (c) future-oriented, promoting medium—and long-term planning; (d) promoting multilateral cooperation between schools, local actors and authorities, scientific communities, the private sector and NGOs, etc.; (e) global connections as part of everyday life, whether in a small village or a large city. These teaching/learning methods promote changes in behaviour and mindsets, and relate not only to knowledge but also to processes, as they teach students how to think—not what to think (Biasutti & Frate, 2017).

Experience-based learning is therefore made up of activities that draw on students' experiences of (in) sustainability in local contexts. Nevertheless, by participating in these experiences, as well as reflecting on and discussing their results, students develop basic sustainability competences. These activities promote understanding of sustainability concepts, principles, and methods, as well as enabling graduates to get involved in solving real-world sustainability problems (Caniglia et al., 2016).

2.3 Learning Activities for Sustainability

Sustainability challenges current paradigms, structures and practices in higher education. Universities are responding to this reality by seeking to make a significant contribution to sustainability. This requires going beyond the integration of key ideas into existing curricula and developing a new way of supporting sustainability action projects, which often take place outside the institution. The sustainability journey requires universities to embrace interdisciplinarity, participatory pedagogy, 'real world' research and to open up institutional boundaries so that the notion of sustainable communities is extended beyond the walls of the university (Tilbury, 2011).

However, pedagogical approaches to developing competences are not as formally developed as systemic thinking competences. An alternative to a real problem-solving experience to develop these competences in an educational context is the use of case studies and group projects in the classroom (Remington-Doucette et al., 2013). To this end, as advocated by Tilbury (2011), a systemic or interconnected vision of sustainability is needed in all institutions to transform the educational experience of students and lead social change towards sustainability.

There are some real examples in the literature. Remington-Doucette et al. (2013) evaluated education for sustainability in a transdisciplinary undergraduate programme in the United States. The authors aimed to present the results of a classroom evaluation designed to determine the extent to which students develop key sustainability competencies during an introductory transdisciplinary sustainability course. To accomplish this goal, the authors asked students to select a real-world sustainability problem and propose a solution to their problem using sustainability science problem-solving methods learned throughout the semester. Based on the evaluation presented by the authors, sustainability competences developed differently in students with different disciplinary affiliations, for example, management graduates did not improve any of the key competences, sustainability graduates improved only the systems thinking competence and sustainability graduates, who were majoring in another traditional discipline, improved all the competences.

Caniglia et al. (2016) aimed to present an experience-based learning framework that provides a bottom-up, student-centred entry point for the development of systems thinking, normative and collaborative competences in sustainability. The authors applied the Phoenix/Tempe and Hamburg/Lüneburg model as part of The Global Classroom, between two universities, Arizona State University in the USA and Leuphana University of Lüneburg in Germany. The application of the experience-based learning framework demonstrates how students have begun to develop the systemic, normative, and collaborative thinking that develops competences in sustainability. They emphasise that experience-based learning contributes to the initial development of sustainability competences in introductory-level courses. It allows students from different disciplinary, social, and cultural backgrounds, for example in international education, to start collaboratively developing these competences.

Molderez and Fonseca (2018) selected two projects from a Belgian university, the Interdisciplinary Assessment Project and the Brookstreet Project. Data was collected through surveys and focus groups with the aim of evaluating what students learnt during two alternative learning activities, i.e. real-world experience, and service learning. The aim of evaluating these learning activities was to assess their effectiveness in developing competences for sustainable development among students. The authors have shown that students perceive that they have developed sustainability competences through these learning activities. The dissimilarity between these projects is an advantage for this research, as comparisons can be made between the different types of learning activities.

Molderez and Fonseca (2018) emphasise that to ensure the development of a broad spectrum of sustainability competences, teachers are advised to combine different learning activities or include a variety of features to promote as many competences as possible, especially due to the interaction between the different competences. Thus, education in this context needs to be reorientated towards a combination of interdisciplinarity, visioning, creativity and fun (Molderez & Fonseca, 2018).

3 Methods

This study, as already mentioned, aims to provide a pedagogical experience in the field of accounting and sustainability in a multicultural context. It reports on the experience of several European higher education institutions with the aim of teaching sustainable development in the field of business, with a particular focus on accounting and management.

However, Remington-Doucette et al. (2013) note that integrating real-world experiences into courses is not so easy for several reasons. Students, particularly those in introductory courses, may lack basic knowledge, skills and understanding of sustainability concepts and methodologies, which limits the learning process. The authors suggest that future research should focus on assessing the development of competencies in sustainability interventions in traditional disciplinary courses and also as students progress to higher level disciplinary courses within transdisciplinary sustainability programmes. The target students for this experiment are those in their second or third year of study, excluding first year students. These are students studying business degrees (management, accounting or related fields). In Portugal, in most editions, the Accounting for Sustainability project has been part of a course unit, "Seminar on Management", where the aim is for students to acquire skills that are transversal to their degree and, in particular, to current issues. In some editions, it has operated as a voluntary, autonomous project in which students have enrolled. Students receive between 2 and 3 ECTS for their participation. In the other countries, one of the two modalities was used.

The following section describes the project, how it works, the partner institutions, the topics covered and the script made available to the students.

4 Pedagogical Experience—Accounting for Sustainability

4.1 Description of the Project

Aim

Before the project starts, the teachers have several online meetings to discuss different issues, both pedagogical and logistical. Such as the dates, the number of students in this edition, the number of teams, the organisation of the teams, the sectors to be used, the questions to be answered (each year all the material is revised and improved) and some logistical questions. The project usually starts in March and ends in April, at the beginning of the second term). During the first week, each country organises sustainability lectures on topics such as Environmental accounting, sustainable growth, company stories, sustainability reports, etc. In the following weeks (almost 3 weeks), the students work on the sustainability accounting cases according to their own law and exchange information with the other foreign members of the team. They have to choose 3 comparable companies in a sector via web conferencing. For example, if we have teams of six students (mixed country teams), 2 students per team read one of the sustainability reports so that they know the answers to the questions before a fixed date.

After almost 4 weeks there was the international week. During this week, all students and teachers travel to the same place, in most editions in Leiria (Portugal). The Portuguese teachers and students welcome the foreign team (teachers and students from Finland, Belgium, Spain and France). During the International Week, the students and teachers have a welcoming session which includes a summary of the main topics to be discussed during the next days of work and an invited speaker. Usually the speaker comes from the sustainability department of a company, or is an academic with research on the topic, or someone from an NGO with social work or environmental projects). In the International week in Leiria, students work in teams (in mixed country teams) comparing sustainability reports from different countries and make their own oral presentations on sustainability. This work preparation began four weeks before the visit to Leiria. The students held several online meetings before meeting in person. In the end of the project, teachers evaluate the students' work in a discussion and oral presentation.

In one of the days of the international week there are a social program for students and teachers. Usually, students like to go to Nazaré Beach and Batalha Monastery, as in all editions we have different students.

The major outcomes of this project are information about sustainability from different international perspectives, working on cases, and communication between students on an international level.

Partners

Most editions of the project have involved four higher education institutions: Polytechnic Institute of Leiria (Portugal), Vives University (Bruges, Belgium), Avans University (Breda, Netherlands) and TAMK (Tampere, Finland). In recent years Avans University (Breda, Netherlands) has been unable to continue in the project and the University of Mondragon (Bilbao, Spain) and the Université Toulouse Capitole (Rodez, France) have joined. The project currently has five partner institutions from Portugal, Belgium, Finland, Spain, and France. The 2023 edition (which took place in March and April) had 85 students (6 Finnish; 19 Belgian; 23 Portuguese; 31 Spanish and 6 French) and 9 teachers (1 Finnish; 2 Belgian; 3 Portuguese; 1 Spanish and 2 French). The number of students per edition (year) has varied between 65 and 100 students.

Focus

This project is an innovative pedagogical proposal that has been running since 2012. After each edition, an evaluation is carried out by the teaching team and the students in order to make improvements for the next edition, both in terms of the way it works and in terms of the educational content and questions in the script.

As mentioned above, the pedagogical proposal focuses on group work developed by teams of students with members of different nationalities. The students have to apply their knowledge to a practical, concrete reality; they have to be able to interpret and critically analyse company reports. As described above, this project culminates in a visit by the partners to the Polytechnic Institute of Leiria, Portugal, where, after virtual communication, the students can meet in person and present their results orally to a jury (Portuguese and foreign professors in the field who accompany the work). The fact that the students come from different cultural, economic and political backgrounds allows them to learn and be open to multidisciplinary approaches. The project also allows the application of methodologies that are more attractive to students, a critical spirit, interdisciplinarity and mobility. On the other hand, it contributes to the teaching of this subject, which needs to be included in higher education curricula.

4.2 Explored Topic—Sustainability Accounting

Initially, theoretical content is taught, and classroom debates are held on the subject of sustainability. These classes take place in each country in the weeks leading up to the international week. Students have to read texts and documents from international accounting organizations concerned with sustainability, such as the European Financial Reporting Advisory Group (EFRAG); Sustainability Accounting Standards Board (SASB), the European Federation of Accountants and Auditors for SMEs (FEAA); Federation of European Accountants (FEE), Global Reporting Initiative (GRI), International Integrated Reporting Council (IIRC), among others. We also suggest reading the document: "BIP Sustainability and reporting Introductory volume", prepared by Belgium teachers. This document includes topics such as: concepts of sustainability and sustainable development; sustainability and business strategy; environmental and social systems used by companies to implementing sustainability; evolution from linear economy to circular economy; legal frameworks related to reporting of non-financial information; sources and benchmarks of sustainability reporting, among others.

In the second phase, students are invited to use concrete case studies of companies that apply sustainability practices and disclosure them by publishing a sustainability report, integrated report, or non-financial information on their website. In the case of Portugal, the Business Council for Sustainable Development (BCSD) publishes a significant collection of "case studies" that students work on. Its website states that "BCSD Portugal is reactivating a project it started in the past to aggregate and share case studies from its member companies, in line with its mission to support member and non-member companies in building more sustainable projects and other stakeholders to have access to the best business practices in the area of sustainability".

(https://bcsdportugal.org/biblioteca-casos-de-estudo/ accessed on 14/12/2023). There are also several useful links where students can find more information. As an example, here are some of the links suggested to Portuguese students:

Some useful links:

Information on GRI: https://www.globalreporting.org/

Sustainability information for companies: https://www.bcsdportugal.org/

Sustainability reports: https://www.sustainability-reports.com/

Case studies about SDGs: https://www.ods.pt/empresas/

Values, Ethics and Responsibility: https://www.ver.pt/

Sustainability reports: www.corporateRegister.com

Finally, students are faced with a practical component of analyzing a sustainability report from a company of their choice from the sector they have been assigned. The possible activity sectors are shown in Table 1.

Generally, each student chooses a company from their own country. They then have to compare their analysis with 2 other reports chosen by students from other countries in the same sector who are part of their group. There are 15 groups (since there are 15 sectors, one group per sector) and each working group can have between 5 and 8 students, depending on the total number of students in that edition. After

Table 1 Activity sectors

Auto construction
Chemicals
Fuel companies
Airline companies
Financial services—banks or insurance companies
Forest and paper products
Telecommunication services
Waste management
Building companies
Production of electrical devices
Food retailers
Food or beverage production
Postal services
Hotel chain
Accounting offices

analyzing the sustainability reports, the students must answer a group of questions as the ones indicated in the script (Table 2).

This script includes various topics previously discussed with the students, such as the SDGs, circular economy, non-financial reporting, auditing the sustainability report, among others. The oral presentation will bring together the main aspects of the script from the 3 companies. Students are given the instructions in Table 3 for preparing the oral presentation.

This experience could be replicated by any Higher Education Institution (HEI) in the Asia-Pacific Region. This experience gives to students and future professionals in the organizations, a conscious of the importance of Sustainable Development working as a driver of prosperity and peace.

5 Conclusions

We conclude that this project fulfils the characteristics of the ESD learning method ology mentioned by Scoullos (2013), as it is interdisciplinary and holistic, student-centred and participatory; the project is value-oriented and promotes critical thinking; it is future-oriented and promotes medium and long-term planning; it promotes multilateral collaborations between schools, local actors and authorities, scientific communities, the private sector and NGOs, either through various invitations to open classes or welcome sessions or through the involvement of the local community. As it is a European project, it also has a regional impact. After each edition, an article is published in the regional newspaper of Leiria, reporting on this multicultural educational experience. Please find attached one of the articles published in the newspaper

Table 2 Example of a questions guide to students

Questions BIP sustainability
Analysis of the sustainability strategy within a company/sector
Based on the analysis of the company's website and the sustainability report (or integrated report) follow and answer the above questions:
1. Start by describing your company. For example, identify the sector of activity, main products and services, company dimension, turnover, number of employees, market share, geographic implementation
2. Identify the main problems in terms of sustainability in the activity sector of your company. Give concrete examples
3. When we talk about sustainability we often talk about the triple P (planet, people and profit)
a. Give few examples of how your company has decreased the negative impact on our planet
b. Another milestone in sustainability is that business is people friendly. Give few examples of people-friendly actions your company has taken
c. Give examples of advertising or other publicity campaigns which relate to the strategy of the company. Justify
4. Give examples of actions that contribute for the achievement of the sustainable development goals (SDG) of the United Nations
5. Is your company member of any organization that pursues sustainability? Explain what reporting system the company is using to communicate its reports sustainability
Policy (ex. GRI-reporting, quality label…)
6. Is there any assurance on what your company reports about sustainability?
7. Give examples from the media that confirm or contradict the social responsibility policy communicated by the company
8. Give example(s) of some actions that your company is already doing, or could be doing, in order to move from a linear economy system to a circular economy

Source: Questions BIP Sustainability, April (2023)

Table 3 Contents of the oral presentation

Oral presentation—instructions
• Very short description of the companies
- What they do? Where they work?
- Size of the company (turnover, employees,…)
• Give some practical examples of how the P of people is incorporated in their sustainability strategy
• Give some practical examples of how the P of planet is incorporated in their sustainability strategy
• Give an example of how sustainability is used in their advertisements or communication
• How do they report on sustainability and if is there any assurance on what they report?
• Did you find some essential differences between what they report and what they do? (greenwashing,…)
• Do you find something in the sustainability strategy about the 17 sustainable development goals of the United Nations? If yes, which goals and how?
• Does your company mention any projects that may contribute to a circular economy?
• How the pandemic affects this sector?
• Conclusion: which of the companies you consider to be the most sustainable and why?

Source: Oral presentation topics BIP Sustainability, April (2023)

"Região de Leiria" in one of the editions of the project. In this way, the community is involved and interest is raised in this issue of sustainability.

The activities of this project, considered an innovative pedagogical experience, promote the understanding of sustainability concepts, principles and methodologies, allowing graduates to participate in solving sustainability problems in the real world, as recommended by Caniglia et al. (2016).

This pedagogical proposal aims to create in students a civic awareness based on ethical values and a critical spirit, providing them with transversal skills that will be very useful in their future professional life as accountants, managers or administrators. This project is in line with the SDGs as well as with various studies and guidelines from the EU and other international organisations such as UNESCO, which provide for the promotion of education for sustainable development. It is a project with a high degree of pedagogical innovation in which the students are the main actors and in which their interaction and problem solving are encouraged with a view to taking sustainability into account.

The fact that the students have different cultural, economic and political backgrounds allows them to learn and be open to multidisciplinarity. The project also allows the application of methodologies that are more attractive to students, a critical spirit, interdisciplinarity and mobility. On the other hand, it contributes to the teaching of this subject, which is essential to be included in university curricula, as stated by various authors such as Leal et al. (2023). As previously mentioned, this experience could be replicated by any higher education institution in the Asia-Pacific region. This experience provides students and future professionals in the organisations with an awareness of the importance of sustainable development as a driver for prosperity and peace.

Future Research Avenues As these teaching/learning methods promote changes in behaviour and mindset (Biasutti & Frate, 2017), it would be interesting in the future to interview students who have participated in previous editions to assess possible changes in behaviour and understand the impact of this project on the daily lives of these students.

Acknowledgements The authos would like to thanks to the team of teachers involved in the Accounting for Sustainability, 2023 project: Fabienne Moeyaert; Ritta Vihuri; Oier Onaindia; Oliver Depite; Daniel Pelissier, Maria João Jorge, Ricardo Marques and Peter Vanwildemeersch.

Appendix

News published in the newspaper "Região de Leiria".

"Business sustainability project brings together four European countries in Leiria

For the 5th year running, the Polytechnic Institute of Leiria (IPLeiria) is hosting around 80 students from three European countries (Belgium, the Netherlands and

Finland), who are meeting at the School of Technology and Management (ESTG, IPLeiria) as part of the "International Week on Sustainability" project and joining students from the degree courses in the Department of Management and Economics, to analyse and compare companies' sustainability practices and their disclosure through sustainability reports. The aim of this comparison is to identify the best corporate practices in terms of sustainability, any signs of greenwashing, as well as to assess any current and future company projects that could contribute to optimising the life cycle of products, i.e. the development of a circular economy.

Vives (Bruges, Belgium), Avans (Breda, Netherlands), TAMK (Tampere, Finland) and IPLeiria (Leiria, Portugal) are the higher education institutions involved in the "International Week on Sustainability" project, which had its first edition in 2012 in Belgium and all the others in Leiria (ESTG, IPLeiria)—the sixth edition of the event was held in 2017."

References

Bebbington, J, Russell, S. and Thomson, I. (2017). Accounting and sustainable development: Reflections and propositions, *Critical Perspectives on Accounting*, 48, p. 21–34.

Bebbington, J. and Unerman, J. (2018). Achieving the United Nations Sustainable Development Goals: An enabling role for accounting research, *Accounting, Auditing & Accountability Journal*, 31 (1), pp.2–24.

Bebbington, J., Brown, J. and Frame, B. (2007). Accounting technologies and sustainability assessment models, *Ecological Economics*, 61(2-3), pp. 224–236.

Biasutti, M., & Frate, S. (2017). A validity and reliability study of the Attitudes toward Sustainable Development scale. *Environmental Education Research*, 23(2), 214–230. https://doi.org/10.1080/13504622.2016.1146660

Brown, J. and Dillard, J. (2019). Accounting education, democracy and sustainability: taking divergent perspectives seriously, Int. J. Pluralism and Economics Education, 10 (1), pp. 24–45.

Brundtland, G. (1993). Sustainability the Global Environment. In Proceedings: Technology and Governance in the 1990s. Papers Presented at the OTA 20th Anniversary Forum on Technology and Governance in the 1990s, April 1993, OTA-A-564 NTIS. Pp. 17–22.

Caniglia, G., John, B., Kohler, M., Bellina, L., Wiek, A., Rojas, C., Laubichler, M. D., & Lang, D. (2016). An experience-based learning framework: Activities for the initial development of sustainability competencies. *International Journal of Sustainability in Higher Education*, 17(6), 827–852. https://doi.org/10.1108/IJSHE-04-2015-0065

Cooper, C.; Taylor, P.; Smith, N. and Catchpowle, L. (2005). A discussion of the political potencial of Social Accounting, *Critical Perspectives on Accounting*, 16 (7), 951–974.

Eugénio, T.; Carreira, P.; Miettinen, N.; and Lourenço, I. (2022), Understanding students' future intention to engage in sustainability accounting in Malaysia and The Philippines. *Journal of Accounting and Emerging Economies*, 12 (4), 695–715. https://doi.org/10.1108/JAEE-10-2020-0277

Frizon, J. and Eugénio, T. (2022). Current developments in Higher Education Sustainability Research in the areas of Accounting and Management. *The International Journal of Management Education,* 20 (3), 100709. https://doi.org/10.1016/j.ijme.2022.100709

Gomes, S.; Jorge, S. and Eugénio, T. (2018). A importância do desenvolvimento sustentável nos currículos académicos, *Revista de la Associacion Española de Contabilidad y Administracion de Empe*sas, pp. 20–24.

Gray, R., Adams, C. and Owen, D. (2014). Accountability, Social Responsibility and Sustainability–Accounting for society and the environment, Pearson Education, ISBN: 978-073-68138-0.
Leal Filho, W., Shiel, C., Paço, A., Mifsud, M., Ávila, L. V., Brandli, L. L., Molthan-Hill, P., Pace, P., Azeiteiro, U. M., Vargas, V. R., & Caeiro, S. (2019). Sustainable Development Goals and sustainability teaching at universities: Falling behind or getting ahead of the pack? *Journal of Cleaner Production*, 232, 285–294. https://doi.org/10.1016/j.jclepro.2019.05.309
Leal Filho, W., Manolas, E., and Pace, P. (2015). The future we want: Key issues on sustainable development in higher education after Rio and the UN decade of education for sustainable development, *International Journal of Sustainability in Higher Education*, 16(1), 112–129.
Leal S., Azeiteiro U., Aleixo A. (2023), Sustainable development in Portuguese higher education institutions from the faculty perspective, *Journal of Cleaner Production*. https://doi.org/10.1016/j.jclepro.2023.139863
Milne, M. and Gray, R. (2013). W(h)ither ecology? The triple bottom line, the global reporting initiative and corporate sustainability reporting, *Journal of Business Ethics*, 118 (1), pp. 13–29.
Molderez, I., & Fonseca, E. (2018). The efficacy of real-world experiences and service learning for fostering competences for sustainable development in higher education. *Journal of Cleaner Production*, 172, 4397–4410. https://doi.org/10.1016/j.jclepro.2017.04.062
Remington-Doucette, S. M., Connell, K. Y. H., Armstrong, C. M., & Musgrove, S. L. (2013). Assessing sustainability education in a transdisciplinary undergraduate course focused on real-world problem solving: A case for disciplinary grounding. International *Journal of Sustainability in Higher Education,* 14(4), 404–433. https://doi.org/10.1108/IJSHE-01-2012-0001
Scoullos, M. (2013). Education for sustainable development in biosphere reserves and other designated areas: a resource book for educators in South-Eastern Europe and the Mediterranean https://unesdoc.unesco.org/ark:/48223/pf0000219946
Sustainable Development Solutions Network (SDSN)—Australia/Pacific. (2017). Getting started with the SDGS in Universities: A guide to for Universities, Higher Education Institutions andthe Academic Sector. Sustainable Development Solutions Network, 56. https://www.miteco.gob.es/es/ceneam/recursos/pag-web/empezar-ods-universidades.aspx
Swaim, J. Maloni, M. Napshin, S. and Henley, A. (2014). Influences on Student Intention and Behaviour Toward Environmental Sustainability. *Journal of Business Ethics*, 465–484.
Tilbury, D. (2011). Higher education for sustainability: a global overview of commitment and progress. Higher Education in the World, 1–21. http://insight.glos.ac.UK/sustainability/Education/Documents/GUNI HE in the World 4 HE's Committment to Sus.pdf
Thomson, I. and Bebbington, J. (2005). Social and environmental reporting in the UK: a pedagogic evaluation, *Critical Perspectives on Accounting*, 16 (5), 507–533
UNESCO. (2014). Shaping the Future We Want UN Decade of Education for Sustainable Development. 2005-2014 Final Report. In United Nations Educational, Scientific and Cultural Organization. http://unesdoc.unesco.org/images/0023/002301/230171e.pdf
United Nations. (2015). Transforming Our World: The 2030 Agenda for Sustainable Development. https://sdgs.un.org/
United Nations Conference on Environment & Development [UNCED]. (1992). Agenda 21, Ch. 36: Promoting Education and Public Awareness and Training. United Nations, New York. https://sustainabledevelopment.un.org/content/documents/Agenda21.pdf
ZILAHY, G.; HUISINGH, D. (2009). The roles of academia in regional sustainability initiatives. *Journal of cleaner production, 17*, (12), 1057–1066.

Sustainability and Universities Operations

Sustainability Maturity: An Analysis of Institutional Development Plans in Brazilian Public Universities

Andrieli de Oliveira Cardoso, Diego Marques Cavalcante, Ana Paula Belizário, Nisrin Naiel Dib Khaled, Débora Londero Kieling, Elda Rodrigues Steinhorst Kraetzig, and Lucas Veiga Ávila

Abstract The aim of this study is to assess, from the perspective of institutional development planning (IDP), the level of maturity of the Sustainable Development Goals (SDGs) at the best-ranked Brazilian public universities in the Times Higher Education (THE) ranking. To this end, a content analysis based was carried out,

---

A. de Oliveira Cardoso (✉) · D. M. Cavalcante · A. P. Belizário · N. N. D. Khaled · D. L. Kieling · E. R. S. Kraetzig · L. V. Ávila
Federal University of Santa Maria—UFSM, Santa Maria, Rio Grande Do Sul, Brazil
e-mail: andrieli@urisantiago.br
URL: http://lattes.cnpq.br/5477568806154554

D. M. Cavalcante
e-mail: diego.cavalcante@acad.ufsm.br
URL: http://lattes.cnpq.br/0511274896694587

A. P. Belizário
URL: https://lattes.cnpq.br/2640940205544007

N. N. D. Khaled
e-mail: nisrin.khaled@acad.ufsm.br
URL: https://lattes.cnpq.br/3868988803586860

D. L. Kieling
URL: http://lattes.cnpq.br/6383849506514480

E. R. S. Kraetzig
e-mail: elda.steinhorst@acad.ufsm.br
URL: http://lattes.cnpq.br/6990391920249058

L. V. Ávila
e-mail: lucas.avila@ufsm.br
URL: http://lattes.cnpq.br/9859670993729250

A. de Oliveira Cardoso
Regional Integrated University of Alto Uruguai and Missões, Santiago Campus of URI, Santiago, Rio Grande Do Sul, Brazil

E. R. S. Kraetzig
Federal University of Pampa—UNIPAMPA, Bage, Rio Grande Do Sul, Brazil

W. Leal Filho et al. (eds.), *Sustainability in the Curriculum and Operations of Universities*, World Sustainability Series,
https://doi.org/10.1007/978-3-032-12678-8_32

considering the IDPs in force at 10 universities and assessing the level of maturity of the 17 SDGs. The results indicate that SDGs 4, 7 and 15 have the highest levels of maturity, while SDGs 5 and 6 are at an intermediate level, and the others at a low level. The study also highlights good practices adopted by universities, such as the explicit inclusion of the SDGs in their IDPs, research and extension initiatives aimed at sustainability, and actions to integrate the SDGs into academic curricula. In addition, the research reveals challenges faced by institutions, such as the scarcity of resources and the lack of clarity on how to operationalize sustainability in the university context, resulting in a partial implementation of the SDGs. By highlighting the role of universities as agents of change, the study demonstrates how these institutions contribute to sustainable development and prepare their students to face global and local challenges related to the SDGs. In addition to fostering discussions and innovative solutions for the implementation of the SDGs, universities stand out for their ability to generate knowledge, invest in research and education, and contribute to meeting the targets set. The results can guide the implementation of more effective policies at other universities and reinforce the importance of sustainability in international rankings and the global impact of higher education.

Keywords Sustainability · Sustainable development goals · THE Ranking · Public Universities · Institutional development plans

1 Introduction

Sustainability is essential in the strategic planning of higher education institutions (HEIs) (Cortese 2003; Stephens et al. 2008). The adoption of the Sustainable Development Goals (SDGs) has driven significant transformations within these institutions. Universities disseminate knowledge and sustainable practices, shaping professionals who are aware of environmental, social, and economic challenges. International rankings and guidelines play a fundamental role in promoting sustainability in higher education, influencing institutional policies and encouraging the adoption of sustainable practices in universities (Leal Filho et al. 2018; Santos and Barbosa 2021).

The incorporation of sustainability in HEIs occurs in different areas: institutional policies, curricular changes, and campus practices. However, the maturity of this implementation varies depending on the institutionalization of initiatives and the commitment of university managers (Lozano 2006; Kapitulčinová et al. 2018). This study assesses the level of sustainability maturity in the Institutional Development Plans (IDPs) of the top-ranked Brazilian public universities according to the Times Higher Education (THE) ranking. The approach is systemic and comparative, analyzing the integration of the SDGs, best practices, challenges, and opportunities.

Federal Institute Farroupilha, Santana Do Livramento, Rio Grande Do Sul, Brazil

Studies indicate that universities adopt the SDGs in their institutional strategies (Lozano et al. 2015; Findler et al. 2019; Viegas and Cabral 2015). Measures include the incorporation of sustainable principles in IDPs, the creation of sustainability offices, and the promotion of research and outreach projects. However, challenges persist, such as a lack of resources, organizational resistance, and difficulties in defining effective impact indicators.

The adopted methodology is qualitative, based on document analysis of IDPs. Key sustainability dimensions were operationalized through the analysis of explicit goals, performance indicators, and the integration of SDGs into institutional plans (Wright 2002; Roorda 2010). Resource management includes policies for energy efficiency, the use of renewable energy, and water conservation. Innovation in teaching and research involves integrating sustainability content into curricula and developing sustainability—focused projects. Social responsibility encompasses outreach programs and sustainable community development. Sustainable governance considers the presence of sustainability committees or offices and the incorporation of sustainability into institutional guidelines.

This research contributes to the theoretical framework by analyzing how universities institutionalize sustainability in their IDPs. It differs from studies that focus on isolated initiatives or general evaluations. It fills the gap in comparative investigations regarding the level of sustainability maturity in Brazilian higher education institutions.

The article is structured as follows: Sect. 2 presents the theoretical framework, exploring sustainability in higher education and sustainability maturity models. Section 3 describes the methodology. Section 4 discusses the results, highlighting best practices and challenges. Finally, Sect. 5 presents the conclusions and suggestions for future research.

2 Theoretical Framework

In 2015, the Member States of the United Nations approved the 2030 Agenda for Sustainable Development, adopting the 17 SDGs. This Agenda is recognized as a global action plan aimed at eradicating poverty and promoting sustainable development in five main areas: people, planet, prosperity, peace and partnerships (United Nations 2015). In this context, Higher Education Institutions (HEIs) occupy a strategic position, contributing to the adoption of the SDGs beyond the boundaries of their campuses and local communities (Mokski et al. 2023). However, the level of implementation of sustainable practices and initiatives varies significantly between developed and developing countries, as well as depending on institutional archetypes and mission (Leal Filho et al. 2023a).

One of the ways in which HEIs can implement the SDGs is through research, which is essential for advancing sustainable development (Leal Filho et al. 2023b). As highlighted by Salvia et al. (2019), incorporating the SDGs into academic research has helped experts deal with both global and local challenges. Another form of contribution is through teaching, where strategies such as creating new courses or modules focused on sustainable development, or revising existing curricula to include topics related to sustainability, have gained prominence (Pallant et al. 2020; Weiss et al. 2021).

University extension is also a powerful tool, allowing universities to develop programs that promote learning beyond formal education (Tasçÿ and Titrek 2020). In addition, universities can boost sustainable development by transforming their internal operations. This includes adopting practices such as using innovative renewable energy systems on campuses (Wang and Guo 2021), energy optimization in buildings, promoting energy conservation (Gui et al. 2021) and efficient solid waste management (Rimantho et al. 2019).

To better understand the effectiveness of these efforts, the concept of sustainability maturity in organizations has been explored in the academic literature. Lozano (2006) proposed an initial model for assessing sustainability maturity, which has since been expanded by more recent studies. These frameworks assess various dimensions, such as institutional governance, stakeholder engagement, policy integration and impact measurement. By analyzing these aspects, universities can identify strengths and areas for improvement in their sustainability initiatives.

Another relevant approach is the concept of sustainable development cycles, which emphasizes that the implementation of sustainability takes place in iterative stages. Understanding these cycles helps institutions to strategically plan and improve their sustainability actions over time.

However, the implementation of these initiatives is not without its challenges. The SDGs cannot simply be applied as a ready-made package to higher education environments, as sustainability-related decisions involve balancing political, moral and epistemic priorities (Chankseliani and McCowan 2021). These authors highlight the need to document the various approaches adopted by universities to promote sustainable development, in order to understand their combined impacts and effects on different institutional areas. In this way, Institutional Development Plans (IDPs) become essential tools for translating the SDGs into concrete practices, strengthening the role of universities as agents of change and leadership in building a more sustainable future.

Universities are key players in sustainability, as they promote economic, social and environmental development through teaching, research and interaction with society, with an emphasis on the region in which they operate (Fiorani and Di Gerio 2022). According to the authors, in this strategic role, the SDGs, integrated with sustainability reports, function as a reference to express the commitment of institutions to socio—environmental issues, with effective communication of impacts and consistency in institutional planning, which strengthens the role of building a fairer, more inclusive and sustainable society.

Although the SDGs have the potential to promote action in universities, such as SDG 5, which calls for gender equality in higher education, implementation still faces challenges such as the lack of teacher training (Merma-Molina et al. 2024). Therefore, factors such as committed institutional policies, awareness-raising actions and training programs can facilitate integration to reach teachers and students. It is therefore essential to invest in quality teacher training, with incentives and specialized support, so that teaching practices are committed to the 2030 Agenda.

The advancement of sustainable development requires the involvement of higher education institutions in the implementation of the SDGs, since they promote education focused on sustainability and adopt policies with this approach (Brusca et al. 2025). Although integrating the SDGs into curricula is the most recurrent way of applying the goals in the university context (Alcántara-Rubio et al. 2022), universities act as agents of transformation on other fronts, such as social impact projects and environmental management practices, contributing directly to the academic community and society (De Iorio et al. 2022). In general, the implementation of the SDGs in institutions has focused especially on SDG 4 and SDG 9, with a predominant emphasis on the social dimension, compared to the economic and environmental spheres (Alcántara-Rubio et al. 2022).

Achieving the goals of the 2030 Agenda is a global challenge, including for universities. Integration is a constant task and the presence of the SDGs in strategic plans, institutional transparency and university reports and financial structures are key factors in effective communication between institutions and the community and a solid commitment to sustainable development (Brusca et al. 2025). Thus, the authors add that universities with a higher level of transparency tend to have a wider dissemination of the SDGs, suggesting that management practices based on transparency and strengthening the financial structure can be fundamental for advancing the goals in higher education, as well as serving as a reference for institutions in different contexts.

3 Methodology

This research proposed a qualitative analysis of the case study type, seeking to describe the possible complexities, the interactions of the variables obtained in the study, as well as their understanding and classification. Due to its qualitative nature, the research focused on developing the study to find answers that would allow us to understand, describe and interpret the facts analyzed, specifically how sustainability was included in the strategic planning, operational practices and academic curricula of the five universities selected for the study.

To assess the level of sustainability maturity, predefined categories were defined based on established frameworks, such as Lozano's maturity model (2006) and recent methodological adaptations. The model used considers dimensions such as governance and policy alignment, stakeholder engagement, curriculum and research integration, operational sustainability and impact assessment. This innovative approach made it possible to classify universities into different stages of maturity, enabling a structured comparative analysis.

A documentary analysis was carried out of the Institutional Development Plans (IDPs) of the five highest ranked Brazilian universities in THE Impact Rankings 2024. The IDP represents university planning and is an essential tool to help HEIs develop their institutional planning. If an institution wants to be recognized for its sustainability practices and stance, these actions must be clearly foreseen and described in its formal planning.

The choice of universities was based not only on their ranking in THE Impact Rankings 2024, but also on geographical representativeness and institutional diversity. The universities selected were: Universidade de São Paulo—USP (101—200), Universidade Federal de Itajubá—UNIFEI (301—400), Universidade Federal do Pará—UFPA (301—400), Universidade Federal de Santa Maria—UFSM (301—400) and Universidade Estadual de Londrina—UEL (301—400). These institutions were chosen to ensure geographical representation across Brazil's five official regions, guaranteeing diversity for the research object of study (Fig. 1).

After selecting the universities, documentary research began, with the aim of extracting and analyzing the information contained in the IDPs as an autonomous method. This research was carried out between January and February 2025, with reference to the 17 SDGs and sustainability. Data collection and analysis were conducted using the content analysis proposed by Bardin (2011), ensuring a systematic method for categorizing information. NVivo software was used to operationalize this analysis, which enabled the identification and grouping of patterns in the documents examined. The evidence was organized in demonstration tables, analyzing aspects such as the organization of the IDPs, their main characteristics, the theoretical

Fig. 1 The top Brazilian universities in THE impact ranking. *Source* Authors, 2025. *Note* University of São Paulo—USP (red icon), Federal University of Itajubá—UNIFEI (green icon), Federal University of Pará—UFPA (blue icon), Federal University of Santa Maria—UFSM (pink icon) and State University of Londrina—UEL (yellow icon)

basis used, their structure and guidelines, always considering how they contemplate or dialogue with the applicability of the 17 SDGs and sustainability.

4 Results and Discussions

When analyzing the institutional development projects of Brazilian universities, we identified different levels of maturity in the integration of sustainability and the SDGs. To facilitate understanding, we present the results in tables and figures, followed by an analytical discussion.

Quantitative Comparison of Universities by SDG Integration

The table below summarizes the level of maturity in sustainability integration at five universities analyzed, based on their Institutional Development Plans (IDPs) and strategic plans (Table 1).

The adoption of sustainable practices by Brazilian universities has grown significantly in recent years. The analysis presented highlights relevant trends, but also reveals structural and strategic gaps that challenge the consolidation of a truly sustainable institutional culture. The survey points to five of the most frequent practices among the universities analyzed, with an absolute emphasis on inclusion and social responsibility programs, which are present in 100% of the institutions. This shows a collective commitment to equity and the social role of the university, even though, paradoxically, this is precisely one of the dimensions that some universities point to as a challenge, which opens up room for reflection.

For example, although UNIFEI is heavily involved in renewable energy and community actions, it admits that it is a challenge to expand its social sustainability initiatives. This suggests that the presence of inclusive programs does not in itself guarantee their effectiveness or scope. This ambivalence is a central point in the debate on university sustainability: the frequency of a practice does not necessarily indicate its maturity or real impact.

Table 1 Maturity level

University	Integration of the SDGs in the mission and vision	Research and innovation	Sustainable infrastructure	Social inclusion and well-being	Partnerships and governance
USP	High	High	Medium	High	High
UNIFEI	Medium	High	High	Medium	High
UFPA	High	Medium	Medium	High	High
UFSM	High	High	High	High	High
UEL	Medium	Medium	High	High	Medium

Source Elaborated by the authors, 2026

On the other hand, waste management and recycling appear in 80% of universities, which shows a relevant and visible operational trend. However, the presence of this practice in high numbers can hide a weakness: it is often limited to selective collection, without involving the academic community in educational or innovative circular economy processes. UFPA and UEL, for example, are mentioned with strengths in this area, but without detailing the degree of community engagement or the existence of structured institutional policies for reduction and reuse.

As for the adoption of renewable energy (60%) and the existence of sustainability research centers (also 60%), we see an interesting correlation between knowledge production and the implementation of sustainable technologies. UNIFEI and USP stand out in this respect. However, the dependence of these practices on research projects can limit their institutional transversality—knowledge is concentrated in academic niches without generating effective changes in university management as a whole.

Perhaps the most worrying statistic is that sustainable mobility initiatives are present in only 40% of institutions. This figure indicates that aspects directly linked to the daily life of the university community, such as transportation and commuting, are still neglected, even though they have a major environmental and social impact. This reveals a mismatch between practices with symbolic visibility (research, energy, social responsibility) and actions that require structural and logistical reorganization, which are generally more complex.

At the level of specific institutions, the case of USP stands out for its broad integration of the SDGs and its strong research apparatus. However, it faces challenges in terms of sustainable infrastructure, revealing that academic excellence does not automatically guarantee an environmentally efficient campus. A similar situation can be seen at UFPA, which has good partnerships and a focus on waste, but also lacks structural investment. UFSM, on the other hand, although it stands out in its articulation of the SDGs and extension programmes, needs more structured governance, something that points to a common organizational weakness in the public sector.

Finally, UEL exemplifies a model of commitment to the social dimension of sustainability, evidenced by its actions of inclusion and international cooperation. However, its challenge lies in consolidating a scientific base and more robust partnerships to sustain these actions in the long term.

Although there has been significant progress in Brazilian universities in terms of sustainability, the frequency of practices should not be confused with depth or effectiveness. There is a clear institutional effort to promote sustainable initiatives, but the challenges are diverse: from poor infrastructure to weak governance mechanisms. In order to evolve, universities need to move beyond the model of one-off, symbolic actions and move towards systemic, cross-sector strategies monitored by sustainable performance indicators. The future of university sustainability depends on their ability to transform good practices into permanent institutional policies.

5 Conclusion

The analysis of the Institutional Development Plans (IDPs) of the five selected Brazilian public universities reveals a promising but also challenging scenario with regard to integrating the SDGs into higher education. Each of the institutions shows a growing commitment to sustainability, but at different levels of maturity and with different approaches.

The University of São Paulo (USP) has established itself as a benchmark for sustainability in Brazil and worldwide. Its initiatives show that it is possible to integrate sustainable practices into large institutions and promote positive impact on multiple fronts. USP not only implements these practices, but also trains responsible professionals to implement these values in their areas of work. The university stands out as a center of innovation, developing technologies and solutions that can be replicated by other institutions and sectors. With measures such as reusing water, using renewable energy and reducing emissions, USP contributes directly to mitigating climate change. Its practices inspire other universities and organizations and show that sustainability is possible and brings economic, social and environmental benefits. In line with the United Nations SDGs, USP emphasizes its commitment to a more sustainable future and demonstrates that universities but also in institutional operations" ou simplesmente remover "but also in practice.

The Federal University of Itajubá (UNIFEI) has made significant efforts to align its teaching, research and extension activities with the SDGs. Initiatives such as the Green Hydrogen Center (CH2V) and the inclusion of sustainable themes in academic curricula stand out. The university has also made progress in implementing sustainable operational practices, such as the efficient management of resources and the promotion of sustainable mobility. With its strong links to the productive sector and history of innovation, UNIFEI has the potential to become a benchmark in renewable energies and environmental management. However, there is still room for greater integration of the SDGs at all institutional levels and for the expansion of partnerships to boost sustainable projects.

The Federal University of Pará (UFPA) plays a key role in promoting sustainability, especially as it is located in a strategic region like the Amazon. The university takes on the responsibility of protecting and studying the Amazon, promoting practices that reconcile sustainable development and environmental preservation. In addition to implementing sustainable actions, UFPA educates the academic and external community, creating a culture of respect for the environment. Its research and projects have a direct impact on the Amazon region and offer solutions that can be applied in other parts of the world. UFPA's actions are aligned with the SDGs, reinforcing its commitment to the global sustainability agenda. The university demonstrates that it is possible to integrate sustainable practices into a public institution, even in the face of economic and social challenges. In this way, UFPA positions itself as a benchmark for sustainability in the Amazon and reaffirms its commitment to transformation towards a more sustainable future.

The Federal University of Santa Maria (UFSM), recognized for its mission to build and disseminate knowledge in a sustainable way, has demonstrated its commitment to the SDGs through the 2016–2026 IDP, strategically integrating them into its goals and actions. Among the main initiatives are the Sustainable UFSM (UMA) program and energy efficiency projects, such as photovoltaic plants and the mini biodiesel plant. In addition, UFSM is seeking advances in basic sanitation, with proposals for collection networks and modern treatment plants, which will also serve as a field of study for research and teaching. Health promotion and the protection of green areas, with the proposal for an Environmental Management System (EMS), are other initiatives that demonstrate UFSM's commitment to community well-being and environmental preservation.

Despite the progress made, there are opportunities to increase UFSM's impact on sustainable development. The effective implementation of the EMS can consolidate the university's environmental management, while the expansion of partnerships with public and private institutions can enable the financing of sustainable projects and innovative initiatives. The cross-cutting integration of the SDGs into academic curricula and sustainability awareness campaigns can increase the engagement of the university community. In addition, UFSM can explore new technologies and innovative methodologies, such as artificial intelligence and circular economy, to optimize its processes and create a permanent committee to monitor and evaluate the adherence of institutional actions to the SDGs, promoting greater involvement of the academic and external community.

The State University of Londrina (UEL) has strategic objectives aimed at sustainability in the university context, and is notable for creating an Environmental Management Plan aligned with the SDGs. This plan encourages environmental actions, leadership and the formation of groups to disseminate environmental education practices. The university is expanding its social impact initiatives, including culture, health and the well-being of the academic community, as well as working on accessibility and reducing physical and methodological barriers to promote an inclusive university. In the environmental context, the Reciclabel projects, the biogas plant and solar energy generation are key to controlling environmental risks and impacts.

Despite the progress made, UEL could give greater prominence to sustainability in the formulation and presentation of the IDP. Including the topic in the university's institutional principles, mission, vision and strategic initiatives could better detail the actions taken and the impacts generated. In addition, the creation, monitoring and dissemination of sustainability indicators could contribute to the achievement of strategic objectives and enable continuous monitoring of institutional progress.

In general, the commitments made, projects and practices already developed in line with the SDGs show that universities are constantly seeking to educate aware citizens and build a fair, inclusive, sustainable and innovative future.

The analysis of the IDPs of the five Brazilian public universities reinforces the importance of the role of higher education institutions as agents of change in promoting sustainability. Although each university has its own particularities and challenges, they all share a commitment to contributing to a more sustainable future, in line with the UN SDGs.

The results of this study highlight the need for universities to strengthen their governance structures in order to institutionalize sustainable practices. A national framework for assessing sustainability maturity could provide standardized benchmarks, allowing for more precise monitoring of progress in the sector. In addition, strengthening public–private partnerships could provide funding and expertise to expand sustainable initiatives at universities.

To facilitate the adoption of similar practices by other universities, it is recommended that a Sustainability Maturity Framework be developed to help institutions assess and improve their integration with the SDGs. Inter-institutional collaboration can strengthen the exchange of knowledge and increase the impact of initiatives. In addition, universities should consider incorporating sustainability indicators into their performance metrics, ensuring more effective institutional planning in line with global environmental and social challenges.

By adopting these measures, Brazilian universities will be able to align themselves even more with global sustainability standards and contribute more effectively to the Sustainable Development Goals.

References

Alcántara-Rubio, L., Valderrama-Hernández, R., Solís-Espallargas, C., & Ruiz-Morales, J. (2022). The implementation of the SDGs in universities: a systematic review. Environmental *Education Research*, 28(11), 1585–1615. https://doi.org/10.1080/13504622.2022.2063798

Alshuwaikhat, H. M.; Abubakar, I (2008). An integrated approach to achieving campus sustainability: assessment of the current campus environmental management practices. *Journal of Cleaner Production*, v. 16, n. 16, p. 1777–1785.

Bardin, Laurence (2011). Content analysis. São Paulo: Edições 70, 229 p.

Brusca, I., Olmo, J., & Pérez-Espés, C. (2025). Are the SDGs embedded in university strategies and reporting practices? Analysing influencing factors. *Public Money & Management*, 1–11. https://doi.org/10.1080/09540962.2025.2477044

Carvalho, J., & Ribeiro, M. (2023). Sustainability in universities: a study on good practices. *Environmental Management Journal*, 15(2), 80–95.

Chankseliani, M., & McCowan, T. (2021). Higher education and the Sustainable Development Goals. *Higher Education, 81*(1), 1–8. https://doi.org/10.1007/s10734-020-00652-w

Cortese, A. D. (2003). The critical role of higher education in creating a sustainable future. *Planning for Higher Education, 31*(3), 15–22.

Costa, J., & Almeida, M. (2022). Sustainability in universities: a comparative study. *Brazilian Journal of Environmental Education*, 15(3), 120–135

De Iorio, S., Zampone, G., & Piccolo, A. (2022). Determinant factors of SDG disclosure in the university context. *Administrative Sciences*,12(1),21. https://doi.org/10.3390/admsci12010021

Findler, F., Schönherr, N., Lozano, R., & Stacherl, B. (2019). Assessing the impacts of higher education institutions on sustainable development—An analysis of tools and indicators. *Sustainability, 11*(1), 59.

Fiorani, G., & Di Gerio, C. (2022). Reporting university performance through the sustainable development goals of the 2030 agenda: lessons learned from Italian case study. *Sustainability*, 14(15), 9006. https://doi.org/10.3390/su14159006

Gui, X., Gou, Z., & Lu, Y. (2021). Reducing university energy use beyond energy retrofitting: The academic calendar impacts. *Energy and Buildings, 231,* 110647. https://doi.org/10.1016/j.enbuild.2020.110647

Kapitulčinová, D. et al (2018). Towards integrated sustainability in higher education: Mapping the use of the Accelerator toolset in all dimensions of university practice. *Journal of Cleaner Production*, v. 172, p. 4367–4382.

Leal Filho, W. et al (2018b). Planning and implementing sustainability in higher education institutions: an overview of the difficulties and potentials. International Journal of Sustainable Development and World Ecology, v. 25, n. 8, p. 712–720.

Leal Filho, W., Skanavis, C., do Paço, A., Rogers, J., Kuznetsova, O., & Castro, P. (2019). The role of planning in implementing sustainable development in a higher education context. *Journal of Cleaner Production, 235*, 678–687.

Leal Filho, W., Aina, Y. A., Dinis, M. A. P., Purcell, W., & Nagy, G. J. (2023a). Climate change: Why higher education matters? *Science of the Total Environment, 892*, 164819. https://doi.org/10.1016/j.scitotenv.2023.164819

Leal Filho, W., Salvia, A. L., & Eustachio, J. H. P. (2023b). An overview of the engagement of higher education institutions in the implementation of the UN Sustainable Development Goals. *Journal of Cleaner Production, 386*, 135694. https://doi.org/10.1016/j.jclepro.2022.135694

Lozano, R (2006). Incorporation and institutionalization of sustainable development into universities: breaking through barriers to change. *Journal of Cleaner Production*, v. 14, n. 9-11, p. 787–796.

Lozano, R. et al (2015). A review of commitment and implementation of sustainable development in higher education: results from a worldwide survey. *Journal of Cleaner Production*, v. 108, p. 1–18.

Merma-Molina, G., Urrea-Solano, M., & Hernández-Amorós, M. J. (2024). The integration of gender equality (SDG 5) into university teaching: the view from the frontline. *Innovative Higher Education*, 49(3), 419–452. https://doi.org/10.1007/s10755-023-09668-3

Mokski, E., Leal Filho, W., Sehnem, S., & Andrade Guerra, J. B. S. O. d. (2023). Education for sustainable development in higher education institutions: An approach for effective interdisciplinarity. *International Journal of Sustainability in Higher Education, 24*(1), 96–117. https://doi.org/10.1108/IJSHE-07-2021-0306

Pallant, E., Choate, B., & Haywood, B. (2020). How do you teach undergraduate university students to contribute to UN SDGs 2030? In W. Leal Filho et al. (Eds.), *Universities as living labs for sustainable development. World Sustainability Series*. Springer, Cham. https://doi.org/10.1007/978-3-030-15604-6_5

Rimantho, D., Hidayah, N., & Herzanita, A. (2019). Analysis of risk failure of solid waste management processes in universities: Case study of Pancasila University Jakarta. In *IOP Conference Series: Earth and Environmental Science* (Vol. 399, No. 1, p. 012045). https://doi.org/10.1088/1755-1315/399/1/012045

Roorda, N (2010). Sailing on the winds of change: The Odyssey to sustainability of the universities of applied sciences in the Netherlands. *Journal of Cleaner Production*, v. 18, n. 7, p. 649–658.

Salvia, A. L., Leal Filho, W., Brandli, L. L., & Griebeler, J. S. (2019). Assessing research trends related to Sustainable Development Goals: Local and global issues. *Journal of Cleaner Production, 208*, 841–849. https://doi.org/10.1016/j.jclepro.2018.09.242

Santos, P. M.; Barbosa, A. M (2021). Sustainability practices in higher education institutions: a bibliometric review. *Revista Sustenere*, v. 3, n. 2, p. 45–62.

Soini, K., & Birkeland, I. (2014). Exploring the scientific discourse on cultural sustainability. *Geoforum, 51*, 213–223.

Stephens, J. C. et al (2008). Higher education as a change agent for sustainability in different cultures and contexts. *International Journal of Sustainability in Higher Education*, v. 9, n. 3, p. 317–338.

Taşçı, G., & Titrek, O. (2020). Evaluation of lifelong learning centers in higher education: A sustainable leadership perspective. *Sustainability, 12*(1), 22. https://doi.org/10.3390/su12010022

Trencher, G. et al (2013). Co-creating sustainability: cross-sector university collaborations for driving sustainable urban transformations. *Journal of Cleaner Production*, v. 50, p. 40–55.

United Nations. (2015). *Transforming our world: The 2030 Agenda for Sustainable Development*. https://sustainabledevelopment.un.org/post2015/transformingourworld

Verhulst, E., & Lambrechts, W. (2015). Fostering the incorporation of sustainable development in higher education: Lessons learned from a change management perspective. *Journal of Cleaner Production, 106*, 189–204.

Viegas, C. V.; Cabral, C. R (2015). Sustainability in Higher Education and Organizational Culture. *Researchgate*, V. 12, N. 1, P. 25–40.

Wang, X., & Guo, L. (2021). How to promote university students to innovative use renewable energy? An inquiry-based learning course model. *Sustainability, 13*(3), 1418 https://doi.org/10.3390/su13031418

Weiss, M., Barth, M., Wiek, A., & von Wehrden, H. (2021). Drivers and barriers of implementing sustainability curricula in higher education: Assumptions and evidence. *Higher Education Studies, 11*(2), 42–64. https://doi.org/10.5539/hes.v11n2p42

Wright, T. S. A (2002). Definitions and frameworks for environmental sustainability in higher education. *Higher Education Policy*, v. 15, n. 2, p. 105–120.

Andrieli de Oliveira Cardoso Master Degree in Accounting Sciences from the Postgraduate Program in Accounting Sciences at the Federal University of Santa Maria—UFSM. Orcid: https://orcid.org/0000-0002-9032-2489. Bachelor in Accounting Sciences at the Regional Integrated University of Alto Uruguai and Missões, Santiago Campus of URI. AV Batista Bonotto Sobrinho n° 733.São Vicente Postcode: 97,711–500—Santiago, RS—Brazil.

Diego Marques Cavalcante PhD student in production engineering in the post-graduate program in Production Engineering at the Federal University of Santa Maria—UFSM. Avenida Roraima, nº 1000. Camobi. 97,105–900—Santa Maria, RS—Brazil.

Ana Paula Belizário Master student in Production Engineering at the Federal University of Santa Maria (UFSM). Santa Maria, Rio Grande do Sul, Brazil. Av. Roraima, nº 1000, 97,105–900, UFSM.

Santa Maria, Rio Grande do sul, Rua Cel Getúlio Mario Zanchi 80.

Nisrin Naiel Dib Khaled Master student in Production Engineering at the Federal University of Santa Maria (UFSM). Santa Maria, Rio Grande do Sul, Brazil. Av. Roraima, nº 1000, 97,105–900, UFSM.

Débora Londero Kieling Master Degree in Accounting Sciences from the Graduate Program in Accounting Sciences at the Federal University of Santa Maria Santa Maria, Rio Grande do Sul, Brazil, Av. Roraima, n. 1000, 97,105–900, UFSM.

Elda Rodrigues Steinhorst Kraetzig Professor and Researcher at Federal University of Pampa—UNIPAMPA

Graduate Program in Production Engineering (PhD candidate) Federal University of Santa Maria—UFSM Master in Production Engineering—UFSM. Bachelor in Business Administration—Federal Institute Farroupilha Santana do Livramento, Rio Grande do Sul, Brazil, Rua. Barão do Triunfo, 1048—Centro, Sant'Ana do Livramento—RS, 97,573–634.

Lucas Veiga Ávila Professor of the Graduate Program in Production Engineering and the Graduate Program in Accounting Sciences. Federal University of Santa Maria.

Sustainable Education: Integrating Walkability and Assistive Technology in University Campuses

Paula Lelis Rabelo Albala, Ana Carolina Cordeiro Correia Lima, and Renan do Nascimento Balzani

Abstract Integrating teaching and research within the university environment enriches both areas. On the one hand, it strengthens students' critical formation by providing practical and up-to-date experiences, thereby enhancing the learning process. On the other hand, it fosters innovation by applying academic knowledge to the development of novel solutions. In this context, the objective of this article is to explore the integration of teaching and research to promote accessibility and inclusive mobility within the university context. The study centers on the elective course "Special Studies in Technology—Walkable Urbanism," offered at the Faculty of Architecture and Urbanism of the University of Brasília (FAU-UnB), which was coordinated with two applied research projects "Acessa UnB" and "Trilhas UnB." Both initiatives aimed to develop Assistive Technologies (AT) to enhance campus accessibility for visually impaired individuals, through two thematic lines: the production of tactile maps and the development of an app for safe and autonomous mobility. The course focused on discussing urban planning oriented towards walkability, adapting its content to align with the research projects. The study applied the Design Science Research (DSR) methodology, combined with active learning principles, and structured in three phases. The first phase involved a literature review conducted with students. The second phase consisted of participatory fieldwork with students and visually impaired volunteers to assess walkability conditions and barriers in the campus, along five predestinated pathways. This phase included the application of the Pedestrian Ergonomic Comfort Index (ICEP). In the final phase, students developed design proposals at both macro and micro scales based on the findings. The project

P. L. R. Albala (✉)
Faculty of Architecture, Urbanism and Design, University of São Paulo, São Paulo, SP, Brazil
e-mail: paula.rabelo@unb.br

P. L. R. Albala · A. C. C. C. Lima · R. do Nascimento Balzani
Faculty of Architecture and Urbanism, University of Brasilia, Campus Darcy Ribeiro, Brasília, DF, Brazil
e-mail: ana.ana@unb.br

R. do Nascimento Balzani
e-mail: renan.balzani@unb.br

W. Leal Filho et al. (eds.), *Sustainability in the Curriculum and Operations of Universities*, World Sustainability Series,
https://doi.org/10.1007/978-3-032-12678-8_33

outcomes included the elaboration of tactile maps, app prototypes, and a collaborative tool to collect georeferenced user feedback. The results demonstrated mutual benefits: students became more engaged through the practical application of their proposals, while the project contributed to advancing knowledge in the development of tactile maps and assistive applications. The initiative highlighted the potential of integrating teaching, research, and innovation to generate practical solutions for real-world challenges. It also reinforced the value of inclusive education, participatory design, and interdisciplinary collaboration. The approach is scalable and replicable, offering a model for other academic and urban environments.

Keywords Accessibility · Walkability · Assistive technology · Inclusive design · Visual impairment

1 Introduction

Sustainability, beyond its environmental dimension, encompasses social and economic aspects. In this context, issues related to inclusion and accessibility play a fundamental role, as they promote environments that aim to accommodate all individuals, regardless of their physical, cognitive, or sensory conditions. Indeed, these themes are related to the Sustainable Development Goals (SDGs), particularly SDG 4, SDG 10, and SDG 11, which address quality education, reduced inequalities, and sustainable cities and communities, respectively (UN 2015).

Within higher education institutions, promoting inclusion and accessibility as part of their transversal policies has become both a challenge and a necessity in the contemporary context. Beyond addressing these topics from a broad perspective, it is essential to raise awareness among faculty and students, enabling them to "see reality" as it is, with its specific issues, challenges, and barriers. To this end, it is fundamental to foster, whether in the classroom or through research, awareness, and reflection on the importance of universal design and inclusive design in educational environments.

In general terms, universal design advocates for a human-centered architecture and design approach that embraces human diversity. It can be understood as the design of products, environments, programs, and services that are conceived to be usable by all people, without the need for adaptation or specialized design, including assistive technologies (ABNT 2021). Inclusive design, in turn, goes beyond accessibility by considering human diversity as an active component in the design process, involving real users in the solutions development (Imrie and Hall 2001). In this debate, it is important to move beyond the physical aspects of the built environment to include, for instance, issues such as digital accessibility.

The creation of reference spaces for accessibility and inclusion on university campuses, both at the architectural and urban scales, can contribute to the physical readjustment of spaces and cultural and behavioral changes within both external and internal environments. In this regard, it is necessary to debate and incorporate pedagogical strategies and examples of best practices that may contribute to the education of professionals who are more sensitive to the demands of accessibility and inclusion, fostering transformation at both academic and professional levels.

Active learning methodologies, such as problem-based learning, can serve as fundamental tools in this process, by engaging students in the identification and resolution of real-world problems. These methods encourage a critical and proactive stance toward architectural and urban barriers that hinder accessibility. Thus, active methodologies reshape the educational process by placing students at the center of learning, fostering autonomy, collaboration, and problem-solving, and promoting greater engagement and meaningful learning (Guarizzo et al. 2024; Diesel et al. 2017). This shift in pedagogy better prepares students for real-world challenges, particularly when applied to global issues such as sustainability (Morán et al. 2015; Guarizzo et al. 2024). This process becomes even more effective when it integrates teaching, research, and community outreach within the university context.

Considering the above, this article aims to explore the integration of teaching and research to promote accessibility and inclusive mobility within the university context, through the elective course "Special Studies in Technology—Walkable Urbanism," offered at the Faculty of Architecture and Urbanism of the University of Brasília (FAU-UnB), developed in coordination with the applied research projects "Acessa UnB" and "Trilhas UnB". These research projects were funded through a public call issued by the Directorate of Accessibility (DACES) at the University of Brasília, dedicated to the development of assistive technologies for UnB's campuses. These projects focused on promoting accessibility through two thematic axes: the production of a campus tactile map and the development of an application to ensure safe and autonomous mobility for visually impaired individuals. This integration was grounded in active teaching methodologies and inclusive design principles, with the objective of contributing to the development of the maps and the app through the provision of technical data and student proposals for the campus.

2 Accessibility and Walkability in the University Context

2.1 Acessibility Overview at the University Campuses

The inclusion of people with disabilities, in any context, must go beyond mere compliance with technical standards and legal requirements. This demand is particularly urgent within the university environment, where conditions of inclusion and accessibility can be determining factors for student retention. Such a perspective requires a profound transformation in how university spaces are conceived, perceived, and appropriated by all users (Martinez 2020). It is therefore essential to integrate diverse fields of knowledge with institutional and pedagogical practices—such as classroom accessibility, universal design in academic units, and the development and incorporation of assistive technologies across various dimensions of institutional life.

Campus urban planning, in this regard, must be approached in a way that ensures equity, autonomy, and safety for the entire academic community, addressing the specific needs and challenges faced by people with disabilities. Particularly in public universities, institutional interventions remain limited and fragmented (Diniz and Silva 2021). Although the Brazilian legal framework guarantees the right to education (Brazil 1988), the lack of incentives and the incomplete implementation of public inclusion policies fail to adequately overcome existing barriers (Ciantelli and Leite 2016).

Architectural, technological, attitudinal, and communicational barriers are particularly significant in hindering full participation in university life (Reginato et al. 2021). These barriers are often invisible, partially mitigated by fragmented inclusion policies, and tend to be perceived only by those directly affected, people with disabilities or others with diverse accessibility needs. Furthermore, qualitative studies indicate a lack of university training in universal accessibility. These obstacles negatively affect the institutional climate and are exacerbated by insufficient investment in support services for people with disabilities (Fahd et al. 2023).

Accessibility measures must, therefore, be tailored to ensure safe and autonomous navigation within the university environment. Architectural obstacles, such as inaccessible or discontinuous pathways, inadequate connections to transportation systems, poorly designed ramps, unsafe pedestrian crossings, and deteriorating sidewalks, represent some of the most critical challenges to creating inclusive and accessible spaces in higher education institutions. Under such conditions, autonomous mobility and orientation become almost impossible for users with disabilities in public universities (Ciantelli and Leite 2016).

Given this scenario, it is crucial to integrate multiple disciplines with institutional and pedagogical practices—such as accessible classrooms, universal design within academic settings, and the development and implementation of Assistive Technologies (AT) in various aspects of institutional life. From this perspective, AT emerges as a fundamental tool for promoting autonomy and the full participation of students, faculty, and staff with disabilities. According to the Brazilian Law of Inclusion (Brazil 2015), AT comprises resources and services aimed at increasing, maintaining, or improving the functional capabilities of individuals with disabilities.

Examples include tactile maps and mobile applications. Tactile maps convey information through raised elements, such as text, symbols, and braille, and are essential for the education, orientation, and mobility of individuals with visual impairments. They help users understand their surroundings and support more autonomous movement within university spaces (Nogueira 2017). Likewise, smartphones and tablets offer new possibilities for orientation and navigation, thanks to real-time information and built-in accessibility features such as zoom, voice assistants, and screen readers (Borges and Mendes 2021).

Recent progress has been made in promoting the inclusion of people with disabilities in Brazilian higher education. These advances are the result of user-centered studies evaluating campus accessibility and the implementation of federal programs.

Notable improvements have been recorded on campuses in cities such as São Carlos, Florianópolis, São Paulo, Rio de Janeiro, Recife, Fortaleza, Cuiabá, and Curitiba (Teixeira and Sumiya 2024).

At the University of Brasília (UnB), accessibility policies have been gradually formalized over recent decades. In the late 1990s, the Support Program for People with Special Needs (PPNE) was created following extensive discussions on the conditions for admission, retention, and graduation of students with disabilities and specific educational needs. In 2019, UnB approved its official Accessibility Policy. Subsequently, in 2020, the institutional structure of the Office of Community Affairs (DAC) was reorganized, leading to the establishment of the Directorate of Accessibility (DACES). Despite this progress, significant challenges remain on UnB campuses regarding the elimination of physical, sensory, informational, and communicational barriers.

Considering the above, this literature review underscores that accessibility on university campuses is a dynamic and evolving field. Its broad and transdisciplinary scope encompasses tools, services, practices, methods, and strategies aimed at fostering autonomy, independence, and quality of life for people with disabilities. Despite institutional advancements and robust legal frameworks, such as those implemented at UnB and other Brazilian universities, persistent architectural, communicational, and attitudinal barriers continue to undermine full inclusion. Consequently, accessibility must be understood as an ongoing and systemic commitment, user-centered and supported by integrated actions in planning, pedagogy, and technology.

2.2 *From Walkability to Accessibility: A Framework for Inclusive Campus Design*

Accessibility on university campuses is also intrinsically connected to walkability, as a well-designed academic setting must ensure safe, comfortable, and inclusive displacements for all users. Walkability is a recent concept, first introduced in the 1990s and generally defined as the "quality of place" (Bradshaw 1993). Over time, the concept has evolved to incorporate multiple dimensions related to walking in urban areas. In a more comprehensive definition, walkability can be described as the extent to which the built environment supports and encourages walking with comfort and safety, connecting people to various destinations within a reasonable timeframe and effort, while offering visual interest along the way (Albala 2022). Walkability for people with disabilities on university campuses is severely hindered by physical and architectural barriers. Therefore, ensuring walkability in the university context is essential for the effective academic and social participation of all campus users.

In addition to this discussion is the notion of universal accessibility, the condition that allows all people, including those with disabilities or reduced mobility, to use urban spaces safely and independently. In the context of university campuses, walkability should therefore be seen not only as part of sustainable urban mobility but also as a fundamental pillar to promote social inclusion and comprehensive campus sustainability. Incorporating walkability, environmental comfort, and universal accessibility into the strategic planning of university spaces can enhance equitable access to educational infrastructure and reinforce the institutional commitment to truly inclusive and welcoming environments. Also, from an urban planning perspective, the notion of "environmental diversity"—which calls for the integration of varied thermal, visual, ergonomic, and auditory conditions to enhance both architectural experience and pedestrian well-being—has been emphasized by several authors (Steane and Steemers 2004; Albala and Kronka 2024).

In this regard, urban ergonomics plays a key role in ensuring walkability, as it addresses how urban space accommodates the dimensions, abilities, and limitations of the human body. For people with disabilities, older adults, children, or with reduced mobility, ergonomic comfort in public space involves not only the removal of physical barriers but also the provision of continuous, adequate surfaces, appropriate inclines, accessible urban furniture, and clear tactile and visual signage. Moreover, additional features such as shaded paths, rest areas, and engaging surroundings further enhance the pedestrian experience. These elements contribute to more inclusive, safe, and functional urban environments, allowing all users to move autonomously and enjoy public space fully. There are many walkability indexes that contribute to evaluating and comparing different urban situations. One of them is the Pedestrian Ergonomic Comfort Index (ICEP), an assessment tool that can be used to evaluate walkability from the perspective of accessibility and well-being (Albala and Kronka 2024). ICEP includes seventeen indicators across five categories: (i) traffic safety, including speed limits, pedestrian crossings, and accident history; (ii) public safety, including lighting and crime rates; (iii) path quality, such as sidewalk width and slope; (iv) attractiveness and diversity, such as active frontages and visual permeability; and (v) mobility and accessibility, including bike infrastructure, transit access, and block dimensions. This framework helps identify gaps and intervention opportunities in the urban built environment, contributing to more inclusive public spaces.

In summary, promoting walkability on university campuses is not merely a matter of improving the quality of walking. It is deeply interrelated to the creation of more accessible and inclusive environments. When walkability is approached from the perspective of architectural and urban barriers, it becomes a guiding principle for rethinking spatial organization and daily experiences across the campus.

3 Methods

The course was developed in coordination with the Acessa UnB and Trilhas UnB projects, which aimed to promote inclusion and accessibility on campus, with a focus on individuals who are blind or have low vision. These initiatives employed participatory methodologies, assistive technologies, and interdisciplinary approaches. The project teams comprised professionals from various fields of knowledge, including architecture, linguistics, pedagogy, engineering, and software development.

Both Acessa UnB and Trilhas UnB adopted the Design Science Research (DSR) methodology, which is particularly suited for problem-solving and system or routine improvement through the development of solutions, including the creation of artifacts (Dresch et al. 2015). Broadly speaking, DSR is a prescriptive research method that, through theoretical and methodological rigor and a trans- and multidisciplinary approach, seeks to generate descriptive, explanatory, and prescriptive knowledge. Its main research methods include case studies, action research, data collection techniques, and validation of the artifacts developed.

Accordingly, the work was structured in three phases. The first phase involved a literature review on accessibility and assistive technologies, with an emphasis on higher education context. The second phase consisted of participatory research (Severino 2016), involving, *in loco,* observations of the main accessibility barriers, both urban and architectural, on campus, with analysis of various routes leading to the university's main buildings. The third and final phase involved the application of the acquired knowledge through the development of assistive technologies, including tactile campus maps and prototypes of a low-cost braille printer, a sensory wristband, and a mobile app designed for users who are blind or have low vision.

The map and app processes development followed four stages: "understand, imagine, design, and prototype," promoting active user participation and applying the principles of Universal Design (Fig. 1).

The course integration into the research projects contributed to knowledge production and application across all three phases. The course scope aimed to provide both theoretical and practical inputs on evaluating walkability on campus, in addition to contributing to the mapping, diagnosis, and proposal of solutions to improve pedestrian routes in predefined areas. The selected routes corresponded to the five

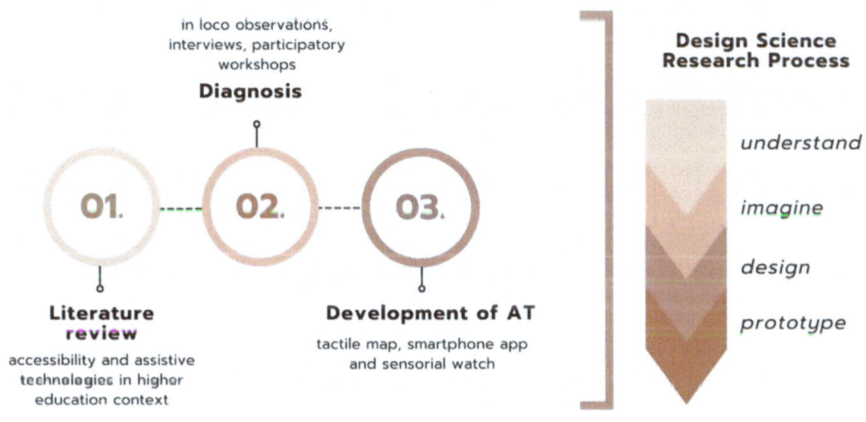

Fig. 1 Research method flowchart. *Source* Author's elaboration (2024)

UnB safety corridors (Fig. 2), which are monitored paths under university security surveillance and benefit from improved public lighting and cleanliness (UnB 2018).

Accordingly, the course's contributions were structured into the following stages:

1. Theoretical and conceptual foundations of walkability and its operative scales;
2. A survey of national and international references on pedestrian-oriented urban design and interventions, conducted through student research;
3. Student participation in a participatory sensory workshop organized by the projects, including members of the academic community who are blind or have low vision;

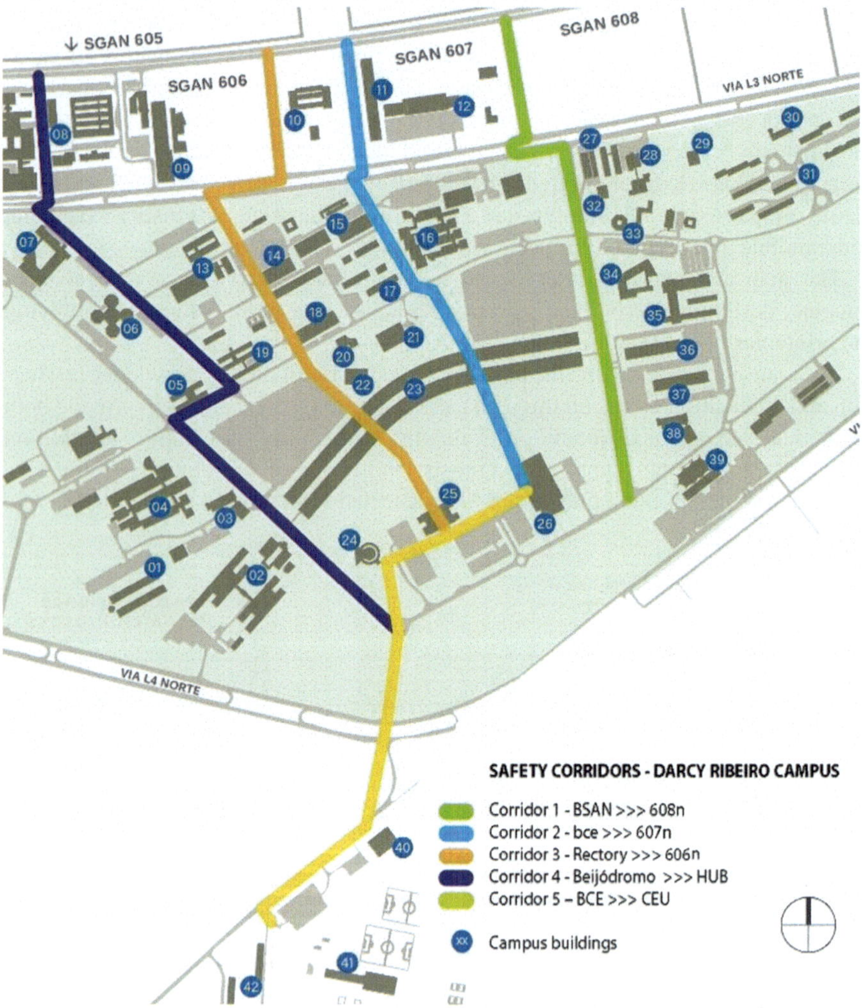

Fig. 2 Safety corridors at UnB. *Source* Author's elaboration, adapted from UnB (2024)

4. Technical diagnosis of the five UnB safety corridors, including on-site measurements and inspections, and application of the ICEP—Pedestrian Ergonomic Comfort Index;
5. Development of macro and micro scale proposals for improving the five routes;
6. GIS-based cartographic mapping of positive and negative walkability conditions across campus, later integrated as a feature of the mobile application;
7. Identification of the most accessible routes between campus buildings to support the development of the tactile map.

4 Results and Discussion

The most recent District Household Sample Survey (PDAD) of the Federal District, where the Darcy Ribeiro campus of the University of Brasília is located, reported 113,642 residents with some form of disability (IPEDF, 2021). Among them, 43.2% have visual impairments, 22.6% have multiple disabilities, and 19.8% have physical impairments. Educational data reveal a stark inequality: only 13.8% of people with disabilities aged twenty-five or older have completed higher education, compared to 35.9% among people without disabilities. These disparities underscore the urgent need for more inclusive educational environments and targeted institutional measures to guarantee access and retention in higher education for this population.

In this context, the integration of the course Special Studies in Technology—Walkable Urbanism with the research projects Acessa UnB and Trilhas UnB provided a rich academic environment, reinforcing the relevance of active methodologies and project-based learning in architectural and urban education. This articulation between theory and practice was central to the development of design proposals. Throughout the semester, students tackled real-world challenges by mobilizing technical knowledge in accessibility, urbanism, ergonomics, universal design, and digital technologies applied to urban space. These collaborative efforts generated concrete outcomes, supported by interdisciplinary dialogue and community engagement.

The development of the project was structured in four main result stages: walkability diagnosis of safety corridors; macro and micro scale design proposals; development of tactile maps; and prototyping of assistive technologies.

4.1 Diagnosis and Proposals for Safety Corridors

The five designated safety corridors at the University of Brasília (UnB) were assessed in terms of their walkability, with a particular focus on the built environment and its urban ergonomics. To conduct this evaluation, the Pedestrian Ergonomic Comfort Index (ICEP) was applied (Fig. 3), which measures walkability across five categories, as previously outlined. Key elements observed included the need for sidewalk

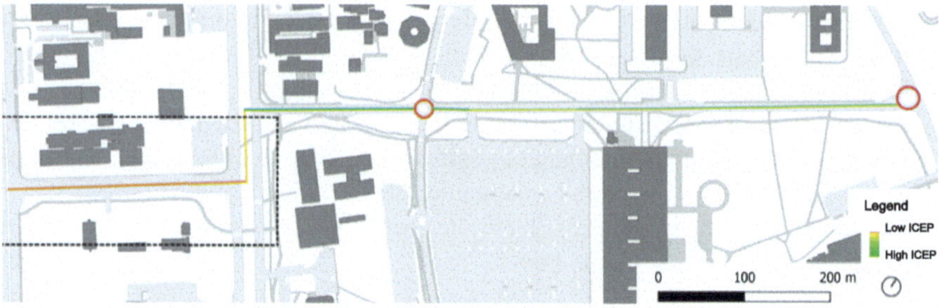

Fig. 3 Evaluation of a part of Corridor 1, developed by student Milana Nogueira. *Source* Authors' personal archive (2024)

upgrades, the installation of tactile and visual signage, the creation of shaded areas, the addition of urban furniture, and the expansion of public lighting infrastructure.

The following summarizes the key findings for each of the corridors:

- Safety Corridor 1 (608N—IREL/IPOL): A bike lane is present but lacks adequate lighting. The sidewalks are narrow, poorly shaded, and include hazardous crossings. Recommendations include improved lighting, horizontal signage, and the addition of urban furniture.
- Safety Corridor 2 (BCE—607N): Characterized by discontinuous sidewalks and low visual and functional appeal. Proposals include improved cycling connectivity, sidewalk upgrades, and the inclusion of benches and vegetation.
- Safety Corridor 3 (606N—Rectory): Marked by blind façades and perceived insecurity. Suggested interventions include activation of building frontages, redesigned pedestrian crossings, and enhanced lighting.
- Safety Corridor 4 (University Hospital—Beijódromo): Primary issues relate to inadequate lighting and sidewalk conditions. Proposed measures involve pavement renewal, improved connections to points of interest, and increased tree planting.
- Safety Corridor 5 (BCE—CEU): Identified as the most critical route, requiring comprehensive restructuring in terms of accessibility, tree cover, signage, and integration with public transportation.

Among the evaluated segments, Corridor 2 (Fig. 4), that starts at the entrance to the Central Library (BCE), stood out as one of the most successful examples of inclusive and walkable campus infrastructure. This segment is fully pedestrian and cyclist-oriented, minimizing conflicts with vehicular traffic. It benefits from adequate lighting, abundant shade providing thermal comfort, and the presence of urban furniture and other points of interest. These features contribute not only to safety and usability but also to the spatial and social quality of the area, making it a model for pedestrian-centered campus design.

In contrast, Corridor 5 (Fig. 4), particularly its peripheral section, exhibited the most critical conditions in terms of walkability and accessibility. This segment suffers

Fig. 4 Segments of safety corridors 2 (left) and 5 (right). *Source* Authors' personal archive (2024)

from a lack of inclusive infrastructure, with fragmented and discontinuous pathways that hinder safe pedestrian flow. The absence of shade leads to high thermal discomfort, especially during the dry season, while poor signage impairs wayfinding, particularly for visually impaired users. Its criticality is accentuated by the fact that it provides access to student housing. These deficiencies underscore the urgent need for targeted interventions to promote equity and safety throughout the campus environment.

Based on field surveys, a georeferenced synthesis map was produced by the students using the Geographic Information System (GIS), consolidating the strengths and weaknesses of all corridors in terms of walkability. The main positive aspects observed included the presence of trees in specific areas, good visibility along certain segments, and connectivity with academic centers. On the other hand, negative aspects were associated with sidewalk discontinuity, lack of tactile signage, insufficient shading, absence of connected bike lanes, unsafe pedestrian crossings, and lack of public lighting (Figs. 5 and 6).

Finally, based on the ICEP evaluation and the diagnostic study, students developed proposals at both macro and micro scales (Figs. 7 and 8).

Overall, the analyses revealed that although the university's safety corridors are equipped with improved surveillance, maintenance, and lighting, they are not necessarily the most walkable or accessible paths for pedestrians. This highlights the need to reassess these corridors to ensure barrier-free mobility for the broadest range of campus users. Furthermore, it is important to revisit the criteria used to define these routes, expanding the concept of "safety" beyond surveillance and control. Urban environments must encourage everyday use and social interaction, fostering a sense of belonging and vitality. Lively and inhabited spaces, whether in cities or on university campuses, are inherently safer. Thus, safety should be conceived not merely as the absence of risk, but as the presence of life, accessibility, diversity, and inclusiveness in public spaces.

Fig. 5 Some aspects concerning walkability in the safety corridors registered during *in loco* diagnosis. *Source* Authors' personal archive (2024)

4.2 Production of Tactile Maps

As an outcome of the *Trilhas UnB* project, a large tactile map of the routes was developed using materials such as thermoreactive paper, 3D printing filament, and resin, following the guidelines of accessibility national legislation, ABNT 9050 (Brasil 2021). These maps were tested by visually impaired individuals during the workshop and throughout the research process, yielding valuable user feedback regarding usability. Consequently, both the materials and textures used as well as the color scheme (black and white) reflect the results of these contributions (Fig. 9). In this case, the course contributed to the identification of accessible paths between academic units, providing the basis for the mapped routes displayed.

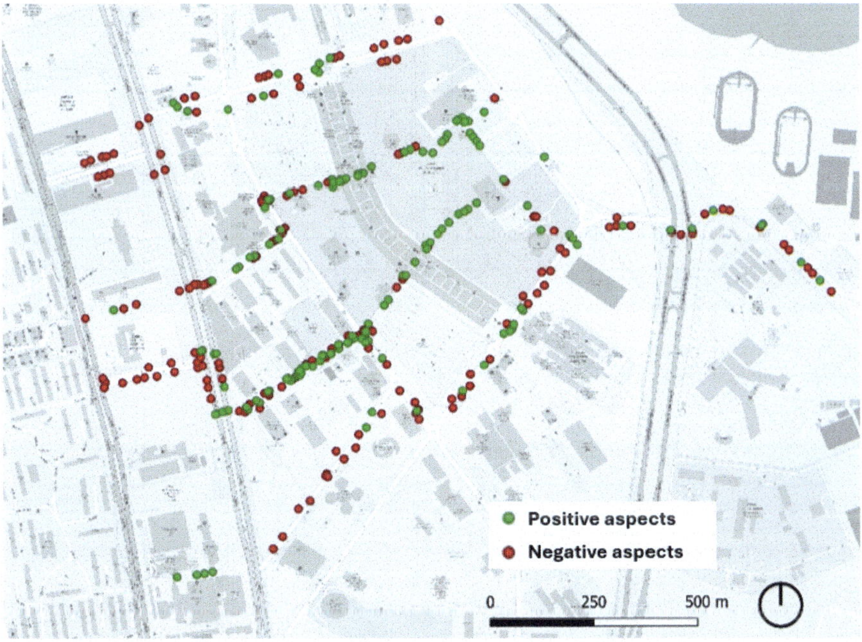

Fig. 6 Positive and negative aspects concerning walkability in the safety corridors. *Source* Authors' personal archive (2024)

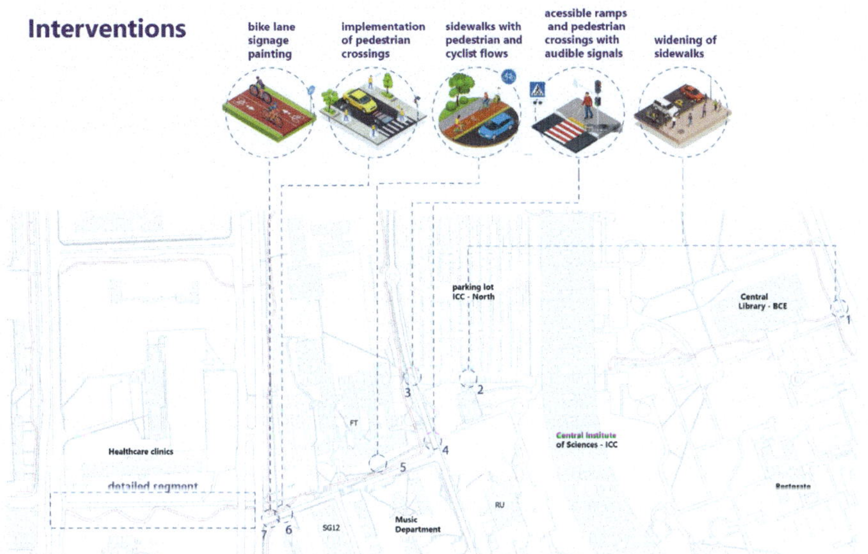

Fig. 7 Macro-scale proposal for Safety Corridor 2, developed by the students Kaio Cardoso and Amanda Macedo. *Source* Authors' personal archive (2024)

Fig. 8 Micro scale proposal for Safety Corridor 2, developed by students Kaio Cardoso and Amanda Macedo. *Source* Authors' personal archive (2024)

Fig. 9 Prototype of tactile map developed for visually impaired users, under development. *Source* Authors' personal archive (2025)

4.3 Integration of Walkability Data with Watch/App

The final phase involved prototyping an integrated assistive technology system comprising a mobile application and a wearable smartwatch device. This system provides sensory feedback (vibration and audio cues) to enhance mobility for visually impaired users on campus. Walkability data collected throughout the study were incorporated into the app's interface design, ensuring language and features were accessible to individuals with visual impairments. Additionally, it featured a collaborative platform that allowed users to contribute georeferenced feedback about the pathways, further enriching the dataset (Fig. 10).

This initiative contributed to advancing sustainability within the curriculum and operational practices of the University of Brasília. Integrating walkability and

Fig. 10 Mobile application interface showcasing collaborative mapping features. *Source* Authors' personal archive (2025)

universal design into academic content not only enhanced students' technical competencies but also strengthened the role of universities as active agents in promoting equitable and sustainable urban environments.

Moreover, this integration demonstrated how universities can lead by example, incorporating sustainability and inclusion principles into teaching, research, operations, and community outreach. The collaborative, interdisciplinary, and transdisciplinary nature of the interventions opened avenues for institutional innovation, positioning assistive technologies as key components in the creation of accessible academic environments.

From a broader perspective, the experience also validated many of the challenges and possibilities previously discussed in the literature. The persistent architectural, technological, and attitudinal barriers found during the fieldwork reaffirm the importance of addressing accessibility from a systemic and user-centered perspective. In parallel, the successful application of tools like ICEP and the development of tactile maps resonate with good practices in inclusive design. Furthermore, by integrating Design Science Research (DSR) and universal design principles into the classroom, the project operationalized pedagogical strategies recommended in the literature, such as active and problem-based learning. The project's results therefore not only reinforce but also expand the academic debate on how accessibility and walkability can be co-developed as strategic pillars for inclusive and sustainable campus planning.

Ultimately, the project effectively contributed to the core objective of this article—the integration between teaching and research to promote accessibility and inclusive mobility in university environments. It translated theoretical concepts into practical applications, involved the academic community in co-creation processes, and produced tangible outputs with direct social impact. In doing so, it reinforced the transformative potential of university education when it actively engages with real-world challenges and adopts interdisciplinary, inclusive, and collaborative approaches.

5 Conclusions

This experience enabled a deeper exploration of issues related to accessibility and pedestrian mobility on the Darcy Ribeiro campus, by combining the concept of walkability with technological innovation. It responded directly to an institutional demand, reinforcing the university's commitment to inclusion while engaging students in socially relevant and practice-oriented learning processes.

The integration of the course *Special Studies in Technology—Walkable Urbanism* with the research projects *Acessa UnB* and *Trilhas UnB* proved to be successful, yielding significant academic and practical outcomes. This collaborative approach enriched the pedagogical experience and contributed to improving the accessibility of the University of Brasília's campus for visually impaired individuals. The synergy between teaching and research fostered a dynamic learning environment in which students confronted real-world challenges and applied theoretical knowledge to propose concrete urban design solutions.

The findings demonstrate the effectiveness of active methodologies—such as problem-based learning and participatory design—in promoting critical thinking, technical innovation, and social responsibility among students. The application of the Pedestrian Ergonomic Comfort Index (ICEP) provided a diagnostic tool that guided the identification of accessibility issues and the formulation of macro and micro scale proposals. The production of tactile maps and the development of a mobile application integrated with sensory feedback represent key milestones, both refined through direct feedback from visually impaired users.

The implications of the project extend beyond the University of Brasília, highlighting the potential for replicability in other academic or urban contexts. By adopting similar interdisciplinary and participatory approaches, other institutions can address a broader range of disabilities—including auditory, motor, or cognitive—and promote more inclusive environments. The experience also demonstrates how integrating assistive technology and inclusive design into university curriculum can operationalize sustainability and equity goals, in alignment with the SDGs.

Nevertheless, there were some limitations. The scope of the student's diagnosis was restricted to five predefined corridors and focused primarily on visual impairments. Expanding the scope to include other disability profiles and additional areas of the campus could further improve the comprehensiveness and impact of the initiative. Additionally, while prototypes were developed, their full-scale implementation and long-term adoption require further institutional commitment and financial investment.

Looking ahead, future research should focus on the scalability and technical refinement of the assistive tools developed, as well as their integration into institutional infrastructure and policy.

In conclusion, the integration of teaching and research not only enhanced the educational process but also contributed to the creation of more inclusive and sustainable university campuses. The project stands as a replicable model of how academic institutions can bridge theory and practice while addressing pressing societal needs.

Acknowledgements We thank the Accessibility Directorate of the University of Brasília (DACES—UnB), especially the then director Sinara Pollom Zardo, for the financial and logistical support in the execution of the projects linked to "Call for Proposals No. 08/2024—Support for the Development of Assistive Technology at UnB". We extend our thanks to the students of the course "Special Studies in Technology—Walkable Urbanism," offered in the second semester of 2024 at FAU-UnB, for their dedication and active participation. We also thank the entire team involved in the projects, as well as the blind and low-vision academic community, whose contribution was essential to the development of the work. Finally, we extend our gratitude to FAU-UnB, especially to "LAPIS—Laboratory for Prototyping, Innovation, and Systems," whose infrastructure and technical support were fundamental for the development of the project, including the printing of tactile maps and the development of the accessible app. We also acknowledge the support of FAPESP through grant number 24/23425-1.

References

Instituto de Pesquisa e Estatística do Distrito Federal (IPEDF). (2021). *Pesquisa Distrital por Amostra de Domicílios—PDAD 2021* [District Sample Household Survey—PDAD 2021] Brasília: IPEDF. https://www.ipe.df.gov.br/pdad/

Albala, P. L. R. (2022). *Percursos de pedestres: caminhabilidade, conforto ambiental e planejamento estratégico* [Pedestrian pathways: walkability, environmental comfort and strategic planning] (Tese de Doutorado). Faculdade de Arquitetura e Urbanismo, Universidade de São Paulo.

Albala, P. L. R., & Kronka, R. C. K. (2024). Pedestrian pathways: an index based on the interrelations between walkability and environmental comfort. *Ambiente Construído, 24*(1). https://doi.org/10.1590/s1678-86212024000100777

Associação Brasileira de Normas Técnicas. (2021). *NBR 9050:2021—Acessibilidade a edificações, mobiliário, espaços e equipamentos urbanos* [Accessibility to buildings, furniture, spaces, and urban equipment]. Rio de Janeiro: ABNT.

Borges, W. F., & Mendes, E. G. (2021). Recursos de acessibilidade e o uso dos dispositivos móveis como tecnologia assistiva por pessoas com baixa visão [Accessibility resources and the use of mobile devices as assistive technology for people with low vision]. *Revista Brasileira de Educação Especial, 27*, e0036.

Bradshaw, C. (1993). Creating a rating system for neighborhood walkability: Towards an agenda for "local heroes." In *14th International Pedestrian Conference*. Boulder: International Pedestrian Conference.

Brazil. (1988). *Constituição da República Federativa do Brasil de 1988* [Constitution of the Federative Republic of Brazil of 1988]. Brasília, DF: Presidência da República.

Brazil. (2015). Lei nº 13.146, de 6 de julho de 2015. Institui a Lei Brasileira de Inclusão da Pessoa com Deficiência (Estatuto da Pessoa com Deficiência) [Brazilian Law of Inclusion of Persons with Disabilities].

Ciantelli, A., & Leite, L. (2016). Ações exercidas pelos núcleos de acessibilidade em universidades federais brasileiras [Actions exercised by accessibility centers in Brazilian federal universities]. *Revista Brasileira de Educação Especial, 22*(3), 385–400.

Diesel, A., Baldez, A. L. S., & Martins, S. N. (2017). Os princípios das metodologias ativas de ensino: uma abordagem teórica [The principles of active learning methodologies: a theoretical approach]. *THEMA, 14*(1), 268–288.

Diniz, E. P. S., & Silva, A. M. (2021). Perspectivas de Estudantes com Deficiências Sobre Facilitadores e Barreiras nas Universidades Públicas de Mato Grosso do Sul [Perspectives of Students with Disabilities about Facilitators and Barriers at Public Universities in Mato Grosso do Sul]. Revista Brasileira de Educação Especial, 27, e0092, 461–476. doi: https://doi.org/10.1590/1980-54702021v27e0092.

Dresch, A., Lacerda, D., & Antunes, J. A. V. Júnior. (2015). *Design Science Research: Método de Pesquisa para Avanço da Ciência e Tecnologia* [Design Science Research: Research Method for the Advancement of Science and Technology] (p. iv). Porto Alegre: Bookman Editora.

Fahd, W. C. B., Santiago-Vieira, S., & Nascimento, I. P. (2023). A política institucional e o núcleo de acessibilidade na universidade: Ações de acesso e permanência dos estudantes público-alvo da Educação Especial [The institutional policy and the accessibility center at the university: Actions for access and permanence of students targeted by Special Education]. *Revista Educação e Políticas em Debate, 12*(3).

Guarizzo, A. B., Silva, A. M. R. da, Araújo, A. F. de, Santos, G. C. dos, Botelho, S. de O., Pereira, S. M. da S., Pereira, T. do R. dos S., & Woodcock, Z. S. P. (2024). Metodologias de aprendizagem ativa: uma mudança de paradigma no ensino [Active learning methodologies: a paradigm shift in teaching]. *Caderno Pedagógico, 21*(5), e4186.

Imrie, R., & Hall, P. (2001). *Inclusive design: Designing and developing accessible environments* (1st ed.). London: Taylor.

Martinez, V. (2020). *Educação, trabalho e evolução: A inclusão social como ferramenta de humanização* [Education, work and evolution: Social inclusion as a tool for humanization]. São Paulo: Editora Agbook.

Morán, J. (2015). Mudando a educação com metodologias ativas [Changing education with active methodologies]. In C. A. de Souza & O. E. T. Morales (Orgs.), *Convergências midiáticas, educação e cidadania: Aproximações jovens* [Media convergences, education and citizenship: Youth approaches]. Ponta Grossa: UEPG.

Nogueira, R. (2017). Mapas táteis padronizados e acessíveis na web [Standardized and accessible tactile maps on the web]. *Revista Brasileira de Cartografia, 67*(5), 1007–1016.

Reginato, V., Zimmermann, C. C., Nappi, J. V. H. V. L., De Quadros, M. M., De Freitas, G., Aliardi, A., Bestetti, B. E., & De Sousa, A. P. R. (2021). Ufsc 100% Acessível: Mapeamento Para a Cidadania / 100% Accessible Ufsc: Mapping for Citizenship. *Brazilian Journal of Development*.

Severino, A. J. (2016). *Metodologia do trabalho científico* (24ª ed., rev. e atual.). São Paulo: Cortez.

Steane, M. A., & Steemers, K. (2004). Environmental diversity in architecture. Routledge.

Teixeira, F. L. L., & Sumiya, L. A. (2024). Barreiras de acessibilidade na educação superior: A percepção dos discentes [Accessibility barriers in higher education: Students' perception]. *Revista Brasileira de Educação em Educação Especial, 11*(3), Artigo e0240029.

Universidade de Brasília (UnB). (2018). *Caminhos pensados para seu bem-estar* [Paths designed for your well-being]. https://noticias.unb.br/publicacoes/76-institucional/2476-caminhos-pensados-para-seu-bem-estar

Universidade de Brasília [UnB]. (2024). Corredores de segurança [Safety corridors]. UnB Notícias. Retrieved February 22, 2026, from https://noticias.unb.br/publicacoes/5926-corredores-de-seguranca

United Nations (UN). (2015). *Transforming our world: the 2030 Agenda for Sustainable Development*. New York: UN. https://sdgs.un.org/2030agenda

Paula Lelis Rabelo Albala Assistant Professor at Faculty of Architecture and Urbanism at the University of Brasilia (FAU-UnB). Postdoctoral researcher at Sciences Center for Development—CCD USP Cidades, University of São Paulo (2025). She holds a PhD in Architecture and Urbanism from FAUUSP (2022), a Master's degree from UPC (2010), and a Bachelor's degree from FAUUSP (2007). Postdoctoral fellow at FAU-UnB (2024). Researcher at LaSUS and Lapis (FAU-UnB), focusing on sustainable architecture and urbanism, environmental comfort, active mobility, accessibility, nature-based solutions, and social inclusion.

Ana Carolina Cordeiro Correia Lima Assistant Professor at Faculty of Architecture and Urbanism—FAU-UnB since 2022. Coordinator of the Evening Architecture and Urbanism Program at FAU-UnB (since 2024). She holds a PhD (2021), a Master's degree (2014), and Bachelor's degree (2011) from FAU-UnB. Member of Reabilita Postgraduate Certificate Course. Former coordinator at UDF. Researcher at LaSUS and Lapis (FAU-UnB), her expertise areas include sustainable design, acoustic comfort, and sensory architecture.

Renan do Nascimento Balzani Assistant Professor at FAU-UnB and Coordinator of the Prototyping, Innovation and Systems Laboratory—Lapis (since 2022). He holds a PhD (2021), Master's degree (2017), and Bachelor's degree (2012) in Architecture and Urbanism from UnB, and a postgraduate certificate in Higher Education Teaching (2015). He is a researcher at LFDC and LaSUS, working on digital fabrication, 3D printing, educational building design, and assistive technologies.

The Implementation of the Green Office UPF: Increased Student Engagement and Sustainability Learning Opportunities

Luciana Londero Brandli, Gabriela Rodrigues, Ana Júlia Loss Marcon, Amanda Lange Salvia, and Bianca Gasparetto Rebelatto

Abstract This study aims to present and analyze the sustainability initiatives carried out by the Green Office at the University of Passo Fundo (UPF) in 2024. Using a participatory action research approach, the study documents the implementation process of the Green Office, which involved organizing workshops, seminars, and collaborative projects focused on waste management, energy efficiency, and sustainable lifestyles. Data were collected through participant observation, interviews, and analysis of project outcomes. The results demonstrate that the Green Office effectively fostered engagement among students and staff, increased sustainability awareness, and facilitated climate change communication both within the university and the local community. The study highlights practical solutions and collaborative dynamics that contributed to embedding a culture of sustainability in the academic environment. These findings provide valuable insights and lessons learned that can support replication and scaling of similar sustainability initiatives in higher education institutions, thereby contributing to broader societal goals of sustainable development.

Keywords Green office · Sustainability · Student engagement · University initiatives · SDGs · Higher education

L. L. Brandli · G. Rodrigues (✉) · A. L. Salvia · B. G. Rebelatto
Graduate Program in Civil and Environmental Engineering, University of Passo Fundo, Passo Fundo, São José, RS, Brazil
e-mail: grodrigues.biol@gmail.com

L. L. Brandli
e-mail: brandli@upf.br

L. L. Brandli · G. Rodrigues · A. J. L. Marcon · A. L. Salvia · B. G. Rebelatto
Green Office UPF, University of Passo Fundo, São José, Passo Fundo, RS, Brazil

W. Leal Filho et al. (eds.), *Sustainability in the Curriculum and Operations of Universities*, World Sustainability Series,
https://doi.org/10.1007/978-3-032-12678-8_34

1 Introduction

Sustainability initiatives in higher education have gained significant attention in recent years, as universities increasingly recognize their pivotal role in addressing global environmental, social, and economic challenges. These institutions are not only centers of knowledge creation through research but also bear the responsibility of shaping the leaders of tomorrow who will drive the global sustainability agenda. As Leal Filho (2020) emphasizes, universities are key in building local capacities for sustainable development, serving as "catalysts for change". The United Nations' 2030 Agenda for Sustainable Development, with its 17 Sustainable Development Goals (SDGs), underscores the importance of education, innovation, and cross-sector collaboration to achieve a more sustainable world by 2030. Within this framework, universities are integrating sustainability into their operations, curricula, research, and community engagement efforts to meet these global goals.

A critical component of advancing sustainability within universities is the establishment of Green Offices—dedicated spaces designed to promote sustainability initiatives and foster collaboration among faculty, staff, and students. These offices aim to ensure sustainability is embedded in all aspects of university life. The Green Office Movement, which originated at Maastricht University in 2010, has become a global model for universities worldwide, emphasizing the importance of involving students, faculty, and staff in sustainability projects to transform institutional operations and culture (Green Office Movement 2010). At Universidade de Passo Fundo (UPF), the Green Office, established in 2021 as part of the international project *Transforming Universities for a Changing Climate—Climate-U* (Salvia et al. 2023), is the first of its kind in Latin America. This Green Office exemplifies the commitment to promoting sustainable practices through educational initiatives and active student engagement, aligning with the broader objectives of the 2030 Agenda and the SDGs.

Student involvement is crucial in sustainability initiatives, as students represent the largest demographic in universities and will be the future leaders tackling global challenges. Universities must equip students with the knowledge and skills needed to contribute to sustainable development. Student-led sustainability initiatives not only empower students to take active roles in environmental and social change but also help foster a culture of sustainability within the academic community. Integrating sustainability into higher education requires overcoming institutional barriers and promoting innovation through collaborative efforts (Brandli et al. 2024). Furthermore, student engagement encourages the adoption of sustainable practices and strengthens the link between theory and practice, enabling students to apply their knowledge to real-world problems. Innovative teaching methods, such as games and competitions, have proven effective in raising awareness and driving behavior change (Blewitt 2010; Gawel et al. 2022).

The Green Office at UPF's initiatives align closely with the SDGs, addressing critical global targets such as climate change mitigation, sustainable consumption, and the conservation of ecosystems. By engaging students, faculty, and staff in sustainability projects, the Green Office serves as a platform for collective action, ensuring that sustainable practices are integrated into the university's daily operations. The concept of universities as *Living Labs* for Sustainable Development (Leal Filho et al. 2021) underscores how universities can act as incubators for innovation, where the SDGs are implemented through interdisciplinary collaboration and practical solutions.

Through the Green Office UPF, participants are involved in activities such as workshops, awareness campaigns, and collaborative projects that tackle critical issues such as energy efficiency, waste reduction, and biodiversity conservation. These initiatives not only make a tangible impact on campus operations but also provide valuable learning opportunities for students and staff. The combination of education, engagement, and practical action ensures that sustainability becomes an integral part of the university's culture.

The objective of this study is to present the sustainability initiatives carried out by the Green Office UPF in 2024, evaluate the collaborative dynamics among participants, and analyze the practical solutions implemented to promote sustainable development within the university. The office was implemented using Participatory Action Research (PAR), and this study aims to provide valuable insights into the impact of these initiatives and contribute to identifying best practices that can be replicated in other academic settings, ultimately supporting the broader achievement of the SDGs and encouraging other institutions to follow the same path.

2 Methodology

This research adopts a case study approach to examine the sustainability initiatives implemented by the Green Office at the University of Passo Fundo (UPF) throughout the year 2024. The primary aim is to assess the effectiveness and impact of these initiatives in promoting sustainable development within the academic environment, as well as their alignment with the Sustainable Development Goals (SDGs). Additionally, the study seeks to understand the level of engagement of the university community in these actions and how these experiences can inform future strategies.

The initiatives analyzed were selected based on their relevance to the institution's sustainability agenda, their demonstrated capacity to engage different segments of the academic and local community, and the tangible outcomes they produced in terms of environmental awareness, behavioral change, and ecological restoration. The analysis focused not only on the activities themselves but also on their contribution to broader sustainability efforts and institutional learning. Each initiative was evaluated in relation to the specific SDGs it addresses, with particular attention to the indicators used to assess its success. These indicators were defined according to the nature and objectives of each action and included both quantitative and qualitative measures. For example, the number of events held, participants involved, waste collected, seedlings planted, and educational materials distributed were considered alongside qualitative feedback from students, teachers, and community members.

To ensure a comprehensive and reliable analysis, the study made use of document review, including internal reports, records, and communication materials; field observations, particularly in activities involving environmental restoration and outreach events; and direct feedback collection through interviews, informal conversations, and testimonials. Table 1 presents an overview of the initiatives, highlighting the associated SDGs, selected indicators, and methods of analysis.

Table 1 Framework for evaluating sustainability initiatives and SDG integration

Initiative	SDG	Indicators	Methods of analysis
Workshops and seminars	4 QUALITY EDUCATION	Number of workshops held; Average number of participants per event; Participant feedback	Document analysis; Feedback collection
Green office sustainable seal	11 SUSTAINABLE CITIES AND COMMUNITIES; 12 RESPONSIBLE CONSUMPTION AND PRODUCTION	Number of departments certified; Criteria: environmental, social, and governance aspects	Seal application report; Interviews with certified departments; Expansion tracking
"Copo consciente"	11 SUSTAINABLE CITIES AND COMMUNITIES; 12 RESPONSIBLE CONSUMPTION AND PRODUCTION	Quantity of disposable cups collected (plastic and Styrofoam); Social media engagement (views, interactions)	Quantitative tracking of collected waste; Social media metrics; Campus feedback
Restoration of degraded areas	15 LIFE ON LAND	Area restored (m^2); Number of seedlings planted; Number of participants involved	Field reports; Observations; Interviews and testimonials from community and students
"Uma noite pelo mundo"	4 QUALITY EDUCATION	Number of students engaged; Interaction with SDG roulette activity; Visibility of Green Office initiatives	Event evaluation; Informal interviews; Visual records of engagement
School visits	4 QUALITY EDUCATION	Number of schools visited; Number of students reached; Educational materials distributed	Feedback from students and teachers; Observations during visits
Participation in "bio week"	4 QUALITY EDUCATION	Number of visitors; Participation in activities (e.g., trails); Engagement with educational materials	Document review; Feedback collection; Evaluation of learning outcomes

3 Results

The results, presented in the following section, reflect a careful assessment of each initiative's contributions to the Sustainable Development Goals, based on the indicators and methods outlined in the methodology. These findings illustrate not only the outcomes achieved—such as reductions in disposable material use, successful ecological restoration efforts, and increased awareness of global sustainability challenges—but also the processes through which these results were obtained. By examining both the quantitative and qualitative dimensions of the initiatives, it was possible to identify key lessons learned, recurring challenges, and factors that facilitated success.

3.1 Sustainable Initiatives: Green Office UPF

One of the key initiatives of the Green Office UPF involves research and outreach projects focused on sustainable development in different fields of knowledge. These projects demonstrate how the academic environment can integrate the SDGs into its activities. With a multidisciplinary approach, these initiatives involve faculty, students, and community members, addressing crucial sustainability challenges (Brandli et al. 2025a). Another innovative initiative is the Sustainable Development Goals Games, designed to enhance student engagement with sustainability. The first edition gathered participants from diverse backgrounds, including students, faculty, staff, NGOs, local government representatives, and businesses. By providing an interactive and collaborative environment, the Games fostered collective engagement in sustainability and climate action (Brandli et al. 2025b).

In addition to these initiatives, other activities focused on education, outreach, and community engagement are regularly conducted to further promote sustainability and environmental awareness. These efforts aim to integrate theoretical knowledge with practical experiences, fostering a deeper understanding of sustainability issues. Below, we present a selection of activities carried out in 2024.

Workshops and seminars on sustainability topics

The workshops covered a wide range of topics, from project presentations and preliminary results to discussions on climate change, sustainable entrepreneurship, and the implementation of the SDGs. In total, 365 registrations were recorded across 10 events. This number highlights the increasing adherence to and importance of these topics within both the academic and professional communities.

In terms of participation, the workshop on the Presentation of the Contingency Plan for Passo Fundo, stood out as the most attended event. This event reflects the growing interest in local issues and the implementation of practical solutions to environmental and social challenges, engaging the community in a meaningful way. Another significant event was the workshop on Climate Change and Sustainable Entrepreneurship, this event highlights the intersection of entrepreneurship and climate change, topics that have become increasingly relevant in the global sustainability context. Each workshop provided an opportunity to deepen knowledge on how the SDGs can be implemented in practical and effective ways, fostering the exchange of experiences and the development of innovative solutions to global challenges.

The growing participation, particularly in events like the Presentation of the Contingency Plan for Passo Fundo, suggests that awareness and community involvement in sustainability are on the rise, and there is a genuine interest in initiatives that provide tangible solutions to environmental, social, and economic problems. This reflects active engagement and a commitment to transformation, aligned with the urgent need for action in favor of the environment and society. These workshops play a crucial role in education and promoting local and global action, significantly contributing to the dissemination of knowledge on sustainability and the SDGs. This action relates to SDG 4 (Quality education), more specifically the 4.7 target "By 2030, ensure that all learners acquire the knowledge and skills needed to promote sustainable development (…)", since the action aims to expose the college students to the SDGs and the sustainable activities developed in campus.

For a more sustainable campus: "Copo Consciente"

Throughout 2024 several initiatives were implemented, the first one being the campaign "*Copo Consciente*". This campaign was designed and executed by the Green Office, the People Management department, the Basic Sanitation department of the university and the Visual Arts department. Its main objective was to reduce the number of plastic cups used in the university campus. To achieve this, plastic cups used by the academic community over 13 days were collected and quantified.

Firstly, our project began promoting via social media, explaining the campaign and highlighting the designated collection points on campus. There were five places that had waste bins distributed with posters explaining our campaign. The collecting process initiated on May 22nd and ended on June 3rd, after this period the Basic Sanitation department gathered, separated, and delivered the collected material to the Green Office on June 5th, where they were cleaned and quantified. A total of 1650 plastic cups were collected together with 170 Styrofoam cups, which were not the focus of the campaign, but still quantified.

Following this, the cups were sent to the Visual Arts department, where a sculpture was designed and created by them. This artwork was then exhibited at the university's community center, the busiest area on campus. The exhibition aimed to raise consciousness among students, encouraging them to adopt more sustainable habits during their time at the university (Fig. 1).

This campaign is related to the SDG 11 (Sustainable cities and communities), since its aim is to reduce the amount of plastic waste on campus, making the academic community more sustainable. It also relates to SDG 12 (Responsible Consumption and Production), since the action aims to promote a more sustainable consumption, without the use of plastic.

Fig. 1 Photos of the waste bins with the poster of the campaign, of the cups collected and of the sculpture exhibited

Green office sustainable seal: UPF + sustainable

Another initiative created aimed to recognize sustainable actions on the academic campus, more specifically in laboratories and restaurants. The seal was requested by one of the laboratories of the Chemical Engineering Department of the university, as they had many sustainable aspects in their laboratory and wanted them to be acknowledged. This request inspired the creation of the seal. To obtain it, it's necessary to answer a questionnaire with indicators based on the ESG (Environmental, social and governance), which were adapted to the university context.

The indicators are divided in three sections, environmental (with 14 indicators), social (with 14 indicators) and governance (6 indicators), in total 34 indicators. The responses to the indicators are measured using the Likert scale in a frequency context, with the following options: (0) does not apply, (1) never, (2) rarely, (3) occasionally, (4) frequently, and (5) very frequently. The number before each response represents the punctuation. Each question has a different weight, chosen by the importance of them. The weight of each question multiplies the score of the response, and all results are added together. For example: the second question of the social section weighs 2 points, hypothetically the laboratory punctuation answer is 5 (really frequently), the result of the punctuation of this question is 10.

Three categories (Fig. 2) were established for the Sustainable Seal, category 1—Tree Seal, category 2—Seedling Seal and category 3—Seed Seal. The symbology of tree, seedling and seed was created by the Environmental Board of the Universidade Federal de Viçosa, in their initiative *UFV + Sustentável*, the idea was used with their consent.

Fig. 2 Image of the design for the three categories of the green office sustainable seal

The category of the seal is determined by the percentage of total possible points scored. Category 1—Tree Seal requires a score between 80 and 100%, Category 2—Seedling Seal requires between 60 and 79%, and Category 3—Seed Seal requires between 40 and 59%. The verification of the indicators was conducted by the Scholarship Holders of the Green Office, who were also responsible for collecting and storing the data.

The first implementation of the *Green Office Sustainable Seal* took place in the Unitary Operations Laboratory, located in the Food Engineering Department and coordinated by Dr. Vandré Barbosa Brião. Among the laboratory's sustainable initiatives are the use of solar energy that covers 100% of its electricity needs and the collection and filtration of rainwater that covers almost 70% of its water needs. Based on the applicable indicators, the maximum score they could achieve was 255, and their total score was 219. That means that they scored 85.8%, qualifying them for Category 1—Tree Seal.

This initiative is related to SDG 11 (Sustainable cities and communities) to the extent that the more sustainable aspects are recognized and encouraged on campus, the more the community will develop in this regard. It is also linked to SDG 12 (Responsible Consumption and Production), since the action aims to recognize and promote more sustainable consumption and production practices in restaurants and laboratories.

Restoration of degraded areas

In 2024 the state of Rio Grande do Sul experienced a flood that devastated environmental areas, including the city of Nova Alvorada, in the area near the Engenho Velho river. In response to this situation, the LAMVIS (Laboratório de Manejo da Vida Silvestre) and Green Office team, along with the school Escola Municipal Edílio Luiz Chesties, planted 120 native trees along the riverbank. LAMVIS cultivated seedling of native trees, such as Araucária, Pitanga, Guabiroba, red and white Camboatá, Uvaia, while the ICMBio (Instituto Chico Mendes de Conservação da Biodiversidade) transported them to Nova Alvorada. This initiative aimed to engage more people in the reforestation of devastated ciliary forests, which are essential to the river water preservation. This action aligns with Goal 15 (Life on Land), as it aims to restore a degraded area, contributing to the conservation of biodiversity, the recovery of native vegetation, and the enhancement of ecosystem services. By promoting ecological restoration, the initiative also supports efforts to combat desertification and land degradation, fostering the sustainable use of terrestrial ecosystems in accordance with the targets established by the 2030 Agenda.

Event "Uma noite pelo mundo"

A further initiative of 2024 was the participation in the event "*Uma noite pelo mundo*", organized by the Department of Languages and Literature, which aims to expose the students to different cultures. The participation of Green Office was invited to expose the students to the Sustainable Development Goals (SDG). To present them in an interactive way, the Green Office team brought roulette wheels featuring all 17 SDGs and when spun, the wheel would land on one of them and the participating student had to answer a question about the SDG data related to the countries whose cultures were featured in the event. This activity allowed for interaction with students, an opportunity to explain the Green Office's work on campus, and a chance to introduce the SDGs in a more engaging way. This action relates to the SDG 4 (Quality education), more specifically the 4.7 target "By 2030, ensure that all learners acquire the knowledge and skills needed to promote sustainable development (…)", since the action aims to expose the college students to the SDGs and the sustainable activities developed in campus.

Participation in "Bio Week"

The participation in "Bio Week" (Fig. 3), an event organized by UPF at Passo Fundo Shopping in celebration of Environment Month. The initiative aimed to promote environmental awareness and stimulate discussions on sustainability, biodiversity, and ecological practices, inviting the local community to reflect on essential issues for the preservation of the planet.

At the space dedicated to the Green Office, materials from the Biological Sciences Program were displayed, highlighting the importance of biodiversity conservation and the various actions developed by the university in support of the environment. Additionally, an interactive perceptual trail was set up, offering visitors an educational and immersive sensory experience about the SDGs. The activity aimed to emphasize

Fig. 3 Image of the bio week's poster and photos of the event

the crucial role each individual plays in building a more sustainable future and in achieving global goals.

The perceptual trail, one of the most impactful and educational attractions of the event, was carefully designed to provide an experience that engaged all the senses. Each station along the trail represented a different SDG, with activities and information related to topics such as environmental conservation, climate change, and sustainable resource use. Throughout the journey, participants were encouraged to reflect on how their daily actions can impact the environment and how adopting more sustainable practices can lead to meaningful change.

To make the experience even more immersive, the trail incorporated sensory elements that fostered a connection with nature. Sounds of nature, images of diverse ecosystems, and tactile and olfactory experiences—such as touching natural materials or identifying forest scents—were integrated into the walk. These elements not only enriched the experience but also helped create a deeper connection with the SDGs and the environmental challenges we face globally.

Thus, the perceptual trail fulfilled its role in informing, sensitizing, and inspiring participants to rethink their daily behaviors towards the environment, encouraging more active engagement in building a more sustainable future. This action also relates to the SDG 4 (Quality education), more specifically the 4.7 target "By 2030, ensure that all learners acquire the knowledge and skills needed to promote sustainable development (…)", since the action aims to expose the local community to the SDGs and the sustainable activities.

School Visits

The regional schools often visit the Green Office space at the university. During these visits, they are shown a PowerPoint presentation that explains what Green Office is, the activities carried out at the university, the importance of a more sustainable lifestyle, the SDGs and their ecological footprint (using the Footprint Calculator website: https://www.footprintcalculator.org/home/en). This action also relates to the SDG 4 (Quality education), more specifically the 4.7 target "By 2030, ensure that all learners acquire the knowledge and skills needed to promote sustainable development (…)", since the action aims to expose the students to the SDGs and the sustainable activities developed in campus.

Table 2 provides a comprehensive summary of the sustainability initiatives implemented under the Green Office program at UPF, highlighting their objectives, observed outcomes, and key insights. The observed results reflect both quantitative outcomes—such as the number of participants, seedlings planted, or materials distributed—and qualitative impacts, such as community engagement, increased awareness, and positive feedback from stakeholders. Collectively, these actions demonstrate the institution's commitment to integrating sustainability into its academic culture and local partnerships.

4 Discussions

The diversity of sustainability initiatives developed by higher education institutions around the world demonstrates the vast array of possibilities for implementing sustainable practices to mitigate environmental impacts. Recognizing that each institution has its own unique needs, challenges, and realities, it is essential to tailor sustainability approaches while pursuing the broader goals of reducing ecological footprints and fostering environmental responsibility. With this in mind, comparing the initiatives developed by the Green Office UPF team in 2024 with similar efforts from other universities provides valuable insights into the effectiveness and potential for scaling sustainability initiatives.

One example of such an initiative at UPF is the "Copo Consciente" campaign, aimed at reducing plastic cup usage on campus. Studying why plastic cups continue to be used despite their well-documented environmental impact is crucial, as demonstrated by the Federal Institute of Paraná/Foz do Iguaçu Campus (Alves 2024). The issue of plastic cup usage is not unique to UPF, as other institutions, like the University of California, Los Angeles (UCLA), have taken significant steps to address similar concerns. UCLA has banned single-use plastic bottles and transitioned to glass and aluminum alternatives, not only reducing plastic pollution but also encouraging more sustainable consumption habits among students and staff.

While UPF's "Copo Consciente" campaign was a temporary awareness-raising initiative, the challenge lies in implementing a long-term strategy that gradually eliminates disposable plastic cups. The experience of the State University of Northern Paraná (Brene et al. 2023) is instructive here, as researchers confirmed the economic feasibility of replacing disposable cups with reusable alternatives. If such a substitution is financially viable, why has it not been widely adopted? Alves (2024) suggests that the convenience of disposable cups, which require no time or resources for cleaning, is a significant barrier to their replacement.

Table 2 Results of sustainability initiatives

Green office sustainability initiatives: results and lessons learned	
	Workshops and seminars Objective: Educate and engage the academic and local community in sustainable practices and the SDG's Observed results: 4 Workshops held; Average of participants per event
	Green office sustainable seal: UPF + sustainable Objective: Promote sustainable actions on campus Observed Results: 2 departments certified; Sustainable criteria divided into environmental, social, and governance
	"Uma noite pelo mundo" Objective: Expose students to different cultures and raise awareness about the SDG's Observed results: 5 schools and 118 students engage positive feedback from students and teachers
	Initiative "Copo consciente" Objective: Reduce plastic cup usage on campus, promoting sustainable consumption practices Observed results: 1650 plastic cups were collected together with 170 Styrofoam cups; On average, each instagram post had 1431 views, 763 reached accounts and 55 interactions
	Restoration of degraded areas Objective: Restore a forest area damaged by flooding, promoting local sustainability Observed results: 120 seedlings ware planted, 57 participants engaged community and student testimonials emphasized hope for the flood-affected region
	School visits Objective: Raise awareness among regional school students about the SDGs and practical sustainability Observed results: At total, 5 schools and classes visited; The total number of students were 118; Educational material delivered
	Participation in "Bio Week" Objective: Raise awareness among regional school students about the SDGs and practical sustainability Observed results: At total, 5 schools and classes visited; The total number of students were 118; Educational material delivered

This issue calls for further research into feasible alternatives and the exploration of behavioral change strategies. Campuses generate a significant amount of plastic waste, which takes centuries to decompose, all the while damaging ecosystems and polluting water and soil (Foresti et al. 2020). Universities must recognize the urgency of addressing plastic consumption, adapting solutions to their unique contexts and working to change the campus culture to embrace sustainable alternatives.

In parallel, universities worldwide are developing initiatives like UPF's "Restoration of Degraded Areas" project. While many institutions focus on tree planting as a means to combat environmental degradation, their approaches vary widely.

The TGBC in Thailand (Komara 2023), for example, implemented a tree planting campaign around its campus, aiming to reduce carbon emissions and improve air quality. However, a key objective of their initiative was to cultivate environmental stewardship and a sense of responsibility among the participating students, fostering cooperation and long-term engagement with sustainability.

The examples of sustainability initiatives across global universities highlight diverse and innovative approaches to mitigating environmental impacts. Despite the varying strategies, these institutions share a common objective: reducing their ecological footprint and promoting a culture of environmental responsibility. For instance, UCLA's comprehensive sustainability approach goes beyond the single-use plastic ban. The university's projects in waste reduction, water conservation, renewable energy, and sustainable food systems provide a holistic model of sustainability. UCLA's policies, such as banning plastic bottles and shifting to more sustainable packaging, are integral to shifting consumption patterns and cultivating a sustainable campus culture (UCLA 2024).

Similarly, Duke University's initiative to install water bottle refill stations, preventing the use of over 400,000 plastic bottles, underscores the importance of making sustainable choices both convenient and accessible. The integration of infrastructure that supports sustainability—such as water refill stations—illustrates how universities can reduce waste and consumption by facilitating environmentally-friendly behavior (Duke University 2024).

Washington State University further exemplifies the potential of circular economy models in higher education with its reusable container program. By diverting 65,000 disposable containers from landfills, the university demonstrates the effectiveness of closed-loop systems and highlights the role of innovation in sustainability efforts. The addition of QR codes to track the containers enhances both functionality and user experience, fostering greater participation and improving the overall efficiency of the program (Washington State University 2024).

These initiatives underscore the pivotal role that universities play in advancing sustainability. As microcosms of society, universities influence student behaviors and broader cultural shifts toward environmental stewardship. Each initiative, whether focused on waste reduction, energy conservation, or sustainable food practices, illustrates that sustainability is not only feasible but can be effectively integrated into university life, regardless of institutional size or resources. Moreover, these examples emphasize the importance of a holistic approach to sustainability, one that engages multiple stakeholders—students, faculty, staff, and the broader community—and integrates sustainability into every facet of university operations. The shared commitment of these institutions to reducing environmental impacts and promoting sustainable practices offers valuable lessons for others looking to create more sustainable.

5 Conclusion

This study provides valuable insights into the sustainability initiatives led by the Green Office at UPF throughout 2024. It highlights the pivotal role of participatory collaboration among students, faculty, and staff in promoting sustainable practices within the university environment. Through active involvement in planning and implementing these initiatives, stakeholders contributed not only to tangible sustainability outcomes but also reinforced the university's commitment to the Sustainable Development Goals (SDGs).

The findings reveal challenges such as engaging diverse stakeholders and maintaining sustained participation over time. Nonetheless, strategies including improved communication, flexible and adaptable planning, and fostering a strong sense of community involvement proved effective in overcoming these obstacles. Lessons learned emphasize the importance of cultivating trust and robust relationships among participants, which are essential for long-term engagement and project success. Additionally, the ability to remain flexible and responsive to feedback was critical in refining initiatives and ensuring their continuous improvement.

For future Green Office initiatives at other universities, adopting a participatory approach with comprehensive stakeholder engagement and consistent monitoring and evaluation is recommended. Institutions should prioritize creating collaborative environments that integrate sustainability across all departments and activities, embedding sustainable practices deeply into their organizational culture. While this study focuses on a single case, which may limit the generalizability of results, the practical experiences and lessons presented offer a useful reference for other higher education institutions aiming to advance sustainability. Future research could explore comparative studies across multiple universities to further validate and expand these findings.

References

Alves, L. F. (2024). Plano de gestão de logística sustentável: Estudo de caso sobre as ações propostas para a redução do consumo de copos descartáveis no IFPR-Foz [Sustainable logistics management plan: Case study on actions proposed to reduce disposable cup consumption at IFPR-Foz]. *Revista Gestão & Sustentabilidade Ambiental*, 13(1), e12554. https://doi.org/10.19177/rgsa.v13e12554

Blewitt, J. (2010). *Sustainability and lifelong learning*. Routledge.

Brandli, L. L., Reginatto, G., Salvia, A. L., & Carretta Diniz, P. H. (2024). *Student engagement on climate learning: What does the academic community say about it? International Journal of Sustainability in Higher Education*. Advance online publication. https://doi.org/10.1108/IJSHE-06-2023-0219

Brandli, L. L., Salvia, A. L., Rodrigues, G., Martins, A. C., Rebelatto, B. G., Diniz, P. H. C., & Mazutti, J. (2025a). Multidisciplinary approaches for advancing the SDGs via outreach and research projects. In *6th World Symposium on Sustainable Development at Universities*.

Brandli, L. L., Salvia, A., Rodrigues, G., Reginatto, G., et al. (2025b). Enhancing student engagement with the SDGs through sustainability games: A case study in Brazil. *Scientia cum Industria*, 14(Special Issue), Article e251427. https://doi.org/10.18226/23185279.e251427

Brene, P. R. A., Silva, B. D. C. C., Debiagi, F., & Oliveira, M. L. M. D. A. (2023). The economic-financial viability of using eco-friendly cups as a substitute for disposable cups at the State University of Northern Paraná. *Journal of Sustainability Perspectives*, 3(2), 170–175. https://doi.org/10.14710/jsp.2023.20478

Climate-U. (2021). *A protocol for participatory action research into climate justice: Principles and tools* (Working Paper Series No. 3). Transforming Universities for a Changing Climate. https://transforminguniversities.org

Duke University. (2024). Duke University sustainability initiatives.

Foresti, T., Oliveira, B. F. D., & Jacques, J. J. D. (2020). Análise comparativa da degradabilidade de copos descartáveis utilizando o método G160 da ASTM. *Revista de gestão & sustentabilidade ambiental. Palhoça, SC. vol. 9, nesp.(ago. 2020), p. 287–303.*

Gawel, E., et al. (2022). Inovação em sustentabilidade e comportamento [Innovation in sustainability and behavior].

Green Office Movement. (2010). *Green Office Model*. Maastricht University.

Komara, C. (2023). Tree planting for carbon reduction at TGBC, Thailand. *Environmental Journal*, 15(4), 45–53.

Leal Filho, W. (2020). *Sustainability in higher education: Developments and perspectives*. Springer.

Leal Filho, W., Brandli, L., & Salvia, A. L. (2021). *Universities as living labs for sustainable development*. Springer.

Salvia, A. L., Mazutti, J., Reginatto, G., Brandli, L. L., Porto, F. R. (2023). *Participatory action research in the implementation of a Green Office: The experience of a Brazilian university*. In W. Leal Filho, A. L. Salvia, E. Pallant, B. Choate & K. Pearce (Eds.), *Educating the Sustainability Leaders of the Future* (World Sustainability Series). Springer, Cham.

University of California, Los Angeles (UCLA). (2024). Sustainability initiatives at UCLA.

Washington State University. (2024). Sustainability at WSU.

Institutionalization Strategies for the 2030 Agenda in the Higher Education Institutions: The Case of the Federal Rural University of Pernambuco

Janielly Mantovani Cravo and Chiara Natércia França Araújo

Abstract The main objective of this article was to conduct a case study on how the Federal Rural University of Pernambuco (UFRPE) had been working to integrate the Sustainable Development Goals (SDGs) into its core areas of activity, taking into account the institutionalization process of the 2030 Agenda. The Higher Education Institutions (HEIs) have an essential educational role, shaping individuals to have skills and knowledge of values and attitudes that take sustainability, among other aspects, into account. In other words, Universities are environments of transformation, with a key role to promote good internal and external practices, exercising their socio-environmental responsibility. The 2030 Agenda guides nations and, more specifically, institutions and people, to promote strategies that target sustainable and resilient actions. It can be observed that this is a global movement and should take place collectively. The 17 Sustainable Development Goals and, thus, the 169 targets display an ambitious plan of integrated and indivisible actions that envision the three pillars of Sustainable Development: economic, social and environmental. The analysis presented in this article addressed the sphere of governance within Public Administration, focusing on the context of HEIs and the instruments employed to institutionalize sustainability. Based on the case study of UFRPE, significant progress was identified, such as the institution's performance of 91.4% in the Integrated Index of Environmental and Social Sustainability, part of the governance and public management assessment conducted by the Federal Court of Accounts (TCU) in 2024, which evidenced the institution's commitment to the Sustainable Development Goals (SDGs). Nevertheless, challenges persisted, particularly of financial and structural nature, along with the need for more robust mechanisms to measure the actual impact of the actions undertaken. The research emphasized the importance

J. M. Cravo (✉) · C. N. F. Araújo
Federal Rural University of Pernambuco (UFRPE), Recife, PE, Brazil
e-mail: janielly.mantovani@ufrpe.br

C. N. F. Araújo
e-mail: chiara.franca@ufrpe.br

W. Leal Filho et al. (eds.), *Sustainability in the Curriculum and Operations of Universities*, World Sustainability Series,
https://doi.org/10.1007/978-3-032-12678-8_35

of integrated strategies across the pillars of teaching, research, extension, management, and innovation, reinforcing the transformative potential of HEIs in promoting a more just, inclusive, and sustainable society.

Keywords Sustainability · 2030 Agenda · SDG · University

1 Introduction

The 2030 Agenda is a guideline for nations and therefore humanity in the sense of orientation towards the implementation of the 169 targets combined into 17 Sustainable Development Goals (SDGs). The SDGs were established by the United Nations Organization (UN) in 2015 and encompass an ambitious and complex agenda to be implemented globally. This agenda contemplates important pillars for the development of the human race and the preservation of the environment, which are: people, planet, prosperity, peace and partnership. Focusing on these pillars, the SDGs seek, cooperatively and in a global scale, to end hunger and poverty, guaranteeing that human beings can live with dignity and equality; to foster sustainable production, consumption and sustainable management in regard to the extraction of natural resources, as well as mitigating the emission of Greenhouse Gases (GHG), aiming to minimize the effects provoked by climate change; to ensure that economic, social and technological developments respect the limits imposed by nature; to implement concepts and practices that promote peaceful, fair and inclusive societies; and at last, although no less challenging, to promote a cooperative environment in a way that nations can prioritize solidarity, specially among most vulnerable/impoverished peoples.

To align Education with the SDGs, a paradigm shift that promotes Education for the Sustainable Development (ESD) is essential. How does one achieve it? It is necessary to create new methodologies, practices and experiences of teaching that promote, in a long run, a balance between human actions and necessities and the limitations imposed by nature when using natural resources. This certainly would enable an educational process that would foster a sustainable development, with human beings more prepared to live in a more responsible and harmonious manner with the environment.

The United Nations Educational, Scientific and Cultural Organization (UNESCO) reaffirms the idea that work involving the SDGs should be continuously practiced within the school environment, thereby engaging both students and teachers in pedagogical practices. It also emphasizes the important role of teachers as transformative agents and facilitators throughout this process (Blasco et al. 2021). Ramos (2022) highlighted in their analyses, among other aspects, that environmental education practices are already familiar to teachers and students in the state of São Paulo and in Brazil, and that these processes are consolidated through the engagement of these actors.

Lima and Feitosa (2024) drew attention to SDG indicators in their research conducted in the city of Natal, RN. They reported that the Brazilian city scored 45.19 out of 100, with eight SDGs showing very low scores, ranging from 0 to 39.99. These include: gender equality (SDG 5), industry, innovation and infrastructure (SDG 9), sustainable cities and communities (SDG 11), responsible consumption and production (SDG 12), life below water (SDG 14), life on land (SDG 15), peace, justice and strong institutions (SDG 16), and partnerships for the goals (SDG 17) (ICS 2023, as cited in Lima and Feitosa 2024). The authors stated that improvements in these indicators are necessary, which depend on social mobilization, investment, and the implementation of effective public policies.

In this context, higher education institutions (HEIs) also play a highly important and transformative role, as they prioritize the education of individuals and the production of knowledge (Corbari et al. 2021). Understanding how these institutions are assimilating the sustainable development process into their operations and governance is necessary. The educational model, as well as production processes, must be aligned with the limits imposed by nature, so that the outcomes reflect environmentally conscious and responsible management. According to Kestin et al. (2017), universities play an essential role in providing students with the knowledge, skills, and motivation to understand and address the Sustainable Development Goals (SDGs), within the framework of "education for sustainable development" (SDG 4). The authors also point out that universities can engage in various activities to support SDG implementation, such as building partnerships, identifying challenges, developing solutions, monitoring practices, and offering advisory and consultancy services to different sectors. Moreover, universities can commit to disseminating and applying knowledge throughout society.

The teaching, research, and outreach strategies provided by the academic environment reflect a privileged setting for pursuing solutions required by the SDGs (Blasco et al. 2021). Souza and Ventura (2024) stated that higher education institutions are catalysts for scientific knowledge and research, with transformative potential when they promote the development of human capital. The authors reached important conclusions in their study, reporting that Brazilian HEIs have the highest number of actions in the areas of administration and governance, followed by activities related to research and teaching. Christ et al. (2024) conducted a bibliometric analysis that identified a predominance of qualitative approaches in research on SDG implementation in Brazilian HEIs, with case studies being the most frequently used strategy.

Thus, the primary objective of this article was to conduct a case study on how the Federal Rural University of Pernambuco (UFRPE) has been working to incorporate the Sustainable Development Goals (SDGs) into its core areas of activity, considering the institutionalization process of the 2030 Agenda. The analysis aimed to identify the main advances, as well as the challenges and barriers encountered, in order to propose strategies that contribute to consolidating the internalization of the SDGs in the operations of Higher Education Institutions (HEIs), recognizing their role as active agents of socio-environmental transformation and promoters of sustainability.

2 Strategies and Challenges for the Institutionalization of the 2030 Agenda in the Higher Education Institutions

The higher education institutions have been protagonists in the promotion of sustainability, answering a global call for action in their local scenarios, in order to achieve the Sustainable Development Goals. Besides acting as agents of transformation through education, the HEIs also take an important role in the local and regional governance, regarding their potential for interaction with various stakeholders (Sedlacek 2013).

A survey conducted by the *Observatorio de la Sustentabilidad en la Educación Superior de América Latina y el Caribe* (OSES-ALC) between 2022 and 2023, with 50 institutions from 6 latin-american countries, from which 18 were brazilian, identified the level of HEIs progress regarding knowledge, commitment and contribution towards SDGs, with 76.7% for brazilian HEIs and 77.4% for the other latin-american HEIs. The study evaluates the HEI contribution with the promotion of sustainability in the different pillars of action (governance and participation, teaching and training, research and technology, extension, and management of campus), identifying that there is a tendency for systemic transformation in the HEIs performance for institutionalizing the 2030 Agenda (Moura-Leite et al. 2024a, b).

Bringing this into the context of Brazilian federal universities, Serafini (2022) identified that only 42% of them incorporate the Sustainable Development Goals in their institutional documents. For Moura-Leite (2023) this incorporation is not a guarantee of the effectiveness of actions carried out by these institutions, even though it represents an effort to contribute with the institutional promotion of SDGs. Therefore, the author defends the adoption of a comprehensive institutional approach of sustainability, focused on incorporating them into the institution's organizational culture in a holistic manner, as presented by Kohl et al. (2022), in which sustainability becomes a part of the learning processes, by being incorporated into the teaching content and methodology, as well as the institution's operations and aspects related to its governance and capacity-building. For this to be achieved, the support of the leadership is necessary, as well as a well-structured strategy of disclosure and capacity-building, involving the institution as a whole (Moura-Leite et al. 2024a).

In this regard, the case of the Federal Rural University of Pernambuco (UFRPE) is presented, focusing on the institutionalization processes of the 2030 Agenda, incorporating SDGs into its actions, including management and governance, extension, teaching, research and innovation, based on the analysis of the institution's official documents.

UFRPE is a Brazilian higher education institution, located in the State of Pernambuco, with campuses in every region of the State and a University population of approximately 20,000 inhabitants (students, professors, technicians and contracted workers). With 112 years of existence, it has a strong tradition in teaching, extension and research, besides a capacity for innovation by seeking to contribute with overcoming socio-environmental problems with projects and research related to sustainable development involving technological, agrarian, human, social and exact sciences (UFRPE 2022a).

At UFRPE, sustainability serves as a cornerstone of institutional strategy, shaping policies, action plans, and initiatives across the areas of research, education, outreach, innovation, and administration. This approach reflects a comprehensive and holistic commitment to sustainability as a guiding principle. Thus, the university made a commitment to institutionalize the 2030 Agenda, integrating it to its institutional planning. The Institutional Development Plan (PDI) 2021–2030,[1] the main planning document for guiding the institutional actions, started to align with the Sustainable Development Goals (UFRPE 2022a). Likewise, the Sustainable Logistic Plan (PLS)[2] also connects its goals and targets with the 2030 Agenda, mainly aiming to implement culture and practices that promote sustainability from the institution's organizational point of view, based on a medium-term planning (UFRPE 2022b). It's worth to highlight that this process of incorporating SDGs in UFRPE institutional documents is being constantly improved, in a continuous process of learning and training of the stakeholders involved in its elaboration.

In 2015, through the Resolution no 51/2015 CONSU/UFRPE, the UFRPE Sustainable Project was created, which is still in effect, aiming to establish an organizational framework and dynamic favorable to the creation and adoption of a sustainable management for the Federal Rural University of Pernambuco (Cravo and Silva 2025). The project is executed by a participative approach, based on two intersectoral committees: (1) Mobilizing Committee: presided over by the Rectorship and formed by all provosts and other strategic sectors of the Institution, which has a deliberative function; (2) Executive Committee: presided over by the Sustainability Coordination of the Provost of Planning and Strategic Management—PROPLAN, and formed by representatives from the Institution's strategic sectors, indicated by the managers. It has the function to study and propose the adoption of better sustainable management practices as well as to elaborate programs and methodologies suitable for UFRPE's reality, among which the Sustainable Logistic Plan (PLS) (Cravo and Silva 2025). The elaboration of the PLS is a legal obligation of every federal institution of teaching and has been the main instrument to guide and boost sustainability at UFRPE since its first edition in 2017. By 2025, UFRPE has been working on the fifth edition of its PLS, as an opportunity to improve the integration of SDGs into the Institution's documents and practices.

As identified by Moura-Leite et al. (2023), despite the legal obligation, only 25% of the federal HEIs had a PLS in effect in 2021. Thus, UFRPE stands out in

[1] UFRPE's Institutional Development Plan (2021–2030), as well as its monitoring dashboard, can be accessed at: https://www.proplan.ufrpe.br/br/content/plano-de-desenvolvimento-institucional.

[2] Every version of the UFRPE's Sustainable Logistic Plan, as well as every respective annual monitoring outcome report can be accessed at: https://www.proplan.ufrpe.br/br/content/documentos-e-relat%C3%B3rios-0.

comparison to the other HEIs, consolidating the PLS as an instrument of institutional sustainability management.

Through the execution of the UFRPE Sustainable Project and the PLS, the Institution implemented actions such as:

- suspension and prohibition of disposable cups purchasing (UFRPE 2020);
- the replacement of physical processes for digital ones, through the implementation of an academic system of integrated management, in 2020 (UFRPE 2021);
- inclusion of courses aligned with the PLS theme in the annual Personal Development Plan (PDP),[3] fostering capacity building of employers in the sustainability field.

In 2019, UFRPE established the Sustainability Coordination (CS) under PROPLAN as part of its institutional framework to support the development and implementation of sustainable management practices. This department is tasked with researching and proposing sustainability initiatives tailored to UFRPE's context, organizing committees focused on Strategic Sustainability Plans, overseeing the execution of the PLS, and supporting organizational units through the systematization, monitoring, and evaluation of relevant data (UFRPE 2025a). These efforts contribute to UFRPE's broader institutional alignment with the goals of the 2030 Agenda.

In the context of socio-environmental governance, UFRPE stood out in the self-assessment questionnaire on governance and public management conducted by the Federal Court of Accounts (TCU) for the year 2024. The institution achieved a score of 91.4% on the Integrated Index of Environmental and Social Sustainability, being classified in the enhanced governance category. This performance exceeded the average of other higher education institutions, which were mostly placed in the intermediate range (40–69.9%). The index assessed aspects such as the existence of environmental and social sustainability management models established by senior leadership; performance monitoring by institutional leadership; the implementation of institutional programs focused on sustainability, diversity, and inclusion; and the promotion of sustainable and inclusive practices both within and beyond the organization's internal environment (TCU 2024).

In addition to the pillar of management and governance, UFRPE has been seeking to integrate SDGs in research and innovation, by requiring on public calls that project and program proposals need to indicate their relation to the Sustainable Development Goals. Besides fostering research projects, through scientific initiation scholarships, for instance, the following projects stand out in the fields of innovation and entrepreneurship: (i) the Institutional Scholarship Program for Initiation in Technological Development and Innovation (PIBITI),[4] which aims to encourage higher education students to develop and transfer new technologies and innovations to society, including solutions addressing environmental and social issues; and (ii)

[3] https://www.progepe.ufrpe.br/plano-pdp.

[4] https://inovacao.ufrpe.br/en/programa-institucional-de-bolsas-de-iniciacao-em-desenvolvimento-tecnologico-E-inovacao-pibiti.

the Institutional Program for Initiation Scholarships in Entrepreneurship (PIEMP),[5] created in 2020 to foster the creation of startups and promote an entrepreneurial and impact-driven culture in society, through innovative solutions focused on the economic, social, and environmental dimensions.

Regarding the pillar of extension, UFRPE has been acting in projects, programs and events that connect university and society, establishing partnerships with governmental and nongovernmental organizations and social movements, contributing to a comprehensive education towards socio-environmental issues (UFRPE 2022a). In this sense, UFRPE published a call for proposals through the Provost of Extension, Culture and Citizenship in 2023 (Public Call No 04/2023—BEXT 2023), to select proposals for university extension programs and projects to be executed in 2023 and 2024, with scholarships directed towards specific themes related to sustainability (Human Rights, Affirmative Actions and Diversity; Earth, life and Sustainability), thus fostering community actions that promote sustainable development internally and externally (UFRPE 2022c). Similarly to research and innovation, the extension call for proposals also required projects and programs to be related to the SDGs. *Source* Management Report, 2023 (UFRPE 2024)

Traditionally linked to the agrarian fields, UFRPE has advanced in contemporary subjects, broadening its thematic range and incorporating emerging fields, such as Environmental Engineering and addressing the socio-environmental issue as a cross-cutting theme in the several courses present in the University (UFRPE 2022a). In the teaching pillar, the incorporation of the "Education for Sustainable Development" concept and the SDGs happen indirectly, with no specific methodology or system implemented for this purpose. However, the Institutional Pedagogic Project (PPI), an autonomous document that is a part of the UFRPE PDI, presents guiding principles for the pedagogical-didactic approach that promotes the development of skills capable of enabling a holistic perspective related to the critical thinking of the impact of each individual's action within their line of work, considering both local and global views, thus contributing to a sustainable development. Following these principles, several Course Pedagogical Projects (PPC) present approaches and curricular components related to education and sustainable development, seeking to shape citizens to be aware of their role to promote sustainability in the several fields of action. The PPCs from the courses of Environmental Engineering, Biological Sciences, Administration, Agronomy, Agroecology, Agricultural and Environmental Engineering, Economic Sciences, Consumption Sciences, Forestry Engineering, among other courses are examples worth mentioning. Not only in the undergraduate courses, but also the postgraduate level, several programs are related to educating for sustainable development. The process of incorporating extension into the UFRPE curriculum (UFRPE 2023), as well as the participation in the Extension Program of Higher

[5] https://empreendedorismo.ufrpe.br/pt-br/participar-do-programa-institucional-de-bolsas-de-iniciacao-ao-empreendedorismo-piemp.

Education in the Postgraduate Level (PROEXT-PG)[6] of the Coordination for the Improvement of Higher Education Personnel (CAPES), are opportunities to guide the incorporation of SDGs in the pillar of teaching, based on the institution's extension experience, while focusing on overcoming social, economic and environmental challenges.

At UFRPE, there is a clear guideline to dedicate efforts in the sense of internalizing socio-environmental commitments when executing guiding policies for academic practices, through dialogical management and the development of teaching, research, and extension policies in a participatory manner (UFRPE 2022a). Besides the guidance provided by the development and implementation of institutional policies, given the cross-cutting nature of the sustainability theme and the complex and wide-ranging scope of the institution's activities, it is important to carry out a more structured mapping, based on the definition of a more robust system, of the practices developed across different areas and spheres of action, in order to measure the actual impact of institutional efforts toward achieving the Sustainable Development Goals, thereby guiding strategies to enhance the institution's performance.

In this sense, UFRPE has been adopting some strategies to support the management and evaluation of the SDG implementation in the institution, with the participation in rankings and processes with external certifications, such as with THE Impact Rankings, which evaluates universities' efforts around the globe in achieving the sustainable development goals, and the SDG Education Seal, a national certification that acknowledges initiatives by brazilian teaching institutions with a real impact in implementing the SDGs. In this regard, in 2024, UFRPE had 47 initiatives contemplated with the SDG Education Seal and achieved 121 social impacts, benefiting a total of 28,025 people (UFRPE 2025b). In the 2023 THE Impact Rankings edition, UFRPE had a global highlight on Goal 2—Zero Hunger and Sustainable Agriculture, achieving the 86º position among the participating universities. Besides, it was placed 8th on Goal 17—Partnerships For The Goals, among the 20 Brazilian Universities participating (THE 2025). These initiatives create an opportunity to identify improvements and points to be improved in the institution's sustainability actions, also assisting in the decision making processes and on planning actions for sustainability, taking into consideration both the internal and external factors of the University.

Furthermore, UFRPE has been strengthening its actions through partnerships, collaborating with and joining networks with several other institutions, from both public and private sectors, aiming to contribute with the construction of a more sustainable society (UFRPE 2025c). Accordingly, UFRPE is part of several groups and networks that act together for the promotion of a sustainable development:

[6] https://www.gov.br/capes/pt-br/acesso-A-informacao/acoes-E-programas/bolsas/programas-estrategicos/desenvolvimento-regional/programa-de-extensao-da-educacao-superior-na-pos-graduacao-proext-pg/programa-de-extensao-da-educacao-superior-da-pos-graduacao-proext-pg.

SDG Brazil Network[7]: coalition of institutions that have a common interest in institutionalizing the 2030 Agenda. This Network has a chair in the National Committee for the Sustainable Development Goals (CNODS);

Brazilian Network of Higher Education Institutions for a Sustainable Development (UniSustentável Network)[8]: aims to promote a cooperative environment with exchange and communication between the higher education institutions participating, in order to contribute to the construction of a more sustainable society. The UniSustentável Network is part of the *Alianza de Redes Iberoamericanas de Universidades por la Sustentabilidad y el Ambiente* (ARIUSA);

SDG HUB Pernambuco/Global Pact[9]: has the objective to accelerate, through regional partnerships, the involvement of the entrepreneurial sector with the UN Sustainable Development Agenda until 2030 along with its 17 Sustainable Development Goals (SDGs);

Forum of Universities for Peace (FOUP)[10]: aims to foster, through the higher education institution's competences and experiences, and in partnership with the governments and other institutions, systematic discussions of social realities relevant to peace and the production of knowledge on this theme englobing its multiple variables." (UFRPE 2025c).

These networks enable the exchange of experiences and learning to improve the institution's action, both internally and externally, by combining different potentials and experiences, in accordance with what Moura-Leite et al. (2024a, b) present, considering the UniSustentável Network as an example, in which, by operating in network, higher education institutions strengthen their capacity to confront the challenges associated with the institutionalization of the 2030 Agenda, through a comprehensive approach to sustainability across their areas of activity.

Serafini (2022) identified the main barriers faced by Brazilian HEIs in implementing the SDGs, including resistance to changes in organizational culture, lack of financial resources, lack of support from the higher administration, absence of a specific organizational structure and leadership for the theme, lack of indicators and tools for measuring progress, and insufficient networks of cooperation among HEIs and partnerships on SDG-related projects. In this context, the main challenges faced by UFRPE are financial difficulties, due to the budget restrictions faced by Brazilian public universities in recent years, as well as other structural factors linked to the restrictions mentioned above, such as an insufficient number of staff. On the other hand, UFRPE has strengths such as the support and involvement of the higher administration in sustainability actions, through the UFRPE Sustainable Project, the engagement of the community through research, extension, and innovation projects, as well as the existence of the Sustainability Coordination as a focal point for a strategic and holistic approach to implementing the SDGs. Moreover, the institution participates actively in networks aimed at promoting sustainability, seeking to strengthen partnerships for the promotion of sustainability both internally and externally.

[7] https://www.redeodsbrasil.org/.

[8] https://redeunisustentavel.com.br/.

[9] https://www.pactoglobal.org.br/hub-ods/.

[10] https://foup.org/.

3 Conclusions

Based on the definition of Sustainable Development (Brundtland Report, published in 1987 by the World Commission on Environment and Development of the UN), nations have been seeking to align their participation and good socio-economic-environmental practices in numerous global conferences. This alignment has been limited by rejections from some developed nations, despite most countries believing that such negotiations are necessary and urgent. While these issues remain unsolved, some countries and institutions are internally promoting actions that display good practices in the three main pillars of sustainable development, in accordance with the goals set by the SDGs and the 2030 Agenda.

In seeking to understand the role of Higher Education Institutions (HEIs) and the effectiveness of their actions in the process of internalizing best practices aligned with the goals of the 2030 Agenda, it is essential to consider several strategic points:

i. The establishment of inclusive governance aligned with the Sustainable Development Goals (SDGs);
ii. The incorporation of specific curricular components, content, practices, and active teaching methodologies that reflect SDG-related actions and contribute to the development of professionals more attuned to labor market demands and the specificities of their respective fields of knowledge;
iii. The encouragement of research projects with applications that align with SDG targets, through partnerships with public and private institutions;
iv. The promotion of outreach projects with social impact, extending benefits beyond the physical boundaries of the university;
v. The implementation of continuous evaluation mechanisms for institutional practices, generating reports that guide potential improvements, strategic adjustments, and the documentation of results.

Section 2 of this article presents an overview of the practices adopted by the Federal Rural University of Pernambuco (UFRPE) to promote the integration of the SDGs into initiatives across the areas of management and governance, outreach, teaching, research, and innovation. As evidence of institutional progress in this process, the result obtained by the university in the 2024 self-assessment questionnaire on governance and public management, conducted by the Federal Court of Accounts (TCU), stands out. UFRPE achieved a score of 91.4% in the Integrated Index of Environmental and Social Sustainability, being classified in the enhanced governance category and surpassing the average of other higher education institutions, which were mostly placed in the intermediate category. This performance reinforces the institution's commitment to consolidating sustainable practices integrated into its academic and administrative mission.

On the other hand, it was observed that, although there have been advances in incorporating the SDGs into institutional policies and projects, it is still necessary to deepen the analysis of ongoing initiatives in order to more accurately identify the real impact of the efforts undertaken and to guide more effective strategies for improving institutional performance.

Considering the points highlighted, it is recognized that UFRPE has been navigating a challenging path, yet one that shows progress in the internalization of best practices aligned with the SDGs. The institution has developed strategic instruments such as the Institutional Development Plan (PDI), the Sustainable Logistics Plan (PLS), resolutions from the University Council, and the creation of a Sustainability Coordination Office, among other managerial determinations that aim to ensure SDG compliance in the medium term. In addition, research and innovation projects are being developed and/or implemented to deliver environmentally sustainable products and solutions. In the area of outreach, projects that have promoted—and continue to promote—the engagement of both internal and external communities are aligned with specific SDG areas. As for teaching, although broader expansion is still needed, some targeted actions have already been implemented, including the creation of new courses focused on sustainability themes.

As a practical implication, this study emphasizes that aligning governance, teaching, research, innovation, and outreach with the principles of the SDGs can enhance the transformative role of HEIs, expanding their social and environmental impact. However, certain limitations should be taken into account, including the focus on a single institution and the absence of quantitative indicators to evaluate impact. As a future direction, it is suggested that comparative studies be conducted across different HEIs, as well as the development of metrics that allow for a more objective assessment of the effectiveness of the implemented actions.

In light of the above, the process of internalizing the SDGs within HEIs presents itself as a complex, long-term challenge—but not an insurmountable one. Along this path, HEIs can play a transformative role that extends beyond their physical boundaries, generating a multiplier effect in society and the world, and acting as active agents in the construction of a more just, inclusive, and sustainable reality.

References

Blasco, N., Brusca, I., & Labrador, M. (2021). Drivers for universities' contribution to the sustainable development goals: An analysis of Spanish public universities. Sustainability, 13(1), 89. https://doi.org/10.3390/su13010089

Christ, G. D., Piffer, M., & Alves, L. R. (2024). Rumo à sustentabilidade: uma análise da implementação dos ODS em Instituições de Ensino Superior [Towards sustainability: An analysis of the implementation of the SDGs in Higher Education Institutions]. REUNIR: Revista de Administração, Contabilidade e Sustentabilidade, 13(4). https://www.reunir.revistas.ufcg.edu.br/index.php/uacc/article/view/1523

Corbari, S. D., Dorado, A., Kniess, C. T., & Freitas, L. (2021). O papel das instituições de ensino superior no alcance dos Objetivos de Desenvolvimento Sustentável (ODS) [The role of higher education institutions in achieving the Sustainable Development Goals (SDGs)]. Instituto de Estudos Avançados da Universidade de São Paulo.http://www.iea.usp.br/pesquisa/projetos-institucionais/usp-cidades-globais/artigos-digitais/o-papel-das-instituicoes-de-ensino-superior-no-alcance-dos-objetivos-de-desenvolvimento-sustentavel-ods

Cravo, J. M., & Silva, J. dos S. (Eds.). (2025). Relatório final de avaliação de desempenho: PLS UFRPE: 2022–2024 [Final performance evaluation report: PLS UFRPE: 2022–2024].

Pró-Reitoria de Planejamento e Gestão Estratégica—Universidade Federal Rural de Pernambuco. https://www.proplan.ufrpe.br/sites/ww2.proplan.ufrpe.br/files/Relat%C3%B3rio_Final_PLS_UFRPE_2022-2024_0.pdf

Kestin, T., van den Belt, M., Denby, L., Ross, K., Thwaites, J., & Hawkes, M. (2017). Getting started with the SDGs in universities. Sustainable Development Solutions Network. https://hdl.handle.net/11541.2/145166

Kohl, K., Hopkins, C., Barth, M., Michelsen, G., Dlouhá, J., Razak, D.A., Abidin Bin Sanusi, Z. & Toman, I. (2022), "A whole-institution approach towards sustainability: a crucial aspect of higher education's individual and collective engagement with the SDGs and beyond", International Journal of Sustainability in Higher Education, Vol. 23 No. 2, pp. 218–236. https://doi.org/10.1108/IJSHE-10-2020-0398.

Lima, M. R. O., & Feitosa, A. A. F. M. (2024). Inserção dos Objetivos do Desenvolvimento Sustentável da Agenda 2030 no currículo escolar [Integration of the Sustainable Development Goals of the 2030 Agenda into the school curriculum]. Anais do Congresso Nacional de Educação—CONEDU. Disponível em: https://editorarealize.com.br/editora/anais/conedu/2024/TRABALHO_COMPLETO_EV200_MD1_ID5474_TB286_01102024164702.pdf

Moura-Leite, R. C. (2023). Abordagem integral institucional da sustentabilidade nas instituições de ensino superior: Como incorporar a sustentabilidade em suas funções e estabelecer parcerias para o fomento aos Objetivos de Desenvolvimento Sustentável [Institutional holistic approach to sustainability in higher education institutions: How to incorporate sustainability into their functions and establish partnerships to promote the Sustainable Development Goals]. Expressa Extensão, 28(1), 34–37. https://doi.org/10.15210/ee.v28i.25562

Moura-Leite, R.C., Lopes, J.C.J. e Yamazaki, C. (2023). Brazilian federal universities and their sustainable practices based on sustainable logistics management plan. International Journal of Sustainability in Higher Education, 24(4), 932–947. https://doi.org/10.1108/IJSHE-02-2022-0057

Moura-Leite, R. C., Carvalho, L. C., & Cravo, J. M. (2024a). Responsabilidade social universitária e o papel das redes: O caso da Rede UniSustentável [University social responsibility and the role of networks: The case of the UniSustentável Network]. In C. G. de Carvalho (Ed.), Responsabilidade Social, 15(11). Brasília: ABMES Editora. ISSN 1809–8754. Retrieved from https://abmes.org.br/editora/detalhe/148

Moura-Leite, R., Sáenz, O., María, A., Holguín, M. T., Callejas, M. M., Bravo, M. T., Velasco, A., Arguedas, M., Hidalgo, C., & Parada, S. (2024b). Avanços no conhecimento, compromisso e contribuições para os Objetivos de Desenvolvimento Sustentável nas instituições de ensino superior da América Latina e do Caribe—Relatório IES brasileiras [Advances in knowledge, commitment and contributions to the Sustainable Development Goals in higher education institutions in Latin America and the Caribbean—Brazilian HEI Report]. Observatorio de la Sustentabilidad en la Educación Superior de América Latina y el Caribe (OSES-ALC). Retrieved from https://oses-alc.net/wp-content/uploads/2024/01/2024-02-22-OSES-ALC-Informe-Brasil.pdf

Ramos, P. de O. (2022). Objetivos do desenvolvimento sustentável (ODS), educação ambiental e o currículo da Cidade de São Paulo [Sustainable Development Goals (SDGs), environmental education, and the curriculum of the City of São Paulo] (Master's thesis, University of São Paulo). USP Theses Repository. https://www.teses.usp.br/teses/disponiveis/106/106132/tde-13122022-100738/

Sedlacek, S. (2013). The role of universities in fostering sustainable development at the regional level. Journal of Cleaner Production, 48, 74-84. https://doi.org/10.1016/j.jclepro.2013.01.029

Serafini, P. G. (2022). Os objetivos de desenvolvimento sustentável nas instituições de ensino superior: Iniciativas e barreiras no processo de implementação [The Sustainable Development Goals in higher education institutions: Initiatives and barriers in the implementation process](Master Thesis). Universidade Federal do Rio Grande do Norte, Centro de Ciências Exatas e de Tecnologia, Programa de Pós-Graduação em Engenharia de Produção. https://repositorio.ufrn.br/handle/123456789/47101

Souza, E. M. R., & Ventura, A. C. (2024). Práticas das universidades federais brasileiras para implementar os ODS: uma revisão sistemática [Practices of Brazilian federal universities to implement the SDGs: A systematic review]. In A. C. Ventura (Ed.), Theory and practice of sustainable development (Chapter 2). Editora Rima. https://editorarima.com.br/wp-content/uploads/2024/04/A-Teoria-e-a-Pratica-do-Desenvolvimento-Sustentavel-Capitulo-2.pdf

Times Higher Education. (2025). Federal Rural University of Pernambuco|World University Rankings. https://www.timeshighereducation.com/world-university-rankings/federal-rural-university-pernambuco

Universidade Federal Rural de Pernambuco. (2020). Relatório anual de acompanhamento do Plano de Gestão de Logística Sustentável—PLS 2019 [Annual Monitoring Report of the Sustainable Logistics Management Plan—PLS 2019]. Pró-Reitoria de Planejamento e Desenvolvimento Institucional. http://ww2.proplan.ufrpe.br/sites/ww2.proplan.ufrpe.br/files/RELAT%C3%93RIO%20PLS%202019%20final.pdf

Universidade Federal Rural de Pernambuco. (2021). Relatório anual de acompanhamento do Plano de Gestão de Logística Sustentável—PLS 2020 [Annual Monitoring Report of the Sustainable Logistics Management Plan—PLS 2020]. Pró-Reitoria de Planejamento e Desenvolvimento Institucional. https://ww2.proplan.ufrpe.br/sites/ww2.proplan.ufrpe.br/files/RELATÓRIO%20DO%20PLANO%20DE%20LOGÍSTICA%20SUSTENTÁVEL%20DE%202020.pdf

Universidade Federal Rural de Pernambuco (UFRPE). (2022a). Plano de Desenvolvimento Institucional da UFRPE—2021–2030 (aditamento). [Institutional Development Plan of UFRPE—2021–2030 (addendum)]. UFRPE. http://ww2.proplan.ufrpe.br/sites/ww2.proplan.ufrpe.br/files/PDI-Final-ADITAMENTO_20_10_2022-1%20%281%29.pdf

Universidade Federal Rural de Pernambuco (UFRPE). (2022b). Plano de Logística Sustentável (PLS). [Sustainable Logistics Plan (PLS)] UFRPE. https://www.proplan.ufrpe.br/sites/ww2.proplan.ufrpe.br/files/PLSfinal%20VERS%C3%83O%20FINAL.pdf. Acesso em: fevereiro, 2025.

Universidade Federal Rural de Pernambuco. (2022c). Regimento interno da Pró-Reitoria de Extensão, Cultura e Cidadania—PROExC [Internal Regulations of the Dean's Office for Extension, Culture and Citizenship—PROExC]. https://www.proexc.ufrpe.br/pt-br/node/225

Universidade Federal Rural de Pernambuco. (2023). Guia de inserção curricular da extensão na UFRPE [UFRPE's Guide for Incorporating Extension into the Curriculum]. Editora Universitária da UFRPE. https://editora.ufrpe.br/sites/editora.ufrpe.br/files/_GUIA%20DE%20INSER%C3%87%C3%83O%20CURRICULAR%20DA%20EXTENS%C3%83O%20NA%20UFRPE.pdf

Universidade Federal Rural de Pernambuco (UFRPE). (2024). Relatório de Gestão do exercício 2023. [Management Report for Fiscal Year 2023]. UFRPE. http://ww2.proplan.ufrpe.br/sites/ww2.proplan.ufrpe.br/files/RELATORIO%2BGESTAO_08%2B02%2B2024_versao%2Bfinal%2B3.pdf

Universidade Federal Rural de Pernambuco. (2025a). Coordenadoria de Sustentabilidade [Sustainability Coordinator] | PROPLAN. https://www.proplan.ufrpe.br/br/content/coordenadoria-de-sustentabilidade.

Universidade Federal Rural de Pernambuco. (2025b). UFRPE ganha Selo ODS Educação com 47 projetos contemplados [UFRPE Receives ODS Education Seal with 47 Certified Projects]. https://ufrpe.br/br/content/ufrpe-ganha-selo-ods-educa%C3%A7%C3%A3o-com-47-projetos-contemplados

Universidade Federal Rural de Pernambuco. (2025c). Sustentabilidade [Sustainability]|PROPLAN. http://ww2.proplan.ufrpe.br/br/content/sustentabilidade

Janielly Mantovani Cravo Master in Environmental Engineering by the Federal Rural University of Pernambuco (UFRPE). Coordinator of Planning and Sustainability within the Directorate of Planning, Results, and Sustainability (DPRS), under the Office of the Pro-Rector for Planning and Administration (PROPLAD) from UFRPE. Associate Coordinator of the Brazilian Network of Higher Education Institutions for Sustainable Development (UniSustentável).

Chiara Natércia França Araújo Associate Professor, PhD, of the Federal Rural University of Pernambuco (UFRPE). Economist, working in more specific areas of business, finance, quantitative methods, and environmental economics. Alternate Representative Member of UFRPE in the Brazilian Network of Higher Education Institutions for Sustainable Development (UniSustentável). Coordinator of the research group "Núcleo Econômico Integrado (NEI)"—dgp.cnpq.br/dgp/espelhogrupo/504537.

Best Practices of Higher Education Institutions in Promoting Gender Equality: A Study on SDG 5 in the Brazilian Context

Rosamaria Moura-Leite, Renata Benedetti Mello Nagy Ramos, Débora Teixeira da Cruz, Jessica Neiva de Figueiredo Leite Araujo, and Cicilia Raquel Maia Leite

Abstract Contemporary Higher Education Institutions (HEIs) play an essential role as agents of social transformation in their territories, contributing to the training of professionals and citizens who are aware of their socio-cultural and environmental responsibilities. This study aims to describe the practices developed by HEIs in relation to gender equity and their connection with the 2030 Agenda. Sustainable Development Goal (SDG) 5 emphasises Gender Equality and is a current issue of extreme relevance in Brazil, where research indicates high rates of psychological, sexual and physical violence against women, as well as inequalities in the labour market, political representation and leadership positions. The methodological design of the research is qualitative, analytical and descriptive. The results emphasise that the 2030 Agenda and the SDGs have driven the integration of actions aimed at sustainable development in the substantive functions of teaching, research and extension in the spheres of governance and management of the institutions. The results show how the three HEIs studied are implementing actions in line with the goals of SDG 5. Among the actions identified are: educational campaigns, welcoming programmes, listening tools and institutional policies to prevent violence. It is considered that by describing good practices, this study helps to encourage other HEIs to promote similar actions, thus expanding the fight against discrimination and violence against women in Brazilian

R. Moura-Leite
Federal University of Mato Grosso Do Sul, Campo Grande, Brazil
e-mail: rosamaria.leite@ufms.br

R. B. M. N. Ramos (✉) · D. T. da Cruz
Unigran Capital University Centre, Campo Grande, Brazil
e-mail: renata.ramos@unigran.br

D. T. da Cruz
e-mail: debora.cruz@unigran.br

J. N. de Figueiredo Leite Araujo · C. R. M. Leite
Rio Grande Do Norte State University, Mossoró, Brazil
e-mail: jessicaneiva@uern.br

W. Leal Filho et al. (eds.), *Sustainability in the Curriculum and Operations of Universities*, World Sustainability Series,
https://doi.org/10.1007/978-3-032-12678-8_36

HEIs. The results reflect the reality of the HEIs analyzed, and it is suggested that future research expand the scope of analysis and assess the effectiveness of the actions undertaken.

Keywords Gender equity · Higher education institutions · Brazilian universities · Sustainable development goal 5 (SDG 5)

1 Introduction

Higher Education Institutions (HEIs) play a strategic role in contemporary society, becoming agents of social transformation and sustainable development (SDSN 2017). In addition to their role of training qualified professionals, HEIs have a responsibility to promote critical reflection and foster actions that contribute to building a fairer and more inclusive society (UNESCO 2022; Moura-Leite 2023). In this context, issues related to gender equality stand out as an urgent and indispensable topic, in line with the Sustainable Development Goals (SDGs) of the 2030 Agenda, especially SDG 5, which aims to achieve gender equality and empower women and girls (Pandit and Paul 2023; Leal Filho et al. 2023).

SDG 5 has nine monitoring targets and 14 global indicators. The Voluntary National Report, published by the Federal Government of Brazil in 2024, presents data for seven indicators relating to five global targets (Merma-Molina et al. 2024). However, only four targets are actually evaluated, due to the difficulty in obtaining monitoring data. The indicators analysed show a reduction in the female homicide mortality rate and in early marriages, but inequalities persist in unpaid domestic work and in women's participation in political and business leadership. Women's access to technology has increased, but there are still disparities, especially among those living in rural areas. In the same document, the Federal Government highlights the priority given to tackling gender-based violence, considering it a central theme in strategic objective 1.7 of the PPA 2024–2027, as well as citing various programmes and policies aimed at achieving SDG 5 (Brazil 2024).

The need to promote gender equality and empower women and girls in Brazil is a historical demand and, despite some progress, there is still a lot to be done. The National Survey on Violence against Women, available on the Federal Senate's website, presents alarming data on the increase in the percentage of women who say they have suffered domestic and family violence, reaching 30 percent by 2023. Another worrying statistic is that 35 percent of women victims are under the age of 19, which reinforces the need to protect girls, a vulnerable group that requires special attention. As for the types of violence suffered, the study reveals a significant increase in moral and psychological violence, as well as an increase in cases of physical, sexual and property violence (Brazil 2024; Muñoz-Mazón 2024).

Organisations, in general, should promote the prevention and combating of the various types of violence against women in the organisational environment, as well as encouraging their empowerment. There are tools to help managers in this process, including ISO 37201—Management Systems to Prevent and Combat Violence Against Women in Organisations, which offers practical guidelines for developing management systems focused on preventing and combating violence against women in the corporate environment (ISO 2024).

Also in the organizational context, a prominent initiative is the UN Global Compact, considered to be the world's largest corporate sustainability initiative, with more than 25,000 participants, including companies and organisations, distributed in more than 67 local networks and operating globally. One of its programmers, called the 'They Lead 2030 Movement', promotes gender parity in the top leadership of organisations by 2030 (Global Compact 2024). In the education sector, the United Nations Educational, Scientific and Cultural Organisations (UNESCO) is carrying out various actions aimed at gender equality in science, technology and innovation.

These include the Organisations for Women in Science for the Developing World (OWSD), which provides research training, career development and networking opportunities for women scientists from developing countries at different stages of their academic and professional careers. Another relevant action, carried out in conjunction with UN Women, was the implementation of the International Day of Women and Girls in Science, celebrated annually on 11 February. In 2025, this date turned 10 years old. UNESCO also makes reports and tools available on its website to support policymakers in implementing and monitoring gender equality in science, technology and innovation.

Considering the role of HEIs as agents of change, this article aims to describe the practices developed by these institutions in relation to gender equity and analyses their connection with the global goals of the 2030 Agenda. Previous studies indicate that the 2030 Agenda has driven the integration of actions aimed at sustainable development into the substantive functions of teaching, research and extension, as well as into institutional governance and management (e.g. Saenz et al. 2024; Moura-Leite et al. 2024a; Serafini et al. 2022; SDSN 2017). And a wide range of studies demonstrate how Brazilian higher education institutions include various aspects of sustainability in their activities (e.g. Evilin et al. 2019; Kihara et al. 2019; Nogueira et al. 2019; Moura-Leite and Pereira 2020; Weber et al. 2020; Campos et al. 2022; Moura-Leite et al. 2023), as well as reporting on the relevant role of collaboration between higher education institutions in developing these actions (Moura-Leite et al. 2024b).

This article is structured in four main sections. The first presents the introduction, which contextualizes the relevance of the topic and sets out the central objective of the research, aimed at analyzing the strategies adopted by Higher Education Institutions (HEIs) to promote gender equity, in line with the Sustainable Development Goals (SDGs), especially SDG 5. The second section is dedicated to the methodological design, which underpins the investigation through a qualitative approach, of a descriptive, exploratory and analytical nature. The third section presents the results of the documentary analysis, based on institutional data available on the official digital platforms of the participating HEIs. Finally, the concluding remarks summaries the main findings and propose avenues for further research into the integration of gender equity into the institutional policies and practices of Brazilian universities, in addition to mentioning the research limitations and suggesting directions for future studies.

2 Methodology

The research adopted a qualitative, descriptive and analytical methodological design, in line with Flick's (2009) assumptions, aimed at understanding social and institutional phenomena in specific contexts. In this sense, it is descriptive in that it reports evidence of institutional actions related to promoting gender equity and analytical in that it critically interprets the data collected, establishing relationships between institutional discourses and actual practices in Higher Education Institutions.

The methodological strategy used was documentary research, based on analyzing public information made available on the websites and institutional platforms of the participating universities. Institutional documents, sustainability reports, management reports, diversity and inclusion policies, among other materials relevant to the scope of the investigation, were examined. The choice of documentary analysis is justified by the institutional nature of the data and the possibility of describing the materialization of the 2030 Agenda guidelines in university policies at different levels of competence.

The study involves the participation of three Higher Education Institutions at different federal, state and private levels, two in the state of Mato Grosso do Sul and one in the state of Rio Grande do Norte, with the aim of correlating their actions with Sustainable Development Goal (SDG) 5—Gender Equality. The research is based on analyzing content available on institutional websites, publications and other evidence publicly disclosed by these institutions. In order to guarantee the ethical compliance of the research, information in the public domain was used, as established by Law No. 12,527 of 18 November 2011 (Access to Information Law).

Data was collected during the month of March 2025. Keywords were searched on the institutional websites of the higher education institutions and institutional documents previously cited were consulted. The keywords used were: women, gender, SDG 5, female. The search for keywords was intended to facilitate access to the institutional actions related to promoting gender equality in each of the three higher education institutions studied, and this study does not intend to describe all the actions related to the issue.

In the results section, the Higher Education Institutions involved in the research were characterized, highlighting their relevance in the context of the study related to Gender Equality. This characterization aims to qualify and support the analysis of the results, providing a more in-depth understanding of the participating institutions.

3 Results

In order to qualify and analyses the results obtained in the research, this section presents a characterization of the participating Higher Education Institutions (HEIs), highlighting their relevance in the context of the study and their inclusion in the actions linked to the United Nations (UN) 2030 Agenda.

The choice to include public and private institutions from different regions of Brazil aims to analyses different institutional approaches to promoting gender equality. This diversity makes it possible to observe the internal policies, structural challenges and impacts of actions aimed at SDG 5 in everyday academic life, contributing to an understanding of the strategies adopted by HEIs to tackle gender inequalities. By analyzing these initiatives in the academic environment, we seek to understand how HEIs develop organizational strategies to tackle gender inequalities, assessing their impacts and potential for replicability.

The Federal University of Mato Grosso do Sul (UFMS) was established in 1962 and became federalized in 1979, making it the largest higher education institution in Mato Grosso do Sul. It has 10 campuses in various cities. In addition to its headquarters in Campo Grande, the capital of the state of Mato Grosso do Sul, the university is home to the Faculty of Pharmaceutical Sciences, Food and Nutrition (Facfan), the Institute of Biosciences (Inbio), the Integrated Health Institute (Inisa), the Faculty of Human Sciences (Fach), the Faculty of Education (Faed),Faculty of Arts, Letters and Communication (Faalc), Faculty of Computing (Facom), Faculty of Law (Fadir), Faculty of Medicine (Famed), Faculty of Veterinary Medicine and Zootechny (Famez), Faculty of Dentistry (Faodo) and Faculty of Engineering, Architecture and Urbanism and Geography (Faeng);Institute of Mathematics (Inma), Institute of Chemistry (Inqui) and Institute of Physics (Infi) UFMS has campuses in Aquidauana, Chapadão do Sul, Corumbá, Coxim, Naviraí, Nova Andradina, Paranaíba, Ponta Porã and Três Lagoas (UFMS 2025).

UFMS has around 40,000 undergraduate and postgraduate students and currently offers 136 undergraduates and 80 postgraduate courses, providing public, free, inclusive and high-quality higher education. UFMS is a higher education institution whose mission is 'to develop and socialize knowledge, training qualified professionals for the transformation of society and the sustainable growth of the country', with the following values: citizenship, effectiveness, ethics, excellence, integrity, interdisciplinary, professionalism, respect, social responsibility, sustainability and transparency. And its vision is 'to be a university accessible to all people and recognized, nationally and internationally, for excellence in teaching, research and extension, entrepreneurship, sustainability, innovation, art and culture, sport and leisure, as well as the popularization of science'. It should be noted that sustainability is present in the mission, values and vision of UFMS (UFMS 2025).

On 20 May 2022, the Board of Trustees approved the UFMS Sustainability Policy by means of Resolution 260.The document institutionalizes the SDGs and the 2030 Agenda at UFMS, since §2 of Art. 1 states that 'The Sustainable Development

Goals—SDGs of the 2030 Agenda of the United Nations—UN, must be observed in the actions promoted by UFMS through the Sustainability Policy' (PROCIDS 2025).

Since 2018, UFMS has been actively integrating the 17 Sustainable Development Goals (SDGs) into its teaching, research and extension projects. As part of a pioneering initiative, the university has adopted the Project Information and Management System (SIGProj) to monitor the alignment of projects with the SDGs. All projects developed at the institution must be registered in this system and, during the registration process, coordinators are instructed to indicate which of the 17 SDGs their projects relate to, and it is mandatory to select at least one goal. As of 2020, it is also compulsory to include a description of the SDGs actually covered in the final reports, allowing for a more accurate assessment of the impacts generated. And as of the second half of 2023, teachers who wish to can record which SDGs their subjects are working on in their teaching plans (PROCIDS 2025).

Every year the university makes publicly available the 'UFMS Sustainable Development Goals Report', which lists the main activities and achievements for each SDG in the year being monitored. This document provides access to information, data and figures related to the SDGs at the institution (PROCIDS 2025). Among the various activities developed by UFMS to meet SDG 5 "I Am Woman (Sou Mulher)" UFMS programmed stands out. It began in 2021 and is aligned with the 'Take Care, I Love You—A UFMS Heart Action' Programmed, bringing together policies, practices and actions aimed at UFMS women (students, civil servants and outsourced employees).

The initiative arose from the actions of the Inclusion, Internationalisation and Affirmative Action Management Committee (CGIIAF), to which the 'I am a Woman' programme is linked. The programme's concept is to work in different areas of service, knowledge and appreciation of women, linking the initiative to SDG 5.To achieve its objectives, the I Am Woman UFMS Programme will encourage specific and cross-cutting actions, based on policies and programmes in Management, Teaching, Research, Extension, Innovation and Entrepreneurship in three strategic axes: Axis 1: Women's entry, permanence and success; Axis 2: Encouraging Teaching, Research, Extension, Entrepreneurship and Innovation led by women; and Axis 3: Promoting an increasingly welcoming and equitable environment for women (Mulher 2024).

In each of the axes, various projects are developed, among which we have chosen a few per axis to describe.

Axis 1 includes projects that encourage the presence of women in STEAM (science, technology, engineering and maths) courses, which have historically been dominated by men, including They're Programming" project, which aims to integrate students as a way of preventing drop-outs, with the participation of around 150 students (Mulher 2024).

In axis 2, one action that has a major impact is the 'Women in Science' call for proposals, which aims to support research and innovation projects led by female professors and administrative technicians. In 2020, R$500,000.00 was made available for this purpose, while in 2021, 2022, 2023 and 2024 the amount invested was R$300,000.00 in each call. In addition, the calls for scientific initiation also received greater investment and, since 2021, have been aligned with both the I Am Woman

UFMS Programme and the Parent in Science® movement, reinforcing the university's commitment to gender equality and the inclusion of mothers in academia (Mulher 2024).

Axis 3 highlights the creation in 2021 of the Policy for Preventing and Confronting Moral and Sexual Harassment, which was approved by University Council Resolution (Coun) No. 127 and is the responsibility of the Inclusion, Internationalisation and Affirmative Action Management Committee (CGIIAF). The actions include: the implementation of eco-technological bus stops with Panic Buttons, connected to campus security, both in the University City and in the regional campuses, as well as the installation of panic buttons in toilets and in the "SOU UFMS APP"; publicising the policy through campaigns on social networks, such as 'UFMS without Harassment', lectures, and lives, as well as the promotion of extension courses to train staff and students on the subject; the creation of a women's ombudsman, a women's ombudsman's office; among other actions (Mulher 2024).

All these actions that make up the 'I am a Woman Programme' show an organisational vision that is sensitive to historical gender inequalities, while at the same time demonstrating active management in reducing these disparities. The adoption of policies such as specific calls for proposals for women scientists, institutional listening channels and safety devices such as the UFMS Alert Button strengthens the organisational culture and contributes to the well-being of students, staff and collaborators.

One of the accolades received for running this programme is the high score it received for SDG 5 in the 2024 edition of the Times Higher Education (THE) Impact international ranking. In the global ranking, which considers 2152 universities from 125 countries, UFMS scored in the 69.9–75.7 range, occupying the 401–600 position for the fourth consecutive year (PROCIDS 2025). The 'I Am a Woman Programme' and its monitoring reports can be consulted on the UFMS website (https://procids.ufms.br/programa-sou-mulher-ufms/). These documents bring together various other actions that have been carried out, which have not been detailed in this article due to space limitations.

The State University of Rio Grande do Norte (UERN), created on 28 September 1968 by Municipal Law No. 20/68, initially under the name of Regional University of Rio Grande do Norte (URRN), is a public higher education institution maintained by the state government, with its administrative headquarters in the municipality of Mossoró. With more than five decades of hi UERN stands out for its territorial capillarity, being present in six municipalities in the state of Rio Grande do Norte Mossoró, Natal, Assú, Caicó, Pau dos Ferros and Patu—through central and advanced campuses (UERN 2024). This strategic distribution allows the institution to reach out significantly to the populations of the interior, with a direct impact on teacher training, regional development and the democratisation of higher education in the semi-arid northeast.

According to the 2024 Management Report (UERN 2024), the university offers 61 undergraduate courses, of which 58 are face-to-face and 3 are distance learning, as well as stricto and lato sensu postgraduate programmes and extension and research projects. UERN's relevance to studies on the 2030 Agenda, particularly with regard to Sustainable Development Goal 5 (SDG 5)—Gender Equality—lies in its commitment to public policies that promote equity, inclusion and female protagonism.

UERN has been recognised nationally for its work in this field, having been awarded the ODS Edu Seal for two consecutive years to institutions committed to the Sustainable Development Goals in the educational context (UERN 2023a, b). Analysing institutional actions on different campuses reveals a diversity of projects aimed at valuing women, critically reflecting on gender and building more inclusive academic spaces.

At the Assú, Caicó and Natal campuses, there are projects with a strong emphasis on critical and cultural education about the role of women in society. In Assú, the 'Gender Relations and Sexuality in Education' project stands out, coordinated by Professor Daiane de Oliveira, which aims to encourage debates on gender equality in the educational environment. In Caicó, the project 'Studies on Women Philosophers and the Epistemicide of Knowledge' proposes a critical approach to the historical erasure of women's intellectual productions. In Natal, the project 'Cinema and Feminisms: An Ethical–Political Interpretation of Gender Relations', coordinated by Prof Maria José Vida, uses audiovisuals as a tool for reflection and debate.

At the Pau dos Ferros campus, activities are concentrated in the area of health, with emphasis on the projects 'Mãe Conectada: Educação em Saúde na Pre gravidez, Parto e Puerpério' (Connected Mother: Health Education in Pregnancy, Childbirth and the Puerperium), coordinated by Prof Giselle dos Santos Costa Oliveira, and 'Assistência de Enfermagem às Mulheres Vítimas de Violência Sexual' (Nursing Assistance to Women Victims of Sexual Violence), by Prof Francisca Adriana Barreto. Both promote reception, information and care for women in vulnerable situations.

In Patu, the 'Ciranda Cirandinha' initiative stands out, coordinated by Prof Dr Cláudia Tomé, which offers a childcare space for the daughters of female students, constituting a policy of permanence and university inclusion. These projects demonstrate UERN's institutional capacity to respond to local realities with transformative and sustainable actions.

In the field of research, training and extension, the work of the 'Simone de Beauvoir' Women's Studies Centre (NEM), linked to the Faculty of Social Work (FASSO), stands out. The centre organises courses, events and workshops on topics related to gender, feminism and human rights, and is a strategic body in the institutionalisation of the gender equality agenda at UERN (Morais 2024).

Among the innovative initiatives is the 'Oi, Meninas' project, created in 2021, which promotes the inclusion of girls in the areas of science and technology, with a focus on Computing. Developed by listening to female students on the Computer Science course, the project offers workshops, mentoring and conversation circles, creating a welcoming space and technical training for young women. It is currently institutionalised as a permanent action with funding from the PIBEX 2025 Programme, which strengthens its sustainability and impact (Gomes et al. 2023; UERN 2025).

In addition to academic actions, UERN is also making progress in terms of institutional representation. In 2023, 17 of the university's 32 leadership positions were held by women, including the Rectorate and most of the pro-rectorates (UERN 2024). These indicators reinforce the university management's commitment to gender equality not just as a discourse, but as an effective practice. The university passed a resolution in its superior council in which it reserves at least 50 percent of the vacancies in senior management positions appointed by the rectors for women. It also approved a policy to combat violence against women.

With this variety of projects and institutional indicators, UERN is consolidating itself as an example of a public university committed to SDG 5, demonstrating that it is possible to integrate teaching, research, extension and management into a transformative agenda for society.

The Unigran Capital University Centre, in turn, represents the private side of the research. Based in Campo Grande—MS, the institution is part of the Unigran Educational group, which has been operating in higher education for over four decades, with its headquarters in Dourados—MS. In 2007, Unigran Capital obtained authorisation from the Ministry of Education (MEC). It began its activities by offering three face-to-face courses—Physical Education, Nursing and Aesthetics. Today there are more than 40 undergraduate and postgraduate courses, including face-to-face, semi-presential and online programmes. All of Unigran Capital's efforts are centred on the student, with the mission of encouraging humanistic, professional and citizen training. The return is realised when students voluntarily take part in social actions or research, teaching and extension activities; when they pass public examinations and selection processes, including postgraduate courses, and also by their results in the National Student Performance Exam (Enade) (Unigran Capital 2023).

UNIGRAN Capital's purpose is to promote higher education aimed at the well-being of society and local, regional and global development. This mission is underpinned by creative, innovative, entrepreneurial pedagogical practices committed to sustainability, based on the principles of democracy, ethics and social justice—values directly aligned with the goals of the 2030 Agenda, especially SDG 5, which aims to achieve gender equality and empower all women and girls (Unigran Capital 2025).

UNIGRAN Capital's institutional vision is to be recognised as a university centre focused on comprehensive training for life, based on promoting self-knowledge, quality management, innovation and entrepreneurship. These pillars favour inclusion, equity and human development, with an emphasis on valuing diversity and gender equity in all dimensions of academic life.

The institutional values guide educational and administrative practices, placing the student at the centre of the training process. This manifests itself in an ongoing commitment to human, professional and citizen training, which values intellectual, emotional, social, cultural, political and professional autonomy. The valorisation of science, the commitment to quality, the search for pedagogical innovation and the practice of ethics strengthen a teaching environment based on transparency, responsibility and credibility. Freedom in teaching, research and the dissemination of culture is understood as an essential principle, ensuring equal access to knowledge, art and quality education. Social responsibility and citizenship are exercised through actions aimed at the sustainable development of local communities, strengthening the institution's role as an agent of social transformation (Unigran Capital 2023).

In this context, entrepreneurial education is promoted as an instrument of personal and collective empowerment, encouraging the formation of autonomous, critical and creative individuals, prepared to make decisions, propose innovative solutions and undertake in a conscious and ethical manner. Life skills training encourages students to reflect critically on their role in society, articulating empirical and scientific knowledge with the construction of an identity committed to democratic values, social justice and gender equality.

The search for institutional information also included the UNIGRAN Capital website, where relevant actions aimed at promoting gender equality were identified, with emphasis on transparency initiatives and compliance with current labour legislation. One of these initiatives was the publication entitled 'UNIGRAN Capital presents the results of the salary equalisation reports for the 1st semester', published on 19 March 2025. The content presents the results for the first half of that year, drawn up in accordance with Law 14.611/2023—known as the 'Equal Pay Law'—which ensures equal pay for men and women who perform the same job. The data presented shows that the pay disparities between the institution's employees are associated with organisational variables such as the gender composition of the workforce, length of service, level of qualification, performance in institutional goals and the degree of responsibility inherent in the functions performed (Unigran Capital 2025).

According to the aforementioned report, the composition of UNIGRAN Capital's workforce is mostly female, accounting for 59.8% of all employees, while men account for 40.2%. In terms of remuneration, it was observed that the average salary for women is 103% of that received by men, while the median value is even higher at 107.2%. These figures point to a scenario of relative gender pay equity, reflecting an institutional policy of valuing women's work and reducing the historical inequalities present in the labour market.

The Unigran Capital University Centre's Inter and Transdisciplinary Extension Centre for Employability and Paid Internships (NEXITEMPER) of the Unigran Capital University Centre, is also a strategic initiative aimed at promoting quality education (SDG 4), gender equality (SDG 5), decent work and economic growth (SDG 8) and reducing inequalities (SDG 10). Created with the aim of centralising and integrating institutional requests, information and communications, the programme adopts an inter- and transdisciplinary approach that broadens the connections between the academic environment and the world of work. NEXITEMPER takes an active role in forwarding CVs, promoting training and managing a database of vacancies, strengthening job placement opportunities for students and graduates. By connecting different areas of knowledge and fostering dialogue between academia, the productive sector and society, the programme contributes to the development of more equitable, inclusive and sustainable career paths (Nexitemper 2022).

In addition to publishing salary reports and the aforementioned Extension project, the UNIGRAN Capital University Centre develops effective actions in line with the principles of Sustainable Development Goal 5 (SDG 5), which aims to promote gender equality and the empowerment of women and girls. Among the most noteworthy initiatives is the Health Work Education Programme (PET-Saúde), 2024–2026 edition, developed under the title 'UNIGRAN Capital in the transversality of equity in the context of health training and work in Campo Grande—MS'. The aim of the project is to integrate a gender perspective into training and professional practices in the field of health, through teaching, research and extension activities aimed at health units in the municipality of Campo Grande (MS) (Pet-Saúde 2024).

Currently in its 11th edition, PET-Saúde includes 150 projects nationwide, all aimed at strengthening integration between teaching, service and the community. The initiative establishes a link between the Unified Health System (SUS) and Higher Education Institutions (HEIs), contributing both to the training of future professionals and to the creation and expansion of conditions that promote the valorisation of workers and future workers within the SUS. The programme adopts an intersectional approach, considering aspects such as gender equity, gender identity, sexuality, race, ethnicity and disabilities, in line with the principles of the National Programme for Gender Equity, Race and Valuing Women Workers in the SUS (Pet-Saúde 2024).

In this context, PET-Health is an important tool for linking academic training and transformative social practices. As well as fostering the production of knowledge and providing practical experience during undergraduate studies, the programme offers grants funded by the Ministry of Health, ensuring fair conditions for students and teachers to carry out their activities in a fair and qualified manner. For the current edition, 40 scholarships have been made available to Unigran Capital, strengthening the academic-professional curricula of the participants and expanding access to training opportunities based on equity, inclusion and social justice (Pet-Saúde 2024).

The project has Thematic Axes of the PET Health Equity Project 2024/2026, namely:

> Axis 1: Valuing workers and future workers within the SUS, Gender, Gender Identity, Sexuality, Race, Ethnicity, Disabilities and intersectionalities in health work; Axis 2: Valuing workers and future workers within the SUS, mental health and violence related to health work; Axis 3: Welcoming and Valuing workers and future workers and health workers in the process of motherhood, welcoming and valuing women, trans men and other people who give birth.

Unigran Capital has also organised awareness-raising and training events on human rights and gender equality. Among the initiatives, the lecture on moral and sexual harassment, which is held continuously, stands out, as does the lecture on violence against women and feminicide, conducted in 2019 by social psychologist Márcia Paulino. In 2024, during an event organised by the Law course in partnership with the Academic League of Criminal Sciences, two highly relevant lectures were held: 'Crimes against life and the Jury Court' and 'History of Domestic Violence against Women with an emphasis on Feminicide', the latter given by Andréa Flores, PhD in Criminal Law. These actions reinforce the institutional commitment to raising the critical awareness of the academic community and tackling gender-based violence. These actions demonstrate an institutional commitment to creating a more aware, respectful and safe academic environment for everyone (Unigran Capital 2019).

4 Conclusion

In a country where gender inequalities persist in multiple spheres, the role of HEIs as catalysts for social transformation and gender justice becomes increasingly crucial. An examination of the initiatives implemented by three higher education institutions—UFMS (federal), UERN (state) and UNIGRAN Capital (private)—reveals the existence of distinct practices to promote gender equality, directly influenced by their governance models, institutional regimes and organisational strategies. Among the various actions described, UFMS demonstrates an alignment with the principles of SDG 5 by structuring an institutional programme, such as 'Sou Mulher' (I am a woman), with policies aimed at female admission, retention and empowerment. UERN, for its part, values the articulation between teaching, research and extension, with a strong community insertion, developing projects that connect gender, education and health. In the private sector, UNIGRAN Capital reveals an approach guided by organisational performance indicators, with emphasis on salary transparency and the promotion of educational actions on gender violence, demonstrating the sector's potential to respond to regulatory frameworks such as Law 14.611/2023.

Higher education institutions are demonstrating that they are working towards a cultural transformation aimed at eradicating violence against women, taking on a leading role in society. By positioning themselves as spaces for critical and civic education, they are guiding society towards building a culture that denaturalises the practices, behaviours and structures responsible for maintaining misogyny and gender violence. By promoting educational actions, awareness campaigns, training events and institutional policies that welcome and protect victims, these institutions reinforce their commitment to human rights and the Sustainable Development Goals (SDGs), especially those relating to gender equality-SDG 5. By documenting replicable institutional practices, this chapter offers strategic insights that can inform policy-making, professional training, and interinstitutional cooperation in addressing SDG 5.

Despite these contributions, this chapter has important methodological limitations. As it is a descriptive study focusing on a limited number of institutions, the results do not allow for generalisations, reflecting only the reality of the HEIs analysed. In order to deepen our knowledge of the subject, we suggest developing comparative studies with a larger universe of institutions in different regions of the country, including both public and private universities. Future research could also look at the effectiveness of the actions undertaken, listening to the various actors involved, such as students, teachers, managers and technical professionals.

It is equally important that these studies result in the development of manuals of good practice, guidelines for recommendations and proposals for pedagogical workshops aimed at continuing training for HEI professionals.

Mapping the institutional actors involved in actions to combat violence and setting up collaborative networks between institutions can strengthen the capillarity and effectiveness of initiatives, promoting more systemic and articulated action.

By describing and systematising good practices, this study contributes not only to recording these actions, but also to encouraging other HEIs to promote similar strategies. In this way, the fight against discrimination and violence against women in the academic sphere is broadened—a topic that, although it has gained greater visibility on the Brazilian scene in recent years, still encounters cultural, institutional and political resistance that makes it difficult to fully address. Therefore, strengthening this debate in higher education institutions is fundamental to consolidating a culture of equity, respect and non-violence.

References

Brazil (2024). National Map of Gender Violence. Data Senado Institute, Consulted on 18/02/2024.https://www.senado.leg.br/institucional/datasenado/mapadaviolencia/#/inicio

Campos, K. C., Moura-Leite, R. C., Lopes, J. C. J., Binotto, E., & Figueiredo, J. C. (2022). Water and sewage, from the perspective of the Sustainable Logistics Management Plan: An analysis of the practices of Brazilian federal universities. Ciência e Natura, 44, 1.

Evilin, F., Moura-Leite, R. C., & Pereira, M. W. G. (2019). Evidence of social responsibility practices of private higher education institutions. Facef Pesquisa, 21, 1.

Federal University of Mato Grosso do Sul (UFMS). (2025). https://www.ufms.br/

Flick, U. (2009). Quality in qualitative research. In: Quality in qualitative research (p. 196). Artmed.

Global Compact (2024) Movement they lead 2030.Global Compact https://www.pactoglobalorg.br/movimentos/movimento-elas-lideram

Gomes, A. F. et al. Hi, Girls: female protagonism in Computing based on listening to female students. Rio Grande do Norte State University, 2023.Available at: http://www.uern.br.Accessed on: 20 Apr. 2025.

ISO (2024). International Organisation for Standardization. https://www.iso.org/home.html

Kihara, B. M. T., Moura-Leite, R., & Lopes, J. C. J. (2019) Sustainable purchasing and contracting at Brazilian federal universities.Revista FSA (Faculdade Santo Agostinho), 16, 27–53.

Leal Filho, W. et al. (2023).Promoting gender equality across the sustainable development goals. Environment, Development and Sustainability, v. 25, n. 12, p. 14177–14198, dec.

Merma-Molina, G.; Urrea-Solano, M. & Hernández-Amorós, M. J. (2024). The Integration of Gender Equality (SDG 5) into University Teaching: the View from the Frontline. Innovative Higher Education, v. 49, n. 3, p. 419–452, June.

Morais, A. (2024).Caring for and valuing women inside and outside the University (With the collaboration of V. Araújo).Portal UERN. https://portal.uern.br/blog/cuidado-e-valorizacao-das-mulheres-dentro-e-fora-da-universidade/

Moura Leite, R. e Pereira, M. (2020). Status of sustainability reports by brazilian higher education institutions. Revista Capital Científico—Eletrônica. 18. 59.

Moura-Leite, R. C. (2023). Integral institutional approach to sustainability in higher education institutions: how to incorporate sustainability into their functions and establish partnerships to promote sustainable development goals. Expressa Extensão, v. 28, p. 34–37, 18 Dec.

Moura-Leite, R. C.; Lopes, J. C. D. J. & Yamazaki, C. (2023). Brazilian federal universities and their sustainable practices based on sustainable logistics management plan. International Journal of Sustainability in Higher Education, v. 24, n. 4, p. 932–947, 15 Mar.

Moura-Leite, R., et al. (2024a).Advances in knowledge, commitment and contributions to the Sustainable Development Goals in higher education institutions in Latin America and the Caribbean—Brazilian HEIs Report. https://doi.org/10.13140/RG.2.2.30547.96804

Moura-Leite, R. C. et al. (2024b). The Brazilian Network of Higher Education Institutions for Sustainable Development (Rede UniSustentável). Journal of Sustainability Perspectives, v. 4, n. 1, p. 1–12, 29 Oct.

Muñoz-Mazón, A. I. et al. (2024). The influence of a gender equality course in tourism on university students' SDG knowledge, attitude and practice (KAP). International Journal of Sustainability in Higher Education, 8 Nov.sustainable logistics management: an analysis of Brazilian federal universities. Journal of Applied Technology, v. 8, n. 1, p. 51–70, 31 May.

Pandit, J. M. & Paul, B. (2023.) Gender Diversity, Sustainable Development Goals and Human Resource Management Practices in Higher Education Indian Journal of Human Development, v. 17, n. 1, p. 111–130, Apr.

Nexitemper (2022) Unigran Capital. Interdisciplinary and Transdisciplinary Extension Centre for Employability and Paid Internships https://www.nexitemper.com.br/ https://www.unigran.br/campogrande/servicosPraVoce/nexitemper/

Nogueira, M. V.; Moura-Leite, R. & Lopes, J. C. D. J. (2019). The quality of life at work the focus of the sustainable logistics management plan.

Pet-Saúde (2024)—Programme that 'exchanges' academics with the SUS is launched at UNIGRAN Capital. https://www.unigran.br/novidades/8248

PROCIDS—Pro-Rectory of Citizenship and Sustainability (2025). https://procids.ufms.br/

Sáenz O., Velasco A., Plata A., Holguin MT, Moura-Leite R, Callejas Restrepo MM, Arguedas-Camacho M, Parada-Guevara S., López C. & Bravo Mercado MT (2024) Advances in knowledge, commitment and contributions to the Sustainable Development Goals in higher education institutions in Latin America and the Caribbean. 013140/RG.2.2.19098.00964. https://oses-alc.net/wp-content/uploads/2024/06/2024-06-1OSES-ALC-informe-a-ministros.pdf

Rio Grande do Norte State University (UERN). (2024). *Relatório de gestão 2023.* https://portal.uern.br/wp-content/uploads/2024/03/Relatorio_Gestao_UERN_2023.pdf

Rio Grande Do Norte State University (UERN). ODS EDU Seal: UERN recognised for the second year running. Mossoró: UERN, 2023 a. Available at: http://portal.uern.br. Accessed on: 20 Apr. 2025.

Rio Grande do Norte State University (UERN). Institutional data and composition of the 2023 management. Mossoró: UERN, 2023b. Available at: http://portal.uern.br. Accessed on: 20 Apr. 2025.

Rio Grande do Norte State University (UERN). Institutional programmes, projects and actions. Mossoró: UERN, 2025. Available at: http://portal.uern.br. Accessed on: 20 Apr. 2025.

Serafini PG, de Moura, JM, de Almeida MR, & de Rezende JFD. (2022). Objetivos de desenvolvimento sustentável em instituições de ensino superior: uma revisão sistemática da literatura. Journal of Cleaner Production, 370, 133473. https://doi.org/10.1016/j.jcle-697pro.2022.133473

Sdsn Austrália/Pacífico (2017): Getting started with the SDGs in universities: Um guia para universidades, instituições de ensino superior e o sector académico. Austrália, Nova Zelândia e edição do Pacífico. Rede de Soluções para o Desenvolvimento Sustentável - Austrália/Pacífico, Melbourne. https://ap-unsdsn.org/wp-content/uploads/University-SDG-Guide_web.pdf

Sou Mulher (2024). UFMS—Pro-Rectory for Citizenship and Sustainability. https://procids.ufms.br/programa-sou-mulher-ufms/

Unigran Capital—(2019) Lecture on Violence against Women and Feminicide at Unigran Capital. https://blog.unigrancapital.com.br/category/eventos/.

Unigran Capital—(2023) Social Responsibility and Social Responsibility and sustainability). https://www.unigran.br/campogrande/institucional/sustentabilidade/responsabilidadeSocial

Unigran Capital (2025) Unigran Capital presents the results of the salary equalisation reports for the 1st semester. https://www.unigran.br/novidades/8371

Unesco. (2022) Contribution of higher education to the Sustainable Development Goals. Development Goals: an analytical framework. Caracas: IESALC, Paris:UNESCO, 50 p. (IESALC).

Working documents, 2). Programme and meeting document, code ED/HE/IESALC/WP/SP/2022/01. Available at: https://unesdoc.unesco.org/. Licence CC BY-SA
Weber, J., Machado, N. S., Moura-Leite, R., & Sehnem, S. (2020).Sustainability indicators for higher education institutions: The case of the Federal University of the Southern Frontier—UFFS.Reunir: Revista de Administração, Ciências Contábeis e Sustentabilidade, 10, 42–55.

Rosamaria Moura-Leite Ph.D. in Administration from the University of Salamanca, Spain. Currently an Associate Professor at the Federal University of Mato Grosso do Sul—School of Business and Administration. Faculty member of the Graduate Program in Administration (PPGAD). Researcher at the Observatory of Sustainability in Higher Education in Latin America and the Caribbean (OSES-ALC). Coordinator of the Brazilian Network of Higher Education Institutions for Sustainable Development (UniSustentável).

Renata Benedetti Mello Nagy Ramos Ph.D. in Education from Dom Bosco Catholic University (UCDB). Master's in Energy Efficiency and Sustainability from the Federal University of Mato Grosso do Sul (UFMS). Coordinator and a Professor of Architecture and Urbanism and Interior Design Technology at the Unigran Capital University Center. Member of the following research groups: Study and Research Group in Educational Technology and Distance Education (GETED/UCDB) and the Brazilian Network of Higher Education Institutions for Sustainable Development (UniSustentável).

Débora Teixeira da Cruz Ph.D. in Health and Development from the Federal University of Mato Grosso do Sul, and a Master's Degree in Bioethics from Universidade do Vale do Sapucaí (UNIVAS). She specialized in Conflict Mediation at the Federal Institute of Southern Minas Gerais. Her diverse academic background includes undergraduate degrees in Radiology (University of Alfenas—UNIFENAS), Psychology (Universidade do Vale do Sapucaí—UNIVAS), Pedagogy (Unigran), and a Bachelor of Laws (Unigran Capital). Dr. Cruz is a full member of Chair 18 of the Brazilian Academy of Radiological Sciences (ABCR). She serves as an Associate Editor for "Psychology and Behavioural Sciences" in the Journal Experimental and Clinical Perspectives, Biomedical Innovations and Health Education (PECIBES) at Maria Aparecida Pedrossian University Hospital (HUMAP) of the Federal University of Mato Grosso do Sul, and is also an Associate Editor for the Brazilian Journal of Radiation Technology Research. She is an Ad hoc reviewer for the Brazilian Journal of Health Research (Federal University of Espírito Santo—UFES) and Interdisciplinary Studies on Ageing (Federal University of Rio Grande do Sul—UFRGS), and an Ad hoc Evaluator for Fundect. Currently, she is the Coordinator of the CST in Radiology and Psychology Internship Supervisor at Unigran Educacional. She also lectures on Health courses at UNGRAN Educacional and for Unigran EAD. Dr. Cruz is the Coordinator of the Inter and Transdisciplinary Extension Center for Paid Internships and Employability (NEXITEMPER) and a member of the Biosafety Committee at Unigran Educacional. Additionally, she lectures on postgraduate courses at the Unigran Capital University Center.

Jessica Neiva de Figueiredo Leite Araujo Ph.D. candidate in Education at the Federal Institute of Rio Grande do Norte (IFRN), with a Master's Degree in Systems and Computing from the Federal University of Rio Grande do Norte (UFRN) and a Bachelor's Degree in Computer Science from the State University of Rio Grande do Norte (UERN). She is a professor at UERN with exclusive dedication and serves as Technical Advisor for Governance and Transparency at the Rector's Office. She is also the institution's Data Protection Officer (DPO). Her academic work focuses on Artificial Intelligence, Educational Technologies, and Teacher Education. She is a member of the Special Committee on Informatics in Education (CEIE) of the Brazilian Computer Society (SBC) and actively participates in academic events and research networks related to sustainability, digital inclusion, and educational innovation. Dr. Neiva coordinates extension projects such as digital literacy for older adults and initiatives to promote girls' participation in technology. She lectures in undergraduate and graduate programs, including the Interinstitutional Doctoral Program in Teaching (RENOEN), and is committed to fostering ethical and inclusive practices in AI adoption in education. She balances her academic and institutional roles with dedication to her family, as a mother of three daughters.

Cicilia Raquel Maia Leite Holds a Ph.D. in Electrical and Computer Engineering from the Federal University of Rio Grande do Norte (UFRN) and completed a postdoctoral fellowship at the Massachusetts Institute of Technology (MIT). Also holds a Master's degree in Electrical Engineering from the Federal University of Campina Grande (UFCG) and a Bachelor's degree in Computer Science from the State University of Rio Grande do Norte (UERN), where she has served as a professor since 2006. Currently serves as Rector of UERN, elected by direct vote in 2021 and re-elected for the 2025–2029 term. Her leadership is characterized by a strong commitment to university autonomy, academic excellence, and social responsibility. Under her administration, UERN achieved full administrative and financial autonomy and expanded its role in advancing the UN 2030 Agenda and the Sustainable Development Goals (SDGs).

Federal Institutes and Regional Sustainable Development: Perspectives on Local Productive Arrangements in Minas Gerais

Luzélia Calegari Santos Moizinho, Marco Antônio Ferreira Inácio, and Carolina Coelho da Silveira

Abstract Local Productive Arrangements (LPAs) consist of companies within the same region that share interdependence, territoriality, and productive specializations. Governance and interaction among stakeholders, such as businesses and educational institutions, can drive endogenous innovation, competitiveness, and sustainable local development. The effectiveness of LPAs depends on coordination and learning among these actors, highlighting the essential role of educational and research institutions. In this context, Federal Institutes (FIs) represent a key public policy for strengthening collective assets and generating positive externalities that benefit LPAs. Based on a qualitative, descriptive, and documentary approach, this study analyzes the alignment of technical, technological, undergraduate, and postgraduate courses offered by three FIs in Minas Gerais with the demands of 36 regional LPAs. The results indicate a strong and direct relationship between the courses offered and local productive needs, demonstrating their contribution to regional sustainable development, in line with Sustainable Development Goal (SDG) No. 9, particularly its target 9.5.

Keywords Local productive arrangements · Federal institutes · Sustainable development · Regional infrastructure

L. C. S. Moizinho
Federal Institute of Education, Science and Technology of Triângulo Mineiro, Uberaba, Brazil
e-mail: luzelia@iftm.edu.br

M. A. F. Inácio · C. C. da Silveira (✉)
School of Management and Business, Federal University of Uberlândia, Uberlândia, Brazil
e-mail: carolinasilveira@ufu.br

M. A. F. Inácio
e-mail: marco.inacio@ufu.br

W. Leal Filho et al. (eds.), *Sustainability in the Curriculum and Operations of Universities*, World Sustainability Series,
https://doi.org/10.1007/978-3-032-12678-8_37

1 Introduction

The establishment of Federal Institutes of Education, Science, and Technology (FIs) by the Law 11.892/2008 is considered a milestone in the advancement of Brazil's educational structure (Otranto 2010; Pacheco 2020). These institutions are innovative, as they do not follow pre-existing national or foreign models, and aim to operate across various teaching modalities, from initial training to postgraduate courses, with a focus on technology (Pacheco 2020). FIs were created by merging earlier professional institutions, resulting in a unique structure (Otranto 2010).

One of the main objectives of FIs is to conduct applied research, fostering the development of technical and technological solutions that benefit local communities (Föeger and Carneiro 2023). Additionally, these institutions promote projects that generate patents and intellectual property, strengthening a culture of innovation and entrepreneurship among their students (Chaves and Cruz 2023). The role of FIs also includes supporting local development, contributing to economic decentralization and the strengthening of local productive and cultural arrangements (Ferreira et al. 2021).

Local Productive Arrangements (LPAs) are potential strategies for fostering regional development (Marini and Silva 2012). Academic research has highlighted correlations between the socioeconomic progress of territories and the existence of productive clusters (Isbasoiu 2007; Lastres and Ferraz 1999; Oliveira and Martinelli 2014; Sforzi et al. 2015). The systemic analysis of the factors and dimensions inherent in local development enables the identification of structural dynamics that were previously underexplored, providing insights to enhance growth strategies for both clusters and the regions that host them (Silva and Martinelli 2021).

Marini and Silva (2012) developed a multidimensional analytical framework encompassing environmental, cultural, economic, spatial, institutional, political, and social aspects that interact within LPA territories. These dimensions reflect the complexity of production territories, characterized as interdisciplinary interfaces that articulate social relations and everyday practices (Silva and Martinelli 2021). Each dimension comprises elements that reflect collective experiences and the actions of stakeholders in regional ecosystems, serving as analytical axes for understanding their dynamics (Marini and Silva 2012).

In Minas Gerais, Brazil, data from the State Secretariat for Economic Development (SEDE, "Secretaria de Desenvolvimento Econômico" in Portuguese) indicate that, as of February 2025, the state had 72 LPAs distributed across its 316 municipalities (SEDE 2025). In 2024, R$10 million was invested in partnership projects involving Scientific, Technological, and Innovation Institutions (ICTs, "Instituições Científicas, Tecnológicas e de Inovação" in Portuguese), businesses, startups, and cooperatives embedded in LPAs (Rodrigues 2024). The goal was to play a key role in boosting the economy, fostering innovation, and strengthening the entrepreneurial environment in Minas Gerais.

Against this backdrop, the following research question arises: do the courses offered by Federal Institutes in Minas Gerais meet the needs and demands of local

LPAs, contributing to regional development? Thus, this study aims to analyze whether the technical, technological, undergraduate, and postgraduate courses offered by Federal Institutes in Minas Gerais align with the demands of LPAs in their regions.

From a social perspective, this research seeks to contribute to sustainable regional development by investigating how technical and higher education can enhance workforce qualification and technological innovation in these territories. The connection between education and the market strengthens the employability of graduates and enhances the competitiveness of businesses within LPAs, reducing socioeconomic inequalities and promoting productive inclusion. Academically, the study fills a gap in the literature by examining the interaction between educational policies and local economic development. It provides empirical insights for improving the curricular guidelines of Federal Institutes and for strengthening public policies focused on regional innovation and entrepreneurship. Finally, analyzing the role of Federal Institutes in training skilled professionals to meet the needs of LPAs contributes to sustainable regional development. This process fosters technological innovation and generates solutions that enhance the competitiveness of local businesses, aligning with SDG 9, especially target 9.5.

To provide context for this analysis, the following section examines the theoretical foundations and evolution of Local Productive Arrangements in Brazil.

2 Local Productive Arrangements in Brazil

Economic literature highlights the role of Marshall (1920) in the discussion of Local Productive Arrangements, which he termed "industrial districts" or "localized industries." Marshall (1920) argued that the geographic concentration of small businesses can generate mutual benefits and allow them to compete with large corporations. This collective efficiency results in positive externalities, as described by Schmitz and Nadvi (1999) and Erber (2008). Cooperation among these businesses is often incidental, arising primarily from geographic proximity (Schmitz 1995). In contrast, internal economies depend on equipment setup, workforce organization, and internal management. Neoclassical theory emphasizes the importance of external economies because internal economies can be easily replicated (Moizinho et al. 2017).

In his study, Becattini (2002) analyzes the evolution of Marshall's arguments from two perspectives: an early phase and a mature phase. In the former, Marshall questions the uncritical acceptance of traditional empirical divisions, such as the nation-state in a specific historical context. This view reflects a pre-Fordist era, where technological development was limited, and geographic location was crucial (Becattini 2002). In Marshall's mature phase, according to Becattini (2002), he is immersed in the dominant Fordist context, where his ideas about internal and external economies are applied not only to industrial districts but also across the entire productive sector.

Subsequent contributions to Marshall's seminal work emphasize both internal and external economies. Becattini (1989) notes that small factories in densely occupied districts have significant advantages over large firms. He exemplifies this with central

Italy, known as the "Third Italy," where small businesses collectively act as a large firm. This specialized organization allows these conglomerates to achieve revenues comparable to those of large corporations and even export Italian products.

According to Cassiolato et al. (2008), LPAs are formed by economic, political, and social agents gathered in a specific territorial space. These agents establish connections among themselves, even if incipient, involving public and private organizations that produce goods or services; institutions dedicated to human resource qualification; research and development centers; and those focused on public policies and funding. Thus, LPAs consist of clusters of nearby businesses, usually belonging to similar or related sectors (Casanueva et al. 2013). These agglomerations generate positive externalities stemming from historical, economic, and social factors, and there are effective mechanisms for knowledge transfer among these businesses (Casanueva et al. 2013; Delgado et al. 2014).

In this sense, LPAs can also be referred to as clusters, innovative milieux, industrial poles, or industrial districts, although there are some conceptual differences (Moizinho et al. 2017), and internationally, the term cluster is more commonly used (Porter 2000).

According to Erber (2008), an externality occurs when collaboration among distinct organizations, such as businesses, universities, and research centers, generates mutual advantages that are not immediately transacted in the market. These externalities, characterized by the absence of direct financial compensation at a given time, may flow in one direction or be reciprocal. However, once the agent responsible for generating these effects assigns them monetary value, the dynamics of the externality change (Moizinho et al. 2017).

From Marshall's perspective, the occurrence of externalities is intrinsically linked to geographic proximity. Thus, the formation of productive clusters, where businesses and their suppliers are located in the same region, fosters an environment conducive to indirect benefits (Becattini 2002). This phenomenon also drives specialization and the enhancement of regional workforce skills, attracted by the opportunities created. Within the field of externality studies, traditional academic analysis tends to emphasize these unintentional or fortuitous effects resulting from the spatial concentration of economic activities (Moizinho et al. 2017).

An often-cited example of a cluster is the designation of origin, which represents a model of productive organization where the geographic region plays a fundamental role (Caldas 2003). In these cases, the reputation and unique characteristics of products such as "Canastra cheese," "Port wine," "Champagne," and "Cerrado coffee" are intrinsically tied to their place of production. The strong specialization of the production process, shaped by specific geographic conditions, fosters the development of skills in the local workforce and stimulates the emergence and improvement of specialized suppliers (Moizinho et al. 2017). This symbiotic relationship between the product and local knowledge contributes to a virtuous cycle of regional development (Crescenzi et al. 2022; Juk and Fuck 2015; Moizinho et al. 2017).

The occurrence of externalities and territorial development depends on establishing multidimensional connections among agents, combining institutional interactions, building trust-based relationships, and forming a collective identity (Carrol and Zeller 2012; Lastres and Cassiolato 2003; Marini and Silva 2012). In this context, synergy arises when cooperation among stakeholders and the existence of shared goals supports the creation of mutual benefits, surpassing individual initiatives (Alvarenga et al. 2013).

In the context of productive agglomerations, local cohesion plays a catalytic role, as the convergence of cultural and ethical values within the community creates an environment conducive to innovation and economic collaboration (Di Giacinto et al. 2014). This dynamic is intrinsically linked to social capital, understood as the set of networks, norms, and reciprocities that facilitate collective action. The relevance of this concept lies in its ability to explain how informal social ties influence economic efficiency and territorial governance (Lastres and Cassiolato 2003; Marini et al. 2012). Furthermore, collaboration among businesses, research institutions, and other local partners strengthens productive capacity and innovation, which are essential for sustainable development (Carvalho 2005).

Relationship networks, marked by bonds of friendship and mutual trust, expedite the dissemination of relevant information, while governance plays a crucial role in stimulating and managing practices (Casanueva et al. 2013; Cassiolato and Szapiro 2003). A broad spectrum of support institutions, both public and private, can optimize the exploration and transfer of information, knowledge, and opportunities, continually enhanced by internal interactions, proximity, and the intensity of exchanges (Molina-Morales and Martínez-Cháfer 2016). In this regard, several authors agree on the need to include the State as a supporting agent for local governance (Carrol and Zeller 2012; Herrerias and Ordoñez 2012; Lastres and Cassiolato 2003; Marini et al. 2012).

Finally, from a social perspective, a fundamental aspect of corporate sustainability, it is noteworthy that cluster initiatives have a robust social foundation and are deeply shaped by context. The close connection with the community enables the integration of socioeconomic goals for businesses and territory, as well as the social networks within their local sphere of influence (Felzensztein et al. 2018). Additionally, LPAs serve as platforms for forming partnerships among businesses, research institutions, and other stakeholders, facilitating access to knowledge and technologies that can drive industrial innovation (Jiménez et al. 2021), which aligns with the targets of Sustainable Development Goal (SDG) 9. Thus, it is evident that LPAs play a key role in promoting smarter and more sustainable economic development (Derlukiewicz et al. 2020; Scheel 2002; Sen and Ongsakul 2018).

Building on this theoretical framework, this study employs a systematic approach to evaluate the alignment between Federal Institute offerings and LPA demands in Minas Gerais.

3 Methodological Procedures

This study aims to analyze whether the technical, technological, undergraduate, and postgraduate courses offered by Federal Institutes in Minas Gerais, Brazil, align with the demands of LPAs in their regions. As such, it can be classified as qualitative and descriptive research, based on bibliographic and documentary analysis.

According to Creswell (2010), qualitative research is an investigative method that seeks to understand how individuals or groups assign meaning to social or human issues. Thus, in terms of its nature, this research is qualitative, as it does not involve quantitative data processing. The focus is on the textual analysis of documents, data, and official information from the government of Minas Gerais.

On the other hand, descriptive research aims to characterize aspects of a specific population and how a particular phenomenon manifest (Sampieri et al. 2013). Therefore, in terms of its objective, this research is descriptive, as in order to verify the alignment of courses offered by Federal Institutes in Minas Gerais with LPAs in the region, we conducted descriptive research using information available on the websites of the Minas Gerais government and educational institutions.

We collected data on LPAs in Minas Gerais from the SEDE's website, under the "Projects" tab, option "LPAs in Minas Gerais" (SEDE 2025). The analysis, in turn, was divided into four stages, as shown in Table 1 Research methodology stages and data sources for analyzing Federal Institute course alignment with Local Productive Arrangements.

In the first stage, the necessary information provided by SEDE for each LPA was compiled in a Microsoft Excel spreadsheet, including: LPA name, sector, main municipality, product/service, specialization, and participating municipalities in the arrangement. As of the completion of this study, 72 LPAs are listed on the Secretariat's website.

In the second stage, the courses offered by FIs in Minas Gerais were compiled. Due to the scope of this study, technical courses (integrated, concurrent, and subsequent), technological, undergraduate, and postgraduate courses were considered. Due to

Table 1 Research methodology stages and data sources for analyzing federal institute course alignment with local productive arrangements

Step	Description	Source/Technique
I	Compilation of recognized LPAs in the state	Secretariat for development of Minas Gerais (SEDE-MG)
II	Compilation of courses offered by FIs in the state	Admission notices for FIs at technical, technological, undergraduate, and postgraduate levels
III	Analysis and cross-referencing of information from stages I and II	Spatial location of reference points Measurement of distances between LPA cities and FI campuses
IV	Analysis of criteria and verification of complementary documentation	Qualitative analysis of the relationship between courses and LPA specializations

the characteristics of Initial and Continuing Education (FIC, "Formação Inicial e Continuada" in Portuguese) courses, such as uncertainty in availability and irregular scheduling—as they depend on demand and campus availability–, these courses were excluded from the scope. Similarly, teacher training courses, whose primary purpose is to train professionals for various areas of education, were also disregarded.

To create the database of offered courses, admission notices for 2025 from all campuses were used. Thus, it is possible that some active courses that did not offer vacancies in the 2025 selection processes were not included in this study.

The third stage involved analyzing each LPA separately. Figure 1 Criteria for classifying course alignment with LPAsillustrates the procedure adopted to classify the alignment of courses offered by FIs with the productive specializations of each arrangement. Initially, the existence of a campus in the city or group of cities of the arrangement was assessed. If present, it was verified whether the course(s) had competencies directly related to those required by the arrangement. When no direct relationship existed, it was analyzed whether the courses had an indirect relationship or no connection at all with the LPA's specializations.

Given the widespread distribution of municipalities comprising LPAs, having campuses in each of them is unfeasible. Thus, the proximity of campuses was also evaluated using the same criteria described above.

Finally, the fourth and last stage involved searching, when necessary, for additional documents such as course pedagogical projects, course offering records, and LPA websites to strengthen the classifications made.

During the research, a need to narrow the study's scope was identified. Considering the specificities of the central and southeast regions of Minas Gerais, which exhibit strong interdependence with federal, state, and private universities, the analysis was focused on the areas covered by the Federal Institute of Education, Science and Technology of the Triângulo Mineiro (IFTM, "Instituto Federal de Educação, Ciência e Tecnologia do Triângulo Mineiro" in Portuguese), the Federal Institute of Education, Science and Technology of Southern Minas Gerais (IFSULDEMINAS, "Instituto Federal de Educação, Ciência e Tecnologia do Sul de Minas" in Portuguese), and the Federal Institute of Northern Minas Gerais (IFNMG, "Instituto Federal do Norte de Minas Gerais" in Portuguese). Each of these institutes encompasses, respectively, 6, 12, and 18 LPAs, totaling 36 LPAs analyzed in Minas Gerais. This geographic delimitation aims to ensure greater depth and robustness in the analytical process.

4 Results and Discussion

State Law 16.296/2006 in Minas Gerais aims to develop a state policy to support and stimulate LPAs, "seeking to strengthen regional economies through the integration and complementarity of local productive chains and the generation of permanent processes of cooperation, diffusion, and innovation" (Minas Gerais 2006).

Through SEDE, the State Government seeks to identify, recognize, and register existing LPAs, "to adapt policies to the characteristics of the LPA and the economic

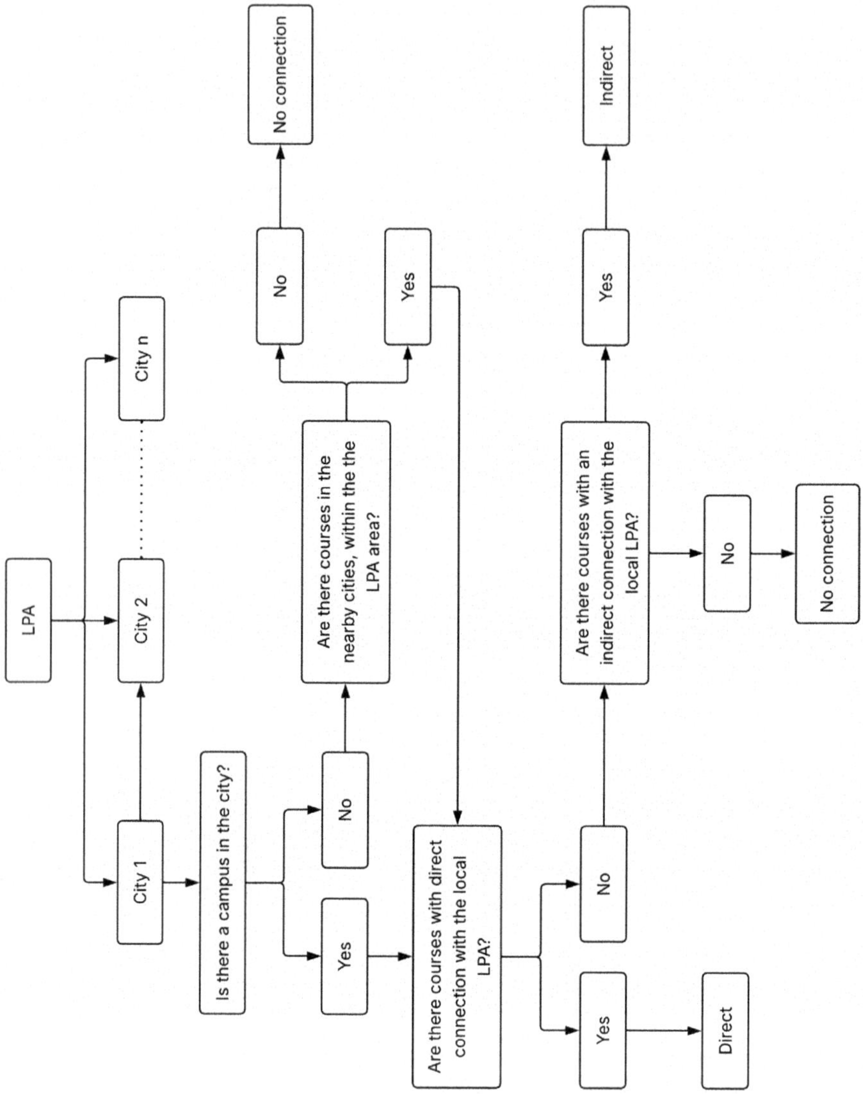

Fig. 1 Criteria for classifying course alignment with LPAs

sector, competitive evolution, integration of public policies for the development of Minas Gerais, and identification of specific demands of each LPA" (SEDE 2025).

The methodology for recognizing LPAs follows a rigorous evaluation described in SEDE Resolution 28/2021, which establishes as requirements measurable parameters for local governance structuring, economic and territorial development, productive linkages, and technological innovation. According to this resolution, an LPA is not recognized when there is no evidence of "local governance, or relevant quantities of sector companies in the municipalities forming the LPA, or a relevant number of sector jobs relative to total sector jobs in the state" (SEDE 2021).

The recognition and governance of an LPA enable access to support policies for the group of agents involved in the arrangement, including capacity-building instruments, research and technological innovation, funding and financing of activities, and investments in logistics and infrastructure (Minas Gerais 2006).

Currently, the SEDE database recognizes 72 LPAs in Minas Gerais. It is noteworthy that the state has 316 municipalities, so not all have an LPA, while some host more than one.

The productive sectors of LPAs are grouped in Table 2. The concentration of arrangements in the apparel sector suggests that cooperation among agents can be a good strategy for competitiveness against international competitors, especially Chinese factories, representing a competitive advantage and strengthening business activities in the market (Porter 2000). Strengthening the local apparel industry generates spillover benefits for the entire sector's productive chain, improving employability and social conditions for those involved.

Table 2 Distribution of local productive arrangements by productive sector in Minas Gerais state (n = 72)

Productive sector	Number of LPAs	Productive sector	Number of LPAs
Processed food	2	Floriculture	1
Beekeeping	3	Fireworks	1
Audiovisual	1	Fruit farming	4
Poultry and swine farming	1	Foundry	2
Mining	1	Gems and jewelry	3
Beverages	6	Dairy	1
Biotechnology	2	Timber/Charcoal	1
Cattle farming/livestock	4	Furniture	5
Coffee farming	4	Seedlings	1
Footwear	3	Plastics	1
Red ceramics	1	Fish farming	1
Apparel	13	ICT	3
Leather	1	Textiles	2
Electronics	1	Total	72
Electrometallurgical	3		

Notable in this distribution are the LPAs in the beverages sector and those linked to agribusiness, the latter being one of the main economic vocations of Minas Gerais due to favorable natural conditions for agriculture and livestock.

After establishing the sectoral distribution of LPAs, the analysis now turns to examining the alignment between Federal Institute courses and these productive arrangements.

Regarding Federal Institutes, Law 11.892/2008 establishes their purposes and characteristics, determining, among other goals, that they must "orient their educational offerings to benefit the consolidation and strengthening of local productive, social, and cultural arrangements, identified based on mapping the potential for socioeconomic and cultural development within the scope of the Federal Institute" (Brazil 2008). To monitor the results achieved by Federal Institutes, and to identify bottlenecks and define priorities for actions, the Federal Court of Accounts conducted an operational audit, the report of which was made available in 2012. This document acknowledges that the Goal and Commitment Agreements signed between SETEC/MEC and Federal Institutes advocate for the "verticalization of education and its integration with research and extension, always for the benefit of society, the construction, dissemination, and democratization of scientific and technological knowledge, as well as support and interaction with local productive, social, and cultural arrangements" (Brazil 2012).

As part of this research, we conducted an analysis of the courses offered by three of the five Federal Institutes in Minas Gerais: the Federal Institute of the Triângulo Mineiro (IFTM), the Federal Institute of Northern Minas Gerais (IFNMG), and the Federal Institute of Southern Minas Gerais (IFSULDEMINAS).

The cross-reference of information about the productive sectors of LPAs located in the geographic regions of the Federal Institutes analyzed with the courses offered at their respective campuses resulted in the summaries presented in Tables 3, 4, and 5.

Table 3 presents the summary of the analysis regarding the courses offered by the Federal Institute of the Triângulo Mineiro (IFTM). It can be observed that this region has six recognized LPAs. In all cases analyzed, there is a direct relationship between the courses offered and the productive sector of the arrangement.

The Coffee Farming in the Cerrado Mineiro LPA covers the largest number of municipalities among all arrangements registered with SEDE, including localities situated in geographic regions served by other Federal Institutes, such as Bambuí, in the coverage area of the IFTM. The courses focused on agricultural sciences, widely offered by the IFTM, directly meet the training demands of this arrangement.

Similar conclusions can be drawn regarding the offering of the Chemistry course in the Ceramics LPA in Ituiutaba, the Electrotechnical course in the Metal-Mechanical Industry LPA in Patos de Minas, and the agricultural courses for the Furniture LPAs in Uberaba and Uberlândia.

Additionally, the more than twenty courses related to IT and computing, distributed across IFTM campuses, can generate significant positive externalities for the Information and Communication Technology (ICT) LPA of the Triângulo Mineiro.

Table 3 Course alignment analysis between federal institute of Triângulo Mineiro (IFTM) and local productive arrangements in its geographic coverage area

LPA name	Productive sector	Main municipality	Number of municipalities	Sector-specific courses	Support courses	Analysis
Coffee farming in the Cerrado Mineiro	Coffee farming	Patrocínio	40	Yes	Yes	Direct
Ceramics of the Triângulo	Red ceramics	Ituiutaba	8	Yes	Yes	Direct
Metal-mechanical industry in Patos de Minas	Metal-mechanical industry	Patos de Minas	1	Yes	N/A	Direct
Furniture of Uberaba	Furniture	Uberaba	1	Yes	N/A	Direct
Furniture of Uberlândia	Furniture	Uberlândia	3	Yes	N/A	Direct
ICT of the Triângulo Mineiro	ICT	Uberlândia	10	Yes	Yes	Direct

Table 4 Course alignment analysis between federal institute of northern Minas Gerais (IFNMG) and local productive arrangements in its geographic coverage area

LPA name	Productive sector	Main municipality	Number of municipalities	Sector-specific courses	Support courses	Analysis
Beekeeping in Taiobeiras	Beekeeping	Taiobeiras	3	Yes	Yes	Direct
Beekeeping in Jequitinhonha	Beekeeping	Turmalina	10	Yes	Yes	Direct
Beekeeping in northern Minas Gerais	Beekeeping	Bocaiúva	17	Yes	Yes	Direct
Artisanal cachaça in Salinas	Beverages	Salinas	6	Yes	Yes	Direct
Coffee farming in the Semi-Arid	Coffee farming	Taiobeiras	6	Yes	Yes	Direct
Fruit farming in Jaíba	Fruit farming	Janaúba	10	No	Yes	Indirect
Eucalyptus in the chapada of Minas	Timber/ Charcoal	Turmalina	9	Yes	Yes	Direct
Gems and jewelry in Três Vales	Gems and jewelry	Teófilo Otoni	12	Yes	Yes	Direct
Furniture in Turmalina	Furniture	Turmalina	3	No	No	None
Pequi in Montes Claros	Fruit farming	Montes Claros	27	Yes	Yes	Direct
Apparel in Espinosa	Apparel	Espinosa	3	No	No	None
Apparel in Taiobeiras	Apparel	Taiobeiras	1	No	Yes	Indirect

Similarly, the analysis of the IFNMG reveals comparable patterns of alignment, though with some regional variations.

Tables 4 and 5 present the summary of the analysis conducted for the geographic region where IFNMG campuses are located. Regarding this region, the concentration of courses related to agriculture and livestock can strengthen mechanisms for training and technological development for most LPAs in the region.

Notable in this region are the synergistic relationships between the Food Engineering course at the Salinas campus and the Artisanal Cachaça LPA in Salinas, classified by SEDE as having consolidated maturity, as well as between the Mining course at the Araçuaí campus and the Gems and Jewelry LPA in Três Vales.

Table 5 Course alignment analysis between the federal institute of southern Minas Gerais (IFSULDEMINAS) and local productive arrangements in its geographic coverage area

LPA name	Productive sector	Main municipality	Number of municipalities	Sector-specific courses	Support courses	Analysis
Stainless steel industry in Lambari	Metal-mechanical industry	Lambari	1	Yes	Yes	Direct
Rotary wings and defense (firearms)	Metal-mechanical industry	Itajubá	2	No	No	None
Cookies in Andradas	Processed food	Andradas	1	Yes	Yes	Direct
Coffee farming in southwest Minas Gerais	Coffee farming	São Pedro da União	21	Yes	Yes	Direct
Footwear in Guaxupé	Footwear	Guaxupé	1	Yes	Yes	Direct
Volcanic beer	Beverages	Poços de Caldas	4	Yes	Yes	Direct
Leather in southern Minas Gerais	Leather	Cristina	1	Yes	Yes	Direct
Lingerie in Monte Belo	Apparel	Monte Belo	3	Yes	Yes	Direct
Electronics in Santa Rita do Sapucaí	Electronics	Santa Rita do Sapucaí	1	Yes	Yes	Direct
Furniture in Passos	Furniture	Passos	1	Yes	Yes	Direct
Ponkan tangerine in Campanha	Fruit farming	Campanha	2	Yes	Yes	Direct
Weaving in Campanha	Textiles	Campanha	1	Yes	Yes	Direct
ICT in Itajubá	ICT	Itajubá	2	Yes	Yes	Direct
Textiles in Guaranésia	Textiles	Guaranésia	1	Yes	Yes	Direct

(continued)

Table 5 (continued)

LPA name	Productive sector	Main municipality	Number of municipalities	Sector-specific courses	Support courses	Analysis
Apparel in Jacutinga	Apparel	Jacutinga	6	No	Yes	Indirect
Apparel in Paraguaçu	Apparel	Paraguaçu	1	No	Yes	Indirect
Apparel in Juruaia	Apparel	Juruaia	1	Yes	Yes	Direct
Wine in southern Minas Gerais	Beverages	Andradas	3	Yes	Yes	Direct

In the municipality of Janaúba, the main municipality of the Fruit Farming LPA in Jaíba, courses related to management and IT are offered, which can indirectly contribute to the development of this arrangement. It is emphasized that this LPA encompasses nine other municipalities, which are close to the Montes Claros State University (UNIMONTES, "Universidade Estadual de Montes Claros" in Portuguese) and the Federal University of Minas Gerais (UFMG, "Universidade Federal de Minas Gerais" in Portuguese), Montes Claros campus.

The productive arrangements for Furniture in Turmalina and Apparel in Espinosa were analyzed as non-aligned with IFNMG courses due to the distance from any campus of the institution. In the case of Turmalina, the closest campus to the municipalities of the arrangement is Araçuaí, located 107 km from Minas Novas and 140 km from Turmalina. According to the methodology adopted in this study, this distance is considered unfeasible, especially due to the transportation challenges faced by many students. A similar analysis was applied to the Apparel LPA in Espinosa, whose nearest campus is 106 km away.

On the other hand, the Apparel LPA in Taiobeiras, which covers a single municipality, is located 50 km from the Salinas campus, where courses in Information Systems are offered. Thus, it can be considered that this campus provides an indirect form of training for the arrangement.

The final regional analysis focuses on IFSULDEMINAS, which encompasses the most diverse range of productive sectors among the three institutes examined.

Table 5 presents the summary of the analysis conducted for the courses offered by IFSULDEMINAS. It was identified that the LPAs in this region of the state encompass a wide variety of productive sectors.

The Coffee Farming LPA in Southwest Minas Gerais brings together the largest number of participating municipalities, totaling 21. In addition to courses in agricultural sciences offered in the region of this arrangement, the Technology in Coffee Farming course, offered at the Muzambinho campus, and the Technological Validation Center (CVT, "Centro de Validação Tecnológica" in Portuguese)—Coffee and Grains, located at the experimental farm of the Guaxupé campus, stand out.

Among its competencies is "establishing partnerships with research entities in the coffee farming sector" (Ministério da Educação 2019). These initiatives represent important training and research development actions specifically for the sector.

The Rotary Wings and Defense LPA focuses on producing solutions and products for aircraft, drones, and the defense sector, including firearms and ammunition (SEDE 2025). Given the specificity of this arrangement, the Federal University of Itajubá (UNIFEI, "Universidade Federal de Itajubá" in Portuguese) exerts predominant influence, being an institutional partner of the LPA (Agência Sebrae Notícias 2022).

Four productive arrangements in the apparel sector were identified. In this context, the Fashion Design course, offered at the Passos campus, is geographically close to the municipalities of Monte Belo (about 60 km) and Juruaia (82 km). The arrangements in Jacutinga and Paraguaçu are farther from the Passos campus but close to other campuses offering courses in management and IT, which can contribute to the development of these arrangements.

The analysis shows that the beverages productive sector receives support from technical courses in Food and Food Engineering, widely offered at campuses in the region or nearby. Additionally, IFSULDEMINAS actively participates in partnerships focused on beer tourism, such as the Volcanic Beer Route project, and in integrating groups for the recognition of the Geographical Indication of Wines from Southern Minas Gerais.

Thus, all LPAs in the region covered by the IFTM have courses directly linked to the corresponding productive sector. In the IFNMG region, of the 12 LPAs analyzed, 8 have courses with direct alignment, while 10 have support courses that aid in the development of the arrangement. Only 2 LPAs were classified as having no courses aligned with the productive sector. In the IFSULDEMINAS region, of the 18 LPAs studied, 15 have courses directly related to their productive areas; only 1 was classified as having indirect alignment, and only 1 has no courses aligned with the productive sector.

The data demonstrate a strong alignment between the courses offered by Federal Institutes and the demands of LPAs, providing qualified training and driving sustainable regional development. This scenario corroborates the studied literature, which indicates that these institutions not only facilitate cooperation between businesses and academic institutions but also foster synergies capable of strengthening productive capacity and stimulating technological and industrial innovations (Carvalho 2005; Derlukiewicz et al. 2020; Jiménez et al. 2021; Scheel 2002).

5 Conclusions

The findings of this research underscore the strategic importance of Federal Institutes as catalysts for regional sustainable development, particularly within the context of Local Productive Arrangements in Minas Gerais.

The results of this research highlight the relevance of Federal Institutes as strategic agents for regional sustainable development, especially in the context of Local Productive Arrangements in Minas Gerais. The analysis of the alignment of technical, technological, undergraduate, and postgraduate courses with the demands of LPAs revealed that the educational offerings of these institutions play a role in workforce training and the promotion of innovation and regional competitiveness, which aligns with SDG 9, particularly target 9.5.

Furthermore, the research emphasizes the importance of public policies that encourage the integration of education, research, and innovation, consolidating FIs as catalysts for sustainable socioeconomic growth. Strengthening these institutions, coupled with fostering strategic partnerships, can expand the positive externalities generated by LPAs, benefiting both local communities and the productive sector.

However, some limitations must be considered. The research relied on secondary data and documentary analysis of LPAs and courses offered by three of the five FIs in Minas Gerais, which may restrict a deeper understanding of local dynamics, and the perceptions of the various stakeholders involved in LPAs. Additionally, the absence of a longitudinal approach prevents the verification of transformations over time, which could provide more robust data on the evolution of the relationship between FIs and LPAs. Therefore, future studies could address these gaps through direct empirical investigations, such as interviews and case studies.

Based on these findings, it is recommended that future research delve deeper into the mechanisms of interaction between FIs and LPAs, considering aspects such as technology transfer, impact on graduate employability, and the inclusion of institutes in regional innovation networks. Thus, it is hoped that this study will contribute to the improvement of educational and economic policies aimed at sustainable development, promoting a more integrated and efficient environment for knowledge generation and innovation.

References

Agência Sebrae Notícias (2022) Itajubá é reconhecida como APL de Asas Rotativas e Defesa [Itajubá is recognized as a Local Productive Arrangement in Rotating Wings and Defense]. https://mg.agenciasebrae.com.br/cultura-empreendedora/itajuba-e-reconhecida-como-apl-de-asas-rotativas-e-defesa

Alvarenga RAM, Matos FRN, Machado DQ, Sobreira MC, Matos LBS (2013) Arranjo produtivo local e desenvolvimento sustentável: uma relação sinérgica no município de Marco (CE) [Local productive arrangement and sustainable development: a synergic relationship in the municipality of Marco (CE)]. Rev Adm Mackenzie 14(5):15–43. https://doi.org/10.1590/S1678-69712013000500002

Becattini G (1989) Sectors and/or districts: some remarks on the conceptual foundations of industrial economics. In: Goodman EJ, Bamford J, Saynor P (eds) Small Firms and Industrial Development in Italy. Routledge, pp 123–135

Becattini G (2002) Del distrito industrial marshalliano a la «teoría del distrito» contemporánea. Una breve reconstrucción crítica [From the Marshallian industrial district to the contemporary

"district theory". A brief critical reconstruction]. Investigaciones Regionales—J Reg Res 1(1):9–32
Brazil (2008) Lei N° 11.892/2008 Institui a Rede Federal de Educação Profissional, Científica e Tecnológica, cria os Institutos Federais de Educação, Ciência e Tecnologia, e dá outras providências [Establishes the Federal Network of Vocational, Scientific and Technological Education, creates the Federal Institutes of Education, Science and Technology, and provides for other matters]. Available from: https://www.planalto.gov.br/ccivil_03/_ato2007-2010/2008/lei/l11892.htm#:~:text=L11892&text=LEI%20N%C2%BA%2011.892%2C%20DE%2029%20DE%20DEZEMBRO%20DE%202008.&text=Institui%20a%20Rede%20Federal%20de,Tecnologia%
Brazil (2012) Relatório Anual de Análise dos Indicadores de Gestão das Instituições Federais de Educação Profissional, Científica e Tecnológica - Exercício 2012 [Annual Report on the Analysis of Management Indicators of Federal Institutions of Professional, Scientific and Technological Education - Fiscal Year 2012]. Available at: https://www.gov.br/mec/pt-br/media/seb-1/pdf/rede_federal/relatorios_publicacoes/setec_analise_indicadores_2012.pdf
Caldas J (2003) As denominações de origem como unidade de planejamento, desenvolvimento local e inclusão social [Designations of origin as units of planning, local development and social inclusion]. Rev Desenv Econômico 5(8):25–32
Carrol MC, Zeller MC (2012) The cognitive limits to economic cluster formation. Urbani Izziv 23(Suppl. 1):553–562. https://doi.org/10.5379/urbani-izziv-en-2012-23-supplement-1-005
Carvalho AMA (2005) Arranjos Produtivos Locais e Sustentabilidade: Políticas públicas promotoras do desenvolvimento regional e da inclusão social [Local Productive Arrangements and Sustainability: Public policies promoting regional development and social inclusion]. Alcance 12(3):317–333
Casanueva C, Castro I, Galán JL (2013) Informational networks and innovation in mature industrial clusters. J Bus Res 66(5):603–613. https://doi.org/10.1016/j.jbusres.2012.02.043
Cassiolato JE, Lastres HMM, Stallivieri F (2008) Introdução: Políticas estaduais e mobilização de atores políticos em arranjos produtivos e inovativos locais [Introduction: State policies and mobilization of political actors in local productive and innovative arrangements]. In: Cassiolato JE, Lastres HMM, Stallivieri F (eds) Arranjos Produtivos Locais: uma alternativa para o desenvolvimento: Experiências de Políticas, vol 2, pp 11–37.
Cassiolato JE, Szapiro M (2003) Uma caracterização de arranjos produtivos locais de micro e pequenas empresas [A characterization of local productive arrangements of micro and small enterprises]. In: Lastres HMM, Cassiolato JE, Maciel ML (eds) Pequena empresa: Cooperação e desenvolvimento local, pp 35–50. Relume Dumará
Chaves AO, Cruz GP (2023) Avaliação das produções tecnológicas e de inovação em Institutos Federais de Educação do Brasil [Assessment of technological and innovation output in Brazilian Federal Institutes of Education]. Rev Gestão Desenvolv 20(1):176–197. https://doi.org/10.25112/rgd.v20i1.3115
Crescenzi R, De Filippis F, Giua M, Vaquero-Piñeiro C (2022) Geographical Indications and local development: the strength of territorial embeddedness. Reg Stud 56(3):381–393. https://doi.org/10.1080/00343404.2021.1946499
Creswell JW (2010) Projeto de Pesquisa. Métodos qualitativa, quantitativo e misto [Research Design: Qualitative, quantitative, and mixed methods], 3rd edn. Artmed
Delgado M, Porter ME, Stern S (2014) Defining clusters of related industries. J Econ Geogr 16(1):1–32. http://www.nber.org/papers/w20375
Derlukiewicz N, Mempel-Śniezyk A, Mankowska D, Dyjakon A, Minta S, Pilawka T (2020) How do clusters foster sustainable development? An analysis of EU policies. Sustain (Switzerland) 12(4). https://doi.org/10.3390/su12041297
Di Giacinto V, Gomellini M, Micucci G, Pagnini M (2014) Mapping local productivity advantages in Italy: Industrial districts, cities or both? J Econ Geogr 14(2):365–394. https://doi.org/10.1093/jeg/lbt021

Erber FS (2008) Eficiência coletiva em arranjos produtivos locais industriais: comentando o conceito [Collective efficiency in local industrial productive arrangements: commenting on the concept]. Rev Nova Econ 18(1):7–32

Felzensztein C, Gimmon E, Deans KR (2018) Coopetition in regional clusters: Keep calm and expect unexpected changes. Ind Mark Manag 69:116–124. https://doi.org/10.1016/j.indmarman.2018.01.013

Ferreira MAM, Felizardo Júnior NÁ, Gava R (2021) A contribuição dos Institutos Federais para o desenvolvimento local: Uma análise comparativa para municípios do estado de Minas Gerais, Brasil [The contribution of Federal Institutes to local development: A comparative analysis for municipalities in the state of Minas Gerais, Brazil]. Res Soc Dev 10(16):e99101623655. https://doi.org/10.33448/rsd-v10i16.23655

Föeger RS, Carneiro FFB (2023) O desenvolvimento científico dos Institutos Federais: crescimento e representatividade na produção científica nacional em periódicos indexados na Web of Science (1970–2020) [The scientific development of Federal Institutes: growth and representativeness in national scientific production indexed in the Web of Science (1970–2020)]. Em Questão 29. https://doi.org/10.1590/1808-5245.29.126063

Herrerias MJ, Ordoñez J (2012) New evidence on the role of regional clusters and convergence in China (1952–2008). China Econ Rev 23(4):1120–1133. https://doi.org/10.1016/j.chieco.2012.08.001

Isbasoiu GM (2007) Industrial clusters and regional development: the case of Timisoara and Montebelluna. MPRA Paper 5037

Jiménez E, de la Cuesta-González M, Boronat-Navarro M (2021) How small and medium-sized enterprises can uptake the sustainable development goals through a cluster management organization: A case study. Sustain (Switzerland) 13(11). https://doi.org/10.3390/su13115939

Juk YV, Fuck MP (2015) Innovations and support institutions: the 1st geographical indication in Brazil. J Technol Manag Innov 10(2):117–126

Lastres HM, Cassiolato JE (2003) Glossário de arranjos e sistemas produtivos e inovativos locais [Glossary of local productive and innovative arrangements and systems]. RedeSist

Lastres HMM, Ferraz JC (1999) Economia da informação, do conhecimento e do aprendizado [The economy of information, knowledge, and learning]. In: Lastres HMM, Albagli S (eds) Informação e globalização na era do conhecimento [Information and globalization in the knowledge age], pp 27–57

Marini MJ, Silva CL, Nascimento DE, Strauhs FR (2012) Avaliação da contribuição de arranjos produtivos locais para o desenvolvimento local [Assessment of the contribution of local productive arrangements to local development]. Rev Bibliográfica Geogr Ciênc Sociais 17(996):1–25

Marini MJ, Silva CL (2012) Desenvolvimento regional e arranjos produtivos locais: uma abordagem sob a ótica interdisciplinar [Regional development and local productive arrangements: an approach from an interdisciplinary perspective]. Rev Bras Gestão Desenvolv Regional 8(2):107–129

Marshall A (1920) Principles of economics, 8th edn. Macmillan.

Minas Gerais (2006). Lei Ordinária Nº. 16.296/2006. Institui a Política Estadual de Apoio aos Arranjos Produtivos Locais e dá Outras Providências [Establishes the State Policy for Support of Local Productive Arrangements and makes other provisions]. Available from: https://www.almg.gov.br/legislacao-mineira/texto/LEI/16296/2006/?cons=1

Ministério da Educação (2019) Fazenda Experimental de Guaxupé [Experimental Farm of Guaxupé]. https://www.muz.ifsuldeminas.edu.br/fazenda-em-guaxupe

Moizinho LCS, Magalhães EN, Braga MPS, Andrade FS (2017) Arranjos Produtivos Locais: a trajetória do conceito e o estado da arte [Local Productive Arrangements: the trajectory of the concept and the state of the art]. Anais Semin Pesq Inov Tecnol—SEPIT 1(1):1–18

Molina-Morales FX, Martínez-Cháfer L (2016) Cluster firms: you'll never walk alone. Reg Stud 50(5):877–893. https://doi.org/10.1080/00343404.2014.952719

Oliveira MF, Martinelli DP (2014) Desenvolvimento local e arranjos produtivos locais: uma revisão sistemática da literatura [Local development and local productive arrangements: a systematic literature review]. Interações 1:47–58. https://doi.org/10.1590/S1518-70122014000100005
Otranto CR (2010) Criação e implantação dos Institutos Federais de Educação, Ciência e Tecnologia—IFETs [Creation and implementation of the Federal Institutes of Education, Science and Technology – IFETs]. Rev Educ Téc Tecnol Ciênc Agrícolas 1(1):89–110
Pacheco E (2020) Desvendando os Institutos Federais: identidade e objetivos [Unveiling the Federal Institutes: identity and objectives]. Educ Prof Tecnol Rev 4(1):1–16. https://doi.org/10.36524/profept.v4i1.575
Porter ME (2000) Location, competition, and economic development: local clusters in a global economy. Econ Dev Q 14(1):15–34
Rodrigues J (2024) Chamada incentiva parcerias entre ICTs e Arranjos Produtivos Locais [Call encourages partnerships between STIs and Local Productive Arrangements]. FAPEMIG Institucional. https://fapemig.br/pt/noticias/1177/
Sampieri RH, Collado CF, Lucio PB (2013) Metodologia de pesquisa [Research methodology], 5th edn. McGraw Hill
Scheel C (2002) Knowledge clusters of technological innovation systems. J Knowl Manag 6(4):356–367. https://doi.org/10.1108/13673270210440866
Schmitz H (1995) Collective efficiency: growth path for small-scale industry. J Dev Stud 31(4):529–566. https://doi.org/10.1080/00220389508422377
Schmitz H, Nadvi K (1999) Clustering and industrialization: introduction. World Dev 28(9):1503–1514
SEDE (2021) Resolução nº 28/2021. Dispõe sobre a definição dos parâmetros e da metodologia de classificação do grau de maturidade dos Arranjos Produtivos Locais – APL [Provides for the definition of parameters and methodology for classifying the degree of maturity of Local Productive Arrangements – LPA]. Available from: https://www.legisweb.com.br/legislacao/?id=414996
SEDE (2025) APL em Minas Gerais [Local Productive Arrangements in Minas Gerais]. Secretaria de Desenvolvimento Econômico. https://desenvolvimento.mg.gov.br/inicio/projetos/projeto/1101
Sen SK, Ongsakul V (2018) Clusters of sustainable development goals: a metric for grassroots implementation. Sustainability 11(3):118–126. https://doi.org/10.1089/sus.2018.0007
Sforzi E, Sforzi F, Boix R (2015) What about industrial district(s) in regional science? Investigaciones Regionales 32:61–73
Silva RG, Martinelli DP (2021) Arranjos Produtivos Locais (APL) e fatores formadores das dimensões do desenvolvimento local [Local Productive Arrangements (LPA) and formative factors of local development dimensions]. Organ Soc 28(96):9–33. https://doi.org/10.1590/1984-92302021v28n9601pt

Luzélia Calegari Santos Moizinho Holds a Ph.D. in Economics, with a focus on economic development and public policies, and a Master's in Administration, specializing in Finance and Controllership, both from the Federal University of Uberlândia. She is the Coordinator of the Management Course and a professor at the Federal Institute of Education, Science and Technology of Triângulo Mineiro. With teaching experience in technical and graduate courses, she covers disciplines such as Accounting and Finance, Operations and Logistics, Statistics, and Economics. She also coordinates research in the areas of Cooperativism, Ecological Economics, and Sustainable Local Development.

Marco Antônio Ferreira Inácio Is a Master's student in the Postgraduate Program in Business Administration at the School of Management and Business of the Federal University of Uberlândia (UFU), Brazil. He holds a Bachelor's degree in Business Administration from the Federal Institute of Triângulo Mineiro (IFTM) and an MBA in ESG and Corporate Sustainability. His research interests include sustainability, climate change, the United Nations Sustainable Development Goals (SDGs), the dark side of organizations, and organizational grief.

Carolina Coelho da Silveira Holds a Bachelor's (2022) and a Master's degree (2024) in Business Administration from the Federal University of Uberlândia. Currently pursuing a Ph.D. in the same program, with a focus on Organizational Management. Research interests include artificial intelligence, text mining (natural language processing and machine learning), sustainability and public administration – with a strong focus on social implications and innovative methodologies to address challenges in public management contexts.

Sustainable Actions at the Federal University of Alfenas—Developments in Teaching, Research, and Extension

Julieta Aparecida Moreira and Elaine Angelina Colagrande

Abstract The development of actions aimed at sustainability within Brazilian Higher Education Institutions (IES) has been highlighted recently, a fact that can be seen from the results of research in different fields of knowledge related to these actions. In this context, this paper aims to identify and reflect on the institution's performance from the set of actions developed in teaching, research and extension over the last four years, which are directly related to environmental and sustainability issues at the Federal University of Alfenas—UNIFAL-MG, basing this discussion on the data recorded by the institution in the Green Metric ranking, in addition to recognizing the main challenges and limitations to these dimensions, about the relationship with sustainability, based on the results presented. Based on a survey carried out by the institutional management, taking into account all the university's academic units, it was possible to obtain information about the subjects offered in graduate and undergraduate teaching, the scientific publications made in research, the projects and programs developed in the field of extension, as well as the events open to the public, which were directly related to sustainability. The results of the survey revealed sectors that have seen an increase in actions, such as publications in the research dimension and extension achievements aimed at activities involving sustainability and environmental education. The teaching dimension presented potential for further development of actions. The set of results can contribute to institutional management to reflect, together with the other members of the university, on future actions to expand the development of actions related to sustainability in the institution. This study also highlights the role of universities in promoting sustainability, highlighting that, although there are institutional advances at UNIFAL-MG regarding the topic, the effective integration of sustainability still demands more articulated and comprehensive efforts at this institution.

J. A. Moreira (✉)
Institute of Natural Sciences, Federal University of Alfenas, Alfenas, MG, Brazil
e-mail: julieta.moreira@unifal-mg.edu.br

E. A. Colagrande
Chemistry Institute, Federal University of Alfenas, Alfenas, MG, Brazil
e-mail: elaine.colagrande@unifal-mg.edu.br

W. Leal Filho et al. (eds.), *Sustainability in the Curriculum and Operations of Universities*, World Sustainability Series,
https://doi.org/10.1007/978-3-032-12678-8_38

Keywords Sustainability · Green metric · Higher education institutions · Sustainable practices

1 Introduction

Sustainability has been establishing itself as one of the foundations for the academic, administrative, and social development of higher education institutions. In this context, universities, as agents of education and social change, play an important role in promoting practices that seek to balance economic progress, social equity, and environmental conservation. Because of this, it is essential that universities implement strategies that connect teaching, research, extension, and university management to contemporary social demands related to sustainability.

In Higher Education Institutions, sustainability must be systematically incorporated into teaching, research, extension, and management activities, ensuring that these principles transcend all academic dimensions. Nevertheless, as Alves and collaborators (2024) point out, some universities have already included sustainability elements in their curricula and pedagogical projects, although the lack of a robust institutional policy limits the adoption of effective practices.

Thus, the integration of sustainability into teaching, through the courses offered, must be complemented by applied research and extension projects developed directly with the community, promoting social engagement and environmental awareness. In this way, the university can become a space for experimentation and innovation in sustainability, seeking to train professionals capable of facing contemporary socio-environmental challenges (Alves et al. 2024).

Higher Education Institutions can play a fundamental role in promoting sustainability and in seeking to build environmental awareness among students. According to Araújo et al. (2024), sustainable initiatives implemented by these institutions generate a favorable impact on the environmental perception of academics, stimulating attitudes of citizenship and activism, which in turn can be reflected in their consumption choices, encouraging the adoption of more sustainable practices.

The adoption of institutional strategies focused on sustainability, such as effective communication, specific educational policies, and the promotion of the expansion of sustainable culture, strengthens students' involvement with environmental issues (Araújo et al. 2024). Therefore, supported by Souza Marques and collaborators (2020), the successful incorporation of sustainability into teaching, research, and university extension can position the institutions as a transformative agent in the search for a more sustainable society, by preparing individuals to add values and attitudes towards benefiting the environment in their daily lives.

Among the responsibilities of higher education is the prerogative of involvement in debates about sustainability, which must go beyond the simple teaching dynamics of the institution, advancing to promoting engagement in extracurricular projects, which involve both the academic community and the external public, including integration into the strategic management of the institution (Souza Marques et al. 2020).

In this sense, it is important to establish means for carrying out periodic assessments in HEIs regarding the development of sustainable practices in these spaces. Among the tools that aim to measure and encourage the adoption of such practices in universities, the *UI Green Metric* stands out as a global ranking that evaluates and classifies institutions based on six thematic axes: infrastructure, energy and climate change, waste, water, transportation, and education (Nogueira dos Santos et al. 2024).

Created by the University of Indonesia (UI) in 2010 (Universitas Indonesia 2024), the *UI Green Metric World University Rankings* is based on data and information collected through a questionnaire sent by the organization to participating universities. This instrument is based on the conceptual framework of Environment, Economy, and Equity, based on indicators and weighting criteria.

This ranking has proven to be effective in promoting, through classification, healthy competition, in addition to the exchange of sustainable experiences and good practices among universities from different regions of the world (Nogueira dos Santos et al. 2024). In addition to certificates to participating institutions, the ranking offers individual results for each of the six thematic axes considered, allowing analysis and institutional decision-making based on the interpretations.

According to Vitoreli and collaborators (2020), the adoption of the Green Metric ranking allows Higher Education Institutions not only to monitor their sustainable performance, but also to identify opportunities for improvement and develop strategies aligned with contemporary sustainability demands. Thus, the use of rankings such as Green Metric can be interpreted as part of a broader movement that aims to evaluate and report on sustainable initiatives in academic institutions. These assessments not only highlight universities that stand out in socio-environmental practices but also encourage higher education institutions to seek improvements in their policies and actions focused on sustainability.

The Federal University of Alfenas (UNIFAL-MG) has participated annually in the *Green Metric ranking* since 2018. It is a Brazilian public, century-old higher education institution located in the south of the state of Minas Gerais, with its headquarters and educational unit in the city of Alfenas, and two university units located in the cities of Poços de Caldas and Varginha. UNIFAL-MG offers a wide range of undergraduate and graduate courses in several areas of knowledge, including Health, Natural Sciences, Exact Sciences, Applied Social Sciences, Humanities, and Literature.

The institution stands out for fulfilling the university tripod, which includes teaching, research, and extension, in addition to its commitment to adequately meet the educational needs of society, engaging in issues and challenges arising from development at local, regional, and national levels. UNIFAL-MG also has a Permanent Commission for the Environment and Sustainability (CPMAS), composed of staff representing different segments, academic units at the headquarters, and other university centers, located in two other municipalities. This commission has a consultative and deliberative nature, intending to promote debates, reflections, and actions on political, economic, social, and environmental issues, contributing to sustainable institutional management.

Given the above, this article aims to present initial reflections on actions related to the sustainability of the Federal University of Alfenas, based on the indicators defined and the results submitted in the Green Metric ranking, considering the years 2020 to 2023, specifically concerning information regarding the dimensions of teaching, research and extension at the university. The objective was to identify and reflect on the institution's performance, based on an analysis of the data reported in the ranking on the Education axis, in addition to recognizing the main challenges and limitations to these dimensions, about the relationship with sustainability, based on the results presented.

2 Methods

Since UNIFAL-MG joined the international ranking in 2018, the institution has committed to annually collecting and submitting the information collected within the deadline established by the aforementioned ranking, based on a pre-defined questionnaire published by the Green Metric organization, containing dozens of indicators for each thematic axis analyzed (infrastructure, energy and climate change, waste, water, transportation and education).

To collect and organize the institutional material submitted to the ranking, the data is requested by the sustainability management from the university's Undergraduate, Graduate and Research, Extension and Culture sectors.

The data was collected for each of the dimensions, including the relationship with the sustainability theme, and the criteria used in this collection were: for the Teaching dimension, the number of Undergraduate and Graduate courses offered; for the Research dimension, the number of scientific productions published and, for the Extension dimension, the sum of the programs, projects and events held annually.

In addition to the institutional information related to the dimensions analyzed in this work, which are directly linked to the theme of sustainability within the university, for better interpretation and analysis of the numerical data, it was necessary to also collect all the disciplines offered, scientific productions published and extension actions developed by UNIFAL-MG in each year of the indicated period (2020–2023), enabling the calculation of percentages for comparisons.

The data collected was processed using descriptive statistics. According to Triola (2024), statistics are essential for organizing, describing, and analyzing data, facilitating the identification of relevant patterns and trends. Descriptive statistics are categories that organize, summarize, and present data in a clear and accessible manner, using tables, graphs, and numerical indicators. Their purpose is to interpret the information collected clearly, offering a comprehensive view of the characteristics of the data under analysis (Martins 2005). Thus, the use of descriptive measures in this work, such as percentage calculation, contributed to a more satisfactory understanding of the implications of the organized information for the discussion of the results presented below.

For the arrangement and analysis of the data, Microsoft Excel software was used, a tool widely used in research due to its practicality and privacy in the processing of numerical information. Thus, it was possible to structure the data collected in a spreadsheet, apply basic functions, and generate graphs, which in turn facilitated the interpretation of the results.

3 Results and Discussion

Based on the data processing, Table 1 was constructed. The results presented were analyzed for the different dimensions: teaching, research and extension, aiming to facilitate understanding, however it is highlighted that these dimensions are developed in an articulated manner, since, to achieve the university's main activities, with full academic training, students are provided with integrated and simultaneous contact with these dimensions.

From the observation of Table 1, it is possible to note that, for the teaching dimension, 2020 had the best result of the analyzed period, considering both the number of subjects related to sustainability offered (91), and their percentage representation (4.86%), based on the total number of subjects made available at the institution that year. In the following years, there was a drop in the values of this dimension, registering in 2023 the percentage of subjects offered that were related to sustainability of 3.94%. Taking into account the results observed in this area, we can infer, for example, that there is a need and room for progress in this dimension, as a consequence of the low number of subjects related to sustainability compared to the total number, however, it is understood that this issue needs to be better addressed and debated within the university.

Table 1 UNIFAL-MG data from 2020 to 2023 submitted to *Green Metric*

Data from green metric—teaching, research, and extension UNIFAL-MG					
Data/year		2020	2021	2022	2023
Teaching	Total subjects offered	1870	2281	1750	1701
	Subjects related to sustainability	91	75	80	67
	Percentage	4,86%	3,28%	4,57%	3,94%
Research	Total publications registered	663	844	868	889
	Publications related to sustainability	68	54	156	123
	Percentage	10,25%	6,39%	17,97%	13,83%
Extension	Total actions developed	318	404	511	556
	Extension actions related to sustainability	19	45	45	62
	Percentage	5,97%	11,13%	8,80%	11,15%

Source The authors (2025)

Regarding the research dimension, there was a decrease in the number of publications in 2021 (6.39%) compared to 2020 (10.25%), but the largest number of scientific productions related to the topic of sustainability were recorded in 2022, comprising 156, which corresponds to 17.97% of the total publications in the year in question. In 2023, however, a small reduction was observed, reaching 13.83% of publications related to sustainability raised in the institution. The analysis of the results of this dimension indicates a trend of continuous improvement, without sudden variations; however, it is observed that there is still room for the expansion of scientific productions related to sustainability in the institution.

In the extension dimension, a reduced number of actions linked to sustainability were identified in 2020 (19), representing 5.97% of the total, highlighting this year with the lowest performance in the dimension, considering the study period. One possible inference about this result is related to the beginning of the pandemic period, when social distancing measures were taken which, in turn, made it impossible to develop part of the extension actions planned for that year, considering that extension actions are characterized by social contact with the community outside the university and the adaptations to carry them out remotely took some time.

In 2021, still in the extension dimension, the percentage of actions carried out was 11.13%, demonstrating, compared to the previous year, a significant advance. In the following year, 2022, although the number of actions developed related to the theme considered in this study remained at 45, the proportional value fell to 8.80%, because the total number of actions carried out showed an increase. In the last year analyzed, 2023, the extension dimension achieved the best result of the period, with the highest number of registered actions associated with sustainability, reaching 62, which represents 11.15% of the total extension actions developed in the year.

Figure 1 shows the percentages of the dimensions analyzed concerning sustainability in the institution, facilitating the view and comparison of the aspects examined in the study. Based on the analysis of the results of the dimensions, it is possible to note that teaching, with regard to the theme of environment and sustainability, presented the lowest percentage index, with little oscillation in the period analyzed, remaining below 5% of the total number of disciplines offered to students at the institution during the period considered in the study, suggesting that this area needs attention to achieve better future results in the institution. One possible inference regarding the results of the teaching dimension is related to the method used in the survey and collection of data regarding the relationship of the disciplines with the theme of sustainability, a factor that may have generated impacts on the results. Situations like this can be reflected upon together with the university's sustainability committee (CPMAS), to identify possible adjustments that can be made to the instruments for future collections.

In the research dimension, in turn, a higher percentage was observed, which reached results above 15% of academic productions related to sustainability in the period in question, although there was a slight drop in the percentage recorded to 13.83%. Based on the results, it was possible to identify, as a possible limitation, the vulnerability in the collection of data referring to the scientific production of the research dimension correlated to sustainability, since, among all the data collected for

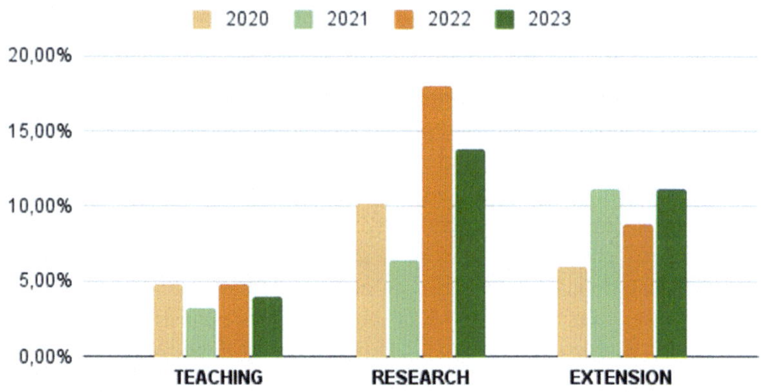

Fig. 1 Percentages related to the sustainability of teaching, research, and extension between 2020 and 2023 at UNIFAL-MG. *Source* the authors (2025)

this research, this sector demonstrated greater difficulty in organizing and providing. This limitation is understood, in part, as a consequence of the lack of human resources in the administrative sectors responsible for managing this information, as well as the lack of a computerized system that makes it possible to filter the data of academic production in an agile and efficient manner.

Intending to contribute to a better organization of institutional information that refers especially to the university's scientific production, aiming to allow agility in the collection of data on specific topics, such as this study, the UNIFAL-MG sustainability committee (CPMAS) has been discussing alternatives to work with the research sector, in the possible creation and provision of a database that provides more efficient and structured access to information on sustainability. This initiative not only facilitates the obtaining of data for research and institutional reports, but can also encourage scientific production in the area, strengthening the visibility and impact of the sustainable actions developed.

Research developed in educational institutions in Brazil reveals a growing concern with the adoption of practices and policies related to sustainability. A study by Barros de Deus et al. (2023) indicates that, despite the considerable increase in publications on the subject in recent years, challenges persist for the implementation of sustainable actions in universities. It was observed, in the aforementioned study, that most of the initiatives are concentrated in federal public institutions, especially in the southern region of the country. These authors emphasize the importance of developing institutional policies that solidify sustainability as a fundamental pillar of academic and administrative practices in Brazilian higher education institutions.

In the same vein, as pointed out by Rohrich and Takahashi (2019), in recent years, research on environmental sustainability in Brazilian HEIs has grown, demonstrating progress in both conceptual understanding and sustainable practices implemented on university campuses. According to these authors, the analyses reveal that sustainability in HEIs unfolds into two main aspects: one related to teaching and research, and the other linked to campus management and the adoption of environmental practices, which, according to them, highlights the urgency of an integrated approach that unites academic training with effective institutional actions, aiming to establish sustainability as a fundamental principle in Brazilian universities.

Thus, UNIFAL-MG has continued to maintain itself as an institution committed to sustainability, constantly seeking to advance in the adoption of sustainable practices and actions on its premises. The university has been striving to integrate sustainability into the academic dimensions of teaching, research and extension, but also into the management and incorporation of environmental practices in the subunits. By aligning these aspects, UNIFAL-MG seeks not only to contribute to the training of professionals sensitive to the issue but also to the construction of a university environment that reflects the urgency of an integrated and effective approach to promoting sustainability, solidifying its performance with distinction in socio-environmental responsibility.

4 Conclusion

The objective of this study was to identify and reflect on the performance of UNIFAL-MG in the dimensions of teaching, research, and extension, considering its relationship with sustainability, based on data submitted to the Green Metric ranking from 2020 to 2023. Additionally, the study sought to identify the main challenges and limitations related to these dimensions, particularly concerning their integration with sustainability. The results indicated that, although the university has incorporated sustainability into its institutional practices, there are still challenges to be overcome for a more systemic integration of this content.

The need was identified to expand the transversality of sustainability in the institution in scientific production, in the academic training of students, and in the development of extension actions, to strengthen the impact of the university in this area beyond its physical boundaries. It is worth noting that this research was conducted exclusively with institutional information reported to Green Metric, which may represent a limitation of the study.

Therefore, it is understood that future research may include qualitative analyses using other instruments to observe, for example, the perception of the academic community on sustainable practices, the performance of organizational management regarding the theme, in addition to the survey and evaluation of other institutional information, with the possibility of applying different sustainability indicators. Furthermore, the teaching dimension was revealed to be in a different condition from the other dimensions discussed, which suggests the need for greater institutional attention to deepen the debate on the incorporation of sustainability into academic curricula, in the search to provide training for students and promote a more enriching reach concerning contemporary issues related to sustainability. Future research may focus on examining the effectiveness of the sustainable institutional policies implemented and assessing the impact of these practices on daily university life and academic development.

References

Alves, M. J. de O., Barbosa, M. de F. N., Vasconcelos, A. C. F. de, & Barbosa, E. M. (2024). Sustentabilidade em instituição de ensino superior público: práticas de ensino, pesquisa, extensão e gestão [Sustainability in a Public Higher Education Institution: Practices in Teaching, Research, Outreach, and Management]. *Cuadernos de Educación y Desarrollo, 16*(9). https://doi.org/10.55905/cuadv16n9-022.

Araujo, C. F., Frio, R. S., Siqueira, C. S. de, Zaluski, F. C., & Hedlund, P. R. (2024). Sustentabilidade nas instituições de ensino superior e os efeitos no comportamento pró-ambiental dos alunos [Sustainability in Higher Education Institutions and Its Effects on Students' Pro-Environmental Behavior]. *Revista Ciências Administrativas, 30*, 1–14. https://ojs.unifor.br/rca/article/view/14477.

Barros de Deus, M. C. da S., Côrtes Lacerda, C., & Costa Radons, M. L. da. (2023). Sustentabilidade ambiental em instituições de ensino brasileiras: estudos realizados de 2015 a 2021 [Environmental Sustainability in Brazilian Educational Institutions: A Review of Studies from 2015 to 2021]. *Revista Sergipana de Educação Ambiental, 10*, 1–20. https://doi.org/10.47401/revisea.v10.18694.

Martins, G. A. (2005). *Estatística geral e aplicada* [General and Applied Statistics] (3ª ed.). Atlas.

Nogueira dos Santos, A., Brasil, M. V. de O., Sumiya, L. A., & De Brelaz, G. (2024). Gestão ambiental sustentável em uma universidade pública federal sob a perspectiva do UI GreenMetric [Sustainable Environmental Management at a Federal Public University: Perspectives from the UI GreenMetric]. *Revista de Gestão Ambiental e Sustentabilidade, 13*(1). https://doi.org/10.5585/2024.23013.

Rohrich, S. S., & Takahashi, A. R. W. (2019). Sustentabilidade ambiental em instituições de ensino superior, um estudo bibliométrico sobre as publicações nacionais [Environmental Sustainability in Higher Education Institutions: A Bibliometric Study of National Publications]. *Gestão & Produção, 26*(2). https://doi.org/10.1590/0104-530X2861-19.

Souza Marques, J. F., Santos, A. V., & Coura Aragão, J. M. (2020). Planejamento e sustentabilidade em instituições de ensino superior à luz dos objetivos do desenvolvimento sustentável [Planning and Sustainability in Higher Education Institutions in Light of the Sustainable Development Goals]. *REUNIR Revista de Administração Contabilidade e Sustentabilidade, 10*(1). https://www.reunir.revistas.ufcg.edu.br/index.php/uacc/article/view/1052.

Triola, M. F. (2024). *Introdução à estatística* [Introduction to Statistics] (14ª ed.). LTC.

Universitas Indonesia. (2024). *Green Metric World University Ranking 2024 Report*. https://greenmetric.ui.ac.id/about/welcome.

Vitoreli, M. C., Pereira, J. A. da S., Guarnetti, R. L., Mariano, E. B., & Gimenes, D. N. V. (2020). Análise do ranking GreenMetric como referencial em ações sustentáveis nas instituições de ensino superior [Analysis of the GreenMetric Ranking as a Framework for Sustainable Actions in Higher Education Institutions]. *Anais do XXVII Simposio de Engenharia de Produção (SIMPEP) UNESP*. https://www.researchgate.net/publication/346237904.

Julieta Aparecida Moreira Master in Ecology and Environmental Technology. Administrative Technician in Education in the Institute of Natural Sciences of the Federal University of Alfenas (UNIFAL-MG), State of Minas Gerais, Brazil. Environment and Sustainable Development Manager of the UNIFAL-MG. Ph.D. student in the Graduate Program in Environmental

Sciences at UNIFAL-MG. Has dedicated himself to the area of Environmental Education, working in university extension.

Elaine Angelina Colagrande Ph.D. in Science. Professor at Chemistry Institute, Federal University of Alfenas (UNIFAL-MG), State of Minas Gerais, Brazil. Professor in the Chemistry Postgraduate Program and the Education Postgraduate Program at Federal University of Alfenas. Research experience: Chemistry teaching, teacher professional development, environmental education.

Post-occupancy Evaluation as a Tool for Promoting Sustainability in the Operation and Maintenance of University Buildings

Gláucia Trindade Pereira and Simone Barbosa Villa

Abstract The adaptive capacity of university buildings is fundamental to prolonging their service life in an efficient and sustainable manner, ensuring that the needs of users are adequately met and contributing to quality in teaching. However, it is possible to assess the performance of buildings, taking into account technical performance features as well as the expectations of their users, making it possible to make adjustments that make buildings more resilient and sustainable. Thus, this article aims to present the development of methodological procedures and standardized Post-Occupancy Evaluation (POE) tools applied to the university context, focusing on the adaptability of buildings and the improvement of environmental, technical, and functional performance throughout their life cycle. This article presents the propositional research stage and part of the empirical research of the ongoing master's thesis entitled "Adaptability in University Buildings: Diagnoses and Strategies from Post-Occupancy Evaluation". The study is based on the hypothesis that systematic, specific, and adaptability-oriented POE can identify weaknesses and potentialities in buildings, contributing to more effective and sustainable interventions. The results of the initial application in the case study of a university building at the Federal University of Uberlândia are discussed, using the Walkthrough tool. The focus is on the adaptive capacity of the built environment, contributing to more efficient and sustainable management of academic spaces. Post-occupancy evaluation (POE) is used for continuous assessments on buildings' performances, taking into account the views of experts and users. This approach provides a broad view of the real and current situation of the building space, providing important information and data for the efficient operation and maintenance of buildings, mitigating environmental impacts, reducing waste and optimizing resources throughout the building's life cycle. It directly aligns with the Sustainable Development Goals, especially SDG

G. T. Pereira · S. B. Villa (✉)
Faculty of Architecture, Urbanism, and Design, Federal University of Uberlândia (FAUED/UFU), Uberlândia, MG, Brazil
e-mail: simonevilla@ufu.br

G. T. Pereira
e-mail: glauciap@ufu.br

W. Leal Filho et al. (eds.), *Sustainability in the Curriculum and Operations of Universities*, World Sustainability Series,
https://doi.org/10.1007/978-3-032-12678-8_39

11, promoting resilient, sustainable and more inclusive communities. As a result, the research led to the creation of standardized Post-Occupancy Evaluation (POE) tools capable of identifying impacts and measuring the adaptive capacity of buildings. The application of the method, with emphasis on walkthrough, allowed for the validation of the tools and the collection of relevant data for space management. It was found that continuous assessments support more efficient maintenance and planning decisions, promoting sustainable and adaptable environments. The integration between technical and administrative aspects and user perception reinforces the importance of participatory and evidence-based practices in university estate management.

Keywords Post · Occupancy evaluation · University building · Evaluation methodology

1 Introduction

Higher education institutions (HEIs) play a central role in promoting quality education and advancing scientific knowledge (Hassanain et al. 2023; Machado et al. 2021; Wang et al. 2022). In this sense, ensuring the quality of the physical environments of these institutions is essential to guaranteeing effective and sustainable teaching, and is one of their responsibilities (Faria et al. 2021). The buildings of Higher Education Institutions (HEIs), especially public ones, must meet a wide variety of demands, considering the complexity and particularities of each environment to meet the specific needs of different users in various areas of knowledge (Oliveira and Oliveira 2019).

Adaptation is a constant aspect in these buildings, whether to extend their durability, improve operational efficiency, accommodate the diversity of users or reduce environmental impacts over time, including variations between the design of the project, its functional use and the agents involved (Heidrich et al. 2017; Saarimaa and Pelsmakers 2020; Spiten et al. 2016). In this way, the adaptive capacity of buildings becomes a fundamental element in meeting social, economic and sustainable demands throughout their existence (Parreira 2020).

Planning in universities presents significant challenges, requiring tools that contribute to projects with better performance and greater sustainability. In this context, the evaluation of buildings in use has emerged as a valuable source of data to improve future interventions (Oliveira and Oliveira 2019). However, there is a lack of specific protocols for evaluation, including maintainability, which directly impacts the sustainability of buildings (Faria et al. 2021). Enabling buildings to be more adaptable, efficient and better maintained can be a determining factor in making buildings more sustainable over time.

Given this scenario, the ongoing master's research entitled "Adaptability in University Buildings: Diagnostics and Strategies from Post-Occupancy Evaluation",[1] aims mainly to develop standardized and systematic methodological procedures and tools for post-occupancy evaluation in university buildings, identifying the impacts and indicators of adaptability of the university-built environment in use. To this end, the study adopts a diversified methodological approach and is characterized as a quantitative–qualitative empirical study, using the hypothetical-deductive scientific method. The Design Science Research method is used to build an artifact for post-occupancy evaluation in university buildings (Dresch et al. 2015), along with the case study method as a research strategy. Thus, the general methodology is structured around the following approaches: (i) Theoretical foundation: systematic literature review (ii) Documentary research: analysis of data, projects and documents from the research case study; (iii) Propositional research: development of a post-occupancy evaluation methodology with a focus on adaptability; (iv) Empirical research: application of the proposed methodology with analysis of the results obtained. This article presents the artifact developed in the proposal stage of the research, consisting of Post-Occupancy Evaluation (POE) instruments, in addition to preliminary results obtained from applying the walkthrough instrument in a pilot project.

2 Adaptability as a Means of Promoting Sustainability

Adaptability can be understood as the ability of the built environment to adapt to a variety of environmental, spatial and social impacts, facilitating its use for multiple purposes and by different users (Pelsmakers and Warwick 2022). Thus, in the context of university buildings, adaptability encompasses a broad capacity to adapt to new demands throughout the life cycle, allowing modifications to both physical and social aspects, where the various user profiles play an active role in satisfying their changing needs over time (Güzelci et al. 2021; Heidrich et al. 2017; Schmidt III and Austin, 2016).

Adaptability in the built environment can be seen as a means of promoting sustainability, since it allows buildings to undergo transformations in a more accessible way, extending the useful life of buildings, while contributing to the reduction of waste produced during adaptations (Hamida et al., 2022).

In this sense, this study relates to the goals of Sustainable Development Goal 11, as it seeks to promote more inclusive, safe, resilient, and sustainable cities and communities (United Nations Brazil, n.d.). Specifically, it contributes to: (i) goal 11.3, by proposing a post-occupancy assessment tool focused on adaptability and resilience, promoting participatory management and the feasibility of identifying possibilities for physical adaptations to buildings in use; (ii) goal 11.6, by providing,

[1] The work is part of a larger research project entitled "[RESILIENT HOUSING] Design strategies for promoting resilience in social housing based on post-occupancy evaluation methods," funded by CNPq, developed at [MORA] Housing Research, Post-graduate Program in Architecture and Urbanism, School of Architecture, Urbanism, and Design, from the Federal University of Uberlândia. https://morahabitacao.com/; http://www.ppgau.faued.ufu.br/.

as a result of applying the method, technical support to optimize the operation and maintenance of buildings, reducing waste and environmental impacts during use and any adjustments; (iii) goal 11.7, by integrating user perception into the analysis of space performance, guiding improvements that ensure more accessible, safe, and inclusive environments for all audiences, especially women, the elderly, and people with disabilities. The research reinforces the importance of management practices that promote planning, resource efficiency, and sustainability in university buildings, contributing to more resilient and sustainable communities.

3 Adaptability in University Buildings

In the scenario of Higher Education Institutions, buildings are largely designed and built without the participation of the end user, which suggests a gap in performance expectations and user behavior during use (Montazami et al. 2015), so adaptability is an important factor in minimizing this gap. Other causes also drive the need to adapt the built environment: technological advances, climate change, changes in regulations, government organization and socio-economic changes, changing user functions and needs, the search for comfort, improved technical performance and energy efficiency (Heidrich et al. 2017; Saarimaa and Pelsmakers 2020; Kamara et al. 2020; Parreira 2020; Memari et al. 2022; Villa et al. 2022). In short, change drivers occur when there is a discrepancy between user expectations and building performance or between technical specifications and actual performance (Kamara et al. 2020). In the university context, pedagogical, social and technological changes require continuous adjustments to ensure functionality and sustainability (Coelho and Krüger 2015). In this way, the adaptive capacity of buildings is essential for dealing with insecurities in the future, being a strategy for protection and guaranteeing the maintenance of adequate functionality (Saleh et al. 2009), as well as being an important mechanism for promoting sustainability.

The adaptability of a building is not only limited to physical changes, but considers how users can adjust to suit it, considering the available resources (Heidrich et al. 2017). The stimuli for change usually result in user adjustments until they become inadequate, requiring changes to the building structure to ensure satisfactory levels of comfort and performance (Kamara et al. 2020).

In this way, the adaptability of buildings occurs on two levels: (i) behavioral adaptations by users and (ii) physical changes to the structure. When user adjustments are no longer sufficient to ensure comfort and performance, building intervention becomes necessary (Heidrich et al. 2017; Kamara et al. 2020). Therefore, it can be understood that adaptability is presented in a cyclical system where, initially, users adapt to the environment until their needs are no longer met, requiring physical modifications (Fig. 1). After the adaptations, the environment is used again until new requirements arise, restarting the cycle throughout the life of the building.

In the context of university buildings, there are some particularities in relation to other types of building. The main differentiating aspects are: (a) use and functions of built environments; (b) users; (c) management of physical infrastructure.

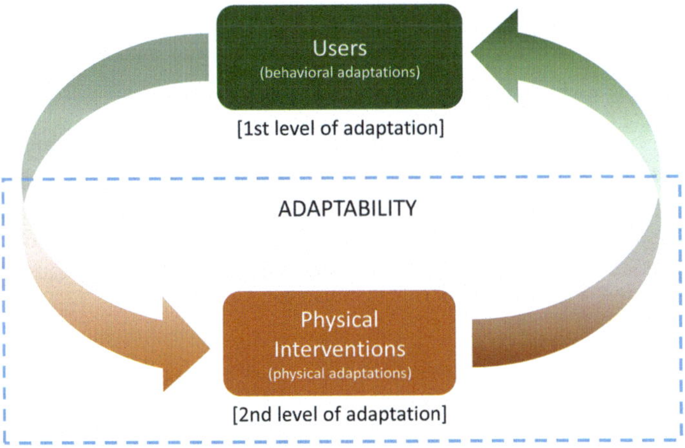

Fig. 1 Adaptability cycle

The most evident feature is the diversity of uses that cover teaching, research and extension activities in different areas of knowledge (Oliveira and Oliveira 2019). Users, on the other hand, can be classified into different categories, depending on their role: students, teachers, administrative staff, outsourced employees, service providers and the external community.

Another important difference refers to physical interventions that cannot be carried out directly by users, requiring formal requests and management approval. This process involves multiple stages, such as registering requests for physical interventions, technical analysis, planning and the availability of institutional resources, making the adaptability cycle more complex (Fig. 2).

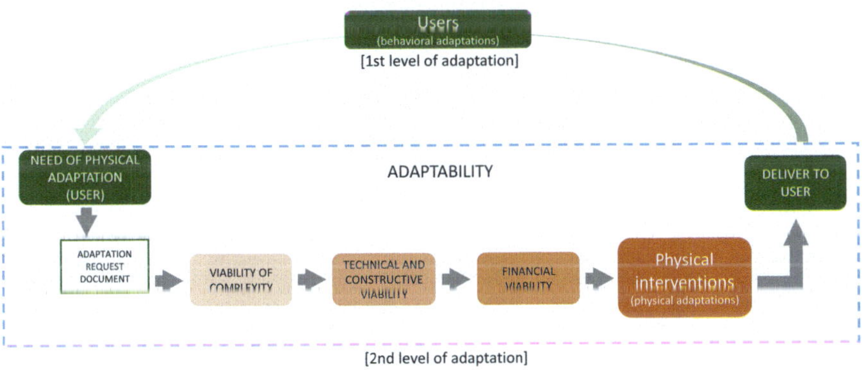

Fig. 2 Adaptability cycle in university buildings

4 Strategies and Evaluation of Adaptability in University Buildings

Adaptability is an intrinsic capacity of the building, but it is difficult to observe and measure, and it is still necessary to measure the adaptation potential of built environments in use (Saleh et al. 2009; Heidrich et al. 2017). It can be seen that in university buildings there are usually elements that contribute to the adaptive capacity of the built space, such as: a modular structure independent of sealing, zoning of the functions of technical and support areas, among others.

To discuss adaptability in the university-built environment, considering its dynamics and diverse users, the question arises of how adaptability can be assessed. In-depth knowledge about the capacity of buildings to adapt, as well as the agents motivating change, is essential for strategic planning and more assertive decision-making to deal with future uncertainties, resulting in better decisions (Buehring and Bishop 2020). Heidrich et al. (2017) highlight the need for tools to evaluate and promote strategies to improve adaptable and resilient projects. From this perspective, this study considers Post-Occupancy Evaluation as a methodology that enables analysis from the perspective of experts as well as users, allowing for comprehensive diagnostics that guide recommendations and necessary interventions to improve building performance (Ono et al., 2018). Considering the need to analyze the specificities of the university setting, indicators were used to identify and analyze the adaptive capacity of buildings.

Adaptability is understood in this research as an attribute of the built environment, which gives the space ample capacity to adapt to the diverse needs of its users, enabling changes in the physical infrastructure and behavioral actions of users (Schmidt III and Austin 2016; Heidrich et al. 2017; Güzelci et al. 2021). Thus, attributes are characteristics of the built environment that provide the desired quality (Villa et al., 2017). These are obtained through indicators and sub-indicators, which are factors that collaborate to achieve the attribute (Bortoli 2023; Oliveira 2023; Parreira 2020; Villa et al. 2022). This results in a relationship between adaptability and the possible means of visualizing the installed capacity of the built environment in use (Fig. 3).

The classification of types of adaptability presented in the studies by Pinder et al. (2017), Heidrich et al. (2017) and Schmidt et al. (2010) are adopted, where six

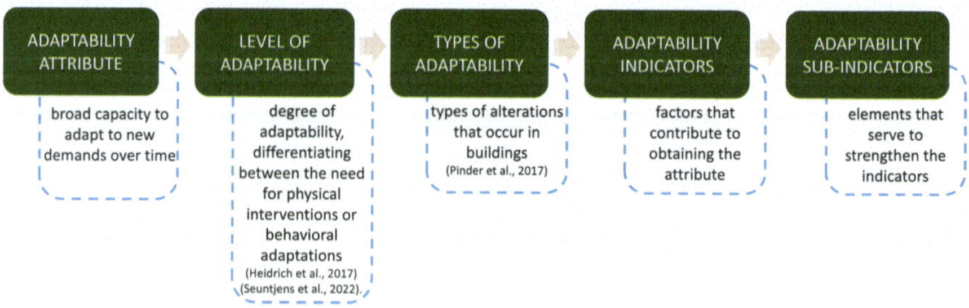

Fig. 3 Relationship between adaptability attribute, level, typology, indicators and sub-indicators

categories are differentiated, considering the level of interventions carried out on the buildings. Based on this classification, an indicator was related to each typology as a means of verifying the various adaptation capacities, and these were the elements that guided the construction of the POE instruments presented in this article (Fig. 4).

The indicators relating to direct user action are: Multifunctionality and Versatility. The indicators referring to the need for physical interventions in the building are: Refit-ability, Conversion, Scalability, Mobility. In addition to the indicators, sub-indicators are adopted for Refit-ability and Scalability. Two sub-indicators were adopted for Refit-ability: Accessibility and Availability. As for Scalability, there are the sub-indicators of: Elasticity and Expansion.

5 Post-Occupancy Evaluation in Higher Education Institution Buildings

Evaluating the built environment involves determining its value and quality to satisfy specific needs, serving as an important tool for understanding decision-making processes (Voordt and Wegen 2013). Considering the specific characteristics and dynamics of university educational buildings, evaluating the built environment in use is an important tool for obtaining feedback and, consequently, optimizing resources in the operation and maintenance of these buildings. The search for adaptability can be an essential factor in improving the quality of buildings used for university teaching, as well as providing a sustainable and more efficient response to future modifications (Hamida et al. 2022).

From this perspective, the POE methodology makes it possible to obtain feedback and helps designers to understand internal environmental factors holistically and systematically, addressing the needs, behaviors and satisfaction of occupants during the use phase (Elsayed et al. 2023; Montazami et al. 2015), while providing a means for creating policies for the continuous improvement of the institutional built environment (Resende et al. 2024).

Post-occupancy evaluation (POE) is a set of methodological procedures that aims to assess whether the objective and subjective needs of users are met during the use of the built environment, throughout its service life (Mallory-Hill et al. 2012; Ono et al. 2018; Saft 2021; Villa et al. 2015). To this end, multiple methods are used to collect and analyze data, including qualitative (subjective) and quantitative (objective) methods, for analyses of physical, functional, and behavioral aspects (Elali 2006). The combination of these methods is essential to ensure a comprehensive understanding of building operation and occupant satisfaction, since meeting the needs of its users is a prerequisite for validating the quality of the building (Strelets et al. 2016). This approach makes it possible to correct inconsistencies in the data, reinforce valid conclusions and discard those that are less representative. Each method has its advantages and disadvantages, so the choice of techniques must consider the size of the sample and the objectives of the study (Bortoli 2023; Elsayed et al. 2023; Voordt and Wegen 2013). Considering that universities may have flaws in their design, implementation, and maintenance processes, resulting in unsatisfactory

ADAPTABILITY DIMENSION	DEFINITION	INDICATOR	DEFINITION OF THE INDICATOR: SUB-INDICATOR	DEFINITION OF THE INDICATOR: DEFINITION	BUILDING LAYERS
ADJUSTABLE	Suitability of tasks for users (Heidrich et al., 2017).	MULTIFUNCTIONALITY		Ability to use the space for different purposes (Abreu; Heitor, 2007) (Parreira, 2020) (Hamida et al.,2022)	stuff
VERSATILE / FLEXIBLE	Suitability of tasks for users (Heidrich et al., 2017).	VERSATILITY		Ability to change the size of the physical space through the existence of mobile elements (Schmidt et al., 2010) (Pinder et al., 2017) (Heidrich et al., 2017).	stuff space plan
REFITABLE	Adequate performance of equipment and facilities (Heidrich et al., 2017).	REFIT-ABILITY		Capacity to improve building performance (Schmidt et al., 2010) (Pinder et al., 2017) (Heidrich et al., 2017) (Hamida, et al., 2023)	space plan services skin
			ACCESSIBILITY	Capacity to equip the building for access and mobility for all users.	
			AVAILABILITY	Ability to access all technical installation systems and their components (Hamida, et al., 2023)	
CONVERTIBLE	Adapting to the new role (Heidrich et al., 2017).	CONVERSION/ CONVERTIBILITY		Ability to change the function of the environment, promoting the spatial reconfiguration of the built space (Abreu; Heitor, 2007) (Hamida, et al., 2023) (Parreira, 2020)	space plan services skin
SCALABLE/ ELASTIC	Changing the dimensions of the rooms and the exterior of the building (Heidrich et al., 2017).	SCALABILITY		Capacity to add new rooms to the building (Parreira, 2020) (Oliveira, 2023)	space plan services skin structure
			ELASTICITY	Capacity to expand the area of the building's internal environments (Parreira, 2020) (Oliveira, 2023) (Tarpio; Huuhka, 2022)	
			EXPANSION/ EXPANDABILITY	Capacity to increase the built area of a building, with an increase in the external limits of the building (Abreu; Heitor, 2007) (Parreira, 2020) (Oliveira, 2023).	
MOVEABLE	Change of building location (Heidrich et al., 2017).	MOBILITY		Ability to change the location of the building by dismantling and reusing part or all of the construction (Schmidt et al., 2010) (Pinder et al., 2017) (Heidrich et al., 2017).	structure site

Fig. 4 Definition of adaptability typologies and indicators

environments, a holistic understanding of the physical space in relation to the behavioral aspects of its users is essential (Del-Roio et al., 2024). In this way, POE results in an up-to-date knowledge base about the analyzed existing buildings (Whittem et al. 2022). This information is essential for developing operational guidelines aimed at improving performance and satisfaction, comfort and productivity of the users (Zhou et al. 2022).

6 Methods

This research adopts the hypothetical-deductive method, with a case study strategy and based on the Design Science Research approach (Dresch et al. 2015), aiming to develop a Post-Occupancy Evaluation (POE) methodology focused on the adaptability of university buildings. It is presented in three major methodological stages: (i) Theoretical research, with a literature review on adaptability, post-occupancy evaluation, and sustainability in university buildings; (ii) Propositional research, with the construction of the POE methodology structured based on the Adaptability Cycle in University Buildings (Fig. 5); (iii) Application of the instrument in a pilot project and preliminary evaluation of the artifact.

This article presents the POE methodology developed, accompanied by the results of the application of the Walkthrough instrument in a case study defined in the master's degree's research.

The methodological proposal for post-occupancy evaluation of university buildings consists of four main instruments: (i) Impact Questionnaire, (ii) Adaptability Questionnaire, (iii) Walkthrough Analysis, (iv) Focus Group. The proposed instruments comprise the Post-Occupancy Evaluation artifact developed for the context of university buildings, integrated into the adaptability cycle in HEIs. These procedures were defined by characterizing the objectives, target audience and techniques based on literature in the area (Fig. 6).

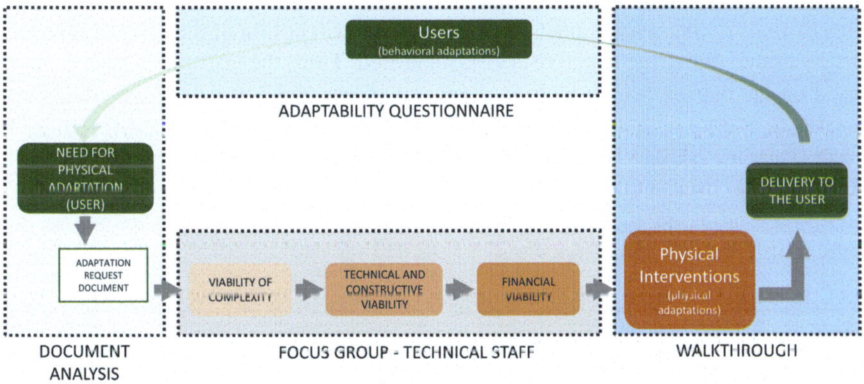

Fig. 5 Relationship between the APO instruments and the adaptability cycle

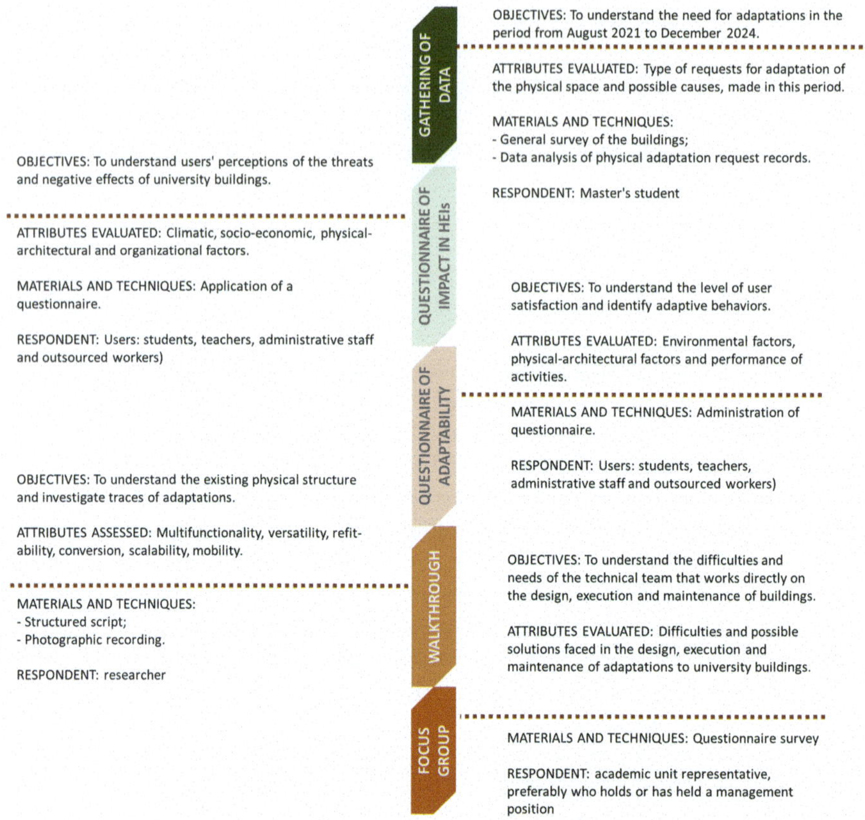

Fig. 6 POE methodological procedures, objectives and assessed attributes

One of the strategies adopted in this research is impact assessment, aiming to understand the factors that can generate negative effects on users of university buildings. Thus, the Questionnaire on Impacts in Higher Education Institutions (QI-HEIs) aims to analyze the main impacts suffered and the perception of the level of discomfort generated due to the negative effects identified. Thus, the instrument is structured based on the causes, which include the threats from which the negative effects are generated, ending with the identification of the perceived level of discomfort of each effect (Parreira 2020; Villa et al. 2017), as illustrated in Fig. 7. In this manner, the Mora Group's Impact Questionnaire[2] was used as a reference for the construction of the IQ-HEI, adjusting the Major Causes, Threats and Negative Effects to the specific context of university buildings.

[2] Grupo [MORA] Pesquisa em Habitação [Research on housing]—research group on housing registered by CNPq and linked to the Postgraduate Program in Architecture and Urbanism of the College of Architecture, Urbanism and Design of the Federal University of Uberlândia. https://morahabitacao.com/; http://www.ppgau.faued.ufu.br.

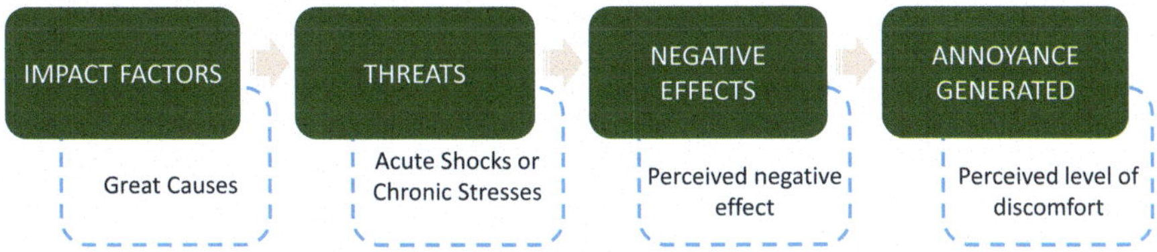

Fig. 7 Basic structure of the impact questionnaire

The instrument has a quantitative approach and is made up of multiple choice closed-ended questions. It is divided into four sections, which correspond to the Major Causes of impacts: (i) Climatic Factors, (ii) Socioeconomic Factors, (iii) Physical-Architectural Factors and (iv) Organizational Factors. The aim of each section is to collect data on the perception of threats and their negative effects, as well as to measure the level of annoyance generated by them (Fig. 8). This questionnaire is aimed at the university community, including teachers, students, technicians and outsourced employees, allowing us to better understand the exposure of the built environment to impacts and, consequently, to analyze the state of vulnerability of buildings and their users.

The second instrument is the Adaptability Questionnaire, which aims to measure users' level of satisfaction with the built environment they use, as well as identifying behavioral adaptation strategies and adaptations to the physical space. The Adaptability questionnaire has a quantitative approach, made up of closed-ended questions, and is aimed at the university community. It is structured around questions that assess user perception and satisfaction regarding environmental, physical-architectural and activity-related aspects, related to academic use (Fig. 9). The instrument seeks to better understand the process of behavioral adaptation of users, and to identify the need for adjustments to the existing physical infrastructure.

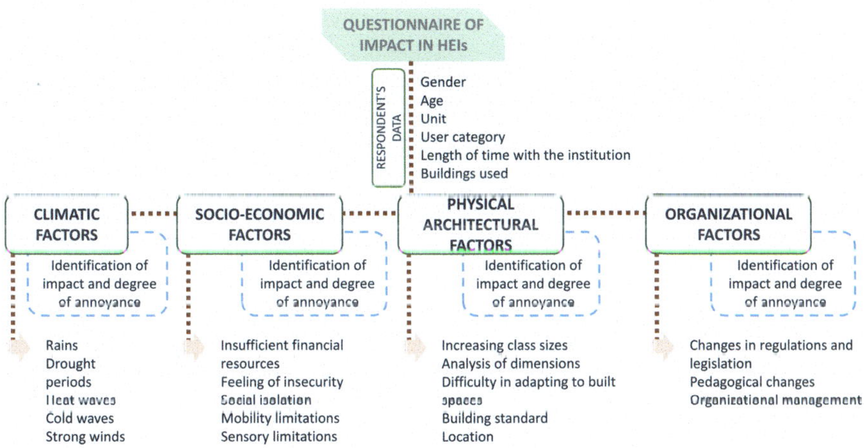

Fig. 8 Breakdown of the sections of the HEI impact questionnaire

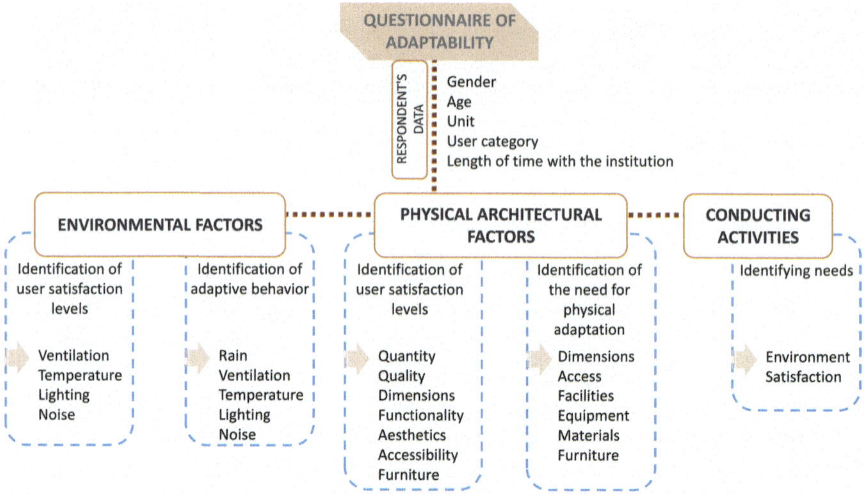

Fig. 9 Details of the sections of the adaptability questionnaire

The walkthrough is an exploratory data collection method that consists of a guided tour of the object of study, preferably with key people who use the building (Ono et al. 2018; Rheingantz et al. 2009). Its main objective is to collect and analyze building data in a short period of time, allowing positive and negative factors of the spaces to be identified (Villa et al. 2015; Villa 2008). In addition, it allows the researcher a direct view of the conditions of the analyzed environment by collecting on-site information, familiarizing themselves with the study site and understanding its physical structure, main changes made, conservation and uses (Ono et al, 2018; Villa et al. 2015).

The proposed walkthrough analysis aims to analyze university buildings with a focus on their present adaptive capacities through identifiable elements. To this end, a script was drawn up based on the attributes and indicators assessed, relating them to the types of adaptability proposed by Schmidt et al. (2010), Pinder et al. (2017) and Heidrich et al. (2017). Each of these typologies has been subdivided according to the adaptability indicators (Fig. 10). Thus, this tool makes it possible to obtain complementary evidence to the analysis of the results of the other tools, from the researcher's point of view.

Finally, as the last instrument, is the qualitative data collection technique, a focus group, which consists of a group activity aimed at discussing a specific topic, to gain a deeper insight into issues related to the research's theme (Fig. 11). According to Ono et al. (2018), the aim of the focus group is to try to reach a consensus opinion among the participants or to identify the opinion of the majority of the participants, using a script to be conducted by a moderator. It is recommended that the group be made up of 6 to 10 people, with a duration that may vary from 90 min to 3 h. This technique makes it possible to capture ideas and feelings, identifying points of consensus and disagreement (Ono et al., 2018; Villa et al. Villa, 2008).

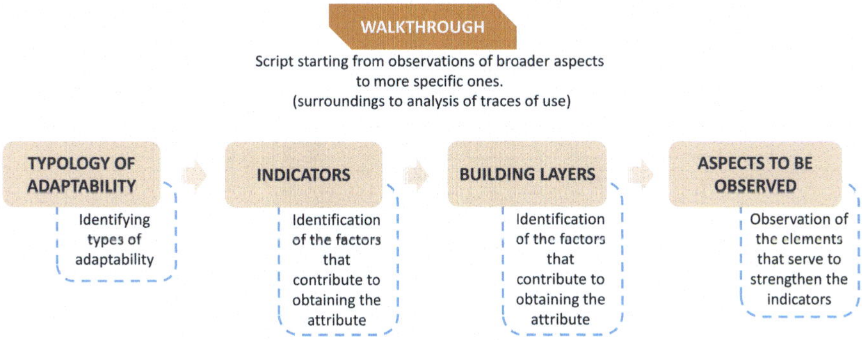

Fig.10 Breakdown of the walkthrough script sections

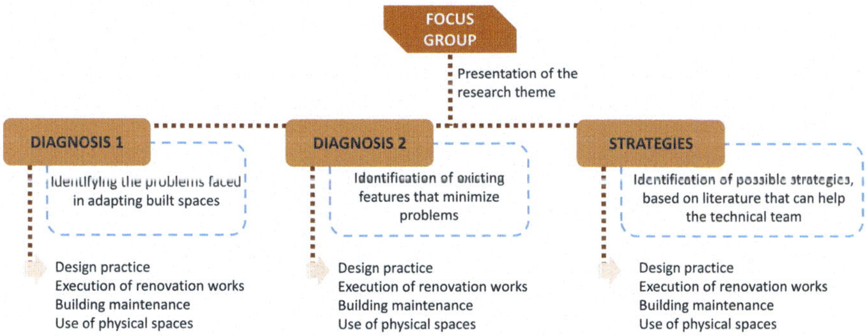

Fig.11 Proposed focus group structure

This research proposes the pilot application of the methodology developed to verify the feasibility and calibration of the instruments. To this end, the FAUeD academic unit was defined as a pilot project, with the object of study being the building known as 5OSM, at the Santa Monica Campus of the Federal University of Uberlândia (UFU), in Uberlândia, Minas Gerais, Brazil.

The 5OSM building (Fig. 12) was selected for its typological and functional representativeness in the context of university buildings, standing out for its multiple uses, academic relevance, and for concentrating most of FAUeD's teaching activities. Located in the central area of the campus, it has separate entrances that facilitate circulation and interaction between users. The building houses 34 classrooms and 8 auditoriums with varying capacities, all equipped with standard furniture and multimedia resources. Of particular note are the auditoriums with movable partitions, which allow for the integration of spaces for academic and cultural events. Wing 5O-B brings together environments with specific furniture geared toward project activities. On the ground floor, the presence of pilotis configured as social and pedagogical support areas expands the potential for institutional use. These characteristics make the building suitable for Post-Occupancy Evaluation, with a focus on technical performance and adaptability.

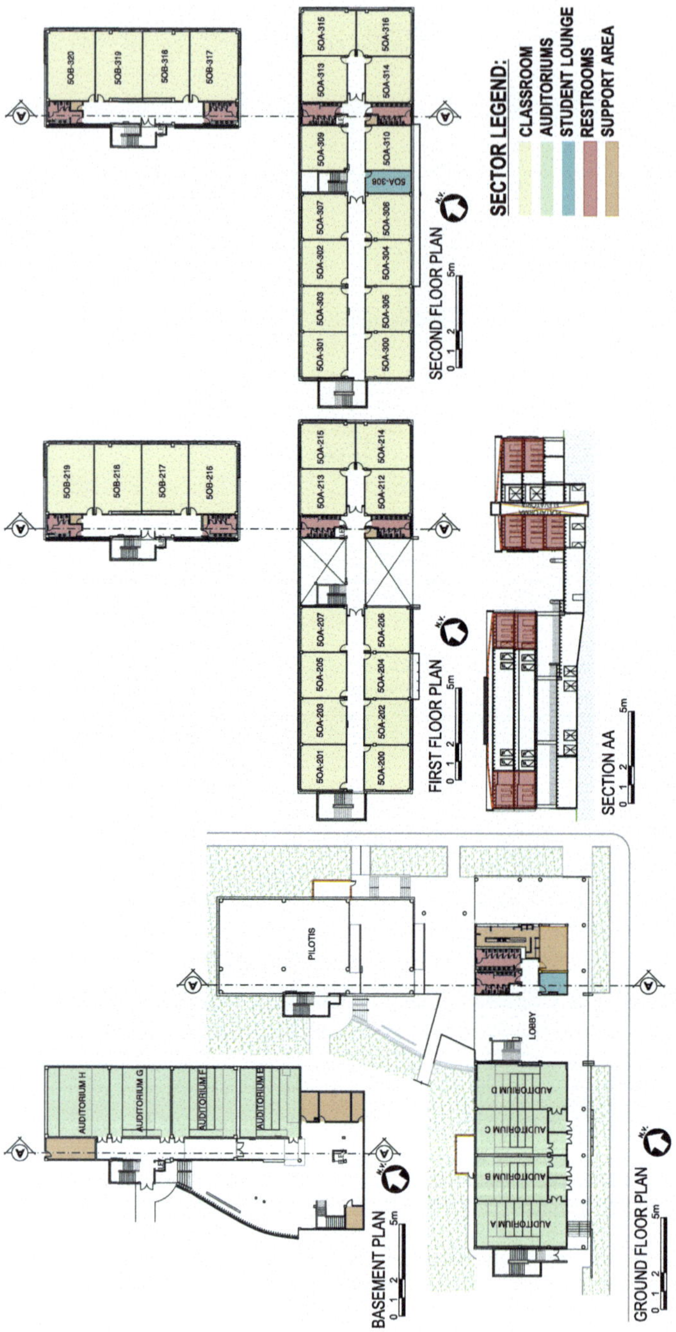

Fig. 12 Sectorization 5OSM building

7 Results and Discussions on the Application of Walkthrough

The application of the walkthrough analysis tool in the 5OSM building allowed the identification of aspects related to the building's adaptive capacities, based on the categories proposed by Schmidt et al. (2010), Pinder et al. (2017), and Heidrich et al. (2017). The data obtained through the on-site survey were initially analyzed based on the building's performance aspects. Next, the aspects indicated in the walkthrough script were observed, based on the theoretical foundation. Finally, the adaptability potential of the different components of the building were assessed, such as: fences, surroundings, circulation, facilities, among others, based on the types of adaptability and their indicators (Fig. 13). This process resulted in a technical diagnosis based on empirical evidence, allowing the identification of weaknesses and potentialities related to the adaptive capacity of the analyzed building.

Mobility, defined as the ability to change the location of a building by dismantling and reusing part or all of the construction (Heidrich et al. 2017; Pinder et al. 2017; Schmidt et al. 2010), is not a recurring feature in university buildings, suggesting its low relevance in the context analyzed. This situation can be observed in the structural systems adopted in these buildings. In the building analyzed, the structure is made of reinforced concrete, therefore not allowing displacement or dismantling.

For the analysis of expandability, the two main sub-indicators were considered, as established in the theoretical framework: (i) expansion, understood as the potential to increase the built area by altering the external boundaries of the building, either horizontally or vertically (Abreu and Heitor 2007; Oliveira 2023; Parreira 2020); (ii) elasticity, which refers to the possibility of expanding the usable area inside the building by integrating or reconfiguring existing environments (Oliveira 2023; Parreira 2020; Tarpio and Huuhka 2022).

With regard to expansion, 5OSM building has a reinforced concrete structural system that is independent of the walls, favoring expansion and integration with other construction systems. The presence of technical ceilings, shafts, and exposed pipes indicates good adaptability of the facilities. However, the layout of the internal circulation and the consolidated occupation of the surrounding area impose restrictions on lateral expansion. Although specific interventions are technically feasible, they require specific solutions and careful compatibility with existing conditions. Thus, the potential for structural adaptability for expansion is classified as moderate to low. On the other hand, the building has demonstrated high potential for elasticity, that is,

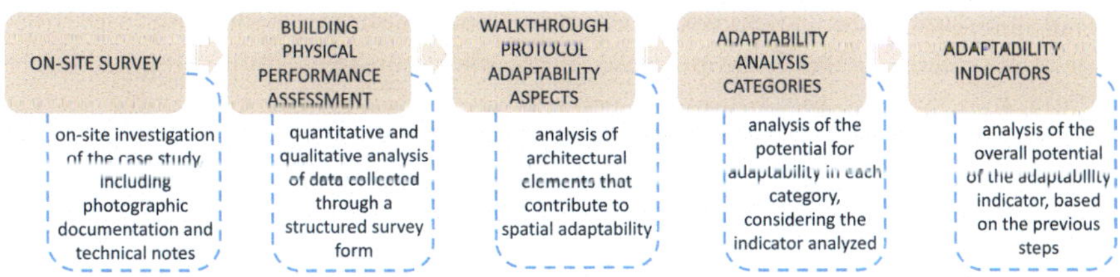

Fig. 13 Structure analysis of the walkthrough script

the ability to expand or redivide the interior of the building. The reinforced concrete structural system, independent of the walls, and the predominant use of internal drywall partitions facilitate the reorganization of spaces with less complex interventions. Added to this is the presence of technical ceilings, shafts, and exposed pipes, contributing to easier interventions in the building's installations. However, conventional ceiling heights limit the vertical division of spaces, while sliding tempered glass frames require compatibility with new partitions.

According to Heidrich et al. (2017), conversion refers to the ability to change the function of an environment. Therefore, the built environment needs to offer favorable conditions for changing the functionality of the space. In the building under analysis, the large dimensions of the classrooms were observed, with the presence of movable furniture and distribution of electrical infrastructure, contributing positively to this indicator. However, the absence of hydraulic points in all environments and the limitations of the drywall sealing system restrict conversion to uses that require such installations. Thus, the classroom spaces have high conversion potential, but the auditoriums, due to their rigid spatial configuration, have low adaptability to new uses, requiring profound structural interventions for functional reconfiguration.

Regarding the refit-ability indicator, understood as the ability to adapt the performance of equipment and facilities to improve building performance (Hamida et al. 2022; Heidrich et al. 2017; Pinder et al. 2017; Schmidt et al. 2010), two sub-indicators were considered: (i) accessibility, such as the possibility of installing equipment and elements to promote universal accessibility for users; (ii) availability, which is the ease of access to all technical installation systems and their components for maintenance, replacement, or readjustment purposes (Hamida et al., 2022). Regarding accessibility for users, the building must provide universal access conditions in its environments, therefore elements that provide mobility must be present, in accordance with current standards. From this perspective, the 5OSM building has operational limitations, since the two vertical platforms were inoperative at the time of the inspection. In addition, tactile signage is restricted to the external areas of the ground floor and basement and is absent from the interior of the other floors. Structurally, the building allows for the installation of external elevators, especially in wing B, where the vertical circulation areas are located next to the front of the building. The insertion of internal elevators would require a technical analysis of the structure to enable cuts in slabs. It is concluded that the building has high potential for adaptation to accessibility requirements, subject to the implementation of physical and operational adjustments in accordance with current regulatory parameters. In terms of availability, there are removable ceilings, accessible shafts, and adequate technical rooms, which favors maintenance and adaptations. However, limitations such as the lack of adequate technical access to the water and sanitation system and the difficulty of accessing the roof compromise future interventions, establishing moderate potential in relation to this sub-indicator.

According to Kamara et al. (2020) and Heidrich et al. (2017), users make behavioral changes and adjustments to adapt to the conditions of the built environment. Thus, there is the indicator of versatility, which is the ability to change the size of the physical space through the existence of movable elements (Heidrich et al.,

2017; Schmidt et al., 2010; Pinder et al., 2017), multifunctionality, defined by the ability to use the space for various purposes (Abreu and Heitor, 2007; Hamida et al., 2022; Parreira, 2020). In terms of versatility, the building analyzed revealed specific solutions in the construction, with moderate potential in this regard. The connection between environments is only possible in the auditoriums, through movable partitions that allow their integration, but this is restricted to this specific use. The classrooms have open floor plans, movable furniture, distributed electrical outlets, and sectorized lighting, favoring subdivision and layout variations. However, the absence of sliding partitions, hydraulic infrastructure, and wired logic network limits more complex functional adaptations.

Finally, the multifunctionality indicator, related to the possibility of diverse use of environments without physical interventions, showed moderate performance in 5OSM building. The classrooms have large floor plans, movable furniture, multimedia resources, Wi-Fi network, sectorized lighting, and distributed electrical outlets, favoring a certain functional flexibility. The presence of large windows and movable blinds contributes to environmental control by the user. However, the fixed compartmentalization between environments and the evidence of exclusive use for teaching activities limit the simultaneous or multiple use of spaces.

Figure 14 presents a summary table of the main findings regarding the building under analysis, establishing the correlation between the physical and construction characteristics observed and the adaptability indicators considered in the methodology.

8 Conclusions

This research is based on the recognition of the importance of the adaptability of university buildings as an essential strategy for sustainability in the operation and maintenance of built spaces. In this context, adaptation strategies aim to extend the useful life of buildings and ensure greater efficiency in the use of available resources, with users playing a central role in identifying demands and implementing the necessary adaptations (Pelsmakers and Warwick 2022).

The methodological proposal of Post-Occupancy Evaluation (POE) with a focus on adaptability proves to be an effective tool for identifying weaknesses, strengths, and opportunities for improvement in university environments. By integrating technical and management aspects with the real needs of users, the approach enables more efficient planning that is aligned with the functionality and usability of spaces. The results obtained with the walkthrough application reinforce this analysis. The instrument enabled the systematic and direct collection of qualitative data in the field, allowing for on-site observation of spatial, structural, and functional conditions. Based on this experience, it will be possible to conduct a more in-depth analysis of the other instruments, providing more robust analyses in the following stages of the study. The walkthrough also proved to be appropriate for the institutional reality, as it is a low-cost tool, applicable in a short time, and capable of generating relevant inputs for building management.

Categories of analysis	Technical observations	Mobility	Ampliability		Conversion	Capability of Refitting		Versatility	Multifunctionality
			expansion	elasticity		accessibility	availability		
Immediate surroundings	Adjacent consolidated areas (roads, parking lots, square)		Limits lateral expansion, requires coordination with existing spaces						
Access and implantation	Building located at an intersection with distances between 3.5 m and 8.3 m		Physical constraints on horizontal expansion						
Structural System	Reinforced concrete structure independent of the enclosures	Does not allow displacement or dismantling of the structure	Allows for constructive continuity or adoption of a complementary structure	Allows for reorganization of environments without structural interference		A feasibility study is required for opening in slab			
Roofing	Thermal and acoustic metal roofing, with access via hatch in the corridors on the 2nd floor						It hinders maintenance and adjustment processes		
Internal sealing	Drywall partitions in classrooms; ceramic tiles in wet areas and stairways			Lightweight, removable partitions facilitate reconfiguration of spaces	Facilitates disassembly, but limits the attachment of heavy equipment without structural reinforcement	Seals independent of the structure allow for reconfiguration of spaces	Promotes insulation between rooms	Does not allow integration of environments through direct user action	
Window frames	Tempered glass sliding windows occupying the space between pillars		They do not prevent expansion, but may require reconfiguration	They require compatibility with partitions in terms of finish and acoustic performance	They provide adequate natural lighting and ventilation without significantly interfering with functional conversion		The type of window frames used presents flaws in terms of thermal and acoustic insulation. Presence of movable brise-soleils, allowing partial control of natural light entry	Adequate natural lighting and ventilation. Requires compatibility with partitions in terms of finish and acoustic performance	Adequate natural lighting and ventilation, with movable brise-soleils in some rooms allowing partial control of natural light
Electrical installations and logics	Presence of shafts, technical ceiling, and exposed pipes; electrical outlets distributed along the walls; Wi-Fi coverage throughout the building		Ease of adapting systems for horizontal or vertical expansion	Good accessibility to hydraulic and electrical networks, facilitating interventions and expansions	Flexibility for use of equipment and different layouts, but with few points in relation to the usable area of the environments	Ease of adapting systems for network expansions and new installations	It allows for easy expansion, maintenance, and adjustments to the systems	Flexibility for use of equipment and different layouts, but with few points in relation to the usable area of the environments	Flexibility for use of equipment and different layouts, but with few points in relation to the usable area of the environments
Hidraulic infrastructure	Absent in rooms; concentrated in specific wet areas				Restricts conversion to uses that require water/sewer connections, such as laboratories		Restricts access to the water tank and hinders maintenance and adjustment processes	Restricts uses that require water/sewer connections	Restricts conversion to uses that require water/sewer connections, such as laboratories
Ceiling height	Standard floor height, with technical ceiling			Ceiling height does not allow for vertical subdivision of spaces (mezzanines, for example)	Does not allow vertical divisions; allows passage of new installations in the technical ceiling	Ceiling height does not restrict the installation of conventional equipment	Allows new installations, but with restrictions in areas where beams are below the ceiling level		
Proportion of the ambients	Classrooms with large, open spaces			Allows spatial subdivision	Large spaces that favor internal reorganization and adaptation for different uses	Allows spatial subdivision	Spacious areas that require air conditioning equipment with greater capacity	Large spaces that favor internal reorganization and adaptation for different uses	Large spaces that favor internal reorganization and adaptation for different uses
Revestimentos e acabamentos	Standard found in various locations throughout the university, serving different purposes.			Use of standardized materials	Use of standardized materials that meet the variety of university uses				
Internal circulation	Layout of toilets and stairs at the ends		Limits direct integration with new volumes in some directions			Lifting platforms not operational at the time of inspection			
Furniture	Movable furniture in classrooms				Allows for easy spatial reorganization	Absence of tactile, visual, and audible signage		Allows for easy spatial reorganization	Allows for easy spatial reorganization
Auditoriums	Rigid configuration with fixed furniture, stage, and stepped floor				Difficulty in converting without major physical and financial interventions			Allows integration of two auditoriums	Difficulty in accommodating different or simultaneous uses without major physical and financial interventions
INSTALLED POTENTIAL		LOW	MODERATE	HIGH	HIGH	HIGH	MODERATE	MODERATE	MODERATE

Fig. 14 Summary board of the walkthrough results

Among the main contributions of the research, we highlight the potential of the methodology to promote the optimization of operational resources, the enhancement of user perception as a basis for more assertive decisions, the strengthening of institutional practices aimed at sustainability, the importance of POE as a tool to support building planning and management, and the feasibility of incorporating

POE results into guidelines for the design, operation, and maintenance of university buildings. In addition, by involving different institutional actors—students, faculty, technicians, and managers—the evaluation process also acts as an instrument of collective awareness, reinforcing the institutional commitment to sustainability and user well-being.

As for limitations, it should be noted that the application of the methodology is still in its early stages, restricted to the implementation of a pilot project. Additional challenges include the typological and functional diversity of university buildings, which may require specific methodological adaptations, as well as the limited time available for the implementation of the research stages, which limits the scope of application. As for future developments, it is proposed to improve the instruments after the completion of the pilot applications, expand the application to different institutions and contexts, and systematically integrate the results into institutional building planning and maintenance policies.

In summary, the approach adopted not only reinforces the commitment to the Sustainable Development Goals (SDGs) but also positions universities as protagonists in promoting more adaptable, efficient, and collective well-being-centered built environments and the well-being of users is reinforced.

By promoting more adaptable, efficient, and user-wellbeing-centered environments, the methodology strengthens HEIs' ability to respond to contemporary challenges and act as a reference in building a more sustainable future.

References

Abreu R, Heitor T (2007) Estratégias de flexibilidade na arquitectura doméstica holandesa: da conversão à multifuncionalidade [Flexibility strategies in Dutch domestic architecture: from conversion to multifunctionality]. Infohabitar—Revista Eletrônica do Grupo Habitar, Lisboa. https://infohabitar.blogspot.com/2007/01/estratgias-de-flexibilidade-na.html. Accessed 22 Apr 2024.

Bortoli K (2023) Resiliência e conforto térmico em habitações de interesse social horizontais em Uberlândia (MG): avaliação para orientação de reformas. [Resilience and thermal comfort in horizontal social housing in Uberlândia (MG): assessment for renovation guidance.] Federal University of Uberlândia.

Buehring J, Bishop PC (2020) Foresight and Design: New Support for Strategic Decision Making. She Ji 6:408–432. https://doi.org/10.1016/j.sheji.2020.07.002.

Coelho C, Krüger M (2015) Towards a methodology to assess adaptability in educational spaces: An entropy approach to space syntax. In: Karimi K, Palaiologou G, Vaughan L, et al. (eds) SSS 2015—10th International Space Syntax Symposium. Space Syntax Laboratory, The Bartlett School of Architecture, University College London.

Del-Roio IG, Fontolan BL, Esperidião AR, Cardoso RAM, Iarozinski-Neto A (2024) Análise fatorial da qualidade do ambiente universitário: Universidade Tecnológica Federal do Paraná campus Curitiba [Factor analysis of university environment quality: Federal University of Technology—Paraná, Curitiba campus]. In: 20th National Meeting on Built Environment Technology—ENTAC 2024, Maceió. Proceedings... Maceio: ANTAC. https://doi.org/10.46421/entac.v20i1.6145.

Dresch A, Lacerda DP, Antunes Jr JAV (2015) Design science research: método de pesquisa para avanço da ciência e tecnologia [Design science research: research method for the advancement of science and technology]. Bookman, Porto Alegre. https://doi.org/10.1007/978-3-319-07374-3.

Elali GA (2006) Uma contribuição da psicologia ambiental à discussão de aspectos comportamentais da avaliação pós-ocupação [A contribution of environmental psychology to the discussion on behavioral aspects of post-occupancy evaluation]. Pós 20, Dec. https://doi.org/10.11606/issn.2317-2762.v0i20p158-169.

Elsayed M, Pelsmakers S, Pistore L, et al (2023) Post-occupancy evaluation in residential buildings: A systematic literature review of current practices in the EU. Build Environ 236:110307. https://doi.org/10.1016/j.buildenv.2023.110307.

Faria RS, Oliveira MF de, Andery PRP (2021) Avaliação de requisitos de manutenção em edificação de instituição federal de ensino.] Assessment of maintenance requirements in federal educational institution buildings.] Even3.

Güzelci OZ, Şen Bayram AK, Alaçam S, et al (2021) Design tactics for enhancing the adaptability of primary and middle schools to the new needs of postpandemic reuse. Archnet-IJAR 15:148–166. https://doi.org/10.1108/ARCH-10-2020-0237.

Hamida MB, Jylh€a T, Remøy H, Gruis V (2022) Circular building adaptability and its determinants—A literature review. https://doi.org/10.1108/IJBPA-11-2021-0150.

Hassanain MA, Sanni-Anibire MO, Mahmoud AS (2023) Design quality assessment of campus facilities through post occupancy evaluation. International Journal of Building Pathology and Adaptation 41:693–712. https://doi.org/10.1108/IJBPA-04-2021-0057.

Heidrich O, Kamara J, Maltese S, et al (2017) A critical review of the developments in building adaptability. International Journal of Building Pathology and Adaptation 35:284–303. https://doi.org/10.1108/IJBPA-03-2017-0018.

Kamara JM, Heidrich O, Tafaro VE, et al (2020) Change factors and the adaptability of buildings. Sustainability (Switzerland) 12. https://doi.org/10.3390/su12166585.

Machado EL, Pasdiora L, Santos A de PL (2021) Avaliação de critérios de desempenho de edificações escolares. [Evaluation of performance criteria for school buildings.] Even3.

Mallory-Hill S, Preiser WFE, Watson C (2012) Enhancing building performance. Wiley-Blackwell, Iowa.

Memari S, Kocaturk T, Lozanovska M, et al (2022) The interdisciplinary conceptualization of future proofing in the context of hospital buildings. Building Research and Information. https://doi.org/10.1080/09613218.2021.2011704.

Montazami A, Gaterell M, Nicol F (2015) A comprehensive review of environmental design in UK schools: History, conflicts and solutions. Renewable and Sustainable Energy Reviews 46:249–264. https://doi.org/10.1016/j.rser.2015.02.012.

Oliveira HS de, Oliveira FL de (2019) Aprendendo com o construído: APO em universidades públicas—um estudo de caso. [Learning from what has been built: POE in public universities—a case study.] EDUFU—Publisher of the Federal University of Uberlandia, pp 1014–1023. https://doi.org/10.14393/sbqp19093.

Oliveira L (2023) Avaliação dos impactos da Covid-19 na habitação social e suas adaptações para um ambiente construído mais resiliente. [Assessment of the impacts of Covid-19 on social housing and its adaptations for a more resilient built environment.] Federal University of Uberlândia.

Ono R, Ornstein SW, Villa SB, França AJGL (2018) Avaliação pós-ocupação na arquitetura, no urbanismo e no design: da teoria à prática [Post-occupancy evaluation in architecture, urbanism, and design: from theory to practice]. Oficina de Textos, São Paulo.

Parreira F (2020) Estratégias de flexibilidade orientadas ao usuário como facilitador da resiliência em habitação de interesse social. [User-oriented flexibility strategies as a facilitator of resilience in social housing.] Federal University of Uberlândia.

Pelsmakers S, Warwick E (2022) Housing adaptability: new research, emerging practices and challenges. Buildings and Cities 3:605–618. https://doi.org/10.5334/bc.266.

Pinder JA, Schmidt R, Austin SA, et al (2017) What is meant by adaptability in buildings? Facilities 35:2–20. https://doi.org/10.1108/F-07-2015-0053.

Resende BPR, Nardi ETSN, Carolino VGC, Saft JBS, Souza TCSS (2024) Avaliação Pós-Ocupação (APO) como instrumento aplicado à gestão de riscos do patrimônio cultural do Arquivo Histórico Municipal (AHM) [Post-Occupancy Evaluation (POE) as an instrument applied to the risk management of the cultural heritage of the Municipal Historical Archive (AHM)]. In: 20th National Meeting on Built Environment Technology—ENTAC 2024, Maceió. Proceedings… Maceió: ANTAC. https://doi.org/10.46421/entac.v20i1.5736

Rheingantz PA, Rocha MMT, França AJGL, Silva LMG (2009) Observando a qualidade do lugar: procedimentos para a avaliação pós-ocupação [Observing place quality: procedures for post-occupancy evaluation]. Coleção PROARQ/FAU/UFRJ, Rio de Janeiro.

Saarimaa B, Pelsmakers S (2020) Better Living Environment Today, More Adaptable Tomorrow? Comparative Analysis of Finnish Apartment Buildings and their Adaptable Scenarios. https://doi.org/10.33357/ys.89676.

Saft JB (2021) Qualidade ambiental na gestão de áreas de guarda de acervos em papel em edifícios históricos na cidade de São Paulo [Environmental quality in the management of archive storage areas in historical buildings in São Paulo]. PhD thesis, Universidade de São Paulo.

Saleh JH, Mark G, Jordan NC (2009) Flexibility: A multi-disciplinary literature review and a research agenda for designing flexible engineering systems. Journal of Engineering Design 20:307–323. https://doi.org/10.1080/09544820701870813.

Schmidt III R, Austin S (2016) Adaptable Architecture: theory and practice. https://doi.org/10.4324/9781315722931.

Schmidt R III, Eguchi T, Austin S, Gibb A (2010) What is the meaning of adaptability in the building industry? In: Proceedings of the 16th International Conference of Open and Sustainable Buildings, pp 233–242.

Spiten TK, Haddadi A, Støre-Valen M, Lohne J (2016) Enhancing value for end users-A case study of end-user involvement. IGLC 2016—24th Annual Conference of the International Group for Lean Construction 153–162.

Strelets K, Perlova E, Platonova M, Pankova A, Romero M, Al-Shabab MS (2016) Post occupancy evaluation (POE) and energy conservation opportunities (ECOs) study for three facilities in SPbPU in Saint Petersburg. Procedia Engineering 165:1568–1578. https://doi.org/10.1016/j.proeng.2016.11.895.

Tarpio J, Huuhka S (2022) Residents' views on adaptable housing: a virtual reality-based study. Buildings and Cities 3(1):93–110. https://doi.org/10.5334/bc.184

United Nations Brazil (n.d.) Objetivo de Desenvolvimento Sustentável 11: Cidades e comunidades sustentáveis [Sustainable Development Goal 11: Sustainable cities and communities]. https://brasil.un.org/pt-br/sdgs/11.

Villa SB, Vasconcellos PB, Bortoli KCR de, Araújo LB de (2022) Lack of adaptability in Brazilian social housing: impacts on residents. Buildings and Cities 3:376–397. https://doi.org/10.5334/bc.180.

Villa SB, Saramago RCP, Garcia LC (2015) Avaliação pós-ocupação no Programa Minha Casa Minha Vida: uma experiência metodológica [Post-occupancy evaluation in the Minha Casa Minha Vida program: a methodological experience]. UFU/PROEX, Uberlândia.

Villa SB, Vasconcellos PB, Araújo LB, Bortoli KCR (2017) Método de análise da resiliência e adaptabilidade em conjuntos habitacionais sociais através da avaliação pós-ocupação e coprodução [Method for analyzing resilience and adaptability in social housing complexes through post-occupancy evaluation and co-production]. Research Report, Federal University of Uberlândia and University of Sheffield, 393 p.

Villa SB (2008) Morar em apartamentos: a produção dos espaços privados e semi-privados nos edifícios ofertados pelo mercado imobiliário no século XXI em São Paulo e seus impactos na cidade de Ribeirão Preto. Critérios para avaliação pós-ocupação [Living in apartments: the production of private and semi-private spaces in 21st century São Paulo residential buildings and their impacts in Ribeirão Preto. Post-occupancy evaluation criteria]. PhD thesis, Universidade de São Paulo.

Voordt TJM, Wegen HBR (2013) Arquitetura sob o olhar do usuário: programa de necessidades, projeto e avaliação de edificações [Architecture through the eyes of the user: requirements program, design, and building evaluation]. Oficina de Textos, São Paulo.

Wang D, Wang X, Gao P (2022) Post-occupancy evaluation of the built environment of college campuses based on students' daily behavior—Evidence from Zijing Road of UJN. Advances in Education, Humanities and Social Science Research 1:277. https://doi.org/10.56028/aehssr.1.1.277.

Whittem V, Roetzel A, Sadick AM, Nakai Kidd A (2022) How comprehensive is post-occupancy feedback on school buildings for architects? A conceptual review based upon Integral Sustainable Design principles. Build Environ 218. https://doi.org/10.1016/j.buildenv.2022.109109.

Zhou H, Zhao Y, Zhang Z, et al (2022) Post occupancy investigation of 40 certified green buildings in Beijing: Results, lessons and policy suggestions. Journal of Building Engineering 60:105153. https://doi.org/10.1016/j.jobe.2022.105153.

Gláucia Trindade Pereira Architect and Urban Planner at the Federal University of Uberlândia (UFU). Holds a Master's degree in Architecture and Urbanism at UFU. Holds a degree in Architecture and Urbanism from UFU (2003) and a postgraduate specialization in Higher Education Teaching (2009). Worked as an Architect and Urban Planner (2003–2006) at Gaia Engenharia e Assessoria Ambiental. Served as Coordinator of the Project Division (2015–2018), Special Advisor (2018–2020), and Director of Projects and Budgets (2020–2024) at the University Administration Office of UFU. Research interests include: Post-Occupancy Evaluation, Building Performance, Accessibility, and Public Management https://lattes.cnpq.br/1032616277543988.

Simone Barbosa Villa Architect and Urban Planner, Full Professor at the Faculty of Architecture, Urbanism, and Design of the Federal University of Uberlândia (FAUED/UFU). Postdoctoral researcher and Visiting Scholar at the University of Cambridge, Department of Architecture. PhD from the Faculty of Architecture and Urbanism at the University of São Paulo. Coordinator of [MORA] Housing Research Group—FAUED/UFU. Faculty member of the Graduate Program in Architecture and Urbanism at FAUED/UFU. Architecture and Urbanism Section Editor for the journal Ambiente Construído (2017–2021). CNPq Productivity Fellowship Holder (PQ2) since 2019. Member of the Advisory Committee of FAPEMIG—Architecture and Engineering Chamber (2020–2023). Author of the books Housing Environmental Quality: Post-Occupancy Evaluation, Post-Occupancy Evaluation in Architecture, Urbanism and Design: from Theory to Practice, and Living in Apartments. Research interests include: Housing, Post-Occupancy Evaluation, Resilience, Building Performance, Design Process, and User-Centered Design http://lattes.cnpq.br/5972522149974965.

Strategic Environmental Assessment and Prospects for a Sustainable University in the Western Region of Para

Jacqueline Bailão Da Silva Lopes and Thiago Almeida Vieira

Abstract The theme of sustainable universities has been researched in several countries and their sustainability policies require strategic analysis using tools with great potential for technical-scientific exploration, such as Strategic Environmental Assessment (SEA). In this case study, the objective is to verify and analyze the strategic sustainable role of the largest federal public university in the western region of Para, in the Brazilian Amazon, through the methodological approach of SEA applied to the university's existing institutional sustainability plan. Among the SEA support techniques used are scenario building, structural analysis/key variable interaction matrix. The results obtained include the prospective sustainability scenario for the university; the main key variables of current influence and direct dependence on the plan and those in potential; the relevant, influential, and dependent social actors identified and related to each other and to the plan's projects. Thus, the elements identified through the SEA and analyzed in the results make up the strategic sustainability analysis for Ufopa and can serve stakeholders, the academic community, and the external public, as inputs for decision-making processes, and good governance. In short, the tested premise is that the SEA approach to the university's sustainability plan focuses on the social, economic, environmental, and governance aspects, prioritizing the integration and harmony between them.

Keywords Sustainable campus · Strategic analysis · Sustainability plans · Higher education

---

J. B. Da Silva Lopes (✉)
Research Society, Nature and Development Graduate Program (PPGSND) of the Federal University of Western Para (Ufopa), Santarem, Para, Brasil
e-mail: jacquelinebailao@hotmail.com

T. A. Vieira
Research and Professor Society, Nature and Development Graduate Program (PPGSND) of the Federal University of Western Para (Ufopa), Santarem, Para, Brasil
e-mail: thiago.vieira@ufopa.edu.br

W. Leal Filho et al. (eds.), *Sustainability in the Curriculum and Operations of Universities*, World Sustainability Series,
https://doi.org/10.1007/978-3-032-12678-8_40

1 Introduction

Investigations into sustainable universities (SUs) have elevated debates and research in several countries, as the interest of higher education institutions (HEIs) in contributing to achieving the 17 Sustainable Development Goals (SDGs) by 2030 also grows (UN, 2015; Avelar et al., 2019; Sanchez-Carrillo et al., 2021). Knowledge about practices and the role of universities for sustainability requires further study (Leal Filho et al., 2019a), since HEIs are key elements in the continuous and integrative process. Further study is also required on systematic and holistic approaches of education for sustainable development (SD) (Lukman & Glavic, 2007; Leal Filho et al., 2019b; Omazic & Zunk, 2021).

In addition to the academic and operational approaches that permeate the work of a SU, whether in Brazil (Brondani et al., 2014; Batista et al., 2019; Rohrich & Takahashi, 2019) or in other countries (Yuan et al., 2013; Leal Filho et al., 2015; BizerriL et al., 2018; Shawe et al., 2019; Ali & Anufriev, 2020; Xiong & Mok, 2020; Şerban et al., 2020; Amaral et al., 2020; Budihardjo et al., 2021) the governance dimension should be considered (Whitehead, 2016; Nwanekezie et al., 2021).

Governance for sustainability in HEIs should be understood here as the management of all the resources the university has, its policies and its involvement with the external community (Bauer et al., 2018; Niedlich et al., 2020). This dimension strongly influences the sustainability of the various HEI configurations (Amaral et al., 2020) and should be inserted as an important element in university models that focus on the multidimensional aspect (co-creative and integrative collaboration) and intra-organizational networks (Giesenbauer & Müller-Christ, 2020).

The aim of this study is to verify and analyze the variables in the strategic role of sustainability at a university with a prominent presence in the Brazilian Amazon, in its main areas of activity which included teaching, research, extension, management and operations, as well as governance.

This case study was based on elements of the strategic environmental assessment (SEA) methodological approach (Partidario, 2000, 2007) (European Commission, 2001a, 2021) to evaluate Ufopa sustainability plan, focusing on the university's strategic impact on local and/or regional development. SEA is an important instrument used in European countries (European Commission, 2001b, 2021) and a valuable tool for sustainability transitions that indicate the necessary adjustments to institutions and their existing governance arrangements and planning (Nwanekezie et al., 2021; Salciccia-Frezza et al., 2025; Shammi et al., 2022).

Ufopa's concern with the premises of sustainable, local, and regional development is contained in institutional macro documents that have existed since the university's creation in 2009. However, the university's sustainability plan, which has been in place since 2014, has reinforced its commitment to the legal precepts of sustainability in force in Brazil. Thus, the existence of a sustainability management instrument at Ufopa led this study to analyze this plan, via the SEA approach, to broaden knowledge about the performance of this HEI and its strategic role in the Brazilian Amazon region in the face of latent environmental issues.

Thus, two scenarios were defined for Ufopa: a pessimistic one (of non-implementation of its sustainability plan); and an optimistic one (looking ahead to Sustainable University status). The SEA is based on the optimistic scenario of implementing the university's sustainability plan and aims to ensure that the social, economic, environmental and governance aspects are integrated and harmonized.

Based on the results obtained, it should be noted that the sustainable strategic focus of the university's plan consists of decision-making criteria focused on governance/policies, teaching/research/extension, innovation, people management, communication and infrastructure/operational (water, energy, waste, transportation, works, purchases).

It is hoped that the results of this study will help shape future sustainable university models suited to the Amazonian reality, relating appropriate strategic aspects (decision criteria, variables, actors, objectives) and the suggestion of indicators to the "Sustainable Ufopa (Sustainable University)" plan, and that this university will advance in evaluations by national and international models.

The sections of this article, in addition to this introduction, consist of are followed by the theoretical framework on environmental management in Brazilian HEIs and the main theme of the SEA approach applied to HEI sustainability plans or other instruments, followed by the methodological procedures adopted, the results and discussion, and the conclusion.

2 Environmental Management in Brazilian Universities

In Brazil, since 2002 there have been reports of sustainable practices in HEIs concerned with the implementation of environmental management systems in universities (Tauchen & Brandli, 2006), sometimes seeking to adapt their campuses to a more sustainable configuration, sometimes implementing environmental practices that permeate teaching, extension, research and the operational axis (service and infrastructure) of Brazilian HEIs (Moura et al., 2019; Rohrich & Takahashi, 2019).

But in 2012 a legislative reinforcement for federal educational institutions in Brazil was launched, through Federal Decree n° 7.746/2012, seeking to regulate sustainable criteria and practices within the Brazilian Federal Public Administration, also determining the elaboration and implementation of sustainable logistics management plans (Brazil, 2012a).

It should be noted that this prerogative of the sustainable plans (SP) in 2012 in Brazil began in 1999 with a project launched by the Ministry of the Environment (MMA), called the Environmental Agenda for Public Administration (A3P), which was voluntary and sought to stimulate a change in the habits of civil servants. In 2008 it gained normative reinforcement through MMA Ordinance No. 61/2008 (Brazil, 2008). Ten years later, the A3P program was instituted in the MMA structure by another ordinance (SECEX/MMA Ordinance No. 28/2018) (Brazil, 2018). From then on, the Ministry began to promote environmental sustainability practices at the national level.

As a result, the Secretariat for Logistics and Information Technology (SLTI), following on from Decree 7.746/2012, published Normative Instruction (IN) No. 10/2012 in the same year, determining rules for the preparation of SP (Brazil, 2012b), for direct, autarchic and foundational bodies or entities and dependent state-owned companies. They now must establish sustainability practices and rationalization of spending and processes in the public administration, from the bidding processes to their operation (functioning) (Brazil, 2014).

These autarchies include all federal higher education institutions, including universities. In view of this, in 2014/2015 Ufopa drew up its SP (SP-Ufopa), which was approved by means of Consun/Ufopa Resolution No. 06/2015 (Ufopa, 2015). This plan contains actions for the adoption of sustainability criteria, and it is hoped that this instrument will be implemented effectively, both in compliance with the legal requirements of the Brazilian context and in the ideal of achieving the SDGs.

3 Strategic Environmental Assessment (SEA) and Its Application to Universities

The environmental impact assessments (EIAs) stipulated since 1969 by the US environmental agency, the National Environmental Policy Act (NEPA) (Egler, 2001), after a meeting of the Council of the European Communities in 1987, came to be strongly encouraged as essential tools for understanding, predicting, and avoiding or minimizing the negative effects of human actions or enhancing the positive effects on the environment (Council Of The European Communities, 1987, p. 9). In the basic text of this event, it was recorded that the Commission's concern would also be extended, "as soon as possible, to cover policies and policy statements, plans and their implementation, procedures, programs" (Council Of The European Communities, 1987, p. 9).

For this reason, the need for methodologies to analyze more strategic instruments such as policies, plans, or programs has led some scholars to highlight the importance of adopting SEA. According to Wood and Djeddour (1992), environmental assessments of policies, plans and programs (PPPs) are more strategic in nature than those applicable to environmental impact assessment (EIA) projects and can differ in several important aspects, including methods and procedures, since the application of SEA must meet the needs of the particular circumstances of each context (Brown & Thérivel, 2000).

As important elements for structuring an SEA, Partidario (2000) highlights four constructive components (3WH) from "Why", to "Who", "What" and "How".

According to IAIA (2002), the criteria for a well-performing SEA process should be integrated, sustainability-oriented, focused on key SD issues, accountable, participatory, and interactive, with repetition to refine the process. Partidario (2013) also recognized these criteria to develop the strategic thinking model, which seeks to promote mental changes, creating a strategic culture in the decision-making process, with institutional cooperation and dialogue, and seeking to avoid conflicts.

From a broader approach to effects and alternatives, the benefits of applying SEA can be seen to be a more systematic approach to the environment, in which it allows consultation and participation to support decision-making towards SD (Fischer, 1999; Tetlow & Hanusch, 2012; Van Doren et al., 2013). And when the costs of the assessment are verified, the benefits of an SEA can outweigh them, by several orders of magnitude and over the long term, for the public and the environment (Chaker, 2006; Therivel & Gonzalez, 2020).

The SEA is a concept with multiple forms that aims to support sustainable decision-making in strategic instruments and not to replace the EIA (Fig. 1). On the contrary, its function is to reinforce the EIA, and the sooner the SEA is used, the more effective its use becomes due to its strategic purpose (Verheem & Tonk, 2000; Partidario, 2000, 2013).

SEA should be integrated and include biophysical, social, economic and institutional issues (Partidario, 2013). Governance is an element to be considered in this process to bring engagement, learning, and dialogues (Monteiro & Partidario, 2017; Baresi et al., 2020), and also overcome sustainability challenges that are complex (Nwanekezie et al., 2021), as each stakeholders tends to have unique expectations, e.g. the Government is more concerned with process; industry with costs, efficiency and certainty; conservation agencies with data (technical rationale) and the public with self-interest and local issues (Cape et al., 2018). Hence, the legitimacy of the SEA process will occur with all stakeholders agreeing that the outcome is fair and acceptable to them (Pope et al., 2018).

Although the understanding of policy, plan, or program may vary in each country, PPPs "can be seen as tools for future planning and for the allocation and distribution of public resources" (Wood & Dejeddour, 1992, p. 6). Sadler and Verheem (1996)

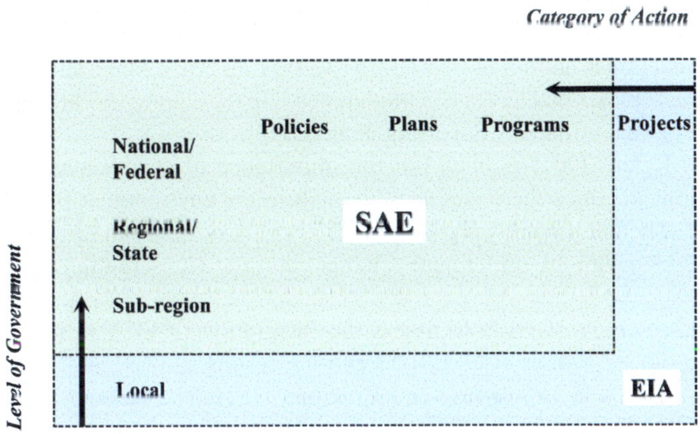

Fig. 1 Applications of environmental impact studies: strategic environmental assessment (SEA) and environmental impact assessment (EIA). *Source* Adapted from Wood and Dejeddour (1992) and Partidario (2000)

define "policy" as a general direction proposed by a government that guides decision-making. The "plan" would be a forward-looking strategy with coordinated priorities, options and measures that elaborate and implement policies, and "program" as a coherent and organized agenda of commitments, instruments, and/or activities that elaborate and implement the plan/policy.

Policies shape subsequent plans, programs and projects and there is a hierarchy of levels in the decision-making of these instruments. The policy is at the top, and there is a difference in the nature of environmental assessment, since at the policy level the assessment deals with more flexible proposals and a wider range of scenarios, the assessment at other levels is gaining more well-defined and ordered specifications (OECD, 2012).

SEA applied to these instruments (PPPs) can better support their formulation or development (Sánchez, 2017), either from a top-down perspective in which SEA principles are incorporated into the formulation of policies and plans, or from a bottom-up perspective in which SEA is applied to plans and programs as an extension of EIA, since the latter has a limited and specific focus and lacks coverage with regard to decisions made at a broader level (Partidario, 2000).

Since 2012, the application of SEA has been reported in at least 60 countries, formally/legally or informally (Tetlow & Hanusch, 2012), and its effectiveness has been pointed out in Portugal, although its potential needs to be further explored (Partidario & Monteiro, 2019). In Germany SEA is applied at all levels of governance and to various types of plans and programs, many of them at the local level (Geißler et al., 2019). In Poland (Tokarczyk-Dorociak et al., 2019) the levels of application are also at local, regional, and national scales, but their effectiveness must be improved to assist in the development of plans/programs, as identified in the Czech Republic (Musil & Smutný, 2019).

In Brazil, EIA has been regulated by law since 1981 and is one of the instruments of the National Environmental Policy (Brazil, 1981), but SEA has not regulated yet. Despite initiatives to establish SEA at the federal level as an important tool for impact studies in the country (Pellin et al., 2011; Almeida et al., 2015), this advance in environmental issues has not been realized (Malvestio, 2013; Malvestio & Montaño, 2019), and the studies that have been implemented in the country so far are voluntary, public, and private initiatives (Oberling et al., 2013).

In 2001, Egler (2001) pointed out the importance of establishing SEA in the Brazilian context, since there is a need to analyze the environment from a broader perspective, whether globally, regionally, locally or sectoral, as indicated in Agenda 21, and the Conventions on Climate Change, and the Conservation of Biological Diversity.

In 2015, a survey of HEIs to assess the integration of SEA in their curricula was carried out by Ramos et al. (2015) and the results showed an unconsolidated integration of SEA in Portuguese and Brazilian HEIs. The authors point out that integrating SEA into academic curricula can lead to significant improvement in higher education practices for SD, as well as the potential to explore it in other areas of research and related approaches.

A study conducted by Siqueira-Gay & Gallardo (2025) in Brazil addresses the importance of teaching SEA in higher education institutions, as well as the difficulty students have in understanding the value of this tool and the lack of national regulation, standardization, and a slow evolution, as cited by Gallardo et al. (2021) in their academic research on SEA for the period from 2000 to 2020.

Other studies on SEA in higher education institutions surveyed are related to teaching and learning in some universities, but no studies have been found that used SEA to verify the strategic role of HEIs, either in the Brazilian (Almeida et al., 2022; Raimundo e Almeida et al., 2024) or global context (Nielsen et al., 2023; Kågström et al., 2023; Ruggerio et al., 2024; González & Therivel, 2020). Given this situation, it is suggested that there may be major challenges in the application and the practice of SEA by managers, including within universities that have proposals focused on sustainability, as in the case of Ufopa.

The study by Budihardjo et al. (2021) on strategies in universities, although not focused on the sustainability instruments of HEIs and the SEA approach, presents important elements for investigating the strategic role of universities and the factors that influence their sustainability.

Universities are important elements in fulfilling the SDGs in the 2030 Agenda and much can be explored about their regional performance, relating it to SEA and SDG instruments (SDG objectives and targets could be incorporated into SEA), in a mutualistic relationship that can be very beneficial (Campo et al., 2020). Operationalizing them is challenging, but it is important to try.

In this sense, SEA is a favorable approach with great potential for evaluating and optimizing sustainability instruments (Tetlow & Hanusch, 2012; White & Noble, 2013) such as sectoral or regional PPPs, which makes this approach appropriate for evaluating, among other instruments, the sustainability plans of universities that operate strategically in considerable regional portions of the Brazilian territory, and where higher education is a driver of transformations in these locations (Vieira, 2017). SEA is a valuable tool for enabling sustainability transitions and is increasingly relevant due to its proactive approach to environmental issues (Nwanekezie et al., 2021; KøRNøV, 2021).

4 Methodological Procedure

This study was based on the reality of a federal public university located in the interior of the Brazilian Amazon, using an inductive approach and a descriptive case study procedure (Marconi & Lakatos, 2003; Yin, 2005).

The methodological steps adopted included a bibliographic search of national and international scientific publication databases (Google Scholar; SciELO, Science Direct, Web of Science, MDPI) to review and update the subject.

Documentary research was carried out in Ufopa's academic and administrative sectors to consult the data and information in the existing sustainability plan. In addition, interviews were conducted with university managers, which allowed the

SEA of this plan to be carried out, according to the methodological guidelines of Partidario (2013).

Ufopa has a multi-campus configuration (seven campuses) in the western part of the Brazilian Amazon. Its headquarters, the Santarem Campus, is in the municipality of Santarem—PA, and the other six campuses are in the interior of the state of Para, which is recognized for its territory with important natural resources (water resources, fauna, flora, minerals). It is in the Amazon biome and recognized for providing raw materials, products and environmental services that favor economies and societies around the world, in addition to the social, ethnic, and cultural wealth that has developed in the territory (Castro et al., 2018; Cortes & Silva Junior, 2021; Rodrigues, 2020).

In addition to the physical facilities in the seven municipalities that Ufopa has, its operations indirectly cover 20 municipalities, which corresponds to 41% of the total area of the state of Para, with a total population of almost 1 million inhabitants (Ufopa, 2019). The Ufopa is the main federal public higher education institution in the west of the state of Para (Ufopa, 2019) and one of the main ones in that part of the Brazilian Amazon. Since its creation, it has played an important role in the social and economic contexts, both locally and regionally (Gomes, 2011; Sousa, 2019; Ramalho, 2020).

Ufopa's internal community comprises around 8921 people: 7831 students (94.3% undergraduates and 5.7% postgraduates) and 1090 civil servants, including 482 teachers and 608 technical administrative staff. Its campuses offer 48 undergraduate and 21 postgraduate courses, around 460 research projects and 153 extension projects and programs. The mission is to produce and socialize knowledge, contributing to citizenship, innovation, and development in the Amazon (Ufopa, 2020).

The object of this evaluation is Ufopa's sustainability plan (SP-Ufopa). Ufopa's macro political and strategic references are contained in institutional documents (statutes, general regulations, institutional development plan, and sustainability plan) and express the understanding of vision, mission, values, and principles, decided jointly by its managers, and which are related to the principles of sustainability. From its creation in 2009 to the present day, the concern to care for the Amazon region and the SD permeates the evolution of these references and macro-policies of Ufopa (2012, 2013, 2014).

Following these guiding references, the SP-Ufopa was drawn up in 2014 by a multidisciplinary team and through public consultation with the internal community, it contains projects for the development of a sustainability policy on all seven campuses and strategic conception at the level of the Brazilian Amazon region (Ufopa, 2015).

The strategic thinking identified in the SP-Ufopa is stated in its general objective, which is to "establish standards and procedures within a project with specific programs for the implementation of a sustainability policy throughout Ufopa's management, covering the campuses" (Ufopa, 2015, p. 22), and is therefore a vision and strategy to be achieved in the long term by the university.

The six projects that make up the SP-Ufopa encompass actions and targets for rationalizing the university's spending and consumption, and a concern to create a

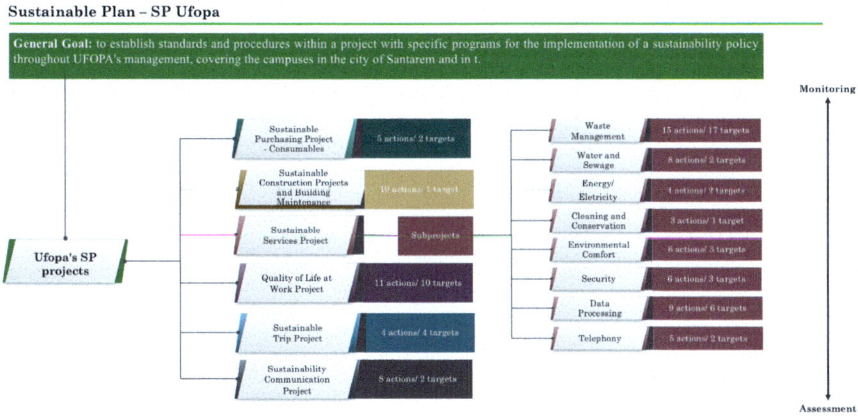

Fig. 2 Projects that are part of the sustainable plan of the Federal University of Western Para. *Source* Authors

culture of sustainability in the academic community. In addition to implementing basic infrastructure, services, and human resources projects that adopt sustainability criteria, the plan provides for continuous monitoring and evaluation of the document (Fig. 2).

To investigate sectoral problems and political priorities in terms of sustainability, as proposed by Partidario (2013), interviews were conducted directly with managers (pro-rectors, directors, and sector coordinators) at the head of Ufopa's administrative sectors (planning, administration, teaching, research, extension, people management, student management, and campus management).

The invitations informed the participants of the purpose and content of the research, and when participation was confirmed, it was up to each manager to indicate the form (online or physical), date, and time of the meeting. With the participant's permission, each interview was recorded, and the answers were recorded and written down. A total of 11 participants were interviewed, and the interview script addressed the following questions about the SP and management for sustainability:

- Are you aware of Ufopa's Sustainable Plan?
- What are the main sustainability issues facing Ufopa (barriers and threats)?
- What opportunities/potential does Ufopa have for sustainability?
- Which environmental issues among the projects foreseen in the SP-Ufopa need to be prioritized?
- Which internal (unit, sub-units, representatives) and/or external (public or private sector entities/companies, representatives) social actors do you consider important for the implementation of the SP and the sustainability process at Ufopa?

The SEA approach for SP-Ufopa is based on the strategic thinking model, as indicated by Partidario (2013), through structuring elements that are integrated and defined according to the context in which they are analyzed in this case Ufopa's SP. The author also points out that this SEA model is not even one, so the identification and constitution of all the elements will depend on the situation and objective of the desired strategic thinking, i.e., the intended sustainability strategy.

In this way, the SEA for the SP-Ufopa, as defined by Partidario (2007, 2013), started with the diagnosis of the plan in its original structure (object of evaluation), supporting the construction of two scenarios sufficient for the situation: a pessimistic scenario, i.e., when the SP would not be implemented, and an optimistic scenario, the desirable prospect that foresees the implementation of the SP for the configuration of Ufopa—Sustainable University.

The SEA of the SP-Ufopa, based on the interviews, was supported by the participants' inductive and practical notes on the university's management aspects. Through the structural analysis of the variables extracted from the SP and others suggested by the managers (variable-variable interaction matrix), the key variables for the plan were identified (Godet, 2007; Godet et al., 2008) with the support of the Micmac software from the Laboratory for Research in Foresight, Strategy and Organization (LIPSOR) (Lipsor, 2004).

5 Results

The scenarios in this case are initially described by two proposals which, in the field of possibilities, indicate a pessimistic and an optimistic possibility for the transition to sustainability at Ufopa. In this sense, the pessimistic scenario shows that the sustainability plan will not be implemented at the university over the next few years, which could lead to situations with unfavorable effects on the development of academic activities and its image, especially in the local and regional context (Table 1).

In contrast, the optimistic scenario is presented as the possibility of the plan for sustainable actions at Ufopa being implemented, in which case this is the best and most favorable scenario, which would imply an increase in the quality of higher education at this HEI and its prestige in the context in which it operates, as well as favoring the balance of the environmental (ecological, social, financial-economic) aspects managed and institutional external (governance). For this reason, scenario 2 is taken as the prospective scenario in this analysis, which is related to the satisfactory implementation of the sustainability plan at Ufopa.

The structural and interaction analysis listed 22 variables identified as priorities by the managers taking part in the survey, both those described in the SP-Ufopa and others that are not included in the plan. They were associated with ten main themes in matrix form: Consumption, Physical Structure, Basic Sanitation, Energy, People, Vegetation, Transportation, University Dissemination, University Education, and Governance (Table 2).

The interaction matrices (variable–variable) Micmac generated (MDI and MDPI) (Tables 3 and 4) describe the relationship of direct influence and potential for direct influence (MDI) between the variables (1: ComprSuste—Sustainable purchasing; 2: ObraManu—Works and Maintenance; 3: ResColeSel—Waste Management

Table 1 Pessimistic and optimistic scenarios proposed in the university's sustainability study

Proposed scenarios	
SCENARIO 1 (pessimistic) Sustainability plan not implemented	SCENARIO 2 (optimistic) Sustainability plan in implementation

• Waste of environmental resources, environmental contamination, and pollution • A demotivated academic community alienated from sustainable university objectives • Fines and sanctions imposed by environmental and other government control bodies • Poorly applied and managed financial resources • Institutional image with low prestige • Reduced levels of teaching quality at the university • Reduced positive influence on the process of sustainable local and regional development	• Rational and efficient use of natural resources, materials and inputs • Environmental management of liabilities (waste, sewage, chemical and biological products, GHG emissions, land occupation, and sealing) • Adoption of eco-efficient technological alternatives • Academic community engaged in environmental practices • Close relationship between the university and external communities (local and regional) • Consolidated partnerships with the various spheres of government and their education, environmental, productive, health, technology, and other sectors • Partnerships with NGOs and the private sector • Active participation in local environmental issues and initiatives • Participation in sustainability networks for higher education, environmental education, etc. • Balanced institutional budget and stable financial aspects • Public university as a reference in higher education for sustainability

Source Authors

and Selective Collection; 4: Agua—Water; 5: Esgoto—Sewerage; 6: EnerEle—Electricity; 7: LimpConser—Cleaning and Conservation; 8: ConforAmb—Environmental comfort; 9: Seguranca—Security; 10: ProceDados—Data processing; 11: Telefonia—Telephony; 12: Saude—Health; 13: Transporte—Transportation; 14: QualidVida—Quality of life; 15: Comunica—Communication; 16: Ensino—Teaching Actions; 17: Pesquisa—Researching Actions; 18: Extensao—Extension activities; 19: IdentCult—Cultural identity; 20: DesenReg—Regional development; 21: DesenLocal—Local development; 22: Financeiro—Financial aspects).

With this, the Micmac program identified the key variables in order of ranking according to the score calculated (Table 5). Among the 10 that exert the greatest

Table 2 Key variables for structural analysis identified for Ufopa

No.	Key variables	Acronym	Relation to SP-Ufopa	Theme
1	Sustainable purchasing	ComprSuste	Sustainable purchasing project	Consumption
2	Works and maintenance	ObraManu	Sustainable works project	Physical structure
3	Waste management and selective collection	ResColeSel	Sustainable services project	Sanitation
4	Water	Agua	Sustainable services project	Sanitation
5	Sewerage	Esgoto	Sustainable services project	Sanitation
6	Electricity	EnerEle	Sustainable services project	Energy
7	Cleaning and conservation	LimpConser	Sustainable services project	People
8	Environmental comfort	ConfAmb	Sustainable Services project	Vegetation
9	Security	Seguranca	Sustainable services Project	People
10	Data processing	ProceDados	Sustainable services project	Consumption
11	Telephony	Telefonia	Sustainable services project	Consumption
12	Health	Saude	Quality of life at work project	People
13	Transportation	Transporte	Sustainable trip project	Transportation
14	Quality of life	QualidVida	Quality of life at work project	People
15	Communication	Comunica	Communication for sustainability project	Dissemination at the university
16	Teaching actions	Ensino	Not related	University education
17	Researching actions	Pesquisa	Not related	University education
18	Extension activities	Extensao	Not related	University education
19	Cultural identity	IdentCult	Not related	People
20	Regional development	DesenReg	Not related	Governance
21	Local development	DesenLocal	Not related	Governance
22	Financial aspects	Financeiro	Not related	Governance

Source Authors

Table 3 Variable–Variable of direct influence (MDI) interaction matrix

	1: ComprSuste	2: ObraManu	3: Res Col Sel	4: Agua	5: Esgoto	6: Ener Ele	7: Limp Conser	8: Confor Amb	9: Seguranca	10: Proce Dados
1: ComprSuste	0	3	3	3	2	3	3	3	3	3
2: ObraManu	3	0	3	3	3	P	3	3	3	3
3: ResColeSel	3	3	0	1	1	P	3	2	2	2
4: Agua	3	3	2	0	3	3	3	3	0	0
5: Esgoto	3	3	3	3	0	3	3	3	0	0
6: EnerEle	3	3	P	3	3	0	3	3	3	3
7: LimpConser	3	3	3	3	3	3	0	3	3	0
8: ConforAmb	3	3	3	3	2	3	3	0	1	0
9: Seguranca	2	3	0	0	0	3	1	0	0	3
10: ProceDados	3	3	3	0	0	3	0	0	3	0
11: Telefonia	3	3	2	0	0	2	0	0	3	2
12: Saude	0	P	2	P	P	P	P	P	P	1
13: Transporte	3	3	3	3	3	1	3	P	3	P
14: QualidVida	3	3	P	3	P	3	3	3	3	P
15: Comunica	3	3	3	3	3	3	3	3	3	3
16: Ensino	3	3	3	3	3	3	3	3	3	3
17: Pesquisa	3	3	3	3	3	3	3	P	3	3
18: Extensao	3	3	3	3	3	3	3	P	3	3
19: IdentCult	P	3	0	P	0	0	0	P	0	0
20: DesenReg	3	3	2	2	P	P	0	0	0	P
21: DesenLocal	3	3	3	3	P	P	P	P	P	P
22: Financeiro	3	3	3	3	3	3	3	3	3	3

	11: Telefonia	12: Saude	13: Transporte	14: Qualid Vida	15: Comunica	16: Ensino	17: Pesquisa	18: Extensão	19: Ident Cult	20: DesenReg	21: Desen Local	22: Financeiro
1: ComprSuste	3	2	3	2	2	3	3	3	P	P	P	3

(continued)

Table 3 (continued)

	11: Telefonia	12: Saude	13: Transporte	14: Qualid Vida	15: Comunica	16: Ensino	17: Pesquisa	18: Extensão	19: Ident Cult	20: DesenReg	21: Desen Local	22: Financeiro
2: ObraManu	2	3	2	2	2	3	3	2	P	P	3	3
3: RasColeSel	1	3	1	2	2	3	3	3	P	1	3	3
4: Agua	0	3	2	3	2	3	3	3	P	P	2	3
5: Esgoto	0	3	0	3	0	3	3	3	0	1	2	3
6: EnerEle	3	3	1	3	3	3	3	2	0	2	3	3
7: LimpConser	0	3	1	3	0	3	3	2	P	0	1	3
8: ConforAmb	0	3	0	3	1	3	3	3	P	1	2	3
9: Seguranca	3	2	3	2	P	P	P	P	P	0	2	3
10: ProceDados	3	2	1	1	3	3	3	3	0	1	2	3
11: Telefonia	0	2	2	2	3	2	2	2	0	0	1	3
12: Saude	1	0	P	3	P	3	3	3	3	P	3	3
13: Transporte	P	3	0	3	3	3	3	3	P	P	2	3
14: QualidVida	P	3	P	0	P	3	3	3	3	2	3	3
15: Comunica	3	3	3	3	0	3	3	3	3	2	3	3
16: Ensino	3	P	3	P	3	0	3	3	3	3	3	3
17: Pesquisa	3	P	3	P	3	3	0	3	3	3	3	3
18: Extensao	3	P	3	P	3	3	3	3	3	3	3	3
19: IdentCult	0	P	0	3	3	3	3	3	0	3	3	2
20: DesenReg	P	P	P	P	3	3	3	3	3	0	3	3
21: DesenLocal	P	P	P	P	3	3	3	3	3	3	0	3
22: Financeiro	3	P	3	P	3	3	3	3	0	P	3	0

Legend 0—No influence; 1—Weak; 2—Moderate influence; 3—Strong influence; P—Potential influences
Source Authors, tables generated by Micmac software © Michel Godet free software (LIPSOR, 2004)

Table 4 Variable–Variable of potential direct influence (MPDI) interaction matrix

	1: Com prSise	2: ObraManu	3: Res Col Sel	4: Agua	5: Esgoto	6: Ener Ele	7: Limp Conser	8: Confor Amb	9: Seguranca	10: Proce Dados
1: ComprSiste	0	3	3	3	2	3	3	3	3	3
2: ObraManu	3	C	3	3	3	3	3	3	3	3
3: RasColeSel	3	3	0	1	1	3	3	2	2	2
4: Agua	3	3	2	0	3	3	3	3	0	0
5: Esgoto	3	3	3	3	0	3	3	3	0	0
6: EnerEle	3	3	3	3	3	0	3	3	3	3
7: LimpConser	3	3	3	3	3	3	0	3	3	0
8: ConforAmb	3	3	3	3	2	3	3	0	1	0
9: Seguranca	2	3	0	0	0	3	1	0	0	3
10: ProceDados	3	3	3	0	0	3	0	0	3	0
11: Telefonia	3	3	2	0	0	2	0	0	3	2
12: Saude	0	3	2	3	3	3	3	3	3	1
13: Transporte	3	3	3	3	3	1	3	3	3	3
14: QualidVida	3	3	3	3	3	3	3	3	3	3
15: Comunica	3	3	3	3	3	3	3	3	3	3
16: Ensino	3	3	3	3	3	3	3	3	3	3
17: Pesquisa	3	3	3	3	3	3	3	3	3	3
18: Extensao	3	3	3	3	3	3	3	3	3	3
19: IdentCult	3	3	0	3	0	0	0	3	0	0
20: DeserReg	3	3	2	2	3	3	0	0	0	3
21: DeserLocal	3	3	3	3	3	3	3	3	3	3
22: Financeiro	3	3	3	3	3	3	3	3	3	3

	11: Telefonia	12: Saude	13: Transporte	14: Quaid Vida	15: Comunica	16: Ensino	17: Pesquisa	18: Extensão	19: Ident Cult	20: DesenReg	21: Desen Local	22: Financeiro
1: ComprSiste	3	2	3	2	2	3	3	3	3	3	3	3

(continued)

Table 4 (continued)

	11: Telefonia	12: Saude	13: Transporte	14: Qualid Vida	15: Comunica	16: Ensino	17: Pesquisa	18: Extensão	19: Ident Cult	20: DesenReg	21: Desen Local	22: Financeiro
2: ObraManu	2	3	2	2	2	3	3	2	3	3	3	3
3: RasColeSel	1	3	1	2	2	3	3	3	3	1	3	3
4: Agua	0	3	2	3	0	3	3	3	3	3	2	3
5: Esgoto	0	3	0	3	0	3	3	3	0	1	2	3
6: EnerEle	3	3	1	3	3	3	3	2	0	2	3	3
7: LimpConser	0	3	1	3	0	3	3	2	3	0	1	3
8: ConforAmb	0	3	0	3	1	3	3	3	3	1	2	3
9: Seguranca	3	2	3	2	3	3	3	3	3	0	2	3
10: ProceDados	3	2	1	1	3	3	3	3	0	1	2	3
11: Telefonia	0	2	2	2	3	2	2	2	0	0	1	3
12: Saude	1	0	3	3	3	3	3	3	3	3	3	3
13: Transporte	3	3	0	3	3	3	3	3	3	3	2	3
14: QualidVida	3	3	3	0	3	3	3	3	3	2	3	3
15: Comunica	3	3	3	3	0	3	3	3	3	2	3	3
16: Ensino	3	3	3	3	3	0	3	3	3	3	3	3
17: Pesquisa	3	3	3	3	3	3	0	3	3	3	3	3
18: Extensao	3	3	3	3	3	3	3	0	3	3	3	3
19: IdentCult	0	3	0	3	3	3	3	3	0	3	3	2
20: DesenReg	3	3	3	3	3	3	3	3	3	0	3	3
21: DesenLocal	3	3	3	3	3	3	3	3	3	3	0	3
22: Financeiro	3	3	3	3	3	3	3	3	0	3	3	0

Legend 0—No influence; 1—Weak; 2—Moderate influence; 3—Strong influence

Source Authors, tables generated by Micmac software © Michel Godet free software (LIPSOR, 2004)

direct influence are the variables linked, in that order, to communication for sustainability, teaching, research, extension, energy, sustainable works, finance, sustainable purchasing, transportation, and water.

On the other hand, in the ranking of the top 10 direct dependency variables, two have dropped off the list, which are communication and transportation (not so dependent, but very influential), and others have entered: waste/selective collection and local development. This suggests that, depending on how the projects of the direct influence variables are implemented (sustainable purchases, sustainable works, finance, communication, etc.), it could change the aspects of waste generation, thus influencing the waste management project at the university.

In the same vein, local development is a variable that suffers high feedback from the effects of the main variables of direct influence, e.g., sustainable works, research/teaching/extension actions, etc. It can boost or slow down this development, given that the socio-economic impact of Ufopa in the region was pointed out by Ramalho (2020).

When it comes to key variables with the potential to directly influence the top 10 ranking, local development and the quality of life of the university's internal community are included. Both in terms of influence and dependency, as well as potential influence, the most related key variables are teaching, research, extension, communication, finance, purchasing and sustainable works.

In this sense, the key variables related to teaching, research, extension, regional development, local development, financial aspects, and cultural identity, which do not have specific projects in the SP-Ufopa, need to be prioritized in order to meet the satisfactory transition process towards sustainability at the university, as stated by Kungl and Hess (2021). This shows that the constitutional principle in higher education in Brazil of the inseparability of teaching, research and extension still contributes to the sustainability of a federal university in the Amazon.

The map of direct influence and dependence between variables (Fig. 3) shows input variables, i.e., those with great influence and little dependence on the system: communication, transportation, environmental comfort, and quality of life.

Among the most unstable (linking) key variables, i.e., those that can change the dynamics of the system at any given moment, are communication with a focus on sustainability, teaching, research, extension, sustainable works and purchases, finance, cleaning and conservation, energy, and electricity. These key points, depending on the greater or lesser priority given to them, can influence sustainability processes and policies at Ufopa.

Waste management and selective collection, as well as local development, are the variables with the highest dependence on the positioning of the previous variables, therefore, they tend to have the greatest effects on these two segments. All the other variables in the fourth quadrant, such as cultural identity, regional development, health, data processing, telephony, security, and sewage do not show considerable effects on the system.

The direct influence graph between the key input, link, and result variables (Fig. 4) predominantly shows very strong influences between them (red lines) showing a

Table 5 Classification of the key variables identified

Ranking	Acronym	Direct influence	Acronym	Direct dependence	Acronym	Direct influence potencial	Acronym	Direct dependency potential
1	Comunica	653	Financeiro	653	Ensino	542	ObraManu	542
2	Ensino	600	Ensino	621	Pesquisa	542	Ensino	533
3	Pesquisa	569	Pesquisa	621	Extensão	542	Pesquisa	533
4	Extensão	569	ObraManu	600	DesenLocal	542	Financeiro	533
5	EnerEle	558	ComprSuste	590	QualidVida	533	ComprSuste	507
6	ObraManu	547	Extensão	590	Comunica	533	Saude	507
7	Financeiro	537	DesenLocal	526	Transporte	516	Extensão	507
8	ComprSuste	526	ResColeSel	495	Financeiro	516	EnerEle	490
9	Transporte	474	Agua	474	ComprSuste	507	QualidVida	481
10	Agua	463	EnerEle	474	ObraManu	499	ResColeSel	456
11	QualidVida	463	LimpConser	453	EnerEle	481	DesenLocal	456
12	LimpConser	453	Seguranca	442	Saude	473	Agua	438
13	ConforAmb	453	Comunica	442	DesenReg	447	Comunica	438
14	ResColeSel	442	Saude	400	Agua	430	ConforAmb	430
15	Esgoto	442	QualidVida	400	ResColeSel	413	LimpConser	421
16	ProceDados	421	Esgoto	368	LimpConser	395	Seguranca	413
17	Telefonia	358	ConforAmb	337	ConforAmb	395	IdentCult	413
18	DesenLocal	347	ProceDados	337	Esgoto	361	Esgoto	404
19	DesenReg	326	Telefonia	326	Seguranca	361	ProceDados	378
20	Seguranca	284	Transporte	326	ProceDados	344	Telefonia	370

(continued)

Table 5 (continued)

Ranking	Acronym	Direct influence	Acronym	Direct dependence	Acronym	Direct influence potencial	Acronym	Direct dependency potential
21	Saude	263	DesenReg	263	IdentCult	327	Transporte	370
22	IdentCult	242	IdentCult	252	Telefonia	292	DesenReg	370

Legend of Acronym: ComprSuste—Sustainable purchasing; ObraManu—Works and Maintenance; ResColeSel—Waste Management and Selective Collection; Agua—Water; Esgoto—Sewerage; EnerEle—Electricity; LimpConser—Cleaning and Conservation; ConforAmb—Environmental comfort; Seguranca—Security; ProceDados—Data processing; Telefonia—Telephony; Saude—Health; Transporte—Transportation; QualidVida—Quality of life; Comunica—Communication; Ensino—Teaching Actions; Pesquisa—Researching Actions; Extensao—Extension activities; IdentCult—Cultural identity; DesenReg—Regional development; DesenLocal—Local development; Financeiro—Financial aspects

Source Authors, table generated by Micmac software © Michel Godet Free software (LIPSOR, 2004)

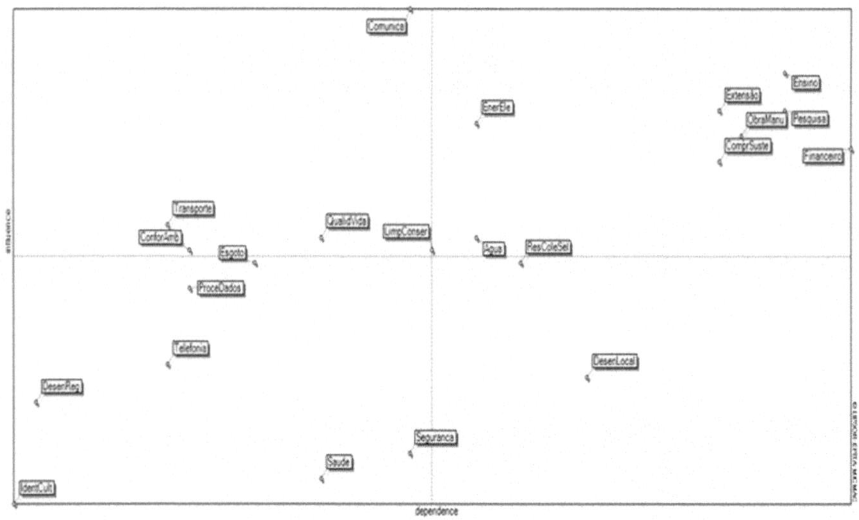

Fig. 3 Direct influence and dependence between the key variables for sustainability at Ufopa. *Legend* ComprSuste—Sustainable purchasing; ObraManu—Works and Maintenance; ResColeSel—Waste Management and Selective Collection; Agua—Water; Esgoto—Sewerage; EnerEle—Electricity; LimpConser—Cleaning and Conservation; ConforAmb—Environmental comfort; Seguranca—Security; ProceDados—Data processing; Telefonia—Telephony; Saude—Health; Transporte—Transportation; QualidVida—Quality of life; Comunica—Communication; Ensino—Teaching Actions; Pesquisa—Researching Actions; Extensao—Extension activities; IdentCult—Cultural identity; DesenReg—Regional development; DesenLocal—Local development; Financeiro—Financial aspects. *Source* Authors, figure generated by Micmac software © Michel Godet Free software (LIPSOR, 2004)

tendency and so will tend to make the whole system more dynamic as more or fewer sustainable actions and initiatives are directed towards one or another variable.

In Brazilian universities, some main themes have been identified in their mission statements, whose highest frequencies of terms were related to development, research, teaching, Training, community, promotion, dissemination, and knowledge (Deus et al., 2016).

The key variables identified in this study enable an interconnection with the variables of initiatives in HEIs studied by Shawe et al. (2019), which cover finance, curriculum, energy, green buildings, research, waste, disclosure, transportation, governance, and water, which in a broader classification are divided into education, disclosure, research, operations, and governance.

In fact, an analysis by Ramos et al. (2018) showed that among the excellent environmental predictors in development analysis are those associated with sanitation, water quality, and vegetation cover, and that these are related to factors such as education, income, quality of life, and longevity. Effective communication is also a key element in establishing more sustainable behavior (Lertpratchya et al., 2017) and a sustainability mindset (Žalėnienė & Pereira, 2021). And academic publications linked to the teaching, research, and extension variables may indicate an increase in

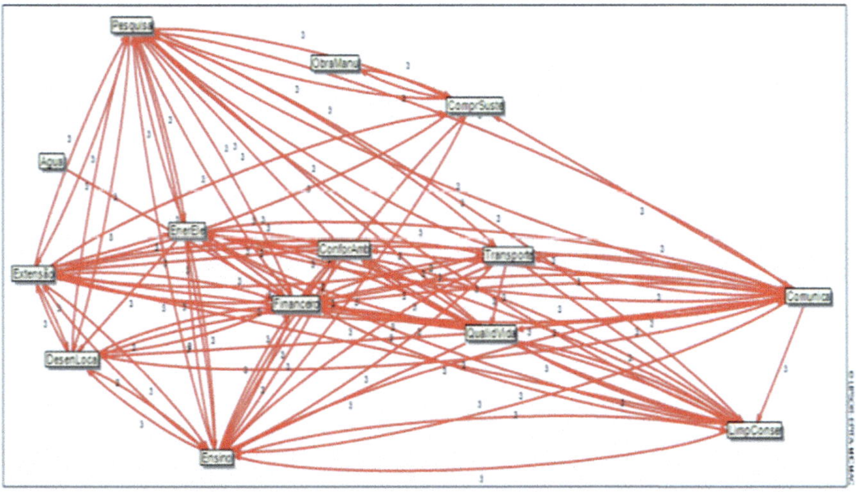

Fig. 4 Direct influence graph between the input, link and result variables. *Legend* ComprSuste—Sustainable purchasing; ObraManu—Works and Maintenance; Agua—Water; EnerEle—Electricity; LimpConser—Cleaning and Conservation; ConforAmb—Environmental comfort; Transporte—Transportation; QualidVida—Quality of life; Comunica—Communication; Ensino—Teaching Actions; Pesquisa—Researching Actions; Extensao—Extension activities; DesenLocal—Local development; Financeiro—Financial aspects. *Source* Authors, figure generated by Micmac software © Michel Godet Free software (LIPSOR, 2004)

environmental practices at HEIs (De Filippo et al., 2019; Leal Filho et al., 2021a; 2021b).

The classification between the direct influence and potential direct influence matrices shows the discrete (hidden) variables that could move up (green lines) or down (red lines) in position (Fig. 5).

In this case, the key variables of teaching, research, and extension will potentially top the ranking, followed by local development and quality of life, which could have a greater direct influence on Ufopa's sustainability. On the other hand, variables such as communication, electricity, and sustainable works will drop down the ranking, most likely to the last two, in response to the future implementation of more sustainable actions at Ufopa.

In any case, these key variables correspond to the social maturity and sustainability criteria for an HEI, as pointed out by Pizzutilo and Venezia (2021). According to these authors, this maturity encompasses three major dimensions (culture, missions and people/community) and 12 criteria: governance, strategy (sustainability policies and plans), inclusive context (location with intense and broad engagement, a criterion that favors university Extension actions), measurement (self-assessments and monitoring), curriculum (integration of courses and content in understanding sustainability, role of teaching), research, dissemination (communication for sustainability), funding (financial aspects in balance), people, consultation (active participation of

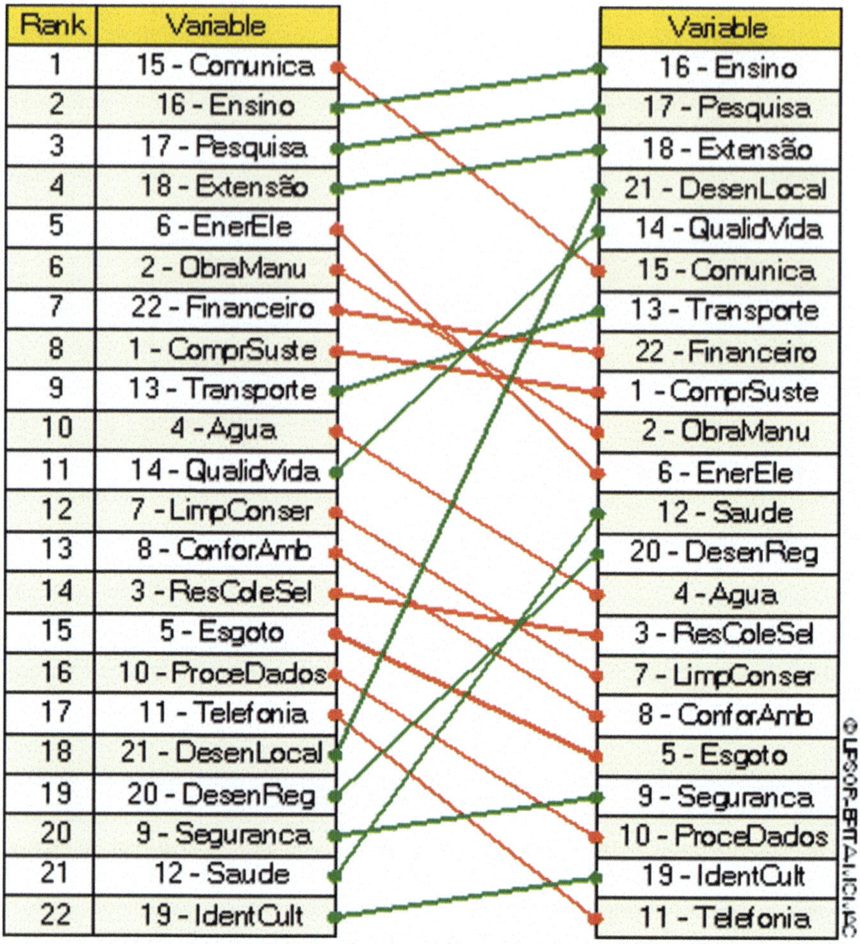

Fig. 5 Ranking of direct influence and potential direct influence matrices. *Note* Rank = position; Variable = variable. Legend: ComprSuste—Sustainable purchasing; ObraManu—Works and Maintenance; ResColeSel—Waste Management and Selective Collection; Agua—Water; Esgoto—Sewerage; EnerEle—Electricity; LimpConser—Cleaning and Conservation; ConforAmb—Environmental comfort; Seguranca—Security; ProceDados—Data processing; Telefonia—Telephony; Saude—Health; Transporte—Transportation; QualidVida—Quality of life; Comunica—Communication; Ensino—Teaching Actions; Pesquisa—Researching Actions; Extensao—Extension activities; IdentCult—Cultural identity; DesenReg—Regional development; DesenLocal—Local development; Financeiro—Financial aspects. *Source* Authors, figure generated by Micmac software © Michel Godet Free software (LIPSOR, 2004)

people), involvement (engagement of people), and co-creation (cooperation with stakeholders, the community and society).

6 Conclusions

In this university, it is confirmed that the social, economic, environmental, and governance aspects seek integration and harmony, with the due implication that the university remains focused on decision-making criteria focused on governance/policies, teaching/research/extension, people and waste management, as well as water, energy, transport, works, purchases, technological innovation, and communication, with effective prioritization of actions that focus on short- and medium-term results.

The findings allowed us to satisfactorily consider the role of SEA in guiding decision-making by stakeholders and in the system, based on factors and variables that deal with the university space in harmony, and operational issues, such as organizational culture, governance, teaching, research and extension, and innovation and technology.

The main limitation of this study was the lack of studies on the role of SEA in HEI—sustainable plans for the purposes of comparative analysis. On the databases checked, no any similar studies in Brazil or other countries were found.

Further research is needed on the sustainable strategic role of HEIs in Brazil and around the world, so that they can support analysis and decision-making, especially among HEIs that have similar contexts. In addition, it is necessary to analyze in greater depth the potential of these institutions to contribute to aspects and factors (governance, infrastructure, economy, culture, etc.) that go beyond their own organizational limits and have an impact at the local, regional, national, and even international levels.

Therefore, it is hoped that this evaluation will stimulate and contribute to the expansion of strategic thinking by decision-makers and the academic scientific communities of higher education in Brazil and around the world, and that research into the strategic role of universities in favor of sustainability will be expanded.

References

Ali EB; Anufriev, V. P. (2020). Towards environmental sustainability in Russia: evidence from green universities. Heliyon, v. 6, n. 8. https://doi.org/10.1016/j.heliyon.2020.e04719

Almeida, M. R. R., Roquetti, D. R., Martins, T. P., & Soto, N. L. C. (2015). Analysis of the Federal Proposal to Implement Strategic Environmental Assessment in Brazil. Rev. Gest. Ambient. Sustentabilidade, v. 4, n. 2, p. 18–32. https://doi.org/10.5585/geas.v4i2.292

Almeida, M. R., Veronez, F. A., & Malvestio, A. C. (2022). Teaching Environmental Impact Assessment in Brazil: is it just a 'Make-believe' approach? Ambiente & Sociedade. https://doi.org/10.1590/1809-4422asoc20210020r1vu2022L2OA

Amaral, A. R. et al. (2020). A review of empirical data of sustainability initiatives in university campus operations. Journal of Cleaner Production, v. 250. https://doi.org/10.1016/j.jclepro.2019.119558

Avelar, A. B. A.; Silva-Oliveira, K. D. D.; Pereira, R. D. S. (2019) Education for advancing the implementation of the Sustainable Development Goals: A systematic approach. The International Journal of Management Education, v. 17, n. 3. https://doi.org/10.1016/j.ijme.2019.100322

Baresi, U.; Vella, K. J.; Sipe, N. G. (2020) A limits-oriented adaptive approach for strategic environmental assessment. Environmental Science & Policy, v. 114, p. 128–139. https://doi.org/10.1016/j.envsci.2020.07.031

Batista, A. S. et al. (2019). Environmental Management in Federal Public Universities: The Appropriation of the Concept of Sustainable Development from the Environmental Agenda in Public Administration (A3P). Id on Line Rev. Mult. Psic., v. 13, n. 44, p. 276–292. https://doi.org/10.14295/idonline.v13i44.1615

Bauer, M. et al. (2018). Sustainability Governance at Universities: Using a Governance Equalizer as a Research Heuristic. Higher Education Policy, v. 31, p. 491–511. https://doi.org/10.1057/s41307-018-0104-x

BizerriL, M. et al. (2018). Sustainability in higher education: A review of contributions from Portuguese Speaking Countries. Journal Of Cleaner Production, v. 171, p. 600–612. https://doi.org/10.1016/j.jclepro.2017.10.048

Brazil. (1981). Law n. 6.938, of August 31, 1981. Provides for the National Environmental Policy, its purposes and mechanisms for formulation and application, and makes other provisions. National Environmental Policy. http://www.planalto.gov.br/ccivil_03/leis/l6938.htm

Brazil. (2008). Ministry of the Environment. Ordinance no. 61 of May 15, 2008. Establishing environmental sustainability practices to be observed by the Ministry of the Environment and its related entities when making sustainable public purchases. http://www.mma.gov.br/estruturas/a3p/_arquivos/36_09102008032817.pdf

Brazil. (2012a) Decree No. 7.746, of June 5, 2012—Regulates Article 3 of Law No. 8.666, of June 21, 1993, to establish criteria and practices for the promotion of sustainable national development, 2012. http://www.planalto.gov.br/ccivil_03/_ato2011-2014/2012/decreto/d7746.htm

Brazil. (2012b). Secretariat of Logistics and Information Technology. Normative Instruction no. 10, of November 12, 2012. Establishes rules for drawing up Sustainable Logistics Management Plans. http://pesquisa.in.gov.br/imprensa/jsp

Brazil. (2014). Secretariat of Logistics and Information Technology. Normative Instruction No. 02, of June 4, 2014. https://www.comprasgovernamentais.gov.br/index.php/legislacao/instrucoes-normativas/304-instrucao-normativa-n-2-de-04-de-junho-de-2014

Brazil. (2018). Ministry of The Environment. Ordinance n. 28, of February 19, 2018. Establishes the Environmental Agenda in Public Administration Program—A3P Program. http://www.mma.gov.br/images/arquivo/80063/Legislacao/Portaria_SECEX_n_28_-_de_19-02-18_-_Institui_o_Programa_A3P.pdf

Brondani, S. C. et al. (2014). Overview of Environment Sustainability in Top Universities in Latin América. Aidis Journal, v. 7, n. 1, p. 1–10. https://www.revistas.unam.mx/index.php/aidis/article/view/46599

Brown, A. L.; Thérivel, R. (2000). Principles to guide the development of strategic environmental assessment methodology. Impact Assessment and Project Appraisal, v. 18, n. 3, p. 183–189. https://doi.org/10.3152/147154600781767385

Budihardjo, M. A. et al. (2021). Towards Sustainability in Higher-Education Institutions: Analysis of Contributing Factors and Appropriate Strategies. Sustainability, v. 13, n. 12. https://doi.org/10.3390/su13126562

Campo, A. G. D.; Gazzola, P.; Onyango, V. (2020). The mutualism of strategic environmental assessment and sustainable development goals. Environmental Impact Assessment Review, v. 82. https://doi.org/10.1016/j.eiar.2020.106383

Cape, L. et al. (2018). Exploring pluralism—Different stakeholder views of the expected and realized value of strategic environmental assessment (SEA). Environmental Impact Assessment Review, v. 69, p. 32–41. https://doi.org/10.1016/j.eiar.2017.11.005

Castro, E. M. R. D. et al. (2018). Critical thinking on the Amazon and the debate on development. NAEA Paper, v. 27, n. 1, p. 29–45. https://doi.org/10.18542/papersnaea.v1i1.11113

Chaker, A. et al. (2006). A review of strategic environmental assessment in 12 selected countries. Environmental Impact Assessment Review, v. 26, n. 1, p. 15–56. https://doi.org/10.1016/j.eiar.2004.09.010

Cortes, J. C. & Silva Junior, R. D. (2021). The Interface Between Deforestation and Urbanization in the Brazilian Amazon. Ambient. soc. 24. https://doi.org/10.1590/1809-4422asoc20190182r1vu2021L1AO

Council Of The European Communities. (1987). Document 41987X1207—Resolution of the Council of the European Communities and of the representatives of the Governments of the Member States (1987–1992). EUR-Lex. https://eur-lex.europa.eu/legal-content/EN/TXT/?uri=CELEX%3A41987X1207

De Filippo, D. et al. (2019). Spanish Universities' Sustainability Performance and Sustainability-Related R&D+I. Sustainability, v. 11, n. 20. https://doi.org/10.3390/su11205570

Deus, R. M.; Battistelle, R. A. G.; Silva, G. H. R. D. (2016). Sustainability insights from the mission statements of leading Brazilian Universities. International Journal of Educational Management, v. 30, n. 3, p. 403–415. https://doi.org/10.1108/IJEM-05-2014-0065

Egler, P. C. G. (2001). Perspectives on the use of the Strategic Environmental Assessment process in Brazil. Strategic Partnerships, v. 6, n. 11. http://seer.cgee.org.br/index.php/parcerias_estrategicas/article/view/166

European Commission. (2001a). Directive 2001/42/EC on the assessment of the effects of certain plans and programs on the environment. https://eur-lex.europa.eu/legal-content/PT/TXT/PDF/?uri=CELEX:32001L0042&from=EN

European Commission. (2001b). EUR-Lex—32001L0042. Directive 2001/42/EC of the European Parliament and of the Council of 27 June 2001on the assessment of the effects of certain plans and programs on the environment. EUR-Lex. https://eur-lex.europa.eu/legal-content/EN/ALL/?uri=CELEX%3A32001L0042

European Commission. (2021). Strategic Environmental Assessment—SEA. https://ec.europa.eu/environment/eia/sea-legalcontext.htm

Fischer, T. B. (1999). Benefits arising from sea application—a comparative review of north west england, noord-holland, and brandenburg-berlin. Environmental Impact Assessment Review, v. 19, n. 2, p. 143–173. https://doi.org/10.1016/S0195-9255(98)00037-7

Gallardo, A. L., Machado, D. M., & Kniess, C. T. (2021). Strategic Environmental Assessment in Brazilian Academic Research. Ambiente & Sociedade. https://doi.org/10.1590/1809-4422asoc20200022r3vu2021L5AO

Geißler, G. et al. (2019). Effectiveness of strategic environmental assessment in Germany?—meta-review of SEA research in the light of effectiveness dimensions. Impact Assessment and Project Appraisal, v. 37, n. 3–4, p. 219–232. https://doi.org/10.1080/14615517.2019.1587944

Giesenbauer, B.; Müller-Christ, G. (2020). University 4.0: Promoting the Transformation of Higher Education Institutions towards Sustainable Development. Sustainability, v. 12, n. 8. https://doi.org/10.3390/su12083371

Godet, M.; Durance, P.; Dias, J. G. (2008). Strategic Foresight for Companies and Territories. La Prospective. http://www.laprospective.fr/dyn/francais/actualites/TOPOSPortugaisV190510.pdf

Godet, M. (2007). Manuel de prospective stratégique Tome 2–2ª ed. La Prospective. http://es.laprospective.fr/dyn/francais/ouvrages/la_prospective_strategique/t1-manuel-de-prospective-strategique-dunod-2007.pdf

Gomes, I. C. C. (2011). The process of creating a university for the Amazon: behind the scenes of the construction of Ufopa. Belem: Dissertation— Federal University of Para, 2011. 152 p. https://repositorio.ufpa.br/jspui/bitstream/2011/2934/1/Dissertacao_ProcessoCriacaoUniversidade.pdf

González, A., & Therivel, R. (2020). Learning support mechanisms in SEA: a review of the potential to optimize outcomes. Impact Assessment and Project Appraisal, 39(1), 51–62. https://doi.org/10.1080/14615517.2020.1830678

International Association For Impact Assessment. (2002). Strategic Environmental Assessment—Performance Criteria. https://iaia.org/uploads/pdf/sp1.pdf

Kågström, M., Kørnøv, L., & Lyhne, I. (2023). Collaboration between researchers and practitioners in environmental assessment: Introduction to the special issue. Environmental Impact Assessment Review, v. 100, https://doi.org/10.1016/j.eiar.2023.107061

KøRNøV, L. (2021). SEA as a change agent: still relevant and how to stay relevant? Impact Assessment and Project Appraisal, v. 39, n. 1, p. 63–66. https://doi.org/10.1080/14615517.2020.1830679

Kungl, G.; Hess, D. J. (2021). Sustainability transitions and strategic action fields: A literature review and discussion. Environmental Innovation and Societal Transitions, v. 38, p. 22–33. https://doi.org/10.1016/j.eist.2020.10.004

Leal Filho, W. et al. (2019a). Sustainable Development Goals and sustainability teaching at universities: Falling behind or getting ahead of the pack? Journal of Cleaner Production, v. 232, p. 285–294. https://doi.org/10.1016/j.jclepro.2019.05.309

Leal Filho, W. et al. (2019b). The role of higher education institutions in sustainability initiatives at the local level. Journal of Cleaner Production, v. 233, p. 1004–1015. https://doi.org/10.1016/j.jclepro.2019.06.059

Leal Filho, W. et al. (2021a). Mapping sustainability initiatives in higher education institutions in Latin America. Journal of Cleaner Production, v. 315. https://doi.org/10.1016/j.jclepro.2021.128093

Leal Filho, W. et al. (2021b). Trends in scientific publishing on sustainability in higher education. Journal of Cleaner Production, v. 296. https://doi.org/10.1016/j.jclepro.2021.126569

Leal Filho, W. et al. (2015). Putting sustainable development in practice: campus greening as a tool for institutional sustainability efforts. In: Leal Filho, W., et al. Sustainability in Higher Education. ed. [S.l.]: Chandos Publishing. Chap. 1, p. 1–19. https://doi.org/10.1016/B978-0-08-100367-1.00001-9

Lertpratchya, A. P. et al. (2017). Assessing the role of college as a sustainability communication channel. International Journal of Sustainability in Higher Education, v. 18, n. 7, p. 1060–1075. https://doi.org/10.1108/IJSHE-09-2016-0172

Lipsor. Laboratoire D'innovation De Prospective Stratégique Et D'organisation. (2004). Micmac Structural Analysis v. 5.1.2. La Prospective, 2004. http://en.laprospective.fr/methods-of-prospective/softwares/59-micmac.html

Lukman, R.; Glavic, P. (2007). What are the key elements of a sustainable university? Clean Technologies and Environmental Policy, v. 9, n. 2, p. 103–114. https://doi.org/10.1007/s10098-006-0070-7

Malvestio, A. C.; Montaño, M. (2019). From medicine to poison: how flexible strategic environmental assessment may be? Lessons from a non-regulated SEA system. mpact Assessment and Project Appraisal, v. 37, n. 5, p. 437–451. https://doi.org/10.1080/14615517.2019.1574390

Malvestio, A. C. (2013). Analysis of the effectiveness of strategic environmental assessment as an environmental policy tool in Brazil. https://www.teses.usp.br/teses/disponiveis/18/18139/tde-10092013-101337/en.php

Marconi, M. D. A.; Lakatos, E. M. (2003). Fundamentals of Scientific Methodology. 5. ed. Sao Paulo: Atlas.

Monteiro, M. B.; Partidário, M. R. (2017). Governance in Strategic Environmental Assessment: Lessons from the Portuguese practice. Environmental Impact Assessment Review, v. 65, p. 125–138. https://doi.org/10.1016/j.eiar.2017.04.007

Moura, M. M. C.; Frankenberger, F.; Tortato, U. (2019). Sustainability in Brazilian HEI: practices overview. International Journal of Sustainability in Higher Education, v. 20, n. 5, p. 832–841. https://doi.org/10.1108/IJSHE-01-2019-0021

Musil, M.; Smutný, M. (2019). Effectiveness of strategic environmental assessment in the Czech Republic,. Impact Assessment and Project Appraisal, v. 37, n. 3–4, p. 199–209. https://doi.org/10.1080/14615517.2019.1578482
Niedlich, S. et al. (2020). Cultures of sustainability governance in higher education institutions: A multi-case study of dimensions and implications. Higher Educ Quarterly, v. 74, n. 4, p. 373–390. https://doi.org/10.1111/hequ.12237
Nielsen, H. N., Larsen, S. V., Puibaraud, I. E., & Kørnøv, L. (2023). Strategic Environmental Assessment Curricula & Problem-Based Learning: A Sourcebook for Capacity Building. Aalborg Universitetsforlag. https://vbn.aau.dk/en/publications/strategic-environmental-assessment-curricula-amp-problem-based-le
Nwanekezie, K.; Noble, B.; Poelzer, G. (2021). Transitions-based strategic environmental assessment. Environmental Impact Assessment Review, v. 91. https://doi.org/10.1016/j.eiar.2021.106643
Oberling, D. F.; Rovere, E. L. L.; Silva, H. V. D. O. (2013). SEA making inroads in land-use planning in Brazil: The case of the Extreme South of Bahia with forestry and biofuels. Land Use Policy, v. 35, p. 341–358. https://doi.org/10.1016/j.landusepol.2013.06.012
OECD, Organization for Economic Cooperation and Development. (2012). Applying strategic environmental assessment: A guide to good practice in development cooperation. Paris: OECD Publishing. https://doi.org/10.1787/9789264026582-en
Omazic, A., & Zunk, B. M. (2021). Semi-Systematic Literature Review on Sustainability and Sustainable Development in Higher Education Institutions. Sustainability, 13 (14), https://doi.org/10.3390/su13147683
Partidario, M. D. R. (2000). Elements of SEA framework: improving the added-value of SEA. Environmental Impact Assessment Review, v. 20, n. 6, p. 647–663. https://doi.org/10.1016/S0195-9255(00)00069-X
Partidario, M. R.; Monteiro, M. B. (2019). Strategic environmental assessment effectiveness in Portugal,. Impact Assessment and Project Appraisal, v. 37, p. 247–265. https://doi.org/10.1080/14615517.2018.1558746
Partidario, M. D. R. (2007). Strategic Environmental Assessment Good Practices Guide: Methodological Guidance. http://content-ext.undp.org/aplaws_publications/1703425/SEA_guide_Portugal.pdf
Partidario, M. D. R. (2013). Portuguese Environment Agency. Best practice guide for Strategic Environmental Assessment—methodological guidelines for strategic thinking in SEA. https://apambiente.pt/sites/default/files/_SNIAMB_Avaliacao_Gestao_Ambiental/AAE/SEA_Guidance_BetterPractices.pdf
Pellin, A. et al. (2011). Strategic environmental assessment in Brazil: considerations on the role of multilateral development agencies. Engenharia Sanitaria e Ambiental, v. 16, n. 1, p. 27–36. https://doi.org/10.1590/S1413-41522011000100006
Pizzutilo, F.; Venezia, E. (2021). On the maturity of social responsibility and sustainability integration in higher education institutions: Descriptive criteria and conceptual framework. The International Journal of Management Education, v. 19, n. 3. https://doi.org/10.1016/j.ijme.2021.100515
Pope, J. Et al. (2018). Are current effectiveness criteria fit for purpose? Using a controversial strategic assessment as a test case. Environmental Impact Assessment Review, v. 70, p. 34–44. https://doi.org/10.1016/j.eiar.2018.01.004
Raimundo e Almeida, M. R., Veronez, F. A., & Malvestio, A. C. (2024). Who teaches, how they teach, and what they think about environmental impact assessment: per spectives from Brazilian university lecturers. Impact Assessment and Project Appraisal. 42(4), 323–337. https://doi.org/10.1080/14615517.2024.2396228
Ramalho, E. E. D. O. (2020). The socioeconomic impact of Ufopa's budget on the municipality of Santarem. Santarém: Dissertation (Master's in Social Sciences)—Postgraduate Program in Social Sciences, Federal University of Western Para. 146 p. https://repositorio.ufopa.edu.br/jspui/handle/123456789/528

Ramos, T. B. et al. (2015). Strategic Environmental Assessment in higher education: Portuguese and Brazilian cases. Journal of Cleaner Production, v. 106, p. 222–228. https://doi.org/10.1016/j.jclepro.2014.12.088

Ramos, S. B. et al. (2018). Prediction of Human Development from Environmental Indicators. Social Indicators Research, v. 138, p. 467–477, 2018. https://doi.org/10.1007/s11205-017-1693-2

Rodrigues, B. S. (2020). Biodiversity and development in the Amazon: a perspective from the technological paradigm of the fourth industrial revolution. Unesp, p. 116–142. https://ieei.unesp.br/index.php/IEEI_MundoeDesenvolvimento/article/view/66

Rohrich, S. S.; Takahashi, A. R. W. (2019). Environmental sustainability in Higher Education Institutions, a bibliometric study on national publications. Gestão & Produção, v. 26, n. 2. https://doi.org/10.1590/0104-530X2861-19

Ruggerio, C. A., Morales-Magaña, M., Paneque-Gálvez, J., & Suárez, F. M. (2024). Teaching–learning environmental conflicts through case studies and experiential immersion: introducing students to transdisciplinary research. Sustain Sci 19, 935–948. https://doi.org/10.1007/s11625-023-01448-4

Sadler, B.; Verheem, R. (1996). Strategic Environmental Assessment: status, challenges and future directions. The Hague: Ministry of Housing, Spatial Planning and the Environment. Netherlands, p. Report n. 53. http://www.jstor.org/stable/enviassepolimana.2.2.203

Salciccia-Frezza, D., Rodríguez-Espinosa, T., & Navarro-Pedreño, J. (2025). Integrating Environmental Sensitivity Analysis into Strategic Environmental Assessment for Sustainable Tourism Planning: A Review. Sustainability, v. 17 (12), 22 pages. https://doi.org/10.3390/su17125439

Sánchez, L. E. (2017). Why isn't strategic environmental assessment advancing in Brazil? Estudos Avançados, v. 31, p. 167–183. https://doi.org/10.1590/s0103-40142017.31890015

Sanchez-Carrillo, J. C.; Cadarso, M. A.; Tobarra, M. A. (2021). Embracing higher education leadership in sustainability: A systematic review. Journal of Cleaner Production, v. 298. https://doi.org/10.1016/j.jclepro.2021.126675

Şerban, E.-C. et al. (2020). Sustainable Universities, from Indifference to Joint Action—A Panel Data Analysis. Amfiteatru Economic, v. 22, n. 54, p. 376–390. https://doi.org/10.24818/EA/2020/54/376

Shammi, M., Halder, P. K., Tareq, S. M., Rahman, M. M., & Kabir, Z. (2022). From environmental impact assessment to strategic environmental assessment in Bangladesh: Evolution, perspective, governance and challenges. Environmental Impact Assessment Review, v. 97. https://doi.org/10.1016/j.eiar.2022.106890

Shawe, R. et al. (2019). Mapping of sustainability policies and initiatives in higher education. Environmental Science and Policy, v. 99, p. 80–88. https://doi.org/10.1016/j.envsci.2019.04.015

Siqueira-Gay, J., & Gallardo, A. L. (2025). Educating on the importance of strategic environmental assessment: an experience from Brazil. Impact Assessment and Project Appraisal. https://doi.org/10.1080/14615517.2025.2493381

Sousa, F. M. L. D. (2019). Public higher education in Santarém: from its origins to the creation of the Federal University of Western Pará (1970–2009). Santarém: Dissertation (Master's in Education)—Postgraduate Program in Education, Federal University of Western Para. 184 p. https://repositorio.ufopa.edu.br/jspui/handle/123456789/538

Tauchen, J.; Brandli, L. L. (2006). Environmental management in higher education institutions: a model for implementation on a university campus. Gestão & Produção, v. 13, n. 3, p. 503–515. https://doi.org/10.1590/S0104-530X2006000300012

Tetlow, M. F.; Hanusch, M. (2012). Strategic environmental assessment: the state of the art. Impact Assessment and Project Appraisal, v. 30, n. 1, p. 15–24. https://doi.org/10.1080/14615517.2012.666400

Therivel, R.; Gonzalez, A. (2020). Is SEA worth it? Short-term costs v. long-term benefits of strategic environmental assessment. Environmental Impact Assessment Review, v. 83. https://doi.org/10.1016/j.eiar.2020.106411

Tokarczyk-Dorociak, K. et al. (2019). Effectiveness of strategic environmental assessment in Poland. Impact Assessment and Project Appraisal, v. 37, n. 3–4, p. 279–291. https://doi.org/10.1080/14615517.2019.1601441

Ufopa. Federal University of Western Para. (2012). Institutional Development Plan (PDI) (2012–2016) of Ufopa. http://www.ufopa.edu.br/proplan/gestao-institucional/pdi/

Ufopa. Federal University of Western Para. (2013). Ufopa Statute. http://www.ufopa.edu.br/ufopa/institucional/regulamento/estatuto/

Ufopa. Federal University of Western Para. (2014). General Regulations. http://www.ufopa.edu.br/ufopa/institucional/regulamento/regimento-geral/

Ufopa. Federal University of Western Para. (2015). Resolution No. 06 of October 20, 2015—Approves the Sustainable Logistics Plan of the Federal University of Western Para. http://www2.ufopa.edu.br/ufopa/arquivo/consun/resolucoes/Resoluon6Consad20.10.15AprovaoPlanodeLogsticaSustentvel.pdf

Ufopa. Federal University of Western Para. (2019). Institutional Development Plan 2019–2023. http://www.ufopa.edu.br/pdi/

Ufopa. Federal University of Western Para. (2020). UFOPA in Numbers Base Year 2019. https://proplan.ufopa.edu.br/proplan/analisa/ufopa-em-nuemros/.

United Nations Organization. Un. (2015). The 17 Goals. Department of Economic and Social Affairs Sustainable Development. https://sdgs.un.org/goals

Van Doren, D. et al. (2013). Evaluating the substantive effectiveness of SEA: Towards a better understanding. Environmental Impact Assessment Review, v. 38, p. 120–130. https://doi.org/10.1016/j.eiar.2012.07.002

Verheem, R. A. A.; Tonk, J. A. M. N. (2000). Strategic environmental assessment: one concept, multiple forms. Impact Assessment and Project Appraisal, v. 18, n. 3, p. 177–182. https://doi.org/10.3152/147154600781767411

Vieira, D. J. (2017). Evolution Of Brazilian Higher Education In Recent Times: New Perspectives For Regional Development? In: Neto, A. M.; Castro, C. N. D.; Brandão, C. A. Regional development in Brazil: policies, strategies, and perspectives. p. 277–305. http://repositorio.ipea.gov.br/handle/11058/7450

White, L.; Noble, B. F. (2013). Strategic environmental assessment for sustainability: A review of a decade of academic research. Environmental Impact Assessment Review, v. 42, p. 60–66. https://doi.org/10.1016/j.eiar.2012.10.003

Whitehead, J. (2016). Prioritizing Sustainability Indicators: Using Materiality Analysis to Guide Sustainability Assessment and Strategy. Business Strategy and the Environment, v. 26, n. 3, p. 399–412. https://doi.org/10.1002/bse.1928

Wood, C.; Dejeddour, M. (1992) Strategic environmental assessment: EA of policies, plans and programs. Impact Assessment, v. 10, n. 1, p. 3–22. https://doi.org/10.1080/07349165.1992.9725728

Xiong, W.; Mok, K. H. (2020). Sustainability Practices of Higher Education Institutions in Hong Kong: A Case Study of a Sustainable Campus Consortium. Sustainability, v. 12. https://doi.org/10.3390/su12020452

Yin, R. K. (2005). Case Study: Planning and Methods. 3. ed. Porto Alegre: Bookman.

Yuan, X., Zuo, J., Huisingh, D. (2013). Green Universities in China—what matters? Journal of Cleaner Production, v. 61, p. 36–45. https://doi.org/10.1016/j.jclepro.2012.12.030

Žalėnienė, I.; Pereira, P. (2021). Higher Education For Sustainability: A Global Perspective. Geography and Sustainability, v. 2, n. 2, p. 99–106. https://doi.org/10.1016/j.geosus.2021.05.001

Jacqueline Bailão Da Silva Lopes She is Ph.D. in Environmental Sciences from the Federal University of Western Pará-UFOPA (2021), Master's degree (2014) and Bachelor's degree (2006) in Environmental Engineering, and also a specialist (2010) in Geoprocessing and Georeferencing. She began teaching exact sciences in 2003 at the high school level, and in 2008 at the university level in environmental subjects. She has experience in the areas of Engineering and Environmental

Sciences, working as a consultant, technician, and researcher on Environmental Planning and Management (Water Resources, Solid Waste and Environmental Sanitation, Energy Resources), Environmental Impact Assessment, Environmental Licensing, Geosciences, working in public and private organizations, in urban and rural areas.

Thiago Almeida Vieira He holds a postdoctoral degree from the Research Centre for Tourism, Sustainability, and Well-being (CinTurs) at the University of Algarve (UAlg). He is an associate professor at the Federal University of Western Pará, where he teaches in the bachelor's degree programs in Agronomy, Forest Engineering, and Agricultural Sciences. He is currently a permanent professor in the Master's program in Society, Environment, and Quality of Life (PPGSAQ/Ufopa), the Doctorate program in Society, Nature, and Development (PPGSND/Ufopa), and the Master's program in Forest Science, Technology, and Innovation (PPGCTIF/Ufopa). He researches quality of life, sustainability, agroecology, and socioeconomic aspects of rural and urban populations.

Promoting Circular Economy in Undergraduate Laboratories: Biomass-Based Filters for Metal Ion Recovery and Waste Valorization

Matheus L. M. Amorim, Vera R. L. Constantino, and Denise F. S. Petri

Abstract This chapter examines various approaches that can be employed to pursue and implement sustainable strategies in undergraduate laboratories, promoting sustainability and a circular economy.

Keyword Undergraduate laboratory · Metal ion waste · Waste valorization

1 Introduction

The circular economy (CE) is emerging as a key approach to sustainable development, with a significant focus on the private sector's role (Stahel 2016). However, higher education lecturers also play a crucial part, though their contributions to CE education remain underexplored (Kirchherr and Piscicelli 2019). Promoting sustainability and CE practices in undergraduate laboratories is crucial not only for reducing waste and environmental impact but also for enhancing the educational experience (Zuin et al. 2019). These practices may include (i) implementing strategies to minimize chemical and material waste through careful planning and efficient use of resources, (ii) developing experiments that are designed to recover or repurpose reagents, thereby reducing reliance on single-use materials, and (iii) fostering a mindset of sustainability and innovation among students by involving them in projects that address real-world environmental challenges. However, implementing these strategies can be challenging as it requires changes in habits and concerted efforts from everyone involved, including students, professors, and laboratory technicians.

M. L. M. Amorim · V. R. L. Constantino · D. F. S. Petri (✉)
Institute of Chemistry, University of São Paulo, São Paulo, Brazil
e-mail: dfsp@iq.usp.br

M. L. M. Amorim
e-mail: matheusmedian@usp.br

V. R. L. Constantino
e-mail: vrlconst@iq.usp.br

W. Leal Filho et al. (eds.), *Sustainability in the Curriculum and Operations of Universities*, World Sustainability Series,
https://doi.org/10.1007/978-3-032-12678-8_41

Waste valorization is crucial for learning and applying principles of the circular economy. Numerous studies have introduced laboratory activities to develop relevant skills, including the valorization of electronic waste (Liu et al. 2023; Gulliani et al 2023), food industry byproducts (Austen et al. 2023), biomass waste (Mackenzie et al. 2019), and plastics (Hudson et al. 2015). These efforts underscore the essential scientific and technical knowledge that students acquire for implementing sustainable processes.

While studies outline liquid waste disposal risks and best practices for laboratory managers (Lara et al. 2017; Ho and Chen 2018; Yang et al. 2022), little research explores its management and the recovery of valuable materials (Furtado et al. 2022; Rai 2023; Braga et al. 2024; Callisaya et al. 2024). In undergraduate Chemistry courses, experiments on acid–base equilibrium, precipitation/solubilization equilibrium, complexation equilibrium, and electrochemistry are commonly conducted in general chemistry laboratory classes. These experiments inherently generate large volumes of aqueous waste containing metal ions. The waste is typically separated into acidic, basic, and metal ion-containing fractions, often comprising Ca^{2+}, Mg^{2+}, Mn^{2+}, Fe^{3+}, Co^{2+}, Ni^{2+}, Zn^{2+}, Pb^{2+}, Cu^{2+}, and chromate ions. Managing such multimetal waste requires careful handling and the development of effective treatment strategies. Traditional methods typically involve precipitating metal ions as hydroxides, oxides, or sulfides, followed by the evaporation of the supernatant and the disposal of the resulting solids in hazardous waste landfills. However, these methods pose significant environmental challenges, particularly for highly toxic metals like lead, which require strict handling protocols to prevent contamination.

On the other hand, some naturally occurring chemical elements are considered endangered due to their rapid depletion by human consumption. For example, lithium (Li), used in batteries, as well as nickel (Ni) and copper (Cu), which are essential for electronic devices, have limited availability. This highlights the urgent need to recycle these elements to ensure their sustainable use and management. Several approaches can be relevant, such as passing the waste through ion-exchange resins that selectively adsorb metal ions for later recovery or disposal. Electrocoagulation, which utilizes sacrificial electrodes to destabilize metal ions and form insoluble hydroxides or oxides, is another effective method. Additionally, metals can be recovered by plating them onto an electrode, a technique particularly effective for Cu^{2+}, Ni^{2+}, and Zn^{2+}. Adsorption methods can also be employed to reduce metal ion concentrations in the aqueous medium and desorb them from the adsorbent for subsequent recovery and reuse.

This study examines alternative strategies for recovering economically valuable metal ions, including copper and nickel, from undergraduate laboratory waste using biomass-based filtration systems. Involving students in designing such adsorbents encourages them to consider both the waste treatment process and the optimization of reagent usage during experiments. Recovery strategies for precious metal ions include ion desorption or thermal treatment to reduce them to their metallic forms, allowing for their reuse in subsequent experiments. These straightforward approaches help introduce students to the concept of a circular economy.

2 Methodology

The production of biomass-based adsorbents for laboratory waste treatment involved the combination of microparticles of sugarcane bagasse (SB) (Furtado et al. 2022; Braga et al. 2024) or *Moringa oleifera* seed husks (MS) (Callisaya et al. 2024) with carboxymethyl cellulose (CMC). The precursor hydrogels, containing 86 wt% SB or 34 wt% MS, CMC, and crosslinkers, were freeze-dried to form cryogels. These cryogels were subsequently heat-treated at 165 °C for 7 min to induce crosslinking. This thermal treatment imparted chemical stability across a pH range of 1.0 to 9.0, as well as high mechanical and dimensional stability. The biomass-based adsorbents (cryogels) presented apparent density (ρ_{ap}) ranging from (70 ± 2) (kg m^{-3}) to (157 ± 3) (kg m^{-3}), porosity larger than 90%, and swelling degree on the order of 20 $g_{water}/g_{cryogel}$.

3 Results: Harnessing Waste Synergies for Sustainability

Utilizing abundant biomass waste to develop adsorbents for treating multi-metal laboratory waste presents a sustainable solution by repurposing two waste streams for environmental benefit. In Brazil, sugarcane bagasse is a plentiful fibrous by product left over after extracting juice from sugarcane. It consists mainly of cellulose, hemicellulose, and lignin, making it a valuable material for biofuel production (Agarwal et al. 2023), paper manufacturing (Azmi and Othman 2023), and adsorbents for wastewater treatment (Furtado et al. 2022; Novaes et al 2022; Ezeonuegbu et al. 2021; Nunes et al. 2023; Braga et al. 2024).

At our institution, we offer a variety of practical classes in both General Chemistry and Analytical Chemistry. During these sessions, students generate waste containing metal ions such as Ca^{2+}, Mg^{2+}, Mn^{2+}, Fe^{3+}, Co^{2+}, Ni^{2+}, Zn^{2+}, Pb^{2+}, Cu^{2+}, and chromate, which are disposed together in the "metal ions" waste container. To prevent precipitation during experiments, they adjust the solution's pH to 1–2 by adding HCl or HNO_3 solutions. By the end of the semester, this process results in the accumulation of plastic canisters filled with an acidic multi-metal mixture. The standard procedure involves adding NaOH solution to precipitate most metal ions as hydroxides or oxides. The resulting slurry then settles at the bottom of the container. The container is left open to allow the complete evaporation of the supernatant, after which the remaining solid residue is sent to an appropriate landfill for metal disposal. Although this is a low-cost procedure, endangered elements such as Mn, Cr, Co, Ni, Zn, Pb, and Cu are permanently buried and do not re-enter any recycling or recovery process.

Polycations can be used to treat municipal wastewater because they promote the flocculation of negatively charged inorganic and organic particles. Many contaminants in wastewater and sewage, such as suspended solids, colloidal particles, and organic matter, carry a negative charge. Polycationic flocculants neutralize these charges, thereby enhancing aggregation and facilitating sedimentation or filtration (Chen et al. 2024). Poly(diallyl dimethyl ammonium chloride) (PDDA) is a polycation commonly used for flocculation (Verezhnikov et al. 2002).

Combining adsorption and flocculation processes enhances efficiency compared to employing either method alone. In the case of laboratory waste, where HCl solution is used to adjust the pH of the medium, the metal ions can exist as negatively charged metal-chloride complexes, such as $[MeCl_3]^-$ and $[MeCl_4]^{2-}$ (Lommelen et al. 2019). Furtado et al. (2022) employed an innovative method to treat laboratory wastewater containing multiple metal ions, as illustrated in Fig. 1. The process involved filtering the waste through columns packed with filters composed of carboxymethyl cellulose (CMC) and sugarcane bagasse (SB) microparticles, a process repeated three times consecutively. After the third filtration, the eluate underwent coagulation using an appropriate amount of PDDA. This combined approach resulted in the supernatant containing less than 10% of the initial metal concentration, with lead (Pb) and mercury (Hg) levels reduced to below 2% of their original concentrations. The treatment method's success lies in its ability to concentrate metal ions from a large initial volume into a significantly smaller volume after treatment. This concentration facilitates the more efficient removal and potential recovery of metals such as lead (Pb) and mercury (Hg).

The strategy depicted in Fig. 1 is noteworthy because it not only reduces the volume of hazardous waste but also facilitates the recycling of valuable metals. In this case, it is necessary to desorb the metal ions by rinsing the adsorbents with a small volume of 1.0 mol L^{-1} HCl. Afterward, the filters can be reused, and the

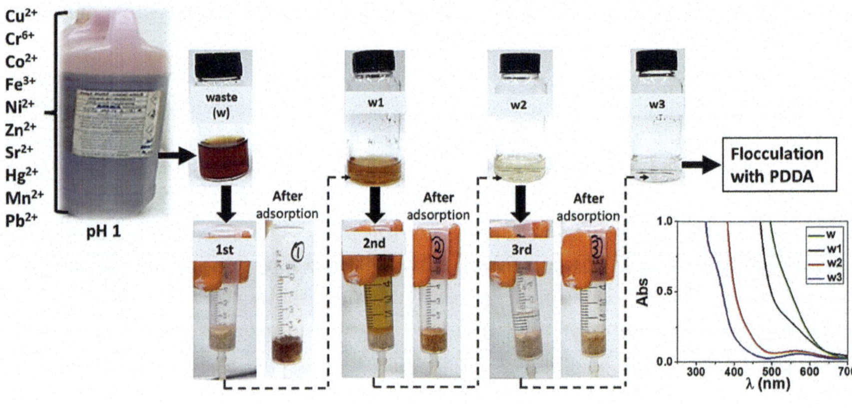

Fig. 1 Photograph of a typical undergraduate laboratory waste (w) containing different metal ions at pH 1. The waste was injected through columns packed with filters composed of CMC/SB three times consecutively. After each filtration, the generated eluate was injected into a fresh filter, yielding w1, w2, and w3, which had much lower absorbance. The remaining metal ions flocculated with PDDA, leaving supernatants with less than 10% of the initial metal concentration, and with lead (Pb) and mercury (Hg) levels lower than 2% of their original concentrations. Reprinted from Furtado et al. (2022) in compliance with Creative Commons CC BY-NC-ND 4.0

eluate will contain a concentrated mixture of multiple metal ions. The advantage of this approach is that the volume of waste to dispose of is significantly reduced. For example, adsorbents prepared with CMC and microparticles derived from *Moringa oleifera* seed husks (MS), a type of biomass, were effective in treating undergraduate laboratory waste (Callisaya et al. 2024). First, aqueous ammonia solution, instead of NaOH, was added to the waste, leading to the precipitation of metal ions as hydroxides or oxides, except for Ni^{2+} and Cu^{2+} ions, which formed soluble complexes, $[Cu(NH_3)_4(H_2O)_2]^{2+}$ (Velásquez-Yévenes and Ram 2022) and $[Ni(NH_3)_4(H_2O)_2]^{2+}$ (Varadwaj et al. 2008). The supernatant was then passed through columns packed with CMC/MS filters, which retained 80% of Cu^{2+} ions. The metal ions were desorbed using a small volume of HCl solution and could be reused up to five times. The recovered Cu^{2+} ions were suitable for other applications or experiments, promoting the circular use of this metal. Notably, the selection of functional groups on the surface of the adsorbents is crucial. CMC chains contain numerous carboxylate and hydroxyl groups, which serve as binding sites for metal ion complexes through electrostatic and ion–dipole interactions, respectively. Figure 2 illustrates the entire process.

Another strategy to treat the undergraduate laboratory waste was based on the partial flocculation of metal ions with PDDA at pH 1 due to the electrostatic interactions between the positively charged PDDA chains and the $[MeCl_3]^-$ and $[MeCl_4]^{2-}$ complexes, followed by the neutralization with aqueous ammonia solution to promote the formation of $[Cu(NH_3)_4(H_2O)_2]^{2+}$ (Velásquez-Yévenes and Ram 2022) and $[Ni(NH_3)_4(H_2O)_2]^{2+}$ (Varadwaj et al. 2008) soluble complexes (Braga et al. 2024), as shown in Fig. 3. The recovery of copper and nickel from the supernatant was achieved through column adsorption, using adsorbents composed either of CMC/SB or chitosan/SB (Braga et al. 2024). The hazardous Pb^{2+} and Hg^{2+}, which had been precipitated as oxides, were further immobilized in cement pastes. This is

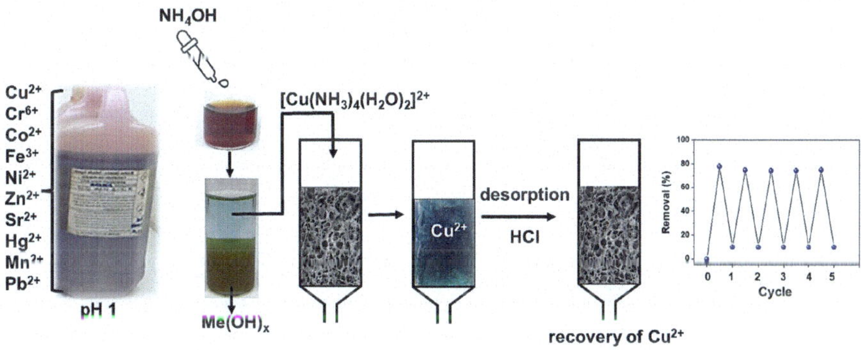

Fig. 2 Photograph of a typical undergraduate laboratory waste containing different metal ions at pH 1. First, aqueous ammonia solution was added to the waste to promote the precipitation of metal ions as hydroxides or oxides and the formation of $[Cu(NH_3)_4(H_2O)_2]^{2+}$ in the supernatant. The supernatant was added to the column packed with CMC/MS adsorbent, which retained 80% of Cu^{2+} ions. The adsorption efficiency remained up to five adsorption/desorption cycles. Reprinted from Callisaya et al. (2024) in compliance with Creative Commons CC BY-NC-ND 4.0

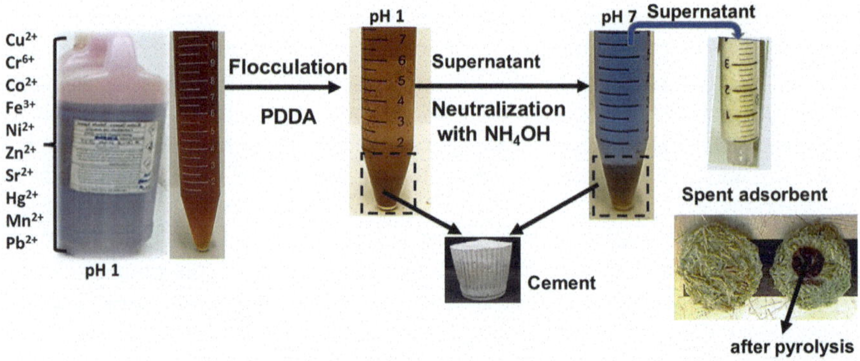

Fig. 3 Photograph of a typical undergraduate laboratory waste containing different metal ions at pH 1. First, PDDA was added to the waste to promote the flocculation of metal ions. The supernatant was then neutralized with aqueous ammonia solution, resulting in the precipitation of metal hydroxides or oxides, while $[Cu(NH_3)_4(H_2O)_2]^{2+}$ was formed in the supernatant. This supernatant was subsequently passed through a column packed with CMC/SB or chitosan/SB adsorbents, which retained 97% of Cu^{2+} ions. The flocculate and precipitate were immobilized in cement pastes, and the spent adsorbents were pyrolyzed using a CO_2 laser. Reprinted from Braga et al. (2024) in compliance with Creative Commons CC BY-NC-ND 4.0

a safe method for avoiding exposure to hazardous metal ions, and no leaching was observed (Braga et al. 2024; Nunes et al. 2023). As an alternative to desorption, the spent adsorbents were pyrolyzed using a CO_2 laser, a method previously employed to create conductive surfaces (Sousa et al. 2024). Preliminary results showed that the electrical resistance decreased from $1 \times 10^6\ \Omega$ m (before pyrolysis) to $2 \times 10^5\ \Omega$ m (after pyrolysis), indicating the potential of this strategy to enhance the conductive properties of copper and nickel elements recovered from laboratory waste.

The strategies outlined above utilize biomass, including sugarcane bagasse and *Moringa oleifera* seed husks. This approach not only facilitates the scalability and feasibility of producing adsorbents but also offers a promising, cost-effective solution for treating complex matrices and recovering high-value metals, such as copper, from undergraduate laboratory waste.

4 Transforming Aluminum Cans and Poly(Ethylene Terephthalate) Bottles into Metal–Organic Frameworks

In Brazil, the reuse of packaging is a vital sector of the economy, ranging from small beverage businesses to large packaging industries. Materials such as glass, aluminum, and cardboard not only generate income for workers but have also driven improvements in working conditions through the formation of cooperatives and the application of recycling technologies. A notable example is aluminum can recycling, where Brazil achieved a 99% rate in 2021, recycling approximately 33 billion cans (https://www.gov.br/pt-br/noticias/meio-ambiente-e-clima/2022/04/indice-de-reciclagem-de-latas-de-aluminio-chega-a-99-e-brasil-se-destaca-como-recordista-mundial). This success in the circular economy is mainly due to the assignment of economic value to waste and the development of practices that enable its reuse.

One such practice is selective waste collection, which is encouraged worldwide. Color-coded bins help separate dry from wet waste, facilitating material reuse. However, this approach is not yet widely applied in chemical laboratories, particularly in educational ones, where waste from wet analysis is often only categorized as "organic" or "metals." This rudimentary classification raises concerns about the recovery of materials indiscriminately discarded, hindering the reuse of technologically valuable elements. Poor waste separation compromises the recovery and reuse of resources. Studies indicate that waste management in academic laboratories remains inefficient, resulting in the loss of valuable chemical substances.

According to Leal Filho et al. (2019), integrating sustainability into higher education necessitates operational and curricular changes, including the introduction of laboratory practices that incorporate waste management into the learning process. The treatment of generated waste should be addressed in experimental protocols to ensure that, from academia onward, work is consistently aligned with the material demands of environmental challenges facing the planet.

The discipline of advanced experimental chemistry, offered to undergraduate students in their final years, proposed a project involving the synthesis of a porous adsorbent from discarded materials. First, terephthalic acid (H_2BDC) was obtained by alkaline hydrolysis of a poly(ethylene terephthalate) (PET) bottle (Ügdüler et al. 2020). H_2BDC was characterized by using structural and spectroscopic techniques to ensure its purity. Afterward, the students were tasked with preparing metal–organic frameworks (MOFs) using terephthalic acid as a ligand or linker. The metal cation was selected based on experiments conducted in the previous Inorganic Chemistry discipline, in which aluminum was extracted from bauxite (Constantino et al. 2002), the primary ore of aluminum, or from a soda can, producing potassium alum, $KAl(SO_4)_2 \cdot 12H_2O$. The 3D porous MOF, known as MIL-53-Al, was synthesized from H_2BDC, potassium alum, and KOH in water at a mild temperature and under atmospheric pressure (Sánchez-Sánchez et al. 2015). After eliminating unreacted linkers from within the pores, MIL-53-Al was tested for the removal of a series of organic dyes from water (Fig. 4), demonstrating the high potential of these adsorbents in pollution remediation.

5 Challenges and Perspectives

Almost 30 years ago, Anastas and Warner (1998) published the well known "Green Chemistry: Theory and Practice" book, in which the 12 principles of Green Chemistry were designed to guide the development of sustainable, environmentally friendly chemical processes. Here's a summary of each principle:

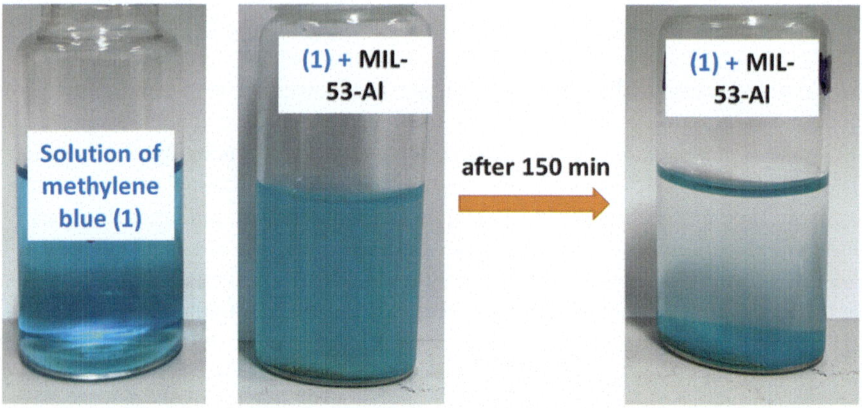

Fig. 4 Photograph of methylene blue dye aqueous solution without (left) and with MOF of Al^{3+} and terephthalate, MIL-53-Al (right). After being stirred for 150 min, the MIL-53-Al powder settled, and two phases could be visualized: a blue solid and an almost colorless supernatant

1. **Prevention**: It's better to prevent waste than to treat or clean it up after it's formed.
2. **Atom Economy**: Design synthetic methods to maximize the incorporation of all materials used in the process into the final product, minimizing waste.
3. **Less Hazardous Chemical Syntheses**: Use substances that are less toxic or hazardous in the chemical synthesis process.
4. **Designing Safer Chemicals**: Design chemicals that are effective but have minimal toxicity to humans and the environment.
5. **Safer Solvents and Auxiliaries**: Avoid the use of solvents, separation agents, and other auxiliary substances when possible, and use safer alternatives when necessary.
6. **Design for Energy Efficiency**: Minimize energy requirements by conducting reactions at ambient temperature and pressure when possible.
7. **Use of Renewable Feedstocks**: Use raw materials that are renewable rather than depleting finite resources.
8. **Reduce Derivatives**: Minimize the use of chemical derivatives (e.g., protecting groups or temporary solvents) that generate additional waste and complexity.
9. **Catalysis**: Use catalytic reactions, which require less energy and generate less waste compared to stoichiometric reactions.
10. **Design for Degradation**: Design chemical products to degrade into non-toxic substances once they are no longer needed.
11. **Real-time Analysis for Pollution Prevention**: Develop analytical tools that enable real-time monitoring of chemical processes to minimize waste and ensure safer operations.
12. **Inherently Safer Chemistry for Accident Prevention**: Design processes and chemicals to minimize the potential for chemical accidents, such as explosions, fires, or releases of toxic materials.

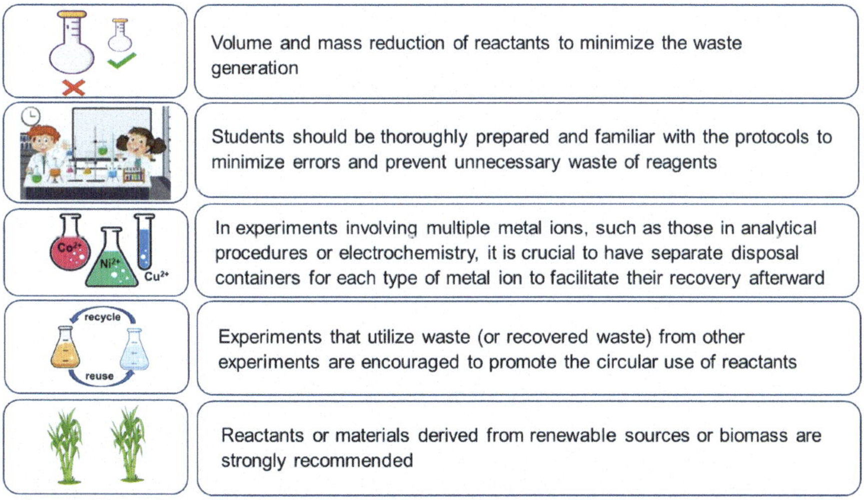

Fig. 5 Suggested practices to be incorporated into undergraduate laboratory classes, whenever possible, to stimulate a collective effort in promoting sustainability and circularity within the curriculum

These principles aim to make chemistry safer, more sustainable, and more efficient by minimizing environmental impact and conserving resources. Figure 5 illustrates several practices that can be incorporated into undergraduate laboratory classes, whenever feasible, to foster a collective effort in promoting sustainability and circularity within the curriculum. These practices require no financial investment and rely primarily on the commitment of the stakeholders. **When metal ions are indiscriminately discarded, recovery becomes unfeasible due to the complexity of the resulting mixtures.** Therefore, it is essential to engage everyone involved in laboratory activities to pursue sustainability and circularity in chemistry actively.

6 Conclusions and Challenges

This chapter presented some strategies for managing waste from teaching laboratories through the production of sustainable adsorbents. These adsorbents were developed from biomass or recyclable household waste, demonstrating the circularity of materials. In this way, undergraduate students can experience a Chemistry education where fundamental chemical concepts are integrated with the principles of sustainability and circularity.

The greatest challenge is engaging professors, technicians, and students to rethink the practices offered to students so that the use and disposal of reagents are carried out sustainably. This means modifying existing experimental protocols or creating new ones. Ideally, the products or by-products generated in experiments from one course would be utilized in experiments from other courses. This would be a didactic way of teaching the principles of sustainability and circularity, while also emphasizing the shared responsibility of everyone involved in addressing these issues.

Acknowledgements D. F. S. Petri thanks FAPESP (Grant 2024/20008-0) and CNPq (Grant 304017/2021) for financial support. M. M. L. Amorim thanks Programa Unificado de Bolsas of the University of São Paulo (Project 733, 2024). V. R. L. Constantino acknowledges CNPq (Grant 314034/2021-8).

References

Agarwal NK, Kumar M, Pattnaik F, Kumari P, Vijay VK, Kumar V (2023) Exploring the Valorization Potential of Sugarcane Bagasse Pith: a Review. Bioenerg Res 16:1280–1295 (2023). https://doi.org/10.1007/s12155-023-10632-4.

Anastas PT, Warner JC "Green Chemistry: Theory and Practice", Oxford University Press: New York, 1998.

Austen LI, Dugmore TI, Matharu AS, Hurst GA (2023) Byproduct Valorization: From Spent Coffee Grounds to Fatty Acid Ethyl Esters. J Chem Educ 100:327–335. https://doi.org/10.1021/acs.jchemed.2c00728.

Azmi AA, Othman SA (2023) Future of sugarcane bagasse paper: A review. AIP Conf Proc 2544:040034. https://doi.org/10.1063/5.0117103.

Braga VHA, Fuentes DP, Oliveira PV, Petri DFS (2024) Biopolymer cryogels for enhanced adsorption of valuable metals from multi-metal waste after flocculation and precipitation. J Environ Chem Eng 12:113306. https://doi.org/10.1016/j.jece.2024.113306.

Callisaya MP, Fuentes DP, Braga VHA, Finzi-Quintão CM, Oliveira PV, Petri DFS (2024) Harnessing carboxymethyl cellulose and *Moringa oleifera* seed husks for sustainable treatment of a multi-metal real waste. Environ Res 252: 118970. https://doi.org/10.1016/j.envres.2024.118970.

Chen M, Xu R, Wu Y, Xiong J, Keleş SZ, Hankins NP (2024) Application of polyelectrolytes for contaminant removal and recovery during water and wastewater treatment: A critical review. J Water Proc Eng 64:105528. https://doi.org/10.1016/j.jwpe.2024.105528.

Constantino VRL, Araki K, Silva DO, Oliveira W (2002) Preparação de compostos de alumínio a partir da bauxita: considerações sobre alguns aspectos envolvidos em um experimento didático. Quim Nova 25:490–498. https://doi.org/10.1590/S0100-40422002000300024.

Ezeonuegbu BA, Machido DA, Whong CMZ, Japhet WS, Alexiou A, Elazab ST, Qusty N, Yaro CA, Batiha GE (2021) Agricultural waste of sugarcane bagasse as efficient adsorbent for lead and nickel removal from untreated wastewater: Biosorption, equilibrium isotherms, kinetics and desorption studies. Biotechnol Rep 26:e00614. https://doi.org/10.1016/j.btre.2021.e00614.

Furtado LM, Fuentes DP, Ando RA, Oliveira PV, Petri DFS (2022) Carboxymethyl cellulose/sugarcane bagasse/polydopamine adsorbents for efficient removal of Pb2+ ions from synthetic and undergraduate laboratory wastes. J Clean Prod 380:134969. https://doi.org/10.1016/j.jclepro.2022.134969.

Gulliani S, Volpe M, Messineo A, Volpe R (2023) Recovery of metals and valuable chemicals from waste electric and electronic materials: a critical review of existing technologies. RSC Sustain 1:1085–1108. https://doi.org/10.1039/D3SU00034F.

Ho CC, Chen MS (2018) Risk assessment and quality improvement of liquid waste management in Taiwan University chemical laboratories. Waste Manag 71:578–588. https://doi.org/10.1016/j.wasman.2017.09.029.

Hudson R, Glaisher S, Bishop A, Katz JL (2015) From lobster shells to plastic objects: a bioplastics activity. J Chem Educ 92:1882–1885. https://doi.org/10.1021/acs.jchemed.5b00108.
Kirchherr J, Piscicelli L (2019) Towards an Education for the Circular Economy (ECE): Five Teaching Principles and a Case Study. Resour Conserv Recycl 150:104406. https://doi.org/10.1016/j.resconrec.2019.104406.
Lara ER, De la Rosa JR, Castillo AIR, de Jesús Cerino-Córdova F, Chuken UJL, Delgadillo SSF, Rivas-García P (2017) A comprehensive hazardous waste management program in a Chemistry School at a Mexican university. J. Clean. Prod. 142:1486–1491. https://doi.org/10.1016/j.jclepro.2016.11.158.
Leal Filho W, Doni F, Vargas VR, Wall T, Hindley A, Rayman-Bacchus L, Emblen-Perry K, Boddy J, Avila LV (2019) The integration of social responsibility and sustainability in practice: Exploring attitudes and practices in Higher Education Institutions. J Clean Prod 220: 152–166. https://doi.org/10.1016/j.jclepro.2019.02.139.
Liu K, Huang S, Jin Y, Lam JCH (2023) Teaching electrometallurgical recycling of metals from waste printed circuit boards via slurry electrolysis using benign chemicals. J Chem Educ 100:782–790. https://doi.org/10.1021/acs.jchemed.2c00637.
Lommelen R, Vander Hoogerstraete T, Onghena B, Billard I, Binnemans K (2019) Model for Metal Extraction from Chloride Media with Basic Extractants: A Coordination Chemistry Approach. Inorg Chem 58:12289–12301. https://doi.org/10.1021/acs.inorgchem.9b01782.
Mackenzie LS, Tyrrell H, Thomas R, Matharu AS, Clark JH, Hurst GA (2019) Valorization of waste orange peel to produce shear-thinning gels. J Chem Educ 96:3025–3029. https://doi.org/10.1021/acs.jchemed.8b01009.
Novaes SD, Oliveira PV, Petri DFS (2022) Hydroxypropyl methylcellulose sugarcane bagasse adsorbents for removal of 17α-ethinylestradiol from aqueous solution and freshwater. Environ Sci Pollut Res 29:63936–63952. https://doi.org/10.1007/s11356-022-20345-4.
Nunes ABSM, Fuentes DP, Mesquita JAFS, Romano RCO, Pileggi RG, Oliveira PV, Petri DFS (2023) Feasibility of sugarcane bagasse/polydopamine as sustainable adsorbents for Cr(VI) with reusability in cement composition. J Hazar Mat Adv 12: 100366. https://doi.org/10.1016/j.hazadv.2023.100366.
Rai R (2023) Copper-based laboratory waste: Its application as a catalyst and recovery. J. Clean. Prod. 432:139775. https://doi.org/10.1016/j.jclepro.2023.139775.
Sánchez-Sánchez M, Getachew N, Díaz K, Díaz-García M, Chebude Y, Díaz I (2015) Synthesis of Metal-Organic Frameworks in water at room temperature: Salts as linker sources. Green Chem 17:1500–1509. https://doi.org/10.1039/C4GC01861C.
Sousa LR, Guinati BGS, Maciel LIL, Baldo TA, Duarte LC, Takeuchi RM, Faria RC, Vaz BG, Paixão TRLC, Coltro WKT (2024) Office paper and laser printing: a versatile and affordable approach for fabricating paper-based analytical devices with multimodal detection capabilities. Lab on a Chip 24:467–479. https://doi.org/10.1039/D3LC00840A.
Stahel W (2016) The circular economy. Nature 531:435–438. https://doi.org/10.1038/531435a.
Ügdüler S, Van Geem KM, Denolf R, Roosen M, Mys N, Ragaert K, Meester SD (2020) Towards closed-loop recycling of multilayer and coloured PET plastic waste by alkaline hydrolysis. Green Chem 22:5376. https://doi.org/10.1039/d0gc00894j.
Varadwaj PR, Cukrowski I, Marques HM (2008) DFT-UX3LYP studies on the coordination chemistry of Ni^{2+}. Part 1: Six coordinate $[Ni(NH_3)n(H_2O)(6\text{-}n)]^{2+}$ complexes. J Phys Chem A 112:10657–66. https://doi.org/10.1021/jp803961s.
Velásquez-Yévenes L, Ram R (2022) The aqueous chemistry of the copper-ammonia system and its implications for the sustainable recovery of copper. Clean Eng Technol 9:100515. https://doi.org/10.1016/j.clet.2022.100515.
Verezhnikov VN, Nikulin SS, Poyarkova TN, Vostrikova YG (2002) Flocculating power of poly-N,N-dimethyl-N,N-diallylammonium chloride of various molecular weights. Russ J Appl Chem 75:461–464. https://doi.org/10.1023/A:1016163309427.

Yang J, Xuan S, Hu Y, Liu X, Bian M, Chen L, Lv S, Wang P, Li R, Zhang J, Shu CM, Dou Z (2022) The framework of safety management on university laboratory. J Loss Prev Process Ind 80:104871. https://doi.org/10.1016/j.jlp.2022.104871.

Zuin VG, Segatto ML, Zandonai DP, Grosseli GM, Stahl A, Zanotti K, Andrade RS (2019) Integrating Green and Sustainable Chemistry into Undergraduate Teaching Laboratories: Closing and Assessing the Loop on the Basis of a Citrus Biorefinery Approach for the Biocircular Economy in Brazil. J Chem Educ 96:2975–2983. https://doi.org/10.1021/acs.jchemed.9b0028.

Matheus L. M. Amorim is an undergraduate Chemistry student at the Chemistry Institute of the Universidade de São Paulo (USP) in Brazil. His research interests include sustainable chemical education, chemical processes, hybrid materials, and green chemistry.

Vera R. L. Constantino is a Full Professor at the Chemistry Institute of the Universidade de São Paulo (USP) in Brazil. She earned her PhD in Chemistry from Universidade Estadual Paulista (Unesp-Araraquara) in 1989 and joined USP in 1990. She conducted her postdoctoral research at Michigan State University in the USA from 1992 to 1994, under the supervision of Professor Thomas J. Pinnavaia, focusing on the reactivity of layered double hydroxides. Additionally, she was an Academic Visitor at the School of Chemistry, University of Bristol, in the United Kingdom, from 2015 to 2016. Vera serves as an Associate Editor for both the New Journal of Chemistry (RSC) and Applied Clay Science (Elsevier). Her research interests encompass the synthesis of layered materials, intercalation chemistry, clay minerals, layered double hydroxides, layered niobates, the exfoliation process, hybrid materials, organic-inorganic nanocomposites, and drug delivery systems.

Denise F. S. Petri holds a Bachelor's degree in Chemistry from the University of São Paulo (1987), a Master's degree in Chemistry from the State University of Campinas (1990), and a Ph.D. in Chemistry from the same institution (1992). She completed postdoctoral research at the Max Planck Institute for Polymer Research in Mainz, Germany (1993–1995), and at the Polymer Institute of the University of Karlsruhe, Germany (1996–1998). She obtained her habilitation from the University of São Paulo in 2004 and has been a Full Professor there since 2014. Her research focuses on Interface Chemistry, particularly in the areas of adsorption, thin films, polysaccharides, hydrogels, and porous materials. She has authored 196 scientific papers and 19 book chapters and is the inventor on three patents. She has supervised 20 Master's students, 11 Ph.D. candidates, and 5 postdoctoral researchers. From June 2022 to December 2024, she served on the editorial board of the Biomacromolecules, Biobased and Biodegradable Polymers section of Polymers (MDPI). Since September 2020, she has served as an Associate Editor of Cellulose (Springer Nature), and since May 2021, she has been a member of the Editorial Board of Carbohydrate Polymers (Elsevier). As of January 2025, she is also a member of the Editorial Advisory Board of ACS Applied Materials & Interfaces. Additionally, since October 2021, she has been member of the Board of Directors of the Brazilian Polymer Association (ABPol).

Solid Waste Management: Achievements and Challenges in Enhancing Sustainability at a Brazilian Higher Education and Research Institution

Ana Maria Maniero Moreira and Wanda Maria Risso Günther

Abstract An effective waste management system poses a significant challenge for cities striving for sustainability. In this context, universities must demonstrate how they implement such a system—not only by incorporating the topic into their curricula but, more importantly, by applying these concepts in their operational practices. Universities generate a substantial quantity and diverse types of waste—some akin to household waste, while others are hazardous. The objective of this study is to present the ongoing efforts of a Higher Education Research Institution located in São Paulo, Brazil, to develop sustainable initiatives aimed at preventing health and environmental impacts, mainly by minimising waste and, most importantly, implementing effective hazardous waste management. This case study employed a three-phase methodological approach. First, a comprehensive review of the scientific literature was conducted to identify the best practices in sustainable waste management within Higher Education and Research Institutions. Second, an in-depth analysis of current regulations and institutional documents was implemented, encompassing internal policies, databases, and operational procedures adopted by the institution under study. Third, the potential for reusing organic waste produced by the institution's university restaurant was assessed. To evaluate both the sustainable actions undertaken and the challenges encountered to date, a comparative analysis was carried out using a framework of 15 best practices identified during the literature review. Of these, ten have already been implemented by the institution. The remaining five—considered highly relevant—will be submitted to the institution's administration as recommendations for future accomplishment. A Solid Waste Management Program does not require significant financial investment but rather relies on trained human resources and continuous community awareness efforts to ensure the permanence of ongoing initiatives and the implementation of new ones.

A. M. M. Moreira (✉) · W. M. R. Günther
School of Public Health, University of São Paulo, São Paulo, Brazil
e-mail: anamaria.fsp@alumni.usp.br; anamariainforme@hotmail.com

W. M. R. Günther
e-mail: wgunther@usp.br

W. Leal Filho et al. (eds.), *Sustainability in the Curriculum and Operations of Universities*, World Sustainability Series,
https://doi.org/10.1007/978-3-032-12678-8_42

Beyond the localised positive impacts, this knowledge could serve as a representative case for similar institutions, encouraging interdisciplinary collaboration and institutional integration.

Keywords Solid waste management · Sustainability · Higher education and research institution · Selective waste collection · Hazardous waste

1 Introduction

Global warming and climate change are not mere illusions but rather pressing environmental challenges that must be addressed immediately (Sugiarto et al. 2022). Greenhouse gases (GHG), including carbon dioxide (CO_2), methane (CH_4), nitrous oxide (N_2O), and fluorinated gases, are widely recognised as the primary drivers of global warming and climate change. The primary sectors contributing to greenhouse gas (GHG) emissions include energy, industrial production, transportation, agriculture, waste management, and construction. Waste disposal, incineration, wastewater treatment, and the decomposition of organic matter in landfills result in the emission of CO_2 from the breakdown of organic materials, and CH_4 from anaerobic fermentation processes in landfills and during waste incineration (Filonchyk et al. 2024).

Within the context of the Post-2015 Development Agenda, solid waste management (SWM) is explicitly incorporated into the Sustainable Development Goals (SDGs), either directly or indirectly linked to more than half of the 17 goals (UNEP 2015). Unfortunately, despite concerted efforts, progress has remained limited. According to UNEP (2024), billions of tons of municipal waste continue to be generated annually, and a significant portion of the global population—possibly billions of people—still lack access to waste collection services.

Solid waste management is not only a fundamental human need but can also be considered a 'basic human right' (UNEP 2024). Proper regulation and management of municipal solid waste structures and systems, including those for hazardous waste, play crucial roles in providing a safe and sustainable environment for all (WHO 2024). If neglected, it can have harmful effects on the environment, society, and the economy. Consequently, a growing number of companies and institutions have recognised their environmental impact and social responsibility, taking proactive measures to enhance their performance and obtain certifications (Disterheft et al. 2012).

As a result, some institutions are recognizing the relevance of the 'Zero Waste Hierarchy' in their operations. Zero Waste (ZW) is an approach centred on resource recovery, consumption reduction, and the design of products with a life-cycle perspective (Singh and Hussain 2021). The authors argue that hierarchies serve as visible strategies for reducing landfill use; however, questions remain regarding their effectiveness in mitigating environmental hazards and conserving natural resources.

The debate on campus sustainability has intensified over the past three decades. Given their significant societal impact, universities are increasingly challenged to take a leadership role in sustainability issues (Disterheft et al. 2012). This responsibility extends beyond merely following a trend or differentiating themselves in the competitive higher education market; it represents a crucial contribution to the transformative changes necessary for the planet's well-being (Bizerril et al. 2017).

University campuses serve as ideal settings for living laboratories (Herth et al. 2024), offering practical learning opportunities for today's students and tomorrow's decision-makers, practitioners, and parents (Haksevenler et al. 2022). A living lab can enhance sustainability education, support research in this field, and promote sustainable behaviour (Amaral et al. 2020).

However, several barriers to sustainable innovation in universities worldwide have been identified in the literature, including lack of organizational commitment, limited management support and environmental committees, insufficient technology, inadequate infrastructure, and the absence of guidelines, policies, and incentives (Ávila et al. 2019).

According to Disterheft et al. (2012), campus sustainability has not yet become a reality in most universities. The authors identified a lack of awareness among students, faculty, and staff. They emphasize the importance of preparing students for contemporary challenges, while at the same time actively involving them in the development of new sustainable strategies. Furthermore, engaging employees and managers can be challenging due to cost and time constraints. Therefore, appointing a team focused on sustainability is essential to ensuring the effective implementation of sustainable development initiatives (Du et al. 2010).

Abubakar et al. (2020) highlight a deficit in assessments and reports, as less than half of Saudi Arabian universities incorporate sustainable development actions into their strategies or development plans.

As mentioned by Dhoulá et al. (2017), campus operations are generally a neglected field in six Central European countries (Czech Republic, Hungary, Poland, Serbia, Slovakia, Slovenia). University managers often do not consider them relevant. Consequently, initiatives in this field, even when they exist, often face indifference or a lack of experience.

Owojori et al. (2020) and Budihardjo et al. (2021) emphasize the importance of an effective Solid Waste Management Plan (SWMP) and underscore the need for a thorough assessment of the characteristics, composition, and quantity of solid waste generated in institutions.

Moreira et al. (2017) state that Brazilian universities often manage solid waste without standardization and, in some cases, fail to meet the legal requirements of the National Solid Waste Policy (Brazil 2010). Common shortcomings include failure to conform with reverse logistics[1] for certain products, insufficient measures to prevent

[1] Reverse logistics is an instrument of economic and social development characterised by a set of actions, procedures, and means aimed at enabling the collection and return of solid waste to the business sector for reuse, either in its own cycle or in other productive cycles, or for another environmentally appropriate final disposal (Brazil 2010, 2020a, 2020b, 2022).

risks to workers and the community, limited participation in reuse programs, and a lack of partnerships with recycling cooperatives. This lack of standardization also leads to uneven quality of solid waste management.

Studies generally emphasize that SWM programs in universities are essential components and key indicators in the development of a green campus. However, they primarily focus on waste reduction or elimination (Osorio et al. 2022). In fact, hazardous waste is only superficially addressed in most global studies, and few have outlined the practical measures implemented to enhance the safety and efficiency of hazardous waste management.

For example, in 2009, a community of students, academic and non-academic staff at a Chemistry School in Mexico successfully implemented a Hazardous Waste Management Programme dealing with significant quantities of hazardous waste, primarily consisting of acidic and toxic inorganic materials with corrosive properties (liquid waste) and glass fragments impregnated with toxic hazardous substances (solid waste) (Lara et al. 2017).

Another exception is the study by Bahçelioğlu et al. (2020), which provided a detailed description of the procedures adopted for hazardous waste management at a university in Turkey. The authors stated that, although each department or unit is responsible for managing its own hazardous waste, the process is also overseen by a central unit, designated as the Hazardous Waste Commission.

This study aims to present the ongoing initiatives undertaken by the School of Public Health (SPH), a Higher Education and Research Institution (HERI) based in São Paulo, Brazil, which has been actively engaged over the past 17 years in efforts to prevent health and environmental impacts—primarily through waste minimization and the effective management of hazardous waste.

2 Methods

This case study employed a three-phase methodological approach. First, a comprehensive review of the scientific literature was conducted to identify the best practices in sustainable waste management within Higher Education and Research Institutions. Second, an in-depth analysis of current regulations and institutional documents was implemented, encompassing internal policies, databases, and operational procedures adopted by the institution under study. Third, the potential for reusing organic waste produced by the institution's university restaurant was assessed.

To assess both the sustainable actions undertaken and the challenges encountered to date, the study carried out an analysis using a framework of 15 best practices identified during the literature review. Of these, ten have already been put into practice by the institution. The remaining five, considered to be highly relevant, will be submitted to the institution's administration as recommendations for future accomplishment.

3 Results and Discussion

Description of the Case Study—School of Public Health

This case study examines the evolution of SWM practices at the School of Public Health (SPH), analysing the challenges, key strategies, and outcomes over time. SPH is a public institution founded in 1918 and incorporated into the University of São Paulo (USP) in 1938. It is located in a central area of São Paulo, Brazil, covering a total area of 22,833 m^2, with a built-up area of 16,191 m^2, according to the institution's administration.

Within its grounds there is a healthcare unit, the Geraldo de Paula Souza Primary Healthcare Centre (GPSPHC), managed by SPH. Figure 1 shows an aerial view of a geographical map highlighting a triangular area (in yellow). The two main buildings are identified as "GPSPHC", "SPH" apart from the surrounding greenery and infraestruture. The Healthcare Centre provides public primary healthcare to residents in its catchment area and involves professors, researchers, and students from SPH, as well as other USP teaching units and partner institutions. In addition, it offers specialised referral services for vulnerable elderly individuals and a sanitary dermatology service for the diagnosis and treatment of sexually transmitted diseases.

The total area includes a main building that houses auditoriums, classrooms, fourteen laboratories, teaching and administrative offices, as well as the related Healthcare Centre, a daycare centre, a reference library, and a university restaurant.

Two undergraduate courses are offered: Nutrition and Public Health, along with six postgraduate programs in the strict sense: Environment, Health and Sustainability; Entomology in Public Health; Epidemiology; Nutrition in Public Health; Global Health and Sustainability; and Public Health. Additionally, the institution holds periodically various cultural and outreach courses, showcasing the diverse educational background of the circulating public.

According to the USP Statistical Yearbook, in 2023, SPH had 68 teaching staff, 203 non-teaching staff, 580 undergraduate students, 744 postgraduate students, and 253 participants in extension courses,[2] with a total population of 1623 people, plus visitors. This large number of individuals is significant for planning awareness initiatives and campaigns to engage in sustainable activities.

Trajectory of the Solid Waste Management at School of Public Health

In 1993, a program called *USP Recicla* was launched, covering all USP units, with the initial objective of reducing and recycling paper. However, since 2009, there has been a significant increase in the frequency and consistency of initiatives at SPH, particularly those related to extending the management to other types of solid waste generated.

A committee comprising students, teachers and other staff members was established to draft and oversee the Solid Waste Management Plan (SWMP), with the

[2] Source: USP Statistical Yearbook 2024. Available in: https://uspdigital.usp.br/anuario/AnuarioControle. Accessed in Jun 2025

Fig. 1 Aerial view of the School of Public Health and Healthcare Centre highlighting the two main buildings. Source: Google Maps

aim of adhering to the prevailing health, environmental and labour regulations at the national, state and municipal levels. This SWMP is currently accessible on the *Quick Access* section of the institution's website.

At that time, some initiatives were initiated with the objective of: (i) preventing the depletion of natural resources; (ii) minimizing waste; (iii) ensuring the proper segregation and disposal of waste; and (iv) disseminating environmental education and raising community awareness.

The committee's primary task was to conduct a comprehensive assessment of the institution's solid waste management situation, including the management of hazardous waste. This assessment revealed problems in the institution's structure, operations and management. A series of participatory discussions were held to address these shortcomings, and the resulting corrective measures were subsequently implemented.

Consequently, since 2009, key sustainable measures have been implemented at SPH and Healthcare Centre, including: (i) identification, characterization, and quantification of all types of waste generated; (ii) continuous supervision of waste handling; (iii) technical support for cleaning and healthcare professionals regarding

Table 1 Checklist of good practices in sustainability in HERIs

	Good practices in HERIs	Studies' references	Adopted by SPH
1	Implementing sustainability by adding environmental subjects in undergraduate and graduate courses	Drahein et al. 2019; Moreira et al. 2018	✔
2	Inviting funding arrangements for sustainability projects	Hugé et al. 2018	
3	Enhancing exchange knowledge between the university managers and the university community	Zen et al. 2016	
4	Encouraging the university community to audit waste management practices	Ebrahimi and North 2017	
5	Hiring professionals to work specifically with waste management	Drahein et al. 2019	
6	Establishing a green office as a model office to the university	Zen et al. 2016	
7	Engaging the community (students, researchers, and staff) in waste management activities based on sustainability principles	Hugé et al. 2018; Moreira et al. 2018	✔
8	Reporting green initiatives and its achieved goals	Tangwanichagapong et al. 2017	✔
9	Disseminating guidelines for proper waste discard	Tangwanichagapong et al. 2017	✔
10	Carrying out management and final disposal of hazardous waste from research laboratories	Moreira et al. 2018	✔
11	Promoting reuse, proper disposal and recycling of obsolete electronic equipment	Alves and Farina 2018	✔
12	Treating the university as a living lab using its capabilities to solve environmental issues	Gallo et al. 2017	✔
13	Creating database on waste management to monitor key performance indicators	Moreira et al. 2018	✔
14	Allocating different types of bins strategically to enhance recycling waste segregation	Oladejo et al. 2018; Zen et al. 2016; Drahein et al. 2019; Ebrahimi and North 2017	✔
15	Raising awareness to the campus community about waste management focusing on recovery, recycling and reuse through campaign, nudging them for a transition to circular economy	Gallo et al. 2017	✔

(*Source* adapted from Vargas and Campos 2020)

occupational safety in solid waste management; (iv) procedures to minimise waste generation and environmental impact; (v) maintenance of the organic waste recovery project; (vi) systematization of selective waste collection; and (vii) promotion of environmental education to enhance awareness and encourage behavioural change.

Achievements and challenges so far

A literature review conducted by Vargas and Campos (2020) identified 15 initiatives aimed at enhancing sustainability in HERIs. These initiatives are presented in Table 1, along with their corresponding studies' references. As the original authors refer to them as "good practices," this terminology has been retained to preserve consistency with the source. The final column of the table indicates which of these practices have already been adopted by SPH.

The comparative analysis indicated that 10 practices (66, 7%) are already adopted in SPH, while the remaining five—specifically items 2 through 6—have not yet been implemented. To facilitate a clearer understanding, each of those practices will be described in detail below.

Good practice 1—Adding environmental subjects in undergraduate and graduate courses (Drahein et al. 2019; Moreira et al. 2018)

USP students may choose courses from any of the 43 HERIs distributed across nine campuses. According to USP platforms *Janus* and *Júpiter*, there are currently 17 undergraduate, and 77 graduate courses related to the environment, sustainability, and solid waste management.

Good practice 2—Inviting funding arrangements for sustainability projects (Hugé et al. 2018)

Some projects receive funding from Brazilian agencies such as the National Council for Scientific and Technological Development (CNPq), the São Paulo Research Foundation (FAPESP), and the Coordination for the Improvement of Higher Education Personnel (CAPES).

However, the ongoing projects require mainly human resources, rather than financial support.

In addition, there are ongoing partnerships with companies (at no cost to SPH) to support the reverse logistics of certain recyclable materials, including batteries, coffee capsules, X-ray films, cleaning sponges, among others.

Good practice 3—Enhancing exchange knowledge between the university managers and the university community (Zen et al. 2016)

Good practice 4—Encouraging the university community to audit waste management practices (Ebrahimi and North 2017)

Good practice 5—Hiring professionals to work specifically with waste management (Drahein et al. 2019)

Good practice 6—Establish a green office as a model office to the university (Zen et al. 2016)

It is evident that the largest challenge in implementing sustainable measures at SPH lies in hiring qualified personnel to establish and maintain the proposed initiatives. Engaging the community to actively participate in these practices remains particularly difficult.

Even projects initiated with financial support rely heavily on teachers, staff, and students volunteering to ensure continuity. This is a relatively common issue, especially in public universities, and underscores the need for stronger institutional commitment to environmental concerns. More importantly, SPH leadership should take responsibility by institutionalizing these activities.

Good practice 7—Engaging the community (students, researchers, and staff) in waste management activities based on sustainability principles (Hugé et al. 2018; Moreira et al. 2018)

At the beginning of each academic year, SPH conducts orientation lectures for new students, introducing key concepts of waste management, explaining the disposal procedures for different types of waste, and addressing related health risks and preventive measures. On this occasion, durable plastic mugs are distributed to encourage students to avoid the use of disposable items while on SPH premises, even at the university restaurant.

Good practice 8—Reporting green initiatives and its achieved goals (Tangwanichagapong et al. 2017)

Both the internal and external communities have access to two key documents available on the SPH website: the ***Solid Waste Management Plan*** **(SWMP)**[3] and the ***Annual Sustainability Report,***[4] which outlines the institution's achievements from the previous year as well as its goals for future improvement.

Good practice 9—Disseminating guidelines for proper waste discard (Hugé et al. 2018; Moreira et al. 2018; Tangwanichagapong et al. 2017)

Each of the 14 research laboratories develops and follows its own Standard Operating Procedures (POP in the Portuguese acronym) and systematically reports to the managers the list of used chemical products ready for collection and incineration.

Periodic training is conducted for cleaning staff, and courses on waste management are offered, along with technical visits to the Healthcare Centre for students. These awareness-raising practices promote knowledge of the implemented system, aim to engage the community, and enhance safety during activities.

Good practice 10—Carrying out management and final disposal of hazardous waste from research laboratories (Moreira et al. 2018)

---

[3] Solid Waste Management Plan *[Plano de Gerenciamento de Resíduos Sólidos]*. Available in https://www.fsp.usp.br/site/. Access in Jun 2025

[4] Annual Sustainability Report 2025 *[Realizações do Programa Socioambiental]*. Available in: https://www.fsp.usp.br/site/institucional/mostra/1477. Access in Jun 2025

Teaching and research activities conducted at SPH, along with clinical assistance provided at the Healthcare Centre, generate hazardous waste, commonly referred to as healthcare waste. This includes infectious materials, animal carcasses, sharps, and chemical substances. Improper management of such waste can pose significant environmental impacts and health risks to those involved, including patients, doctors, technicians, students, researchers, cleaning staff, external waste collectors, and others.

The segregation, packaging, collection, storage, internal and external transportation, and proper treatment of infectious and chemical waste are strictly monitored at SPH and at the Healthcare Centre. Compliance with current regulations is ensured, and measures to reduce waste generation are continuously evaluated.

In the city of São Paulo, the treatment of healthcare waste—including autoclaving for infectious agents and incineration for chemicals and animal carcasses—generally occurs outside the institutions. The municipal sanitation system, funded through a monthly fee, is responsible for the external collection and treatment of waste generated by all healthcare facilities in the city, ensuring universal and safe service delivery. To access this service, institutions must register in the system, report data on the quantities of waste generated, and issue a document to accompany the transportation of each shipment. In return, waste generators receive a certificate from the concessionaire confirming the proper disposal of the healthcare waste sent.

Good practice 11—Promoting reuse, proper disposal and recycling of obsolete electronic equipment (Alves and Farina 2018)

In 2024, SPH collected 48 kg of printer ink cartridges and toners, alongside 64 kg of electronic waste (e-waste). E-waste collected at SPH, as well as at other institutions within USP, is sent to the Centre for the Disposal and Reuse of Computer Waste (CEDIR), located on a different USP campus. There, the waste is assessed for return to suppliers, reuse, or, when feasible, donation to underprivileged communities outside USP.

Good practice 12—Treating the university as a living lab using its capabilities to solve environmental issues (Gallo et al. 2017)

Theoretical concepts such as ***Reverse Logistics Systems*** and the ***Circular Economy*** have begun to be applied in practice, as exemplified by two projects developed at SPH: *Ecopoint* and *Reuse of organic waste from University Restaurant.*

1. ***Results of the implementation and maintenance of an Ecopoint for voluntary delivery of special waste***

In 2023, an *Ecopoint* (Fig. 2) was installed to facilitate community participation in the reverse logistics system for special materials. It consists in a row of black recycling bins lined up against a wall indoors. Each bin has a label indicating the type of waste it is meant for. Above the bins, a sign reads "Ecoponto" with recycling symbols and additional text.

The *Ecopoint* is installed in an accessible and busy area in SPH, where several collection containers are provided for the community, in order to voluntarily dispose

Fig. 2 Ecopoint of the School of Public Health, showing indoor recycling bins identified by waste type

of special waste, which is currently included in the reverse logistics system. Depending on demand, external collection and appropriate transportation for treatment or recycling are carried out, funded by the suppliers of each material following their specific routes.

In 2024, the community delivered the following quantities into the *Ecopoint* and other SPH collection points: 160 kg of used batteries; 142 kg of coffee capsules; 20 kg of X-ray plates; 16 kg of plastic packaging lids; 13 kg of pens and other writing devices; 6.5 kg of cleaning sponges; and 226 kg of expired medicines. By providing users with the convenience of an *Ecopoint*, it is anticipated that, over time, the volume of materials completing the cycle will increase, and additional types of materials will be incorporated into the reverse logistics system.

2. ***Results of the study on the reuse of organic waste generated at the university restaurant through composting***

Biodegradable organic matter is typically the largest component of the total waste generated at HERIs (Torrijos et al. 2021). If this waste is not properly separated and directed for recovery, it will end up in landfills or even dumps, potentially leading to soil and water contamination as well as the production of gases such as methane. Uncontrolled methane emission contributes to the greenhouse effect. Organic waste is recognised as a pollutant because it decomposes rapidly in natural environments, generates odours, attracts vectors, and can often cause diseases due to its high levels of degradable organic compounds and moisture content (Yaser et al. 2022).

Institutions implementing a composting project can divert compostable waste generated in their restaurants and other sectors from landfills. Additionally, an equivalent quantity of leaves, pruning, and grass clippings collected from the institution's gardens can be integrated into the process. This approach offers multiple environmental benefits, including reducing greenhouse gas emissions from off-site waste transportation, extending the lifespan of landfills, and producing organic compost that can be utilised in the facility's vegetable garden and landscaping, thereby closing an environmentally and economically responsible cycle.

Recovery of waste through composting is considered a key in achieving and adopting the circular economy model in the coming years. However, composting food waste might be challenging due to its inconsistent characteristics, as it depends on its sources and the consumers' eating habits (Yaser et al. 2022). Teixeira et al. (2020) proposed to carry out extension activities with students, with the aim of nudging the restaurant users towards the reduction of waste, as well as raising employees' awareness about proper segregation.

SPH's university restaurant mainly serves its internal public of undergraduate and postgraduate students but also accepts other USP students and staff. According to the restaurant administration, an average of 1085 meals are served daily, including 75 for breakfast, 580 for lunch and 430 for dinner. The menu includes animal or vegetarian protein, side dishes, raw and/or cooked salad, dessert and juice. The delivery system is not self-service; therefore, the servings are portioned by the staff, except for the salads, which are served at will.

It is important to highlight that the use of disposable plastic cups was banned in the restaurant some years ago. Each user is required to bring their own reusable cup, which they receive upon entering the institution.

A gravimetric study was conducted for one day in September 2024 by a group of undergraduate students. First, the bags of waste from the food processing sector (kitchen), service sector, and leftovers and scraps from service trays (food stations) were weighed. The bags were then emptied onto a tarpaulin to assess the effectiveness of sorting into recyclables, rejects, and organic waste. Once the waste was properly separated, each category was weighed individually.

According to the data, approximately 211 kg/day are routinely disposed of as ordinary waste by the university restaurant. The results from waste segregation based on material characteristics were as follows: 140 kg of plate leftovers and remnants of prepared food from the service trays; 46 kg of compostable waste (such as vegetable and fruit peelings, scraps, and vegetable stalks); 19 kg of recyclable materials, including 13 kg of cardboard and 6.0 kg of plastic containers; and 6 kg of rejects.

Based on an estimate of 20 working days per month, the restaurant generates approximately 920 kg of compostable waste monthly, calculated as 46 kg per day. In practice, none of the compostable material was directed to composting, and only 13 kg of cardboard was pre-sorted for recycling (Fig. 3).

Since 2009, a composting area (Fig. 4) has existed within SPH. It is currently receiving approximately 226 kg/month of compostable waste generated by various sectors of the institution, excluding the university restaurant. Therefore, the compostable fraction produced during the restaurant's food preparation process (approximately 920 kg/month) should be separated and redirected to this facility. If combined with an equivalent amount of green waste collected from the gardens, the total amount diverted from the landfill could reach 2292 kg/month.

As a result of this study, the measures that will be proposed for restaurant managers to adopt a more sustainable approach include: conducting surveys to assess consumer preferences regarding the menu; incorporating waste reduction strategies into future service contracts; providing adequate training to employees on waste segregation

Fig. 3 Cart in the restaurant storage area containing plastic bags filled with cardboard separated to be recycled

Fig. 4 Garden composting area with piles of organic material at different stages of decomposition

by category; supplying appropriate and clearly labelled equipment (such as waste containers); and promoting awareness through notices and posters with clear disposal instructions for both staff and consumers. Additionally, Yaser et al. (2022) recommend educating communities through on-site demonstrations, awareness campaigns, and workshops.

Good practice 13—Creating a database on waste management to monitor key performance indicators (Moreira et al. 2018)

Universities generate a large quantity and variety of waste, similar to household waste, such as recyclables, non-recyclables, and organic materials. However, they also produce hazardous waste, such as infectious, sharps, radioactive, and chemical substances, because of teaching and research activities. Additionally, other types of waste, such as used batteries, fluorescent and Light Emitting Diode (LED) lamps, tires, cooking oil, lubricants, construction debris, and e-waste, require specialised management based on their type, characteristics, and level of hazardousness.

Rodríguez-Guerreiro et al. (2024) calculated that the average waste generation rate in Higher Education Institutions (HEIs) worldwide is 0.19 ± 0.21 kg/day per person, with a median of 0.093 kg/day per person. On average, organic waste accounted for 30 ± 19% of total waste, followed by paper and cardboard at 23 ± 13%, and plastics at 18 ± 11%. Glass and metals comprised 3 ± 3% each. These quantities vary significantly based on the specificities of the institutions and the research methodology. In the study by Rodríguez-Guerreiro et al. (2024), for example, only household waste was considered. Garden waste, hazardous, and electronics waste were only reported by some institutions.

HERIs, such as SPH—which generate a wide variety of waste types, including hazardous waste such as healthcare waste—are legally required (Brazil, 2010) to implement, regularly update, and maintain accessible their respective Solid Waste Management Plans (SWMPs), which must include waste characterisation, its quantification, and all ongoing procedures. The quantification of almost all types of waste generated is a practice that was implemented several years ago. According to 2024 data, the average monthly waste generation at SPH and the Healthcare Centre was 8818.5 kg. Of this total, 7920 kg (89.9%) was non-recyclable, 626.2 kg (7.1%) recyclable, 250 kg (2.8%) infectious waste, and 22.3 kg (0.3%) hazardous chemical waste. It is important to consider that the organic fraction of the waste generated at SPH—both compostable and non-compostable—continues to be discarded as ordinary waste, despite the existence of an on-site composting area.

Figure 5 illustrates a slight increase in the generation of the four main types of waste—recyclable, non-recyclable, infectious, and chemical—over the past three years. This increasement is likely attributable to the rise in the number of users and activities following the resumption of operations after the COVID-19 pandemic. Nevertheless, further studies are necessary to better understand this trend and to support the development of the necessary corrective measures.

(Source: provided by the institution's administration)

Good practice 14—Allocating different types of bins strategically to enhance recycling waste segregation (Oladejo et al. 2018; Zen et al. 2016; Drahein et al. 2019; Ebrahimi and North 2017)

It can be also observed in Fig. 5 that the percentage of recyclables collected relative to the total waste generated remained low over the last three years, despite the measures taken so far. A significant portion of recyclable materials continues to be discarded as ordinary waste, while some non-recyclable materials are mistakenly included in the recyclable stream. In addition, it has often been observed that recyclable materials, even when placed in the appropriate containers, were wet and/or contaminated, compromising the quality of the remaining recyclables and leading to their misclassification as garbage. As a result, it has been determined that the waste segregation process needs to be reassessed. The inconsistency appears to be cultural, indicating the need for behavioural changes concerning environmental issues. According to Hu et al. (2021), environmental knowledge and community regulations are key factors

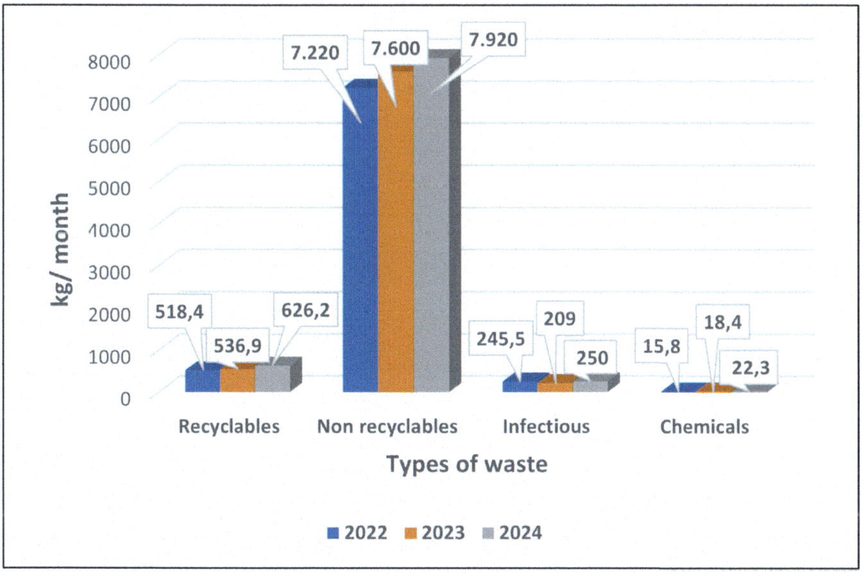

Fig. 5 Bar chart comparing monthly waste generation (kg) for the years 2022, 2023, and 2024

in behavioural interventions aimed at reinforcing the constructs underlying individuals' intentions. The authors define a construct as an individual's conscious plan or decision to exert effort in order to engage in a particular behaviour.

The selective collection of recyclables is crucial for mitigating the negative environmental impact of solid waste. Increasing the rate of collected recyclables would help to reduce the volume of waste sent to landfills, lower disposal costs, and potentially generate income and profit through recycling (Ferreira et al. 2023). Most likely, the problem is still a lack of awareness and engagement within the campus community. To address this issue, actions were proposed to enhance waste separation. The strategic distribution and labelling of containers for the two waste categories (Fig. 6) were adjusted. In the halls, different container formats and sizes were placed side by side, the smaller one for non-recyclables and the larger one for recyclables, while different colours were assigned to the bags (respectively black and blue) to facilitate proper segregation.

Good practice 15—Raising awareness among the campus community about waste management focusing on recovery, recycling and reuse through campaigns, nudging them to a transition towards circular economy (Gallo et al. 2017)

Projects, research, lectures, workshops, campaigns, announcements through digital media and electronic newsletters, as well as events related to environmental and occupational health, have become key strategies for disseminating information, raising awareness within the community, and ensuring the continuity of the program.

Fig. 6 Containers available for waste segregation, side by side, and labelled as "Não recicláveis" (Non recyclable) and "Recicláveis" (Recyclable)

As an example of the institution's community engagement in the Circular Economy System, the 'Book Exchange Fair,' traditionally held every June 5 (World Environment Day), became a permanent event (Fig. 7) since 2024, following a request from students. Currently, a table filled with stacks of books is permanently available. Above the table, a banner reads "Feira de Troca de Livros Usados na FSP/USP". The banner encourages participation and sustainability, inviting people to exchange their used books. This simple initiative, at no additional costs, increased the circulation of textbooks and other literary works, bringing both financial and environmental benefits.

Then, an important goal for SPH is to inspire students to develop projects that utilize the campus as a living classroom. Universities and colleges worldwide are

Fig. 7 Permanent used-book exchange fair at the School of Public Health, promoting reuse and sustainability

increasingly using their campuses as living laboratories for sustainability problem-solving and engagement. The projects should focus on experiential learning and campus sustainability, integrating research with campus operations (Rivera and Savage 2020). According to Rooney and McMillin (2010), this learning experience is shaped by much more than traditional classroom instruction.

4 Conclusion

Until recently, sustainable solid waste management was primarily basic, involving the separation of recyclable materials such as plastic, paper, glass, and metal packaging, which were then sent to private recycling centres or waste picker cooperatives. The recycling industry continues to expand, and even daily consumers encounter materials that are no longer considered refuse, with new products increasingly being manufactured from recycled materials. Additionally, drop-off points and collection companies are proliferating, facilitating proper disposal and the closure of the recycling loop.

On the other hand, the population is gradually adopting more sustainable habits and recognizing the importance of their actions toward sustainability for the economy, the environment, and society. In Brazil, regulations have been implemented to expand the range of products that must be included in the reverse logistics system, imposing obligations on consumers, retailers, distributors, and manufacturers.

Of course, that sorting, sanitising, packaging, and transporting each material collected to different collection points, sometimes located far away, requires additional effort on the part of the consumer. These efforts are even greater when it comes to recycling organic waste, as individuals may need to create their own composting area at home or within their communities, given that there is still no widespread collection program for this third fraction of waste in Brazilian cities.

SPH, as an exemplary measure, demonstrates its commitment to the proper disposal of each type of waste generated on its premises and creates opportunities to educate students and staff on the correct methods for doing so. Students and staff learn how to operate their own composting systems, or alternatively, they can bring organic waste from their homes to contribute to the composting system they helped create and are actively maintaining. It is expected that the community involved will continue to support the projects, which is essential for their sustainability.

There are limited number of studies, both in developed and developing countries, which have explored the management of hazardous waste generated in research laboratories, regardless of whether the treatment occurs on or off campus.

All individuals utilizing university research laboratories should be well-versed, both theoretically and practically, in proper procedures to prevent accidents when handling, packaging, and disposing of hazardous waste, including infectious, chemical, sharps, and radioactive materials In addition to proper segregation, they should also learn strategies to minimize waste generation, such as neutralizing certain chemicals before discarding them into the sewage system to mitigate risks and reduce treatment costs. These represent the new challenges for SPH.

Solid Waste Management Programs bring together professionals and curricula from multiple disciplines, fostering a comprehensive approach to sustainability. The main advantage is their ability to attract new projects and supporters, as they do not require significant financial investment but rather rely on trained human resources to ensure the practical continuity of ongoing initiatives and the implementation of new ones.

Acknowledgements The authors would like to commend the undergraduate nursing students for selecting the topic and for the excellent development of the study entitled *"Analysis on Waste Generation in University Restaurants"*, presented as the final project for the course *Fundamentals of Environmental Health/2024*.

They also extend their gratitude to the staff of the institution under study for their continuous and invaluable collaboration throughout the years in the composting project.

References

Abubakar IR, Aina YA, Alshuwaikhat, HM (2020) Sustainable Development at Saudi Arabian Universities: An Overview of Institutional Frameworks. Sustainability 12 (19):8008. https://doi.org/10.3390/su12198008

Alves DS, Farina MC (2018) Disposal and reuse of the information technology waste: a case study in a Brazilian university. European Business Review, 30(6): 720–734, 2018. https://doi.org/10.1108/EBR-08-2016-0117

Amaral AR, Rodrigues E, Rodrigues-Gaspar A, Gomes A (2020) A review of empirical data of sustainability initiatives in university campus operations. J. Clean. Prod. 250:119558. https://doi.org/10.1016/j.jclepro.2019.119558

Ávila LV, Beuron TA Damke, L.I., Pereira RS, Klein LL (2019) Barriers to innovation and sustainability in universities: an international comparison. International Journal of Sustainability in Higher Education 20(5): 805–821. https://doi.org/10.1108/IJSHE-02-2019-0067

Bahçelioğlu E, Buğdaycı E, Doğan N, Şimşek N, Kaya SO, Alp E (2020) Integrated solid waste management strategy of a large campus: A comprehensive study on METU campus, Turkey. Journal of Cleaner Production 265: 121715. https://doi.org/10.1016/j.jclepro.2020.121715

Bizerril M, Rosa M, Carvalho T, Jesus J (2017). Sustainability in Higher Education: a review of contributions from Portuguese speaking countries. Journal of Cleaner Production 171:600–612. https://doi.org/10.1016/j.jclepro.2017.10.048

Brazil. Law n.12,305 of August 2nd (2010). Institutes the National Policy on Solid Waste; alters law n. 9,605 of February 12th, 1998; and makes other provisions. [Institui a Política Nacional de Resíduos Sólidos; altera a Lei nº 9.605, de 12 de fevereiro de 1998; e dá outras providências]. https://www.planalto.gov.br/ccivil_03/_ato2007-2010/2010/lei/l12305.htm

Brazil. Decree n. 10,240 of February 12 (2020a) Regulates item VI of the caput of Article 33 and Article 56 of Law n. 12,305, of August 2, 2010, and complements Decree n. 9,177, of October 23, 2017, regarding the implementation of a reverse logistics system for household electrical and electronic products and their components. [Regulamenta o inciso VI do **caput** do art. 33 e o art. 56 da Lei n° 12.305, de 2 de agosto de 2010, e complementa o Decreto n° 9.177, de 23 de outubro de 2017, quanto à implementação de sistema de logística reversa de produtos eletroeletrônicos e seus componentes de uso doméstico]. Retrieved from https://www.planalto.gov.br/ccivil_03/_ato2019-2022/2020/decreto/d10240.htm

Brazil. Decree n. 10,388 of June 5 (2020b) Regulates paragraph 1 of Article 33 of Law n. 12,305/2010 (National Policy on Solid Waste–PNRS) and establishes the Reverse Logistics System for Expired or Unused Household Medications. [Regulamenta o § 1° do **caput** do art. 33 da Lei n° 12.305, de 2 de agosto de 2010, e institui o sistema de logística reversa de medicamentos domiciliares vencidos ou em desuso, de uso humano, industrializados e manipulados, e de suas embalagens após o descarte pelos consumidores]. Retrieved from https://www.planalto.gov.br/ccivil_03/_Ato2019-2022/2020/Decreto/D10388.htm

Budihardjo M, HumairaN, Putri S, RamadanB, Syafrudin, S, Yohana E (2021) Sustainable Solid Waste Management Strategies for Higher Education Institutions: Diponegoro University, Indonesia Case Study. Sustainability 13(23):13242. https://doi.org/10.3390/su132313242

Disterheft A, Caeiro SSFS, Ramos MR, Azeiteiro UMM (2012) Environmental Management Systems (EMS) implementation processes and practices in European higher education institutions e Top-down versus participatory approaches. Journal of Cleaner Production 31: 80–90. https://doi.org/10.1016/j.jclepro.2012.02.034

Drahein AD, De Lima EP, Da Costa SEG (2019) Sustainability assessment of the service operations at seven higher education institutions in Brazil. Journal of Cleaner Production 212: 527–536. https://doi.org/10.1016/j.jclepro.2018.11.293

Du S, Bhattacharya CB, Sen S (2010) Maximizing business returns to corporate social responsibility (CSR): the role of CSR communication. Int. J. Manag. Rev 12 (1): 8–19. https://doi.org/10.1111/j.1468-2370.2009.00276.x

Ebrahimi K, North LA (2017) Effective strategies for enhancing waste management at university campuses. International Journal of Sustainability in Higher Education 18 (7)1123–1141. https://doi.org/10.1108/IJSHE-01-2016-0017

Ferreira CM, Alves AG, Portela FA, Freire LL, Moreira Junior FA (2023) Avaliação gravimétrica de resíduos orgânicos como ferramenta para tomada de decisões: estudo de caso em um restaurante acadêmico de uma instituição pública federal. Revista Brasileira de Meio Ambiente, 11 (3) 33–36. https://revistabrasileirademeioambiente.com/index.php/RVBMA/article/view/1480/399

Filonchyk M, Peterson MP, Zhang L, Hurynovich V, He Y (2024) Greenhouse gases emissions and global climate change: Examining the influence of CO2, CH4, and N2O. Science of the Total Environment 935: 173359. https://doi.org/10.1016/j.scitotenv.2024.173359

Gallo M, Marotta V, Magrassi F, Taramasso AC, del Borghi A (2017) University campus waste prevention and reduction: A circular-economy approach. Economics and Policy of Energy and the Environment 2017 (1): 235–252. https://doi.org/10.3280/EFE2017 001012

Haksevenler BH, Kavak FF, Akpinar S (2022) Separate waste collection in higher education institutions with its technical and social aspects: A case study for a university campus. **Journal of Cleaner Production** 367: 133022. https://doi.org/10.1016/j.jclepro.2022.133022

Herth A, Verburg R, Blok K (2024) The Innovation Power of Living Labs to EnableSustainability Transitions: Challenges and Opportunities of On-Campus Initiatives. Creativity and Innovation Management 34: 255–518. https://doi.org/10.1111/caim.12649

Hu J, Tang K, Qian X, Sun F, Zhou W (2021) Behavioral change in waste separation at source in an international community. An application of the theory of planned behavior. Waste Management, 135: 397–408. https://doi.org/10.1016/j.wasman.2021.09.028

Hugé J, Mac-Lean C, Vargas (2018). Maturation of sustainability in engineering faculties – From emerging issue to strategy? Journal of Cleaner Production 172: 4277–4285. https://doi.org/10.1016/j.jclepro.2017.07.143

Lara ER, De la Rosa JR, Castillo AIR, Cerino-Cordova FJ, Chuken UJL, Delgadillo SSF, Rivas-Garcia P (2017) A comprehensive hazardous waste management program in a Chemistry School at a Mexican university. Journal of Cleaner Production 142:1486-1491. https://doi.org/10.1016/j.jclepro.2016.11.158

Moreira RM, Cetrulo TB, Mendizábal–Cortes AD, Cetrulo NM, Malheiros TF (2017) Brazilian National Solid Waste Policy applied as a tool to enhance an University Campus Waste Management. Journal of Management and Sustainability 7 (3):101–111. https://doi.org/10.5539/jms.v7n3p101.

Moreira R, Malheiros TF, Alfaro JF, Cetrulo T B, Ávila LV (2018) Solid waste management index for Brazilian Higher Education Institutions. Waste Management 80:292–298. https://doi.org/10.1016/j.wasman.2018.09.025

Oladejo OS, Auta AM, Ibikunle PD, Omamofo EK (2018) Solid waste generation, characteristics and material recovery potentials for Landmark University Campus. International Journal of Civil Engineering and Technology 9 (9): 1071–1082. http://www.iaeme.com/ijciet/issues.asp?JType=IJCIET&VType=9&IType=9

Osorio A, Úsuga L, Vásquez R, Nieto-Londoño C, Rinaudo M, Martínez J, Filho W (2022) Towards carbon neutrality in Higher Education Institutions: Case of two private universities in Colombia. Sustainability (Switzerland) 14(3). https://doi.org/10.3390/su14031774

Owojori O, Edokpayi J, Ratshalingwa M, Odiyo J (2020) Characterisation, recovery and recycling potential of solid waste in a university of a developing economy. Sustainability 12: 5111. https://doi.org/10.3390/su12125111

Rivera CJ, Savage C (2020) Campuses as living labs for sustainability problem-solving: trends, triumphs, and traps. J Environ Stud Sci 10, 334–340. https://doi.org/10.1007/s13412-020-00620-x

Rodríguez-Guerreiro M-J, Torrijos V, Soto M A (2024) Review of Waste Management in Higher **Education Institutions: The Road to Zero Waste and Sustainability. Environments, 11: 293.** https://doi.org/10.3390/environments11120293

Rooney M, McMillin J (2010) The Campus as a Classroom: Integrating People, Place, and Performance for Communicating Climate Change. In: Leal Filho W (eds) Universities and Climate Change. Climate Change Management. Springer, Berlin, Heidelberg. pp117–136 https://doi.org/10.1007/978-3-642-10751-1_10

Singh S, Hussain CM (2021). Zero waste hierarchy for sustainable development. In: Hussain CM (ed) Concepts of Advanced Zero Waste Tools. Chapter Six. 123–142. https://doi.org/10.1016/B978-0-12-822183-9.00006-4

Sugiarto A, Lee CW, Huruta AD (2022) A Systematic Review of the Sustainable Campus Concept. Behav Sci 12 (5): 130. https://doi.org/10.3390/bs12050130

Tangwanichagapong S, Nitivattananon V, Mohanty B, Visvanathan C (2017) Greening of a campus through waste management initiatives: Experience from a higher education institution in Thailand. International Journal of Sustainability in Higher Education 18 (2): 203–217. https://doi.org/10.1108/IJSHE-10-2015-0175

Teixeira RA, Nicolau Korres AM, Borges RM, Rabello LL, Ribeiro IC, Bringhenti JR (2020) Sustainable Practices for the Organic Waste Management Generated in an Educational Institution Restaurant. In: Leal Filho W et al. Universities as Living Labs for Sustainable Development. World Sustainability Series. Springer, Cham. https://doi.org/10.1007/978-3-030-15604-6_49

Torrijos V, Dopico DC, Soto M (2021) Integration of food waste composting and vegetable gardens in a university campus. Journal of Cleaner Production 315 https://doi.org/10.1016/j.jclepro.2021.128175

Vargas DB, Campos LMS (2020). Waste Management in Higher Education Institutions: A State-of-the-art Overview. Proceedings of the 5th NA International Conference on Industrial Engineering and Operations Management Detroit, Michigan, USA. IEOM Society International. https://www.ieomsociety.org/detroit2020/papers/16.pdf

UNEP. United Nations Environment Programme. Global Waste Management Outlook (2015). UNEP/ISWA. ISBN: 978-92-807-3479-9. Global Waste Management Outlook |UNEP - UN Environment Programme

UNEP. United Nations Environment Programme. Beyond an age of waste Turning rubbish into a resource. (2024) Global Waste Management Outlook 2024.UNEP/ ISWA. Global Waste Management Outlook 2024 | UNEP - UN Environment Programme

WHO. World Health Organization (2024) Chapter 4. Solid waste. In: Compendium of WHO and other UN guidance in health and environment, 2024 update. Geneva: World Health Organization. https://iris.who.int/handle/10665/378095. https://cdn.who.int/media/docs/default-source/who-compendium-on-health-and-environment/who_compendium_chapter4.pdf

Yaser AZ, Lamaming J, Suali E, Rajin M, Saalah S, Kamin Z, Safie NN, Aji NAS, Wid N (2022) Composting and Anaerobic Digestion of Food Waste and Sewage Sludge for Campus Sustainability: A Review. International Journal of Chemical Engineering 2022, Article ID 6455889, https://doi.org/10.1155/2022/6455889

Zen I S, Subramaniam D, Sulaiman H, Saleh AL, Omar W, Salim MR (2016) Institutionalize waste minimization governance towards campus sustainability: A case study of Green Office initiatives in Universiti Teknologi Malaysia. Journal of Cleaner Production 135: 1407–1422. https://doi.org/10.1016/j.jclepro.2016.07.053

Ana Maria Maniero Moreira is a doctor with Master, PhD and Postdoctoral degrees in Science from the School of Public Health, University of Sao Paulo (USP). She is a Visiting Professor at the School of Public Health, USP, where she teaches and conducts research in Environmental Health. With experience in Environmental Management and Public Policy, her work and projects focus on solid waste and healthcare waste management.

Wanda Maria Risso Günther is a Civil Engineer and Sociologist with a PhD in Public Health from the University of São Paulo (USP) and a postdoctoral degree from the Autonomous University of Madrid. She is a Full Professor at the Faculty of Public Health, USP, and serves as the Coordinator of the USP Global Cities Synthesis Centre and Coordinator of the USP Sustainable Municipalities Program, both based at the Institute of Advanced Studies (IEA/USP).

Smart Campus for Environmental Sustainability: A Case Study of ICTS-UNESP Focused on Photovoltaic Generation and Energy Consumption Forecasting

Bruno Oliveira Zarpellon, Eduardo Paciência Godoy, Fernando Pinhabel Marafão, Luis De Oro Arenas, and Helmo K. Morales Paredes

Abstract A smart campus incorporates advanced technologies and data-driven decision-making processes to optimize various operations within a university setting. A fundamental pillar of a smart campus is environmental sustainability, exemplified by many applications designed to curtail environmental impact. These applications include the generation of photovoltaic (PV) energy, the management and reduction of energy consumption, the monitoring of water usage and air quality, and initiatives to encourage recycling, among others. Since 2022, the Institute of Science and Technology of Sorocaba (ICTS) at the São Paulo State University (UNESP) in Brazil has been actively implementing smart campus technologies, with a particular focus on energy consumption monitoring, meteorological data analysis, PV energy generation, and air quality monitoring. This work presents the ICTS smart campus, highlighting the significant potential of smart campuses in reducing the environmental impact of universities. It presents empirical evidence on the environmental benefits of such initiatives from two key perspectives. Firstly, the study focuses on the role of PV generation in reducing campus-wide greenhouse gas (GHG)

B. O. Zarpellon (✉) · E. P. Godoy · F. P. Marafão · L. De Oro Arenas · H. K. M. Paredes
Institute of Science and Technology, Sorocaba, São Paulo State University (UNESP), Sorocaba, Brazil
e-mail: bruno.o.zarpellon@unesp.br

E. P. Godoy
e-mail: eduardo.godoy@unesp.br

F. P. Marafão
e-mail: fernando.marafao@unesp.br

L. De Oro Arenas
e-mail: luis.arenas@unesp.br

H. K. M. Paredes
e-mail: helmo.paredes@unesp.br

W. Leal Filho et al. (eds.), *Sustainability in the Curriculum and Operations of Universities*, World Sustainability Series,
https://doi.org/10.1007/978-3-032-12678-8_43

emissions, evaluating the effectiveness of the PV system in mitigating these emissions. Secondly, it employs a predictive statistical tool, Autoregressive Integrated Moving Average (ARIMA), to forecast future energy consumption, offering valuable insights for campus energy planning. The ARIMA also helps determine the amount of PV generation required to achieve a complete transition to a 100% clean energy campus. The study's promising results serve as a practical case study, demonstrating actionable strategies for enhancing sustainability on smart campuses.

Keywords Smart campus · Data integration · Environmental sustainability · Energy consumption monitoring · Internet of things · ARIMA · PV generation

1 Introduction

According to the Cambridge Dictionary, sustainability is "*the quality of causing little or no damage to the environment and therefore being able to continue for a long time*". Following this definition, in 2015, the United Nations (UN) and its members adopted the 2030 Agenda for Sustainable Development to improve the quality of life for all people worldwide and reduce the impact of humanity on nature by 2030 (United Nations 2015). This agenda comprises 17 Sustainable Development Goals intended to guide the countries that have adopted them. The main goals include the eradication of poverty and hunger, the provision of quality education, the sustainable management of water and sanitation for all, access to affordable, reliable, sustainable and modern energy for all, the promotion of responsible consumption and production patterns, and taking urgent action to combat climate change and its impacts, among others (United Nations 2020).

Whilst the term sustainability is mostly associated with environmental issues, the goals defined in the UN's 2030 Agenda show that the sustainability concept extends beyond the environmental domain, encompassing aspects such as human well-being, education, and economic prosperity. The smart city concept was first coined in the 1990s and is now widely accepted as a technological solution to several problems faced by urban areas, including the ones related to sustainability. Several contributions associated with this subject can be found in the literature (Cui et al. 2017; Kumar et al. 2018; Supangkat et al. 2018; Yang et al. 2021; Kurniawan et al. 2022). Despite the evolution of this field of study, the implementation of smart cities can be burdensome due to several factors, such as the size of the cities, the presence of several management organizations and the lack of interest among inhabitants in actively participating and collaborating in the implementation of changes. It is evident that there are numerous similarities between university campuses and cities (Torres-Sospedra et al. 2015). In addition, the challenges associated with the implementation of a smart city are more easily overcome in a campus context, and thus the smart city concept can be applied to a university campus, thereby originating the smart campus concept.

A smart campus is a university campus that employs advanced and/or intelligent technologies and initiatives in several areas of the campus. The objective is to improve campus operation and enhance the quality of life of its members (Zarpellon et al. 2024). The concept of a smart campus can be considered as a tool to improve the sustainability of universities curriculum and operations. Some of the applications which are commonly implemented on a smart campus directly relate to reducing its environmental impact, such as air quality monitoring, water and energy consumption monitoring and control, and photovoltaic (PV) energy generation. In contrast, other applications pertain to the human element, encompassing improved education quality, optimized campus expenses and promoted well-being.

In the current Brazilian context, several smart campuses have been implemented, including those of Facens (2019), Unicamp (Universidade Estadual de Campinas 2017), Instituto Mauá de Tecnologia (IMT) (Instituto Mauá de Tecnologia 2018), and São Paulo State University (UNESP) in Sorocaba (Institute of Science and Technology Sorocaba 2022). Even though there are several smart campus initiatives in Brazil, most of them are practical implementations, with a lack of theoretical works to support them and further promote the concept. Additionally, several works (Bandeira et al. 2022; Brito de Jesus and Gomes Camelo 2024) point out that most models implemented in the Brazilian context address only the technological aspect of the smart campus, focusing on using technologies to monitor campus processes and the energy and environmental aspects of the campus, but lacking applications and initiatives focused on management, education, research, and the human aspect of the campus. Therefore, applications of energy consumption monitoring and implementation of PV generation systems on these campuses are a common trend, with most campuses conducting the former and a small number conducting the latter.

Although there is a growing literature interest in the smart campus subject (Chagnon-Lessard et al. 2021), there is a paucity of works that have addressed the smart campus concept with a focus on its impact on the sustainability of the university in which it is implemented, even in the Brazilian context, where several applications directly related to energy sustainability are well developed. This work addresses this gap by examining the smart campus implemented at the Institute of Science and Technology (ICTS), a campus of the São Paulo State University (UNESP) located in Sorocaba, Brazil. The study builds upon the previously presented architecture of the ICTS's smart campus (Zarpellon et al. 2024), expanding the original work to demonstrate how the applications developed at the campus contribute to UNESP's sustainability. To accomplish this objective, the present work evaluates the quantity of greenhouse gas (GHG) emissions that have been averted since the implementation of the PV generation system on the campus. Furthermore, it employs a statistical tool to predict campus consumption based on historical consumption data. This is done to demonstrate how the application of these tools can assist in a more effective planning of energy usage and, therefore, optimize it.

2 The ICTS Smart Campus

The ICTS campus is a relatively small university campus, regarding its size and programs offered. It has a total area of 16,172 m^2, with a constructed area of 7,600 m^2. The ICTS proudly offers two graduation programs and five post-graduation programs, encompassing two areas of study and has approximately 700 students. It is essential to understand these numbers to grasp the campus's consumption figures outlined in Sect. 4.2.

The smart campus implemented at the ICTS campus is based on monitoring variables pertinent to environmental sustainability, such as water and energy consumption, PV energy generation, climate conditions, air quality, and additional variables, namely compressed air monitoring and ambient monitoring. The infrastructure for monitoring these variables is based on a flexible, scalable, and low-cost IoT architecture, built based on a layered structure to easily accommodate any new application when designed according to the predetermined layers (Zarpellon et al. 2024). Figure 1 presents the layers of the architecture. The description of each layer is further presented.

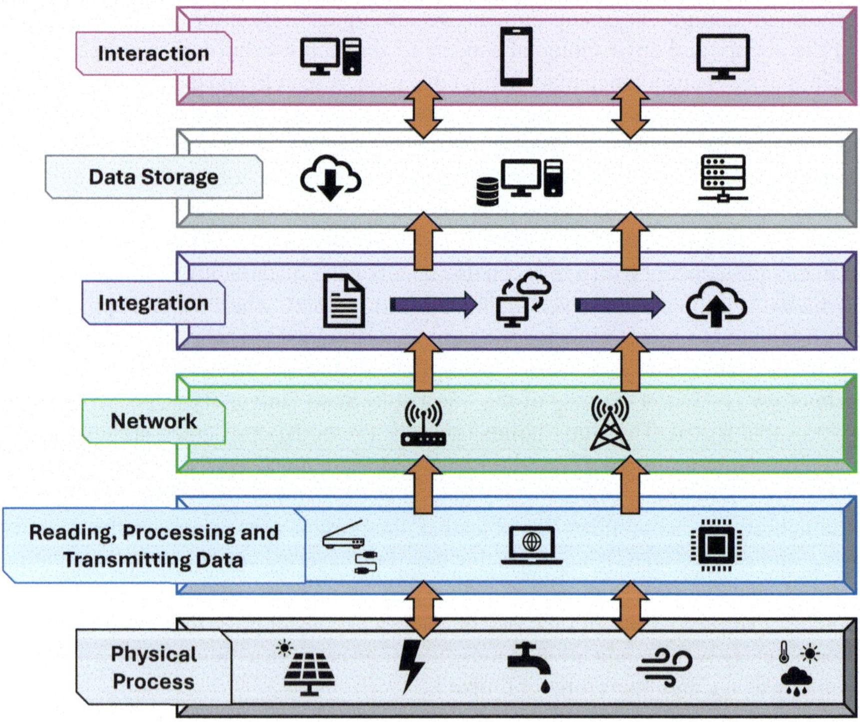

Fig. 1 Layered architecture adopted in the ICTS smart campus (Zarpellon et al. 2024)

- Physical Process: all physical processes monitored in the smart campus are located on this layer, along with the sensors utilized to measure their associated variables. Regarding the ICTS, some of the processes and their respective sensors are energy consumption monitoring (energy meters), water consumption monitoring (water meters, flow and pressure sensors), and climate variables (meteorological station), among others.
- Reading, Processing and Transmitting Data: on this layer are located the devices used to read data from the sensors installed at the physical process layer, pre-processing this data and forwarding it via the network to the upper layers of the architecture. At the ICTS some of the adopted devices are ESP32s, Raspberry Pi and a notebook.
- Network: this layer represents the communication network responsible for transmitting data between the reading, processing and transmitting data layer and the integration layer. At UNESP the current networks utilized are Wi-Fi and LoRaWAN.
- Integration: the integration layer receives data read from the physical process and sent via the network, formats it to the format expected at the data storage and stores it in the utilized databases. At the ICTS the Node-RED platform was adopted for the integration of data.
- Data Storage: the layer where data are stored for later use, consisting of databases. For the ICTS architecture, the databases utilized are InfluxDB and MySQL.
- Interaction: the highest logical level of the architecture consists of an interface for retrieving and visualization of the data stored in the database. The interface in the context of this work is built in Grafana.

The architecture proposed by Zarpellon et al. (2024) is generic and easily replicable with a low-cost implementation in any campus aiming to become a smart one. All data collected at the ICTS is freely available for download and use for case studies based on a real scenario at the link https://smartcampus.sorocaba.unesp.br/d/WusCo658k9/main-screen?orgId=1&kiosk. Further details about the architecture, its implementation at the ICTS and replication steps at other campuses can be found in Zarpellon et al. (2024).

3 Methodology for Evaluating the ICTS Smart Campus Sustainability

Based on the two metrics defined previously to assess the sustainability brought to the ICTS by the implementation of the smart campus, this section presents the theoretical foundation associated with each one of the metrics adopted. The first half of this section presents the necessary concepts to understand how much GHG emissions are avoided by the PV generation installed at the campus since its deployment at the ICTS. It presents a quantification of how much GHG is avoided by each kWh generated by the PV generation according to previous studies, while also presenting the historical generation since the start of its production. The statistical tool chosen for predicting future consumption is presented in the second half of the section. It is called Autoregressive Integrated Moving Average (ARIMA). This section also presents the advantages of using it for time-series prediction, the definition of its parametrization, general information about the consumption data of the campus that integrates the evaluated dataset and information regarding the preprocessing steps necessary for utilizing this dataset.

3.1 Greenhouse Gas Emissions Avoided by PV Implementation

The main components of GHG are CO_2, CH_4, N_2O and fluorinated gases, such as hydrofluorocarbons, perfluorocarbons, sulfur hexafluoride, and nitrogen trifluoride (U.S. Environmental Protection Agency 2025a). Depending on its abundance, duration of presence and how strong it is, each gas has its impact on global warming, causing an improvement in Earth's temperature by absorbing energy and slowing the rate at which it is released to space (U.S. Environmental Protection Agency 2025b). The emission of these gases is caused mainly by the use of fossil fuels (Mohibullah et al. 2006). This has caused an exponential growth in the importance of using renewable energy resources as an alternative energy source.

Several studies in the literature have evaluated how the use of renewable energy resources can decrease the amount of GHG emissions (Mohibullah et al. 2006; Akinyele et al. 2014; Amponsah et al. 2014; Breyer et al. 2015; Hudy et al. 2018; Rojas-Hernández 2022), mainly focusing on CO_2 due to it being the most prevalent gas present in GHG and addressing case studies from the emissions of specific countries, reviewing prior studies related to GHG emissions or addressing the economic savings from the utilization of renewable energy systems.

In this context, one of the most relevant studies is conducted in Amponsah et al. (2014), where the authors address not only how much GHG emissions are associated with different non-renewable sources, but also the emissions associated with the complete life-cycle of several renewable generation sources. The life cycle of these systems includes all or some of: (i) fuel production (i.e., to account for the non-consumable portion of the produced fuel) and transportation to the plant, (ii) facility construction, (iii) facility operation and maintenance, and (iv) dismantling (Mann and Spath 2001). Therefore, even though renewable energy power plants do not have GHG emissions associated with their process of generating energy, there are emissions indirectly associated with it, and these must also be considered when evaluating how much GHG emissions were avoided by implementing the renewable energy resource.

The amount of GHG emissions associated with the PV generation life cycle can vary from 9 to 300 gCO_{2eq}/kWh, according to the study conducted in Amponsah et al. (2014). As a comparison, another reference addressed by this study points out to

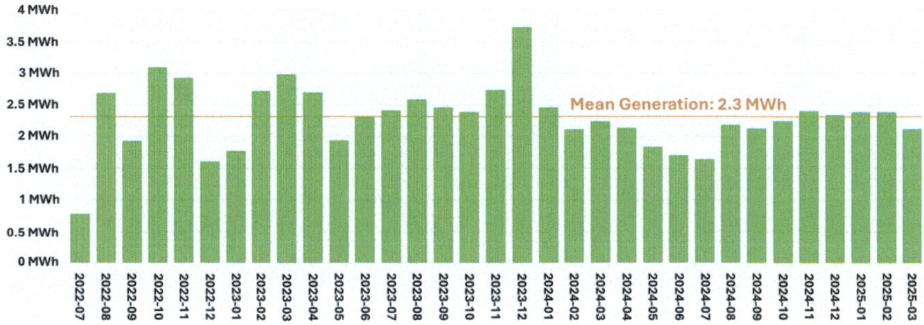

Fig. 2 PV generation history graph

46 gCO_{2eq}/kWh. Regarding non-renewable energy sources, Amponsah et al. evaluate the GHG emissions of natural gas, oil, coal and lignite, which are, respectively, 499, 733, 888 and 1054 gCO_{2eq}/kWh. The utilized unity identifies the total CO_2 emissions equivalent based on all GHG emissions (Peng et al. 2013; Ritchie et al. 2020) throughout the life cycle of the generation source, divided by a multiplication of the annual power generation by the total lifetime of the generation source, in years. In Breyer et al. (2015), the authors evaluate seven different PV generation scenarios and point out that the GHG emissions avoided can range from 618 to 820 gCO_{2eq}/kWh, while except for PV power plants and rooftop systems in Germany, all remaining scenarios evaluated have additional financial benefits.

Regarding the specific case study addressed in the present paper, the PV generation system installed at the ICTS has a 20 kW nominal capacity. The data stored from it is measured at the inverter connected to the PV generation panels in 5-min intervals during the daily hours where sunlight is available and dates back to July of 2022. Figure 2 and Table 1 present, respectively, the bar graph corresponding to the month-by-month and mean monthly generation, and the monthly generation values since the installation of the PV system.

As seen in Table 1, the PV generation system has generated a total of 76.05 MWh since its installation 2 years and 9 months ago. This amount of energy is not the real amount generated, since due to problems associated with the devices reading data, network issues or server issues, sometimes the architecture failed in registering data in the database. However, even with these gaps in the data, with the available data we can still evaluate how much the generated energy accounted for in avoiding GHG emissions.

Based on the GHG emissions associated with the PV generation life cycle reported by Amponsah et al. (2014), we can estimate the minimum and maximum GHG emissions resulting from PV energy generation utilizing Eq. (1), where GHG_e is the amount of GHG emissions by the energy source (in gCO_{2eq}), GHG_{cpk} is the emissions associated with the generation of a single kWh by that source (in gCO_{2eq}/kWh) and EG is the total amount of energy (in kWh) produced by the generation source. Since all energy generated by the PV panels is consumed on campus, Eq. (1) can be applied to calculate the GHG emissions that would have been produced if

Table 1 PV generation historical values

Month and year	PV generation (MWh)	Month and year	PV generation (MWh)
July, 2022	0.78	December, 2023	3.73
August, 2022	2.69	January, 2024	2.46
September, 2022	1.93	February, 2024	2.11
October, 2022	3.10	March, 2024	2.24
November, 2022	2.93	April, 2024	2.13
December, 2022	1.61	May, 2024	1.84
January, 2023	1.78	June, 2024	1.70
February, 2023	2.72	July, 2024	1.64
March, 2023	2.98	August, 2024	2.18
April, 2023	2.70	September, 2024	2.12
May, 2023	1.94	October, 2024	2.24
June, 2023	2.31	November, 2024	2.40
July, 2023	2.41	December, 2024	2.34
August, 2023	2.58	January, 2025	2.39
September, 2023	2.46	February, 2025	2.38
October, 2023	2.39	March, 2025 (up to the 25th)	2.11
November, 2023	2.73		
Total generation		**76.05 MWh**	
Mean monthly generation		**2.30 MWh**	

the equivalent energy was supplied by a non-renewable source. This allows us to quantify the GHG emissions avoided by utilizing renewable PV energy instead of fossil-based generation.

$$GHG_e = GHG_{epk} * EG \tag{1}$$

3.2 *Autoregressive Integrated Moving Average*

Time series data are a series of observations or measurements taken in chronological order. Considering that the information of a time series follows trends, and its historical data can be useful for predicting future values, the study and prediction of time series has been gaining more attention and rapidly developing. Statistical modelling refers to using mathematical models and statistical assumptions to evaluate and make predictions about real-world data (Heavy.AI 2022). By using statistical models, we can further study historical data, evaluating hypotheses made about it and transforming it into more valuable and understandable information (Breiman 2011).

The most common type of statistical model used to evaluate time series data is the Autoregressive Integrated Moving Average (ARIMA). ARIMA evaluates the existing dependence between a time series observations over time, using past values of the series to predict future values. It is a linear regression formulation where the predictors are dependent on lagged values of the dependent variable and prediction errors, and can be specified based on three parameters: p, which is the degree of autoregression that identifies the number of lags of Y to be used as predictors; d, the degree of differencing needed to make the series stationary; and q, the moving average degree, referring to the number of lagged forecast errors that should go into the ARIMA Model (Shahriari et al. 2025). The advantage of using ARIMA is that it works better with non-stationary data, requiring a smaller number of parameters for estimation and being competent in dealing with linearity present in data (Bhaskar and Singh 2012; Gupta et al. 2019). Additionally, ARIMA is one of the most widely adopted predictor tools to work with time series data throughout the literature and is very well suited for stationary time series data, which is the case for the ICTS smart campus consumption data.

Regarding the specific case study addressed in this paper, the energy consumption monitoring system consists of eleven Sirax BM1200 meters from Camille Bauer located at different places on the campus. Nine of these meters are responsible for registering the energy consumption of specific sectors or buildings of the campus, while the remaining two register the consumption at the two main switchboards of the campus, accounting for the total consumption of the campus. The consumption data are read each minute sequentially by a single notebook connected to the meters by an RS-485 connection cable, and the data collection started in early September 2022. The evaluated data ranges from late October 2022 to early January 2025.

Energy consumption prediction has several applications, such as better understanding consumption patterns, planning utility purchase and studying the amount of renewable energy and batteries necessary to make the campus fully independent in terms of energy production. The initial predictions conducted in this work have the main objective of evaluating and using the collected data for the first time since the implementation of the smart campus. They also demonstrate how monitoring this data and saving it for later usage can improve the sustainability of a smart campus. Discussions related to the other applications of data are also conducted.

To predict energy consumption based on the historical data gathered from the database, two situations were evaluated. In the first one, the data was resampled to hourly consumption, resulting in 13,565 observations. The original data consists of more than 1 million observations. In the second case, energy consumption was resampled to gather data in a daily basis, resulting in 652 observations. Both these scenarios were tested to identify which one of the sampling rates is better to predict future energy consumption values, with the minute sampling rate not being used due to the computational burden necessary to train and test the model, while also avoiding a problem existent only in this sample rate, where the energy consumption has no visible change from some measurements to the next ones.

For both cases of resampled data, after the appropriate pre-processing of the data (such as resampling and removing null values and outliers), the series are tested for stationarity, autocorrelation and partial autocorrelation. Based on the tests conducted, the starting set of parameters adopted for the p, d and q orders of the ARIMA model are equal to 1.0, 0.0 and 1.0, respectively. Additionally, for the daily sampled data, the prediction algorithm is also tested using $p = 7.0$, meaning a one-week data window. Approximately two-thirds of the data are used as training data (66% of the dataset), while the remaining data are utilized as testing data. All simulations are conducted at the Google Colab web environment.

The models are tested using walk-forward validation. This procedure involves separating the data into train and test data. The model is trained using the train data. Then, the model is used to predict the first value of test data. Based on the result, the model is then retrained, incorporating the feedback from its first iteration and then tries to predict the next value. The model is retrained with the new error identified, attempting to predict the next test value after, and so on. Therefore, it is a good approach to train an ARIMA model, since it captures evolving patterns and can detect concept drifts, but it is computationally intensive.

To evaluate models' results, the following classical metrics are adopted: Mean Absolute Percentage Error (MAPE), Median Absolute Percentage Error (MDAPE), Mean Absolute Error (MAE), Root Mean Squared Error (RMSE), alongside the visual similarity between the curves. Furthermore, we compare the mean consumption value from the test datasets and the predicted values. This is done because in the context of the present case study the most important information is to know in advance the mean energy consumption on a predetermined time window (daily or hourly) and not necessarily predicting the exact amount of energy consumption for a given timestep.

4 Results and Discussions

In this section the results of the paper are presented and discussed based on the two sustainability metrics characterized previously. In the first part of the section, based on the theory presented in Sect. 3.1, the amount of GHG emissions associated to the use of energy coming from PV generation is compared to the emissions which would be associated to using the same amount of energy while coming from four different non-renewable generation sources (natural gas, oil, coal and lignite) and, based on that, the amount of GHG emissions avoided in the ICTS since the installation of the PV generation is quantified. These absolute numbers are also translated into yearly GHG emissions reductions and compared to the average yearly emission of a single person in Brazil, for a better contextualization of the overall positive impact of installing this renewable generation at the campus. Further, in the second part of this section, the consumption predicted data are presented graphically, the metrics adopted for the evaluation of the predictions conducted are presented and a discussion is conducted regarding a comparison between each one of the three ARIMA models adopted. Furthermore, both subsections present comparisons between the results of this paper and the results found by other papers in literature which have conducted similar studies.

4.1 Greenhouse Gas Emissions Avoided by PV Implementation

Utilizing the ICTS PV generation data presented in Sect. 3.1 and Eq. (1), the minimum and maximum amount of GHG emissions associated with the PV generation at the ICTS since its implementation is calculated. This data, the number of emissions associated with the same amount of energy generated by each non-renewable source and the comparison between these emissions associated with non-renewable generation sources with worst- and best-case scenarios of PV generation are presented in Table 2. A comparison between minimum and maximum emissions associated with the PV generation is not conducted.

Table 2 shows that the PV generation system at the ICTS has a possible range of kilograms of CO_2 equivalent emissions avoided. In the worst-case scenario, the amount was 15,133.95 $kgCO_{2eq}$ (approximately 15 tonnes of CO_{2eq}). In the best-case scenario, it was 79,472.25 $kgCO_{2eq}$ (approximately 80 tonnes of CO_{2eq}). Considering the total period in which the PV generation was installed (2 years and 9 months), the amount of CO_2 equivalent emissions avoided per year can vary from 5,503.25 $kgCO_{2eq}$ to 28,899.00 $kgCO_{2eq}$, in worst- and best-case scenarios, respectively.

Table 2 Comparison of GHG emissions associated with the amount of energy generated by the PV generation system at the ICTS vs the amount of GHG emissions associated with the generation of the same amount of energy by a non-renewable generation source

Energy source	GHG emissions associated with source lifecycle considering the amount of energy generated at the ICTS in the timespan considered ($kgCO_{2eq}$)	Emissions avoided with minimum GHG emissions of PV generation ($kgCO_{2eq}$)	Emissions avoided with maximum GHG emissions of PV generation ($kgCO_{2eq}$)
PV generation (minimum emissions)	684.45		
PV generation (maximum emissions)	22,815.00	–	–
Natural gas	37,948.95	37,264.50	**15,133.95**
Oil	55,744.65	55,060.20	32,929.65
Coal	67,532.40	66,847.95	44,717.40
Lignite	80,156.70	**79,472.25**	57,341.70

Even though Brazil has most of its energy matrix coming from renewable sources, according to Ritchie et al. (2020), in 2023 Brazil emitted 2.38 billion tonnes of CO_2 equivalents, translating into 11.3 tonnes of CO_2 equivalents per person in the country. Therefore, the PV energy generation installed at the ICTS accounted for diminishing half of the yearly CO_{2eq} emissions of a single person in the worst-case scenario and almost the annual emissions of three persons in the best-case scenario. When considering the mean of both cases, the usage of PV generation compensates for approximately the amount of yearly emissions equivalent to 1.5 persons. It is a modest number compared to the total emissions of the country. However, it is already a significant reduction in emissions and environmental impact, while the generation can be further expanded to contribute even more to avoidance of emissions.

Meanwhile, it is a great case study to assess the implementation of renewable energy initiatives at a growing number of university campuses and smart campuses, and how these implementation can exponentially reduce the amount of CO_{2eq} emissions of these campuses, helping in the avoidance of overall GHG emissions of any country. The ICTS is a small campus when compared to other universities, which also indicates that the installation of renewable generation on other campuses can have a greater impact on GHG emissions reduction. In addition, any university with renewable energy systems implementation and sustainability in its curriculum could also work in developing initiatives to better raise awareness of people to avoid emissions and have less impactful ways of life and consumption.

The results presented can also be compared to other studies in literature which evaluate the GHG emissions avoided by the implementation of PV generation systems, even though none of these studies address these avoidances in a university campus context. Hudy et al. (2018) indicate that in 2014 the per capita emissions in Europe were 7.11 tonnes of CO_{2eq}, while Poland in the same year had an avoidance factor of 0.96 $kgCO_{2eq}/kWh$ for its PV generation. Considering these numbers, authors identify that if the non-renewable generation coming from coal is replaced by PV generation, panels with the area of 1 km^2 can compensate for the emissions of 5,700 to 7,900 EU citizens depending on the location where they are installed. These numbers are even better than the results presented here, since they are calculated considering a much larger PV generation scale, smaller per capita emissions and the $kgCO_{2eq}$ emissions avoided per kWh generated are more than triple the amount considered in our best-case scenario.

Rojas-Hernández (2022) presents a study of CO_2 emissions payback time (the time during operation stage that a PV system will take to avoid the number of gases emitted during the rest of its life cycle) based on PV generations installed in Costa Rica, considering a life cycle of 25 years of three different PV power plants. The results show that for all case studies, payback time exceeds the total lifecycle of the PV generation plants by approximately 10 years, demonstrating that even in simpler cases, such as the one addressed in this paper, this must be considered when analyzing the CO_2 emissions impact of PV generation.

Mohibullah et al. (2006) evaluated the CO_2 mitigation potential of several renewable energy sources per year in India, with the PV potential being 2,051.09 ×

10^6 kgCO_2/year. Finally, Akinyele et al. (2014) investigated the costs and CO_2 emissions avoided by implementing PV generation in four case studies in Nigeria, indicating that in a near future PV generation systems will be a financial competitive option over diesel generation and that in developing countries the cost of renewable generation is cheaper than in developed ones, achieving emissions reductions at a relatively lower cost.

All the presented cases are similar in scope to what is conducted in the present paper. However, due to the scale considered in these studies, it is difficult to compare their results with the ones presented here, something which also highlights the importance of evaluating the CO_{2eq} emissions avoided in smaller case studies and smart campuses scenarios.

It is also important to note that by adopting the methodology proposed by Amponsah et al. (2014), which considers the GHG emissions associated with the PV renewable energy life cycle, we have also accounted for any emissions associated with the PV generation process, which is not a common practice throughout general literature. Although this diminishes the amount of GHG emissions avoided when compared to situations where no emissions are associated with the renewable energy process, it is a more realistic approach. Additionally, the estimated emissions avoided are undervalued, since the amount of data available in the database does not fully contain the data of the 2 years and 9 months of functioning of the PV generation system, as stated previously.

4.2 *Energy Consumption Prediction Using ARIMA*

Figure 3 presents the results from the trained ARIMA model when predicting future values using only one previous sample in the prediction ($p = 1.0$) for the hourly sampled data. Similarly, Fig. 4 presents the results for the model trained with the daily sampled data and $p = 1.0$, while Fig. 5 refers to the results obtained when the model is trained with the daily sampled data and $p = 7.0$. Table 3 presents the most relevant error metrics for each one of the predictions, while some of the main descriptive statistics of the complete dataset, the test dataset and the model's predicted data are presented in Table 4.

Each error metric presented in Table 3 is now discussed. Firstly, the MAPE indicates the average magnitude of the absolute error produced by the model's predictions as percentage of the actual values. The values obtained indicate that, on average, the forecasts deviate by about 14.34%, 21.31% or 19.66%, respectively, with the results obtained with the hourly data prediction being the best ones and the results obtained with the daily data prediction using $p = 1.0$ the worst. 20% is not a small error, and even considering the best-case scenario, 14% is also not neglectable. However, for the application presented here, which addresses mainly an initial better understanding of the consumption patterns of the campus and evaluating the amount of consumption of the campus, while also relating this to the PV generation (how much of the consumption comes from it and how much more generation would be necessary for

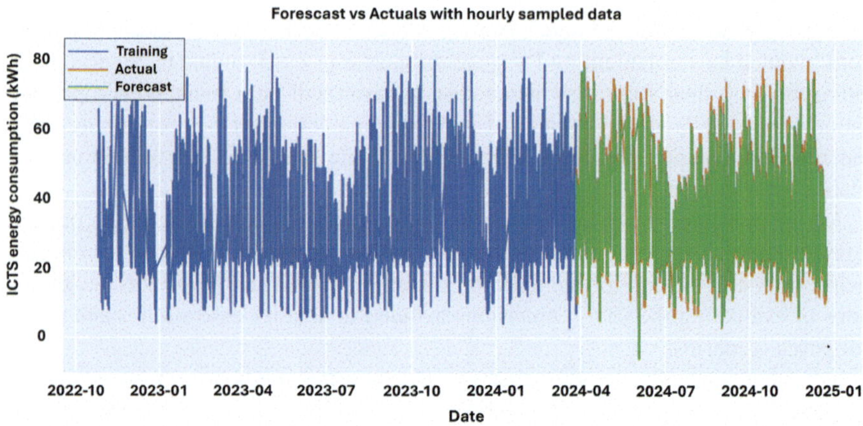

Fig. 3 ICTS energy consumption prediction using ARIMA with $p = 1.0$, $d = 0.0$ and $q = 1.0$ and hourly sampled data

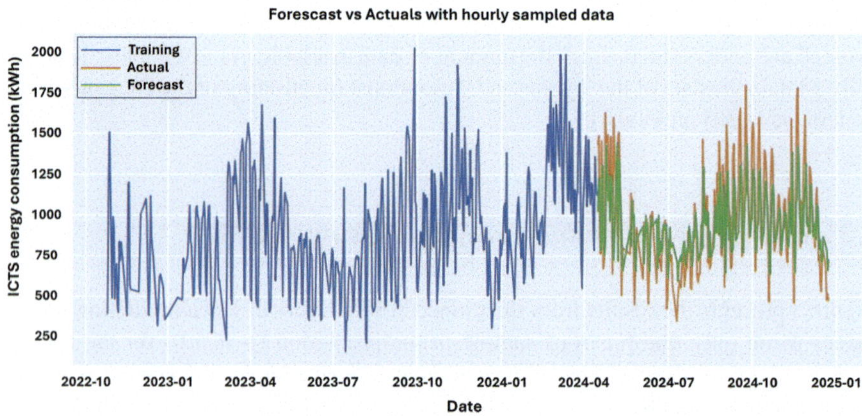

Fig. 4 ICTS energy consumption prediction using ARIMA with $p = 1.0$, $d = 0.0$ and $q = 1.0$ and daily sampled data

the campus to become fully energy auto-sustainable), this MAPE can be considered as not so relevant.

The MDAPE is similar to MAPE, but instead of taking the mean of the absolute percentage errors, it takes the median. Therefore, it is less affected by extreme values and gives a good measure of the central tendency of the prediction errors. From Table 3, it is seen that for each model, it is equal to 8.97%, 17.21% and 14.87%, respectively. These values being smaller than their respective MAPE indicate that the predicted data probably are not evenly distributed. Therefore, most of the time the prediction errors are smaller than what is indicated by MAPE, which is a good sign for the precision of the trained models when considering the number of predictions

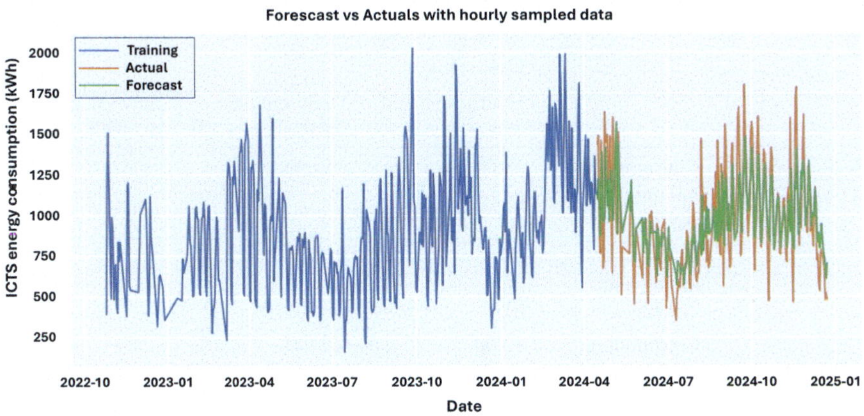

Fig. 5 ICTS energy consumption prediction using ARIMA with $p = 7.0$, $d = 0.0$ and $q = 1.0$ and daily sampled data

Table 3 Error metrics associated with each prediction case evaluated

	Case study		
	Hourly sampled data with $p = 1.0$	Daily sampled data with $p = 1.0$	Daily sampled data with $p = 7.0$
MAPE (%)	14.34	21.31	19.66
MDAPE (%)	8.97	17.21	14.87
MAE (kWh)	4.00	196.20	175.54
RMSE (kWh)	6.29	254.72	231.1

closer to the actual values in a batch of predictions. Once again, the hourly data model yields the smallest error, followed by the daily data with $p = 7.0$.

The MAE evaluates the accuracy of predictions or forecasts using the average of the absolute differences between the real values and the predicted ones. The values obtained from the models are 4.00 kWh, 196.20 kWh and 175.54 kWh. For this metric, it is more difficult to compare the error between the hourly data case and the daily data case since the absolute values of the test data have different magnitudes. However, if we take a ratio from these values based on the mean values of the test values presented for each case in Table 4, the MAE values are translated into 0.12, 0.20 and 0.18. Once again, the hourly consumption data has the best results, with the daily data with $p = 1.0$ having the worst results.

RMSE quantifies the average magnitude of errors between the predicted or forecasted values and the actual values but gives more weight to large errors than MAE, due to calculating the square root of the squared difference between real and predicted values. From the values in Table 3, one can see that when giving more weight to the extreme prediction values, the error of each model increases. Although when considering the ratio from RMSE value to the test data mean value the hourly data keeps

Table 4 Descriptive statistics of the original dataset, the test dataset and the predicted data

		Case study		
		Hourly sampled data with $p = 1.0$	Daily sampled data with $p = 1.0$	Daily sampled data with $p = 7.0$
Statistics of the complete dataset	Consumption minimum value (kWh)	1.00	159.00	
	Consumption mean value (kWh)	30.64	920.75	
	Consumption maximum value (kWh)	82.00	202300	
Statistics related to test and predicted data	Test data mean consumption value (kWh)	32.04	971.20	
	Predicted data mean consumption value (kWh)	31.83	941.06	968.73
	Difference between test data mean and predicted data mean values (kWh)	0.21	30.14	2.47

the best prediction results (0.20, 0.26 and 0.24, respectively), the absolute values increase relatively more for the hourly data than for the daily data when compared to the MAE, which indicates that more extreme values are present for the hourly data prediction.

Regarding the difference between the mean values of the test data and the predictions data presented in Table 4, for hourly data it is 0.21, for daily data with $p = 1.0$ it is 30.14, and for daily data with $p = 7.0$ it is 2.47, with the predicted data mean being smaller than the actual values for all cases. Although the absolute value obtained for hourly data is the smallest among the three, when comparing these values to their respective test data mean values, the results from the predictions using daily data and $p = 7.0$ yield the best values. This is explained since when considering the normalization of these values regarding the mean values of the test data, they represent a smaller magnitude (respectively 0.006, 0.031 and 0.002). The results from both the hourly model and the daily data model with $p = 7.0$ are satisfactory, providing a value considerably close to the actual one, and allowing for better insights into the mean consumption for the campus.

Finally, when analyzing the graphics presented comparing predictions and real test values, all three models have achieved good precision, following the trends observed in test data. However, for both models using daily data for training, the amplitude range has a clearer difference, with the predictions not achieving the highest and lowest values observed in test data.

After evaluating the three models' predictions based on six metrics, we can say that the model trained with hourly sampled data has the best results among the three. The better results found for this model can be attributed to the amount of data available for its training, with the hourly data having a total of 13,565 rows of data, which falls to 652 after resampling it to be daily. However, this improved precision comes with a much higher computational cost, with the model requiring almost 6 h to train, in contrast to the 5 min required to train the daily consumption model with $p = 7.0$. Additionally, the computational burden necessary to train and test the model with hourly data also prevented testing predictions of hourly consumption using $p = 24.0$ (hourly data from the previous day), since the amount of time necessary to train and validate the model overcomes the maximum connection time allowed at Google Colab, crashing the algorithm execution.

It is important to point out that for energy consumption management and energy dispatch, daily predictions are customarily more utilized, which means that the hourly data predictions must be further evaluated and adapted to be better used in practical case scenarios. Additionally, a more thorough data preprocessing and evaluation of different parametrizations for the ARIMA models can enhance the overall precision of the presented results.

Similarly to what was done in the previous subsection, the results of the consumption predictions conducted here can also be compared to other studies which utilized ARIMA models for time series prediction.

In their work, Sheoran et al. (2022) adopted ARIMA and an improved version of it (window sliding ARIMA, or WSARIMA) to predict wind speed and solar irradiance for wind and PV generation. In this context, data from two different locations in India were used, with the prediction of values sampled in a daily and weekly sampling rate. Considering only the ARIMA predictions conducted with different parametrizations depending on the sampling rate and location, the results for wind speed were: RMSE of 0.91 m/s for daily sampled data and 0.60 m/s for weekly sampled data (mean value of the original datasets is 2.64 m/s for both cases) in Dhanora; and RMSE of 0.97 m/s for daily sampled data and 0.68 m/s for weekly sampled data (mean values of the original datasets are 2.75 m/s and 2.74 m/s, respectively) in Nowlaipalle. And for the global irradiance, the ARIMA results are: RMSE of 100.39 W/m^2 for daily sampled data and 68.35 W/m^2 for weekly sampled data (mean values of the original datasets are 331.46 W/m^2 and 331.48 W/m^2, respectively) in Dhanora; and RMSE of 114.63 W/m^2 for daily sampled data and 64.88 W/m^2 for weekly sampled data (mean values of the original datasets are 341.90 W/m^2 and 341.93 W/m^2, respectively) in Nowlaipalle.

Gupta et al. (2019) also utilize ARIMA and an improved version of it (in this case ARIMA + Generalized Autoregressive Conditional Heteroskedasticity, or GARCH) to predict wind power with data from three different locations. The metrics presented within the results are the RMSE and MAE, with the values achieved being, respectively: 3.24 MW and 2.88 MW for data from Lake Bonney, 2.58 MW and 2.21 MW for data from Mount Millar and 1.27 MW and 1.09 MW for data from Starfish Hill. The real values to be predicted range from 5.36 to 9.38 MW for data from Lake Bonney, 4.66–8.15 MW for data from Mount Millar and 2.30–4.02 MW for data from Starfish Hill. Lastly, similarly to the present work, Chen et al. (2022) also utilize ARIMA and LSTM to load forecast, achieving an RMSE of 0.46 (absolute values of consumption data range from approximately 2.6–6.5), a MAPE of 9.37% and a worst relative error (WRE) of 24.12% when predicting consumption using data from a city. The model is also applied to three different customers' consumption data, with the results for RMSE, MAE and WRE being, respectively: 2.00, 1.71 and 16.68% for User2; 13.75, 7.67 and 736.26% for User4; and 12.42, 8.03 and 712.50% for User5. The mean consumption value for User2 is 26.27, for User4 is 53.72 and for User5 is 40.79.

When comparing the results of the references presented with the ones presented in this paper, it is possible to determine that these results are quite similar, indicating that even if there are issues associated with the predictions conducted here, the results are within the expected range for an ARIMA prediction model. Additionally, all references present at best only three different numeric metrics and one graphical metric for evaluating their results, while in the present paper four different numeric metrics and two qualitative metrics are adopted, enriching the analysis conducted. Finally, all references compared the ARIMA model results with another prediction methods, being them improved versions of ARIMA or totally different tools, having better results for these alternative methods than when utilizing ARIMA, which is a strong indicative that even though ARIMA is widely used in time series prediction and its results are satisfactory, it is possible to refine time series predictions even more when evaluating different predictions methodologies.

The results obtained in this work can be utilized for better planning the energy consumption coming from the utility, while the trends observed in the graphics can indicate in which months or periods of the day the consumption is higher. This information is crucial not only because it can bring economic advantages to the campus, but if studied more can also aid in the optimization of energy usage. A lower consumption yields a smaller number of GHG emissions associated with the campus. On the other hand, the mean daily consumption observed from the complete dataset (920.75 kWh/day) can be compared to the mean monthly energy generated from the PV generation panels (2.30 MWh), indicating that the current PV monthly generation can only provide energy for approximately two and a half days. Even though this is an initial estimation, it already indicates that for the ICTS to become fully maintained by this renewable energy source (therefore having the smallest amount of GHG emissions possible), the generation capacity would have to be increased by at least fifteen times its current capacity, while also having the employment of batteries to store surplus energy produced in the moments of the day with peak irradiance for later utilization.

Furthermore, all energy consumption registered can also be presented using CO_{2eq} as unity, something that shows a more comprehensible and straightforward status of the campus regarding emissions and is a better metric to help raise awareness of the campus members of its environmental impact. Converting the consumption unity of the mean values of the complete consumption history presented in Table 4, the GHG emissions associated with the campus hourly consumption can be 15,289.36 gCO_{2eq}, 22,459.12 gCO_{2eq}, 27,208.32 gCO_{2eq}, and 32,294.56 gCO_{2eq} if coming from non-renewable sources, while it would vary from 275.76 to 9,192.00 gCO_{2eq} if coming from PV generation. If this evaluation is conducted regarding daily sampled data, the emissions related to non-renewable sources would be 459,454.25 gCO_{2eq}, 674,909.75 gCO_{2eq}, 817,626 gCO_{2eq}, and 970,470.5 gCO_{2eq}, with the numbers varying from 8,286.75 to 276,225.00 gCO_{2eq} if coming completely from PV generation. When utilizing these numbers, the study conducted on energy consumption not only shows most of the GHG emissions associated with the campus but can also be directly related to the study about emissions avoided by PV generation, increasing the sustainability addressed by the present study.

Additionally, the work conducted here complements what is done in Zarpellon et al. (2024), providing a vision of how implementing a smart campus in a university can bring environmental value to the university campus. Such works are important to increase the promotion of sustainability initiatives in universities, guiding how to implement smart campuses and showing practical results that evidence its importance and the sustainability brought by its implementation.

5 Conclusion

This work evaluates the implementation of a smart campus at a university in Brazil (ICTS, UNESP), exploring the potential of renewable energy generation and energy consumption monitoring to enhance the sustainability of the campus. To this end, an evaluation of the photovoltaic (PV) generation system's contribution to the mitigation of greenhouse gas (GHG) emissions since its implementation was conducted. Additionally, the energy consumption historical data are utilized to train Autoregressive Integrated Moving Average (ARIMA) models to predict energy consumption, thereby facilitating a more effective consumption planning and management.

Regarding the PV generation, the findings of the study indicate that, given the current capacity, the ICTS has the potential to circumvent annual GHG emissions that are equivalent to half of the yearly CO_{2eq} emissions of a single person in Brazil in the worst-case scenario and the yearly emissions of roughly three persons in the best-case scenario. Even though these numbers may appear modest, they are noteworthy considering the underestimation of the evaluated data, which was influenced by instances of failure in the architecture connection, resulting in the non-registration of some of the data from the 2 years and 9 months since the installation of the PV generation system. Furthermore, the capacity of the generation installed can be expanded to contribute even more to the ICTS' GHG emissions reduction, while concurrently serving as a compelling case study for other universities with PV generation installed. This will facilitate understanding the extent to which emissions are being averted and the way it can contribute to the sustainability of their respective campuses.

For energy consumption prediction, three models of ARIMA were trained using different sampling rates and parametrization, namely: hourly sampled data with $p = 1.0$, $d = 0.0$, $q = 1.0$; daily sampled data with $p = 1.0$, $d = 0.0$, $q = 1.0$; and daily sampled data with $p = 7.0$, $d = 0.0$, $q = 1.0$. The best results were achieved with the model trained with hourly sampled data. This model returned a mean absolute percentage error (MAPE) of 14.34%, a median absolute percentage error (MDAPE) of 8.97%, a mean absolute error (MAE) of 4.00 kWh, a root mean squared error (RMSE) of 6.90 kWh, a difference between the mean value of the test data and the predictions of 0.21 kWh and a greater similarity between the plots of real values of the test data and the predicted ones, but with a considerably higher demand for computational resources when training and testing the model. While the results obtained can be further refined with enhanced preprocessing of the initial data, optimization of the model's parameters, use of computers with a better processing power and assessment of alternative prediction methodologies (such as long short-term memory, or LSTM), they successfully achieved the main objective of this work, which is to initially utilize and evaluate the utility of the monitored data. Additionally, these predictions, alongside the GHG emissions study, contribute to the state of the art of how smart campuses can contribute to their campus sustainability.

For future studies, it is recommended to improve the prediction models presented for a more effective energy consumption planning. It is also interesting to conduct studies that apply this predicted information to the campus' annual planning, with the aim of evaluating its practical impact on campus sustainability. Furthermore, as stated in the introduction, sustainability encompasses not only environmental impact but also encompasses other subjects, including people's well-being, education and the economy. Further studies regarding the smart campus implementation at the ICTS could also evaluate the sustainability contributions and impact of the campus on these other fields of sustainability.

Acknowledgements This work was supported in part by the São Paulo Research Foundation (FAPESP) under Grant numbers 2022/15423-3 and 2025/08820- 4, in part by the National Council for Scientific and Technological Development (CNPq) under Grant 304018/2025-2, and in part by the Coordination for the Improvement of Higher Education Personnel (CAPES) under Grant 001.

References

Akinyele DO, Nair NKC, Rayudu RK, Seah WKG (2014) Clean development mechanism projects for developing countries: Potential for carbon emissions mitigation and sustainable development. In: 2014 18th National Power Systems Conference, NPSC. IEEE, p 1–6

Amponsah NY, Troldborg M, Kington B, et al (2014) Greenhouse gas emissions from renewable energy sources: A review of lifecycle considerations. Renewable and Sustainable Energy Reviews 39:461–475. https://doi.org/10.1016/j.rser.2014.07.087

Bandeira L, Bandeira B, Neto M de SA, Casimiro A (2022) SMART CAMPUS NO BRASIL: um estudo exploratório [SMART CAMPUS IN BRAZIL: an exploratory study]. In: 11° IFBAE

Congresso do Instituto Franco-Brasileiro de Administração de Empresas. Rennes, France, p 1–16
Bhaskar K, Singh SN (2012) AWNN-Assisted Wind Power Forecasting Using Feed-Forward Neural Network. IEEE Transactions on Sustainable Energy 3:306–315. https://doi.org/10.1109/TSTE.2011.2182215
Breiman L (2011) Statistical Modeling: The Two Cultures. Statistical Science 16:199–215. https://doi.org/10.1214/ss/1009213726
Breyer C, Koskinen O, Blechinger P (2015) Profitable climate change mitigation: The case of greenhouse gas emission reduction benefits enabled by solar photovoltaic systems. Renewable and Sustainable Energy Reviews 49:610–628. https://doi.org/10.1016/j.rser.2015.04.061
Brito de Jesus RG, Gomes Camelo AK (2024) SMART CAMPUS : caminhos possíveis para universidades brasileiras [SMART CAMPUS: possible paths for Brazilian universities]. Revista Brasileira de Aprendizagem Aberta e a Distância 24:1–19. https://doi.org/10.17143/rbaad.v24i3.785
Chagnon-Lessard N, Gosselin L, Barnabe S, et al (2021) Smart campuses: Extensive review of the last decade of research and current challenges. IEEE Access 9:124200–124234. https://doi.org/10.1109/ACCESS.2021.3109516
Chen S, Lin R, Zeng W (2022) Short-Term Load Forecasting Method Based on ARIMA and LSTM. In: 2022 IEEE 22nd International Conference on Communication Technology (ICCT). IEEE, p 1913–1917
Cui Q, Kan H, Zhao J, et al (2017) Current status of smart city researches in China using the co-word analysis method. In: 4th International Conference on Smart and Sustainable City, ICSSC 2017. p 4–9
Facens (2019) Home - Smart Campus Facens. https://smartcampus.facens.br/en/. Accessed 29 jun 2025
Gupta A, Sharma KC, Vijayvargia A, Bhakar R (2019) Very Short term Wind Power Prediction Using Hybrid Univariate ARIMA-GARCH Model. In: 2019 8th International Conference on Power Systems: Transition towards Sustainable, Smart and Flexible Grids, ICPS 2019. IEEE, p 1–6
Heavy.AI (2022) Statistical Modeling. https://www.heavy.ai/technical-glossary/statistical-modeling. Accessed 1 mar 2025
Hudy W, Piaskowska-Silarska M, Noga H, et al (2018) Analysis of the possibility of reducing the amount of air pollution using photovoltaic systems. In: Proceedings of the 2018 19th International Carpathian Control Conference, ICCC 2018. IEEE, p 565–569
Institute of Science and Technology Sorocaba (2022) SMART CAMPUS ICTS. https://smartcampus.sorocaba.unesp.br/d/WusCo658k9/main-screen?orgId=1&kiosk. Accessed 29 jun 2025
Instituto Mauá de Tecnologia (2018) Home - Smart Campus Mauá. https://smartcampus.maua.br/. Accessed 29 jun 2025
Kumar NM, Goel S, Mallick PK (2018) Smart cities in India: Features, policies, current status, and challenges. In: International Conference on Technologies for Smart City Energy Security and Power: Smart Solutions for Smart Cities, ICSESP 2018 - Proceedings. IEEE, Bhubaneswar, India, p 1–4
Kurniawan B, Supangkat S, Lumban Gaol F, Ranti B (2022) Framework for Developing Smart City Models in Indonesian Cities (Based On Garuda Smart City Framework). In: 2022 International Conference on ICT for Smart Society. IEEE, p 1–5
Mann MK, Spath PL (2001) A life cycle assessment of biomass cofiring in a coal-fired power plant. Clean Products and Processes 3:81–91. https://doi.org/10.1007/s100980100109
Mohibullah, Imdadullah, Ashraf I (2006) Estimation of CO2 mitigation potential through renewable energy generation. In: First International Power and Energy Conference, (PECon 2006). Putra Jaya, Malaysia, p 24–29
Peng J, Lu L, Yang H (2013) Review on life cycle assessment of energy payback and greenhouse gas emission of solar photovoltaic systems. Renewable and Sustainable Energy Reviews 19:255–274. https://doi.org/10.1016/j.rser.2012.11.035

Ritchie H, Rosado P, Roser M (2020) Greenhouse gas emissions. In: OurWorldinData.org. https://ourworldindata.org/greenhouse-gas-emissions. Accessed 1 mar 2025

Rojas-Hernández I (2022) CO2 Emission Payback Time of c-Si Photovoltaic Systems in Costa Rica. In: 2022 IEEE 40th Central America and Panama Convention, CONCAPAN 2022. p 1–5

Shahriari S, Sisson SA, Rashidi T (2025) Modelling time series with temporal and spatial correlations in transport planning using hierarchical ARIMA-copula Model: A Bayesian approach. Expert Systems with Applications 274:1–19. https://doi.org/10.1016/j.eswa.2025.126977

Sheoran S, Pasari S, Shukla S (2022) Application of Window Sliding ARIMA in Wind Speed and Solar Irradiance Forecasting. In: 2022 IEEE 7th International Conference on Recent Advances and Innovations in Engineering (ICRAIE). IEEE, p 321–325

Supangkat SH, Arman AA, Nugraha RA, Fatimah YA (2018) The implementation of Garuda Smart City framework for smart city readiness mapping in Indonesia. Journal of Asia-Pacific Studies 32:169–176

Torres-Sospedra J, Avariento J, Rambla D, et al (2015) Enhancing integrated indoor/outdoor mobility in a smart campus. International Journal of Geographical Information Science 29:1955–1968. https://doi.org/10.1080/13658816.2015.1049541

U.S. Environmental Protection Agency (2025a) Overview of Greenhouse Gases. https://www.epa.gov/ghgemissions/overview-greenhouse-gases. Accessed 1 mar 2025

U.S. Environmental Protection Agency (2025b) Understanding Global Warming Potentials. https://www.epa.gov/ghgemissions/understanding-global-warming-potentials. Accessed 1 mar 2025

United Nations (2015) The Sustainable Development Agenda. https://www.un.org/sustainabledevelopment/development-agenda/. Accessed 3 nov 2023

United Nations (2020) The 2030 Agenda for Sustainable Development's 17 Sustainable Development Goals (SDGs). In: 4th Sustainable Development Goals Youth Summer Camp

Universidade Estadual de Campinas (2017) Smart Campus - Unicamp. https://smartcampus.prefeitura.unicamp.br/. Accessed 29 jun 2025

Yang J, Kwon Y, Kim D (2021) Regional smart city development focus: The South Korean National Strategic Smart City Program. IEEE Access 9:7193–7210. https://doi.org/10.1109/ACCESS.2020.3047139

Zarpellon BO, Arenas L de O, Godoy EP, et al (2024) Design and implementation of a smart campus flexible Internet of Things architecture on a Brazilian university. IEEE Access 12:113705–113725. https://doi.org/10.1109/ACCESS.2024.3444471

Bruno Oliveira Zarpellon received the B.S. degree in automation engineering from FACENS, Sorocaba, in 2020, and the M.Sc. degree in electrical engineering from São Paulo State University (UNESP), Sorocaba, São Paulo, in 2024. Currently is a Ph.D. student at the São Paulo State University (UNESP), Bauru. He did exchange studies with the University of Wisconsin–Madison, in 2015 and 2016, and did an internship exchange with the Fraunhofer Institute for Production Technology, Aachen, in 2018 and 2019. His research interests include smart campuses, the Internet of Things, PCB design, smart metering devices and microgrids.

Eduardo Paciência Godoy received the B.Eng. degree in control and automation engineering from the Federal University of Itajubá, Minas Gerais, Brazil, in 2003, and the M.Sc. and Ph.D. degrees in mechanical engineering from the University of São Paulo, São Paulo, Brazil, in 2007 and 2011, respectively. Currently, he is an Associate Professor with São Paulo State University (UNESP), São Paulo. His research interests include automation and networked systems, wireless networks and telemetry, the Internet of Things, and Industry 4.0.

Fernando Pinhabel Marafão received the B.S. degree in electrical engineering from São Paulo State University (UNESP), Brazil, in 1998, and the M.Sc. and Ph.D. degrees from the University of Campinas (UNICAMP), Brazil, in 2000 and 2004, respectively. In 2002, he joined the

Power Electronics Group, University of Padua, Italy, as a Visiting Student. In 2013, he joined Colorado School of Mines, Golden, CO, USA, as a Visiting Scholar on autonomous and intelligent distributed energy systems. Since 2005, he has been an Associate Professor with the Group of Automation and Integrating Systems, UNESP. His current research interests include smart grid technologies, renewable energies, energy management, and power theories. He is a member of Brazilian Power Electronics Society (SOBRAEP) and Brazilian Automation Society (SBA).

Luis de Oro Arenas received the B.S. degree in electronic engineering from the National University of Colombia, Bogotá, Colombia, in 2012, and the M.Sc. and Ph.D. degrees in electrical engineering from São Paulo State University (UNESP), Sorocaba, Brazil, in 2014 and 2019, respectively. He conducted his postdoctoral studies with the Power Electronics Laboratory, UNESP/FEIS, in 2019, and the Automation and Integrated Systems Group (GASI), UNESP, from 2020 to 2022. He is currently an Assistant Professor with the Department of Environmental Engineering, ICTS, UNESP. His research interests include electronic instrumentation and data science applied to environmental systems, power electronics, power quality, and digital and analog signal processing.

Helmo K. Morales Paredes received the B.Sc. degree in electrical engineering from San Agustin National University, Arequipa, Peru, in 2002, and the M.Sc. and Ph.D. degrees in electrical engineering from the University of Campinas, Campinas, Brazil, in 2006 and 2011, respectively. In 2009, he was a Visiting Student with the University of Padova, Padua, Italy. In 2014 and 2018, he was a Visiting Scholar with the University of Nottingham, Nottingham, U.K., and the Colorado School of Mines, Golden, CO, USA, respectively. Since December 2011, he has been an Associate Professor with São Paulo State University (UNESP), Sorocaba, Brazil. From 2017 to 2023, he led the Automation and Integrated Systems Group, UNESP. His research interests include power quality, harmonics and unbalance propagation, grid-connected converters for renewable energy systems, and microgrid control. Dr. Morales Paredes is a Senior Member, IEEE and a Member of the Brazilian Power Electronics Society and the Brazilian Automation Society. He received the Prize Paper Award from the IEEE Transactions on Power Electronics in 2011, Best Paper Awards from the Brazilian Power Quality Society in 2021 and 2023, and second place in the Typhoon HIL Industrial Challenge at SEPOC 2025.

Putting the Green Campus Initiatives of Higher Education Institutions' (HEIs) into Practice: Their Impact on Students' Proactivity Towards Sustainable Development

Jucelia Appio Frizon, Teresa Eugénio, Isadora Flessak, and Nelson Natalino Frizon

Abstract Education is a powerful tool for behaviour change, and therefore higher education institutions (HEIs) have a fundamental role to play. This study examines the relationship between the green campus initiatives of a Brazilian public university in Paraná, Brazil, and students' proactivity towards sustainable development (SD). An online survey, with a quantitative-descriptive approach, was distributed through Google Forms to the students of the university, obtaining a sample of 310 respondents. The survey questions were pre-tested and scored on a 5-point Likert scale. The results of the multiple regression analysis, adjusted R^2 25% ($p < 0.001$), show that the independent constructs are: Extracurricular activities on SD, approach to SD in the classroom and green campus operations have a positive and significant effect on students' proactivity towards SD. An underlying argument is that students engage in SD by taking the best university initiatives as a precedent, whereby university initiatives can lead people to proactive behaviour towards SD. These findings can be useful for those involved in planning SD-oriented actions in HEIs in any part of the world, such as the Asia-Pacific region. It reinforces the role of HEIs as co-creators of

J. A. Frizon
Center for Applied Social Sciences, LeiriaWestern Paraná Sate University (UNIOESTE), Francisco Beltrão-PR, Brazil
e-mail: jucelia.frizon@unioeste.br

J. A. Frizon · T. Eugénio (✉)
Center for Applied Research in Management and Economics (CARME), Polytechnic Institute of Leiria, Leiria, Portugal
e-mail: teresa.eugenio@ipleiria.pt

I. Flessak · N. N. Frizon
Center for Applied Social Sciences, Western Paraná State University (UNIOESTE), Francisco Beltrão-PR, Brazil
e-mail: isadora@flessak.com.br

N. N. Frizon
University Institute of Lisboa (ISCTE), Lisboa, Portugal

W. Leal Filho et al. (eds.), *Sustainability in the Curriculum and Operations of Universities*, World Sustainability Series,
https://doi.org/10.1007/978-3-032-12678-8_44

change by promoting the principles of SD among future professionals, seeking ways to expand the dialogue on SD in HEIs, especially when it comes to achieving the SDGs, intertwined with the idea of increasing student participation and engagement, and contributing to prosperity and peace.

Keywords Green campus · Higher education institutions · Sustainable development · Students · Brazil

1 Introduction

The increasing demand for sustainable practices requires higher education institutions (HEIs) to be agents of environmental change (Cho, 2019), taking a leadership role in the development and implementation of sustainable practices and initiatives (Aleixo et al., 2018; Berchin et al., 2021). The leadership role of higher education institutions should be linked to their actions and become spaces of sustainability (Monavvarifard et al., 2019).

Achieving the Sustainable Development Goals (SDGs) of the United Nations UN) 2030 SD Agenda (United Nations, 2015) is a challenge, and HEIs are consistently seen as contributing to the pursuit of sustainability initiatives. For example, through knowledge creation and dissemination, research, extension, management activities and operations (Aleixo et al., 2021; Leal Filho et al., 2023; Leal Filho et al., 2019b; Leal Filho et al., 2019f; León-Fernández et al., 2018; Monavvarifard et al., 2019; Ribeiro et al., 2021), promoting education for sustainable development (Leal Filho et al., 2019c), teaching and learning for sustainability (Leal Filho et al., 2023), carrying out the management of environmental impacts caused by activities (Leal Filho et al., 2019e; León-Fernández et al., 2018), introducing active policies and promoting sustainable experiences (Leal Filho et al., 2019d). Moreover, the application of these policies and practices in universities can be interesting for developing students' environmental awareness in their academic education (Moqbel et al., 2020).

While it is recognised that sustainability research in higher education demonstrates the academic excellence of HEIs, there is a paucity of literature that specifically evaluates how this work is being undertaken (Leal Filho et al., 2023), yet further investigation into the changes within the university is required to maximise its role in climate action (Brandli et al., 2023). This is in line with the argument that sustainability is a social construct and that it is only through the involvement of multiple stakeholders that a holistic definition can be constructed. This study examines the relationship between the Green Campus Initiatives of in this study examines the relationship between the Green Campus Initiatives of a Brazilian public university in Paraná, Brazil, and students' proactivity towards SD. To this end, a cross-sectional survey with a descriptive design and quantitative approach was chosen, addressed to students associated in a university.

The novelty of this research is based on exploring the perception of students in a Brazilian public university regarding green campus practices that affect their proactivity, providing quantitative information. Therefore, to what extent are green campus practices perceived by students?

To the best of our knowledge, this study is one of the first efforts to investigate students' perceptions of environmentally sustainable practices at a Brazilian public university in Paraná, Brazil. This study provides suggestions for sustainable paths towards greener universities.

This study adopts a bottom-up approach (León-Fernández et al., 2018) when studying students. The participatory approach can be seen as both a requirement and a benefit for the global paradigm shift towards SD, contributing to the integration of academic culture (Disterheft et al., 2015). Homer and Khor (2022) recommend activities that enhance the adoption of sustainable development by making universities a 'living lab' for sustainability.

It is sought to broaden the dialogue on SD in universities and the idea of participation, particularly from the perspective of student engagement, as students are the future leaders. It is assumed that Green Campus initiatives are important for the dissemination of SD concepts in universities. This concept supports the argument that "dialogue transforms those who engage in it" (UNESCO, 2022, p. 58).

After this introduction, environmental education in HEIs and Green Campus issues are presented and the hypothesis is established. The methodology of the study is then described and the results and conclusions are presented and discussed.

2 Environmental Educations in Higher Education Institutions

In recent decades, there has been a remarkable increase in the international debate on integrating sustainability in higher education. Despite considerable efforts, there are still many challenges (Leal Filho et al., 2023) as the issue of sustainability in higher education has become an urgent need rather than a choice (Cho et al., 2020).

Environmental education in the academic field received greater international attention in the 1970s through attempts to highlight the importance of the environment and promote environmental awareness (Berchin et al., 2021; Lozano et al., 2015) in order to increase the sense of responsibility and solidarity between people and nations, thus enabling the resolution of environmental challenges at local and global levels (UNESCO-UNEP, 1977). Since then, environmental education has evolved from a strictly environmental perspective to a more comprehensive and holistic concept that includes social and economic issues (Berchin et al., 2021).

Since the Stockholm Declaration of 1972, there has been a steady development of national and international sustainability declarations relevant to higher education. The Stockholm Declaration was the first declaration to refer to sustainability in higher education, albeit indirectly, by discussing inter- and intra-generational equity. The Principle 19 states that education "broaden the basis for enlightened opinions and responsible conduct by individuals, enterprises and communities in protecting and improving the environment in its full human dimension" (UNESCO, 1972, p. 5).

The Conference on Environmental Education in Tbilisi, of the 1977, has been one of the most important moments in the evolution declarations related to environmental education, arguing that should be provided to people of all ages, all levels of ability and taught in formal and non-formal settings (UNESCO-UNEP, 1977).

Since the Talloires Declaration of 1990, HEIs have been guided to set an example of environmental responsibility by establishing institutional policies and practices for ecology, resource conservation, recycling, waste reduction, and environmentally sound operations (UNESCO, 1990).

The Halifax Declaration was launched in December 1991 at the conference on university action for SD in Halifax, Canada, highlights the role that universities could play, at local, national and international levels, in improving environmental and development issues such as the continuing widespread degradation of the environment, poverty and the devastating effects of unsustainable environmental practices in order to contribute to SD (International Institute for Sustainable Development (IISD), 2003).

In 1992, Agenda 21 was the result of the United Nations Conference on Environment and Development held in Rio de Janeiro, Brazil, which became known as Rio-92. Chap. 36 deals with Education, Awareness and Training, specifically on issues related to sustainability in education, including initiatives that individuals, governments and nations can take to ensure Sustainable Development, recognizing that various countries must develop their own programs according to their specific needs, policies and responsibilities. The three main thrusts are: (a) reorienting education towards Sustainable Development; (b) increasing public awareness of environmental issues; and (c) promoting environmental training for educators (United Nations Conference on Environment & Development [UNCED], 1992).

2.1 Green Campus

The lack of change structures can be seen as an important reason why sustainability remains a niche topic in most higher education organisations. Adomßent et al. (2019) highlight the crucial importance of self-efficacy approaches, in particular the Green Office Model, which aims to strengthen collaboration between all members of the higher education community.

The Green Campus concept is considered to be in a process of continuous improvement and guided by the precepts of sustainability, with particular attention to sustainable education and the integration of campus infrastructure, with the aim of continuously integrating SD concepts into all aspects of the campus (Zhu & Dewancker, 2021).

Green campus policies promote environmentally friendly practices in university operations and infrastructure, leading to an increase in initiatives related to waste reduction, energy efficient buildings, sustainable transport and organic food (Sima et al., 2019). This requires appropriate institutional structures capable of dealing with such complexity, and an active community with a high level of commitment. For example, if management systems for SD in higher education institutions are only implemented through a top-down process, it may achieve improvements within universities but exclude the educational aspect (Disterheft et al., 2012).

The participatory approach can be seen as both a requirement and a benefit for the global paradigm shift towards SD, contributing to the integration of academic culture (Disterheft et al., 2015). This call is supported by Disterheft et al. (2012), who studied universities across Europe. The authors emphasise a top-down approach, stating that it can be less time consuming, more focused on improving environmental and operational performance, and oriented towards compliance with regulations and administrative requirements, while a participatory (bottom-up) approach, in a more systemic perspective, fosters the new development of teaching and learning environments. Furthermore, a participatory approach, while more time and personnel intensive, focuses on empowerment, combining technical and analytical skills with the creation of understanding and awareness to achieve not only better environmental performance, but also a better embedding of SD as a whole.

León-Fernández et al. (2018) state that a process of improvement in green campus management can only be guaranteed if there is social commitment, measured in terms of actual participation. This relationship is confirmed by Dagiliūtė et al. (2018) comparative study on students' perceptions of green and non-green universities in Lithuania. The results showed that students from a green university are more proactive and have a better understanding of the concept of SD. In the same vein, Eugénio et al.'s (2021) study with students in Malaysia and the Philippines also found that the higher the university's concern for sustainability, the higher the students' attitudes, subjective norms and behavioural control towards environmental sustainability and, indirectly, the higher the students' intention to engage in sustainability.

Several studies have concluded that students consider SD essential in the meantime there seems to be a gap between knowing the concept and putting it into action (Dagiliute et al., 2018; Ribeiro et al., 2021), as it can be challenging for students to associate concepts and put them into practice (Rampasso et al., 2019), more than that, students are more likely to be proactive when they have SD knowledge, while HEIs are the mirror that provides them with opportunities to participate in pro-SD efforts in collaborative systemic thinking. In this line of thinking, HEIs are profoundly responsible for transferring knowledge to create a more sustainable future (Shawe et al., 2019). Indeed, this view is present in the Lüneburg Declaration (2001), which emphasizes that HEIs have a catalytic role in providing knowledge for SD. The study of Lee et al. (2017) further validates this observation by reporting that student knowledge related to SD results in significant changes in attitudes, intentions, and behaviour.

Based on the studies revisited, the research hypothesis is presented.

H1. Green Campus initiatives of a Brasileira HEIs positively affect students' Proactivity towards SD.

The study in a Spanish HEI by León-Fernández et al. (2018) substantiates this discussion by arguing that developing a participatory action plan with a set of goals, objectives, indicators, and promoters can be a good starting point for other HEIs that wish to improve their environmental management and involve community members in their progress toward SD, thereby contributing to the environmental awareness and education of the actors involved. For this purpose, as part of university education for sustainable development, teachers and students alike have a responsibility in reaching these goals (Adomßent et al., 2019).

3 Materials and Methods

This study examines the relationship between the green campus initiatives of a Brazilian public university in Paraná, Brazil, and students' proactivity towards SD. For this purpose, the students' perceptions were collected in February 2023 through a structured questionnaire with an online application via Google Forms. Therefore, it is characterised as a descriptive study supported by the quantitative approach.

3.1 Survey Procedures

A questionnaire survey was developed to assess students' perceptions of green campus initiatives on the part of the HEI, and students' proactivity towards SD. We based our categories on a literature review of empirical studies such as Aleixo et al. (2021), who attempted to analyse students' perceptions of SD in Portuguese public HEIs; Ribeiro et al. (2021), who analysed whether the Green Campus initiatives of four Brazilian universities are a viable strategy to disseminate SD concepts in HEIs, increase students' knowledge and proactivity about SD; Eugénio et al. (2021), who investigated whether the level of sustainability concern of HEIs in Malaysia and the Philippines is positively associated with students' intention to engage in sustainability; Monavvarifard et al. (2019), who explored the factors of Iranian students' engagement in the process of value co-creation for SD; and Dagiliūtė et al. (2018), who analysed students' perceptions of differences in sustainability performance between two Lithuanian higher education institutions, Vytautas Magnus University (a non-green university) and Kaunas University of Technology (a green university).

The questionnaires, published in English, were translated into Portuguese. In order to assess the content validity, a consensus was sought among teachers and researchers who are experts in the content area (Alexandre & Coluci, 2011). The questionnaire had two sections. The first concerned the profile of the respondents, while the second section sought to measure students' perceptions of the university's green campus initiatives (Appendix 1). All variables were measured using a 5-point Likert scale (1 strongly disagree; 2 disagree; 3 undecided; 4 agree; 5 strongly agree).

Appendix 1 Green Campus initiatives and student proactivity regarding SD scale
Please indicate the extent to which you disagree/agree with the statements by using the following scale:

	1. Strongly Disagree	2. Disagree	3. Undecided	4. Agree	5. Strongly Agree
[1] level of approach to sustainable development (SD) in the classroom					
At least once a semester, I have classroom activities related to economic, environmental and/or social sustainability					
Teachers usually address SD issues during their lessons					
[2] extracurricular activities on SD					
The university where I study promotes pro-sustainability actions					
My university promotes research activities related to SD					
My university promotes outreach activities related to sustainability					
My university promotes quizzes/challenges/games related to SD					
My university promotes lectures/workshops on SD					
[3] communication on SD					
There are communications in my university that help promote SD					
The university website presents information on its pro-sustainability actions					

(continued)

Appendix 1 (continued)

	1. Strongly Disagree	2. Disagree	3. Undecided	4. Agree	5. Strongly Agree
I have read strategic documents of my university on SD					
My university clarifies my questions on SD					
[4] green infrastructure on campus					
I consider that the campus infrastructure adheres to the concepts of sustainability					
My university has actions to promote the use of renewable energy					
My university has actions to promote energy efficiency					
My university has actions to promote water consumption efficiency					
There is a possibility of recycling waste at my university					
My university has actions to promote waste management					
My university encourages sustainable transport					
[5] level of proactivity in relation to SD					
Initiatives related to waste sorting are included in my daily life					

(continued)

Appendix 1 (continued)

	1. Strongly Disagree	2. Disagree	3. Undecided	4. Agree	5. Strongly Agree
Pro-sustainability initiatives related to saving electricity are included in my daily life					
Initiatives related to the use of sustainable transport are included in my daily activities					
Pro-sustainability initiatives related to water use are included in my daily life					
I reuse things that can be useful to me or to other people					
I buy energy-efficient appliances					
Initiatives related to social activities are included in my daily life					
I take part in events or activities that support sustainability					

3.2 *Data Collection and Sampling*

The sample selection strategy was non-probabilistic by convenience. Invitation letters were sent to the most diverse sectors of the HEI to disseminate the survey to the greatest number of students. The population studied was the students enrolled in 2023 at there were 1,387 students regularly enrolled in undergraduate courses in Administration, Economic Sciences, Law, Social Work, Nutrition, Medicine, Bachelor's Geography, Bachelor's Geography and Pedagogy, and they were invited to take part in the survey. To this end, considering a tolerable sampling error of 5% (BARBETTA, 2004), a sample of 310 students was required. We also followed the recommendation of Hair Jr et al. (2005) that the n should be at least five respondents per research item (26 variables × 5 = 130). Students voluntarily participated in the study. The valid sample consisted of 301 students.

3.3 Data Analysis

The data collected via Google Forms were analysed with IBM Statistical Package for the Social Sciences (SPSS) Statistics 21. Initially, descriptive statistics were carried out to present the profile of the students who responded to the survey. Next, the variables relating to student perceptions of the HEI's green campus initiatives and student productivity in relation to the SD were analysed by mean, followed by Spearman correlation test and multiple linear regression analysis.

4 Results

4.1 The University

The University, which is being analysed, has some green campus initiatives in 2023, such as: Selective waste collection; Medicinal Garden; Use of renewable energy sources; Gardening projects and planting native trees; Encouraging the use of recycled paper; Encouraging car sharing, a system to control drinking water consumption. As far as education is concerned, the curricular units deal with the subject, for example, in the Administration Course: Environmental Management; Economics Course: Environmental Economics; Law Course: Environmental Law; Geography Course: Fundamentals of Environmental Planning and Management; Applied Environmental Planning and Management (among others); Medicine Course: Collective Health II (there is a specific point in the syllabus for environmental education policies); Nutrition Course: Food and Nutrition Security (a point in the syllabus deals with Products, production and consumption: impact on the environment and health); Nutrition course: Nutrition in Collective Feeding (a point in the syllabus deals with Waste management and sustainability in collective feeding); Pedagogy course: Geography in school teaching (two points in the syllabus deal with environmental education. National Curriculum Guidelines for Environmental Education); Social Work course: Agrarian Question, Rurality's and Regional Development (in general, the subject deals with The Green Revolution. Private appropriation of natural resources and environmental impacts. The process of agricultural modernisation and environmental education).

At the campus, extension projects such as the "Assessment of the quality of the waters of the Marrecas River in the Municipality of Francisco Beltrão, Paraná", linked to the Health Sciences Centre; the first Agroecology Meeting linked to the Administration course; the gardening project entitled green campus and student proactivity. The university was awarded the SESI ODS 2021 label in the Universities category. The award-winning project is aligned with the goals of health and well-being, quality education, gender equality, drinking water and sanitation, decent work and economic growth. This project evaluates the impact of exposure to pesticides on the health of women with breast cancer.

The Regional Sustainable Development and Innovation Agency (AGEUNI) was set up on the campus. This agency acts as a gateway to society's demands, both from the private sector and the municipality. The university is also mentioned in the fourth edition of the Impact Rankings 2022 by Times Higher Education (THE), a magazine linked to the British newspaper The Times. It is ranked 29th in the country and 801-1000th in the world. It stands out for its activities that contribute to achieving SDGs 2, 3, 8 and 17.

4.2 Sample Profile and Descriptive Analysis

With regard to gender, the majority of respondents identified themselves as female, accounting for 72% of responses. This discrepancy may be indicative of a greater willingness on the part of women to participate in research, or perhaps there is a more significant presence of women in certain courses or areas of activity. It is important to note that the analysis of these variables can provide valuable insights for future research investigations.

In terms of age, the vast majority of respondents are aged between 15 and 25, accounting for 84.51% of the total, followed by 12.90% aged between 26 and 35, 1.95% aged between 36 and 45 and 0.64% aged between 46 and 55.

With regard to the courses with the highest percentage of participation, it is interesting to note that administration and medicine had 17.41% each, followed by pedagogy with 17.10%. Economic sciences accounted for 14.20% of the total; law 13.87% of the survey; geography and nutrition both with 8.71% and social work 2.59%.

Based on the results of green campus initiatives by HEIs and student proactivity, some results stand out. Students seem to consider it important: "I reuse things that can be useful to me or to other people" (Mean: 3.69); "Pro-sustainability initiatives related to water use are included in my daily life" (Mean: 3.37); "Pro-sustainability initiatives related to saving electricity are included in my daily life" (Mean: 3.27); "I buy energy-efficient appliances" (Mean: 3.33); "Initiatives related to waste sorting are included in my daily life" (Mean: 3.18); "There is a possibility of recycling waste at my university" (Mean: 3.16).

The survey also pointed to teaching activities that are little perceived by students, highlighting: "My university promotes quizzes/challenges/games related to SD" (Mean: 2.06); "I have read strategic documents of my university on SD" (Mean: 2.13); "My HEI has actions to promote the use of renewable energy" (Mean: 2.23); "My university has actions to promote energy efficiency" (Mean: 2.28); and "At least once a semester, I have classroom activities related to economic, environmental and/or social sustainability" (Mean: 2.32).

As pointed out in the study by Aleixo et al. (2021) students believe that higher education institutions can and should have more initiatives in the area of sustainable development. This indicates that there is room for improvement in the dissemination and promotion of sustainable actions and initiatives at the educational institution analyzed, both in terms of offering sustainability-related activities in the classroom and in communicating the information and resources available. In addition, teachers' commitment to addressing sustainability issues in their lessons does not seem to be widely perceived by the students surveyed. These results are in line with the literature (Dagiliūtė et al., 2018; Ribeiro et al., 2021).

The results corroborate the findings of the study conducted by Aleixo et al. (2021) in which the majority of students are concerned about climate change, the majority talk about some reuse, reduction and recycling practices, but fewer students contribute to SD by participating in activities (e.g. environmental or community volunteering) organized by environmental organizations or the HEI. This also corroborates the study by Ribeiro et al. (2021) indicating a more utilitarian view, based on obtaining a benefit, i.e. saving resources. In the social sphere, there is low adherence to daily social activity initiatives and participation in SD-related events or activities compared to the environmental sphere. A similar result was found in other studies (Rampasso et al., 2019).

4.3 Hypothesis Test

Since the studies used as a basis for the questionnaire were Aleixo et al. (2021), Ribeiro et al. (2021), Eugénio et al. (2021), Monavvarifard et al. (2019), Dagiliūtė et al. (2018) and these validated the research instruments, exploratory factor analysis was not carried out in this study. Based on this, there are five constructs explored in this research: [1] Level of approach to Sustainable Development (SD) in the classroom; [2] Extracurricular activities on SD; [3] Communication on SD; [4] Green infrastructure on campus; [5] Level of proactivity in relation to SD.

The constructs were administered by the average value (arithmetic mean) of the variables that made up each construct, followed by Spearman correlation test and multiple linear regression analysis.

Bivariate correlation analysis is a statistical technique that measures the relationship between two variables. To this end, in this research we opted for Spearman correlation, which is a non-parametric correlation also known as rank-based correlation coefficients. It can be used to assess the strength and direction of the association between two variables and can be expressed in terms of a correlation coefficient. The correlation coefficient is a number that ranges from −1 to 1, where −1 indicates a perfect negative correlation (when one variable increases, the other decreases), 0 indicates no correlation and 1 indicates a perfect positive correlation (when one variable increases, the other also increases) (Dancey & Reidy, 2008).

Table 1 shows the *Spermann's r* correlation between the constructs.

The constructs with the highest correlation were [3] Communication on SD and [2] Extracurricular activities on SD (0.682***), followed by [4] Green infrastructure on campus and [3] Communication on SD (0.667***), [3] Communication on SD and [1] Level of approach to Sustainable Development (SD) in the classroom (0.475***) and [2] Extracurricular activities on SD and [5] Level of proactivity in relation to SD (0.463***). The correlations were moderate and positive.

Table 1 Spearman correlation test

	1	2	3	4	5
[1] Level of approach to sustainable development (SD) in the classroom	1				
[2] Extracurricular activities on SD	0.460***	1			
[3] Communication on SD	**0.475*****	**0.682*****	1		
[4] Green infrastructure on campus	0.359***	0.521***	**0.667*****	1	
[5] Level of proactivity in relation to SD	0.377***	**0.463*****	0.382***	0.395***	1

Note correlation is significant at the ***$p < 0.005$ level (2 ends)

Communication on SD allows for dissemination and participation of campus users, contributing to an ethical and social learning perspective. It should enable the development of strategies with clear and tangible objectives for participants. In general, the process should be geared towards training and capacity building, enabling the different activities to be carried out on campus and in society in general. This construct is strongly correlated with the others, as it is assumed that without communication there is no knowledge (Dagiliūtė et al., 2018; Ribeiro et al., 2021).

Next, multiple regression analysis was used to test the hypothesis. According to Hair Jr. et al. (2005), this statistical technique can be used to analyse the relationship between a single dependent variable (criterion) and several independent variables (predictors). The aim of multiple regression analysis is to predict changes in the dependent variable in response to changes in the independent variables. Table 2 shows multiple regression analysis.

In the summary of the multiple linear regression models and the overall fit statistics, it was found that the adjusted R^2 were significant (*p-value* of ANOVA <.001). This result means that the linear regression explained 25%. It is advisable to include the Durbin-Watson test for autocorrelations. The Durbin-Watson of 2.078 are results that corroborate the literature that calls for values between 0 and 4 (Gujarati & Porter, 2011) assuming that there is no first-order linear autocorrelation in the multiple linear regression data.

The result of the regression analysis in the model reveals that the construct [5] Level of proactivity in relation to SD is explained by the constructs [2] Extracurricular activities on SD ($\beta = 0.245$***), followed by the construct [1] Level of approach to Sustainable Development (SD) ($\beta = 0.185$***) and [4] Green infrastructure on campus ($\beta = 0.134$***). These results support Hypothesis *H1. Green Campus initiatives of a Brasileira HEIs positively affect students' Proactivity towards SD*

Table 2 Multiple regression analysis

Independents variables (IV)	Dependent variable [5] Level of proactivity in relation to SD
[1] Level of approach to sustainable development (SD) in the classroom	**β = 0,185***** t = 3231 (0,052)
[2] Extracurricular activities on SD	**β = 0,245***** t = 3084 (0,101)
[3] Communication on SD	β = 0,060 t = 0,864 (0,085)
[4] Green infrastructure on campus	**β = 0,134**** t = 2016 (0,093)
Constant (unstandardised)	t = 3247*** (0,205)
Z	26,694
Standard error	101,170
R^2	.259
Adjusted R^2	**.250**
p-value (ANOVA)	**<.001**
Durbin-Watson	**2078**

Note Number of observations = 310
Probabilistic notes: non-standardised standard error in parenthesis; *** $p < 0.010$; ** $p < 0.050$

5 Discussion

Several studies have highlighted the importance of extracurricular activities. Guided reflection on these experiences contributes to the integral human and professional formation of participants in sustainable development (Leal Filho et al., 2019a; Leal Filho et al., 2019c). In addition, all programmes must enable learners to consider the issues raised by sustainability, not only in the context of their disciplines, but also at a more general level, as citizens who will be affected and have an impact. Certain disciplines may make a stronger contribution, but all objectives require some consideration by all staff and students if progress is to be made (Leal Filho et al., 2019c).

It is also possible to emphasise the importance of encouraging sustainability initiatives in relation to everyday social activities and events or activities that support sustainability. These arguments support those of Ribeiro et al. (2021), according to which co-curricular activities have a positive impact on increasing students' perceptions of the importance of SD.

The question of how universities can be agents of change by putting sustainability education into practice is crucial and deserves the attention of both practitioners and academics (Leal Filho et al., 2019c). Therefore, universities face a great challenge as they take on the mission of training professionals, opening horizons to the future and new paradigms, and helping to balance teaching and SD (Disterheft et al., 2015).

The academic community has shown an interest in learning about climate change, although the university has not yet institutionalised this topic, which shows an opportunity for action (Brandli et al., 2023). However, according to the results of this study, teaching activities are still not very well perceived by students. Higher education institutions still do not seem to be very oriented towards teaching this topic, confirming Beddewela et al. (2017) and Aleixo et al. (2021). Part of the problem may be the lack of teachers who understand and link the SD topic in their classes, and even their difficulty in adapting current disciplines to SD concepts and discussions, confirming the need to promote capacity building and continuous training (Berchin et al., 2021).

Our findings corroborate the study by (Homer & Khor, 2022) in Malaysia. The respondents rated the importance of sustainable resource use, technology and campus facilities to appeal to students. In this sense, energy efficiency initiatives in higher education institutions can strongly contribute to creating and promoting habits and attitudes that support SD (Leal Filho et al., 2019b).

6 Conclusion

The role of higher education institutions (HEIs) in sustainable development (SD), and in particular in achieving the SDGs by 2015 (United Nations, 2015), is undisputed. In fact, the fourth SDG (Education), among others, highlights the importance of preparing students to consider the impact of SD in their actions as citizens and professionals; and the seventh goal (Ensure access to clean energy), in turn, is linked to the generation of affordable, reliable and sustainable energy for all (Leal Filho et al., 2019b).

Higher education institutions should play a transformative role in education, acting as co-creators of change and promoting the principles of SD in future professionals. Education is a powerful tool for behaviour change, so HEIs have a fundamental role to play.

This study confirms that stakeholder engagement is particularly important for SD with regard to the educational and institutional dimensions (Disterheft et al., 2015). Indeed, understanding the potential role that HEIs play in students' proactivity could help address global challenges (Leal Filho et al., 2019e). In this line of argument, the younger generations (as observed in the sample of this research) can indeed play an important role in addressing sustainability by understanding and implementing holistic and transdisciplinary approaches that address the dimensions of sustainability and their interrelationships (Sonetti et al., 2016).

An underlying argument is that students are engaged in SD, taking the best HEI initiatives as a precedent, whereby HEI initiatives can lead people towards proactive behaviour in relation to SD. These findings can be useful for those involved in planning SD-oriented actions in HEIs and strengthen the role of HEIs as co-creators of change, promoting the principles of SD in future professionals and looking for ways to broaden the dialogue on SD in HEIs, especially when it comes to achieving the SDGs, intertwined with the idea of increasing student participation and engagement.

One of the possible contributions of the research is to identify critical areas where interventions are needed to improve campus sustainability. For example, surveys could be conducted to understand students' perceptions of the availability of infrastructure for cycling, public transport and recycling on campus. Based on this data, the university can create policies and invest in infrastructure to encourage the use of sustainable transport, reduce the carbon footprint and promote environmental awareness.

Another possible contribution of the research is to identify existing sustainable practices on campus that can be expanded and improved. For example, research could be conducted to understand how the campus manages its waste and how this could be improved. This research can lead to the creation of more sustainable waste management policies and practices on campus.

It can also help identify barriers to the adoption of sustainable practices on campus and develop strategies to overcome them. For example, research can be conducted to understand why students do not adopt sustainable practices, such as recycling, and develop strategies to encourage the adoption of these practices.

These and other research findings, as highlighted by Leal Filho et al. (2021), can help to make better use of the many opportunities offered by sustainability publications.

References

Adomßent, M., Grahl, A., & Spira, F. (2019). Putting sustainable campuses into force: Empowering students, staff and academics by the self-efficacy Green Office Model. *International Journal of Sustainability in Higher Education, 20*(3), 470–481. https://doi.org/10.1108/IJSHE-02-2019-0072

Aleixo, A. M., Leal, S., & Azeiteiro, U. M. (2018). Conceptualization of sustainable higher education institutions, roles, barriers, and challenges for sustainability: An exploratory study in Portugal. *Journal of Cleaner Production, 172*, 1664–1673. https://doi.org/10.1016/j.jclepro.2016.11.010

Aleixo, A. M., Leal, S., & Azeiteiro, U. M. (2021). Higher education students' perceptions of sustainable development in Portugal. *Journal of Cleaner Production, 327*(9), 129429. https://doi.org/10.1016/j.jclepro.2021.129429

Alexandre, N. M. C., & Coluci, M. Z. O. (2011). Validade de conteúdo nos processos de construção e adaptação de instrumentos de medidas. *Ciencia e Saude Coletiva, 16*(7), 3061–3068. https://doi.org/10.1590/S1413-81232011000800006

Beddewela, E., Warin, C., Hesselden, F., & Coslet, A. (2017). Embedding responsible management education—Staff, student and institutional perspectives. *International Journal of Management Education, 15*(2), 263–279. https://doi.org/10.1016/j.ijme.2017.03.013

Berchin, I. I., Dutra, A. R. de A., & Guerra, J. B. S. O. de A. (2021). How do higher education institutions promote sustainable development? A literature review. *Sustainable Development, 29*(6), 1204–1222. https://doi.org/10.1002/sd.2219

Brandli, L. L., Reginatto, G., Salvia, A. L., & Mazutti, J. (2023). *Academic Community Expectations on Climate Change Learning and Engagement: A Case Study at University of Passo Fundo* (pp. 111–124). In: Leal Filho, W., Lange Salvia, A., Pallant, E., Choate, B., Pearce, K. (eds) Educating the Sustainability Leaders of the Future. World Sustainability Series. Springer, Cham. https://doi.org/10.1007/978-3-031-22856-8_7

Cho, C. H., Kim, A., Rodrigue, M., & Schneider, T. (2020). Towards a better understanding of sustainability accounting and management research and teaching in North America: a look at the community. *Sustainability Accounting, Management and Policy Journal, 11*(6), 985–1007. https://doi.org/10.1108/SAMPJ-08-2019-0311

Cho, M. (2019). Campus sustainability: An integrated model of college students' recycling behavior on campus. *International Journal of Sustainability in Higher Education, 20*(6), 1042–1060. https://doi.org/10.1108/IJSHE-06-2018-0107

Dagiliūtė, R., Liobikienė, G., & Minelgaitė, A. (2018). Sustainability at universities: Students' perceptions from Green and Non-Green universities. *Journal of Cleaner Production, 181*, 473–482. https://doi.org/10.1016/j.jclepro.2018.01.213

Dancey, C. P., & Reidy, J. (2008). *Estatística sem matemática para psicologia usando SPSS para Windows*. Porto Alegre: Artes Médicas.

Disterheft, A., Caeiro, S., Azeiteiro, U. M., & Leal Filho, W. (2015). Sustainable universities - A study of critical success factors for participatory approaches. *Journal of Cleaner Production, 106*, 11–21. https://doi.org/10.1016/j.jclepro.2014.01.030

Disterheft, A., Ferreira Da Silva Caeiro, S. S., Ramos, M. R., & De Miranda Azeiteiro, U. M. (2012). Environmental Management Systems (EMS) implementation processes and practices in European higher education institutions - Top-down versus participatory approaches. *Journal of Cleaner Production, 31*, 80–90. https://doi.org/10.1016/j.jclepro.2012.02.034

Eugénio, T., Carreira, P., Miettinen, N., & Lourenço, I. M. E. C. (2021). Understanding students' future intention to engage in sustainability accounting: the case of Malaysia and Philippines. *Journal of Accounting in Emerging Economies, Vol ahead*-(351). https://doi.org/10.1108/JAEE-10-2020-0277

Gujarati, D. N., & Porter, D. C. (2011). *Econometria básica* (5th ed.). Porto Alegre: AMGH.

Hair Jr., J. F., Anderson, R, E., Tatham, R. L., & Black, W. C. (2005). *Multivariate data analysis*. Pearson Education Limited.

Homer, S. T., & Khor, K. S. (2022). Exploring the perceptions of Malaysian Gen Z towards the impact of COVID-19 on sustainable development. *Environmental Science and Pollution Research, 29*(57), 85700–85716. https://doi.org/10.1007/s11356-021-17763-1

International Institute for Sustainable Development (IISD). (2003). *The Halifax Declaration*. http://www.iisd.org/educate/declarat/halifax.htm

Leal Filho, W., Doni, F., Vargas, V. R., Wall, T., Hindley, A., Rayman-Bacchus, L., Emblen-Perry, K., Boddy, J., & Avila, L. V. (2019a). The integration of social responsibility and sustainability in practice: Exploring attitudes and practices in Higher Education Institutions. *Journal of Cleaner Production, 220*, 152–166. https://doi.org/10.1016/j.jclepro.2019.02.139

Leal Filho, W., Salvia, A. L., Paço, A. do, Anholon, R., Gonçalves Quelhas, O. L., Rampasso, I. S., Ng, A., Balogun, A. L., Kondev, B., & Brandli, L. L. (2019b). A comparative study of approaches towards energy efficiency and renewable energy use at higher education institutions. *Journal of Cleaner Production, 237*, 117728. https://doi.org/10.1016/j.jclepro.2019.117728

Leal Filho, W., Shiel, C., Paço, A., Mifsud, M., Ávila, L. V., Brandli, L. L., Molthan-Hill, P., Pace, P., Azeiteiro, U. M., Vargas, V. R., & Caeiro, S. (2019c). Sustainable Development Goals and sustainability teaching at universities: Falling behind or getting ahead of the pack? *Journal of Cleaner Production, 232*, 285–294. https://doi.org/10.1016/j.jclepro.2019.05.309

Leal Filho, W., Skanavis, C., Kounani, A., Brandli, L. L., Shiel, C., Paço, A. do, Pace, P., Mifsud, M., Beynaghi, A., Price, E., Salvia, A. L., Will, M., & Shula, K. (2019d). The role of planning in implementing sustainable development in a higher education context. *Journal of Cleaner Production, 235*, 678–687. https://doi.org/10.1016/j.jclepro.2019.06.322

Leal Filho, W., Vargas, V. R., Salvia, A. L., Brandli, L. L., Pallant, E., Klavins, M., Ray, S., Moggi, S., Maruna, M., Conticelli, E., Ayanore, M. A., Radovic, V., Gupta, B., Sen, S., Paço, A., Michalopoulou, E., Saikim, F. H., Koh, H. L., Frankenberger, F., … Vaccari, M. (2019e). The role of higher education institutions in sustainability initiatives at the local level. *Journal of Cleaner Production, 233*, 1004–1015. https://doi.org/10.1016/j.jclepro.2019.06.059

Leal Filho, W., Will, M., Salvia, A. L., Adomßent, M., Grahl, A., & Spira, F. (2019f). The role of green and Sustainability Offices in fostering sustainability efforts at higher education institutions. *Journal of Cleaner Production, 232*, 1394–1401. https://doi.org/10.1016/j.jclepro.2019.05.273

Leal Filho, W., et al. (2021). Sustainability reporting in higher education institutions: An overview of recent developments. International Journal of Sustainability in Higher Education, 22(3), 561–579. https://doi.org/10.1108/IJSHE-02-2020-0079

Leal Filho, Walter, Frankenberger Silva, F., Salvia, A., Shiel, C., Paço, A., Price, L., Brandli, L. L., Rampasso, I. S., Anholon, R., Quelhas, O. L. G., & Pretorius, R. W. (2023). An overview of research trends on sustainability in higher education—an exploratory study. *International Journal of Sustainability in Higher Education, 24*(5), 1161–1175. https://doi.org/10.1108/IJSHE-08-2022-0252

Lee, W. E., Birkey, R. N., & Patten, D. M. (2017). Exposing Students to Environmental Sustainability in Accounting: An Analysis of Its Impacts in a US Setting. *Social and Environmental Accountability Journal, 37*(2), 81–96. https://doi.org/10.1080/0969160X.2016.1270225

León-Fernández, Y., Gomera, A., Antúnez, M., Martínez-Escrich, B., Villamandos, F., & Vaquero, M. (2018). Enhancing environmental management in universities through participation: the case of the University of Córdoba. *Journal of Cleaner Production, 172*, 4328–4337. https://doi.org/10.1016/j.jclepro.2017.06.103

Lozano, R., Ceulemans, K., Alonso-Almeida, M., Huisingh, D., Lozano, F. J., Waas, T., Lambrechts, W., Lukman, R., & Hugé, J. (2015). A review of commitment and implementation of sustainable development in higher education: Results from a worldwide survey. *Journal of Cleaner Production, 108*, 1–18. https://doi.org/10.1016/j.jclepro.2014.09.048

Monavvarifard, F., Baradaran, M., & Khosravipour, B. (2019). Increasing the sustainability level in agriculture and Natural Resources Universities of Iran through students' engagement in the value Co-creation process. *Journal of Cleaner Production, 234*, 353–365. https://doi.org/10.1016/j.jclepro.2019.06.175

Moqbel, S., Abu-Zurayk, R., Bozeya, A., Alsisan, R., & Al Bawab, A. (2020). Assessment of sustainable recycling at The University of Jordan. *International Journal of Sustainability in Higher Education, 21*(6), 1111–1129. https://doi.org/10.1108/IJSHE-11-2019-0334

Rampasso, I. S., Anholon, R., Silva, D., Cooper Ordoñez, R. E., Santa-Eulalia, L. A., Quelhas, O. L. G., Leal Filho, W., & Granada Aguirre, L. F. (2019). Analysis of the perception of engineering students regarding sustainability. *Journal of Cleaner Production, 233*, 461–467. https://doi.org/10.1016/j.jclepro.2019.06.105

Ribeiro, J. P. M., Hoeckesfeld, L., Dal Magro, C. B., Favretto, J., Barichello, R., Lenzi, F. C., Secchi, L., Montenegro de Lima, C. R., & Salgueirinho Osório de Andrade Guerra, J. B. (2021). Green Campus Initiatives as sustainable development dissemination at higher education institutions: Students' perceptions. *Journal of Cleaner Production, 312*(June). https://doi.org/10.1016/j.jclepro.2021.127671

Shawe, R., Horan, W., Moles, R., & O'Regan, B. (2019). Mapping of sustainability policies and initiatives in higher education institutes. *Environmental Science and Policy, 99*(June), 80–88. https://doi.org/10.1016/j.envsci.2019.04.015

Sima, M., Grigorescu, I., & Bălteanu, D. (2019). An overview of campus greening initiatives at universities in Romania. *International Journal of Sustainability in Higher Education, 20*(3), 410–422. https://doi.org/10.1108/IJSHE-01-2019-0036

Sonetti, G., Lombardi, P., & Chelleri, L. (2016). True green and sustainable university campuses? Toward a clusters approach. *Sustainability (Switzerland), 8*(1), 1–23. https://doi.org/10.3390/su8010083

The Lüneburg Declaration. (2001). *Lüneburg Declaration presented to United Nations Secretary-General Kofi Annan.* https://www.iau-hesd.net/sites/default/files/documents/luneburgdeclaration_0.pdf

UNESCO-UNEP. (1977). *Recommendations of the Intergovernmental Conference on Environmental Education Tbilisi.* USSR. France: UNESCO Paris. https://kykpee.org/wp-content/uploads/2015/08/the_declaration_of_tbilisi.pdf

UNESCO. (1972). *Declaration Of The United Nations Conference On The Human Environment.* Stockholm: UNESCO. https://undocs.org/en/A/CONF.48/14/Rev.1

UNESCO. (1990). *The Talloires Declaration.* Gland: UNESCO. http://ulsf.org/wp-content/uploads/2015/06/TD.pdf

UNESCO. (2022). *Knowledge-driven actions : Transforming higher education for global sustainability Independent Expert Group on the Universities and the 2030 Agenda.* https://unesdoc.unesco.org/ark:/48223/pf0000380519

United Nations. (2015). *Transforming Our World: The 2030 Agenda for Sustainable Development.* https://sdgs.un.org/

United Nations Conference on Environment & Development [UNCED]. (1992). *Agenda 21, Ch. 36: Promoting Education and Public Awareness and Training.* United Nations, New York. https://sustainabledevelopment.un.org/content/documents/Agenda21.pdf

Zhu, B., & Dewancker, B. (2021). A case study on the suitability of STARS for green campus in China. *Evaluation and Program Planning, 84*(August 2019), 101893. https://doi.org/10.1016/j.evalprogplan.2020.101893

Exploring Students' Attitudes Toward a Paperless Campus in Selected Educational Institutions in Pabna, Bangladesh: An Empirical Investigation on the Ground of Epistemic Emotions and Cognitions

Md. Hafiz Iqbal, Md. Amzad Hossain, and Ibrahim Habib

Abstract Current assessments of demand for paperless curricula in higher education exhibit significant ambiguity, marked by evidential gaps and methodological challenges. This study explores the spectrum of teacher and student perspectives and to develop perception-based teaching and learning strategies for paperless environments. To address this, our study investigates the perceptions of 385 students towards a paperless campus, employing a Q-methodology framework with 20 statements. This approach reveals three distinct discourses: a positive outlook, a stance of uncertainty, and a fatalistic perspective. Our analysis identifies common ground among these diverse viewpoints, demonstrating a consensus that the successful adoption of paperless campuses hinges on social mobilization, technological infrastructure, and robust curriculum support from educational institutions and government bodies. This research contributes to the ongoing discourse on campus greening and sustainable operations by providing a nuanced understanding of student perspectives, thereby informing the development of targeted educational policies.

Keywords Campus greening · Climate action · Q-methodology · Sustainability · Epistemology

---

Md. H. Iqbal (✉)
ICT Division, Bangladesh National Commission for UNESCO (BNCU), Dhaka, Bangladesh
e-mail: ict@bncu.gov.bd

Md. A. Hossain
Social Innovation Science, Graduate School of Innovation and Practice for Smart Society, IDEC Institute, Hiroshima University, Higashi-Hiroshima, Japan
e-mail: m236224@hiroshima-u.ac.jp

I. Habib
Department of Economics, School of Business and Economics, North South University, Dhaka, Bangladesh
e-mail: ibrahim.habib.242@northsouth.edu

W. Leal Filho et al. (eds.), *Sustainability in the Curriculum and Operations of Universities*, World Sustainability Series,
https://doi.org/10.1007/978-3-032-12678-8_45

1 Introduction

The escalating global concern over carbon emissions, driven by deforestation and fossil fuel consumption, underscores the urgent need for sustainable practices. Observational studies consistently demonstrate a significant link between paper consumption and deforestation (Prandi et al. 2021; Rehmani et al. 2022), with global paper use rising by an estimated 3% annually (Lackner et al. 2022). This surge directly impacts forests, as the paper industry fells over 900 million trees each year (Del Rio et al. 2022), and is identified as a major contributor to deforestation (Asamoah et al. 2020).

In response to this environmental challenge, the concept of a 'paperless campus,' aligned with the Sustainable Development Goals (SDGs), has gained considerable traction (Prandi et al. 2021). A paperless campus minimizes paper usage through digital tools and technologies for learning, communication, and documentation (Kareem et al. 2020). Key components include the 3R principles (reuse, recycle, reduce), digital administrative processes, electronic course materials, e-libraries, distance learning, online submissions and grading, and electronic record keeping (Iqbal and Farid, 2015; Kang et al. 2021; Silvis, 2022). Notably, transitioning to a paperless campus can reduce educational expenses and mitigate the demand for logging, promoting environmental stewardship and contributing to SDG achievement.

The Graduate School for International Development and Cooperation (IDEC) at Hiroshima University's Higashi-Hiroshima Campus exemplifies a successful paperless graduate school (Iqbal and Farid 2015). IDEC has implemented practices like the 3R principles, online examinations, and digital platforms for document sharing, demonstrating the practical application of paperless strategies in higher education.

However, the transition to paperless campuses is not without challenges. In Bangladesh, for example, there's a lack of clarity regarding benefits and potential disparities. Field visits to Pabna University of Science and Technology (PUST) and Pabna Government Edward College (GEC) revealed mixed student perceptions, with concerns about unequal access to technology versus the potential for sustainable learning. This ambiguity underscores the need for a nuanced understanding of paperless campus implementation, particularly in contexts like Bangladesh.

Therefore, this study aims to contribute to existing research on paperless campuses and the SDGs in low-income countries. Specifically, we employ Q methodology to explore teacher and student attitudes towards paperless campuses and provide an overview of curriculum and assessment techniques in this context within Bangladesh. Our objectives are to map the spectrum of perspectives and develop perception-based teaching and learning strategies.

The remainder of this paper is structured as follows: Section 2 reviews the literature, Sect. 3 outlines the methodology, Sect. 4 presents the results and discussion, and Sect. 5 concludes with policy implications and limitations.

2 Literature Review

While the invention of paper has undeniably contributed to societal advancement (Chin et al. 2019), its production carries significant environmental and social costs. The pulp and paper industry, a major consumer of virgin wood, ranks among the most energy-intensive sectors globally (Andersson and Thollander, 2019). This industry accounts for approximately 6% of global industrial energy use and 2% of direct industrial CO_2 emissions (Del Rio et al. 2022). Furthermore, the industry's toxic waste poses a threat to aquatic ecosystems, human health, and biodiversity (Singh et al. 2022).

In response to these concerns, the concept of a paperless campus has emerged as a promising initiative aligned with several SDGs, including responsible consumption and production, climate action, and quality education (Ramaswamy et al. 2021). Paperless campuses can significantly reduce paper consumption and waste, thereby contributing to SDG 12 (Degirmenci and Recker, 2023). By minimizing the carbon footprint associated with paper production and disposal, they also support SDG 13 (Prandi et al. 2021). Moreover, the provision of digital learning materials and the promotion of digital literacy skills enhance access to quality education (SDG 4) (Tsiligkiris and Ilieva, 2022). Paperless platforms also facilitate collaboration and communication (Tarteer et al. 2022).

However, the successful implementation of paperless campuses presents several challenges. Adequate technological infrastructure, including reliable internet connectivity and access to digital devices, is essential (Kareem et al. 2020). Ensuring equitable access is crucial to prevent a digital divide among students. Moreover, the transition requires significant changes in workflows and processes, along with substantial initial investment and training for staff. Data security and privacy concerns also necessitate robust cybersecurity measures and data protection policies (Habib et al. 2021; Mishra et al. 2022).

Despite these challenges, numerous universities worldwide, including those in the USA, UK, Hong Kong, and Malaysia (Iqbal and Farid, 2015), have adopted paperless campus strategies. While many institutions in low—and middle-income countries express interest in this transition, implementation gaps remain.

Notably, while existing research has assessed the effectiveness of paperless campuses, studies focusing on students' attitudes are lacking. This study aims to address this gap, particularly from the perspective of a low-income country, providing valuable insights into the student experience and contributing to a more comprehensive understanding of paperless campus implementation.

3 Materials and Methods

3.1 Present Study

This study was conducted at PUST and GEC, both located in Bangladesh. These institutions, with a combined student population of approximately 30,000 are committed to achieving the SDGs by 2030 and adapting to the Fourth Industrial Revolution (4IR). However, traditional teaching practices remain prevalent. These institutions were selected due to the presence of three distinct student groups: those with positive attitudes towards paperless campuses, those with uncertain views due to technological challenges, and those with a fatalistic perspective. Bridging these diverse viewpoints requires an epistemological understanding of their underlying beliefs.

3.2 Epistemological Approach

An epistemological approach was employed to address the contrasting student perspectives on paperless campuses. Epistemology, the theory of knowledge, provides a framework for examining how individuals acquire, understand, and justify knowledge (Blaikie, 1993; Greene et al. 2016). In this context, it facilitated the exploration of respondents' epistemic beliefs about paperless campuses and their critical thinking when confronted with conflicting information (Muis et al. 2021). This approach allowed us to move beyond simply documenting opinions to understanding the foundations of those opinions.

3.3 Mixed Method Research (MMR)

A mixed methods research (MMR) design was utilized to provide a comprehensive understanding of student attitudes. MMR combines quantitative and qualitative data collection and analysis, offering a more complete picture than either approach alone (Creswell, 2014). This study employed a pragmatic approach, integrating survey data (quantitative) with interview data and open-ended responses (qualitative). Q-methodology was used as the primary tool within the MMR framework.

3.4 Q-Methodology

To investigate the subjective attitudes of students towards paperless campuses, Q-methodology was employed, a method involving participants ranking a set of statements (Q-sorts) based on their personal perspectives (Webler et al. 2007). The process began with site selection and participant exploration, where PUST and GEC were chosen due to their diverse student perspectives. Subsequently, a concourse of statements related to paperless campuses was developed through document reviews and student interviews. From an initial collection of 33 statements, 20 were selected as the final Q-sorts (Table 1), with approval from education experts and policymakers. Participants then ranked these 20 statements based on their agreement or disagreement, and statistical factor analysis was used to identify common viewpoints (factors), which were subsequently interpreted and described. Notably, the concourse was developed using student interviews that revealed diverse perspectives on the benefits, drawbacks, and implementation of paperless campuses; however, these interviewees were not included in the final Q-sort participant pool.

The statement sorting exercise, conducted between June and August 2023 at PUST and GEC, involved a purposive sample of students. Prior to the sorting activity, participants were briefed on the potential impacts of a paperless campus. During the survey, each of the 385 respondents received: (1) 20 cards, each displaying a statement and its corresponding number; (2) an instruction bar representing a quasi-normal distribution; and (3) an answer sheet for recording their rankings. Participants were instructed to read all statements, sort the cards according to their level of agreement or disagreement using a standard Likert scale (−3: strongly disagree to +3: strongly agree), and place the cards on the instruction bar, as depicted Fig. 1. In addition to the sorting task, respondents were interviewed to gather insights into their personal experiences with paperless campuses, any challenges they encountered, and their overall opinions on the concept.

Table 1 Matrix for categorization of statement

Statement topic	Academic aspect	SDG aspect	Environmental aspect
Responsible consumption	1	1	1
Climate action	1	1	1
Quality teaching and learning	3	1	1
Collaboration and communication	1	1	1
Miscellaneous views	3	2	1

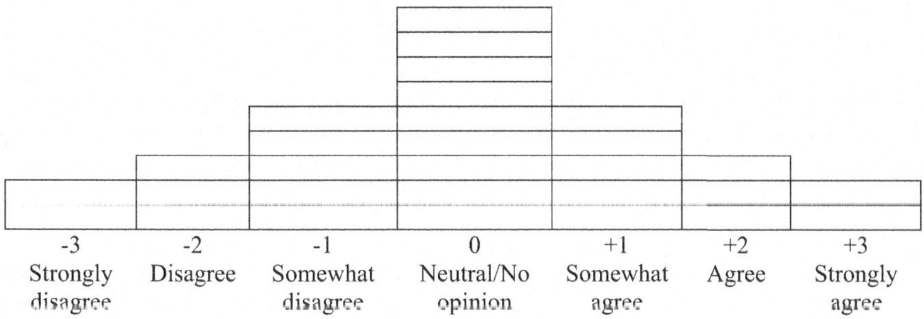

Fig. 1 Guide bar with quasi-normal distribution

Following the sorting exercise, the data was analyzed using PQ Method, a standard tool in Q-methodology. This analysis, specifically factor analysis, aims to identify distinct attitude groups, or 'defining types,' within the broader discourse surrounding the research topic. Initially, Principal Component Analysis (PCA) was employed to extract four unrotated factors with eigenvalues exceeding 1. Subsequently, Varimax rotation was applied to explore two-, three-, four-, and five-factor solutions. Ultimately, the three-factor solution was deemed statistically significant, justifiable, and provided the most coherent interpretation of the sorted data, thus revealing three distinct discourses among the participants.

To determine the most pertinent factor for each Q-sort, we followed the methodology outlined by Mahlalela et al. (2022), considering factor loadings at a significance level of $p < 0.10$, with a threshold of 2.58 times the standard error. The standard error for factor loadings was calculated using the formula $1/\sqrt{N}$, where N represents the number of statements (Christensen and Golino, 2021). In our study, with 20 statements, the standard error was estimated to be 0.577 $(2.58 \times 1/\sqrt{20})$. Consequently, factor loadings exceeding 0.577 were considered significantly loaded on the respective factor. Statements that did not significantly load on any of the three identified factors were excluded from further analysis.

4 Results and Discussion

Factor interpretation relies on statement scores, which are initially calculated as z-scores and subsequently transformed back into the original Q-sort value format (ranging from −3 to +3, as detailed in Table 2). Positive scores indicate a student's favorable perception of a given statement, while negative scores reflect an unfavorable perception. Based on these scores, we identified three distinct student groups: 'Benefit Believers,' who hold positive views regarding paperless campuses; 'Risk-Averse Individuals,' who express negative perceptions; and 'Traditionalists,' who maintain a more neutral stance.

Factor A-Benefit Believers

The findings shown in Table 2 imply that factor A encompasses a viewpoint that is predisposed to be favorably incline toward the idea of a paperless campus. Due to its role in popularizing paperless campus, the factor raises some questions concerning social mobilization (statement 1, converted factor score + 3). Students are likely to adopt a paperless campus because of the curriculum design and technological supports (10, +3) from their educational institutions and government, which are significantly associated with the paperless campus. Upload course materials (4, + 3), e-book and digital library (5, +3), distance learning (6, +2), online submission and grading (9, +2), environmental benefits (11, +1), cost effectiveness of learning (12, +2), waste disposal (14, −1), physical storage (16, −2), formative assessment (17, +2), data protection (18, +1), and education policy (20, + 1) are all necessary

Table 2 Q-sorts value for each statement

No	Statement	A	B	C
1	Social mobilization is necessary for paperless campus	3^{*}	1^{**}	3^{*}
2	Reuse, recycle and reduce practice can promote paperless campus	2	2	1^{***}
3	Digital administrative process is corelated with paperless campus	3	−1	1^{***}
4	Paperless campus is significant to upload course materials	3^{**}	-2^{**}	−1
5	E-book and digital library are the potential outcome of paperless	3^{***}	0	0
6	campus	2^{**}	-3^{*}	0
7	Distance learning is performed well under the provision of paperless	3	1	−2
8	campus	1	0	3^{*}
9	Video-conference is one of the major components of paperless campus	2^{*}	1^{**}	−2
10	Web-mail is the necessary factor of paperless campus	3^{**}	3^{*}	1^{***}
11	Paperless campus facilitates online submission and grading	1^{**}	1^{***}	0
12	Curriculum design & technological supports are essential for paperless	2^{**}	0	−3
13	campus	3	−3	−2
14	Environmental benefits are the outcome of paperless campus	-1^{*}	−3	−3
15	Paperless campus ensures cost effectiveness of learning	−3	-1^{**}	0
16	Students from rural areas are not preferred to paperless campus	-2^{*}	1^{*}	2
17	Paperless campus removes paper waste disposal in a larger amount	2^{**}	2	3^{**}
18	Paperless campus reduces printing cost	1^{***}	1^{***}	−2
19	Physical storage is not required under the provision of paperless	2	1	3
20	campus	1^{**}	2^{**}	-1^{**}
	Formative assessment should be highly reflected in paperless campus			
	Data protection is an urgent matter under the provision of paperless campus.			
	Data security and privacy may be hampered under paperless campus.			
	Guideline of paperless campus should be included in the education policy			

Note Factor A = benefit believers, B = risk averse, C = Traditionalist. $^{*}p < 0.01$, $^{**}p < 0.05$, $^{***}p < 0.10$

for components of a paperless campus. All these components are positively related to the paperless campus except waste disposal and physical storage because of their negative signs. The negative sign of waste disposal and physical storage implies that respondents in this group are not in favor of waste disposal and physical storage on paperless campus. But we cannot say reuse, recycle, and reduce practices, digital administrative process, video conference, web-mail, students from rural areas, reduce printing cost, and data security and privacy because of their insignificant relationship with the paperless campus. Benefit believers advocate paperless campus practices because they value learning gain over student life. The students in this group are generally more interested in paperless campus.

Factor B- Risk-Averse Individuals or Risk Perceivers

Factor B represents a viewpoint that, while not entirely opposed to paperless campuses, is significantly less enthusiastic than Factor A. This group's perspective is strongly influenced by social mobilization (statement 1, +1), uploading course materials (4, −2), distance learning (6, −3), online submissions and grading (9, + 1), curriculum and technological support (10, +3), environmental benefits (11, +1), reduced printing costs (15, −1), physical storage (16, +1), data protection (18, +1), and education policy (20, +2). Notably, their negative views on uploading course materials, distance learning, and reduced printing costs suggest a reluctance towards these specific aspects of a paperless environment. Conversely, statements related to e-books and digital libraries (5, 0), webmail (8, 0), and the cost-effectiveness of learning (12, 0) elicited neutral responses from this group, indicating a lack of strong opinion in these areas.

Factor C- Traditionalists or Fatalist

Factor C reveals a somewhat fatalistic attitude towards paperless campuses. This group strongly emphasizes the importance of social mobilization (statement 1, + 3), reuse, recycle, and reduce practices (2, +1), digital administrative processes (3, +1), webmail (8, +3), curriculum design and technological support (10, +1), and formative assessment (17, +3). However, they express reservations about education policy (20, −1), suggesting a perception that current policies do not adequately support the implementation of paperless campuses. Furthermore, they hold neutral views on e-books (5, 0), digital libraries (6, 0), environmental benefits (11, 0), and printing costs (15, 0), indicating a lack of strong opinions in these areas.

This group's fatalistic perspective likely stems from a limited understanding of the benefits and drawbacks of paperless campuses. Despite this, they actively advocate for social mobilization, curriculum design, and technological support as crucial drivers for advancing paperless campus initiatives. Interviews suggest that they perceive subsidies for social mobilization, curriculum design, and technological support as particularly important contributors.

Consensus Statement

Despite the distinct perspectives represented by each factor, several areas of consensus emerged. Notably, all three groups agreed that social mobilization, curriculum design, and technological support from educational institutions and the government are crucial for promoting paperless campuses in Bangladesh. This consensus, reflected in the statistically significant positive rankings for statements 1 and 10 in Table 2, underscores the students' shared need for awareness programs and subsidized access to essential gadgets and internet packages.

However, each factor also exhibited a degree of caution regarding the implementation of paperless campuses. The differences lie in the nature and intensity of their concerns. Factor A, representing 'Benefit Believers,' expressed an urgent desire for paperless campus adoption, driven by a strong belief in its environmental benefits. Factor B, the 'Risk-Averse Individuals,' were more hesitant, expressing concerns about the limited perceived benefits and potential learning disparities arising from unreliable support. Factor C, the 'Traditionalists,' displayed a fatalistic outlook, acknowledging both the potential advantages and disadvantages of paperless campuses.

The converted factor scores reveal distinct characteristics for each group. In factor A, students demonstrated pragmatism, a mild progressive mindset, and environmental awareness, advocating for increased social mobilization, tailored curriculum design, and institutional and governmental support. In contrast, in factor B, students were apprehensive about the benefits of paperless campuses and worried about potential learning discrimination. Furthermore, in factor C, students influenced by traditional learning preferences, shared concerns about the quality of education in a paperless environment.

5 Conclusions

This study provides an initial understanding of student perceptions regarding paperless campuses in the Pabna district, focusing on their potential to protect the environment, enhance learning, facilitate digital interaction, promote responsible consumption, and foster collaboration. The analysis revealed three distinct student groups, each exhibiting varying attitudes towards the implementation of paperless practices, particularly concerning reuse, recycling, digital administration, digital resources, distance learning, and data security. However, a significant consensus emerged regarding the pivotal role of social mobilization, curriculum design, and technological support from government and educational institutions. Students across all groups emphasized the need for reliable energy supplies, affordable internet packages, and subsidized gadgets to facilitate a smooth transition.

Notably, the students in this study, as key stakeholders, did not exhibit extreme positions, instead expressing a willingness to adopt paperless campuses given adequate support and benefits derived from collaborative efforts involving government, NGOs, civil society, media, and technology companies. This nuanced perspective suggests that persuading students to shift from traditional to paperless practices may be more complex than anticipated.

While the non-random selection of participants limits the generalizability of our findings, Q-methodology effectively highlights critical factors influencing the adoption of paperless campuses (Sneegas et al. 2021). The identified group-specific motivations and interests suggest that tailored policy interventions, employing diverse instruments, will be necessary for successful implementation (Karlen et al. 2019).

A key limitation of this study is the restricted scope of respondent discussions, which could have yielded a broader range of relevant statements. Future research should address this by conducting more extensive need-based discussions across a wider study area to gain a more comprehensive understanding of the factors influencing paperless campus adoption.

References

Andersson, E., & Thollander, P. (2019). Key performance indicators for energy management in the Swedish pulp and paper industry. *Energy Strategy Reviews*, *24*, 229–235. https://doi.org/10.1016/j.esr.2019.03.004

Asamoah, O., Kuittinen, S., Abrefa Danquah, J., Quartey, E.T., Bamwesigye, D., Mario Boateng, C., & Pappinen, A. (2020). Assessing wood waste by timber industry as a contributing factor to deforestation in Ghana. *Forests, 11*(9), 939. https://doi.org/10.3390/f11090939

Blaike, N. (1993). *Approaches to Social Inquiry*. Cambridge: Polity Press

Chin, T., Yang, Y., Zhang, P., Yu, X., & Cao, L. (2019). Co-creation of social innovation: Corporate universities as innovative strategies for Chinese firms to engage with society. *Sustainability, 11*(5), 1438. https://doi.org/10.3390/su11051438

Christensen, A.P., & Golino, H. (2021). On the equivalency of factor and network loadings. *Behavior Research Methods, 53*(4), 1563–1580. https://doi.org/10.3758/s13428-020-01500-6

Creswell, J.W. (2014). *Research Design: Qualitative, Quantitative and Mixed Methods Approaches*. California: SAGE Publication

Degirmenci, K., & Recker, J. (2023). Breaking bad habits: a field experiment about how routinized work practices can be made more eco-efficient through IS for sense making. *Information and Management, 60*(4), 103778. https://doi.org/10.1016/j.im.2023.103778

Del Rio, D.D.F., Sovacool, B.K., Griffiths, S., Bazilian, M., Kim, J., Foley, A.M., & Rooney, D. (2022). Decarbonizing the pulp and paper industry: a critical and systematic review of sociotechnical developments and policy options. *Renewable and Sustainable Energy Reviews, 167*, 112706. https://doi.org/10.1016/j.rser.2022.112706

Greene, J.A., Sandoval, W.A., & Braten, I. (2016). The role of epistemic beliefs in students' self-regulated learning with computer-based learning environment: conceptual and methodological issues. *Education Psychology, 45*, 245–257. http://doi:10.1080/00461520.2010.515932

Habib, M.N., Jamal, W., Khalil, U., & Khan, Z. (2021). Transforming universities in interactive digital platform: case of city university of science and information technology. *Education and Information Technologies, 26*, 517–541. https://doi.org/10.1007/s10639-020-10237-w

Iqbal, M.H., & Ahmed, F. (2015). Paperless campus: the real contribution towards a sustainable low carbon society. *IOSR Journal of Environmental Science, Toxicology and Food Technology, 9*(8), 10–17. https://www.iosrjournals.org/iosr-jestft/papers/vol9-issue8/Version-1/C09811017.pdf

Kang, Q., Lu, J., & Xu, J. (2021). Is e-reading environmentally more sustainable than conventional reading? Evidence from a systematic literature review. *Library and Information Science Research, 43*(3): 101105. https://doi.org/10.1016/j.lisr.2021.101105

Kareem, A.K., Bello, M.L., Ishola, A.A., & Jimoh, L.A. (2020). E-Government, information and communications technology support and paperless environment in Nigerian public universities: issues and challenges. *Journal of Technology Management and Business, 7*(1), 65–74. https://publisher.uthm.edu.my/ojs/index.php/jtmb/article/view/5384

Karlen, Y., Suter, F., Hirt, C., & Merki, K.M. (2019). The role of implicit theories in students' grit, achievement goals, intrinsic and extrinsic motivation, and achievement in the context of a long-term challenging task. *Learning and Individual Differences, 74*, 101757. https://doi.org/10.1016/j.lindif.2019.101757

Lakner, C., Mahler, D.G., Negre, M., & Prydz, E.B. (2022). How much does reducing inequality matter for global poverty? *The Journal of Economic Inequality, 20*(3), 559–585. https://doi.org/10.1007/s10888-021-09510-w

Mahlalela, L.S., Jourdain, D., Mungatana, E.D., & Lundhede, T.H. (2022). Diverse stakeholder perspectives and ecosystem services ranking: application of the Q-methodology to Hawane Dam and nature reserve in Eswatini. *Ecological Economics, 197*, 107439. https://doi.org/10.1016/j.ecolecon.2022.107439

Mishra, A., Alzoubi, Y.I., Gill, A.Q., & Anwar, M.J. (2022). Cybersecurity enterprises policies: a comparative study. *Sensors, 22*(2), 538. https://doi.org/10.3390/s22020538

Muis, K.R., Chevrier, M., Denton, C.A., & Losenno, K.M. (2021). Epistemic emotions and epistemic cognition predict critical thinking about socio-scientific issues. *Frontiers in Education, 6*, 669908. http://doi:10.3389/feduc.2021.669908

Prandi, C., Ceccarini, C., Nisi, V., & Salomoni, P. (2021). Designing interactive infographics to stimulate environmental awareness: an exploration with a university community. *Multimedia Tools and Applications*, *80*, 12951–12968. https://doi.org/10.1007/s11042-020-09140-w

Ramaswamy, M., Marciniuk, D.D., Csonka, V., Colò, L., & Saso, L. (2021). Reimagining internationalization in higher education through the United Nations sustainable development goals for the betterment of society. *Journal of Studies in International Education*, *25*(4), 388–406. https://doi.org/10.1177/10283153211031046

Rehmani, M., Iftikhar, U., Khokhar, M.N., Rehman, H.I.U., Irshad, H., & Anwer, N. (2022). Does the traditional recruitment procedure of higher education institutions cause paper waste? *Cogent Business and Management*, *9*(1): 2151679. https://doi.org/10.1080/23311975.2022.2151679

Silvis, D. (2022). Renewing learning: reimagining the newness of families' sociotechnical practices towards ecologically sustainable media engagement. *Learning, Culture and Social Interaction*, *35*, 100644. https://doi.org/10.1016/j.lcsi.2022.100644

Singh, A.K., Kumar, A., & Chandra, R. (2022). Environmental pollutants of paper industry wastewater and their toxic effects on human health and ecosystem. *Bioresource Technology Reports*, *20*, 101250. https://doi.org/10.1016/j.biteb.2022.101250

Sneegas, G., Beckner, S., Brannstrom, C., Jepson, W., Lee, K., & Seghezzo, L. (2021). Using Q-methodology in environmental sustainability research: a bibliometric analysis and systematic review. *Ecological Economics*, 180, 106864. https://doi.org/10.1016/j.ecolecon.2020.106864

Tarteer, S., Badah, A., Khlaif, Z.N. (2022). Employing google classroom to teach female students during the covid-19 pandemic. *Computers in the Schools*, *38*(4), 300–321. https://doi.org/10.1080/07380569.2021.1988318

Tsiligkiris,V., & Ilieva, J. (2022). Global engagement in the post-pandemic world: challenges and responses. Perspective from the UK. *Higher Education Quarterly*, *76*(2), 343–366. https://doi.org/10.1111/hequ.12390

Webler, T., Danielson, S., & Tuler, S. (2007). Guidance on the use of Q method for evaluation of public involvement programs at contaminated sites. Greenfield, MA: Massachusetts: Social and Environmental Research Institute.

Žalėnienė, I., & Pereira, P. (2021). Higher education for sustainability: a global perspective. *Geography and Sustainability*, *2*(2), 99–106. https://doi.org/10.1016/j.geosus.2021.05.001

The manufacturer's authorised representative in the EU is Springer Nature Customer Service Centre GmbH, Europaplatz 3, 69115 Heidelberg, Germany. If you have any concerns regarding our products, please contact ProductSafety@springernature.com

Printed and bound by CPI Group (UK) Ltd, Croydon, CR0 4YY
21/09/2026
02220009-0019